Plant Ecology: Principles and Practices

Plant Ecology: Principles and Practices

Edited by **Molly Ismay**

R CALLISTO REFERENCE

New York

Published by Callisto Reference,
106 Park Avenue, Suite 200,
New York, NY 10016, USA
www.callistoreference.com

Plant Ecology: Principles and Practices
Edited by Molly Ismay

International Standard Book Number: 978-1-63239-659-4 (Hardback)

Printed in the United States of America.

Contents

Preface

The main aim of this book is to educate learners and enhance their research focus by presenting diverse topics covering this vast field. This is an advanced book which compiles significant studies by distinguished experts. This book addresses successive solutions to the challenges arising in the area of application, along with it; the book provides scope for future developments.

Ecology is the science which deals with the relation of an organism to its surrounding environment that is, every condition required for its existence. Plant ecology can be further divided into plant ecophysiology, landscape, ecology, plant population ecology, etc. It covers both, basic science and applied science and focuses on water, carbon, and nutrient cycling; physical stress; plant functional traits; and trait diversity, covering ecosystem effects and functional growth analysis in an environmental context. This book presents researches and studies performed by experts across the globe. It contains some path-breaking studies in the field of plant ecology. This book includes contributions of experts and scientists which will provide innovative insights into the subject. It will prove to be immensely beneficial for ecologists, botanists, agriculturists, environmentalists, researchers and students alike.

It was a great honour to edit this book, though there were challenges, as it involved a lot of communication and networking between me and the editorial team. However, the end result was this all-inclusive book covering diverse themes in the field.

Finally, it is important to acknowledge the efforts of the contributors for their excellent chapters, through which a wide variety of issues have been addressed. I would also like to thank my colleagues for their valuable feedback during the making of this book.

Editor

Live substrate positively affects root growth and stolon direction in the woodland strawberry, *Fragaria vesca*

Erica M. Waters* and Maxine A. Watson

Department of Biology, Indiana University, Bloomington, IN, USA

Studies of clonal plant foraging generally focus on growth responses to patch quality once rooted. Here we explore the possibility of true plant foraging; the ability to detect and respond to patch resource status **prior to rooting**. Two greenhouse experiments were conducted to investigate the morphological changes that occur when individual daughter ramets of *Fragaria vesca* (woodland strawberry) were exposed to air above live (non-sterilized) or dead (sterilized) substrates. Contact between daughter ramets and substrate was prohibited. Daughter ramet root biomass was significantly larger over live versus dead substrate. Root:shoot ratio also increased over live substrate, a morphological response we interpret as indicative of active nutrient foraging. Daughter ramet root biomass was positively correlated with mother ramet size over live but not dead substrate. Given the choice between a live versus a dead substrate, primary stolons extended preferentially toward live substrates. We conclude that exposure to live substrate drives positive nutrient foraging responses in *F. vesca*. We propose that volatiles emitted from the substrates might be effecting the morphological changes that occur during true nutrient foraging.

Keywords: **plant nutrient foraging, clonal plants, woodland strawberry, *Fragaria vesca*, root biomass, stolon trajectory**

Edited by:
Sergio Roiloa,
University of A Coruña, Spain

Reviewed by:
Rubén Retuerto,
Universidad de Santiago
de Compostela, Spain
Zhenzhu Xu,
Chinese Academy of Sciences, China
Beáta Oborny,
Eötvös Loránd University, Hungary

*Correspondence:
Erica M. Waters,
Department of Biology, Indiana
University, Bloomington, IN 47405,
USA
emwaters@umail.iu.edu

Introduction

Optimal foraging theory (OFT) proposes that organisms forage for nutrients in a way that maximizes energy intake per unit time (MacArthur and Pianka, 1966; Charnov, 1976; Norberg, 1977; Oaten, 1977). Resources often occur in patches within an environment and the theory predicts that there is an optimum pattern of visitation that provides an organism with maximum benefits for minimum output of energy. Application of the theory requires two conditions: (1) that individuals can move through and explore an environment and (2) that individuals can distinguish between and respond to patches of varying quality. The theory includes factors regarding both within-patch ("exploitation", i.e., how long to remain, how to efficiently capture resources) and between-patch ("true foraging", i.e., patterns of locating resources, time spent searching) behavior (Charnov, 1976; Oaten, 1977). Optimal adjustment of these factors results in an increased uptake of energy, and thus improves fitness.

Optimal foraging theory was originally used as a means of understanding and predicting animal behavior. It posited that animals adjust both foraging time and patch visitation order to maximize energy acquisition (MacArthur and Pianka, 1966; Charnov, 1976; Pyke et al., 1977; Pyke, 1984; McNamara et al., 2006). However, complex thought in animals involving predation risk

(Brown, 1988; Higginson et al., 2012), food choice (Houston et al., 2011; Cressman et al., 2014), and memory (Freidin and Kacelnik, 2011) confounds the theory (Perry and Pianka, 1997).

The theory has also been applied to clonal plants (Slade and Hutchings, 1987; Birch and Hutchings, 1994; Cain, 1994; de Kroon and Hutchings, 1995; Grime and Mackey, 2002). Clonal plants are unique in the plant world for the ability of genets (aggregates of plants that are the products of a single seed) to change their location over time, and therefore explore and effectively exploit a heterogeneous environment for light and nutrients. They do this through the production of ramets, or potentially physiologically independent genetically identical units (**Figure 1**). Compared to most animals, plant movement is slow; it occurs via growth processes and benefits accrue due to maintenance of connections between sister ramets. As new ramets are produced and extend into the environment, older ramets die, essentially moving the genet through space. Ramets remain connected via stolons or rhizomes for variable lengths of time (Jónsdóttir and Watson, 1997) and these connections allow for transport of nutrients and hormones between the mother and daughter ramets (Alpert and Mooney, 1986; Jónsdóttir and Watson, 1997; Hutchings, 1999). Thus, clonal plants fulfill the first requirement for application of OFT through their ability to move. But, can they do this in a selective way? Can they distinguish between and respond to patches of differing quality in a heterogeneous environment?

For plants to forage for light or nutrients, they must be able to sense, interpret, and respond to environmental signals that specify habitat quality in a way that results in the non-random placement of individual ramets in appropriate patches. Thus, clonal plant foraging can be said to occur if placement of daughter ramets occurs more frequently in high quality than in low quality patches (Cain, 1994). In stoloniferous plants (those with above-ground connections between ramets), the means of sensing and responding to light patches in a heterogeneous environment has been well studied (Slade and Hutchings, 1987; Methy et al., 1990; Kemball et al., 1992; Dong, 1993, 1995;

Dong and Pierdominici, 1995; Stoll and Schmid, 1998; Grime and Mackey, 2002; Lepik et al., 2004; Dauzat et al., 2008). Detection of differences in red/far-red ratios via phytochromes and other photoreceptors induces plant morphological responses such as enhanced elongation rates or increased leaf area in response to low photosynthetically active radiation (PAR; Ballare et al., 1997; Smith, 2000; Franklin and Quail, 2010). These responses assist daughter ramets in locating and then occupying high light patches. Similarly, plants utilize red/far-red radiation ratios to assess areas of high vs. low density (i.e., neighbor sensing) (Schmitt and Wulff, 1993; Ballare, 1999; Marcuvitz and Turkington, 2000; Smith, 2000; Franklin and Quail, 2010). The majority of studies focused on morphological changes in leaf area and shoot biomass to differing light conditions, while others suggest that plasticity of spacers in length and branching intensity play a more critical role in light foraging, particularly in keeping ramets in light-rich patches (Kemball et al., 1992; Oborny, 1994; de Kroon and Hutchings, 1995; Dong, 1995; Dauzat et al., 2008).

Far less is known about the capacity of plants to detect nutrient-rich environments. Early studies focused on the proliferation of roots **after** plant establishment ("exploitation") (Birch and Hutchings, 1994; Cain, 1994; de Kroon and Hutchings, 1995; van Kleunen and Fischer, 2001). Studies found that lateral root elongation is highly responsive to the presence of nitrates (Zhang and Forde, 2000), and results in an abundance of root mass in richer patches (Leyser and Fitter, 1998; Jansen et al., 2006). Connected ramets in complementary environments increase the size of organs that obtain the most abundant resource (Stuefer et al., 1996). In a light-rich environment clonal ramets increase the mass of shoots and leaves, whereas in a nutrient-rich environment, root growth is increased. While an overall increase in biomass indicates that ramets are situated in an abundance of resources, evidence of clonal ramet foraging arises when the ratio of root:shoot biomass increases or decreases in response to an increase in nutrients or light, respectively. These morphological changes indicate a preferential allocation of resources to specific organs specialized for the capture of the abundant resource (Tuomi et al., 1983). These studies mirror those related to light in that they indicate that once plants enter a rich environment they alter their morphology in ways that enhance exploitation. Evidence of between-patch foraging – the ability of a developing stolon to distinguish between nutrient-rich or nutrient-poor patches – also exists.

Precision of foraging depends on "the ability of a species to perceive the heterogeneity and respond to it" (Wijesinghe et al., 2001) and there is evidence of this ability in many clonal plants. Salzman (1985) demonstrated that when given a choice between saline or non-saline soils, *Ambrosia psilocstachya* placed 67% of its rhizomes in non-saline soils. While it may be argued that the salinity suppressed plant growth, similar patterns of nutrient patch detection also have been found in stoloniferous plants. *Cuscuta subinclusa* exhibited coiling responses prior to physiological connection and exploitation of its host, indicating an ability to survey and interpret its surroundings and adjust development appropriately (Kelly, 1990). To date, the most striking example of patch recognition and differentiation was reported by Roiloa and Retuerto (2006). They found that

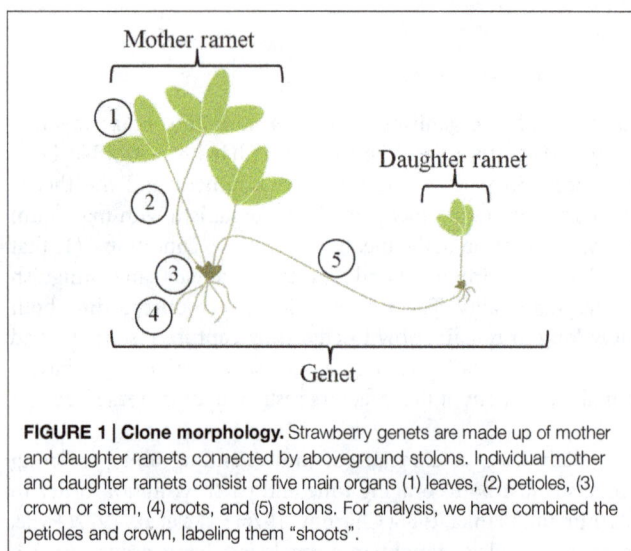

FIGURE 1 | Clone morphology. Strawberry genets are made up of mother and daughter ramets connected by aboveground stolons. Individual mother and daughter ramets consist of five main organs (1) leaves, (2) petioles, (3) crown or stem, (4) roots, and (5) stolons. For analysis, we have combined the petioles and crown, labeling them "shoots".

offspring ramets of *Fragaria vesca*, when given a choice of six soils of varying quality, preferentially grew into higher quality soils first. Only after these higher quality soils were colonized did the newest formed ramets colonize lower quality patches. These findings were in stark contrast to the homogeneous control, where daughter ramet placement was random. This experiment was particularly interesting because unlike *A. psilocstachya*, *F. vesca* is stoloniferous, demonstrating that clonal plants are capable of precision foraging aboveground. *While the study demonstrated that F. vesca are able to detect and respond to nutrient-rich patches, it did not investigate the morphological changes that occur when the ramet encounters a nutrient-rich patch, and raises the question: are there changes and, if so, are they consistent with optimal nutrient foraging?*

We conducted a series of experiments designed to investigate the morphological changes that occur in developing ramets **prior to rooting** in response to unsterilized (live) versus sterilized (dead) field substrate. Our goal was to determine whether airborne signals are able to alter development such that newly developing ramets can be placed into favorable nutrient patches.

First, we examined local root growth and development of *F. vesca* daughter ramets exposed to live versus dead, nutrient-rich field substrates. Once a ramet roots, it can no longer move, so if true foraging is to occur, there must be air-borne clues signaling soil quality. Therefore, during the experiments, rooting into the substrate was prevented. We hypothesized that **unrooted** ramets exposed to air above live substrates would exhibit an increase in root biomass and root:shoot ratio compared to dead controls.

Because prior studies demonstrated that ramets of *F. vesca* are placed into higher quality soils first (Roiloa and Retuerto, 2006), we also developed two experiments to look at the trajectory of stolon extension when given the choice of nutrient-rich versus nutrient-poor patches. We hypothesized that there will be a positive response in the direction of growth of the extending stolon, such that it grows toward the rich substrate. Both parameters, a positive alteration of stolon direction and an increase in the root biomass and root:shoot ratio of **unrooted** ramets, would be taken as a positive indicators of nutrient foraging.

Materials and Methods

Experimental Species

Fragaria vesca, the woodland strawberry, is an herbaceous perennial native to the northern hemisphere. Growth occurs clonally via production of stolons; distribution of ramets within a colony indicates a guerrilla growth form (Angevine, 1983).

Substrate Collection

In order to determine responsiveness to airborne signals from live substrate, we collected soil and litter from a site colonized by *F. vesca* in Aurora, IN (N 39 05.225 W 084 55.663) in March and September 2012. Prior to collection, all strawberries were removed from the substrate. Leaf and stick litter was harvested and placed in plastic bags. Field soil no more than three inches deep was collected and stored in plastic bags. Any residual root

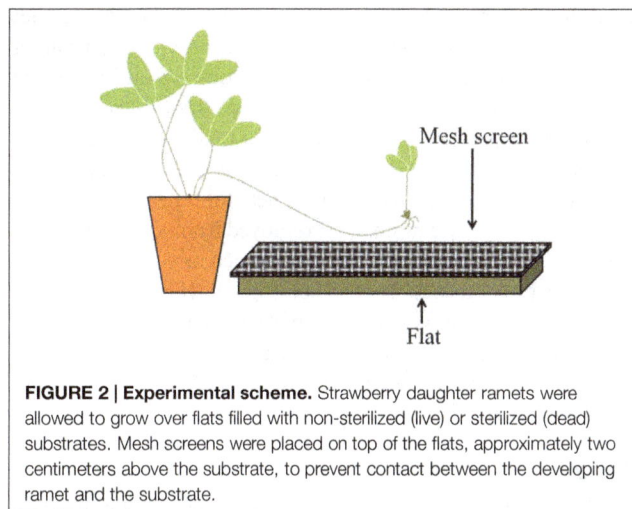

FIGURE 2 | Experimental scheme. Strawberry daughter ramets were allowed to grow over flats filled with non-sterilized (live) or sterilized (dead) substrates. Mesh screens were placed on top of the flats, approximately two centimeters above the substrate, to prevent contact between the developing ramet and the substrate.

mass in the soil was removed prior to storage. Soil and litter were stored in cool, dry, dark conditions until the experiment was initiated. At the beginning of each experiment, half of the collected field soil and litter was autoclaved, the other half was not.

Fragaria Vesca Propagation

Woodland strawberries were propagated from a single clone in the greenhouse at Indiana University, Bloomington, IN, USA in the spring of 2012, the fall of 2012, the fall of 2013 and again in the spring of 2014. These genetically identical ramets were individually potted in 12 cm diam. pots filled with SunGro Metro-Mix and watered daily.

Root Growth Experimental Design and Plant Data Collection

In the spring of 2012, 15 flats (27.3 cm × 54.0 cm × 6.1 cm) each were filled with 1L of field soil and covered with 0.5 L of leaf litter; an additional 15 contained dead (autoclave-sterilized) field substrate (soil plus litter). In order to prevent direct contact between developing ramets and the substrate, a sheet of aluminum screen mesh was placed over each flat; the mesh was situated approximately two centimeters above the substrate (**Figure 2**). Treatments were randomly placed along both sides of a bench in the greenhouse so that half faced east and half faced west.

Thirty potted strawberries with new daughter stolons at least 35 cm in length were randomly assigned to a treatment (live or dead) and were placed at the short edge of each flat, one per flat. The daughter stolon was directed toward and allowed to extend over the mesh-covered substrate. Developing stolons and ramets were not allowed to root into or come into direct contact with the substrate. Both the mother plants and substrate flats were watered daily with tap water and subjected to a 16-h light cycle. The identical experiment was repeated in the fall of 2012 with 80 individual ramets of the same genotype.

Primary stolon length was measured daily. All other newly emerging stolons were clipped from the mother over the duration of the experiment. Dates of initiation of daughter ramet development and root formation were recorded. Initiation of ramet development was identified by the upward curling of the stolon tip, accompanied by leaf production. Root formation was defined by the presence of primordial root hairs extending from the base of the developing daughter ramet. Three days after root formation on the daughter ramet, the entire assemblage (mother and daughter ramet) was removed from the experiment, harvested and separated into organs; for each ramet, the stem and petioles were combined and labeled "shoot" (**Figure 1**). Leaves were scanned into tif files and Image J (U.S. National Institutes of Health, Bethesda, MD, USA) was used to measure leaf area. All plant organs, including leaves were dried at 60°C for 3 days and weighed to mean ± 1 mg.

Statistical Analysis

The statistical analyses were designed to determine if there was an effect of substrate type (live versus dead) on organs of the daughter ramet. Because previous studies indicated that the size of the mother ramet can affect growth responses (e.g., Cain, 1990), we also examined the effect of mother ramet size. Data sets were analyzed for normality based on QQ plots and Kolmogorov–Smirnov test values: for the spring 2012 data, all factors were log transformed to establish normality; for the fall 2012 data only daughter ramet aboveground dry mass, daughter ramet leaf dry mass, daughter ramet root dry mass and stolon length were log transformed, while daughter shoot dry mass, stolon dry mass and total mother dry mass were not. We performed a series of ANCOVAS on the daughter ramet leaf dry mass, daughter ramet shoot dry mass (later combined as aboveground dry mass), daughter ramet root dry mass, stolon dry mass and stolon length. Mother ramet total dry mass (maternal effect) was analyzed as a covariate for all factors. We used partial Eta squared (η^2) to estimate the effect size. All analyses were performed using SPSS (IBM Corp., Released 2011. IBM SPSS Statistics for Windows, Version 20.0. Armonk, NY: IBM Corp.,).

Stolon Trajectory Experimental Design and Plant Data Collection

In the fall of 2013, forty-three flats (27.3 cm × 54.0 cm × 6.1 cm) each were filled with 1 L of live field soil and covered with 0.5 L of leaf litter (live substrate). An additional 43 flats (27.3 cm × 54.0 cm × 6.1 cm) each were filled with autoclaved play sand (Hardscapes by Quikrete). Treatments consisted of a runway made of three glass blocks (6 in. × 8 in. × 4 in., Pittsburgh Corning Premiere). Each runway was flanked by one flat filled with live substrate and one flat filled with sand; distribution of live substrate on the left versus right side was randomized (**Figure 3**). Treatments were randomly placed along both sides of a bench in the greenhouse so that approximately half faced east and half faced west. Forty-three potted strawberries with new daughter stolons at least 35 cm in length were randomly assigned to treatments and placed at the short edge of the runway, one per runway. The daughter stolon was directed toward and allowed to extend along the glass runway. Both mother plants and substrate flats were watered daily with tap water and subjected to a 16-h light cycle. The identical experiment was repeated in the spring

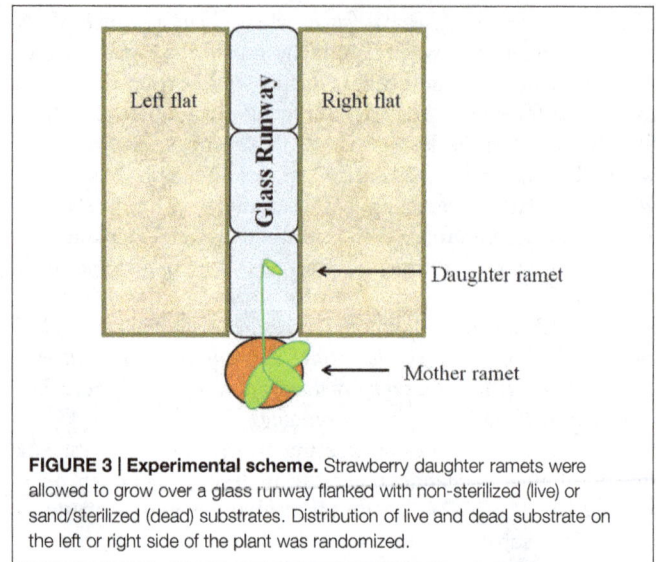

FIGURE 3 | Experimental scheme. Strawberry daughter ramets were allowed to grow over a glass runway flanked with non-sterilized (live) or sand/sterilized (dead) substrates. Distribution of live and dead substrate on the left or right side of the plant was randomized.

of 2014 with the following changes: (1) there were 46 individual strawberries (23 per treatment) and (2) the sand treatment was replaced with sterilized field substrate and litter (dead substrate).

Primary stolon growth was monitored daily. Once a developing stolon extended beyond the edge of the glass blocks, either in the direction of the live or the dead substrate, or off the end of the runway, the date was recorded and the individual was removed from the experiment. All other newly developing stolons were clipped from the mother over the duration of the experiment.

Statistical Analysis

These statistical analyses were designed to determine if there was an effect of substrate type (live versus dead/sand) on the direction of stolon extension. Because there was an equal probability of the strawberry stolon extending into the live substrate, the dead substrate, or growing past the end of the glass runway (no choice), we performed a series of chi-square analyses with the final choice as the categorical variable. All analyses were performed using SPSS (IBM Corp., Released 2011. IBM SPSS Statistics for Windows, Version 20.0. Armonk, NY: IBM Corp.,).

Results

Root Experiment: Spring (Table 1)

Exposure to live substrate affected individual plant organs to different degrees. Most notably, average root dry mass was nearly three times greater on daughter ramets exposed to live versus dead substrates ($p < 0.001$). There was no significant difference in size of leaves, shoots or stolons in plants growing over live versus dead substrate. Because we considered changes in the root:shoot ratio as indicative of nutrient foraging, we analyzed the difference in root:shoot ratio between ramets exposed to the two substrate treatments. We found a significantly higher root:shoot ratio over live (0.02) versus dead substrate (0.01) [$F_{(1,28)} = 42.56$, $p \ll 0.001$].

TABLE 1 | Analysis from the spring experiment.

	Mean		Treatment effect		Maternal effect		
Factor	Live substrate treatment	Dead substrate treatment	$F(1,27)$	P	$F(1,27)$	P	Effect size (η^2)
Leaf dry mass (mg)	139.56 ± 15.32	150.40 ± 12.59	0.55	0.46	12.82	**0.001**	0.32
Shoot dry mass (mg)	49.88 ± 5.27	54.64 ± 4.07	1.03	0.32	22.39	**<0.001**	0.45
Root dry mass (mg)	2.94 ± 0.24	1.06 ± 0.08	68.05	**<0.001**	2.93	0.10	0.10
Stolon dry mass (mg)	218.73 ± 28.44	242.69 ± 32.16	0.17	0.69	28.95	**<0.001**	0.52
Stolon length (mm)	296.47 ± 16.23	302.53 ± 20.47	0.00	0.96	8.05	**0.01**	0.23

Effects of live versus dead substrate (treatment effect) and total dry mass of the mother ramet (maternal effect) on the organs of the daughter ramet. Bold face highlights effects significant at p < 0.05.

Maternal effect differed between substrate treatments. Daughter ramet root biomass over dead substrate was independent of maternal size, whereas over live substrate, there was a strong correlation between the two ($r^2 = 0.561$; $n = 15$; $p = 0.015$; **Figure 4A**) In contrast, aboveground dry mass (leaf + shoot) was significantly correlated with mother ramet size over both live ($r^2 = 0.656$, $n = 15$; $p = 0.004$) and dead ($r^2 = 0.599$; $n = 15$; $p = 0.009$) substrates (**Figure 4B**).

Root Experiment: Fall (Table 2)

Similar to the results from the spring experiment, root dry mass was significantly greater when daughter ramets were exposed to live substrate ($p < 0.001$). Substrate treatment had a significant effect on shoot dry mass ($p = 0.03$). Leaf dry mass was only marginally affected ($p = 0.06$) although daughter ramets produced more over live than dead substrate. Neither stolon dry mass nor stolon length was affected by substrate treatment. These results differ from those obtained in the spring, when only root dry mass was significantly affected by substrate type. As before, root:shoot ratio was significantly higher in daughter ramets exposed to live (0.02) versus dead substrate (0.01) [$F(1,76) = 29.74$, $p \ll 0.001$]. Interestingly, the root:shoot ratios were similar between spring and fall.

Consistent with results from the spring experiment, daughter ramet root biomass over dead substrate was independent of maternal size, while over live soils the two factors were marginally correlated ($r^2 = 0.242$; $n = 38$; $p = 0.072$; **Figure 4C**). Mother ramet size was significantly correlated with aboveground dry mass over dead ($r^2 = 0.179$; $n = 40$; $p = 0.003$) but not live substrate (**Figure 4D**).

Stolon Trajectory Experiment

In order to rule out any developmentally predetermined directional growth of the stolon, we analyzed the frequency of the stolon extending to the left versus the right of the glass blocks and found no statistically significant preference for growth direction ($\chi^2 = 2.78$; df = 1; $p = 0.096$).

When *F. vesca* stolons were given the choice of live substrate or sand, 67.4% grew into the live substrate, 16.3% grew into the sand, and 16.3% extended beyond the glass runway ($\chi^2 = 22.52$; df = 2; $p < 0.001$; **Figure 5**). When the experiment was repeated using live versus dead substrate, 56.5% grew into live substrate, 21.7% grew into dead substrate, and 21.7% extended beyond the glass runway ($\chi^2 = 11.13$; df = 2; $p < 0.05$; **Figure 6**).

Discussion

We hypothesized that in order to grow into nutrient rich patches, the developing daughter ramet must be able to: (1) detect the patch and (2) respond to the presence of the patch. We suggest two types of responses indicative of nutrient foraging: (1) an increase in root biomass and root:shoot ratio prior to rooting and (2) an alteration of stolon growth trajectory in the direction of a nutrient rich patch. As a first step in testing this hypothesis, we asked whether root growth differs when a ramet extends over live versus dead substrate.

We found a consistent effect of substrate (live versus dead) on the daughter ramet root dry mass; daughter ramets produced more root biomass over live substrate (**Tables 1** and **2**). We also saw an increase in root:shoot ratio, an indicator of the relative allocation of biomass. Not only were our results consistent with our hypothesis, but also ratios in the spring and fall for both treatments were nearly identical. Because ramets were not allowed to come into contact with the substrate, this suggests that the consistency in root:shoot ratio is a programmed response of the daughter ramet to the presence of volatiles emitted from the substrate. The increase in root biomass correlated with maternal size over live (but not dead) substrate (**Figure 3**). One possible explanation for this pattern is that exposure to live substrates initiates a cascade of events increasing the distribution of resources from mother to daughter ramet.

In terms of stolon trajectory, we expected to see a higher frequency of growth toward nutrient rich versus nutrient poor patches. In both experiments, our results were consistent with our hypothesis, in that the majority of *F. vesca* individuals extended stolons into the live substrate (**Figures 4** and **5**). Furthermore, in both experiments, the frequency of stolon extension into nutrient-poor flats (sand or dead substrate) occurred equally, suggesting no differential influence from the less ideal patch.

The experiments in this study strongly demonstrate that ramets of *F. vesca* can identify and respond to nutrient-rich substrate patches, however, the mechanisms behind this capacity are less clear. Because the ability of a new ramet to explore an environment ends once rooting occurs, it is fundamentally important for the plant to be able to predict (based on environmental cues) the quality of the surrounding substrate. Thus, we designed our experiments in such a way to highlight morphological responses to nutrient-rich

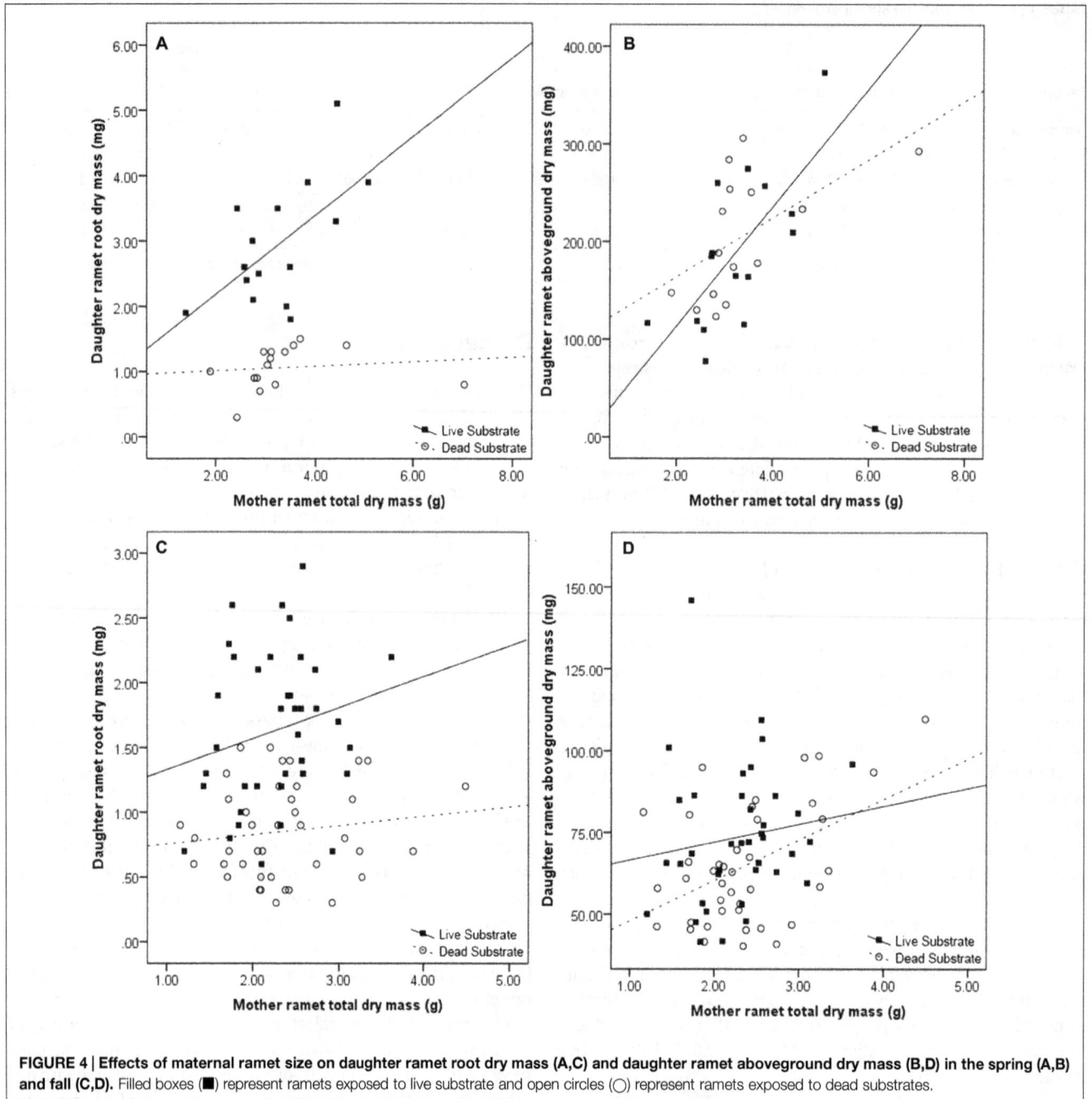

FIGURE 4 | Effects of maternal ramet size on daughter ramet root dry mass (A,C) and daughter ramet aboveground dry mass (B,D) in the spring (A,B) and fall (C,D). Filled boxes (■) represent ramets exposed to live substrate and open circles (○) represent ramets exposed to dead substrates.

TABLE 2 | Analysis from the fall experiment.

	Mean		Treatment effect		Maternal effect		
Factor	Live substrate treatment	Dead substrate treatment	$F(1,75)$	P	$F(1,75)$	P	Effect size (η^2)
Leaf dry mass (mg)	49.79 ± 2.72	43.75 ± 2.12	3.56	0.06	6.22	**0.01**	0.08
Shoot dry mass (mg)	23.77 ± 0.94	21.15 ± 0.91	4.70	**0.03**	5.04	**0.03**	0.06
Root dry mass (mg)	1.63 ± 0.09	0.85 ± 0.06	48.74	**<0.001**	1.96	0.17	0.03
Stolon dry mass (mg)	285.56 ± 10.67	267.75 ± 11.22	2.37	0.13	21.97	**<0.001**	0.28
Stolon length (mm)	469.18 ± 11.87	464.43 ± 12.59	0.19	0.66	1.99	0.16	0.03

Effects of live versus dead substrate (treatment effect) and total dry mass of the mother ramet (maternal effect) on the organs of the daughter ramet. Bold face highlights effects significant at p < 0.05.

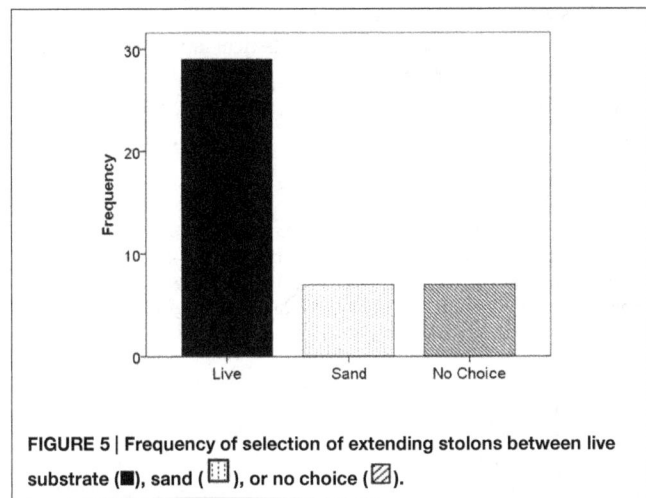

FIGURE 5 | Frequency of selection of extending stolons between live substrate (■), sand (▦), or no choice (▨).

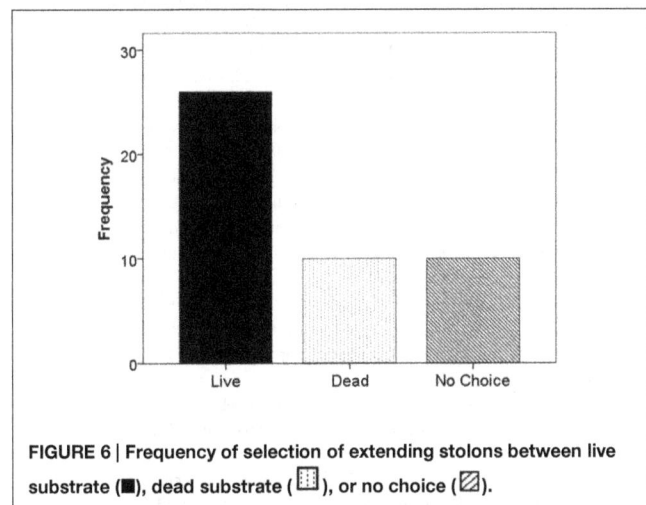

FIGURE 6 | Frequency of selection of extending stolons between live substrate (■), dead substrate (▦), or no choice (▨).

and nutrient-poor substrates independent of soil contact. Our positive results, specifically the increase in root:shoot ratio of developing ramets and the alteration of stolon trajectory, have led us to propose the following mechanism of patch detection.

The soil environment is highly heterogeneous, and the nature of the soil environment is primarily determined by its inhabitants; a nutrient-rich environment also is a substrate environment rich in microflora, microinvertebrates or larger fauna (Chaparro et al., 2012). Belowground volatile emission can influence the community (Wenke et al., 2010; Tumlinson, 2014) by controlling the bacterial and fungal population (Fiddaman and Rossall, 1993; Mackie and Wheatley, 1999; Kai et al., 2007; Vespermann et al., 2007), attracting herbivores (Neveu et al., 2002; Rasmann et al., 2005; Johnson and Gregory, 2006; Ali et al., 2010), and moderating plant growth (Ryu et al., 2003; Splivallo et al., 2007). It is highly likely that the soil inhabitants produce volatile organic compounds (VOCs) that could be detected by a foraging clonal plant. In this scenario, a developing ramet at the terminal end of an extending stolon would have an opportunity to effectively sample the nutrient

environment without the morphological commitment to rooting. This would increase the likelihood of the plant establishing roots in a nutrient-rich/high-quality patch and would explain how a stoloniferous clonal plant might identify and grow into these patches.

Volatile organic compounds are naturally produced chemicals that are critical in influencing ecological interactions both above and belowground (Hughes and Sperandio, 2008; Faure et al., 2009; Kai et al., 2009; Insam and Seewald, 2010; Wenke et al., 2010; Tumlinson, 2014). They are produced by a large variety of organisms, including microbes (Zhang et al., 2007; Kai et al., 2009; Ortiz-Castro et al., 2009), fungi (Splivallo et al., 2007; Tarkka and Piechulla, 2007; Morath et al., 2012; Hung et al., 2015), and plants (Niinemets et al., 2004; Kant et al., 2009; Zhao et al., 2011). Along with mediating communication between different species, they also are byproducts released in response to temperature changes (Asensio et al., 2007; Zhao et al., 2011; Hartikainen et al., 2012), herbivory (Farmer, 2001; Rasmann et al., 2005; Poelman et al., 2013), pathogens (Huang et al., 2012; Panka et al., 2013), and drought (Asensio et al., 2012; Bourtsoukidis et al., 2014; Copolovici et al., 2014).

Volatile organic compounds are often implicated in the promotion of secondary responses, including plant growth. One highly cited example demonstrated that compounds emitted from *Bacilus subtilis* GB03 and *B. amyloliquefaciens* IN937a significantly increased the growth of *Arabidopsis* seedlings as compared to a non-growth promoting strain of *E. coli* and water controls (Ryu et al., 2003). In a similar study, Kai and Piechulla (2009) looked at the effects of *Serratia odorifera* volatiles on growth in *Arabidopsis* in an open system; they concluded that presence of volatiles significantly increased plant growth and a possible role of bacterially emitted CO_2 was suggested. More recently, Minerdi et al. (2011) found that volatiles emitted from *Fusarium oxysporum*, specifically β-Caryophyllene, increased root and shoot length as well as fresh biomass of *Lactuca sativa*. They concluded that this increased growth was the result of the upregulation of seven expansin proteins.

Increase in root growth as a result of exposure to VOCs has been widely discussed in the literature (Zhang et al., 2007; Minerdi et al., 2011; Zamioudis et al., 2013) and is a likely explanation of the results in the current study. VOCs may also explain how *F. vesca* were able to locate nutrient-rich patches in past experiments (Roiloa and Retuerto, 2006). While not directly addressed in the current study, this mechanism of patch-detection might also contribute to the increased root-foraging plasticity in aggressive invaders (Keser et al., 2014), which are perhaps more sensitive or more responsive to volatile clues of nutrient availability. Our on-going studies seek to elucidate mechanisms governing plant foraging by examining the ability of individuals to respond to specific volatile cues emitted from substrates and how these volatiles might elicit specific responses in plant foraging and invasion. We want to determine whether specific growth promoting volatiles are emitted by live *versus* dead substrates and whether they affect stolon trajectory given that establishing a growth trajectory toward a nutrient-rich patch is a necessary precursor to colonization and successful nutrient foraging.

Acknowledgments

Funding was provided by the Indiana University Graduate School. Assistance in greenhouse maintenance was provided by the Indiana University Greenhouse Staff. Beate Henschel and Wesley Beaulieu from the Indiana Statistical Consulting Center provided assistance with statistical analysis. A special thanks to Ross Wilkerson of University Information Technology Services for assistance with data input and graphical editing.

References

Ali, J. G., Alborn, H. T., and Stelinski, L. L. (2010). Subterranean herbivore-induced volatiles released by citrus roots upon feeding by *Diaprepes abbreviatus* recruit endomopathogenic nematodes. *J. Chem. Ecol.* 36, 361–368. doi: 10.1007/s10886-010-9773-7

Alpert, P., and Mooney, H. A. (1986). Resource sharing among ramets in the clonal herb, Fragaria chiloensis. *Oecologia* 70, 227–233. doi: 10.1007/BF00379244

Angevine, M. W. (1983). Variations in the demography of natural populations of the wild strawberries *Fragaria vesca* and *F. virginiana. J. Ecol.* 71, 959–974. doi: 10.2307/2259605

Asensio, D., Filella, I., and Llusia, J. (2007). On-line screening of soil VOCs exchange responses to moisture, temperature and root presence. *Plant Soil* 291, 249–261. doi: 10.1007/s11104-006-9190-4

Asensio, D., Yuste, J. C., Mattana, S., Ribas, A., Llusia, J., and Penuelas, J. (2012). Litter VOCs induce changes in soil microbial biomass C and N and largely increase soil CO_2 efflux. *Plant Soil* 360, 163–174. doi: 10.1007/s11104-012-1220-9

Ballare, C. L. (1999). Keeping up with the neighbours: phytochrome sensing and other signalling mechanisms. *Trends Plant Sci.* 4, 97–102. doi: 10.1016/S1360-1385(99)01383-7

Ballare, C. L., Scopel, A. L., and Sanchez, R. A. (1997). Foraging for light: photosensory ecology and agricultural implications. *Plant Cell Environ.* 20, 820–825. doi: 10.1046/j.1365-3040.1997.d01-112.x

Birch, C. B. D., and Hutchings, M. J. (1994). Exploitation of patchily distributed soil resources by the clonal herb Glechoma hederacea. *J. Ecol.* 82, 653–664. doi: 10.2307/2261272

Bourtsoukidis, E., Kawaletz, H., Radacki, D., Schutz, S., Hakola, H., Hellen, H., et al. (2014). Impact of flooding and drought conditions on the emission of volatile organic compounds of *Quercus robur* and *Prunus serotina. Trees* 28, 193–204. doi: 10.1007/s00468-013-0942-5

Brown, J. S. (1988). Patch use as an indicator of habitat preference, predation risk, and competition. *Behav. Ecol. Sociobiol.* 22, 37–47. doi: 10.1007/BF00395696

Cain, M. L. (1990). Patterns of Solidago altissima ramet growth and mortality: the role of below-ground ramet connections. *Oecologia* 82, 201–209. doi: 10.1007/BF00323536

Cain, M. L. (1994). Consequences of foraging in clonal plant species. *Ecology* 75, 933–944. doi: 10.2307/1939417

Chaparro, J. M., Sheflin, A. M., Manter, D. K., and Vivanco, J. M. (2012). Manipulating the soil microbiome to increase soil health and plant fertility. *Biol. Fertil. Soils* 48, 489–499. doi: 10.1007/s00374-012-0691-4

Charnov, E. L. (1976). Optimal foraging, the marginal value theorem. *Theor. Popul. Biol.* 9, 129–136. doi: 10.1016/0040-5809(76)90040-X

Copolovici, L., Kannaste, A., Remmel, T., and Niinements, U. (2014). Volatile organic compound emissions from *Alnus glutinosa* under interacting drought and herbivory stresses. *Environ. Exp. Bot.* 100, 55–63. doi: 10.1016/j.envexpbot.2013.12.011

Cressman, R., Krivan, V., Brown, J., and Garay, J. (2014). Game theoretic methods for functional response and optimal foraging behavior. *PLoS ONE* 9:e88773. doi: 10.1371/journal.pone.0088773

Dauzat, J., Clouvel, P., Luquet, D., and Martin, P. (2008). Using virtual plants to analyze the light-foraging efficiency of a low-density cotton crop. *Ann. Bot.* 101, 1153–1166. doi: 10.1093/aob/mcm316

de Kroon, H., and Hutchings, M. J. (1995). Morphological plasticity in clonal plants - the foraging concept reconsidered. *Ecology* 83, 143–152. doi: 10.2307/2261158

Dong, M. (1993). Morphological plasticity of the clonal herb *Lamiastrum galeobdolon* (L.) Ehrend and Polatschek in response to partial shading. *New Phytol.* 124, 291–300. doi: 10.1111/j.1469-8137.1993.tb03819.x

Dong, M. (1995). Morphological responses to local light conditions in clonal herbs from contrasting habitats, and their modification due to physiological integration. *Oecologia* 101, 282–288. doi: 10.1007/BF00328813

Dong, M., and Pierdominici, M. G. (1995). Morphology and growth of stolons and rhizomes in 3 clonal grasses, as affected by different light supply. *Vegetatio* 116, 25–32.

Farmer, E. E. (2001). Surface-to-air signals. *Nature* 411, 854–856. doi: 10.1038/35081189

Faure, D., Vereecke, D., and Leveau, J. H. J. (2009). Molecular communication in the rhizosphere. *Plant Soil* 321, 279–303. doi: 10.1007/s11104-008-9839-2

Fiddaman, P. J., and Rossall, S. (1993). The production of antifungal volatiles by *Bacillus subtilis. J. Appl. Microbiol.* 74, 119–126.

Franklin, K. A., and Quail, P. H. (2010). Phytochrome functions in *Arabidopsis* development. *J. Exp. Bot.* 61, 11–24. doi: 10.1093/jxb/erp304

Freidin, E., and Kacelnik, A. (2011). Rational choice, context dependence, and the value of information in European starlings (*Sturnus vulgaris*). *Science* 334, 1000–1002. doi: 10.1126/science.1209626

Grime, J. P., and Mackey, J. M. L. (2002). The role of plasticity in resource capture by plants. *Evolut. Ecol.* 16, 299–307. doi: 10.1023/A:1019640813676

Hartikainen, K., Riikonen, J., Nerg, A.-M., Kivimaenpaa, M., Ahonen, V., Tervahauta, A., et al. (2012). Impact of elevated temperature and ozone on the emission of volatile organic compounds and gas exchange of silver birch (*Betula pendula* Roth). *Environ. Exp. Bot.* 84, 33–43. doi: 10.1016/j.envexpbot.2012.04.014

Higginson, A. D., Fawcett, T. W., Trimmer, P. C., McNamara, J. M., and Houston, A. I. (2012). Generalized optimal risk allocation: foraging and antipredator behavior in a fluctuating environment. *Am. Nat.* 180, 589–603. doi: 10.1086/667885

Houston, A. I., Higginson, A. D., and McNamara, J. M. (2011). Optimal foraging for multiple nutrients in an unpredictable environment. *Ecol. Lett.* 14, 1101–1107. doi: 10.1111/j.1461-0248.2011.01678.x

Huang, M., Sanchez-Moreiras, A. M., Abel, C., Sohrabi, R., Lee, S., Gershenzon, J., et al. (2012). The major volatile organic compound emitted from *Arabidopsis thaliana* flowers, the sesquiterpene (E)-β-caryophyllene, is a defense against a bacterial pathogen. *New Phytologist* 193, 997–1008.

Hughes, D. T., and Sperandio, V. (2008). Inter-kingdom signaling: communication between bacteria and their hosts. *Nat. Rev. Microbiol.* 6, 111–120. doi: 10.1038/nrmicro1836

Hung, R., Lee, S., and Bennett, J. W. (2015). Fungal volatile organic compounds and their role in ecosystems. *Appl. Microbiol. Biotechnol.* 99, 3395–3405. doi: 10.1007/s00253-015-6494-4

Hutchings, M. J. (1999). Clonal plants as cooperative systems: benefits in heterogeneous environments. *Plant Species Biol.* 14, 1–10. doi: 10.1046/j.1442-1984.1999.00001.x

Insam, H., and Seewald, M. S. A. (2010). Volatile organic compounds (VOCs) in soils. *Biol. Fertil. Soils* 46, 199–213. doi: 10.1007/s00374-010-0442-3

Jansen, C., van Kempen, M. M., Bogemann, G. M., Bouma, T. J., and de Kroon, H. (2006). Limited costs of wrong root placement in *Rumex palustris* in heterogeneous soils. *New Phytol.* 171, 117–126. doi: 10.1111/j.1469-8137.2006.01733.x

Johnson, S. N., and Gregory, P. J. (2006). Chemically mediated host-plant location and selection by root-feeding insects. *Physiol. Entomol.* 31, 1–13. doi: 10.1111/j.1365-3032.2005.00487.x

Jónsdóttir, L., and Watson, M. A. (1997). "Extensive physiological integration: an adaptive trait in resource-poor environments? Pages 109-136," in *The Ecology and Evolution of Clonal Plants*, eds H. de Kroon and J. van Groenendael (Leiden: Backbuys Publishers).

Kai, M., Effmert, U., Berg, G., and Piechulla, B. (2007). Volatiles of bacterial antagonists inhibit mycelial growth of the plant pathogen *Rhizoctonia solani. Arch. Microbiol.* 187, 351–360. doi: 10.1007/s00203-006-0199-0

Kai, M., Haustein, M., Molina, F., Petri, A., Scholz, B., and Piechulla, B. (2009). Bacterial volatiles and their action potential. *Appl. Microbiol. Biotechnol.* 81, 1001–1012. doi: 10.1007/s00253-008-1760-3

Kai, M., and Piechulla, B. (2009). Plant growth promotion due to rhizobacterial volatiles–an effect of CO_2? *FEBS Lett.* 583, 3473–3477. doi: 10.1016/j.febslet.2009.09.053

Kant, M. R., Bleeker, P. M., Van Wijk, M., Schuurink, R. C., and Haring, M. A. (2009). Plant volatiles in defence. *Adv. Bot. Res.* 51, 613–666. doi: 10.1016/S0065-2296(09)51014-2

Kelly, C. K. (1990). Plant foraging - a marginal value model and coiling response in *Cuscuta subinclusa*. *Ecology* 71, 1916–1925. doi: 10.2307/1937599

Kemball, W. D., Palmer, M. J., and Marshall, C. (1992). The effect of local shading and darkening on branch growth, development and survival in *Trifolium repens* and *Galium aparine*. *Oikos* 63, 366–375. doi: 10.2307/3544962

Keser, L. H., Dawson, W., Song, Y.-B., Yu, F.-H., Fischer, M., Dong, M., et al. (2014). Invasive clonal plant species have a greater root-foraging plasticity than non-invasive ones. *Oecologia* 174, 1055–1064. doi: 10.1007/s00442-013-2829-y

Lepik, M., Liira, J., and Zobel, K. (2004). The space-use strategy of plants with different growth forms, in a field experiment with manipulated nutrients and light. *Folia Geobot.* 39, 113–127. doi: 10.1007/BF02805241

Leyser, O., and Fitter, A. (1998). Roots are branching out in patches. *Trends Plant Sci.* 3, 203–204. doi: 10.1016/S1360-1385(98)01253-9

MacArthur, R. H., and Pianka, E. R. (1966). On optimal use of a patchy environment. *Am. Nat.* 100, 603–609. doi: 10.1086/282454

Mackie, A., and Wheatley, R. E. (1999). Effects and incidence of volatile organic compound interactions between soil bacterial and fungal isolates. *Soil Biol. Biochem.* 31, 375–385. doi: 10.1016/S0038-0717(98)00140-0

Marcuvitz, S., and Turkington, R. (2000). Differential effects of light quality, provided by different grass neighbors, on the growth and morphology of *Trifolium repens* L. (white clover). *Oecologia* 125, 293–300. doi: 10.1007/s004420000453

McNamara, J. M., Green, R. F., and Olsson, O. (2006). Bayes' theorem and its applications in animal behavior. *Oikos* 112, 243–251. doi: 10.1111/j.0030-1299.2006.14228.x

Methy, M., Alpert, P., and Roy, J. (1990). Effects of light quality and quantity on growth of the clonal plant *Eichhornia crassipes*. *Oecologia* 84, 265–271. doi: 10.1007/BF00318283

Minerdi, D., Bossi, S., Maffei, M. E., Gullino, M. L., and Garibaldi, A. (2011). *Fusarium oxysporum* and its bacterial consortium promote lettuce growth and expansin A5 gene expression through microbial volatile organic compound (MVOC) emission. *FEMS Microbiol. Ecol.* 76, 342–351. doi: 10.1111/j.1574-6941.2011.01051.x

Morath, S. U., Hung, R., and Bennett, J. W. (2012). Fungal volatile organic compounds: a review with emphasis on their biotechnological potential. *Fungal Biol. Rev.* 26, 73–83. doi: 10.1016/j.fbr.2012.07.001

Neveu, N., Grandgirard, J., Nenon, J. P., and Cortesero, A. M. (2002). Systemic release of herbivore-induced plant volatiles by turnips infested by concealed root-feeding larvae *Delia radicum* L. *J. Chem. Ecol.* 28, 1717–1732. doi: 10.1023/A:1020500915728

Niinemets, U., Loreto, F., and Reichstein, M. (2004). Physiological and physicochemical controls on foliar volatile organic compound emissions. *Trends Plant Sci.* 9, 180–186. doi: 10.1016/j.tplants.2004.02.006

Norberg, R. A. (1977). An ecological theory on foraging time and energetics and choice of optimal food-searching method. *J. Anim. Ecol.* 46, 511–529. doi: 10.2307/3827

Oaten, A. (1977). Optimal foraging in patches: a case for stochasticity. *Theor. Popul. Biol.* 12, 263–285. doi: 10.1016/0040-5809(77)90046-6

Oborny, B. (1994). Growth rules in clonal plants and environmental predictability - a simulation study. *J. Ecol.* 82, 341–351. doi: 10.2307/2261302

Ortiz-Castro, R., Contreras-Cornejo, H. A., Macias-Rodriguez, L., and Lopez-Bucio, J. (2009). The role of microbial signals in plant growth and development. *Plant Signal. Behav.* 4, 701–712. doi: 10.4161/psb.4.8.9047

Panka, D., Piesik, D., Jeske, M., and Baturo-Ciesniewska, A. (2013). Production of phenolics and the emission of volatile organic compounds by perennial ryegrass (*Lolium perenne* L.) I *Neotyphodium lolii* association as a response to infection by *Fusarium poae*. *J. Plant Physiol.* 170, 1010–1019. doi: 10.1016/j.jplph.2013.02.009

Perry, G., and Pianka, E. R. (1997). Animal foraging: past, present and future. *Trends Ecol. Evol.* 12, 360–364. doi: 10.1016/S0169-5347(97)01097-5

Poelman, E. H., Harvey, J. A., van Loon, J. J. A., Vet, L. E. M., and Dicke, M. (2013). Variation in herbivore-induced plant volatiles corresponds with spatial heterogeneity in the level of parasitoid competition and parasitoid exposure to hyperparasitism. *Funct. Ecol.* 27, 1107–1116. doi: 10.1111/1365-2435.12114

Pyke, G. H. (1984). Optimal foraging theory - a critical review. *Annu. Rev. Ecol. Syst.* 15, 523–575. doi: 10.1016/j.tree.2011.03.003

Pyke, G. H., Pulliam, H. R., and Charnov, E. L. (1977). Optimal foraging - selective review of theory and tests. *Q. Rev. Biol.* 52, 137–154. doi: 10.1146/annurev.es.15.110184.002515

Rasmann, S., Kollner, T. G., Hiltpold, I., Toepfer, S., Kuhlmann, U., Gershenzon, J., et al. (2005). Recruitment of entomopathogenic nematodes by insect-damaged maize roots. *Nature* 434, 732–737. doi: 10.1038/nature03451

Roiloa, S. R., and Retuerto, R. (2006). Small-scale heterogeneity in soil quality influences photosynthetic efficiency and habitat selection in a clonal plant. *Ann. Bot.* 98, 1043–1052. doi: 10.1093/aob/mcl185

Ryu, C. M., Farag, M. A., Hu, C. H., Reddy, M. S., Wei, H. X., Pare, P. W., et al. (2003). Bacterial volatiles promote growth in *Arabidopsis*. *Proc. Natl. Acad. Sci. U.S.A.* 100, 4927–4932. doi: 10.1073/pnas.0730845100

Salzman, A. G. (1985). Habitat selection in a clonal plant. *Science* 228, 603–604. doi: 10.1126/science.3983647

Schmitt, J., and Wulff, R. D. (1993). Light spectral quality, phytochrome and plant competition. *Trends Ecol. Evol.* 8, 47–51. doi: 10.1016/0169-5347(93)90157-K

Slade, A. J., and Hutchings, M. J. (1987). The effects of light intensity on foraging in the clonal herb *Glechoma hederacea*. *J. Ecol.* 75, 639–650. doi: 10.2307/2260196

Smith, H. (2000). Phytochromes and light signal perception by plants - an emerging synthesis. *Nature* 407, 585–591. doi: 10.1038/35036500

Splivallo, R., Novero, M., Bertea, C. M., Bossi, S., and Bonfante, P. (2007). Truffle volatiles inhibit growth and induce an oxidative burst in *Arabidopsis thaliana*. *New Phytol.* 175, 417–424. doi: 10.1111/j.1469-8137.2007.02141.x

Stoll, P., and Schmid, B. (1998). Plant foraging and dynamic competition between branches of *Pinus sylvestris* in contrasting light environments. *J. Ecol.* 86, 934–945. doi: 10.1046/j.1365-2745.1998.00313.x

Stuefer, J., de Kroon, H., and During, H. (1996). Exploitation of environmental heterogeneity by spatial division of labour in a clonal plant. *Funct. Ecol.* 10, 328–334. doi: 10.2307/2390280

Tarkka, M. T., and Piechulla, B. (2007). Aromatic weapons: truffles attack plants by the production of volatiles. *New Phytol.* 175, 381–383. doi: 10.1111/j.1469-8137.2007.02165.x

Tumlinson, J. H. (2014). The importance of volatile organic compounds in ecosystem functioning. *J. Chem. Ecol.* 40, 212–213. doi: 10.1007/s10886-014-0399-z

Tuomi, J., Hakala, T., and Haukioja, E. (1983). Alternative concepts of reproductive effort, costs of reproduction, and selection in life history evolution. *Am. Zool.* 23, 25–34.

van Kleunen, M., and Fischer, M. (2001). Adaptive evolution of plastic foraging responses in a clonal plant. *Ecology* 82, 3309–3319. doi: 10.2307/2680154

Vespermann, A., Kai, M., and Piechulla, B. (2007). Rhizobacterial volatiles affect the growth of fungi and *Arabidopsis thaliana*. *Appl. Environ. Microbiol.* 73, 5639–5641. doi: 10.1128/AEM.01078-07

Wenke, K., Kai, M., and Piechulla, B. (2010). Belowground volatiles facilitate interactions between plant roots and soil organisms. *Planta* 231, 499–506. doi: 10.1007/s00425-009-1076-2

Wijesinghe, D. K., John, E. A., Beurskens, S., and Hutchings, M. J. (2001). Root system size and precision in nutrient foraging: responses to spatial pattern of nutrient supply in six herbaceous species. *Ecology* 89, 972–983. doi: 10.1111/j.1365-2745.2001.00618.x

Zamioudis, C., Mastranesti, P., Blilou, I., and Pieterse, C. M. J. (2013). Unraveling root developmental programs initiated by beneficial *Pseudomonas* spp. Bacteria. *Plant Physiol.* 162, 304–318. doi: 10.1104/pp.112.212597

Zhang, H., and Forde, B. G. (2000). Regulation of *Arabidopsis* root development by nitrate availability. *J. Exp. Bot.* 51, 51–59. doi: 10.1093/jexbot/51.342.51

Zhang, H., Kim, M. S., Krishnamachari, V., Payton, P., Sun, Y., Grimson, M., et al. (2007). Rhizobacterial volatile emissions regulate auxin homeostasis and cell expansion in *Arabidopsis. Planta* 226, 839–851. doi: 10.1007/s00425-007-0530-2

Zhao, F.-J., Shu, L.-F., Wang, Q.-H., Wang, M.-Y., and Tian, X.-R. (2011). Emissions of volatile organic compounds from heated needles and twigs of *Pinus pumila. J. For. Res.* 22, 243–248. doi: 10.1007/s11676-011-0157-9

Conflict of Interest Statement: The authors declare that the research was conducted in the absence of any commercial or financial relationships that could be construed as a potential conflict of interest.

2

Species identity and neighbor size surpass the impact of tree species diversity on productivity in experimental broad-leaved tree sapling assemblages under dry and moist conditions

Torben Lübbe *, Bernhard Schuldt and Christoph Leuschner

Plant Ecology and Ecosystems Research, Albrecht von Haller Institute for Plant Sciences, University of Göttingen, Göttingen, Germany

Edited by:
Ülo Niinemets,
Estonian University of Life Sciences,
Estonia

Reviewed by:
Thomas Spiegelberger,
Institut National de Recherche en
Sciences et Technologies pour
l'Environnement et l'Agriculture,
France
Bartosz Adamczyk,
Luke, Finland

*Correspondence:
Torben Lübbe
tluebbe@gwdg.de

Species diversity may increase the productivity of tree communities through complementarity (CE) and/or selection effects (SE), but it is not well known how this relationship changes under water limitation. We tested the stress-gradient hypothesis, which predicts that resource use complementarity and facilitation are more important under water-limited conditions. We conducted a growth experiment with saplings of five temperate broad-leaved tree species that were grown in assemblages of variable diversity (1, 3, or 5 species) and species composition under ample and limited water supply to examine effects of species richness and species identity on stand- and tree-level productivity. Special attention was paid to effects of neighbor identity on the growth of target trees in mixture as compared to growth in monoculture. Stand productivity was strongly influenced by species identity while a net biodiversity effect (NE) was significant in the moist treatment (mostly assignable to CE) but of minor importance. The growth performance of some of the species in the mixtures was affected by tree neighborhood characteristics with neighbor size likely being more important than neighbor species identity. Diversity and neighbor identity effects visible in the moist treatment mostly disappeared in the dry treatment, disproving the stress-gradient hypothesis. The mixtures were similarly sensitive to drought-induced growth reduction as the monocultures, which may relate to the decreased CE on growth upon drought in the mixtures.

Keywords: aboveground productivity, belowground productivity, complementarity effect, drought sensitivity, interspecific competition, neighbor effect, selection effect

INTRODUCTION

Recent findings from several biodiversity experiments with planted young trees and observational studies in forests suggest that forest productivity is often enhanced by higher tree diversity (e.g., Zhang et al., 2012; Scherer-Lorenzen, 2014). Contradicting evidence does also exist, however, showing no or even a negative relationship of diversity to forest productivity in diversity experiments (Lang et al., 2012; Grossiord et al., 2013; Li et al., 2014) or in forests (Firn et al., 2007; Szwagrzyk and Gazda, 2007; Jacob et al., 2010; Von Oheimb et al., 2011). Theory predicts that three mechanisms may lead to a positive diversity effect on stand productivity, a selection effect (SE) (the probability of including productive species in the sample increases with increasing species richness), greater complementarity in resource consumption at the stand level, and facilitative interactions that enhance growth (Vandermeer, 1992; Loreau and Hector, 2001). A key process in the diversity–function relationship in forests is competition, which is underlying the selection process in mixed forests, but which is also important for the complementarity effect (CE) as complementary resource use should reduce competition intensity. Increasing diversity should lead to increasingly asymmetric competitive interactions in a stand. Species identity influences stand productivity not only through the traits of the occurring species, but also via neighbor effects on the growth of target trees; the latter effects may be species-specific.

Only few experiments with planted young trees are able to separate between true diversity effects on productivity as caused by resource use complementarity and/or facilitation, and SE, which are driven by the presence of certain species with specific properties (Potvin and Gotelli, 2008; Lang et al., 2012; Grossiord et al., 2013). This is also true for effects of tree neighbor composition on growth. In dependence of their competitive strength, neighbors may decrease or increase the growth of target trees in relation to growth in monoculture. Consequently, these effects should differ between pure stands and mixtures and vary with neighbor species identity (Stoll and Newbery, 2005; Pretzsch and Schütze, 2009; Mölder et al., 2011; Lang et al., 2012). The size and density of neighbors are known as key factors influencing the competitive ability and performance of target plants (e.g., Weiner, 1990). However, their effect has been found difficult to separate from tree identity effects, i.e., neighbor properties other than plant size and density acting on target plants. Several studies showed that neighbor identity effects can be modified or even masked by crowding or tree size effects (Uriarte et al., 2004; Potvin and Dutilleul, 2009; Von Oheimb et al., 2011; Lang et al., 2012).

The interplay between species identity and diversity effects on forest productivity and the relative importance of neighbor effects on tree growth are not well understood. Even less is known about the environmental dependence of these processes on forest ecosystem functioning. The stress-gradient hypothesis applied to forests predicts that resource use complementarity and facilitation are of greater significance in stressful environments (Callaway and Walker, 1997), i.e., in forests exposed to dry, cold, or nutrient-poor conditions, which seems to be supported by empirical studies (e.g., Vilà et al., 2007; Pretzsch et al., 2010; Paquette and Messier, 2011). If positive diversity effects on productivity were indeed larger under stressful conditions, tree species richness could serve to enhance community resistance against environmental hazards. However, it is not well known whether more diverse forests capture resources more rigorously under limiting conditions compared to monocultures (Forrester, 2014). Functional biodiversity research in forests would also benefit from deeper insights into the role of species identity and associated SE on productivity and other ecosystem functions.

Recent comprehensive observational studies along a natural diversity gradient in an old-growth temperate deciduous forest with decreasing abundance of European beech (*Fagus sylvatica* L.) in Hainich National Park (Thuringia, Germany) showed that tree species identity exerted a large influence on various ecosystem functions, while diversity itself seemed to be only of secondary importance (Gebauer et al., 2012; Jacob et al., 2013). Three- and five-species stands were not more productive aboveground than monospecific beech stands (Jacob et al., 2010) but had a higher fine root productivity in ingrowth cores (Meinen et al., 2009). In addition, the stem wood production of beech was higher and its sensitivity to environmental fluctuation lower in more diverse neighborhoods on clay-rich soils, highlighting the role of tree neighborhood effects (Mölder et al., 2011; Mölder and Leuschner, 2014).

Here, we present the results of a tree diversity experiment with potted sapling assemblages, designed to complement the findings obtained from the observational studies in the Hainich mixed forest. The five temperate broad-leaved tree species used in the study (*Fraxinus excelsior* L., *Acer pseudoplatanus* L., *Carpinus betulus* L., *Tilia cordata* L., *F. sylvatica* L.) are also the most abundant species in the Hainich forest; they differ in important morphological and functional traits (Köcher et al., 2009, 2012; Legner et al., 2013). We established three diversity levels (1-, 3-, and 5-species) with all possible monocultures (5) and 3-species combinations (10) and cultivated the plants for 16 months at both ample and water-limited conditions. Study goal was to disentangle the effects of tree diversity and tree species identity on the productivity at the stand level (five trees each) and the tree level under both favorable and resource-limited conditions. Special emphasis was put on neighborhood effects on tree growth and their alteration with increasing diversity.

We tested the hypotheses that (i) stand productivity increases with diversity, but species identity is a more influential factor, (ii) the growth performance of target trees is significantly influenced by the species composition of the neighborhood, (iii) the neighborhood effect is mainly a tree size effect rather than a species identity effect, and (iv) diverse stands reduce their productivity under drought less than monocultures because they reach a higher resource use complementarity.

MATERIALS AND METHODS

Experimental Design

A replicated diversity experiment with 1- to 2-yr-old saplings of the five common Central European broad-leaved tree species [*F. excelsior* (European ash), *A. pseudoplatanus* (sycamore maple), *C. betulus* (European hornbeam), *T. cordata* (small-leaved lime), and *F. sylvatica* (European beech)] was established in April 2011 in the Experimental Botanical Garden of the University of Göttingen (coordinates: 51°33′ N, 9°57′ E, 177 m a.s.l.) and conducted for two vegetation periods until harvest in August 2012 (duration: 15 months, ~450 days). Five saplings were planted together each in a pot of 0.05 m^3 volume (height: 0.30

m, diameter: 0.58 m) filled with coarse-grained sand (98% sand, 1.8% silt, 0.2% clay). The plants were arranged systematically at equal distances to each other to expose them to similar competition intensities. We established three diversity levels (1, 3, or 5 species per pot) and grew all five species in monoculture (all five plants of the same species; five types of monocultures), in 3-species mixture (ten possible combinations with three out of five species) and in 5-species mixture (all plants of different species identity). Thus, 16 different species combinations were investigated. The experiment was further conducted with two different soil moisture treatments (moist, dry), which allowed us to test for diversity and species identity effects on growth under optimal and resource-limited conditions. Due to limitations in plant material and work force, the dry treatment could not be carried out with the full set of species combinations used in the moist treatment. The 10 possible 3-species mixtures were reduced in the dry treatment to five representing each species in three different combinations (**Table 1**). We defined target values of maximal volumetric soil water content (SWC) for the moist (~21%) and the dry treatment (~12%), equivalent to 95%, and 57% of field capacity, respectively. In total, 185 pots with 925 tree individuals were monitored. For details see Lübbe et al. (2015).

The pots were installed under a light-transmitting roof, which excluded all precipitation. The pots were set up at random position in a grid pattern for minimizing the impact of possible environmental gradients.

During July-September 2011 and May-August 2012, mean SWC content varied between 12 and 20% in the moist and 7 and 12% in the dry treatment. Accordingly, lowest soil matrix potentials reached −84 kPa in the moist and −869 kPa in the dry treatment, respectively (Lübbe et al., 2015). Soil moisture content and the amount of required irrigation water were determined gravimetrically. For details on plant care and soil moisture control see Lübbe et al. (2015). Details on climatic conditions are provided in Figure A1.

Measurement of Productivity, Allocation Patterns, and Plant Morphology

The final harvest of all plants took place within a 7-week period in July/August 2012, i.e., up to 16 months after the onset of the experiment. By applying a rotating harvesting scheme, one replicate pot per treatment and species combination was collected every week, thereby avoiding different experimental durations of the treatments. The roots were washed out from the substrate under flowing water. Shoot length (L_{Shoot}) and maximum root length (L_{Root}) were determined and the stem diameter at ground level was measured in two directions perpendicular to each other for calculating basal area (BA). Leaf, stem and root mass were oven-dried (70°C, 72 h) and weighed at a precision of 10 mg. The specific leaf area (SLA) of fully expanded leaves in the upper crown was determined for a subset of trees using WinFolia software (Régent, Quebec, Canada); it served for calculating the total leaf area (LA) of the trees. Besides metrics related to tree size, biomass, and biomass partitioning, we calculated the root-to-shoot ratio (RS) and the relative increment in BA (BAI), shoot length (LI_{Shoot}), and root length (LI_{Root}) for the entire growth period of 450 d by subtracting initial from final size or biomass (initial plant metrics are given in Table A1). Furthermore, relative growth rates (RGRs) were calculated considering aboveground, below-ground and total biomass (RGR, in $g\,g^{-1}\,450\,d^{-1}$). Growth was measured with the aim (i) to compare the productivity of a tree assemblage in a pot among different species combinations, diversity levels, and soil moisture levels, and (ii) to analyze the productivity of the five species on the tree individual level in its dependence on diversity, neighborhood, and soil moisture level. Net biodiversity effects (NE), selection effects (SE), and complementarity effects (CE) on stand productivity were calculated after Loreau and Hector (2001) using Equation (1):

$$NE = CE + SE = N\,\overline{\Delta RY}\,\overline{Y_M} + N\,cov\,(\Delta RY, Y_M) \qquad (1)$$

where N is the number of species in mixture and ΔRY is the difference between observed and expected relative yield (the latter being derived from the species' relative abundance in the mixture upon planting). Y_M is the yield of a species in monoculture. Horizontal bars above terms symbolize average values across the species in mixture. COV is the covariance of the two variables in parentheses. Neighborhood effects on the growth performance of a target species were investigated in the 3-species mixtures, where for every species all six possible neighborhood

TABLE 1 | Design of the experiment with five tree species, three diversity levels (mono, monocultures; mix 3, 3-species mixtures; mix 5, 5-species mixtures) and moist and dry treatments with the number of replicates.

Thee diversity	Species combination	Replication (n)	
		moist	dry
MONO			
	A. pseudoplatanus	7	7
	C. betulus	7	7
	F. sylvatica	7	7
	F. excelsior	7	7
	T. cordata	7	7
MIX-3			
	A.p. – C.b. – F.s.	7	
	A.p. – C.b. – F.e	7	6
	A.p. – C.b. – T.c	7	6
	A.p. – F.s. – F.e.	7	6
	A.p. – F.s. – T.c.	7	
	A.p. – F.e. – T.c.	7	
	C.b. – F.s. – F.c.	7	
	C.b. – F.s. – F.e.	7	6
	C.b. – F.e. – T.c.	7	
	F.s. – F.e. – T.c.	7	6
MIX-5			
	A.c. – C.b. – F.s.		
	–F.e. – T.c.	8	7

In the dry 3-species mixtures, only five of the 10 possible combinations were realized. A.p., Acer pseudoplatanus; C.b., Carpinus betulus; F.e., Fraxinus excelsior; F.s., Fagus sylvatica; T.c., Tilia cordata.

constellations with the four other species were realized in the moist treatment. In the dry treatment, only three of the six possible combinations were available (**Table 1**). We calculated the competitive ability index (CA) after Grace (1995), which compares the growth performance of a target species in mixture with that in monoculture (Equation 2).

$$CA = \frac{(RGR_{mix} - RGR_{mono})}{RGR_{mono}} \qquad (2)$$

We did this for all six neighborhood constellations of a species in the 3-species mixtures (moist treatment; only three in the dry treatment). To disentangle the effect of the each two neighbors in the 3-species mixtures, we also calculated the CA of a target species for all species combinations where one of the four possible neighbors was present.

Statistical Analysis

To avoid pseudo-replication, we used the pots as replicate units in samples consisting of the different individuals of a species. We thus averaged over all individuals of a species in a pot. Statistical analyses were done with R software, version 3.0.0 (R Development Core Team, 2014). We conducted Two-way ANOVAs to test for effects of the factors species composition (Type II SS considering incomplete data) or tree diversity (Type III SS for unbalanced designs) in assumed interaction with soil moisture treatment on parameters characterizing productivity and biomass partitioning at the pot level (*car* package). For the additive partitioning procedure of biodiversity effects, grand means of the NE, SE, and CE were tested against zero by one-sample *t*-tests. We further tested the effect size of species richness (three or five species; *t*-test) or species composition (ANOVA) on the variance of the three diversity effects. At the tree-individual level, Three-way ANOVAs were conducted for analyzing effects of species identity, diversity level, and moisture treatment on various growth-related parameters. The effect of the neighbor constellation in 3-species mixtures on the RGR and CA of the five species was tested individually by One-way ANOVAs in the moist and dry treatment. To test for the influence of certain neighbor species on the RGR_{total} and CA of a target species in 3-species mixtures, we applied generalized linear models (glm), where the presence of heterospecific neighbors was introduced through dummy variables (yes/no). For separating between effects of neighbor identity and crowding on the growth of target species in the mixed pots, we further conducted ANCOVA analyses with the species composition of the neighborhood as predictor variable and several parameters characterizing the size of the neighbors (biomass, LA, shoot length, root length) introduced separately as co-variables. The residuals of all models were tested for violation of the normality (Shapiro-Wilk test) and homoscedasticity assumptions (Levene's test). Multiple comparisons among the means of different species, species combinations or diversity levels were performed with Tukey contrasts (*glht ()*, *multcomp* package). Pairwise comparisons among the two moisture treatments were done with Student's *t*-test, Welch's *t*-test, or the Mann-Whitney U-test, depending on data structure.

RESULTS

Stand Productivity and Biomass Partitioning

While average phytomass production and RGR of the sapling assemblages tended to increase slightly from the monospecific to the 3-species and the 5-species mixtures for most studied parameters (phytomass, LA, BA, RGR_{above}, RGR_{below}, RGR_{total}), the increase was significant only for LA (in the moist treatment), and LI_{Root} (**Table 2**). In contrast, L_{Shoot} and root:shoot ratio (RS) were not affected. These results are consistent with those from Two-way ANOVA, which showed a significant diversity effect only for LA [$F_{(2, 183)} = 3.78, p < 0.05$] but not for the other productivity parameters including RGR_{total} [$F_{(2, 183)} = 1.10, p > 0.10$].

Additive partitioning of biodiversity effects after Loreau and Hector (2001) showed, for the moist treatment only, a significant NE on biomass ($t = 3.87, p < 0.01$), which was mainly due to a positive CE ($t = 3.67, p < 0.01$; **Figure 1**). Across all 11 mixtures, a significant SE on biomass production was not detected. CE, SE and NE were not influenced by species richness (3-species vs. 5-species mixtures), and the species composition of the mixtures influenced only the size of the SE significantly ($F = 3.34, p < 0.01$). Similar patterns for NE and CE were detected for various other growth-related parameters with strongest effects for LA and L_{Root} (Table A3). Significant SE co-occurred in case of below-ground biomass, LA and BA.

All 12 productivity-related parameters except RS were significantly affected by the moisture treatment (**Table 2**). In contrast to the moist treatment, significant NE and CE occurred in the dry treatment only by exception (above-ground biomass and LA, Table A3). The reduction in RGR from the moist to the dry treatment tended to increase with diversity and it was more conspicuous in root growth than in shoot growth (RGR_{below}: 19, 26, and 28% reduction in the monospecific, mix 3 and mix 5 category, respectively).

Species Identity Effects on Stand Productivity and Biomass Partitioning

Comparing the pot-level productivity of the 16 (moist treatment) or 11 species combinations (dry treatment) with Two-way ANOVA revealed highly significant effects of the species combination [$F_{(15, 169)} = 3.75, p < 0.001$] and of the moisture treatment [$F_{(1, 183)} = 38.28, p < 0.001$] on RGR_{total}. The largest productivity differences existed among the five monocultures (3.7–6.0 g g^{-1} 450 d^{-1} in the moist treatment, 2.6–5.3 g g^{-1} 450 d^{-1} in the dry treatment) with highest RGR_{total} in *F. excelsior* and lowest in *A. pseudoplatanus* (difference significant in both treatments; **Figure 2**). Except for one 3-species mixture (*Fagus-Fraxinus-Tilia*: 6.3 g g^{-1} 450 d^{-1} in the moist treatment), the RGR of all 3- and 5-species mixtures remained in the productivity range set by the five monocultures, and transgressive over-yielding was restricted to this single mixture.

Variation in pot-level RGR_{total} among the different mixtures was smaller in the dry than in the moist treatment, and

TABLE 2 | Various parameters characterizing productivity and plant-internal biomass partitioning (pot-level data: 5 plants each) averaged over the three diversity levels in the moist and dry treatments.

Moisture treatment	Diversity level	No. of replicates (n)	Phytomass (g)	RS (g g^{-1})	LA (m^2)	BA (cm^2)
Moist	mono	35	511.90 ± 27.88 a	1.08 ± 0.06 a	1.46 ± 0.07 a	10.32 ± 0.81 a
	mix3	70	547.70 ± 15.37 a	1.12 ± 003 a	1.65 ± 0.04 b	10.99 ± 0.31 a
	mix5	8	554.88 ± 42.04 a	1.11 ± 0.06 a	1.55 ± 0.08 ab	11.81 ± 0.68 a
Dry	mono	35	425.70 ± 22.70 a*	1.11 ± 0.06 a	1.29 ± 0.07 a°	8.60 ± 0.70 a*
	mix3	30	434.59 ± 13.69 a***	1.10 ± 0.04 a	1.41 ± 0.05 a***	8.75 ± 0.38 a***
	mix5	7	445.49 ± 23.88 a	1.02 ± 0.02 a	1.45 ± 0.05 a	9.36 ± 0.29 a**

Moisture treatment	Diversity level	No. of replicates (n)	L$_{Shoot}$ (cm)	L$_{Root}$ (cm)	LI$_{Shoot}$ (%)	LI$_{Root}$ (%)
Moist	mono	35	100.37 ± 4.06 a	70.94 ± 3.27 a	121.51 ± 12.18 a	184.89 ± 8.37 a
	mix3	70	97.87 ± 1.36 a	76.67 ± 167 a	108.46 ± 4.13 a	205.78 ± 5.43 b
	mix5	8	98.45 ± 3.45 a	82.75 ± 3.43 a	107.36 ± 7.28 a	230.15 ± 13.70 b
Dry	mono	35	80.08 ± 2.02 a††	61.66 ± 2.18 a	99.31 ± 9.79 a*	162.98 ± 9.39 a*
	mix3	30	85.44 ± 1.72 a***	65.75 ± 1.48 a***	82.78 ± 5.98 a***	164.64 ± 6.45 a***
	mix5	7	87.98 ± 1.18 a*	69.52 ± 3.87 a*	85.33 ± 2.49 a*	177.36 ± 15.43 a*

Moisture treatment	Diversity level	No. of replicates (n)	BAI (%)	RGR$_{above}$	RGR$_{below}$	RGR$_{total}$
Moist	mono	35	337.15 ± 22.90 a	6.10 ± 0.43 a	3.77 ± 0.24 a	4.68 ± 0.28 a
	mix3	70	329.00 ± 11.76 a	6.38 ± 0.19 a	4.23 ± 0.18 a	5.11 ± 0.17 a
	mix5	8	341.58 ± 25.37 a	6.44 ± 0.52 a	4.33 ± 0.47 a	5.19 ± 0.47 a
Dry	mono	35	258.73 ± 16.61 a**	4.70 ± 0.30 a**	3.06 ± 0.20 a*	3.70 ± 0.21 a**
	mix3	30	240.75 ± 11.02 a***	4.96 ± 0.22 a***	3.14 ± 0.13 a***	3.85 ± 0.15 a***
	mix5	7	250.03 ± 11.03 a**	5.22 ± 0.33 a°	3.11 ± 0.22 a*	3.97 ± 0.27 a*

*For phytomass, leaf area (LA) and basal area (BA), cumulative values for the five plants are given, for root:shoot ratio (RS), shoot length (L$_{shoot}$), root length (L$_{root}$), shoot and root length increment (LI$_{shoot}$, LI$_{root}$, in percent of initial value), basal area increment (BAI, in percent), and RGR, averages over the five plants are presented. Relative growth rates (RGR) are given in g g^{-1} 450 d^{-1}. Different small letters indicate significant differences between diversity levels (p < 0.05); asterisks in the dry treatment indicate significant differences between moisture treatments in a diversity level (°p < 0.10; *p < 0.05; **p < 0.01; ***p < 0.001). Note different no. of replicates in the diversity levels.*

significantly different productivities of the various 3-species combinations appeared only in the moist treatment.

Growth of the Five Species as Dependent on Neighborhood Diversity and Composition

The individual-based RGR$_{total}$ analysis allows comparing the growth performance of the species in defined neighborhood constellations. Three-way ANOVA indicated for all growth-related parameters highly significant effects of species identity [RGR$_{total}$: $F_{(4, 420)}$ = 20.30, p < 0.001] and also of the moisture treatment [RGR$_{total}$: $F_{(1,423)}$ = 21.57, p < 0.001], except for RS. Diversity effects were significant only for L$_{Root}$ [$F_{(2, 422)}$ = 8.17, p < 0.001] and LI$_{Root}$ [$F_{(2, 422)}$ = 8.53, p < 0.001].

When all individuals of a species from all species combinations were pooled in the analysis, productivity (RGR$_{total}$) decreased in the sequence *Fraxinus > Tilia > Carpinus > Fagus > Acer* in the moist and the dry treatment (**Figure 3**: first bars of the species blocs). For the other productivity parameters, the species ranking differed in some cases (Table A2).

When comparing a species' RGR$_{total}$ in monoculture, 3-species mixture, and 5-species mixture (**Figure 3**), RGR$_{total}$ of *T. cordata* was significantly higher in 5-species mixture than

in monoculture (5.05, 6.71, and 8.61 g g^{-1} 450 d^{-1} in 1-, 3-, and 5-species assemblages), which was reflected in the significant increase in LA of *Tilia* plants from 1- to 3-species assemblages (Table A7). A non-significant tendency for higher growth rates with increasing diversity was also observed in *C. betulus* (Table A6), while *F. sylvatica*, *A. pseudoplatanus* and *F. excelsior* (Table A4) showed no productivity trend across the three diversity levels. However, *A. pseudoplatanus* increased both L$_{Root}$ and LI$_{Root}$ in 5-species mixture compared to the monoculture (moist treatment), but decreased RS in 5-species mixture in the dry treatment (Table A5). In contrast, *F. sylvatica* saplings tended to grow better in monoculture than in 3-species mixtures, which was also visible in higher LA, BA, RGR$_{above}$, and a smaller RS (Table A8).

In *T. cordata*, the drought-induced reduction in RGR$_{total}$ increased with diversity (monoculture: −18%, 3-species mixtures: −23%, 5-species mixtures: −50%). For *F. excelsior* (−17, −24, −4%), *A. pseudoplatanus* (−30, −37, −2%), *C. betulus* (−23, −17, −22%), and *F. sylvatica* (−26, −21, −38%), no consistent trends with increasing diversity were visible. In the 5-species mixture, *A. pseudoplatanus* and *F. excelsior* reduced growth only marginally compared to the moist treatment, while *F. sylvatica* and *T. cordata* suffered larger reductions.

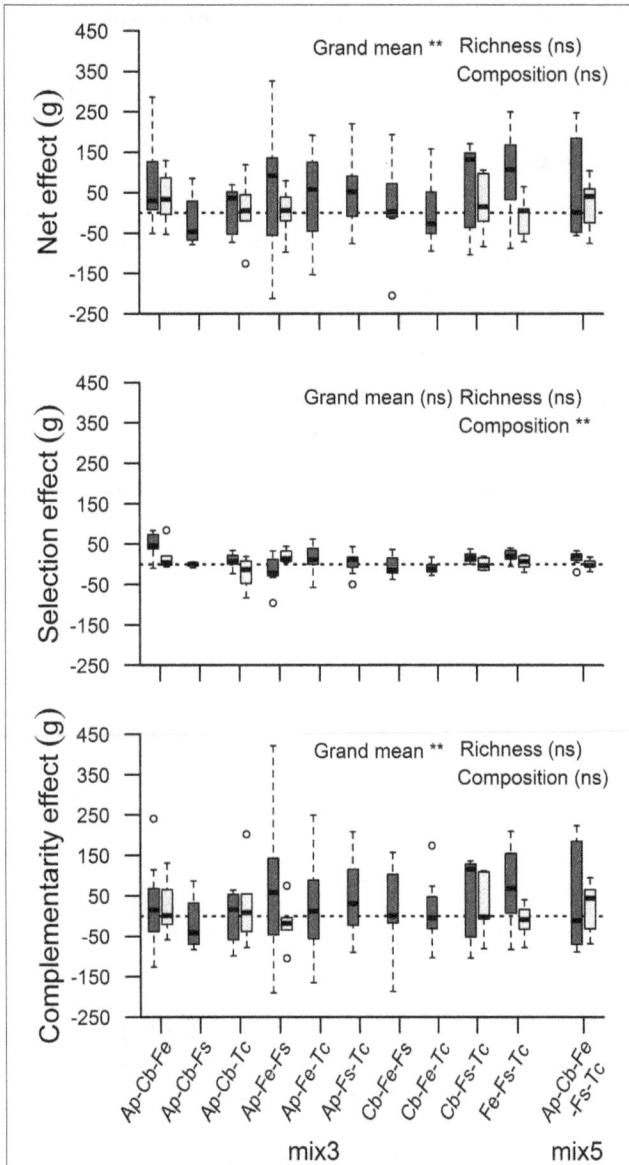

FIGURE 1 | Additive partitioning of biodiversity effects on accumulated biomass of mixed tree assemblages: Net diversity effect, selection effect, and complementarity effect in their dependence on species richness (three vs. five species) and species composition. Asterisks indicate significant effects for the moist treatment (dark boxes; $**p < 0.01$; ns, non-significant). In the dry treatment, no significant effects were detected (bright boxes). Circles above and below boxplots show outlier values.

Importance of Neighbor Species Identity

Analysis of variance indicated significant neighborhood effects on the growth response of target species. The superior growth of *F. excelsior* and *T. cordata* in certain 3-species constellations of the moist treatment is reflected in significantly higher CA of the target species in the respective mixtures [**Figure 4**; ANOVA: $F_{(5, 35)} = 3.72$, $p < 0.01$ for *F. excelsior*; $F_{(5, 34)} = 2.24$, $p < 0.1$ for *T. cordata*]. RGR$_{total}$ of *F. excelsior* was remarkably high in coexistence with *Acer* and *Carpinus* (Figure A2, upper panel) in the moist treatment and the corresponding CA

indices were significantly higher than for mixtures with *Acer—Fagus*, *Carpinus—Fagus,* and also *Carpinus—Tilia* (**Figure 4**). However, the outstanding performance of *F. excelsior* in combination with *Acer—Carpinus* was not observed under dry conditions (**Figure 4**, lower panel). All CA scores for *T. cordata* were positive indicating better growth in mixture than monoculture with highest values for the coexistence with *Fagus—Fraxinus*. In contrast, *F. sylvatica* reached highest growth rates in monoculture resulting in negative CA scores across all heterospecific constellations. Species-specific neighbor effects were less important in the dry treatment. The RGR of *C. betulus* was higher in monoculture than in mixture with *Fagus—Tilia* (Figure A2, lower panel), but the CA scores of the different 3-species constellations did not differ (**Figure 4**).

The explicit analysis of pairwise neighbor interactions on the growth performance of target species in the moist treatment showed *A. pseudoplatanus* and *F. sylvatica* to grow fastest in conspecific neighborhood (negative CA scores; **Figure 5**; GLM, *glht*), while *F. excelsior*, *C. betulus*, and *T. cordata* performed better in mixture. Three of the five species did not show significantly different competitive abilities in response to different neighbor species. Only *T. cordata* achieved a significantly higher CA score in neighborhood to *Fagus* than in vicinity to *Acer* ($p < 0.05$). *F. excelsior* showed highest CA scores in coexistence with *Acer*, which tended to be higher than the scores for *Tilia* or *Fagus* as neighbors ($p < 0.10$).

The Importance of Neighbor Size for Competitive Interactions

Effects of neighbor size on the CA score of the target species were tested by introducing either neighbor biomass, LA or plant size as co-variable in ANCOVA runs (**Table 3**). For *F. excelsior* and *C. betulus* in the moist treatment, the models explaining CA were significantly improved when the neighbor's LA was included as co-variable while the interaction term of biomass × neighborhood species composition was the most important covariate for *T. cordata*. The species identity of the neighbors (factor SpecComp) was, however, only influential for the CA of *F. excelsior* (secondary to LA) and *T. cordata*, where it was the dominant factor. In the two species with negative CA scores in interspecific interaction (*F. sylvatica* and *A. pseudoplatanus*), variation neither in neighbor size nor neighbor species identity influenced CA. In the dry treatment, neighbor size effects on productivity were much smaller (significant effect of LA in *T. cordata*, marginally significant effect in *A. pseudoplatanus*) (**Table 3**).

DISCUSSION

Tree Diversity and Identity Effects on Productivity

We found a significant NE on total (above- and below-ground) biomass production and growth-determining parameters such as LA in the moist treatment in support of our first hypothesis. Additive partitioning of biodiversity effects after Loreau and Hector (2001) showed that the diversity effect was mainly caused

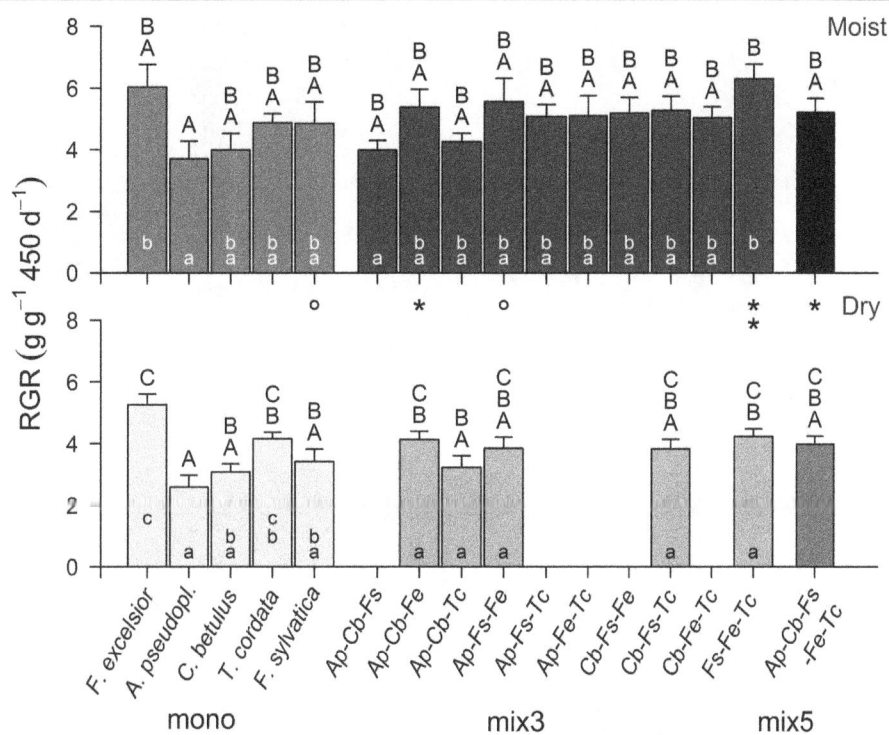

FIGURE 2 | Average relative growth rate (RGR; above- and below-ground) of tree assemblages differing in species composition and diversity in the moist (upper panel) and dry (lower panel) treatment (mean ± SE of 6–8 replicate pots). Different capital letters indicate significant differences ($p < 0.05$) between the species combinations in the full sample (moist: 16, dry: 11 combinations), different small letters indicate significant differences between the species combinations within a diversity level. Asterisks indicate significant differences between the moisture treatments for a species combination ($^{o}p < 0.10$; $^{*}p < 0.05$; $^{**}p < 0.01$). For species abbreviations see **Table 1**.

by a CE and not by a SE; the latter refers to a replacement process in which more productive species achieve dominance in the assemblage. This result meets the assumptions for an experiment with tree saplings because a positive SE could only result from canopy expansion of the more productive species, but not from competition-induced alteration of species abundances in the assemblages, as may take place in communities of more short-lived plants.

The resulting net diversity effect increased RGR_{total} in the mixtures by ~10% compared to the average of the monocultures and thus was relatively small. Moreover, a productivity increase occurred only from the monospecific to the 3-species mixtures but not from the 3- to the 5-species mixture. Thus, a diversity increase from one to three species seems to enhance resource use complementarity, but not a further diversity increase from three to five species. This matches the stand transpiration data from this experiment, which show a comparable net diversity effect on water consumption but no difference in transpiration rate between 3-species and 5-species mixtures (Lübbe et al., 2015). Due to the large contribution of water spending species (*F. excelsior* and *T. cordata*) to stand transpiration in the mixed tree assemblages in the moist treatment, the net diversity effect was interpreted mainly as a SE. The observed LA increase with diversity, which is a main determinant of plant water loss, was assigned to both complementarity and SE (Table A3). In contrast

to earlier studies (e.g., Forrester et al., 2010), water use efficiency of productivity was not different between the diversity levels (Table A9), i.e., the productivity increase was not greater than the transpiration increase with growing species diversity.

The small diversity effect in our experiment might in part be a consequence of the young age of the saplings and the short duration of the experiment. Complementarity in resource use could increase with the development of structurally more complex canopies and root systems, and the manifestation of a substantial SE in tree assemblages might take years or decades. A meta-analysis of plant diversity experiments indeed found that CE on productivity increase over time (Cardinale et al., 2007). However, diversity effects on forest productivity do not seem to be a universal phenomenon (Forrester, 2014). Diversity experiments with planted trees produced mixed results with either positive (e.g., Erskine et al., 2006; Healy et al., 2008) or lacking diversity effects on productivity or biomass (e.g., Nguyen et al., 2012; Grossiord et al., 2013). Further, a sapling experiment with tropical tree species also did not show diversity effects on tree growth (Lang et al., 2012; Li et al., 2014), even though positive interactions were observed.

Various explanations for only small or lacking diversity effects on stand productivity have been proposed including a low potential for growth stimulation under non-limiting conditions, young tree age, and not fully developed tree interactions, low

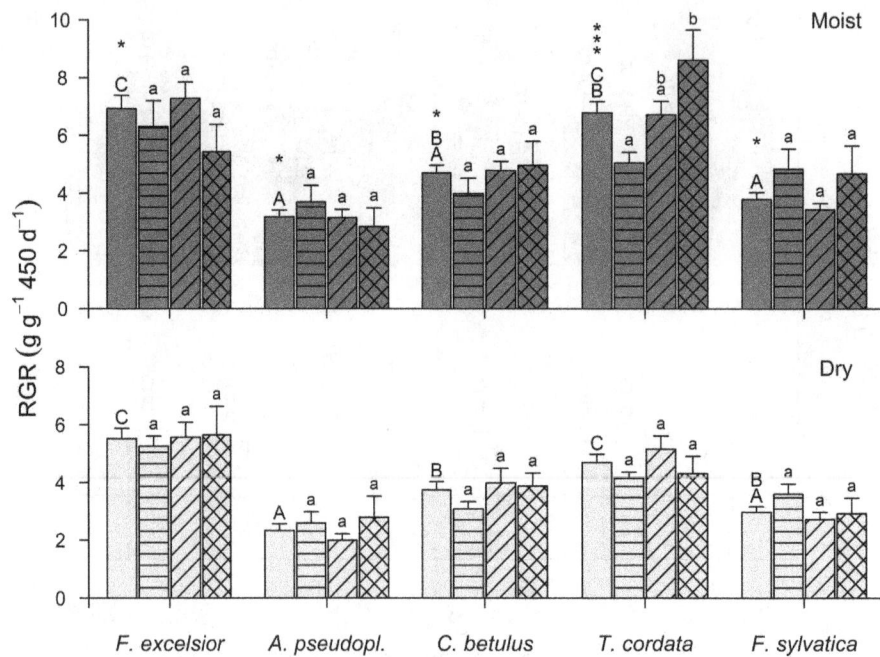

FIGURE 3 | Relative growth rate (above- and below-ground) of the five species in the moist (upper panel) and dry treatment (lower panel) in monoculture (second bar of a group), 3-species mixture (3rd bar), 5-species mixture (4th bar), and as average of all constellations (first bar, no hatching) (means ± SE). Different capital letters indicate significantly different species averages ($p < 0.05$), different small letters significant differences between the three diversity levels within a species. The number of asterisks gives the level of significance for the growth reduction from the moist to the dry treatment of a species (*$p < 0.05$; ***$p < 0.001$).

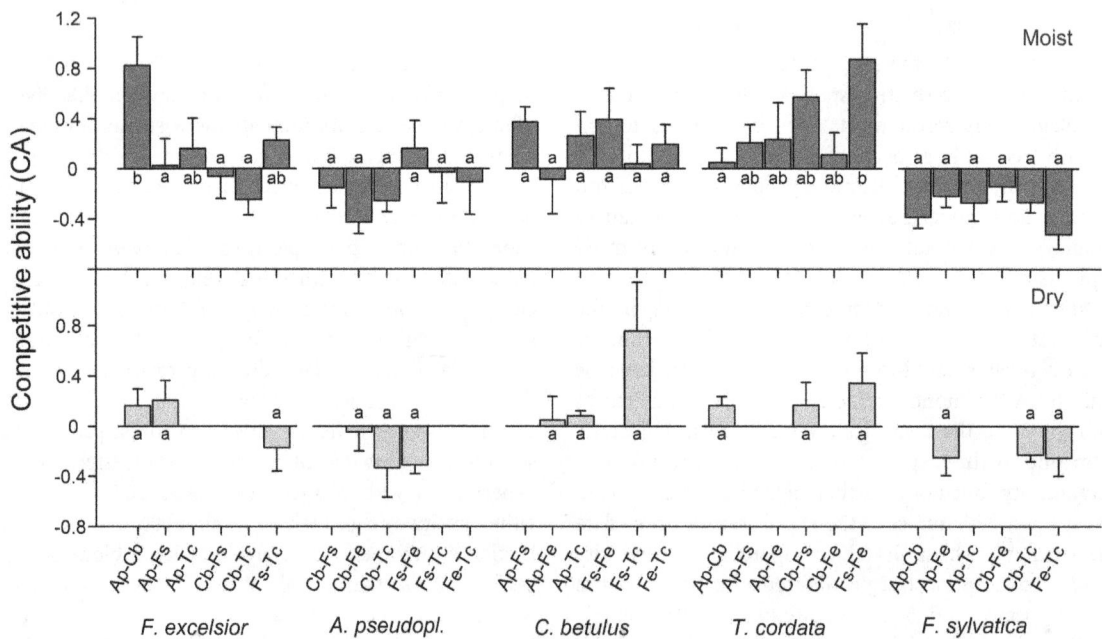

FIGURE 4 | Competitive ability (expressed as CA index) of the five species when grown in six (moist treatment) or three (dry treatment) different 3-species neighborhood constellations (means ± SE of 6–7 replicate pots). For species abbreviations see **Table 1**. Different small letters indicate significant ($p < 0.05$) differences in CA of the target species between different neighborhood constellations. A positive CA indicates better growth in mixture than in monoculture.

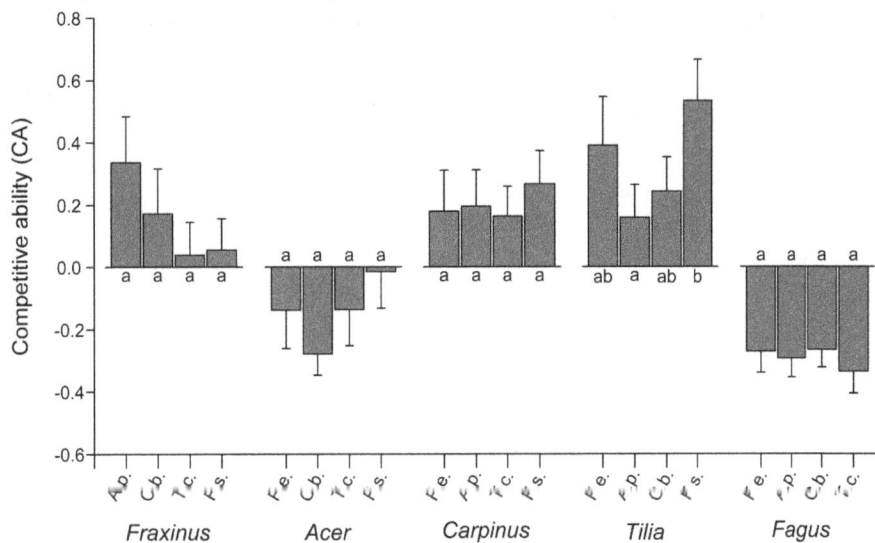

FIGURE 5 | Competitive ability of the five species in the moist treatment when grown in neighborhood of the respective four other species (means ± SE of 21 neighborhood replicates). CA was calculated by pooling the each three 3-species neighborhood constellations in which that neighbor species occurred. Different small letters indicate significant differences ($p < 0.05$) in CA between neighbor constellations of a target species. A.p., *Acer pseudoplatanus*; C.b., *Carpinus betulus*; F.e., *Fraxinus excelsior*; F.s., *Fagus sylvatica*; T.c., *Tilia cordata*.

species numbers, and more or less symmetric competition due to missing functional differentiation among the tree species (Von Oheimb et al., 2011; Lang et al., 2012; Grossiord et al., 2013; Li et al., 2014). Niche differentiation certainly requires the presence of species with sufficient functional dissimilarity as given for instance in case of *F. sylvatica* and *Picea abies* (Pretzsch and Schütze, 2009) or *Eucalyptus globulus* and *Acacia mearnsii* (Forrester et al., 2004), for which complementary resource use and overyielding were observed. Our five broad-leaved species differ in important morphological and physiological traits, but they are functionally more similar than these species pairs, in particular at young age.

The most and the least productive monocultures (*F. excelsior* and *A. pseudoplatanus*) differed nearly by a factor of two in their biomass production in the moist treatment. Similarly large interspecific differences were found for the water consumption of the trees, as the most productive species also transpired most (Lübbe et al., 2015). The majority of other tree diversity experiments also reported a prominent tree identity effect on productivity (e.g., Lang et al., 2012; Grossiord et al., 2013). Our experimental results match observations in the Hainich mixed forest in that species identity was much more influential than diversity. However, the diversity effect on above-ground productivity in the sapling experiment, even though weak, was not detected in the mature stands with 1, 3, or 5 species (Jacob et al., 2010).

Is the Neighbor Identity Effect Mainly a Size Effect?

Loreau and Hector (2001) quantified the SE by the covariance between the monoculture yield of the species and the change in relative yield of the species in the mixtures. Species that profit from the mixture will expand their canopies and root systems at the expense of inferior species and will eventually dominate the mixture by numbers. Our detailed analysis of neighborhood effects on the species' growth in mixture and monoculture allows insights into the mechanisms underlying selection and species identity effects on productivity. Accordingly, the large observed variation in productivity among the different mixture types is only in part caused by the species constellations and species-specific differences in yield; specific neighbor effects (positive or negative) on the productivity of a target species in mixture add to the variation in yield, thus supporting our second hypothesis. This result is in accordance with other studies demonstrating effects of neighborhood composition on tree growth (Massey et al., 2006; Mölder et al., 2011; Von Oheimb et al., 2011; Lang et al., 2012). Most neighborhood interactions in our study were markedly asymmetric as has been found for other tree mixtures as well (Canham et al., 2004, 2006; Potvin and Dutilleul, 2009; Mölder and Leuschner, 2014).

We found considerable differences in the CA of the five species; the species' CA scores depended largely on neighbor identity. While the fast-growing species generally were better competitors in mixture, slower-growing species (*A. pseudoplatanus* and *F. sylvatica*) were inferior competitors. Fast-growing species (in particular *F. excelsior* and *T. cordata*) were more sensitive to the specificity of the neighborhood constellation than the less productive trees.

A neighbor's tree height and biomass are properties likely influencing the growth of a target species, as these attributes typically correlate with the consumption of light, water and nutrients. Our ANCOVA results indicate that neighbor size is a dominant factor, supporting our third hypothesis. In four of the five species (moist or dry treatment), determinants of light

TABLE 3 | ANCOVA results for the five species on the dependence of competitive ability index (CA) on the predictor variable species composition of the neighborhood (six or three constellations in the moist or dry treatments, respectively) and the most influential parameter characterizing neighbor plant size (leaf area LA, basal area BA, or biomass) as covariate.

Species		Moist treatment				Dry treatment			
	Predictor	df	SS	F	p	Predictor	df	SS	F	p
F. excelsior	LA	1	3.18	**17.76**	**<0.001**	LA	1	0.45	2.84	0.116
	SpecComp	5	0.39	**2.18**	**0.081**	SpecComp	2	0.10	0.30	0.743
	Error	32	0.18			Error	13	2.08		
A. pseudopl	LA	1	0.65	2.70	0.111	LA	1	0.30	**3.57**	**0.085**
	SpecComp	5	0.66	0.55	0.736	SpecComp	2	0.30	1.78	0.214
	Error	32	7.68			Error	11	0.92		
C. betulus	LA	1	1.86	**8.14**	**0.007**	LA	1	0.22	0.56	0.469
	SpecComp	5	0.62	0.54	0.744	SpecComp	2	1.43	1.87	0.201
	Error	34	7.76			Error	11	4.20		
T. cordata	Biomass	1	0.19	0.84	0.366	LA	1	0.92	**6.02**	**0.032**
	SpecComp	5	3.70	**3.34**	**0.018**	SpecComp	2	0.00	0.00	0.999
	Bm × SpecComp	5	3.11	**2.81**	**0.036**	Error	11	1.69		
	Error	27	5.79							
F. sylvatica	BA	1	0.17	2.16	0.151	LA	1	0.08	1.09	0.315
	SpecComp	5	0.39	0.97	0.448	SpecComp	2	0.32	2.04	0.170
	Error	33	2.63			Error	13	1.01		

Bold F and p-values highlight significant effects on CA scores (p < 0.10).

interception and canopy space occupation (LA or biomass) were detected as influential variables affecting the neighbor's CA. Due to fixed plant numbers and distances in the pots, differences in plant size are the main determinant of variation in neighbor crowding. The dominant effect of neighbor size on the growth of target trees is in agreement with results obtained in other tree mixing studies (Uriarte et al., 2004; Potvin and Dutilleul, 2009), which found a larger effect of crowding than neighbor species identity. Our results meet the expectation that neighbor effects on the target tree's growth rate are mainly resource depletion effects controlled by the size of the neighbors, while traits unrelated to size (leaf and root physiological properties, direct chemical and mechanical interactions, indirect biotic interactions, etc.) must be of secondary importance.

No Support for the Resource Gradient Hypothesis

From the resource gradient hypothesis, we had expected stronger resource CE in the dry than the moist treatment and a less pronounced growth decline in the mixtures than the monocultures (He et al., 2013; Forrester, 2014). However, we obtained no clear indication that more diverse stands were more resistant against drought-induced productivity reduction, disproving our fourth hypothesis. This finding is in agreement with the results of a quantification of stand water consumption in our experiment revealing a smaller net diversity effect with respect to transpiration in the dry than in the moist treatment (Lübbe et al., 2015). It also concurs with findings on radial

growth in mixed coniferous mountain forests, in which species composition, but not species richness, determined community resistance against drought (DeClerck et al., 2006). In fact, species richness may increase drought exposure in mixed forests when more diverse stands exploit soil water reserves more completely than monospecific stands do (e.g., Grossiord et al., 2014). Beneficial effects of mixed stands with respect to drought resistance have been demonstrated in the form of reduced drought sensitivity of growth in certain tree species (Lebourgeois et al., 2013; Pretzsch et al., 2013; Mölder and Leuschner, 2014). In our study, none of the species showed clear improvement in growth performance in mixture than in monoculture in the dry treatment. The lacking CE with respect to transpiration (Lübbe et al., 2015) and growth in the mixtures of the dry treatment might also be related to the restrictions set by a pot trial, when limited soil volume does not allow for distinct root space partitioning. In the moist treatment of our experiment, in contrast, canopy space partitioning between different species likely has taken place which may have reduced competition for light. This would fit to the prediction of reduced competition for light driving mixture effects in stands with high resource supply (Forrester, 2014), matching findings from other tree diversity experiments (Potvin and Dutilleul, 2009; Lang et al., 2012).

CONCLUSIONS

This sapling study was conducted in conjunction with an observational study in an old-growth mixed forest containing the same species composition. The setting allows some careful extrapolation of the experimental results to real world systems. A CE on productivity existed but it was relatively small and less influential than species identity. Moreover, neighbor effects were found to strongly determine the individual growth performance of tree saplings.

Under drought, the observed CE was smaller than for ample water supply. Contradicting the insurance hypothesis of biodiversity, diverse tree assemblages showed no higher resistance to drought than monocultures. Future biodiversity experiments with trees should search for both positive and negative diversity effects in other water-limited mixed stands and assess the evidence for the proposed insurance function of tree diversity in forests under drought.

ACKNOWLEDGMENTS

We thank the Deutsche Forschungsgemeinschaft (DFG) for funding GRK 1086 ("The role of biodiversity for biogeochemical cycles and biotic interactions in temperate deciduous forests"), as well as Sebastian Fuchs, Eva Akomeah, and the gardener's team of the Experimental Botanical Garden in Göttingen for plant care and support with experimental procedures. We further acknowledge support by the Open Access Publication Funds of the University of Goettingen.

REFERENCES

Callaway, R., and Walker, L. (1997). Competition and facilitation: a synthetic approach to interactions in plant communities. *Ecology* 78, 1958–1965. doi: 10.1890/0012-9658(1997)078[1958:CAFASA]2.0.CO;2

Canham, C., LePage, P. T., and Coates, K. D. (2004). A neighborhood analysis of canopy tree competition: effects of shading versus crowding. *Can. J. For. Res.* 787, 778–787. doi: 10.1139/x03-232

Canham, C. D., Papaik, M. J., Uriarte, M., McWilliams, W. H., Jenkins, J. C., and Twery, M. J. (2006). Neighborhood analyses of canopy tree competition along environmental gradients in New England forests. *Ecol. Appl.* 16, 540–554. doi: 10.1890/1051-0761(2006)016[0540:NAOCTC]2.0.CO;2

Cardinale, B. J., Wright, J. P., Cadotte, M. W., Carroll, I. T., Hector, A., Srivastava, D. S., et al. (2007). Impacts of plant diversity on biomass production increase through time because of species complementarity. *Proc. Natl. Acad. Sci. U.S.A.* 104, 18123–18128. doi: 10.1073/pnas.0709069104

DeClerck, F. A., Barbour, M., and Sawyer, J. (2006). Species richness and stand stability in conifer forests of the Sierra Nevada. *Ecology* 87, 2787–2799. doi: 10.1890/0012-9658(2006)87[2787:SRASSI]2.0.CO;2

Erskine, P. D., Lamb, D., and Bristow, M. (2006). Tree species diversity and ecosystem function: can tropical multi-species plantations generate greater productivity? *For. Ecol. Manage.* 233, 205–210. doi: 10.1016/j.foreco.2006.05.013

Firn, J., Erskine, P. D., and Lamb, D. (2007). Woody species diversity influences productivity and soil nutrient availability in tropical plantations. *Oecologia* 154, 521–533. doi: 10.1007/s00442-007-0850-8

Forrester, D. I. (2014). The spatial and temporal dynamics of species interactions in mixed-species forests: from pattern to process. *For. Ecol. Manage.* 312, 282–292. doi: 10.1016/j.foreco.2013.10.003

Forrester, D. I., Bauhus, J., and Khanna, P. K. (2004). Growth dynamics in a mixed-species plantation of *Eucalyptus globulus* and *Acacia mearnsii. For. Ecol. Manage.* 193, 81–95. doi: 10.1016/j.foreco.2004.01.024

Forrester, D. I., Theiveyanathan, S., Collopy, J. J., and Marcar, N. E. (2010). Enhanced water use efficiency in a mixed *Eucalyptus globulus* and *Acacia mearnsii* plantation. *For. Ecol. Manage.* 259, 1761–1770. doi: 10.1016/j.foreco.2009.07.036

Gebauer, T., Horna, V., and Leuschner, C. (2012). Canopy transpiration of pure and mixed forest stands with variable abundance of European beech. *J. Hydrol.* 442–443, 2–14. doi: 10.1016/j.jhydrol.2012.03.009

Grace, J. B. (1995). On the measurement of plant competition intensity. *Ecology* 76, 305–308. doi: 10.2307/1940651

Grossiord, C., Granier, A., Gessler, A., Jucker, T., and Bonal, D. (2014). Does drought influence the relationship between biodiversity and ecosystem functioning in boreal forests? *Ecosystems* 17, 394–404. doi: 10.1007/s10021-013-9729-1

Grossiord, C., Granier, A., Gessler, A., Pollastrini, M., and Bonal, D. (2013). The influence of tree species mixture on ecosystem-level carbon accumulation and water use in a mixed boreal plantation. *For. Ecol. Manage.* 298, 82–92. doi: 10.1016/j.foreco.2013.03.001

He, Q., Bertness, M. D., and Altieri, A. H. (2013). Global shifts towards positive species interactions with increasing environmental stress. *Ecol. Lett.* 16, 695–706. doi: 10.1111/ele.12080

Healy, C., Gotelli, N. J., and Potvin, C. (2008). Partitioning the effects of biodiversity and environmental heterogeneity for productivity and mortality in a tropical tree plantation. *J. Ecol.* 96, 903–913. doi: 10.1111/j.1365-2745.2008.01419.x

Jacob, A., Hertel, D., and Leuschner, C. (2013). On the significance of belowground overyielding in temperate mixed forests: separating species identity and species diversity effects. *Oikos* 122, 463–473. doi: 10.1111/j.1600-0706.2012.20476.x

Jacob, M., Leuschner, C., and Thomas, F. (2010). Productivity of temperate broad-leaved forest stands differing in tree species diversity. *Ann. For. Sci.* 67, 503–512. doi: 10.1051/forest/2010005

Köcher, P., Gebauer, T., Horna, V., and Leuschner, C. (2009). Leaf water status and stem xylem flux in relation to soil drought in five temperate broad-leaved tree species with contrasting water use strategies. *Ann. For. Sci.* 66, 1. doi: 10.1051/forest/2008076

Köcher, P., Horna, V., Beckmeyer, I., and Leuschner, C. (2012). Hydraulic properties and embolism in small-diameter roots of five temperate broad-leaved tree species with contrasting drought tolerance. *Ann. For. Sci.* 69, 693–703. doi: 10.1007/s13595-012-0189-0

Lang, A. C., Härdtle, W., Baruffol, M., Böhnke, M., Bruelheide, H., Schmid, B., et al. (2012). Mechanisms promoting tree species co-existence: experimental evidence with saplings of subtropical forest ecosystems of China. *J. Veg. Sci.* 23, 837–846. doi: 10.1111/j.1654-1103.2012.01403.x

Lebourgeois, F., Gomez, N., Pinto, P., and Mérian, P. (2013). Mixed stands reduce Abies alba tree-ring sensitivity to summer drought in the Vosges mountains, western Europe. *For. Ecol. Manage.* 303, 61–71. doi: 10.1016/j.foreco.2013.04.003

Legner, N., Fleck, S., and Leuschner, C. (2013). Within-canopy variation in photosynthetic capacity, SLA and foliar N in temperate broad-leaved trees with contrasting shade tolerance. *Trees* 28, 263–280. doi: 10.1007/s00468-013-0947-0

Li, Y., Härdtle, W., Bruelheide, H., Nadrowski, K., Scholten, T., von Wehrden, H., et al. (2014). Site and neighborhood effects on growth of tree saplings in subtropical plantations (China). *For. Ecol. Manage.* 327, 118–127. doi: 10.1016/j.foreco.2014.04.039

Loreau, M., and Hector, A. (2001). Partitioning selection and complementarity in biodiversity experiments. *Nature* 412, 72–76. doi: 10.1038/35083573

Lübbe, T., Schuldt, B., Coners, H., and Leuschner, C. (2015). Species diversity and identity effects on the water consumption of tree sapling assemblages under ample and limited water supply. *Oikos.* doi: 10.1111/oik.02367. [Epub ahead of print].

Massey, F. P., Massey, K., Press, M. C., and Hartley, S. E. (2006). Neighbourhood composition determines growth, architecture and herbivory in tropical rain forest tree seedlings. *J. Ecol.* 94, 646–655. doi: 10.1111/j.1365-2745.2006.01127.x

Meinen, C., Hertel, D., and Leuschner, C. (2009). Root growth and recovery in temperate broad-leaved forest stands differing in tree species diversity. *Ecosystems* 12, 1103–1116. doi: 10.1007/s10021-009-9271-3

Mölder, I., and Leuschner, C. (2014). European beech grows better and is less drought sensitive in mixed than in pure stands: tree neighbourhood effects on radial increment. *Trees* 28, 777–792. doi: 10.1007/s00468-014-0991-4

Mölder, I., Leuschner, C., and Leuschner, H. H. (2011). δ13C signature of tree rings and radial increment of *Fagus sylvatica* trees as dependent on tree neighborhood and climate. *Trees* 25, 215–229. doi: 10.1007/s00468-010-0499-5

Nguyen, H., Herbohn, J., Firn, J., and Lamb, D. (2012). Biodiversity–productivity relationships in small-scale mixed-species plantations using native species in Leyte province, Philippines. *For. Ecol. Manage.* 274, 81–90. doi: 10.1016/j.foreco.2012.02.022

Paquette, A., and Messier, C. (2011). The effect of biodiversity on tree productivity: from temperate to boreal forests. *Glob. Ecol. Biogeogr.* 20, 170–180. doi: 10.1111/j.1466-8238.2010.00592.x

Potvin, C., and Dutilleul, P. (2009). Neighborhood effects and size-asymmetric competition in a tree plantation varying in diversity. *Ecology* 90, 321–327. doi: 10.1890/08-0353.1

Potvin, C., and Gotelli, N. J. (2008). Biodiversity enhances individual performance but does not affect survivorship in tropical trees. *Ecol. Lett.* 11, 217–223. doi: 10.1111/j.1461-0248.2007.01148.x

Pretzsch, H., Block, J., Dieler, J., Dong, P. H., Kohnle, U., Nagel, J., et al. (2010). Comparison between the productivity of pure and mixed stands of Norway spruce and European beech along an ecological gradient. *Ann. For. Sci.* 67, 712–724. doi: 10.1051/forest/2010037

Pretzsch, H., and Schütze, G. (2009). Transgressive overyielding in mixed compared with pure stands of Norway spruce and European beech in Central Europe: evidence on stand level and explanation on individual tree level. *Eur. J. For. Res.* 128, 183–204. doi: 10.1007/s10342-008-0215-9

Pretzsch, H., Schütze, G., and Uhl, E. (2013). Resistance of European tree species to drought stress in mixed versus pure forests: evidence of stress release by inter-specific facilitation. *Plant Biol.* 15, 483–495. doi: 10.1111/j.1438-8677.2012.00670.x

R Development Core Team (2014). *R: A Language and Environment for Statistical Computing.* Vienna: R Foundation for Statistical Computing. Available online at: http://www.R-project.org

Scherer-Lorenzen, M. (2014). "The functional role of biodiversity in the context of global change," in *Forests and Global Change*, eds D. Coomes, D. Burslem, and W. Simonson (Cambridge, UK: Cambridge University Press), 195–237.

Stoll, P., and Newbery, D. (2005). Evidence of species-specific neighborhood effects in the Dipterocarpaceae of a Bornean rain forest. *Ecology* 86, 3048–3062. doi: 10.1890/04-1540

Szwagrzyk, J., and Gazda, A. (2007). Aboveground standing biomass and tree species diversity in natural stands of Central Europe. *J. Veg. Sci.* 18, 555–562. doi: 10.1111/j.1654-1103.2007.tb02569.x

Uriarte, M., Condit, R., Canham, C. D., and Hubbell, S. P. (2004). A spatially explicit model of sapling growth in a tropical forest: does the identity of neighbours matter? *J. Ecol.* 92, 348–360. doi: 10.1111/j.0022-0477.2004.00867.x

Vandermeer, J. (1992). *The Ecology of Intercropping.* Cambridge University Press.

Vilà, M., Vayreda, J., Comas, L., Ibáñez, J. J., Mata, T., and Obón, B. (2007). Species richness and wood production: a positive association in Mediterranean forests. *Ecol. Lett.* 10, 241–250. doi: 10.1111/j.1461-0248.2007.01016.x

Von Oheimb, G., Lang, A. C., Bruelheide, H., Forrester, D. I., Wäsche, I., Yu, M., et al. (2011). Individual-tree radial growth in a subtropical broad-leaved forest: the role of local neighbourhood competition. *For. Ecol. Manage.* 261, 499–507. doi: 10.1016/j.foreco.2010.10.035

Weiner, J. (1990). Asymmetric competition in plant populations. *Trends Ecol. Evol.* 5, 360–364. doi: 10.1016/0169-5347(90)90095-U

Zhang, Y., Chen, H. Y. H., and Reich, P. B. (2012). Forest productivity increases with evenness, species richness and trait variation: a global meta-analysis. *J. Ecol.* 100, 742–749. doi: 10.1111/j.1365-2745.2011.01944.x

Conflict of Interest Statement: The authors declare that the research was conducted in the absence of any commercial or financial relationships that could be construed as a potential conflict of interest.

3

A 6-Year-Long Manipulation with Soil Warming and Canopy Nitrogen Additions does not Affect Xylem Phenology and Cell Production of Mature Black Spruce

Madjelia C. E. Dao[1]*, Sergio Rossi[2], Denis Walsh[2], Hubert Morin[2] and Daniel Houle[3,4]

[1] *Département Productions Forestières, Institut de l'Environnement et de Recherches Agricoles, Ouagadougou, Burkina Faso,* [2] *Département des Sciences Fondamentales, Université du Québec à Chicoutimi, Chicoutimi, QC, Canada,* [3] *Direction de la Recherche Forestière, Forêt Québec, Ministère des Forêts de la Faune et des Parcs, Québec, QC, Canada,* [4] *Ouranos, Consortium Sur la Climatologie Régionale et l'Adaptation aux Changements Climatiques, Montréal, QC, Canada*

Edited by:
Boris Rewald,
University of Natural Resources
and Life Sciences, Vienna, Austria

Reviewed by:
Zhenzhu Xu,
Chinese Academy of Sciences, China
Gerhard Wieser,
Bundesforschungs- und
Ausbildungszentrum für Wald, Austria

*Correspondence:
Madjelia C. E. Dao
dao.ebou@gmail.com

The predicted climate warming and increased atmospheric inorganic nitrogen deposition are expected to have dramatic impacts on plant growth. However, the extent of these effects and their interactions remains unclear for boreal forest trees. The aim of this experiment was to investigate the effects of increased soil temperature and nitrogen (N) depositions on stem intra-annual growth of two mature stands of black spruce [*Picea mariana* (Mill.) BSP] in Québec, QC, Canada. During 2008–2013, the soil around mature trees was warmed up by 4°C with heating cables during the growing season and precipitations containing three times the current inorganic N concentration were added by frequent canopy applications. Xylem phenology and cell production were monitored weekly from April to October. The 6-year-long experiment performed in two sites at different altitude showed no substantial effect of warming and N-depositions on xylem phenological phases of cell enlargement, wall thickening and lignification. Cell production, in terms of number of tracheids along the radius, also did not differ significantly and followed the same patterns in control and treated trees. These findings allowed the hypothesis of a medium-term effect of soil warming and N depositions on the growth of mature black spruce to be rejected.

Keywords: boreal forest, cambial activity, climate change, N deposition, intra-annual growth, increased soil temperature, wood anatomy, xylogenesis

INTRODUCTION

Surface temperature is projected to rise over the 21st century under all assessed emission scenarios (IPCC, 2014). Recent forecasts for the boreal forest of eastern Canada estimate increases of 3°C in mean annual temperature for the year 2050 (Plummer et al., 2006). Plant phenology, the timings of plant development and growth, is one of the traits sensitive to regional climate warming (Schwartz et al., 2006; Cleland et al., 2007). Apart from higher average temperatures, plants may also have to cope with other effects of global change, especially enhanced atmospheric nitrogen (N) deposition. Anthropogenic N depositions have greatly altered the N cycle and plant nutrition in the last two centuries and are projected to increase in the future (Thomas et al., 2010). In the boreal forest, plant growth is often considered to be limited by low temperatures and the availability of N (Reich et al., 2006). Thus, understanding the combined effects of warming and increased N depositions on xylem phenology, tree growth and the amount of cells produced, is critical for improving the prediction of tree responses to future climate.

The key role of soil and air temperatures in xylem phenology and cell production has recently been demonstrated (Rossi et al., 2008; Dufour and Morin, 2013). Soil temperature <6°C strongly inhibits xylem activity and water uptake in various conifers (Alvarez-Uria and Körner, 2007). Also, observations at the northern treeline showed no xylogenesis activity when soil temperature was < 3–5°C (Körner, 2003). Many studies underline the importance of soil temperature in defining the growing season (Körner, 2003; Alvarez-Uria and Körner, 2007). Kilpeläinen et al. (2007) reported that increased temperature resulted in thicker cell walls and higher wood density in Scots pine (*Pinus sylvestris* L.).

Studies on the effect of N deposition on plant growth revealed increased impact of N deposition on plant growth but decreased wood density and cell wall thickness in conifers (Hättenschwiler et al., 1996; Kostiainen et al., 2004).

The combined effect of warming and N fertilization can also be observed in xylem anatomy (Kostiainen et al., 2004; Kilpeläinen et al., 2007). Zhao and Liu (2009), by combining treatments of infrared warming and N deposition in China, obtained further increased performance of *P. tabulaeformis* seedlings but reduced that of *P. asperata*.

Most experimental investigations of N deposition and warming on plant growth have examined juvenile wood or used unrealistically high rates of N addition, so it is not possible to extrapolate the results to the conditions occurring in natural forest ecosystems. Few studies have used an accurate quantity of the additional N inputs that are expected in boreal forest ecosystems in the future, and together with soil warming or singly, the results indicated limited effects on the N status and growth rate after 3-year studies (Lupi et al., 2012; D'Orangeville et al., 2013).

As many factors affect tree growth patterns, short-term studies might be influenced by the confounding effect of several interacting environmental variables on plant growth (Battipaglia et al., 2015). Although these studies indicated no immediate effects after 3 years, we expected that cumulative effects of soil warming and increased N deposition would produce significant modifications in tree growth in the medium term. Few experiments have been done with concomitant variation of soil temperature and N deposition. Thus, the effects in the medium and long term on these environmental factors on trees, especially on cambium phenology and xylem cell production, remain largely unknown. To make realistic predictions of the comprehensive effects of climate change based on experiments a more complete understanding of the relationships between environmental factors (in terms of N deposition and increased temperature) and plant growth requires more years of manipulations.

In this study, we investigated if and how soil temperature and inorganic N deposition influence xylem phenology. We hypothesized that, in the medium term, increased soil temperature and N addition will enhance xylem production and cell differentiation and, in turn, increase the growth rate in boreal forests. To test this hypothesis, during 2008–2013, we used a unique experimental design in the field where inorganic N was repeatedly applied through artificial rain events on the tree canopy and the soil was warmed by 4°C with buried heating cables in two sites of the boreal forest of Québec, QC, Canada.

MATERIALS AND METHODS

Study Site and Tree Selection

The study took place in two mature even-aged stands of black spruce [*Picea mariana* (Mill.) BSP] located at different latitudes and altitudes in the boreal forest of Quebec, Canada. The more northern site Bernatchez (abbreviated as BER) is located near Lac Bernatchez, in the Monts-Valin (48°51′N, 70°20′W, 611 m a.s.l.) while the other Simoncouche (SIM) is in the Laurentides Wildlife Reserve, within the Simoncouche research station (48°13′N, 71°15′W, 350 m a.s.l.). Both regions are included in the balsam fir-white birch ecological domain (Saucier et al., 1998), with an understorey vegetation mainly composed of *Kalmia angustifolia* L., *Ledum groenlandicum* Oeder, *Cornus canadensis* L., *Vaccinium myrtilloides* Michx., and soil vegetation of *Sphagnum* sp. and mosses [*Hylocomium splendens* (Hedw.), *Pleurozium schreberi* (Brid.), *Ptilium crista-castrensis* (Hedw.) De Not.]. The soil in both regions is podzol with a mor-type humus (Rossi et al., 2015). The mean annual temperature is 0.3 and 2.0°C at BER and SIM. From May–September mean annual rainfall is 401.8 and 425.4 mm, at SIM and BER, respectively. SIM derived from a forest fire in 1922, while the forest fire at the origin of the stand in BER has been estimated to have occurred between 1865 and 1870. The stands are growing on gentle slopes (8–17%) and drained glacial tills.

In each site, six co-dominant trees were chosen with upright stem, healthy overall appearance and similar growth patterns. The homogeneity in growth rates was assessed during a preliminary investigation by extracting wood cores and counting the number of tracheids along three previous tree rings (Rossi et al., 2007). The average diameter at breast height and the average height of sampled trees were 17 ± 2 and 21 ± 4 cm, and 15 ± 2 and 14 ± 2 m, at BER and SIM, respectively.

Experimental Design

In each site, two treatments were combined: an increase in soil temperature (H-treatment) and a canopy application of artificial rain enriched with nitrogen (N-treatment). The combination of the treatments resulted in four experimental groups: heated only trees (H), N-enriched only trees (N), heated, and N-enriched trees (NH) and control trees, for which the soil was not heated and that received no N-enrichment (C). The two treatments were attributed randomly to experimental trees resulting in a random split plot design with three replications.

For the H-treatment, heating cables were installed during autumn 2007 between the organic and mineral soil layers, at about 20 cm depth, where the majority of the root system of black spruce is localized (Ruess et al., 2003), following a spiral pattern at a distance of 90–200 cm from the stem collar. Cables were laid by cutting the soil vertically with a shovel or a knife and manually inserting the cable in the resulting narrow "trench", which was then rapidly reclosed. To account for potential root damage and soil disturbance during cable laying, non-heating cables were also installed around non-heated trees (C and N).

Power was supplied by a diesel generator located at 200 m from the site. H treatment consisted of increasing the soil temperature by 4°C during the first part of the growing season. This led to an earlier snowmelt and an increase in annual soil temperature in agreement with the estimates for 2050 by the FORSTEM climatic model developed for the boreal forest of eastern Canada (Houle et al., 2012). Heating started on different dates according to year and site, usually with a 2 weeks delay between SIM and BER to reflect the difference in temperature between the two sites (Lupi et al., 2010) and achieve a 1–2 weeks earlier snowmelt in heated plots. Soil temperature was measured between the coils of the cables around three heated and three control trees.

The temperature differential between control and treated trees was maintained during April–July, the period in which most cambial division takes place (Thibeault-Martel et al., 2008), to reproduce an earlier snow melt and a longer snow free period. Soil temperature was measured every 15 min and data were stored as hourly averages in CR1000 dataloggers (Campbell Scientific Corporation, Canada). Volumetric water content of heated and non-heated plots was measured in July 2009 using a CS-616 probe (Campbell Scientific Corporation, Canada) mounted on a portable device to check for differences in soil moisture content. No significant difference was found between heated (H and NH) and non-heated trees (C and N) (Lupi et al., 2012).

Artificial rain was produced by sprinklers installed above the canopy of each tree. Each week, the equivalent of 2 mm rainfall was applied to the canopy, during the frost-safe period (June to September), for a number of weeks varying between 12 and 14. Rain was applied over a circular 3-m radius area centered on the stem of each experimental tree, which allowed the canopy area to be covered. Non-N-enriched trees (C and H) were irrigated with a water solution reproducing the chemical composition of natural rainfall at the studied sites (Duchesne and Houle, 2006, 2008), while for N-enriched trees (N and NH), a threefold increase in ammonium nitrate (NH4NO3) concentration was used (**Table 1**).

It is expected that frequent artificial rain additions directly to the canopy, with relatively low inorganic N concentration, better imitate the way anthropogenic derived N depositions are reaching boreal forest ecosystems than massive soil applications do.

Meteorological Data

At each site, a standard weather station was installed in a forest gap to measure air, soil temperature and snow depth. The soil temperature was measured both on the mineral and organic layers, at 20–30 and 5–10 cm depth, respectively, and air

TABLE 1 | Ion concentration in the artificial rain.

Ion	Control (μequiv. L^{-1})	N-treatment (μequiv. L^{-1})
Na$^+$	2.24	2.24
Ca^{+2}	5.00	5.00
Mg^{+2}	1.66	1.66
K$^+$	0.76	0.76
H$^+$	16.18	16.18
Cl$^-$	2.24	2.24
SO4^{-2}	23.69	23.69
NH4$^+$	14.93	44.78
NO3$^-$	14.93	44.78

temperature at a height of 2 m. Snow depth was measured with an acoustic distance sensor that quantifies the elapsed time between emission and return of an ultrasonic pulse and automatically corrects for variations of the speed of sound during the year using the measurements of air temperature. Data were collected every 15 min and stored as hourly averages in CR10X dataloggers (Campbell Scientific Corporation, Logan, UT, USA). Daily mean values were later calculated for further analysis.

Sample Collection and Preparation

During 2008–2013, microcores (2.5 mm diameter and 25 mm long) were collected weekly from the stem from April to October with a Trephor (Rossi et al., 2006a) following a counterclockwise-elevating spiral centered at breast height. In SIM and BER, samples were collected weekly in early summer (May–June) and every 2 weeks during July–October. Micro-cores usually contained the previous 5–10 tree rings and the developing annual layer with the cambial zone and adjacent phloem tissues. Wood samples were always taken at 5–10 cm intervals to avoid the formation of resin ducts as a reaction to disturbance. The micro-cores were placed in Eppendorf micro-tubes containing a water:ethanol solution (1:1). Microcores were dehydrated through successive immersions in ethanol and Histosol and embedded in paraffin (Rossi et al., 2006b). Transverse sections 6–10 mm thick were cut with a rotary microtome, stained with cresyl violet acetate (0.16% in water) and observed within 20–30 min under visible and polarized light at magnifications of 400–500× to differentiate the cambium and developing xylem cells. The cambial zone and cells in radial enlargement showed only a primary wall, which, unlike the secondary wall, did not shine under polarized light (Gričar et al., 2006). Cambial cells were characterized by thin cell walls and small radial diameters, while enlarging cells had a radial diameter at least twice that of a cambial cell. Cells in wall thickening shone under polarized light and during the maturation process showed a coloration varying from light to deep violet. As lignification advanced, a blue coloration starting from the cell corners spread into the secondary walls. Since lignin deposition may persist after the end of cell wall thickening (Gindl et al., 2000), cells were considered mature when the violet was completely replaced by blue (Rossi et al., 2006b; Rossi et al., 2014).

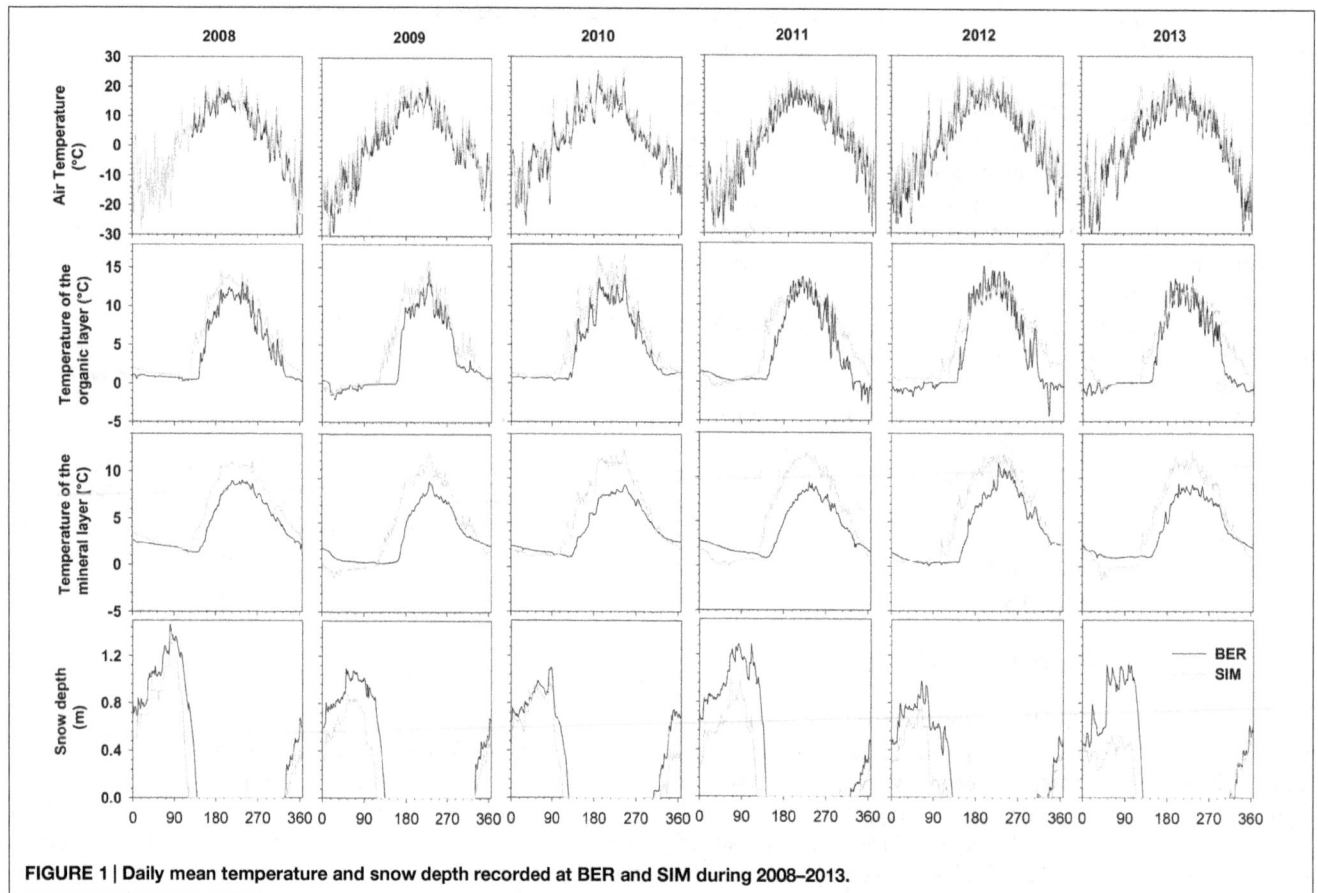

FIGURE 1 | Daily mean temperature and snow depth recorded at BER and SIM during 2008–2013.

The number of cells in each phase was counted along three radial rows and the total number of xylem cells was calculated as the sum of cells in radial enlargement and wall thickening and lignification and mature cells. In spring, xylem formation was considered to have begun when the average number of cells in enlarging phase between the three radial rows was more than one. In late summer, when no further cells were observed undergoing wall thickening and lignification, xylem formation was considered complete.

The phenology of xylem development was assessed for each tree. Four phenophases, computed in days of the year (DOY), were considered, including onset and ending of both cell enlargement and wall thickening and lignification. Duration of xylem formation was calculated as the difference between the onset of cell enlargement and the ending of lignification.

Statistical Analysis

Analysis of variance for repeated measures (ANOVAR) in a split-plot mixed design was used for all variables of phenology and cell production, with site as block factor, treatments N × H as main plots, and Year as repeated factor. As sampling times were correlated, the selection of the covariance structure was based on the lower Akaike's information criterion (AIC). The first-order autoregressive [AR(1)] provided the suitable correlation structure (Wolfinger, 1993). ANOVAR was performed with the MIXED procedure of JMP Pro 11.1.1 (SAS Institute, Cary, NC,

USA). Normality and homoscedasticity were graphically verified on residual plots of the linear models (Quinn and Keough, 2002).

RESULTS

Temperature and Snow Depth

The sites were characterized by long winters with temperatures close to or below zero from October to April. BER, located at the higher latitude and altitude, was the coldest site in winter with the absolute minimum temperatures being measured in 2013 reaching −41.63°C (**Figure 1**). The average temperature for the period ranging between DOY 122–273 (May–September) in BER was two degrees lower than in SIM (11.8°C vs. 13.8°C). Summers were short, with absolute maximum temperatures reaching 26°C in 2010 (**Figure 1**).

During 2008–2013, mean temperatures in the organic and mineral layers varied between 3.4°C and 5.5°C. During winter, soil temperature remained below 3°C and was always lower at BER (**Figure 1**). Summer temperatures in the organic and mineral soil reached 10–15°C, starting to increase only after snowmelt (**Figure 1**). Maximum absolute snow depth varied between 1.0 and 1.5 m, with 2012 being the year with the least snow depth (**Figure 1**). The moment of snowmelt varied between years, but, on average, occurred 15 days later in BER.

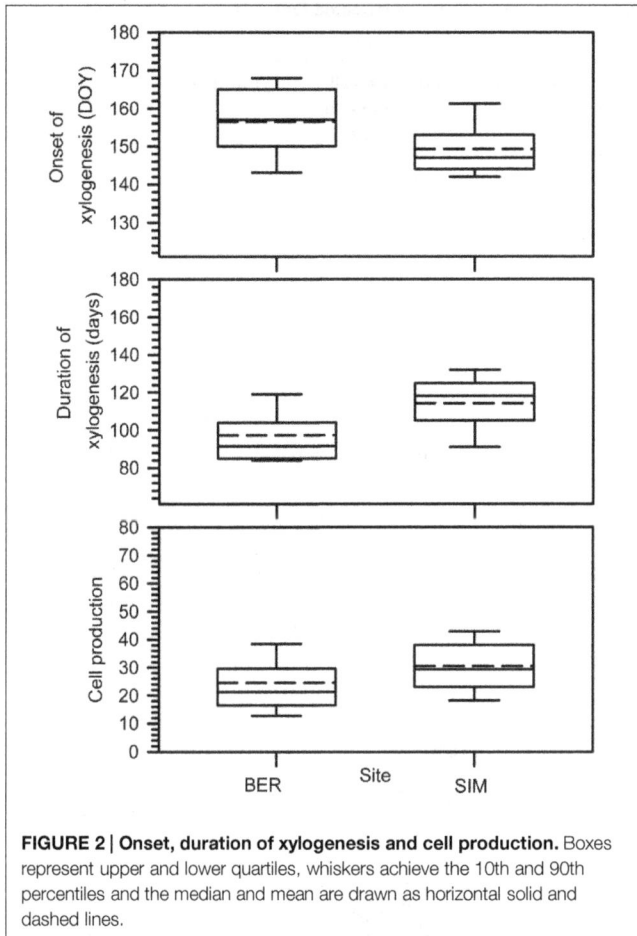

FIGURE 2 | Onset, duration of xylogenesis and cell production. Boxes represent upper and lower quartiles, whiskers achieve the 10th and 90th percentiles and the median and mean are drawn as horizontal solid and dashed lines.

Xylogenesis

The onset of xylogenesis started from late May to mid-June (DOY 146–166) (**Figure 2**). The onset of xylem growth was observed later in BER, the colder site. Xylogenesis lasted 95–120 days on average; SIM clearly had the longest duration of xylogenesis, reaching 121 days. Cell production, corresponding to the number of cells produced along the tree ring, differed between the two sites (**Figure 2**). On average, SIM had the highest cell productions and the longest period of growth, with 39 cells.

Differences between Sites

The ANOVAR test detected a significant difference in onset of cell enlargement ($P < 0.05$) between sites (**Table 2**). The onset occurred between DOY 140 and 170 and between DOY 146 and 160 at BER and SIM, respectively (**Figure 3**). It began 1 week earlier in BER, but the latest onset of cell enlargement and the greatest inter-tree variability during the 6 study years were also observed in BER (**Figure 3**). The dynamics of the onset of cell wall thickening was similar in both sites for all treatments but the model showed a significant effect of site ($P < 0.001$), with trees starting wall thickening 11 days earlier in SIM than in BER (DOY 148 and 159, respectively) (**Figure 3**). The onset of cell maturation was significantly different between sites ($P < 0.01$) (**Table 2**) with SIM being the earliest (**Figure 3**). Even if, on

average, cell enlargement ended 10 days earlier at BER (DOY 198) than at SIM (DOY 208), no statistically significant difference was found between sites ($P > 0.05$) (**Table 2**). The duration of xylogenesis was significant between sites ($P < 0.01$) (**Table 2**), with the whole process being completed in 35 days (between DOY 82 and 117) in BER and 32 days (between DOY 100 and 132) in SIM (**Figure 3**). There was no significant difference for cell production between sites ($P > 0.05$) (**Table 2**). On average, trees at the SIM site produced more cells in the tree ring, with the highest and lowest values of 48 and 20 cells observed in 2008 and 2012, respectively (**Figure 3**).

Differences between Years

Years had highly significant effects on xylem phenology and cell production ($P < 0.001$) and significant effects of the interactions between site and years were also observed for the onset of wall thickening, the mature cell and cell production ($P < 0.05$) (**Table 2**). The dynamics of xylem formation remained similar in each xylem phenological phase. No significant effect on phenological phase was found with the N treatment but the interaction of N × year was significant ($P < 0.01$). An exception in the year 2010 was observed in the onset of cell wall thickening, which occurred markedly earlier, in late May (DOY 148) (**Figure 3**). The onset of cells maturation also appeared earlier in 2010 (**Figure 3**).

The ending of cell enlargement was observed between mid-July and late August (DOY 198–242) in all sites excepted in 2013 when it occurred later at SIM (mid-September, DOY 252) (**Figure 3**). The ending of cell thickening occurred in late September for all treatments in both sites during 2008–2013, except in 2009 in SIM, where the highest variations between treatments were observed (40 days between control and the other treatments) (**Figure 3**).

During 2008–2013, important variations occurred in the number of cells produced, with the highest variability and the greatest number of cells produced in BER and SIM in 2012 and 2013 (**Figure 3**). In 2013, xylogenesis started 18 days earlier in BER on DOY 82, and ended 15 days earlier on DOY 117, thus the duration of growth was markedly longer in BER (**Figure 3**).

Effects of the Treatments

No significant effect of the treatments was found (**Table 2**). The onset of cell wall thickening occurred between mid- (DOY 170) to late- (DOY 181) June for all treatments, with no significant treatment effects (**Table 2**). The treatments also had no significant effect on the moment at which the first mature cells were formed (**Table 2**). The ending of cell enlargement occurred at the same time for all treatments and ANOVAR showed no significant difference between treatments (**Table 2**). It was observed between mid-July and late August (DOY 198–242) in both sites (**Figure 3**). The last cells in wall lignification, which corresponded to the ending of xylem differentiation, occurred between late August and mid-October, with SIM being the later site to complete differentiation (**Figure 3**). ANOVAR performed on the ending of cell wall thickening and lignification showed no significant effect of the treatments (**Table 2**). It was not possible to find significant

TABLE 2 | *F*-values of the mixed procedure with repeated measurements using Site as block factor, Year as repeated factor for the different phases of xylem phenology and cell production.

Source of variation	Onset of cell enlargement	Onset of wall thickening and lignification	Onset of cell mature	Ending of cell enlargement	Ending of wall thickening and lignification	Duration of xylogenesis	Total cell number
Site	7.83*	27.08***	26.11***	2.27	12.06**	11.73**	2.29
N	0.21	0.84	0.89	0.01	1.24	0.16	0.02
Site × N	0.04	0.00	0.41	0.11	1.50	0.63	0.04
T	2.18	1.59	1.49	1.39	0.43	1.30	0.20
Site × T	0.40	0.09	0.00	1.76	0.90	0.04	0.00
N × T	0.27	2.53	3.44	1.78	0.67	0.54	0.59
Site × N × T	0.40	0.47	0.06	2.04	0.07	0.03	0.14
Year	10.93***	57.88***	51.51***	17.37***	14.60***	7.46***	11.36***
Site × Year	2.07	2.74*	2.97*	0.64	1.42	1.88	2.73*
N × Year	0.60	0.24	1.49	0.66	1.60	1.37	4.33**
Site × N × Year	0.73	0.41	0.92	0.40	0.14	0.19	1.63
T × Year	1.26	0.20	0.24	0.73	2.16	1.87	0.47
Site × T × Year	0.82	0.16	0.83	0.33	0.52	0.87	1.38
N × T × Year	0.26	1.47	1.13	2.25	1.03	0.70	2.32
Site × N × T × Year	0.44	0.93	0.36	0.61	0.39	0.63	1.05

The treatments are reported as N (nitrogen), and H (soil warming), and N × H (interaction between N and H).
*p < 0.05; **p < 0.01, and ***p < 0.001.

effects of the treatments on the overall period for completing process of xylogenesis (**Table 2**).

DISCUSSION

This study conducted in two matured black spruce stands of the boreal forest of Quebec, Canada tested the hypothesis that xylem phenology and cell production were affected by increased soil temperature and inorganic N availability in precipitation. Soil temperature was increased by 4°C during the first part of the growing season and precipitations containing three times the current inorganic N concentration in ambient precipitation were repeatedly applied during the growing season from 2008 to 2013. The experiment consisted of frequent canopy applications of inorganic N at realistic concentrations with the aim of simulating future rain composition. After a 6-year experiment, our results showed no substantial change between treatments in xylem phenology and cell production, so our hypothesis had to be rejected. However, a potential effect of soil warming and increased N deposition could still occur under a longer period of experimentation.

In previous studies, localized warming of the stem often increased cell productions but only in the zone of treatment application (Gričar et al., 2006, 2007). Soil warming significantly enhanced diameter growth of woody individuals, especially shrubs (Farnsworth et al., 1995). McWhirter (2013) investigated the responses of *Malus coronaria* (Crap apple) seed germination and seedling growth to warming and nitrogen in old temperate forests and in greenhouse and the results suggested direct effects on germination and establishment of seedlings. In northern China, the warming response of plant phenology (including flowering and fruiting date as well as reproductive duration)

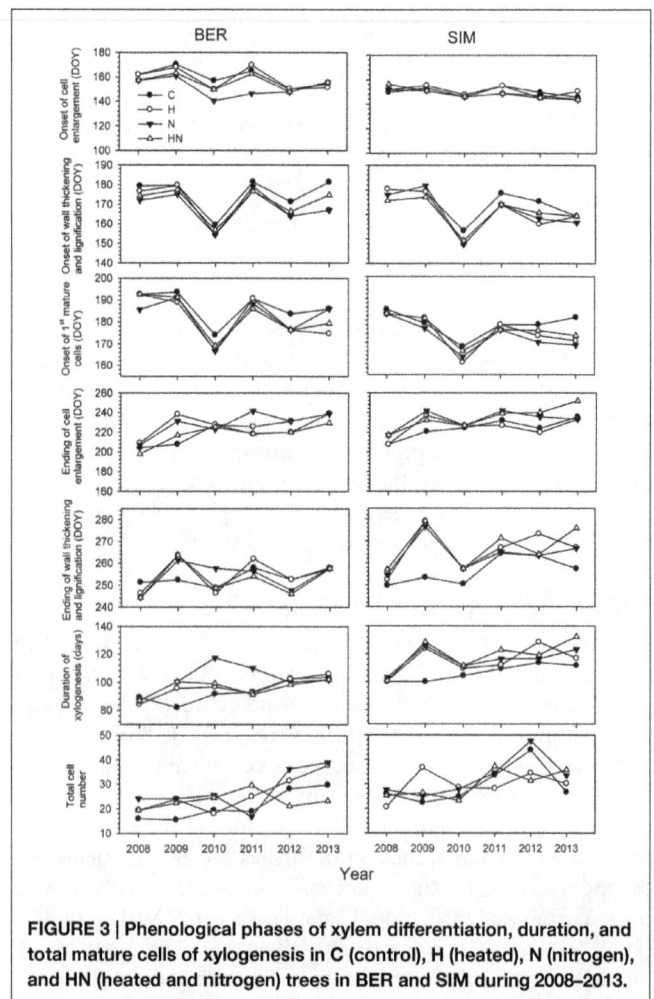

FIGURE 3 | Phenological phases of xylem differentiation, duration, and total mature cells of xylogenesis in C (control), H (heated), N (nitrogen), and HN (heated and nitrogen) trees in BER and SIM during 2008–2013.

is larger in earlier than later flowering species in temperate grassland systems, but no interactive effect between warming and N addition was found on any phenological event (Xia and Wan, 2013).

In contrast to our results, Lupi et al. (2012) by combining early season soil warming with canopy applications of water containing N at a concentration three times higher than the ambient precipitation in the same sites, detected in the short term, that soil warming resulted in earlier onset and extended duration of xylogenesis in the root and along the stem. The results of Lupi et al. (2012) were not confirmed on our longer period of observations. Melillo et al. (2011) reported carbon gains in the woody tissues of trees in a 7-year soil warming study in a mixed hardwood forest ecosystem. (Overdieck et al., 2007) also recorded increased stem diameter, stem height and stem mass for beech seedlings grown for 2.5 years at increased air temperature.

The absence of growth stimulation in our study, which used realistic increased concentrations of inorganic N concentration in precipitation, suggests that growth stimulation (due to increasing N availability) is not to be expected in the future for the boreal forest of eastern Canada. However, given the experimental conditions used and the gaps in our understanding of N foliar uptake, it would be premature to definitely conclude an absence of effects. For instance, in a labeling experiment with ^{15}N, less than 5% of the label was recovered in live foliage and wood after 2 years of N addition to the canopy with a helicopter (Dail et al., 2009). The majority of the label was recovered in twigs and branch materials. Thus, most of the N was retained on plant surfaces, branches and main-stem bark, with little being assimilated into foliage that could then be transported to new forming cells in the stem. In good agreement with the latter study, another recent study involving canopy application of inorganic ^{15}N in a coniferous stand (Gélinas-Pouliot, 2013) has shown that small twigs, not needles, was the main sink for the added N.

It is, however, possible that the N scavenged by the twigs may take a certain time to reach the stem where cell divisions occur and that a N effect could be observed with a longer period of experimentation. It has also been suggested that changes in nutrient cycling due to increased N deposition and its potential effect on tree growth, may become significant only in the medium and long term, since trees seem less receptive than other plants and microorganisms to the uptake of inorganic and organic N in the short term (Näsholm et al., 2009). Thus, although a direct effect of earlier snowmelt and higher soil temperature on tree growth does not appear likely based on our results, an effect of N addition could potentially appear with longer term addition.

Our results revealed the impact of time scale on xylem phenology and cell production. As shown in **Figure 3** and **Table 1**, the onset of cell wall thickening and the first mature cell had already started earlier in 2010 at the end of May when snowmelt had just finished and air temperature reached values above zero. This led to the conclusion that the onset of wood formation can be affected by snowmelt and temperatures (Rossi et al., 2011a; Dufour and Morin, 2013). Based on studies of conifers across a wide range of different geographical locations, Rossi et al. (2008) found that air temperature is also a critical factor limiting the differentiation of xylem cells.

During the 6 years of treatments, the maximum amount of 48 cells produced occurred in 2012, the least snow depth and warmest year. Several authors had found that the onset and ending increments are affected by an air temperature threshold (Deslauriers et al., 2008; Dufour and Morin, 2010) as well as snowmelt and soil temperature (Rossi et al., 2011b; Lupi et al., 2012). Kalliokoski et al. (2012) indicated that weather variation induced differences of up to 28 days in the onset of tracheid formation between years for Norway spruce. Mäkinen et al. (2003) also revealed that interannual variations in increment onset can be important. The investigations on montane Mediterranean tree species (*Cedrus libani)* at different altitudes reported differences in onset, duration and end of cambial activity and xylogenesis as well as growth rates with respect to temperature, especially daily means of air and stem temperature (Guney et al., 2015).

In the medium term, despite some observed significant effects, the results showed similar dynamics between sites for xylem phenology for all treatments. The differences often observed between sites may therefore rather indicate that xylem phenology and cell production are controlled by temperature and snowmelt. Thus the differences in timing can be explained by the lower average air temperature at the BER site.

CONCLUSION

The two boreal forest sites studied showed significant differences for xylem phenology of black spruce, except for the ending of cell enlargement and cell production, which could be attributed to their difference in average annual air temperature.

It was, however, found that a 6-year experiment of soil warming and increased inorganic N additions applied directly to the canopy failed to induce significant differences in xylem phenology and cell production at both sites, which allowed our hypothesis to be rejected. Different results could be expected with a longer-term experiment. For instance, it is possible that the N added to the canopy could be slowly translocated from the twigs to the dividing cells and that could lead to an increase in growth in the longer term.

FUNDING

This work was supported by the Natural Sciences and Engineering Research Council of Canada, the Ouranos consortium and the "Ministère des Forêts de la Faune et des Parcs du Québec". MD received additional financial support through a scholarship from the Program Canadien des Bourses de la Francophonie (PCBF).

ACKNOWLEDGMENTS

The authors want to acknowledge all persons who contributed to data collection, installations in the field and laboratory assistance, especially F. Gionest, D. Laprise, M. Perrin, L. Balducci, H. A. Bouzidi, M. Montoro, Jannie Trambley.

REFERENCES

Alvarez-Uria, P., and Körner, C. (2007). Low temperature limits of root growth in deciduous and evergreen temperate tree species. *Funct. Ecol.* 21, 211–218. doi: 10.1111/j.1365-2435.2007.01231.x

Battipaglia, G., Zalloni, E., Castaldi, S., Marzaioli, F., Cazzolla-Gatti, R., Lasserre, B., et al. (2015). Long tree-ring chronologies provide evidence of recent tree growth decrease in a central African tropical forest. *PLoS ONE* 10:e0120962. doi: 10.1371/journal.pone.0120962

Cleland, E. E., Chuine, I., Menzel, A., Mooney, H. A., and Schwartz, M. D. (2007). Shifting plant phenology in response to global change. *Trends Ecol. Evol.* 22, 357–365. doi: 10.1016/j.tree.2007.04.003

Dail, D. B., Hollinger, D. Y., Davidson, E. A., Fernandez, I., Sievering, H. C., Scott, N. A., et al. (2009). Distribution of nitrogen-15 tracers applied to the canopy of a mature spruce-hemlock stand, Howland, Maine, USA. *Oecologia* 160, 589–599. doi: 10.1007/s00442-009-1325-x

Deslauriers, A., Rossi, S., Anfodillo, T., and Saracino, A. (2008). Cambial phenology, wood formation and temperature thresholds in two contrasting years at high altitude in southern Italy. *Tree Physiol.* 28, 863–871. doi: 10.1093/treephys/28.6.863

D'Orangeville, L., D. Houle, B., Côté, Duchesne, L., and Morin, H. (2013). Increased soil temperature and atmospheric N deposition have no effect on the N status and growth of a mature balsam fir forest. *Biogeosciences* 10, 4627–4639. doi: 10.5194/bg-10-4627-2013

Duchesne, L., and Houle, D. (2006). Base cation cycling in a pristine watershed of the canadian boreal forest. *Biogeochemistry* 78, 195–216. doi: 10.1007/s10533-005-4174-7

Duchesne, L., and Houle, D. (2008). Impact of nutrient removal through harvesting on the sustainaibility of the boreal forest. *Ecol. Appl.* 18, 1642–1651. doi: 10.1890/07-1035.1

Dufour, B., and Morin, H. (2010). Tracheid production phenology of *Picea mariana* and its relationship with climatic fluctuations and bud development using multivariate analysis. *Tree Physiol.* 30, 853–865. doi: 10.1093/treephys/tpq046

Dufour, B., and Morin, H. (2013). Climatic control of tracheid production of black spruce in dense mesic stands of eastern Canada. *Tree Physiol.* 33, 175–186. doi: 10.1093/treephys/tps126

Farnsworth, E. J., Nunez-Farfan, J., Careaga, S. A., and Bazzaz, F. A. (1995). Phenology and growth of three temperate forest life forms in response to artificial soil warming. *J. Ecol.* 83, 967–977. doi: 10.2307/2261178

Gélinas-Pouliot, M. (2013). *The Fate of* ^{15}*N-Labeled Ammonium Nitrate Applied on Trees Canopy in a Mature Balsam-Fir Stand*, Québec. Master thesis, Université du Quebec à Chicoutimi, Chicoutimi, QC.

Gindl, W., Grabner, M., and Wimmer, R. (2000). The influence of temperature on latewood lignin content in treeline Norway spruce compared with maximum density and ring width. *Trees* 14, 409–414. doi: 10.1007/s004680000057

Gričar, J., Zupančič, M., Čufar, K., Gerald, K., Uwe, S., and Oven, P. (2006). Effect of local heating and cooling on cambial activity and cell differentiation in the stem of norway spruce (*Picea abies*). *Ann. Bot.* 97, 943–951. doi: 10.1093/aob/mcl050

Gričar, J., Zupančič, M., Čufar, K., and Oven, P. (2007). Regular cambial activity and xylem and phloem formation in locally heated and cooled stem portions of Norway spruce. *Wood Sci. Technol.* 41, 463–475. doi: 10.1007/s00226-006-0109-2

Guney, A., Kerr, D., Sokucu, A., Zimmermann, R., and Kuppers, M. (2015). Cambial activity and xylogenesis in stems of *Cedrus libani* A. Rich at different altitudes. *Bot. Stud.* 56:20.

Hättenschwiler, S., Schweingruber, F. H., and Körner, C. H. (1996). Tree ring responses to elevated CO_2 and increased N deposition in *Picea abies*. *Plant Cell Environ.* 19, 1369–1378. doi: 10.1111/j.1365-3040.1996.tb00015.x

Houle, D., Ariane, B., Duchesne, L., Travis, L., and Richard, H. (2012). Projections of future soil temperature and water content for three southern Quebec forested sites. *J. Climate* 25, 7690–7701.

IPCC (2014). *Climate Change 2014: Synthesis Report Contribution of Working Groups I, II and III to the Fifth Assessment Report of the Intergovernmental Panel on Climate Change*, eds Core Writing Team, R. K. Pachauri, and L. A. Meyer (Geneva: IPCC), 151.

Kalliokoski, T., Reza, M., Jyske, T., Mäkinen, H., and Nöjd, P. (2012). Intra-annual tracheid formation of Norway spruce provenances in southern Finland. *Trees* 26, 543–555. doi: 10.1007/s00468-011-0616-0

Kilpeläinen, A., Zubizarreta-gerendiain, A., Luostarinen, K., Peltola, H., and Kellomäki, S. (2007). Elevated temperature and CO_2 concentration effects on xylem anatomy of Scots pine. *Tree Physiol.* 27, 1329–1338. doi: 10.1093/treephys/27.9.1329

Körner, C. (2003). Carbon limitation in trees. *J. Ecol.* 91, 4–17. doi: 10.1046/j.1365-2745.2003.00742.x

Kostiainen, K., Kaakinen, S., Saranpää, P., Sigurdsson, B. D., Linder, S., and Vapaavuori, E. (2004). Effect of elevated [CO_2] on stem wood properties of mature Norway spruce grown at different soil nutrient availability. *Glob. Change Biol.* 10, 1526–1538. doi: 10.1111/j.1365-2486.2004.00821.x

Lupi, C., Morin, H., Deslauriers, A., and Rossi, S. (2010). Xylem phenology and wood production: resolving the chicken-or-egg dilemma. *Plant Cell Environ.* 33, 1721–1730. doi: 10.1111/j.1365-3040.2010.02176.x

Lupi, C., Morin, H., Deslauriers, A., Rossi, S., and Houle, D. (2012). Increasing nitrogen availability and soil temperature: effects on xylem phenology and anatomy of mature black spruce1This article is one of a selection of papers from the 7th International Conference on Disturbance Dynamics in Boreal Forests. *Can. J. For. Res.* 42, 1277–1288. doi: 10.1139/x2012-055

Mäkinen, H., Nöjd, P., Kahle, H.-P., Neumann, U., Tveite, B., Mielikäinen, K., et al. (2003). Large-scale climatic variability and radial increment variation of *Picea abies* (L.) Karst. in central and northern Europe. *Trees* 17, 173–184. doi: 10.1007/s00468-002-0220-4

McWhirter, B. D. (2013). *Tree Seedling Establishment in Response to Warming and Nitrogen Addition in a Temperate Old Field.* thesis, The University of Western Ontario, London, ON.

Melillo, J. M., Butler, S., Johnson, J., Mohan, J., Steudler, P., Lux, H., et al. (2011). Soil warming, carbon–nitrogen interactions, and forest carbon budgets. *Proc. Natl. Acad. Sci. U.S.A.* 108, 9508–9512. doi: 10.1073/pnas.1018189108

Näsholm, T., Kielland, K., and Ganeteg, U. (2009). Uptake of organic nitrogen by plants. *New Phytol.* 182, 31–48. doi: 10.1111/j.1469-8137.2008.02751.x

Overdieck, D., Ziche, D., and Böttcher-Jungclaus, K. (2007). Temperature responses of growth and wood anatomy in European beech saplings grown in different carbon dioxide concentrations. *Tree Physiol.* 27, 261–268. doi: 10.1093/treephys/27.2.261

Plummer, D. A., Caya, D., Frigon, A., Côté, H., Giguère, M., Paquin, D., et al. (2006). Climate and climate change over north America as simulated by the Canadian RCM. *J. Climate* 19, 3112–3132. doi: 10.1175/JCLI3769.1

Quinn, G. P., and Keough, M. J. (2002). *Experimental Design and Data Analysis for Biologists*. Cambridge, UK: University Press.

Reich, P. B., Hobbie, S. E., Lee, T., Ellsworth, D. S., West, J. B., Tilman, D., et al. (2006). Nitrogen limitation constrains sustainability of ecosystem response to CO_2. *Nature* 440, 922–925.

Rossi, S., Anfodillo, T., and Menardi, R. (2006a). Trephor: a new tool for sampling microcores from tree stems. *IAWA J.* 27, 89–97. doi: 10.1163/22941932-90000139

Rossi, S., Deslauriers, A., and Anfodillo, T. (2006b). Assessment of cambial activity and xylogenesis by microsampling tree species: an example at the alpine timberline. *Iawa J.* 27, 383–394.

Rossi, S., Cairo, E., Krause, C., and Deslauriers, A. (2015). Growth and basic wood properties of black spruce along an alti-latitudinal gradient in Quebec, Canada. *Ann. For. Sci.* 72, 77–87. doi: 10.1007/s13595-014-0399-8

Rossi, S., Deslauriers, A., Anfodillo, T., and Carraro, V. (2007). Evidence of threshold temperatures for xylogenesis in conifers at high altitudes. *Oecologia* 152, 1–12. doi: 10.1007/s00442-006-0625-7

Rossi, S., Deslauriers, A., Griçar, J., Seo, J.-W., Rathgeber, C. B. K., Anfodillo, T., et al. (2008). Critical temperatures for xylogenesis in conifers of cold climates. *Glob. Ecol. Biogeogr.* 17, 696–707. doi: 10.1111/j.1466-8238.2008.00417.x

Rossi, S., Girard, M. J., and Morin, H. (2014). Lengthening of the duration of xylogenesis engenders disproportionate increases in xylem production. *Glob. Chang Biol.* 20, 2261–2271. doi: 10.1111/gcb.12470

Rossi, S., Morin, H., and Deslauriers, A. (2011a). Multi-scale influence of snowmelt on xylogenesis of black spruce. *Arct. Antarct. Alp. Res.* 43, 457–464. doi: 10.1657/1938-4246-43.3.457

Rossi, S., Morin, H., Deslauriers, A., and Plourde, P. Y. (2011b). Predicting xylem phenology in black spruce under climate warming. *Glob. Change Biol.* 17, 614–625. doi: 10.1111/j.1365-2486.2010.02191.x

Ruess, R. W., Hendrick, R. L., Burton, A. J., Pregitzer, K. S., Sveinbjornssön, B., Allen, M. F., et al. (2003). Coupling fine root dynamics with ecosystem carbon cycling in black spruce forests of interior alaska. *Ecol. Monogr.* 73, 643–662. doi: 10.1890/02-4032

Saucier, J. P., Bergeron, J. F., Grondin, P., and Robitaille, A. (1998). Les régions écologiques du Québec méridional (3ᵉ version), un des éléments du système hiérarchique de classification écologique du territoire mis au point par le ministère des ressources naturelles. *L'Aubelle* 124, s1–s12.

Schwartz, M. D., Ahas, R., and Aasa, A. (2006). Onset of spring starting earlier across the Northern Hemisphere. *Glob. Change Biol.* 12, 343–351. doi: 10.1111/j.1365-2486.2005.01097.x

Thibeault-Martel, M., Krause, C., Morin, H., and Rossi, S. (2008). Cambial activity and intra-annual xylem formation in roots and stems of abies balsamea and *Picea mariana. Ann. Bot.* 102, 667–674.

Thomas, R. Q., Canham, C. D., Weathers, K. C., and Goodale, C. L. (2010). Increased tree carbon storage in response to nitrogen deposition in the US. *Nat. Geosci.* 3, 13–17.

Wolfinger, R. (1993). Covariance structure selection in general mixed models. *Commun. Stat. Simul. Comput.* 22, 1079–1106.

Xia, J., and Wan, S. (2013). Independent effects of warming and nitrogen addition on plant phenology in the Inner Mongolian steppe. *Ann. Bot.* 111, 1207–1217. doi: 10.1093/aob/mct079

Zhao, C., and Liu, Q. (2009). Growth and photosynthetic responses of two coniferous species to experimental warming and nitrogen fertilization. *Can. J. For. Res.* 39, 1–11. doi: 10.1139/X08-152

Conflict of Interest Statement: The authors declare that the research was conducted in the absence of any commercial or financial relationships that could be construed as a potential conflict of interest.

4

Capturing spiral radial growth of conifers using the superellipse to model tree-ring geometric shape

Pei-Jian Shi[1,2†], *Jian-Guo Huang*[2,3*†], *Cang Hui*[4,5], *Henri D. Grissino-Mayer*[6], *Jacques C. Tardif*[7], *Li-Hong Zhai*[2,3], *Fu-Sheng Wang*[1] and *Bai-Lian Li*[8]

[1] Co-Innovation Centre for Sustainable Forestry in Southern China, Bamboo Research Institute, Nanjing Forestry University, Nanjing, China, [2] Key Laboratory of Vegetation Restoration and Management of Degraded Ecosystems, South China Botanical Garden, Chinese Academy of Sciences, Guangzhou, China, [3] Provincial Key Laboratory of Applied Botany, South China Botanical Garden, Chinese Academy of Sciences, Guangzhou, China, [4] Department of Mathematical Sciences, Centre for Invasion Biology, Stellenbosch University, Matieland, South Africa, [5] Mathematical and Physical Biosciences, African Institute for Mathematical Sciences, Cape Town, South Africa, [6] Department of Geography, The University of Tennessee, Knoxville, TN, USA, [7] Centre for Forest Interdisciplinary Research, University of Winnipeg, Winnipeg, MB, Canada, [8] Ecological Complexity and Modelling Laboratory, Department of Botany and Plant Sciences, University of California, Riverside, Riverside, CA, USA

Edited by:
Sergio Rossi,
Université du Québec à Chicoutimi,
Canada

Reviewed by:
Martin De Luis,
University of Zaragoza, Spain
Fabio Lombardi,
Università del Molise, Italy

*Correspondence:
Jian-Guo Huang
huangjg@scbg.ac.cn

[†] These authors have contributed equally to this work.

Tree-rings are often assumed to approximate a circular shape when estimating forest productivity and carbon dynamics. However, tree rings are rarely, if ever, circular, thereby possibly resulting in under- or over-estimation in forest productivity and carbon sequestration. Given the crucial role played by tree ring data in assessing forest productivity and carbon storage within a context of global change, it is particularly important that mathematical models adequately render cross-sectional area increment derived from tree rings. We modeled the geometric shape of tree rings using the superellipse equation and checked its validation based on the theoretical simulation and six actual cross sections collected from three conifers. We found that the superellipse better describes the geometric shape of tree rings than the circle commonly used. We showed that a spiral growth trend exists on the radial section over time, which might be closely related to spiral grain along the longitudinal axis. The superellipse generally had higher accuracy than the circle in predicting the basal area increment, resulting in an improved estimate for the basal area. The superellipse may allow better assessing forest productivity and carbon storage in terrestrial forest ecosystems.

Keywords: basal area, cross section, major semi-axis, polar coordinate, rotation, tree-rings

INTRODUCTION

Tree rings are natural archives of environmental changes and they have long been used in exploring the effects of endogenous (e.g., competition) and exogenous (e.g., climate, disturbances) factors on tree growth (Fritts, 1976). For example, tree-ring data have been widely used in climate reconstructions (Cook et al., 2002; Frank and Esper, 2005), disturbance reconstructions (Bergeron et al., 2004; Stoffel and Corona, 2014), investigation of species competition and succession (Callaway, 1998; Linares et al., 2010; Huang et al., 2013), and assessments of forest carbon storage and equilibrium (Guyette et al., 2008; Davis et al., 2009; Ma et al., 2012).

Traditionally forest sciences including tree-ring techniques often have assumed that tree rings on a cross section approximate a series of concentric circles (Biondi and Qeadan, 2008; West, 2009).

Based on this assumption, mean annual ring-width and basal area increment are usually obtained and commonly used as two basic parameters for investigating environmental effects on growth and for assessing forest growth, productivity, and carbon sequestration. Mean annual ring-width is often calculated from two radial growth measurements along two directions with an angle of between $90°$ and $180°$ on a cross section collected at diameter at breast height (DBH). Basal area increment is calculated from the difference in area encircled by two adjacent rings (e.g., Biondi and Qeadan, 2008; Huang et al., 2013, 2014).

In many cases, however, tree rings can be better depicted by ellipses rather than circles, or are more inclined to be elliptical around a common centre. This geometric shape of the tree-ring boundary (thereafter referred to as tree-ring shape) has been ignored in practice given that the bias between the circular and ellipse is assumed to be small (West, 2009) although tree rings have long been used in multidisciplinary ecological research for nearly 75 years. Gielis (2003a,b) introduced a superellipse equation that can capture a wide range of geometric shapes in nature. The superellipse equation is a generalization of the traditional ellipse equation and can even produce the outline of a rectangle under special parameter values. If a tree ring can be better fitted by a superellipse function which bears a major axis and a minor axis, the following hypotheses then need to be further tested. First, because the tree trunk does not often look perfectly round, radial growth of trees may follow a spiral growth pattern over time on the cross section in contrast to spiral grain over the longitudinal axis (see the following section below), such that the direction of the major axis of the superellipse may vary with age. Second, basal area increment can be better estimated using the superellipse equation than the circle equation.

Spiral grain is a growth phenomenon in trees characterized by a helical structure of fibers around the pith rather than a longitudinal structure of fibers along the stem axis (Skatter and Kucera, 1998). Spiral grain along the longitudinal axis has been widely observed and reported in many coniferous and some broadleaf species (Harris, 1989; Kubler, 1991; Skatter and Kucera, 1998; Wing et al., 2014). Supplementary Figure 1 exhibits an example of spiral grain of dragon juniper [*Sabina chinensis* (L.) Ant. cv. Kaizuca]. A general agreement is that many tree species (particularly conifers) usually develop a left-handed (L) spirality (when viewed from below) while young. The grain angle then shifts gradually toward right-handed (R) spirality, ending up with a remarkable right-handed grain angle during the mature stage (Skatter and Kucera, 1998). This is the "LR" pattern commonly observed (Harris, 1989) while the opposite pattern ("RL") has also been proposed but only for fewer tree species (Balodis, 1972; Harris, 1989; Harding and Woolaston, 1991). The LR or RL pattern is widely believed to be controlled strongly by genetic factors and less by environmental factors, such as strong wind or water shortage that dominates on one side of the tree (Kubler, 1991; Gapare et al., 2009; Wing et al., 2014). For example, Wing et al. (2014) found no correlation between spiral grain in bristlecone pines (*Pinus longaeva* D.K. Bailey) and environmental factors. In contrast, spiral growth on the radial section was previously acknowledged (Kubler, 1991) but has never been thoroughly investigated and understood. Knowledge on spiral growth on the radial section obtained through model fit with the superellipse may help better understand the long-term debate on spiral growth over the longitudinal section mentioned above, which is closely related to wood quality and forest productivity. Consequently they may together contribute to an improved estimation of growth of trees and forests, as well as for carbon storage and equilibrium of terrestrial forest ecosystems, and ultimately sustainable forest management within the context of global change.

In this study, we attempt to: (1) use the superellipse equation to model tree-ring shapes of conifers which usually bear clear annual ring growth pattern; and (2) explore whether any spiral growth exists along the radial section over time and, if it does, determine whether it is related to spiral grain over the longitudinal axis.

MATERIALS AND METHODS

Superellipse Equation
The superellipse equation is a generalized ellipse equation that can produce the circle, ellipse, square, and rectangle (Gielis, 2003a,b):

$$\left|\frac{x}{a}\right|^n + \left|\frac{y}{b}\right|^n = 1 \tag{1}$$

where x and y represent the Cartesian coordinates; a represents the major semi-axis radius; b ($0 < b \leq a$) represents the minor semi-axis radius; and n is a power. It can also be formulated using the polar coordinates (a transformation using $x = r\cos\varphi$ and $y = r\sin\varphi$; Gielis, 2003a,b):

$$r = \left(\left|\frac{\cos\phi}{a}\right|^n + \left|\frac{\sin\phi}{b}\right|^n\right)^{-1/n} \tag{2}$$

where r represents the radial distance between the pole and a point on the boundary, and φ the angle of the radial vector. The superellipse equation becomes a typical ellipse equation when $n = 2$. Let $k = b/a \leq 1$, and Equation (2) can be rewritten as:

$$r = a \cdot \left(|\cos\phi|^n + \left|\frac{\sin\phi}{k}\right|^n\right)^{-1/n} \tag{3}$$

Examples for different n ranging from 0.2 to 4 are illustrated in **Figure 1**.

Parametric Fitting
We define a standard superellipse equation: the origin coordinate is (0, 0), and the major axis is aligned with the horizontal axis. However, the planar coordinates of a tree ring are usually extracted from a scanned image, with the centre not exactly in the origin and the major axis not aligned with the horizontal axis (**Figure 2**). In this case, we refer to such a shape as a non-standard superellipse and its equation as the non-standard superellipse equation. To fit the parameters of a non-standard superellipse equation, we need first to transform the boundary coordinates to the standard format of the superellipse. Let $x_1 = x' - x_0$, $y_1 =$

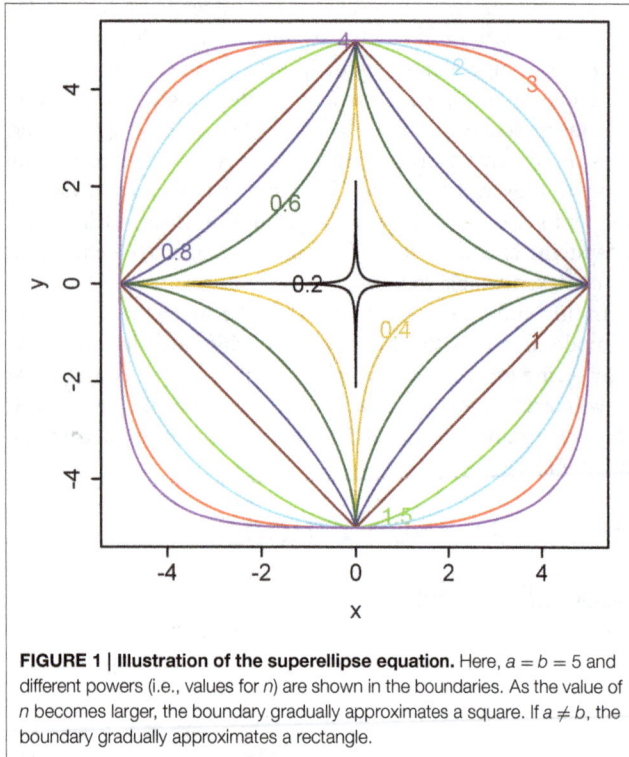

FIGURE 1 | Illustration of the superellipse equation. Here, $a = b = 5$ and different powers (i.e., values for n) are shown in the boundaries. As the value of n becomes larger, the boundary gradually approximates a square. If $a \neq b$, the boundary gradually approximates a rectangle.

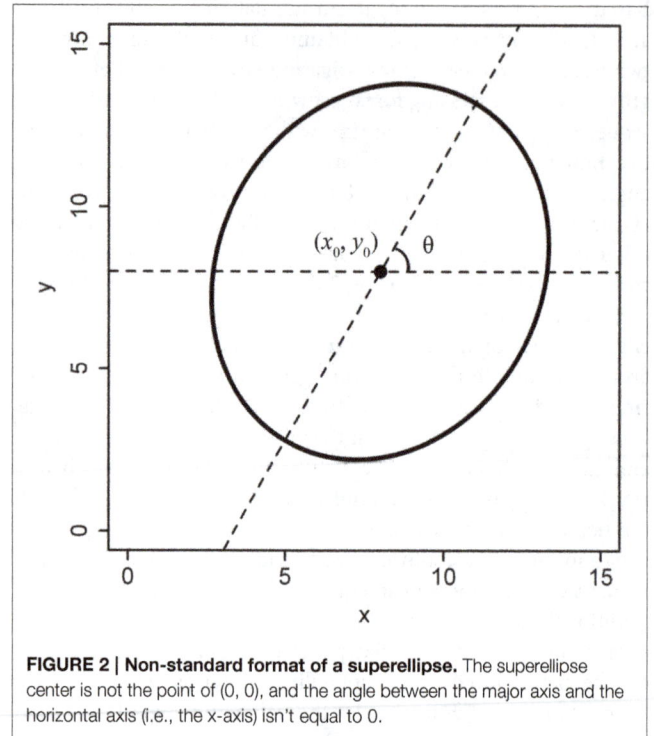

FIGURE 2 | Non-standard format of a superellipse. The superellipse center is not the point of (0, 0), and the angle between the major axis and the horizontal axis (i.e., the x-axis) isn't equal to 0.

$y' - y_0$. Here, x' and y' are the x- and y-coordinates extracted from a non-standard format of the superellipse; (x_0, y_0) is the coordinate of the superellipse centre (i.e., the pole). Let φ' be the angle coordinate corresponding to the point of (x_1, y_1) in a non-standard superellipse boundary. Obviously, $\varphi' = \arctan(y_1/x_1)$. Let θ be the angle between the major axis in the non-standard superellipse equation and the horizontal axis. Then the angle coordinate (φ) in the standard superellipse equation is $\varphi = \varphi' - \theta$. Then we have:

$$\begin{cases} x = x_1 \cos\theta + y_1 \sin\theta \\ y = y_1 \cos\theta - x_1 \sin\theta \end{cases} \quad (4)$$

where x and y are the x- and y-coordinates in the standard superellipse equation. We can fit the parameters of x_0, y_0, and θ together with the three original model parameters a, k, and n, using the optimization algorithm of Nelder and Mead (1965) [see the function "optim" in R software (R Development Core Team, 2013)]. This optimization algorithm has proven effective for estimating the parameters of a non-linear model (Shi et al., 2013). In Appendices S1-S2, we provided a MATLAB function "profile" (M-file; see Appendices S1-S1 in Supplementary Material) for extracting the planar coordinates from a tree-ring image and two R functions (i.e., "optim.sf" and "fit.sf" R-files; also see Appendices S1-S1 in Supplementary Material) for fitting the model parameters of a transformed superellipse equation (i.e., a non-standard superellipse equation). The estimated angle between the major axis and the horizontal axis for these two R functions was defined in the range of $(-2\pi, 2\pi)$.

Evaluation

To verify the validity of the superellipse equation and relevant R functions, we developed a "simu.sf" function (R-file; see Appendices S1-S2 in Supplementary Material) for simulating the planar coordinates based on the given model parameter values ($x_0 = y_0 = 200$, $\theta = \pi/4$, $a = 50$, $k = 0.95$, and $n = 1.9$). Because the actual tree-ring shape can slightly deviate from a standard superellipse, the effects of the variation in a tree-ring boundary on the parameter estimation were considered during the simulation. Thus, the "simu.sf" function was designed to permit a variation in the radial coordinate (i.e., r) by setting the optional value of coefficient of variation ("CV") in any direction. The effects of different CVs on the parameter estimation was investigated by the goodness-of-fit when CV = 4, 3, 2, 1, 0.5, and 0%.

To check how the number of the points extracted from a tree-ring image affects the parameter estimation and the model fit, data points of 200, 400, 800, 1600, 3200, and 6400 were randomly sampled from a simulated tree ring when CV = 1%. When the model parameters were given, the area encircled by a tree ring was actually fixed. For these two types of simulations, we compared the area calculated from the real model parameters and from the fitted model parameters. In general, the width of the confidence interval (CI) of a parameter estimate becomes narrower when the sample size increases. The relevant R functions to calculate the area encircled by a tree ring and the CIs of the model parameters are provided in Appendices S1 and S2.

Although these simulation methods can provide a reasonable result for evaluating model performance, the simulated data only followed a pre-defined superellipse. To evaluate whether the actual tree rings of conifers can follow the superellipse equation (i.e., whether this precondition of superellipse shape holds), it is necessary to explore the spiral growth of conifers using actual tree rings. We examined six actual cross sections collected at DBH from three tree species, including cross-sections from four white spruce (*Picea glauca* (Moench) Voss.) trees (Huang et al., 2013), one from black spruce [*Picea mariana* (Mill.) B.S.P.] (Tardif et al., 2001), and one from Douglas-fir [*Pseudotsuga menziesii* (Mirbel) Franco] (Grissino-Mayer, 1996) (see Supplementary Table 1, Supplementary Figure 2).

For each of the cross sections, we examined whether the angle (θ) between the major axis and the horizontal axis changes when tree ages over time. We defined the angle of the horizontal axis as 0, then defined the angle change due to the rotation of the major axis in an anti-clockwise direction and in a clockwise direction as a positive number and a negative number, respectively. If the angle of θ_{i+1} in the $(i+1)$th year was larger than the angle of θ_i in the ith year, an anti-clockwise spiral growth in the increments of $(\theta_{i+1} - \theta_i)$ was observed; otherwise, a clockwise spiral growth in the increments of $(\theta_i - \theta_{i+1})$ was observed. Obviously, the clockwise rotation results in a left-handed spiral grain while the anti-clockwise rotation leads to a right-handed spiral grain. As the shape of a superellipse is symmetrical around the major axis or the minor axis, the produced tree-ring shapes for the angle θ and $\theta \pm \pi$ should be the same in theory. Assume that the real angle of a tree ring is θ_2. Whether its estimate is $\widehat{\theta}_2$ or $\widehat{\theta}_2 \pm \pi$ will not affect the description for tree-ring shape. However, comparison of the angles from tree rings at different ages can be negatively affected. Assume that the real angles for two adjacent tree rings are θ_1 and θ_3 and their corresponding estimates are $\widehat{\theta}_1$ and $\widehat{\theta}_3$, respectively. If the real angle θ_2 is incorrectly estimated to be $\widehat{\theta}_2 - \pi$, tree-ring angle at the middle age can then be largely underestimated compared to the angles of the neighbors. Therefore, the R function "angle.corr" was developed to automatically correct the angles that have been overestimated or underestimated to make them rank in a normal order (see Appendices S1 and S2). This function can also detect abnormal angle estimates from bad fitting. To test whether a general trend of spiral growth within species exists, we compared the corrected angles of the four log cross sections of white spruce.

To test whether a tree ring still follows a circle equation or a pure ellipse equation rather than a superellipse equation, we further tested whether the ratios of minor to major semi-axis (values for k) and the powers (values for n) were different among the four cross sections of white spruce. If tree rings follow a circle equation, the ratios should be identical (= 1), and the powers should also be identical (= 2). If the tree rings follow a pure ellipse equation, the ratios should be smaller than 1 and the powers should be identical (= 2). Because the frequency distributions for the parameters k and n were unknown, the Kruskal-Wallis test was used to compare the difference among the four log sections (Hollander and Wolfe, 1973).

To compare the validity and complexity of the model when using the superellipse equation and the circle equation in describing tree-ring shapes, we used the Akaike Information Criterion (AIC; Burnham and Anderson, 2004), which can reflect the trade-off between the goodness-of-fit and the model complexity. Model comparisons with the AIC were performed

for both the simulated and real tree rings. The simulated tree rings were produced by the superellipse equation with $0.75 < k < 1$ and $1.7 < n < 2.3$ for our focal species. The effect of the number of data points on a tree ring (100, 200, 400, and 800, respectively) on model performance was also checked using the AIC. Five real tree rings per cross-section were randomly chosen for white spruce, black spruce and Douglas fir, and one ring for jack pine, red pine, tamarack, and white cedar (see **Table 1**, Supplementary Table 2 for details).

To test if the superellipse equation performs better than the traditional circle equation in estimating the basal area, we compared the goodness-of-fit (with the χ^2 value) of the areas that were calculated using the superellipse equation and using the circle equation, when the radius is exactly equal to the major axis, the minor axis, and the average of both, respectively. The χ^2 value was calculated by

$$\chi^2 = \sum_{i=1}^{q} \frac{\left(A_i - \hat{A}_i\right)^2}{\hat{A}_i} \qquad (5)$$

where A_i and $\widehat{A}_i$ represent the actual basal area and the predicted basal area encircled by tree ring in year i, respectively; q represents the total number of years in a cross section. The lower the χ^2 value, the better the fit of the model. The circle equation might often overestimate or underestimate the basal area increment in any given time of period when using the major axis or the minor axis as the radius, respectively. Therefore, average error proportion (%) in annual basal areal increment, which was defined as mean ratios of the absolute values of differences between the predicted and actual annual areal increments to the actual annual areal increments, was also calculated. An improved accuracy, as expressed by the difference of average error proportion obtained by the superellipse and the circle-3, was further calculated to assess the predicative capacity of the model.

RESULTS

Tree-ring shapes can be fitted by the superellipse equation with high precision. The simulations confirmed that the optimization method could produce reliable parameter estimates. The goodness-of-fit, indicated by adjusted R^2 and χ^2, declined with the increase of CV, yet the parameter estimates were still reliable even for CV = 4% (**Table 1** and Supplementary Figure 3). The same pattern emerged using different numbers of data points, yet the coefficients of determination remained almost unchanged (**Table 2** and Supplementary Figure 4). A narrower 95% CI for any parameters was found when the number of data points increased. The 95% CI of the parameter θ only had an absolute difference of 0.03 (< 5% of the real value) for a sample size ≥ 800.

When actual tree-ring samples were used, the superellipse model was able to fit the planar coordinates. **Figure 3** displays the fitted results for the six actual tree cross sections. For any tree cross section, each of the predicted tree rings is symmetrical

TABLE 1 | Comparison between actual and estimated parameters under different coefficients of variation (CV) (number of data points = 1000).

Parameters	Actual values	Estimates					
		CV = 4%	CV = 3%	CV = 2%	CV = 1%	CV = 0.5%	CV = 0%
x_0	200	200.24	200.13	200.01	199.96	199.99	199.92
y_0	200	199.93	199.98	199.95	199.99	199.99	200.05
θ	0.7854	0.7881	0.7493	0.8042	0.7729	0.7828	0.7722
a	50	50.32	50.05	49.99	50.04	50.03	49.97
k	0.95	0.9383	0.9480	0.9517	0.9498	0.9498	0.9508
n	1.9	1.9105	1.9037	1.8931	1.8967	1.8997	1.9008
R^2	–	0.2039	0.3101	0.4712	0.7900	0.9350	0.9927
χ^2	–	75.8	41.3	18.6	4.7	1.2	0.1
Area (cm^2)	7309	7328	7313	7308	7314	7315	7307

TABLE 2 | Comparison between actual and estimated parameters under different numbers of data points (CV = 1%).

Parameter	Real value	Estimate (with 95% CI)		
		Size = 200	Size = 400	Size = 800
x_0	200	199.93 (199.82, 200.02)	200.01 (199.94, 200.08)	199.98 (199.94, 200.02)
y_0	200	200.08 (199.97, 200.16)	200.04 (199.97, 200.11)	200 (199.95, 200.04)
θ	0.7854	0.7935 (0.7619, 0.8296)	0.7921 (0.7705, 0.8166)	0.7803 (0.7647, 0.7938)
a	50	50.03 (49.85, 50.15)	50.14 (50.04, 50.26)	49.99 (49.92, 50.06)
k	0.95	0.9479 (0.9449, 0.9519)	0.9476 (0.9446, 0.9505)	0.95 (0.948, 0.9515)
n	1.9	1.9134 (1.895, 1.9388)	1.8898 (1.8733, 1.9034)	1.9028 (1.8951, 1.9125)
R^2	–	0.7937 (0.7501, 0.841)	0.7884 (0.7583, 0.8168)	0.7888 (0.7665, 0.8135)
χ^2	–	0.96 (0.77, 1.11)	1.97 (1.71, 2.2)	3.75 (3.3, 4.09)
Area (cm^2)	7309	7324 (7301, 7341)	7316 (7301, 7331)	7312 (7303, 7320)

Parameter	Real value	Estimate (with 95% CI)		
		Size = 1600	Size = 3200	Size = 6400
x_0	200	199.99 (199.96, 200.02)	200 (199.96, 200.13)	199.99 (199.97, 200.02)
y_0	200	200 (199.97, 200.04)	199.98 (199.96, 200.13)	200 (199.98, 200.04)
θ	0.7854	0.7912 (0.7812, 0.8027)	0.7844 (0.7711, 0.8017)	0.7891 (0.7817, 0.7967)
a	50	50.04 (49.97, 50.09)	49.99 (49.89, 50.24)	50.01 (49.91, 50.01)
k	0.95	0.9491 (0.9477, 0.9506)	0.9493 (0.945, 0.95)	0.9497 (0.9491, 0.9509)
n	1.9	1.8992 (1.8929, 1.906)	1.9052 (1.8855, 1.9205)	1.9017 (1.9002, 1.913)
R^2	–	0.79 (0.7736, 0.8049)	0.7764 (0.7569, 0.7863)	0.7815 (0.7725, 0.7887)
χ^2	–	7.64 (7.08, 8.15)	16.64 (16.02, 17.97)	31.67 (30.7, 32.86)
Area (cm^2)	7309	7311 (7303, 7315)	7308 (7296, 7328)	7311 (7302, 7314)

around its major axis; however, such symmetry is difficult to be observed intuitively when all tree rings are superposed together due to the angle rotation.

For white spruce (**Figures 4A–D**), a general pattern of angle rotation was not found, even for WS-1 and WS-2 collected from the same site but exhibited different trends in the angle change (**Figures 4A,B**). As shown in **Figure 4A**, an obvious low point appeared in the 10th ring for this white spruce. Therefore, tree rings at age ≤ 10 years rotated in a clockwise direction, resulting in a left-hand spiral grain over the longitudinal axis. Tree rings at age > 10 years rotated in a reverse clockwise direction, leading to a right-hand spiral grain over the longitudinal axis. In contrast

to WS-1, white spruce WS-2 showed an inverse trend, with a peak point at age of 15 years (**Figure 4B**). The reverse clockwise rotation was observed at age ≤ 15 and the clockwise rotation was found at age > 15. Correspondingly, the right-hand and left-hand spiral grain over the longitudinal axis was expected, respectively. The angle change trend in WS-3 was more or less similar to that of WS-1, but its lowest point appeared in the 25th ring (**Figure 4C**). Interestingly, white spruce sample WS-4 exhibited a completely different trend from those of the first three white spruce trees. As shown in **Figure 4D**, a peak point occurred in the 19th year and a lowest point appeared in the 36th year. Therefore, tree rings first rotated in an anti-clockwise direction at age ≤ 19,

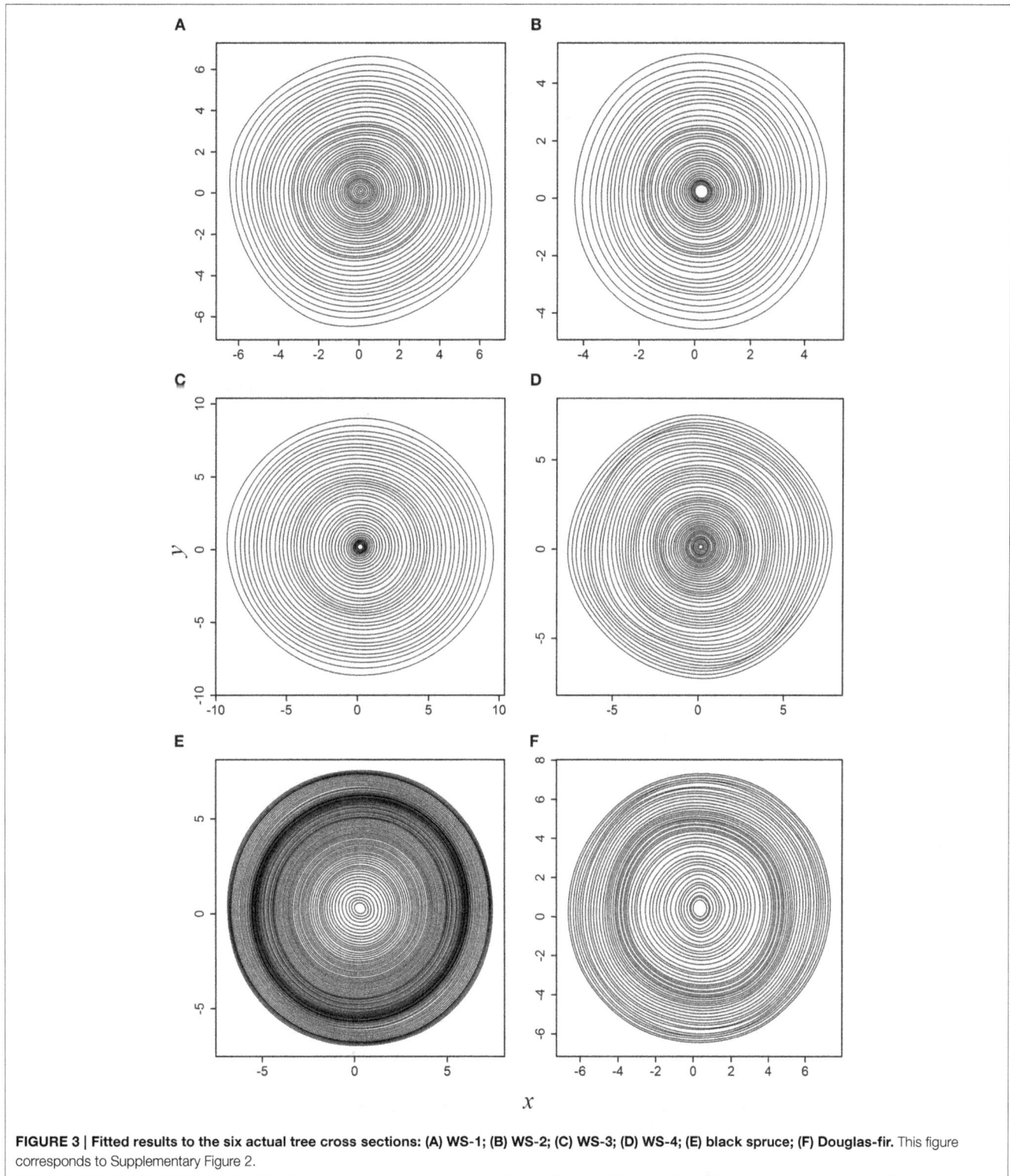

FIGURE 3 | Fitted results to the six actual tree cross sections: (A) WS-1; (B) WS-2; (C) WS-3; (D) WS-4; (E) black spruce; (F) Douglas-fir. This figure corresponds to Supplementary Figure 2.

then rotated in a clockwise direction at $19 < \text{age} \leq 36$, and finally rotated in an anti-clockwise direction at age > 36 again.

Tree-ring angles of black spruce kept a continuous decrease during the first 30 years, indicating a clockwise rotation (**Figure 4E**). However, the angle changes were not very substantial from the 30th year to the 70th year. Afterwards, tree-ring angles began to become larger and larger, which means an anti-clockwise rotation. Compared to tree species mentioned

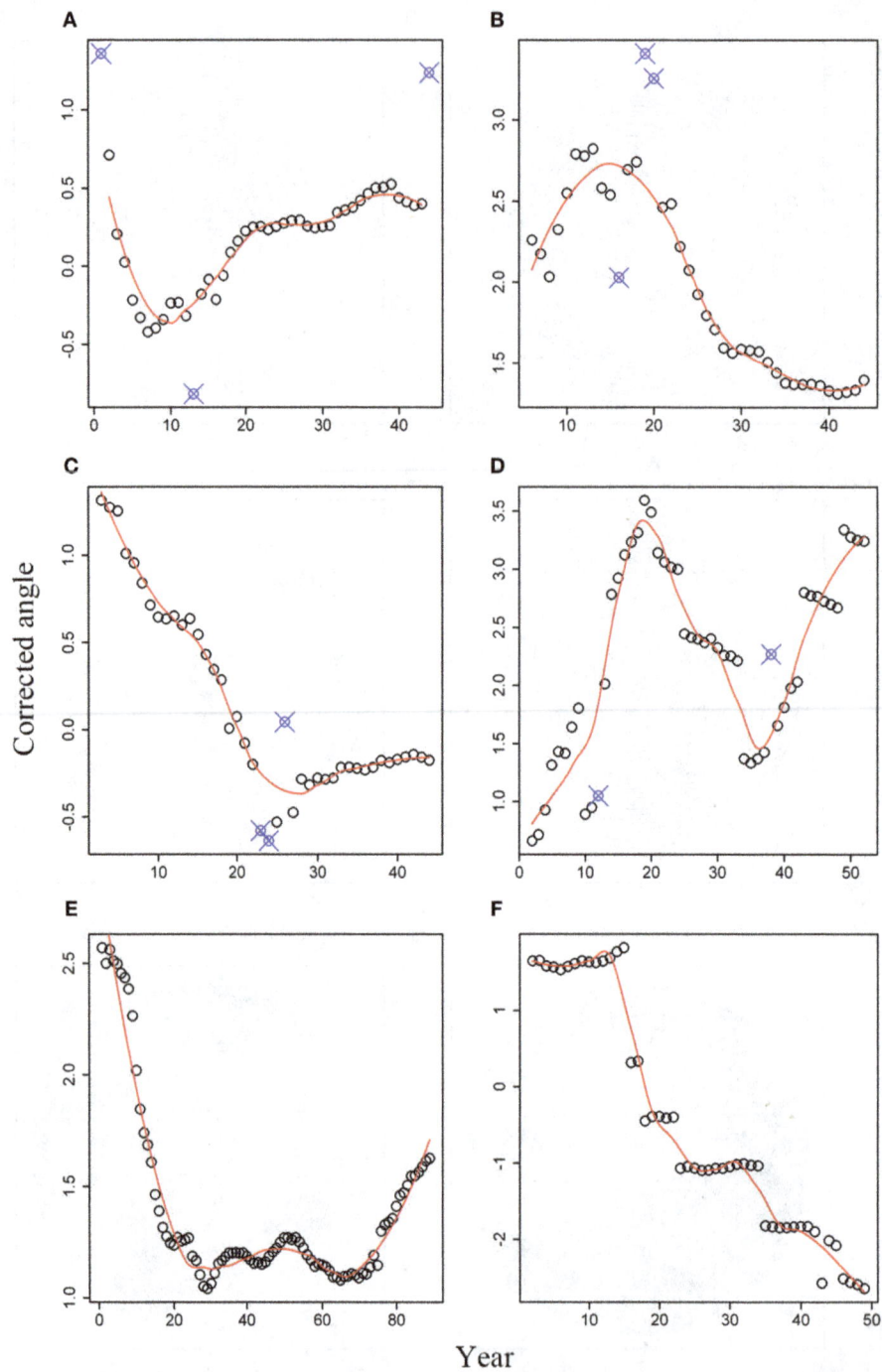

FIGURE 4 | Corrected angles for six tree cross sections: (A) WS-1; (B) WS-2; (C) WS-3; (D) WS-4; (E) black spruce; (F) Douglas-fir. Small open circles represent the corrected angles; Solid line represents the predicted values based on the local regression method; Small open circles with signs of "X" represent abnormal data that were not used.

above, tree-ring angles for Douglas-fir have its distinct rotation, which was characterized by several plateaus and an overall decreasing trend over time (**Figure 4F**).

The Kruskal-Wallis test showed a significant difference in the ratios of minor to major semi-axis ($\chi^2 = 12.1$, $df = 3$, $P =$

0.007; **Figure 5A**), and a significant difference in the powers n ($\chi^2 = 23.2$, $df = 3$, $P < 0.01$; **Figure 5B**) among the four samples of white spruce. The pairwise test for the median of k showed insignificant difference between any pair of WS-1, WS-2, and WS-3 ($P > 0.05$), except for WS-4 which showed significant

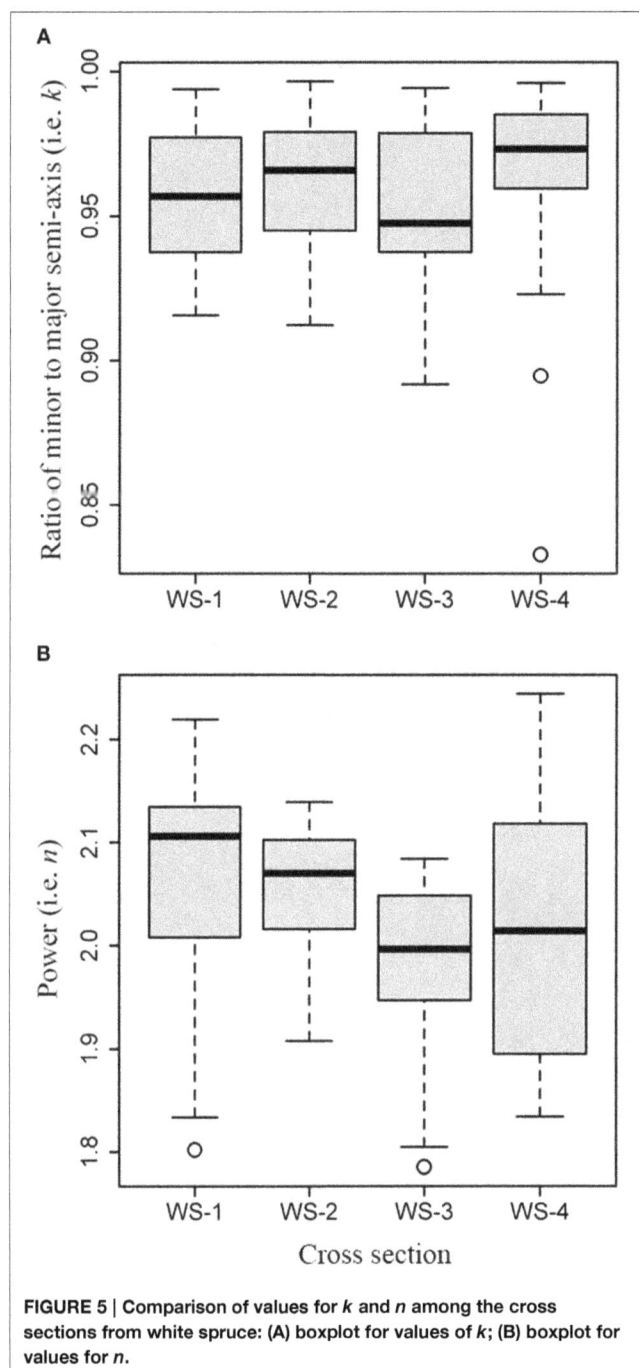

FIGURE 5 | Comparison of values for *k* and *n* among the cross sections from white spruce: (A) boxplot for values of *k*; (B) boxplot for values for *n*.

equation in describing the real tree-ring shapes of conifers. The AIC values obtained from the superellipse equation were lower than those obtained from the circle equation when $k < 1$, $n < 2$ and $n > 2$ (**Figure 6**). It implicates that the superellipse equation is generally better than the circle equation in describing the simulated tree-ring shapes except when the tree-ring shape is perfectly round. The results also showed a decline of the AIC score with the increase of data points on a simulated tree ring. The same conclusion of the superellipse equation superior to the circle equation was also drawn from using the AICs for real tree rings (see Supplementary Table 2) as the estimated k is usually smaller than 1 and n unequal to 2 for any real tree rings.

In addition, the results showed that the superellipse equation had the lowest χ^2 values which mean the best goodness-of-fit compared to the other three calculations using the circle equation, as shown in **Table 3**. Average error proportion calculated by the superellipse was much lower than that calculated by the circle. The results of improved predicative accuracy between the superellipse and the circle-3 showed that the superellipse generally had higher accuracy than the circle in predicting the basal area increment, ranging from 2.31 to 12.57% for our focal species (**Table 3**).

DISCUSSION

Tree-Ring Shape and Improved Estimates of the Basal Area

Our modeling results showed that tree-ring shape can be best fitted by the superellipse equation with high precision. This suggests that tree rings do not follow either a circle equation or a pure ellipse equation, but somewhere in-between, i.e., a superellipse equation. Theoretically, tree-ring shape should be a circle under genetic influences only because cambium cell differentiation of a healthy tree is assumed to be at the same rate along the circumference during the growing season (Lupi et al., 2014). However, due to the influences of external factors such as environmental stresses (e.g., light availability, water stress) and biogeophysical factors (e.g., position, slope), trees have to physiologically adjust the rate of cambium cell differentiation along the circumference during the growing seasons to survive in or adapt to the local environmental conditions. Consequently, a "superelliptical" tree ring is produced, as widely observed in terrestrial forest ecosystems. Local conditions in the cambium that influence wood formation at any given instant are believed to be unique because the immediate environment of a cambial initial (weather and nutrient factors, growth regulators, physical stresses) varies continuously over time (Downes et al., 2009). A recent micro-sampling based study investigated the process of cambium cell differentiation of black spruce along the circumference during the growing season, and found that the onset of xylogenesis along the circumference varies within an individual tree (Lupi et al., 2014). More earlywood cells (corresponding to a wider annual ring) were often found in a warm year than a cold year (Rossi et al., 2007; Deslauriers et al., 2008; Huang et al., 2011), suggesting that xylem cell number is mainly determined by environmental factors such as

differences with WS-1 and with WS-3 ($P < 0.05$), while significant difference in the median of n between any pair of WS-1, WS-2, and WS-3 ($P < 0.05$) was found, except for WS-4 which showed significant difference with WS-1 ($P < 0.05$) only. Tree-ring shape is not a standard ellipse because the values for n from the first two cross sections are higher than 2 (**Figure 5B**). It also showed that the ratios and powers in the superellipse equation can be different even for the four samples from the same species.

The results of model comparison using the AIC showed that the superellipse equation performed better than the circle

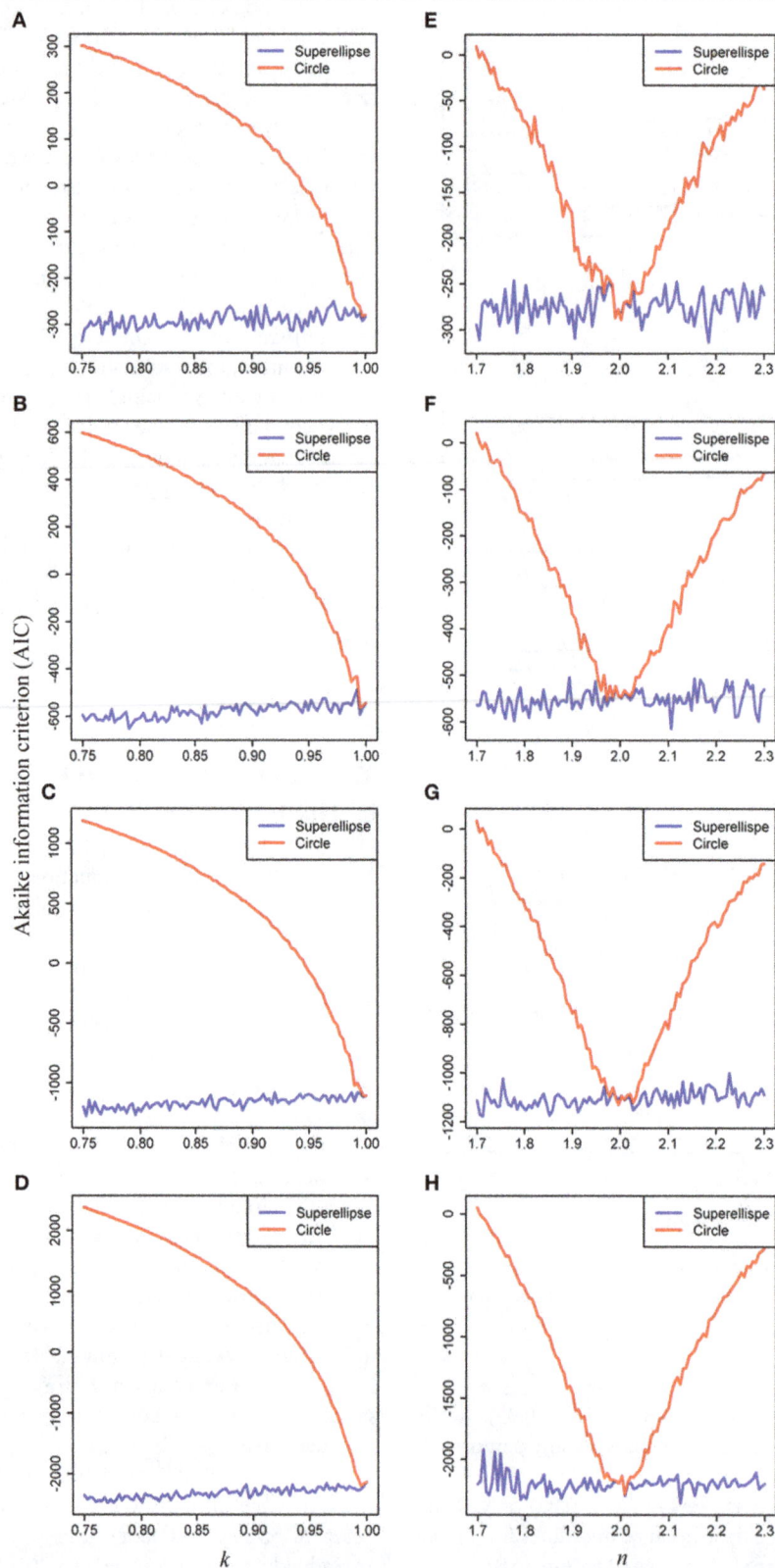

FIGURE 6 | Effects of the parameters _k_ and _n_ on the Akaike Information Criterion (AIC) when simulating tree rings by the superellipse and circle equations with 100 (A,E), 200 (B,F), 400 (C,G), and 800 (D,H) data points, respectively. The simulated tree rings were produced by the superellipse with $x_0 = y_0 = 200$, $a = 50$, $n = 2$, and $0.75 < k < 1$ for panels **(A–D)**, with $x_0 = y_0 = 200$, $a = 50$, $k = 1$, and $1.7 < n < 2.3$ for panels **(E–H)**. We permitted 5% coefficient of variation in the distances of data points on a simulated tree ring from the corresponding pole.

TABLE 3 | Comparison of the goodness-of-fit (χ^2) among the predicted basal areas by different equations.

Sample	Superellipse[a]($\times 10^{-4}$)	Circle-1[b]	Circle-2[c]	Circle-3[d]	AEP_S (%)	AEP_C3 (%)	Accuracy (%)
WS-1	<0.1	2.06	5.57	1.09	0.15	7.18	7.03
WS-2	1.5	1.00	3.21	0.19	0.14	4.06	3.92
WS-3	5.2	7.92	1.55	0.25	0.17	3.38	3.21
WS-4	5.2	4.82	2.88	2.31	0.25	12.82	12.57
Black spruce	3.8	5.24	2.98	0.41	0.57	4.18	3.61
Douglas fir	2.1	3.36	1.97	1.18	0.24	12.55	2.31

[a]Value represents the χ^2 value between the actual areas at different ages and the predicted areas using the superellipse equation.
[b]Value represents the χ^2 value using the circle equation when the radius equals the major semi-axis (Circle-1).
[c]Value represents the χ^2 value using the circle equation when the radius equals the minor semi-axis (Circle-2).
[d]Value represents the χ^2 value using the circle equation when the radius equals the mean of both the major semi-axis and the minor semi-axis. AEP indicates average error proportion (%) in predicted basal areal increment between two adjacent rings by the superellipse (AEP_S) and the circle-3 (AEP_C3). Because the Circle-3 can get the lowest χ^2 value among the three circle equations, we only used the Circle-3 here. WS1 to WS4 were from white spruce. An improved accuracy (Accuracy) between the superellipse and the circle-3 was calculated as the difference between AEP_C3 and AEP-S.

temperature. The importance of various drivers of xylogenesis may shift from factors mainly varying at the site level (e.g., climate) at the beginning of the growing season to factors varying at the individual tree level (e.g., possibly genetic variability) at the end of the growing season (Lupi et al., 2014).

The results of comparison of goodness-of-fit (χ^2) among the areas predicted by different equations showed that tree-ring shape can be best fitted by the superellipse equation compared to by the circle equation, i.e., by the major semi-axis or the minor semi-axis or the average from both. Basal area is one of the basic parameters in forestry and forest ecology and has been widely used in estimates of forest growth and productivity as well as carbon storage and equilibrium (Phillips et al., 1998; Ma et al., 2012). Our results might indicate that traditional circle-based estimates of the basal area might be more or less over- or under-estimated in practice because traditional calculations of the basal area were, directly or indirectly, based on the DBH. DBH values measured in practice may range from the length of the major axis to that of the minor axis due to its feature of spiral growth of the cross-section over age (West, 2009; Torres and Lovett, 2013). The improved accuracy obtained by the superellipse than the other commonly used circle-based approaches might also indicate an improved estimate of the basal area through the superellipse fit. This may contribute to a better understanding of forest growth and productivity, as well as carbon balance and dynamics in terrestrial forest ecosystems given that forests cover 31% of the world's land surface (FAO, 2010).

All evidence together suggests that tree-ring geometric shape can be better depicted by the superellipse than the circle commonly used in practice. Although our study focused on coniferous species only due to its clear annual ring pattern, this geometric shape of tree rings is believed to be universal and common in deciduous species as well.

Spiral Growth on the Cross Section

Our study first quantitatively confirmed that radial growth follows a spiral growth pattern over time despite that the angle change between successive years being much smaller than reported previously in spiral grain-based studies, such as 30–50° in ponderosa pines (*Pinus ponderosa*) (Leelavanichkul

and Cherkaev, 2004; Wing et al., 2014). It is widely accepted that spiral grain originates from the cambium and spiral grain formation can be coupled to cell divisions taking place in the cambial region (Harris, 1989; Eklund and Säll, 2000). Therefore, spiral growth over the cross section might be closely related to spiral grain over the longitudinal axis although this potential relationship needs to be further quantified. Previous studies have attempted to explore the spiral grain angles over time along the longitudinal axis (Danborg, 1994; Gjerdrum et al., 2002; Watt et al., 2013), but the potential relationship between the rotation angles of tree rings on the cross-section over time and the grain angles along the longitudinal axis over time was not quantified. Although spiral grain pattern over the longitudinal axis for many tree species is less visible or even undetected due to a minor change in the shifting angle, we conclude that spiral growth over the cross section might be common in many tree species, including both coniferous and deciduous species.

Previous studies often claim that coniferous trees shift the direction of spiral grain pattern from LR to RL or vice versa, but rarely clearly confirm the exact year when this shift occurred (Kubler, 1991). One or two extreme points observed in the angle change over time in our study might indicate the exact or critical year or years when the direction of spiral grain over the longitudinal axis shifted from LR to RL or from RL to LR. However, the potential mathematical link between the spiral growth over the cross section and spiral grain over the longitudinal axis has not been elucidated yet and thus merits further investigation. In addition, the extreme points found in early years or late years in our study also suggest that the direction shift in spiral grain, either LR to RL or RL to LR pattern, may occur in both young and old stages of tree growth. These findings are counter to the results from previous studies that found a LR pattern normally formed in early stages of growth but gradually shifted to a RL pattern in the older stages of growth for coniferous trees in the northern hemisphere. Further, our results counter the suggestion that two shifts in spiral grain pattern are rarely observed (Skatter and Kucera, 1998). These differences might be partly attributed to different perspectives of investigation, i.e., longitudinal axis vs. cross section. In addition, spiral grain over the longitudinal axis or spiral growth over the cross section is a time-dependent growth phenomenon and is difficult to monitor over time. Most of the previous studies were based on field observations with a quantitative analysis over time lacking. Consequently, the complexity of the mechanisms behind spiral grain over the longitudinal axis might be underestimated.

Fluctuations in the corrected angles of the radial section over time differed within a site and across sites and species. This finding indicates that spiral growth on the radial section is determined less by population and species genetics, but more by macro- and micro-environmental factors. Our findings are contrary to some of the previous studies that claimed that spiral grain pattern in conifers is strongly genetic (Kubler, 1991; Skatter and Kucera, 1998) such as Sitka spruce [*Picea sitchensis* (Bong.) Carr.] (Hansen and Roulund, 1997) and loblolly pine (*Pinus taeda* L.) (Zobel et al., 1968). The results from our study are in general agreement with other studies that claimed that spiral grain is strongly affected by environmental factors such as wind (Koizumi et al., 2007). Overall, various hypothetical reasons for spiral grain have been proposed, such as the Earth rotation, optimal structure for even distribution of sap between the roots and the crown, wind torque, and the relief of growth stresses in the cambial zone (Wing et al., 2014). Still, a consensus has not been reached. From the perspectives of ecology and evolution, we infer that spiral growth over the cross section, which is closely related to spiral grain over the longitudinal axis, is controlled by both genetic and environmental factors, but the importance of both factors in affecting spiral growth or spiral grain might be shifting over time, i.e., site and species- specific, even individually.

Model Advantages and Limitations

We have justified that the superellipse equation is superior to the traditional circle equation in modeling tree-ring boundary as according to the AIC. We are confident that the superellipse equation not only fits tree-ring shapes of coniferous species, but also can be widely applied in fitting tree-ring shape of many tree species. Supplementary Figures 5, 6 exhibited the application of the superellipse equation on additional four species of conifers. The most important advantage is that estimates of forest growth and productivity as well as carbon storage can be improved when the superellipse equation is employed in the future, in contrast to the ill-fitting of tree-ring shapes by the circle equation.

The standard superellipse includes three parameters (i.e., a, k, and n) whereas the circle equation only has a single parameter (radius), reflecting a more complex structure of the former than the latter. In general, for model selection, the adjusted coefficient of determination (R^2_{adj}), AIC, corrected AIC, Bayesian information criterion (BIC), deviance information criterion (DIC), and residual information criterion (RIC) are better than the RSS, R^2, and χ^2 given the trade-off between the goodness-of-fit of the model and the complexity of the model (see Shi and Ge, 2010 and references therein). In practice, forest growth and productivity as well as carbon storage might have often been under- or over-estimated due to the ill-fitting of tree-ring shapes by the circle equation. To better estimate forest productivity and carbon dynamics in forest ecosystems, investigators are usually concerned less about the

model structural complexity, but more on the goodness-of-fit of the model. That means, the better the fitting is, the better a model would be. Given that the original input parameters of our model are obtained from the geometric shape of tree-ring boundary rather than ring-width measurements from a single core or two cores or cross sections commonly collected at DBH from the field, our model is convenient to be applied in practice if the cross-section of trees can be collected in the field and then the parameters of tree-ring shape can be obtained using our method (see Appendices S1–S2 in Supplementary Material). Otherwise currently it is less convenient to obtain these parameters due to a lack of such a tool to automatically measure tree-ring shape in the field, which however deserves to be further invented for the practitioners.

Although our study showed that the suprellipse equation could fit the tree-ring shapes of conifers very well, it is only a descriptive model. A better model for describing tree-ring shape and growth should be based on the dynamics of growth for trees, especially necessarily considering the biophysical mechanism of tree stem formation accompanied with growth stresses (Archer, 1989). Although no convincing evidence has demonstrated that the macrocosmic spiral growth in tree trunk could be related to microcosmic microfibril angle in fiber (Barnett and Bonham, 2004; Jordan et al., 2007), there might be a certain relationship between them. Thus, a mechanical model that could link the effects of growth stresses on the microfibril angle in fiber to the spiral growth in conifers merits further investigation.

AUTHOR CONTRIBUTIONS

PS and JH designed the study. PS analyzed the data. JH and PS wrote the paper. All the coauthors discussed and commented the paper.

ACKNOWLEDGMENTS

This work was supported by the 100 Talents Program of the Chinese Academy of Sciences (Grant number: Y421081001) and National Natural Science Foundation of China (Grant number: 31570584) to JH; Special Fund for Public Welfare Projects of Forestry (Grant number 201204106), and National Natural Science Foundation for Young Scholars of China (Grant number 31400348), the Priority Academic Program Development of Jiangsu Higher Education Institutions, and the Startup Foundation of Nanjing Forestry University (Grant number GXL038) to PS; the National Research Foundations of South Africa (Grant number 76912 and 81825) to CH. We thank the reviewers and the associate editor for their helpful comments and suggestions on the previous versions. We also thank Y.L. Ding, Z.Y. He, Z.Y. Cao, Heand F. Conciatori, and J. Gielis for their valuable help during the preparation of this manuscript.

REFERENCES

Archer, R. R. (1989). On the origin of growth stresses in trees. *Wood Sci. Technol.* 23, 311–322. doi: 10.1007/BF00353247

Balodis, V. (1972). Influence of grain angle on twist in seasoned boards. *Wood Sci.* 5, 44–50.

Barnett, J. R., and Bonham, V. A. (2004). Cellulose microfibril angle in the cell wall of wood fibres. *Biol. Rev. Camb. Philos. Soc.* 79, 461–472. doi: 10.1017/S1464793103006377

Bergeron, Y., Gauthier, S., Flannigan, M., and Kafka, V. (2004). Fire regimes at the transition between mixedwood and coniferous boreal forest in Northwestern Quebec. *Ecology* 85, 1916–1932. doi: 10.1890/02-0716

Biondi, F., and Qeadan, F. (2008). A theory-driven approach to tree-ring standardization: defining the biological trend from expected basal area increment. *Tree-Ring Res.* 64, 81–96. doi: 10.3959/2008-6.1

Burnham, K. P., and Anderson, D. R. (2004). Multimodel inference: understanding AIC and BIC in model selection. *Sociol. Methods Res.* 33, 261–304. doi: 10.1177/0049124104268644

Callaway, R. M. (1998). Competition and facilitation on elevation gradients in subalpine forests of the northern Rocky Mountains, USA. *Oikos* 82, 561–573. doi: 10.2307/3546376

Cook, E. R., Palmer, J. G., and D'Arrigo, R. D. (2002). Evidence for a 'Medieval Warm Period' in a 1,100 year tree-ring reconstruction of past austral summer temperatures in New Zealand. *Geophys. Res. Lett.* 29, 1–4. doi: 10.1029/2001GL014580

Danborg, F. (1994). Spiral grain in plantation trees of Picea abies. *Can. J. For. Res.* 24, 1662–1671. doi: 10.1139/x94-215

Davis, S. C., Hessl, A. E., Scott, C. J., Adams, M. B., and Thomas, R. B. (2009). Forest carbon sequestration changes in response to timber harvest. *Forest Ecol. Manage.* 258, 2101–2109. doi: 10.1016/j.foreco.2009.08.009

Deslauriers, A., Rossi, S., Anfodillo, T., and Saracino, A. (2008). Cambial phenology, wood formation and temperature thresholds in two contrasting years at high altitude in southern Italy. *Tree Physiol.* 28, 863–871. doi: 10.1093/treephys/28.6.863

Downes, G. M., Drew, D., Battaglia, M., and Schulze, D. (2009). Measuring and modelling stem growth and wood formation: an overview. *Dendrochronologia* 27, 147–157. doi: 10.1016/j.dendro.2009.06.006

Eklund, L., and Säll, H. (2000). The influence of wind on spiral grain formation in conifer trees. *Trees Struct. Funct.* 14, 324–328. doi: 10.1007/s004680050225

FAO. (2010). *Global Forest Resources Assessment 2010.* Food and Agriculture Organization of the United Nations. Available online at: http://www.fao.org/forestry/fra/fra2010/en/

Frank, D., and Esper, J. (2005). Temperature reconstructions and comparisons with instrumental data from a tree-ring network for the European Alps. *Int. J. Climatol.* 25, 1437–1454. doi: 10.1002/joc.1210

Fritts, H. C. (1976). *Tree Rings and Climate.* New York, NY: Academic Press.

Gapare, W. J., Baltunis, B. S., Ivkovic, M., and Wu, H. X. (2009). Genetic correlations among juvenile wood quality and growth traits and implications for selection strategy in *Pinus radiata* D. Don. *Ann. Forest Sci.* 66, 606p1–606p9. doi: 10.1051/forest/2009044

Gielis, J. (2003a). A generic geometric transformation that unifies a wide range of natural and abstract shapes. *Am. J. Bot.* 90, 333–338. doi: 10.3732/ajb.90.3.333

Gielis, J. (2003b). *Inventing the Circle: The Geometry of Nature.* Antwerpen: Geniaal Press.

Gjerdrum, P., Säll, H., and Storø, H. M. (2002). Spiral grain in Norway spruce: constant change rate in grain angle in Scandinavian sawlogs. *Forestry* 75, 163–170. doi: 10.1093/forestry/75.2.163

Grissino-Mayer, H. D. (1996). "A 2129-year reconstruction of precipitation for northwestern New Mexico, USA," in *Tree Rings, Environment, and Humanity,* eds J. S. Dean, D. M. Meko, and T. W. Swetnam (Tucson, AZ: Radiocarbon), 191–204.

Guyette, R. P., Dey, D. C., and Stambaugh, M. C. (2008). The temporal distribution and carbon storage of large oak wood in streams and floodplain deposits. *Ecosystems* 11, 643–653. doi: 10.1007/s10021-008-9149-9

Hansen, J. K., and Roulund, H. (1997). Genetic parameters for spiral grain, stem form, pilodyn and growth in 13 years old clones of Sitka spruce (*Picea sitchensis* (Bong.) Carr.). *Silvae Genet.* 46, 107–113.

Harding, K. J., and Woolaston, R. R. (1991). Genetic-parameters for wood and growth-properties in Araucaria-Cunninghamii. *Silvae Genet.* 40, 232–237.

Harris, J. M. (1989). *Spiral Grain and Wave Phenomena in Wood Formation.* Berlin: Springer-Verlag. doi: 10.1007/978-3-642-73779-4

Hollander, M., and Wolfe, D. A. (1973). *Nonparametric Statistical Methods.* New York, NY: John Wiley & Sons.

Huang, J. G., Bergeron, Y., Zhai, L. H., and Denneler, B. (2011). Variation in intra-annual radial growth (xylem formation) of *Picea mariana* (Pinaceae) along a latitudinal gradient in western Quebec, Canada. *Am. J. Bot.* 98, 792–800. doi: 10.3732/ajb.1000074

Huang, J. G., Deslauriers, A., and Rossi, S. (2014). Xylem formation can be modeled statistically as a function of primary growth and cambium activity. *New Phytol.* 203, 831–841. doi: 10.1111/nph.12859

Huang, J. G., Stadt, K. J., Dawson, A., and Comeau, P. G. (2013). Modelling growth-competition relationships in trembling aspen and white spruce mixed boreal forests of Western Canada. *PLoS ONE* 8:e77607. doi: 10.1371/journal.pone.0077607

Jordan, L., He, R. C., Hall, D. B., Clark, A., and Daniels, R. F. (2007). Variation in loblolly pine ring microfibril angle in the Southeastern United States. *Wood Fiber Sci.* 39, 352–363.

Koizumi, A., Oonuma, N., Sasaki, Y., and Takahashi, K. (2007). Difference in uprooting resistance among coniferous species planted in soils of volcanic origin. *J. Forest Res.-Jpn.* 12, 237–242. doi: 10.1007/s10310-007-0001-4

Kubler, H. (1991). Function of spiral grain in trees. *Trees-Struct Funct.* 5, 125–135. doi: 10.1007/BF00204333

Leelavanichkul, S., and Cherkaev, A. (2004). Why the grain in tree trunks spirals: a mechanical perspective. *Struct. Multidiscip. O.* 28, 127–135. doi: 10.1007/s00158-003-0311-x

Linares, J. C., Camarero, J. J., and Carreira, J. A. (2010). Competition modulates the adaptation capacity of forests to climatic stress: insights from recent growth decline and death in relict stands of the Mediterranean fir *Abies pinsapo*. *J. Ecol.* 98, 592–603. doi: 10.1111/j.1365-2745.2010.01645.x

Lupi, C., Rossi, S., Vieira, J., Morin, H., and Deslauriers, A. (2014). Assessment of xylem phenology: a first attempt to verify its accuracy and precision. *Tree Physiol.* 34, 87–93. doi: 10.1093/treephys/tpt108

Ma, Z. H., Peng, C. H., Zhu, Q. A., Chen, H., Yu, G. R., Li, W. Z., et al. (2012). Regional drought-induced reduction in the biomass carbon sink of Canada's boreal forests. *Proc. Natl. Acad. Sci. U.S.A.* 109, 2423–2427. doi: 10.1073/pnas.1111576109

Nelder, J. A., and Mead, R. (1965). A simplex method for function minimization. *Comput. J.* 7, 308–313. doi: 10.1093/comjnl/7.4.308

Phillips, O. L., Malhi, Y., Higuchi, N., Laurance, W. F., Nunez, P. V., Vasquez, R. M., et al. (1998). Changes in the carbon balance of tropical forests: evidence from long-term plots. *Science* 282, 439–442. doi: 10.1126/science.282.5388.439

R Development Core Team (2013). *R: A Language and Environment for Statistical Computing.* Vienna: R Foundation for Statistical Computing. Available online at: http://www.R-project.org

Rossi, S., Deslauriers, A., Anfodillo, T., and Carraro, V. (2007). Evidence of threshold temperatures for xylogenesis in conifers at high altitudes. *Oecologia* 152, 1–12. doi: 10.1007/s00442-006-0625-7

Shi, P. J., and Ge, F. (2010). A comparison of different thermal performance functions describing temperature-dependent development rates. *J. Therm. Biol.* 35, 225–231. doi: 10.1016/j.jtherbio.2010.05.005

Shi, P. J., Men, X. Y., Sandhu, H. S., Chakraborty, A., Li, B. L., Ou-Yang, F., et al. (2013). The "general" ontogenetic growth model is inapplicable to crop growth. *Ecol. Model.* 266, 1–9. doi: 10.1016/j.ecolmodel.2013.06.025

Skatter, S., and Kucera, B. (1998). The cause of the prevalent directions of the spiral grain patterns in conifers. *Trees Struct. Funct.* 12, 265–273. doi: 10.1007/s004680050150

Stoffel, M., and Corona, C. (2014). Dendroecological dating of geomorphic disturbance in trees. *Tree Ring Res.* 70, 3–20. doi: 10.3959/1536-1098-70.1.3

Tardif, J., Conciatori, F., and Bergeron, Y. (2001). Comparative analysis of the climatic response of seven boreal tree species from northwestern Québec, Canada. *Tree-Ring Res.* 57, 169–181.

Torres, A. B., and Lovett, J. C. (2013). Using basal area to estimate aboveground carbon stocks in forests: La Primavera Biosphere's Reserve, Mexico. *Forestry* 86, 267–281. doi: 10.1093/forestry/cps084

Watt, M. S., Kimberley, M. O., Harrington, J. J., Riddell, M. J. C., Cown, D. J., and Moore, J. R. (2013). Differences in intra-tree variation in spiral grain angle for radiata pine. *N. Zeal. J. For. Sci.* 43, 12. doi: 10.1186/1179-5395-43-12

West, P. W. (2009). *Tree and Forest Measurement.* Berlin: Springer-Verlag. doi: 10.1007/978-3-540-95966-3

Wing, M. R., Knowles, A. J., Melbostad, S. R., and Jones, A. K. (2014). Spiral grain in bristlecone pines (*Pinus longaeva*) exhibits no correlation with environmental factors. *Trees Struct. Funct.* 28, 487–491. doi: 10.1007/s00468-013-0965-y

Zobel, B., Stonecypher, R. W., and Browne, C. (1968). Inheritance of spiral grain in young Loblolly pine. *Forest Sci.* 14, 376–379.

Conflict of Interest Statement: The authors declare that the research was conducted in the absence of any commercial or financial relationships that could be construed as a potential conflict of interest.

5

Beyond ectomycorrhizal bipartite networks: projected networks demonstrate contrasted patterns between early- and late-successional plants in Corsica

Adrien Taudiere[1]*, François Munoz[2,3], Annick Lesne[4,5], Anne-Christine Monnet[1], Jean-Michel Bellanger[1], Marc-André Selosse[6], Pierre-Arthur Moreau[7] and Franck Richard[1]

[1] UMR 5175, CEFE – CNRS – Université de Montpellier – Université Paul Valéry Montpellier – EPHE – INSERM, Montpellier, France, [2] UM2, UMR AMAP, Montpellier, France, [3] French Institute of Pondicherry, Pondicherry, India, [4] CNRS, LPTMC UMR 7600, Université Pierre et Marie Curie-Paris 6, Sorbonne Universités, Paris, France, [5] CNRS, IGMM UMR 5535, Université de Montpellier, Montpellier, France, [6] CNRS, Muséum National d'Histoire Naturelle, UMR 7205, Origine, Structure et Evolution de la Biodiversité, Paris, France, [7] Département de Botanique, Faculté des Sciences Pharmaceutiques et Biologiques, Université Lille, Lille, France

Edited by:
Sergio Rossi,
Université du Québec à Chicoutimi,
Canada

Reviewed by:
Vicky Martine Temperton,
Leuphana University Lüneburg,
Germany
Raffaella Balestrini,
Consiglio Nazionale delle Ricerche,
Italy

*Correspondence:
Adrien Taudiere
adrien.taudiere@cefe.cnrs.fr

The ectomycorrhizal (ECM) symbiosis connects mutualistic plants and fungal species into bipartite networks. While links between one focal ECM plant and its fungal symbionts have been widely documented, systemic views of ECM networks are lacking, in particular, concerning the ability of fungal species to mediate indirect ecological interactions between ECM plant species (projected-ECM networks). We assembled a large dataset of plant–fungi associations at the species level and at the scale of Corsica using molecular data and unambiguously host-assigned records to: (i) examine the correlation between the number of fungal symbionts of a plant species and the average specialization of these fungal species, (ii) explore the structure of the plant–plant projected network and (iii) compare plant association patterns in regard to their position along the ecological succession. Our analysis reveals no trade-off between specialization of plants and specialization of their partners and a saturation of the plant projected network. Moreover, there is a significantly lower-than-expected sharing of partners between early- and late-successional plant species, with fewer fungal partners for early-successional ones and similar average specialization of symbionts of early- and late-successional plants. Our work paves the way for ecological readings of Mediterranean landscapes that include the astonishing diversity of below-ground interactions.

Keywords: bipartite networks, projected networks, host-specificity, ecological strategies, Mediterranean forests, ectomycorrhiza, ecological mycorrhizal network

INTRODUCTION

Evaluating the extent and functions of the ecological links that soil biota create among terrestrial plants is a fascinating challenge in ecology. In temperate and boreal forests, ectomycorrhizal (ECM) symbiosis ecologically binds together 3% of terrestrial plant species and more than 6000 filamentous fungal species. Below-ground mycelia directly connect short roots to soil resources, and provide pathways for reciprocal nutrient fluxes and water exchanges among associated individuals (Smith and Read, 2008). ECM symbiosis contributes to indirect interactions among trees through shared fungus partners (Bingham and Simard, 2012), and these interactions facilitate seedling establishment (Richard et al., 2009) and species coexistence (Selosse et al., 2006) in plant communities. ECM symbiosis thereby contributes to plant community dynamics during both primary (Nara, 2006) and secondary ecological succession (Simard et al., 1997; Richard et al., 2009; van der Heijden and Horton, 2009; Bingham and Simard, 2012).

One of the most striking and consistently observed properties of ECM symbioses is the variation over three orders of magnitude of the number of ECM fungal species associating with a plant species, ranging from a few fungal species associating with *Sarcodes sanguinea* and *Neottia nidus-avis* (Kretzer et al., 2000; Selosse et al., 2002) to over 1800 for *Pseudotsuga menziesii* (Molina et al., 1992).

Ectomycorrhizal fungal species also display large variation in the number of associated plant species, e.g., *Cenococcum geophilum* and *Laccaria amethystina* have been found on most European ECM trees (Horton and Bruns, 2001; Roy et al., 2008) whereas *Alpova alpestris* only associates with one tree species in the genus *Alnus* (Moreau et al., 2011). Phylogenetically constrained interactions have been shown in some specialized lineages, e.g., *Suillus* sp. associate only with Pinaceae, *Leccinum* with Betulaceae/Salicaceae, and *Alnicola* with *Alnus* (Wu, 2000; den Bakker et al., 2004; Rochet et al., 2011). Hereafter, specialism and generalism are used *sensu* Öpik and Moora (2012), i.e., referring to the ability of organisms to associate with large (interaction generalists; see Glossary in **Box 1**) or small (interaction specialists) numbers of partners.

Demonstrations of the ecological and evolutionary advantages of specialism vs. generalism in ECM symbiosis are lacking, for several reasons: most ECM fungal lineages are refractive to *in vitro* cultivation, fungal species may appear less host-specific than they really are due to cryptic diversity, and determinants of fungal host-specialization are still controversial (Hawksworth, 2001; Bruns et al., 2002). It is commonly accepted that generalism may enable ECM plants to extend the habitats in which they can establish as seedlings (Bruns et al., 2002; Botnen et al., 2014). Thus, generalist plants may gain access to a greater reservoir of compatible ECM inoculum and may therefore establish more easily than specialist plants. The ability of ECM plants to colonize new areas more rapidly may result from plant ability either to associate with numerous distantly related partners (direct plant generalism), or to associate with a set of fungi that do so (indirect plant generalism through fungal generalism). During the first steps of ecological succession, generalism may primarily drive vegetation dynamics. Early-successional plants first establish in newly available habitats through fungi-mediated facilitation processes (e.g., Nara, 2006), while late-successional tree species colonize (e.g., Selosse et al., 2006; Richard et al., 2009) and outcompete (Bruns et al., 2002) in pioneer vegetation through facultative epiparasitism. At the end of the succession, ECM communities are hyper-diverse assemblages classically associated with long-lived late-successional tree species in old stands (Visser, 1995; Dahlberg et al., 1997; Smith et al., 2002; Richard et al., 2005; Walker et al., 2005). It has been hypothesized that this pattern of fungal diversity enrichment with forest aging may be driven by a late accumulation of host-specific fungi (Last et al., 1983; Smith et al., 2002; Twieg et al., 2007).

Two questions arise. First, is there a trade-off between the number of fungal partners and their specialization at the plant species level? Second, do early- and late-successional plant species differ in their association patterns, with late-successional species accumulating more specialized fungal species than early-successional plants? We here explore these questions by analyzing the correlation between the number of symbionts of a plant species and the average specialization of these symbionts, in regard to the successional status of the plant species in Corsica.

The analysis of bipartite interaction networks has provided important insights into ecological (Bascompte and Jordano, 2007; Chagnon et al., 2014) and evolutionary questions (Ives and Godfray, 2006; Krasnov et al., 2012). Analyses have demonstrated that the topology of the network depends on whether the interactions are mutualistic or antagonistic (Thébault and Fontaine, 2010). To date, most research on bipartite ecological networks has focused on plant-pollinator or plant–herbivore interactions (Thébault and Fontaine, 2010). The structure of local plant–fungi bipartite networks has been investigated at the species level in the context of endomycorrhizal (Chagnon et al., 2012; Montesinos-Navarro et al., 2012; Öpik and Moora, 2012) and orchid mycorrhizal associations (Martos et al., 2012). In the context of the ECM symbiosis, ever-increasing information is available on below-ground associations to inform the links between ECM fungal and plant species. However, studies examining these interactions in a network perspective are lacking (but see Bahram et al., 2014). Moreover, most studies have considered the ECM community as a static biotic component of the plant's ecological niche.

We here argue that investigating the structure of ECM interspecific networks is a path to explore the ecological role of ECM symbiosis at a systemic level. We have analyzed an ECM ecological bipartite network (see Glossary) constructed from a unique qualitative dataset that exhaustively assembled field and molecular records on ECM plant–fungi associations in the island of Corsica, France. ECM links between plant and fungal species are considered at the scale of the whole island. More specifically, our aim was to understand the potential of ECM symbiosis for creating fungi-mediated ecological interactions (e.g., facilitation or competition) between plant species, and reciprocally between ECM fungal species by means of shared host plants. These indirect interactions among species of the same kind (plants vs. fungi) mediated by species of the other kind constitute *projected networks* (see Glossary; Latapy et al., 2008; Nacher and Akutsu, 2011).

Our study investigated three main questions. First, whether there is a trade-off between the number and the specialization of fungal species associated with a plant species; second, whether plant species associated with fewer fungal species in the bipartite network are linked to fewer plant species in the projected

---

BOX 1 | Glossary.

   Bipartite network: a network linking two distinct sets of nodes (with no links among nodes of the same set). **Figures 1** and **2A** illustrate the ectomycorrhizal (ECM) ecological network linking plant species with fungal species. Bipartite networks are also termed 2-mode networks or bimodal networks.

   Common mycorrhizal network (CMN): below-ground network where fungal mycelia physically connect roots of different plant individuals.

   Degree (k): in network terminology the degree k_n of a node n is the number of links it has established with other nodes. Here, the degree of a species (either a fungal or a plant species) corresponds to the number of its symbiotic partners. It measures its interaction specialization.

   Ecological mycorrhizal network: mutualistic interaction network linking together plant and fungal entities (e.g., individuals, populations, species) able to establish a mycorrhizal connection in at least one ecological context and during one ontological stage. The ecological ECM network studied here at the species level (**Figure 1**) is qualitative (binary links) and only informs on the potentiality of two species to interact.

   Interaction specialization: tendency to interact with few or lot of partners. A species that interacts with many species (high degree k) is termed an **interaction generalist** and a species that interacts with few species (low degree k) is termed an **interaction specialist**. In the case of the ECM ecological network, fungal interaction specialization is often called host-specificity of the fungal species in mycological literature.

   Modular network: a modular network is made of subsets of nodes highly connected between them and poorly connected to others. These subsets are called modules or clusters. Modularity in an ecological network may reflect ecological (e.g., spatial or successional position for the plant species) and evolutionary (e.g., coevolution) processes.

   Partner specialization (c): for a focal node n, we defined c_n as the mean degree of its partners (its direct neighbors in the bipartite network), where the mean is taken over the set of its partners. Here, the partner specialization of a focal plant species is the mean number of host plant species of its fungal symbionts (**Figure 2A**). The average of c_n over plant species of same bipartite degree recovers the standard bipartite degree correlation (sometimes termed connectivity correlation).

   Projected degree (l): the projected degree l_n of a node n of a bipartite network is its degree in the corresponding projected network, that is, the number of nodes of the same set sharing at least one neighbor in the bipartite network (**Figure 2B**). Here, the projected degree of a focal plant species is the number of other plant species sharing fungal partners with it.

   Projected network: 1-mode network built from a bipartite network, by considering only nodes of one set, and linking two nodes if they share at least one neighbor in the bipartite network (**Figure 2B**); the links of a projected network are also termed indirect links, mediated by nodes of the other set. A bipartite network is associated with two projected networks. Here, the plant projected network links ECM plants through their shared fungal partners (Supplementary Figure S10).

   Projected weight (s): we defined the projected weight s_n of a focal node n as the total number of indirect, two-step connections to its neighbors in the projected network in its projected network (**Figure 2B**). Here, the projected weight of a focal plant species is the number of fungal species shared with other plants, where each fungal species is counted as many times as it indirectly links the focal species to another plant species. The projected weight of a focal plant species is in general different from its projected degree l_n, but related to its partner specialization c_n and its bipartite degree k_n according to $c_n = 1+(s_n/k_n)$.

---

network; and finally, when the ecological strategies of plants are considered, whether early- and late-successional species display different interaction patterns, in either the bipartite network or the projected plant network.

MATERIALS AND METHODS

Study Area

We assembled and analyzed a database on plant species and their associated ECM fungal species all over the Mediterranean island of Corsica (Conservatoire Botanique National de Corse, unpublished data; data are given in Supplementary Table S1, references in Supplementary References S2, detailed description of the methods in Supplementary Methods S3, and the workflow of data filtering in Supplementary Figure S4). The island covers 8681 km^2 of mountainous territory. It is the sole island in the Mediterranean basin presenting large surfaces of well-preserved native ECM forests (Quézel and Médail, 2003; Richard et al., 2005). The distribution and numbers of plant species and ECM fungal species are exceptionally well-known after over a century of intensive botanical and mycological surveys (Jeanmonod and Gamisans, 2007).

Plant Species Included in the Analysis

We constructed a network based on all known ECM plant species present in Corsica (Gamisans, 1991; Jeanmonod and Gamisans, 2007) excluding recently introduced species (*Eucalyptus globulus, Larix decidua, P. menziesii,* and *Populus nigra*). We also excluded mycoheterotrophic and mixotrophic plants (orchids and Pyroleae; Selosse and Roy,

2009) from the analysis because of the atypical physiology of their relationship with ECM fungi. Eleven additional plant species were excluded from the analysis because of their very low frequency (*Pinus pinea, Populus canescens, Pinus halepensis,* and *Quercus robur* subsp. *robur*) or because their ECM fungal communities are insufficiently documented (*Arbutus unedo, Fumana laevipes, Fumana thymifolia, Ostrya carpinifolia, Populus tremula, Quercus humilis* subsp. *Humilis,* and *Quercus petraea* subsp. *petraea*; cf. Supplementary Methods S3).

Salix species cover large areas in many locations of the island, where the different species co-occur and their local ECM communities have not been critically analyzed using molecular tools. The same is true for *Cistus* species. Therefore, we conservatively considered each of these genera as a single taxon in the analysis (*Salix* sp. on one hand and *Cistus* sp. one the other hand; cf. Supplementary References S2 and Supplementary Methods S3). The 16 resulting plant taxa (hereafter called plant species) included all the tree species dominating the forest stages of plant ecological successions in Corsica (Gamisans, 1991). They represented all forest ecosystems from sea level (sclerophyllous oak forests) to the upper altitudinal tree limit.

Fungal Species Included in the Analysis

For macromycetes, 411 species out of 610 ECM fungal species recorded in Corsica (67%, Supplementary Table S1; Supplementary Methods S3) were included in the network analysis. We retained only taxa for which were available (i) a well-resolved taxonomic treatment, and (ii) reliable data on host association within Corsica based on published records (Supplementary References S2).

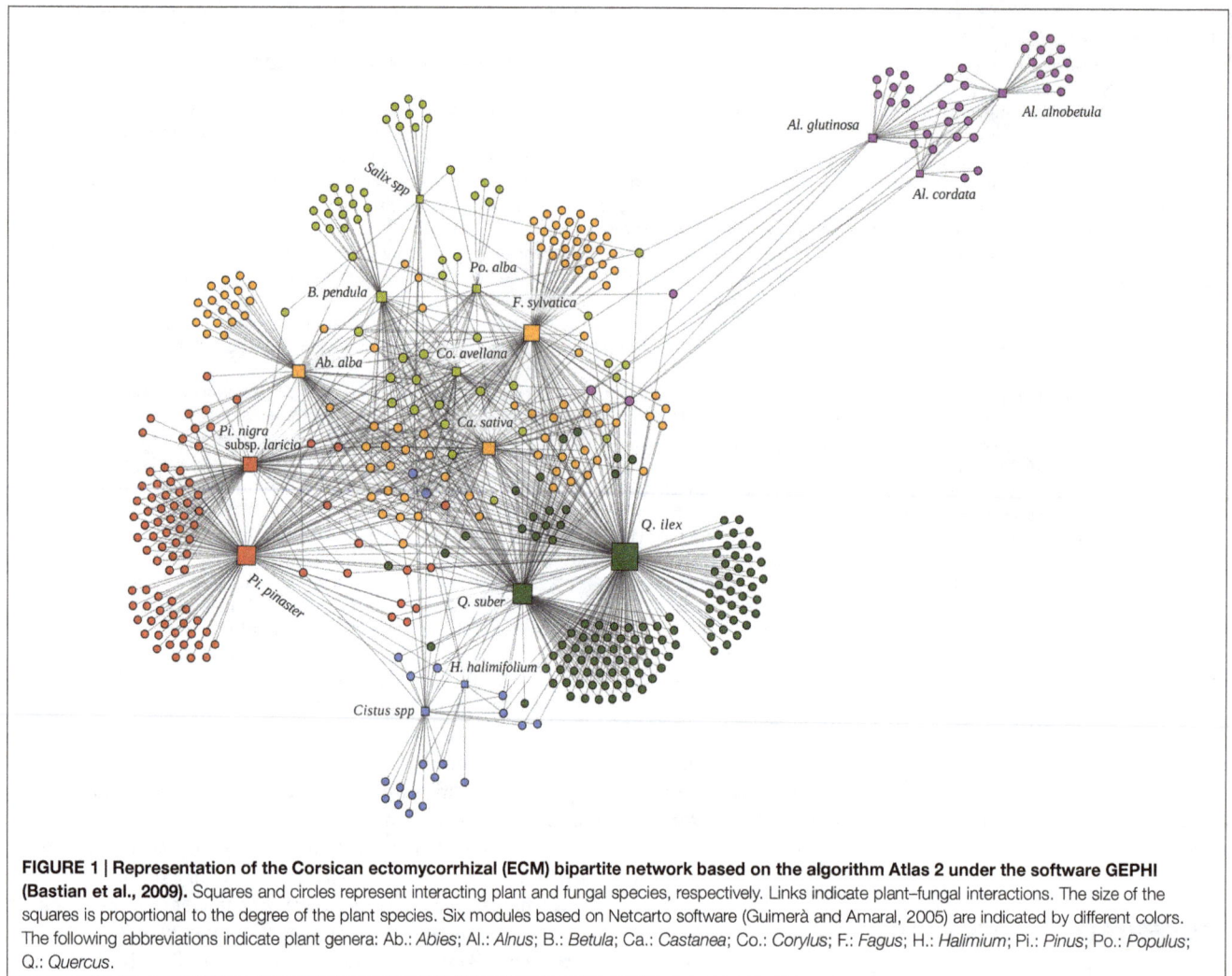

FIGURE 1 | Representation of the Corsican ectomycorrhizal (ECM) bipartite network based on the algorithm Atlas 2 under the software GEPHI (Bastian et al., 2009). Squares and circles represent interacting plant and fungal species, respectively. Links indicate plant–fungal interactions. The size of the squares is proportional to the degree of the plant species. Six modules based on Netcarto software (Guimerà and Amaral, 2005) are indicated by different colors. The following abbreviations indicate plant genera: Ab.: *Abies*; Al.: *Alnus*; B.: *Betula*; Ca.: *Castanea*; Co.: *Corylus*; F.: *Fagus*; H.: *Halimium*; Pi.: *Pinus*; Po.: *Populus*; Q.: *Quercus*.

Validation of the Interactions Included in the Analysis

In total, we defined 993 interactions (**Figure 1**) from 61 informative analyzed datasets extracted from either fruitbody surveys or DNA belowground studies (Supplementary References S2). This two sources of information are two complementary views of plant–fungal ECM links (Horton and Bruns, 2001). Briefly, we compiled a large set of publications (both peer-reviewed and 'gray literature'), books and records from expert field mycologists. The associations of the 411 selected fungal species with the 16 plant species were compiled as follows. To be validated, each association between plants and fungi was ascertained by molecular studies (Richard et al., 2004; Moreau et al., 2011; Rochet et al., 2011; Roy et al., 2013), or by validated observations within a monospecific forest/chaparral stand (one single ECM host) in Corsica, or by both methods. Molecular data were obtained from mycorrhizal sequencing in published studies. Only mycorrhizal root tips that were (i) taxonomically assigned to a given fungal species (blast) and (ii) related to a given plant host were included in the dataset. Herbarium collections (published or not; most came from the Lille University herbarium LIP, France) associated with field observations were

systematically used to validate the taxonomic identity of fungal species.

Analysis of Bipartite and Projected Networks

We built a bipartite network including the 16 plant species and 411 ECM fungal species (**Figure 1**). The corresponding 16×411 binary matrix of association has value 1 at position (p, f) if fungal species f has been reported to be an ECM symbiont of plant species p, and 0 otherwise (Supplementary Table S1). Following network theory terminology, the number of ECM fungal species that are linked with a given plant species p is called its bipartite degree k_p (**Figure 2**). The number of plant species that are linked with a given fungal species f is called its bipartite degree k_f. Low k values characterize interaction-specialist species (*sensu* Öpik and Moora, 2012), while high k values characterize interaction-generalist species. Hereafter, plant specialization (generalism vs. specialism) will be used to refer to the realized biotic part of plant niche, and not to any theoretical niche of ECM plants in the Corsican region. The fungal species linked with the plant species p may also be linked with other plant species, and the number of plant species that are linked through one or more

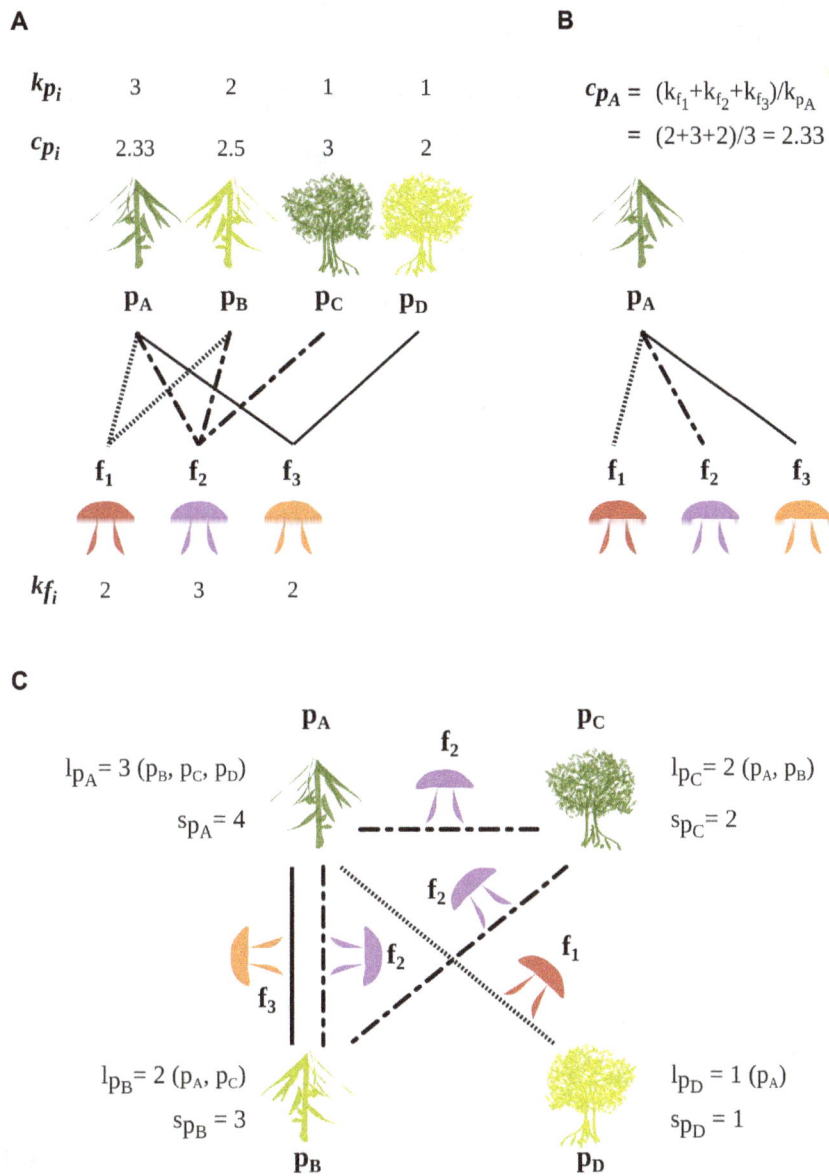

FIGURE 2 | A schematic bipartite network of four ECM plant species (p_A, p_B, p_C, p_D) and their three ECM fungal symbionts (f_1, f_2, f_3). The degree in the bipartite network is k_{p_i} for a plant species p_i and k_{f_j} for a fungal species f_j. The average degree of the fungal partners of a given plant species p_i is c_{p_i}. **(A)** Displays the degrees of plant and fungal species and the average degree of the fungal partners of each plant species. **(B)** Illustrates the detailed calculation of c_{p_A}. **(C)** Illustrates the plant projected network composed of indirect (two-step) links through fungi, i.e., plant–plant links mediated by shared fungal symbionts. The number of plant species linked to a given plant species through s_{pi} fungi-mediated links is l_{p_i}. The numbers of indirect (two-step) links between a given plant species p_i and the other plant species is s_{pi}. In all panels, dotted, broken and continuous lines show plant–plant links made by the fungal species f_1, f_2, and f_3, respectively. The roles played by plants and fungi can be inverted to analyze the network from a fungal perspective.

of these shared fungal partners is called the projected degree of the plant species, l_p. These links constitute the projected network of plant species (or "plant projected network"). We likewise defined the projected degree of a fungal species, l_f, as the number of associated fungal species in the projected network, i.e., linked through plant species to the focal fungal species f (**Figure 2**).

A plant species can interact with another plant species *via* a single fungal species or *via* several ones. The number of fungal

species linking two plant species is a measure of the strength of their association in the projected network. The total number of indirect (two-steps) links of the plant species p to other plant species is called its weight s_p in the plant projected network (links between plant species in **Figure 2C**). Likewise, the total number of plant-mediated links of a fungal species f to other fungal species is its weight s_f in the fungal projected network.

While l_p represents the number of plant species to which p is linked in the plant projected network, s_p represents the total number of links established *via* fungal species with these plant species. The contribution to s_p of a plant species is simply the number of ECM fungal species that it shares with p. Therefore, the ratio of s_p to l_p measures in an integrated way the redundancy of fungal species in establishing links between p and other plant species.

If the number of links of a given plant species to other plant species is selected for, then we may expect a trade-off for the plant between its specialization (value of k_p) and the specialization of the fungi (value of k_f) with which it associates. Indeed, a plant species associated with many specialized fungal species (high k_p and partners with low k_f) may be as well-linked (same s_p) with as many other plant species (same l_p) as a plant species associated with few generalist fungal species (low k_p and partners with high k_f). To explore this question, we introduced an additional quantity, c_p, measuring the average specialization of the fungal partners of a given plant species p. It is defined as the mean degree (in the bipartite network) of the fungal partners of the plant species, p, where the average runs over the set of these fungal partners. We then devised two null models to analyze the relationships between c_p and k_p (see below). A summary of the different network-related notions and their notation is given in the Glossary (**Box 1**).

Null Models

Null models were used to simulate random situations sharing some minimal constraints with the actual network by randomizing links among plant and fungi. Deviations of observed network statistics from the statistics measured in these random situations inform on whether species are more generalist or specialist than expected by chance, or whether they are associated with partners that are themselves more generalist or specialist than expected by chance, given the specified constraints.

First we devised a simple random model (null model 1) considering independently each pair of plant and fungal species and drawing a link with a probability equal to the density of links (actual number of links divided by the maximal number of links given the number of plant and fungal species) observed in the real network. As a consequence, the mean values of plant and fungal bipartite degrees are the same in this null model 1 and the real network.

The distributions and relationships between k_p and c_p values obtained with this null model were compared with the distribution and relationships obtained in another null model, in which the bipartite degree of each plant or fungal species is exactly the same as in the real network, but the links between species were randomized (null model 2). Both null models were generated using the *permat* functions within the R package Vegan with (model 2; method *quasiswap* of the function *permatswap*; Patefield, 1981) or without (model 1; function *permatfull*) preserving column and row sums (Oksanen et al., 2015). Using these two null models, it is possible to identify the correlations between k_p and c_p values which are simply due to the degree distributions in the dataset, i.e., to the distribution of specialism vs. generalism. Comparison of the data with null model 1 allowed detecting whether species bipartite degrees were distributed

randomly within the network (Gotelli and Entsminger, 2003). We present our results on partner specialization as a scatter plot, keeping track of the plant species identity and the variance of their partner specialization (**Figure 3A**). Finally, observed values of c_p and k_p were compared with their statistics obtained in the null model 2 in order to detect whether the actual network differs from a random network in which levels of species specialization (values of k_p and k_f) were identical to those observed in the real data.

Given that the association matrix is highly asymmetric (16 plants * 411 fungi), our analysis focused on a plant perspective as more information (more potentially associated partners) is available for each plant, allowing a more powerful statistical analysis. The fungal-perspective analysis is presented in Supplementary Figure S5.

ECM Association Patterns and Ecological Strategy of Plant Hosts

Early- stage vs. late-stage status of each plant species was determined according to reference work on their ecology in Corsica (Jeanmonod and Gamisans, 2007; Rameau et al., 2008). Accordingly, all species belonging to the Salicaceae, Betulaceae, and Cistaceae were classified as early-stage plants, while all species belonging to the Pinaceae and Fagaceae were classified as late-stage plants. We compared the degrees of plant species in the bipartite network (k_p), their weights in the projected network (s_p) and average numbers of hosts of their fungal symbiotic species (c_p) in the sets of early-stage and late-stage plant species, using Mann–Whitney non-parametric tests. We measured modularity using the simulated annealing algorithm implemented in Netcarto (Guimerà and Amaral, 2005). This software also simulates random graphs with the same degree distribution as the original network to test for modularity significance (Guimerà et al., 2004).

RESULTS

Relationships between the Number of ECM Partners and their Specialization

Among the 16 plant species, association patterns varied from species-poor (low k_p values) fungal communities associated with Betulaceae (*Alnus, Betula, Corylus*, whose number of associated ECM fungal species ranged from 14 to 46) and Cistaceae (*Cistus* and *Halimium* with $k_p = 26$ and $k_p = 12$, respectively) to rich ECM fungal assemblages (high k_p values) associated with *Quercus suber* and *Quercus ilex* evergreen oaks (respectively 133 and 197 ECM linked fungal species; **Figure 3A**). There was no significant correlation between the bipartite degree k_p of a plant species and the interaction specialism c_p (average number of host species) of its associated fungal species (Spearman non-parametric rank correlation test, $p = 0.71$; **Figure 3A** green points).

The variances of the plant (k_p) and fungal (k_f) bipartite degrees were much higher in the real network than in the simple random model (null model 1; **Figure 3A** and Supplementary Figure S5 for k_f). We then constrained the null model so that the degree of each plant (k_p) and fungal species (k_p) is the same

A

Average number of hosts of the symbionts (c_p)

- Early stage plants
- Late stage plants
- Null model 2 means with 95% confidence interval
- Confidence ellipse for Null model 1 (95%)

Co. avellana
Ca. sativa
Po. alba
B. pendula
Ab. alba
H. halimifolium
Pi. nigra subsp. laricio
F. sylvatica
Q. suber
Salix spp
Cistus spp
Pi. pinaster
Al. cordata
Q. ilex
Al. glutinosa
Al. alnobetula

Plant degree (k_p) in the bipartite network = Number of ECM fungal associates

B

Difference with a random distribution of symbionts' specialization

Towards symbionts generalism

Ca. sativa
Q. suber
Co. avellana
F. sylvatica
Q. ilex
B. pendula
Pi. nigra subsp. *laricio*
Ab. alba
Po. alba
H. halimifolium
Salix spp
Cistus spp
Pi. pinaster
Al. cordata
Al. glutinosa
Al. alnobetula

Towards symbionts specialism

Plant degree (k_p) in the bipartite network = Number of ECM fungal associates

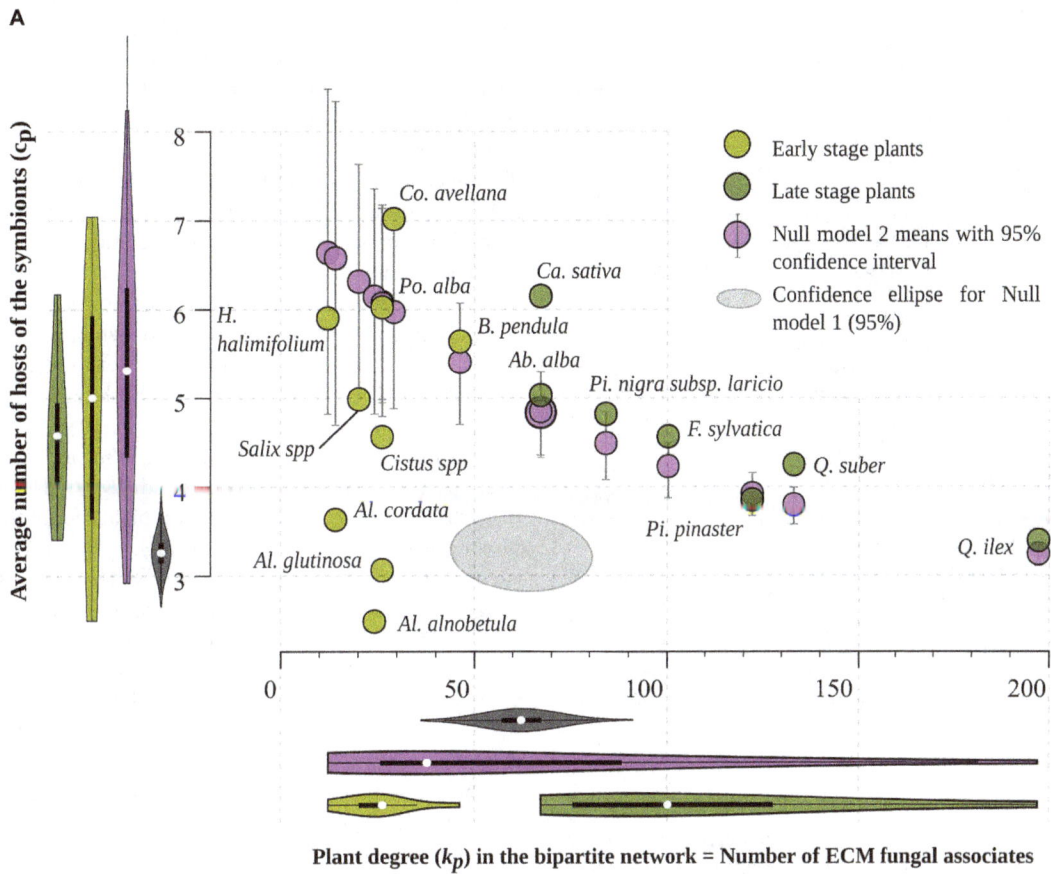

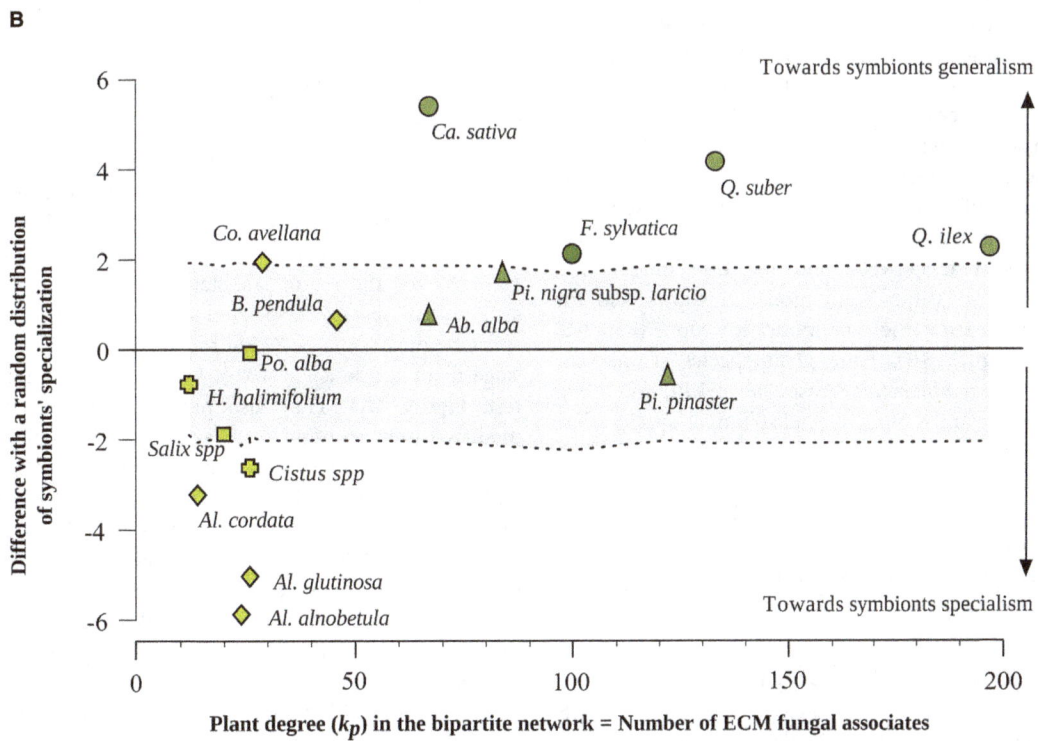

FIGURE 3 | Continued

as in the real network (null model 2, 999 permutations). The simulation revealed a strong negative correlation between plant degree (k_p) and mean interaction specialism of fungal partners (c_p) (**Figure 3A** purple points). Hence, k_p and k_f values impose a constraint on the structure of the network which *a priori* results in a strong negative correlation between number of associated fungal species and degree of specialization of these fungal species (**Figure 3A** purple points). In the real network, the absence of correlation between the bipartite degree k_p of a plant species and the average number of hosts c_p of its associated fungal species is unexpected given the distribution of k_p and k_f values.

Comparing the real network and the null model with constrained k_p and k_f values (null model 2) shows that the value of c_p deviates from the null model for more than half of the species (**Figure 3B**). Four plant species (*Cistus* sp., *Alnus glutinosa*, *Alnus alnobetula* subsp. *suaveolens*, and *Alnus cordata*) host significantly more specialist fungal species than would be expected by chance (given the distributions of degrees; p-values < 0.025 using the null model 2 for these species) while the four Fagaceae species (*Q. ilex*, *Q. suber*, *Fagus sylvatica*, and *Castanea sativa*) and *Corylus avellana* host significantly more generalist fungal species than expected by chance (**Figure 3B**; p-values < 0.025 using the null model 2 for these species).

Projected Networks of ECM Plants

Ten plant species are linked to all other plant species ($l_p = 15$; **Table 1**). Indeed, the number of plant species to which a plant species is linked in the projected network varies from 11 (*A. cordata* and *A. alnobetula*) to all other 15

species (*A. glutinosa* and all species belonging to Cistaceae, Pinaceae, or Fagaceae; **Table 1**). Behind the tendency of all plant species to saturate their projected network, there is wide variation in the number of ECM-mediated links (two-step links) that a given plant species makes with all other plant species (s_p ranging from 36 ECM-mediated links for *A. alnobetula* to 474 links for *Q. ilex*; Supplementary Figure S6). For instance, 12 ECM partners link *Halimium halimifolium* with all plant species (altogether mediating 59 two-step links), so that the neighborhood of *Halimium halimifolium* in the projected network includes all the plant species of the present survey. In contrast, *Betula pendula* symbiotic links do not saturate the plant–plant projected network even though its 46 associated ECM fungal species establish 214 fungi-mediated links.

Endemic alders, *A. cordata* and *A. alnobetula*, with $l_p = 11$ plant partners through $s_p = 37$ and 36 fungi-mediated links respectively, interact with slightly less plant species than the wide-ranging *A. glutinosa* ($l_p = 15$ through $s_p = 54$ links; **Table 1**). In our analysis, these two endemic alders establish no ECM fungi-mediated links with *Betula*, *Corylus*, *Populus*, or *Salix*.

There is wide variation in the number of ECM-mediated links (two-step links) that a given plant species makes with all other plant species (s_p values, taken as the weight of the plant species in the projected network), ranging from 36 ECM-mediated links for *A. alnobetula* to 474 links for *Q. ilex* (Supplementary Figure S6).

The redundancy l_p/s_p (ratio of the projected degree l_p to the number of comprised links s_p) ranges from $11/36 = 0.306$ for *A. alnobetula*, to $15/474 = 0.032$ for *Q. ilex* (**Table 1**). These results indicate a lower redundancy of the ECM-mediated links of *A. alnobetula* to other plant species as compared to *Q. ilex*.

ECM Associations along Ecological Succession

Early-stage vs. late-stage plant species displayed contrasted distributions of their numbers of associated fungal species (**Figure 3** and **Table 1**, with a maximal degree $k_p = 46$ for the set of early-stage species and a minimal degree $k_p = 67$ for the set of late-stage species). On average, early-stage plant species associated with five times fewer fungal species (mean $k_p = 24.78$ for early-stage vs. 110 for late stage plant species; $p < 10^{-4}$, Mann–Whitney non-parametric test; **Figure 4A**). They also present five times fewer fungi-mediated links to other plant species in the projected network (mean $s_p = 97.67$ vs. 365; Mann–Whitney $p < 10^{-4}$; **Table 1** and **Figure 4B**) and five times higher l_p/s_p ratios (mean ratio = 0.38 vs. 0.083; Mann–Whitney $p < 10^{-3}$). Nevertheless, their associated fungal species did not differ in their partner specialization c_p (average number of host species of its symbionts; mean $c_p = 4.83$ vs. 4.59; Mann–Whitney $p = 0.84$) from those associated with late-stage plant species (**Figures 3A** and **4C**).

We compared the sharing of ECM symbionts observed in the real system (i) between early- and late-stage plant species, (ii) among early-stage plant species, and (iii) among late-stage plant species, against the sharing calculated in the null model

TABLE 1 | Network properties of the 16 ECM plant taxa.

	H. halimifolium	*Al. cordata*	*Salix sp.*	*Al. Alnobetula*	*Cistus sp.*	*Al. glutinosa*	*Po. alba*	*Co. avellana*	*B. pendula*	*Ca. sativa*	*Ab. alba*	*Pi. nigra subsp. laricio*	*F. sylvatica*	*Pi. pinaster*	*Q. suber*	*Q. ilex*
k_p	12	14	20	24	26	26	26	29	46	67	67	84	100	122	133	197
c_p	5.92	3.64	5	2.5	4.58	3.08	6.04	7.03	5.65	6.16	5.04	4.83	4.58	3.87	4.26	3.41
l_p	15	11	13	11	15	15	13	13	13	15	15	15	15	15	15	15
s_p	59	37	80	36	93	54	131	175	214	346	271	322	358	350	434	474
n_{hs}	1	2	8	13	7	8	18	3	14	1	18	2	29	27	1	39

k_p: degree in the bipartite network, i.e., the number of their fungal partner species; c_p: average number of host species of their symbionts; l_p: degree of plant species in their projected network; s_p: total number of links of a plant species within its projected network; n_{hs}: number of hyper-specialists, i.e., strictly associated fungal taxa. See Glossary for more complete definitions of network metrics.
The following abbreviations were used to indicate plant genera. Ab.: Abies; Al.: Alnus; B.: Betula; Ca.: Castanea; Co.: Corylus; F.: Fagus; H.: Halimium; Pi.: Pinus; Po.: Populus; Q.: Quercus.

with constrained k_p and k_f values (null model 2). Fewer ECM fungal species (62) than expected by chance (108.52 ± 4.95 under null model 2) interact with both early- and late-stage plant species ($p < 10^{-3}$). More ECM fungal species (37) are exclusively associated with early-stage plant species than expected by chance (25.54 ± 4.21 under null model 2; $p < 10^{-3}$) given the number of partner species. On the contrary, the number of ECM fungal species (272) exclusively associated with late-stage plant species is close to the value expected by chance (277.21 ± 4.33 using null model 2; $p = 0.09$). These results show that during ecological successions, late-successional vegetation stages accumulate rich ECM communities comprised of specialist fungi that differ in composition compared to the communities associated with early-successional stages (Supplementary Figures S7–S9).

We further used the modularity of the whole network to characterize the sharing of ECM symbionts among early- and late-stage plant species. We detected six modules

in a significantly modular system ($M = 0.458$, null model $= 0.362 ± 0.002$). Three modules included only early-stage plant species (represented in light green, blue and purple in **Figure 1**) and the three others included the late-stage species (represented in orange, red and dark green in **Figure 1**).

Comparing *A. glutinosa* (early-stage) and *Q. ilex* (late-stage) illustrates the contrast between early- and late-stage plant species association patterns (**Figure 5**). Fungi-mediated interactions of *Q. ilex* are (i) quantitatively highly variable (from two links toward *A. cordata* and *A. alnobetula* to 131 links toward *Q. suber*) and (ii) qualitatively more numerous toward Fagaceae (*Q. suber*, *C. sativa*, and *F. sylvatica* accounted for 52.95% of the 474 indirect links of *Q. ilex*). All but two fungal species associating with *Q. suber* also associate with *Q. ilex* in Corsica (*Inocybe fibrosoides* is strictly associated with *Q. suber* and *Boletus pulverulentus* is strictly associated with *Q. suber* and *C. avellana*).

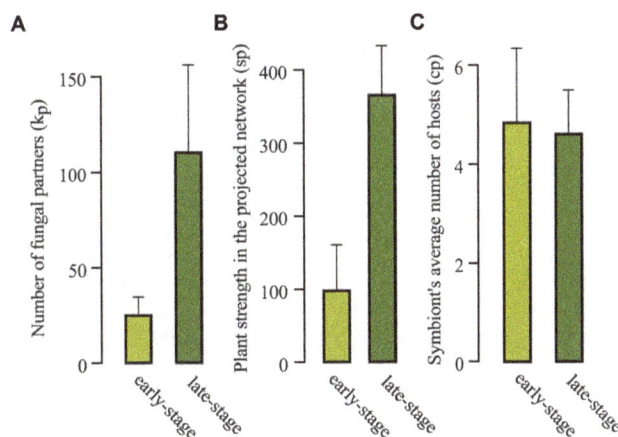

FIGURE 4 | (A) Comparison of the average plant bipartite degree (average of s_p), **(B)** average number of fungi-mediated links (average of s_p), and **(C)** average mean number of host species of fungal symbionts (average of c_p) over the set of early-stage plant species (light green) and the set of late-stage (dark green) plant species. Bars indicate standard errors.

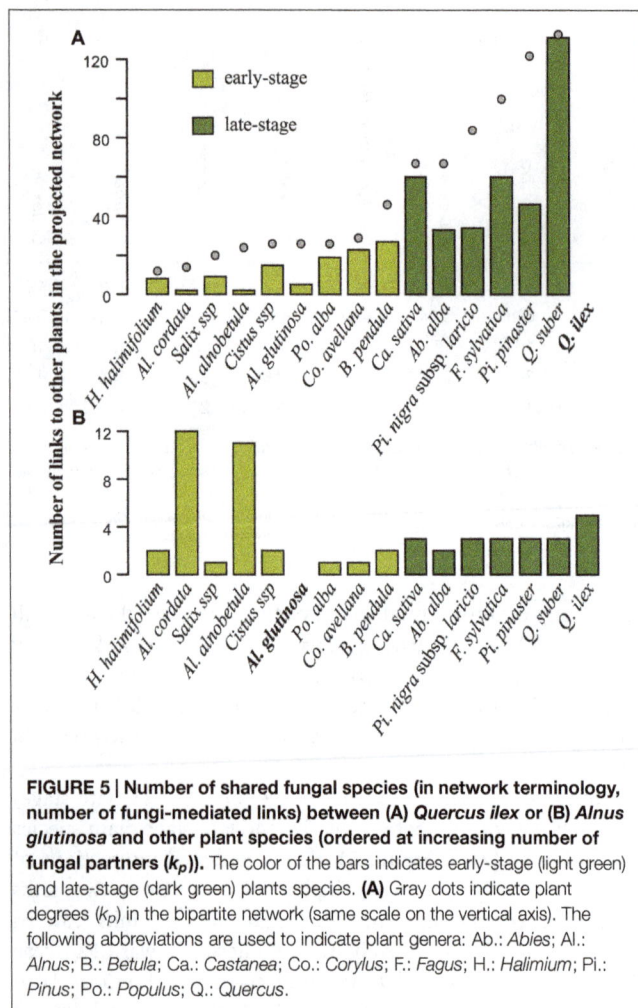

FIGURE 5 | Number of shared fungal species (in network terminology, number of fungi-mediated links) between (A) *Quercus ilex* or (B) *Alnus glutinosa* and other plant species (ordered at increasing number of fungal partners (k_p)). The color of the bars indicates early-stage (light green) and late-stage (dark green) plants species. (A) Gray dots indicate plant degrees (k_p) in the bipartite network (same scale on the vertical axis). The following abbreviations are used to indicate plant genera: Ab.: *Abies*; Al.: *Alnus*; B.: *Betula*; Ca.: *Castanea*; Co.: *Corylus*; F.: *Fagus*; H.: *Halimium*; Pi.: *Pinus*; Po.: *Populus*; Q.: *Quercus*.

late-successional plant species share fewer fungal symbionts than expected by chance, with early-successional species presenting fewer links caused by fewer fungal partners than late-successional plant species, and no difference in average specialization of fungal symbionts of early- and late-successional plant species.

Relationship between the Number and the Specialization of ECM Symbionts

In our analysis, plant species hosting the richest ECM fungal assemblages associate with fungi whose host ranges are similar to those of plants with few fungal partners (**Figure 3A**). In other words, the analysis of the whole dataset shows no trade-off between plant specialization and the specialization of its symbionts. As a consequence, the number of fungal partners of a plant species (k_p) poorly reflects the number of plant species with which it has fungi-mediated indirect interactions (l_p).

Insights into the Hyper-diversity of Oak Symbionts

The positions of *Quercus* species in both **Figures 3A,B** confirm the well-known hyper-diversity of ECM communities of evergreen oaks (Richard et al., 2004; Morris et al., 2008). Here, we show that the high diversity of fungal symbionts dissimulates contrasted specialization degrees, with few hyper-generalists occurring among a disproportionately high number of interaction-specialist fungal species. Both oak species deviate from the null model by associating with fungi that are less specialized on average than expected by chance (**Figure 3B**). Interestingly, all but two fungal symbionts of *Q. suber* are also associates of *Q. ilex*. Only two of the 133 fungal symbionts of *Q. suber* are host-specific, while 39 species out of the 197 fungal associates of *Q. ilex* are host-specific (**Table 1**). This pattern may result from a mix of phylogenetic and ecological proximity. Indeed, the distribution and the ecological range (including soil chemical conditions and climatic requirements) of *Q. suber* in Corsica tend to be included within those of *Q. ilex* (Gamisans, 1991). This extensive sharing of fungal symbionts between the two oak species results in a low specialization of oak symbionts (low number of single-host fungal species). However, in these hyper-diverse communities, the symbionts are mostly oak specialists that establish few links to tree species belonging to other genera: 57 of the 131 fungal species associated with both oak species are only associated with oaks (Supplementary Figure S10).

Similarly, *A. glutinosa* presents a high variation in the number of fungi-mediated links toward other plant species, ranging from one (*C. avellana, P. alba,* and *Salix* sp.) to 12 (*A. cordata*). Both *A. glutinosa* and *Q. ilex* show higher numbers of fungi-mediated links toward plant species belonging to their own genus (*A. alnobetula* and *A. cordata* for *A. glutinosa*, and *Q. suber* for *Q. ilex*) and, for *Q. ilex*, to its family (*C. sativa* and *F. sylvatica* for *Q. ilex*; **Figures 1** and **5**).

DISCUSSION

In this paper we analyzed the main properties of the ECM bipartite network linking host plants and fungal species in Corsica and deciphered ECM network structure in an ecological perspective. We did not detect significant correlation between the specialization of plant species and the specialization of their symbionts across the whole plant dataset, but rather observed decoupled patterns of symbiont diversity (number of fungal partners associated with a given plant species) and the specialization of these symbionts (average number of plant partners of these symbionts). Our results show that early- and

Contrasted Structures of Bipartite and Plant Projected Networks

Our ECM network presents a modular structure (**Figure 1**) due to abundant plant–plant connections at genus (e.g., within *Alnus*, **Figure 5B**), family (e.g., within Fagaceae, **Figure 5A**; see also Supplementary Figure S7) and ecological (early- vs. late-stage, **Figure 4** and Supplementary Figure S8) levels. Our results complement those of previous studies that suggested modularity of below-ground mycorrhizal networks (Chagnon et al., 2012; Martos et al., 2012; Montesinos-Navarro et al., 2012).

The analysis of the structure of plant–plant interaction patterns (Supplementary Figure S10) through compatible fungal species reveals that generalism prevails in the plant projected network. Among the 16 species, 10 of them interact with all the other plant species and only two interact with fewer than 13 plant species (**Figure 3A**; **Table 1**). This tendency to saturate the projected network (Supplementary Figure S10) contrasts with the high heterogeneity in plant degree found in the bipartite network (**Figure 3A**). As a consequence, similar numbers of neighbors in the plant projected network hide highly variable numbers of symbionts (**Table 1**; **Figure 5**). The contrasted patterns observed for *A. glutinosa* and for *Q. ilex* illustrate this point. In our analysis, *A. glutinosa* saturates the plant–plant network based on only 54 fungal-mediated links whereas *Q. ilex* use 474 such links to do so (**Figure 3A**). This result points out the composite nature of *Alnus*-associated communities, which are comprised of (i) fungal symbionts belonging to *Alnus*-associated fungal lineages (e.g., *Alpova*, *Alnicola*), (ii) interaction specialists scattered across distant fungal lineages and (iii) a few generalist fungal species (Rochet et al., 2011; Roy et al., 2013).

Four species, including the three alder species and the *Cistus* sp. group, host more specialized symbionts on average in their projected network than expected by chance. Two of them, the endemic *A. alnobetula* subsp. *suaveolens* and *A. cordata*, displaying the lowest numbers of fungi-mediated links with other plant species in the whole dataset, share no fungal species with four other genera, *Betula*, *Corylus*, *Popula*, and *Salix*. This result may be surprising when we consider the convergent affinity of alders and the four latter genera for hygrophylic habitats in Corsica. These patterns may be partially explained by ecological requirements of *A. alnobetula* subsp. *suaveolens* and *A. cordata*. The first species highly dominates shrubby vegetation above the current altitudinal limit of forests, at high elevation (Jeanmonod and Gamisans, 2007). In these ecosystems, only endomycorrhizal trees (*Acer*, *Sorbus*) and scattered montane ECM species (*Pinus*, *Abies*, and *Fagus*), but no *Betula*, *Corylus*, *Populus* and *Salix*, remain in the landscape (Jeanmonod and Gamisans, 2007). The second *Alnus* species, *A. cordata*, displays (i) a low dependence on water compared to other alders, *Salix* and *Populus*, and (ii) an ability to establish under *Pinus*, *Abies*, *Fagus*, and *Quercus* during secondary successions (Gamisans, 1991; Jeanmonod and Gamisans, 2007). For these two endemic alders, their positions in the projected network indicate an unusual below-ground ecology.

Below-ground Ecological Strategies of ECM Plants

In previously published literature, both the number of associated fungal species and the number of shared fungal species between plant species have been used for evaluating the potential of plants to interact with other plants through ECM fungal networks (Richard et al., 2005; Nara, 2006; Bingham and Simard, 2012). In our study, we did not hypothesize that the existence of a link between two plant species in the projected network induces any effect of one plant species on the local establishment of another. In this regard, our approach strongly differs from individual plant-centered studies investigating physical networks (Common Mycorrhizal Networks; e.g., Nara, 2006). The links we study here encompass a wide spectrum of plant–plant interactions, from the simple ability for a given plant to provide suitable habitats for

other plant species through dispersed fungal propagules, to the possibility that roots of co-occurring species are inter-connected through shared mycelia.

A high plant degree (high number of fungal partners) has been hypothesized to increase physical networking and facilitate seedling establishment under pre-established trees in either conspecific (Dickie et al., 2005; Walker et al., 2005; Bingham and Simard, 2012) or mixed species populations (Amaranthus and Perry, 1989; Horton et al., 1999; Selosse et al., 2006) during secondary successions. The number of shared symbionts has been widely used as a proxy of the strength of the fungi-mediated association between pairs of host plants. Based on these ECM community overlaps, previous studies have suggested for instance facilitation of the establishment of oak forests in a Corsican succession process (*Quercus* – *Arbutus*; Richard et al., 2009), facilitation of coexistence of *Quercus* species (Walker et al., 2005), facilitation of presence of a species of *Helianthemum* at the edge of *Quercus* forests (Dickie et al., 2004) or the capacity of *Betula papyrifera* to constitute a nurse plant for *Pseudotsuga menziesii* (Simard et al., 1997; Simard and Durall, 2004; Bingham and Simard, 2012).

Beyond such fungi-mediated interactions between plant species, the ecological and evolutionary implications of the ability of an ECM plant species to interact with a high number of plant species remain undocumented. A saturated ECM projected network may increase fungal inoculum availability by maintaining compatible reservoirs on alternative hosts. This tendency to share symbionts with many other plants may enlarge the biotic component of the plant niche. Specifically, a high degree in the projected network may facilitate seedling establishment in vegetation of various stages and composition. Such plant-mediated facilitation mechanisms may have been particularly favored in Mediterranean ecosystems where summer drought drastically impacts tree recruitment (Bruno et al., 2003; Gómez-Aparicio et al., 2004). Further studies are required to ascertain the place of ECM inoculum-driven processes in facilitation mechanisms (Richard et al., 2009), which may potentially counterbalance local accumulation of pathogens on plant species (the so-called Janzen–Connell effect, already reported for Mediterranean tree species; Steinitz et al., 2011).

For fungal species, the ECM projected network provides a view of the pool of potentially interacting fungal species. As in plant-centered approaches, calculating the relative overlap in host plants between pairs of fungal species (for instance, using Jaccard distances) would provide fungal species attributes. Mediterranean forests are generally dominated by locally monospecific tree stands (Quézel and Médail, 2003). In this context, a high fungal degree (high number of hosts) may be important for broadening the range of forest types where a fungus can establish. Alternatively, constraining environmental conditions in Mediterranean ecosystems may act as primary abiotic filters selecting for both plant and fungal specialists. We note here that *Salix* and *Alnus* species are adapted to hydromorphic soils, where their host-specific fungal partners may exhibit adaptations to this specific abiotic environment.

In our study, ECM association patterns significantly differed depending on the ecological strategy of their plant host. On average, early-stage plants had five times fewer fungal associates than late-stage ones (**Figure 4**). Our data do not support the hypothesis that early-stage plants differ in the specialization of their symbionts from late successional species, but suggest that early-stage plants associate with significantly fewer ECM fungal species than late-stage plants. Additionally, long-lived late-successional plant species that dominate forest ecosystems for centuries allow a lasting fungal recruitment, which may entail the accumulation of pioneer ECM species. Interestingly, our dataset encompasses mostly species belonging to late-successional fungal genera (*Russula, Boletus, Amanita,* various "Aphyllophorales" including Bankeraceae and Hydnaceae; Last et al., 1983) that increase in relative abundance with tree aging (Horton and Bruns, 2001). Our results thus suggest that the cumulated effect of late-successional host-plant establishment (above-ground dynamics driven by plant ecological strategies) and host aging-related accumulation of late-successional stage fungal species (below-ground dynamics driven by fungal ecological strategies) lead to the hyper-diversity of ECM communities in mature forests, with no effect on the average specialization of the below-ground ECM diversity (Twieg et al., 2007).

CONCLUSION

To the best of our knowledge, our work documents for the first time the below-ground fungal counterpart of the ecological strategies of ECM host plants. Despite the unprecedented rhythm of research describing ECM communities worldwide, their role in plant community dynamics in general, and in host coexistence processes in particular, remains largely unexplored, mainly because most ECM fungi are not cultivable. We assembled a large qualitative dataset to propose a systemic view of the interactions established by the tremendous ECM fungal diversity at the scale of a large Mediterranean island. The ECM symbiosis shapes ecological and evolutionary-based interaction modules, and can help us understand the below-ground niche differences between early- and late-stage plants. The huge diversity of symbionts contributes to saturate plant–plant networks through highly variable numbers of fungi-mediated interactions between plants. Assembling quantitative data of below-ground plant–fungi interaction is crucial to grasp the ecological dimension of these contrasted patterns.

AUTHOR CONTRIBUTIONS

AT, FM, AL, FR originally formulated the idea, developed the methodology and performed statistical analyses. A-CM, J-MB, P-AM, FR generated data. AT, FM, P-AM, AL, FR wrote the initial manuscript. AT, FM, AL, A-CM, J-MB, M-AS, P-AM, and FR contributed to the final manuscript.

ACKNOWLEDGMENTS

This study was funded by the Conservatoire National Botanique de Corse. We are grateful to L. Hugot and J. Alesandri for helping us in assembling field data, and to Doyle McKey for checking the English. All experiments performed in this study comply with the current laws of France.

REFERENCES

Amaranthus, M. P., and Perry, D. A. (1989). Interaction effects of vegetation type and Pacific madrone soil inocula on survival, growth, and mycorrhiza formation of Douglas-fir. *Can. J. For. Res.* 19, 550–556. doi: 10.1139/x89-087

Bahram, M., Harend, H., and Tedersoo, L. (2014). Network perspectives of ectomycorrhizal associations. *Fungal Ecol.* 7, 70–77. doi: 10.1016/j.funeco.2013.10.003

Bascompte, J., and Jordano, P. (2007). Plant-animal mutualistic networks: the architecture of biodiversity. *Annu. Rev. Ecol. Evol. Syst.* 38, 567–593. doi: 10.1146/annurev.ecolsys.38.091206.095818

Bastian, M., Heymann, S., and Jacomy, M. (2009). Gephi: an open source software for exploring and manipulating networks. *ICWSM* 8, 361–362.

Bingham, M. A., and Simard, S. W. (2012). Mycorrhizal networks affect ectomycorrhizal fungal community similarity between conspecific trees and seedlings. *Mycorrhiza* 22, 317–326. doi: 10.1007/s00572-011-0406-y

Botnen, S., Unni, V., Carlsen, T., Eidesen, P. B., Davey, M. L., and Kauserud, H. (2014). Low host specificity of root-associated fungi at an Arctic site. *Mol. Ecol.* 23, 975–985. doi: 10.1111/mec.12646

Bruno, J. F., Stachowicz, J. J., and Bertness, M. D. (2003). Inclusion of facilitation into ecological theory. *Trends Ecol. Evol.* 18, 119–125. doi: 10.1016/S0169-5347(02)00045-9

Bruns, T. D., Bidartondo, M. I., and Taylor, D. L. (2002). Host specificity in ectomycorrhizal communities: what do the exceptions tell us? *Integr. Comp. Biol.* 42, 352–359. doi: 10.1093/icb/42.2.352

Chagnon, P.-L., Bradley, R. L., and Klironomos, J. N. (2012). Using ecological network theory to evaluate the causes and consequences of arbuscular mycorrhizal community structure. *New Phytol.* 194, 307–312. doi: 10.1111/j.1469-8137.2011.04044.x

Chagnon, P.-L., Bradley, R. L., and Klironomos, J. N. (2014). Plant–fungal symbioses as ecological networks: the need to characterize more than just interaction patterns. *Fungal Ecol.* 12, 10–13. doi: 10.1016/j.funeco.2014.05.002

Dahlberg, A., Jonsson, L., and Nylund, J.-E. (1997). Species diversity and distribution of biomass above and below ground among ectomycorrhizal fungi in an old-growth Norway spruce forest in south Sweden. *Can. J. Bot.* 75, 1323–1335. doi: 10.1139/b97-844

den Bakker, H. C., Zuccarello, G. C., Kuyper, T. W., and Noordeloos, M. E. (2004). Evolution and host specificity in the ectomycorrhizal genus *Leccinum*. *New Phytol.* 163, 201–215. doi: 10.1111/j.1469-8137.2004.01090.x

Dickie, I. A., Guza, R. C., Krazewski, S. E., and Reich, P. B. (2004). Shared ectomycorrhizal fungi between a herbaceous perennial (*Helianthemum bicknellii*) and oak (*Quercus*) seedlings. *New Phytol.* 164, 375–382. doi: 10.1111/j.1469-8137.2004.01177.x

Dickie, I. A., Schnitzer, S. A., Reich, P. B., and Hobbie, S. E. (2005). Spatially disjunct effects of co-occurring competition and facilitation. *Ecol. Lett.* 8, 1191–1200. doi: 10.1111/j.1461-0248.2005.00 822.x

Gamisans, J. (1991). *La Végétation de la Corse*. Aix-en-Provence: Édisud.

Gómez-Aparicio, L., Zamora, R., Gómez, J. M., Hódar, J. A., Castro, J., and Baraza, E. (2004). Applying plant facilitation to forest restoration: a meta-analysis of the use of shrubs as nurse plants. *Ecol. Appl.* 14, 1128–1138. doi: 10.1890/03-5084

Gotelli, N. J., and Entsminger, G. L. (2003). Swap algorithms in null model analysis. *Ecology* 84, 532–535. doi: 10.1890/0012-9658(2003)084[0532:SAINMA]2.0.CO;2

Guimerà, R., and Amaral, L. A. N. (2005). Functional cartography of complex metabolic networks. *Nature* 433, 895–900. doi: 10.1038/nature 03288

Guimerà, R., Sales-Pardo, M., and Nunes Amaral, L. A. (2004). Modularity from fluctuations in random graphs and complex networks. *Phys. Rev.* 70:25101. doi: 10.1103/PhysRevE.70.025101

Hawksworth, D. L. (2001). The magnitude of fungal diversity: the 1.5 million species estimate revisited. *Mycol. Res.* 105, 1422–1432. doi: 10.1017/S0953756201004725

Horton, T. R., and Bruns, T. D. (2001). The molecular revolution in ectomycorrhizal ecology: peeking into the black-box. *Mol. Ecol.* 10, 1855–1871. doi: 10.1046/j.0962-1083.2001.01333.x

Horton, T. R., Bruns, T. D., and Parker, V. T. (1999). Ectomycorrhizal fungi associated with Arctostaphylos contribute to *Pseudotsuga menziesii* establishment. *Can. J. Bot.* 77, 93–102. doi: 10.1139/b98-208

Ives, A. R., and Godfray, H. C. J. (2006). Phylogenetic analysis of trophic associations. *Am. Nat.* 168, E1–E14. doi: 10.1086/505157

Jeanmonod, D., and Gamisans, J. (2007). *Flora Corsica*. Aix-en-Provence: EdiSud.

Krasnov, B. R., Fortuna, M. A., Mouillot, D., Khokhlova, I. S., Shenbrot, G. I., and Poulin, R. (2012). Phylogenetic signal in module composition and species connectivity in compartmentalized host-parasite networks. *Am. Nat.* 179, 501–511. doi: 10.1086/664612

Kretzer, A. M., Bidartondo, M. I., Grubisha, L. C., Spatafora, J. W., Szaro, T. M., and Bruns, T. D. (2000). Regional specialization of *Sarcodes sanguinea* (Ericaceae) on a single fungal symbiont from the *Rhizopogon ellenae* (Rhizopogonaceae) species complex. *Am. J. Bot.* 87, 1778. doi: 10.2307/26 56828

Last, F. T., Mason, P. A., Wilson, J., and Deacon, J. W. (1983). Fine roots and sheathing mycorrhizas: their formation, function and dynamics. *Plant Soil* 71, 9–21. doi: 10.1007/BF02182637

Latapy, M., Magnien, C., and Vecchio, N. D. (2008). Basic notions for the analysis of large two-mode networks. *Soc. Netw.* 30, 31–48. doi: 10.1016/j.socnet.2007.04.006

Martos, F., Munoz, F., Pailler, T., Kottke, I., Gonneau, C., and Selosse, M.-A. (2012). The role of epiphytism in architecture and evolutionary constraint within mycorrhizal networks of tropical orchids. *Mol. Ecol.* 21, 5098–5109. doi: 10.1111/j.1365-294X.2012.05692.x

Molina, R., Massicotte, H. B., and Trappe, J. M. (1992). "Specificity phenomena in mycorrhizal symbioses: community-ecological consequences and practical implications," in *Mycorrhizal Functioning: An Integrated Plant Fungal Process*, ed. M. F. Allen (London: Chapman and Hall), 357–423.

Montesinos-Navarro, A., Segarra-Moragues, J. G., Valiente-Banuet, A., and Verdú, M. (2012). The network structure of plant-arbuscular mycorrhizal fungi. *New Phytol.* 194, 536–547. doi: 10.1111/j.1469-8137.2011.04045.x

Moreau, P.-A., Rochet, J., Richard, F., Chassagne, F., Manzi, S., and Gardes, M. (2011). Taxonomy of Alnus-associated hypogeous species of *Alpova* and *Melanogaster* (Basidiomycota, Paxillaceae) in Europe. *Cryptogam. Mycol.* 32, 33–62. doi: 10.7872/crym.v32.iss1.2012.033

Morris, M. H., Smith, M. E., Rizzo, D. M., Rejmánek, M., and Bledsoe, C. S. (2008). Contrasting ectomycorrhizal fungal communities on the roots of co-occurring oaks (*Quercus* spp.) in a California woodland. *New Phytol.* 178, 167–176. doi: 10.1111/j.1469-8137.2007.02348.x

Nacher, J. C., and Akutsu, T. (2011). On the degree distribution of projected networks mapped from bipartite networks. *Physica A* 390, 4636–4651. doi: 10.1016/j.physa.2011.06.073

Nara, K. (2006). Ectomycorrhizal networks and seedling establishment during early primary succession. *New Phytol.* 169, 169–178. doi: 10.1111/j.1469-8137.2005.01545.x

Oksanen, J., Blanchet, F. G., Kindt, R., Legendre, P., Minchin, P. R., O'Hara, R. B., et al. (2015). *R Package Version, 2-3-0*.

Öpik, M., and Moora, M. (2012). Missing nodes and links in mycorrhizal networks. *New Phytol.* 194, 304–306. doi: 10.1111/j.1469-8137.2012.04121.x

Patefield, W. M. (1981). Algorithm AS159. An efficient method of generating r x c tables with given row and column totals. *Appl. Stat.* 30, 91–97. doi: 10.2307/2346669

Quézel, P., and Médail, F. (2003). *Ecologie et Biogéographie des Forêts du Bassin Méditerranéen*. Paris: Elsevier.

Rameau, J.-C., Mansion, D., Dumé, G., and Albert, C. H. (2008). *Flore Forestière Française Tome 3*. Paris: Institut pour le développement forestier.

Richard, F., Millot, S., Gardes, M., and Selosse, M.-A. (2005). Diversity and specificity of ectomycorrhizal fungi retrieved from an old-growth Mediterranean forest dominated by *Quercus ilex*. *New Phytol.* 166, 1011–1023. doi: 10.1111/j.1469-8137.2005.01382.x

Richard, F., Moreau, P.-A., Selosse, M.-A., and Gardes, M. (2004). Diversity and fruiting patterns of ectomycorrhizal and saprobic fungi in an old-growth Mediterranean forest dominated by *Quercus ilex* L. *Can. J. Bot.* 82, 1711–1729. doi: 10.1139/b04-128

Richard, F., Selosse, M.-A., and Gardes, M. (2009). Facilitated establishment of *Quercus ilex* in shrub-dominated communities within a Mediterranean ecosystem: do mycorrhizal partners matter? *FEMS Microbiol. Ecol.* 68, 14–24. doi: 10.1111/j.1574-6941.2009.00646.x

Rochet, J., Moreau, P.-A., Manzi, S., and Gardes, M. (2011). Comparative phylogenies and host specialization in the alder ectomycorrhizal fungi *Alnicola*, *Alpova* and *Lactarius* (Basidiomycota) in Europe. *BMC Evol. Biol.* 11:40. doi: 10.1186/1471-2148-11-40

Roy, M., Dubois, M.-P., Proffit, M., Vincenot, L., Desmarais, E., and Selosse, M.-A. (2008). Evidence from population genetics that the ectomycorrhizal basidiomycete *Laccaria amethystina* is an actual multihost symbiont. *Mol. Ecol.* 17, 2825–2838. doi: 10.1111/j.1365-294X.2008.03 790.x

Roy, M., Rochet, J., Manzi, S., Jargeat, P., Gryta, H., Moreau, P.-A., et al. (2013). What determines Alnus-associated ectomycorrhizal community diversity and specificity? A comparison of host and habitat effects at a regional scale. *New Phytol.* 198, 1228–1238. doi: 10.1111/nph.12212

Selosse, M.-A., Richard, F., He, X., and Simard, S. W. (2006). Mycorrhizal networks: des liaisons dangereuses? *Trends Ecol. Evol.* 21, 621–628. doi: 10.1016/j.tree.2006.07.003

Selosse, M.-A., and Roy, M. (2009). Green plants that feed on fungi: facts and questions about mixotrophy. *Trends Plant Sci.* 14, 64–70. doi: 10.1016/j.tplants.2008.11.004

Selosse, M.-A., WEIss, M., Jany, J.-L., and Tillier, A. (2002). Communities and populations of sebacinoid basidiomycetes associated with the achlorophyllous orchid *Neottia nidus-avis* (L.) LCM Rich. and neighbouring tree ectomycorrhizae. *Mol. Ecol.* 11, 1831–1844. doi: 10.1046/j.1365-294X.2002.01553.x

Simard, S. W., and Durall, D. M. (2004). Mycorrhizal networks: a review of their extent, function, and importance. *Can. J. Bot.* 82, 1140–1165. doi: 10.1139/b04-116

Simard, S. W., Perry, D. A., Jones, M. D., Myrold, D. D., Durall, D. M., and Molina, R. (1997). Net transfer of carbon between ectomycorrhizal tree species in the field. *Nature* 388, 579–582. doi: 10.1038/41557

Smith, J. E., Molina, R., Huso, M. M., Luoma, D. L., McKay, D., Castellano, M. A., et al. (2002). Species richness, abundance, and composition of hypogeous and epigeous ectomycorrhizal fungal sporocarps in young, rotation-age, and old-growth stands of Douglas-fir (*Pseudotsuga menziesii*) in the Cascade Range of Oregon U.S.A. *Can. J. Bot.* 80, 186–204. doi: 10.1139/b02-003

Smith, S. E., and Read, D. J. (2008). *Mycorrhizal Symbiosis*. New York, NY: Elsevier/Academic Press.

Steinitz, O., Troupin, D., Vendramin, G. G., and Nathan, R. (2011). Genetic evidence for a Janzen-Connell recruitment pattern in reproductive offspring of *Pinus halepensis* trees. *Mol. Ecol.* 20, 4152–4164. doi: 10.1111/j.1365-294X.2011.05203.x

Thébault, E., and Fontaine, C. (2010). Stability of ecological communities and the architecture of mutualistic and trophic networks. *Science* 329, 853–856. doi: 10.1126/science.1188321

Twieg, B. D., Durall, D. M., and Simard, S. W. (2007). Ectomycorrhizal fungal succession in mixed temperate forests. *New Phytol.* 176, 437–447. doi: 10.1111/j.1469-8137.2007.02173.x

van der Heijden, M. G. A., and Horton, T. R. (2009). Socialism in soil? The importance of mycorrhizal fungal networks for facilitation in natural ecosystems. *J. Ecol.* 97, 1139–1150. doi: 10.1111/j.1365-2745.2009.01570.x

Visser, S. (1995). Ectomycorrhizal fungal succession in jack pine stands following wildfire. *New Phytol.* 129, 389–401. doi: 10.1111/j.1469-8137.1995.tb04309.x

Walker, J. F., Miller, O. K., and Horton, J. L. (2005). Hyperdiversity of ectomycorrhizal fungus assemblages on oak seedlings in mixed forests in the southern Appalachian Mountains. *Mol. Ecol.* 14, 829–838. doi: 10.1111/j.1365-294X.2005.02455.x

Wu, Q. (2000). Phylogenetic and biogeographic relationships of eastern Asian and eastern North American disjunct *Suillus* species (fungi) as inferred from nuclear sibosomal RNA ITS Sequences. *Mol. Phylogenet. Evol.* 17, 37–47. doi: 10.1006/mpev.2000.0812

Conflict of Interest Statement: The authors declare that the research was conducted in the absence of any commercial or financial relationships that could be construed as a potential conflict of interest.

6

Influence of low- and high-elevation plant genomes on the regulation of autumn cold acclimation in *Abies sachalinensis*

Wataru Ishizuka[1]*, Kiyomi Ono[2], Toshihiko Hara[2] and Susumu Goto[3]

[1] Forestry Research Institute, Hokkaido Research Organization, Dibai, Japan, [2] Institute of Low Temperature Science, Hokkaido University, Sapporo, Japan, [3] Graduate School of Agricultural and Life Sciences, The University of Tokyo, Tokyo, Japan

Edited by:
Glenn Thomas Howe,
Oregon State University, USA

Reviewed by:
Jason Holliday,
Virginia Polytechnic Institute and State
University, USA
Anne Y. Fennell,
South Dakota State University, USA

*Correspondence:
Wataru Ishizuka
wataru.ishi@gmail.com

Boreal coniferous species with wide geographic distributions show substantial variation in autumn cold acclimation among populations. To determine how this variation is inherited across generations, we conducted a progeny test and examined the development of cold hardening in open-pollinated second-generation (F_2) progeny of *Abies sachalinensis*. The F_1 parents had different genetic backgrounds resulting from reciprocal interpopulational crosses between low-elevation (L) and high-elevation (H) populations: L × L, L × H, H × L, and H × H. Paternity analysis of the F_2 progeny using molecular genetic markers showed that 91.3% of the fathers were located in surrounding stands of the F_1 planting site (i.e., not in the F_1 test population). The remaining fathers were assigned to F_1 parents of the L × L cross-type. This indicates that the high-elevation genome in the F_1 parents was not inherited by the F_2 population via pollen flow. The timing of autumn cold acclimation in the F_2 progeny depended on the cross-type of the F_1 mother. The progeny of H × H mothers showed less damage in freezing tests than the progeny of other cross-types. Statistical modeling supported a linear effect of genome origin. In the best model, variation in freezing damage was explained by the proportion of maternally inherited high-elevation genome. These results suggest that autumn cold acclimation was partly explained by the additive effect of the responsible maternal genome. Thus, the offspring that inherited a greater proportion of the high-elevation genome developed cold hardiness earlier. Genome-based variation in the regulation of autumn cold acclimation matched the local climatic conditions, which may be a key factor in elevation-dependent adaptation.

Keywords: cold acclimation, interpopulation variation, elavational cline, phenology, *Abies sachalinensis*, modeling, paternity analysis

INTRODUCTION

Seasonal growth cycles are well described for plant taxa distributed in boreal, sub-boreal, and temperate climates. In particular, for evergreen coniferous species, cold acclimation often occurs before temperatures drop below freezing (Aitken and Hannerz, 2001; Chuine, 2010). During this acclimation, evergreen plants shift their physiological condition from an active growth phase to a hardening phase that involves some degree of freezing tolerance. This physiological shift is responsible for a series of mechanical and biochemical changes in the cell: for example, intracellular accumulation of compatible osmolytes such as soluble sugars, accumulation of polypeptides and/or proteins, and the vesiculation of vacuoles (Fujikawa et al., 2006). The timing of cold acclimation affects fitness. A long duration of the hardy phase may help plants avoid autumn and spring freezing damage. However, it can also limit annual photosynthetic activity and growth. Therefore, because of a trade-off between growth and risk of freezing damage, the timing of phenological events can be a key driver for adaptation to cold climates. During their long evolutionary histories, conifers have acquired an adaptive schedule of cold acclimation to survive under their local environmental conditions, using changes in daylength and/or temperature as regulatory signals (Kozlowski and Pallardy, 2002). Population differences in the acclimation schedule can be observed within species that inhabit wide geographic distributions spanning diverse climatic regions. Interpopulational genetic variation in phenological traits is associated with climatic differences (Rehfeldt, 1989; Oleksyn et al., 1998). Furthermore, interpopulational genetic variation for many boreal conifers is greater during autumn cold acclimation than during spring dehardening (Aitken and Hannerz, 2001; Howe et al., 2003). It is a consistent trend that populations from high latitudes and elevations exhibit earlier development of cold hardening than those from low latitudes or elevations (Rehfeldt, 1989; Skrøppa and Magnussen, 1993; Oleksyn et al., 1998; Savolainen et al., 2004; Notivol et al., 2007; Mimura and Aitken, 2010).

Common garden trials and reciprocal transplantation experiments are powerful tools for detecting adaptive variation in plants and the drivers of natural selection. A previous study using transplanted materials of the sub-boreal conifer *Abies sachalinensis* Mast. (Sakhalin fir) has revealed an adaptive cline in the timing of autumn cold acclimation along an elevational gradient (Ishizuka et al., 2015). In that study, earlier development of autumn cold acclimation in trees derived from high-elevation populations than in those from low-elevation populations was detected at all transplantation sites. Modeling analysis was then conducted to examine the environmental conditions responsible for the detected physiological variation. The results have shown that the genetic variation in response to the temperature change might be an important driver of elevation-dependent adaptation (Ishizuka et al., 2015). However, the mechanisms by which the genome generates quantitative adaptive differences are not clearly understood. Quantitative traits resulting as a consequence of genome inheritance and genome effects are complex (Sykes et al., 2006; Cané-Retamales et al., 2011). Furthermore, some studies have reported epigenetic phenomena which have regulated phenological traits (Johnsen et al., 2005; Kvaalen and Johnsen, 2008). Perennial plants have a "memory" of the climate they experienced during embryogenesis. These epigenetic signals may be transmitted through several mechanisms, including cytosine methylation and histone modification, and may enhance adaptation at a population level (Johnsen et al., 2005; Kvaalen and Johnsen, 2008). Controlled crosses using mother plants that have been exposed to the same environmental conditions can be used to test the effects of genome inheritance on the regulation

of the autumn cold acclimation. However, previous ecological studies have not extensively evaluated the complex effects of genome inheritance on the development of cold acclimation in trees.

We investigated the genetic basis of variation in the regulation of autumn cold acclimation using Sakhalin fir. This conifer plays an important role in the timber production in northern Japan. It is a dominant climax species in natural sub-boreal forests of the Far East, thus forming an important component of forest ecosystems (Kato, 1952). The geographic range of this species extends from eastern Siberia to Hokkaido, the northernmost island of the Japanese archipelago (**Figure 1A**). The elevational distribution extends from 100 to 1600 m a.s.l. This species survived in a glacial refugium in Hokkaido, and had become a dominant species by the time of the last glacial maximum, 20000 years ago (Igarashi, 1996; Tsumura and Suyama, 1998). Interpopulational variation in several traits has been observed within this elevational range, including variation in growth and survival (Goto et al., 2011; Ishizuka and Goto, 2012), resistance to biological stresses such as disease or rats (Kurahashi and Hamaya, 1981; Saho et al., 1994), and cold hardiness (Eiga and Sakai, 1984). In our recent study, we reported that clinal variation in the timing of cold acclimation appears to result from adaptation to the local climate at elevations ranging from 230 to 1200 m (Ishizuka et al., 2015).

In the present study, we used open pollination to obtain second-generation (F_2) progeny for testing the genetic basis of variation in phenological traits. The maternal population (F_1) was produced by reciprocal crosses between two distinct ecotypes: a population inhabiting low elevations and a population inhabiting high elevations. The genetic background of the maternal F_1 trees varied among cross-types (low × low, hybrids between elevations, high × high), and the trees were planted at a single site in order to ensure their exposure to the same environment. Thus, we could evaluate the effects of genome inheritance by examining the traits of the F_2 population. Using the F_2 progeny, we conducted (1) paternity analysis based on microsatellite markers and (2) statistical modeling of freezing damage. Our objectives were to (1) measure variation in the timing of autumn cold acclimation of the F_2 progeny and (2) establish whether this variation could be explained by the genomic background inherited from the low-elevation or high-elevation populations.

MATERIALS AND METHODS

F₁ Planting Site

We conducted this study in the University of Tokyo Hokkaido Forest (UTHF), located in central Hokkaido, northern Japan (43°20′N, 142°30′E; **Figure 1A**). In the UTHF, natural sub-boreal forests cover about 150 km² of the south-western slope of Mt. Dairoku (1459 m a.s.l.), with an elevational gradient of more than 1200 m. With increasing elevation, dramatic changes occur in soil type, snow depth, and forest vegetation, including changes in the prevalence of the bamboo species, *Sasa senanensis* and *Stigmella kurilensis* (Kato, 1952; Toyooka et al., 1983; Nakata et al., 1994). However, the primary factor changing

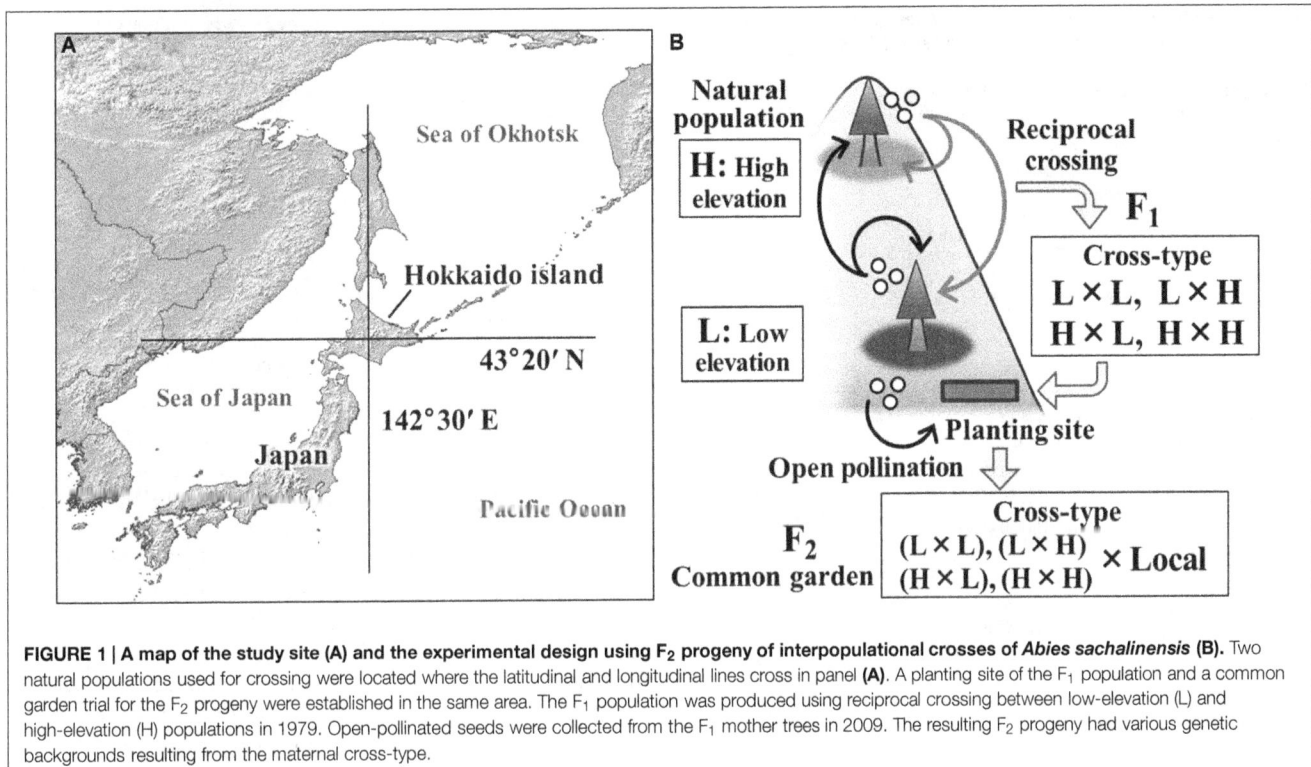

FIGURE 1 | A map of the study site (A) and the experimental design using F$_2$ progeny of interpopulational crosses of *Abies sachalinensis* (B). Two natural populations used for crossing were located where the latitudinal and longitudinal lines cross in panel **(A)**. A planting site of the F$_1$ population and a common garden trial for the F$_2$ progeny were established in the same area. The F$_1$ population was produced using reciprocal crossing between low-elevation (L) and high-elevation (H) populations in 1979. Open-pollinated seeds were collected from the F$_1$ mother trees in 2009. The resulting F$_2$ progeny had various genetic backgrounds resulting from the maternal cross-type.

with elevation is air temperature. Based on monitoring data recorded hourly from 2009 to 2010 at 230 m, the annual and winter (October–March) mean temperatures were 6.53°C and −1.78°C, respectively. Temperatures at 1100 m were 1.75°C and −6.43°C, respectively.

Sakhalin fir, *A. sachalinensis*, grows from the lowland (200 m) to the high-elevation zone (1200 m) in this area. In 1979, natural populations at 400–530 m and 1100–1200 m were selected as low-elevation (L) and high-elevation (H) populations, respectively, and artificial reciprocal crossing was performed by Kurahashi (Kurahashi, 1995; Goto et al., 2011). Within each population, five and three adult trees were selected as mother (seed parents) and father trees (pollen donors), respectively (Kurahashi and Hamaya, 1981). Pollen from the three father trees in each population was pooled and used for the crosses. The generated progeny were categorized into four cross-types (female × male): L × L, L × H, H × L, and H × H (**Figure 1B**). We refer to these progeny as the F$_1$ population. Seedlings of the F$_1$ population were grown in an outdoor nursery at a lowland site of the UTHF. In 1986, the F$_1$ individuals were transplanted to the lowland planting site (220 m; **Figure 1B**). At the planting site, the progeny were arrayed in grids at a spacing of 1.2 m, assigning one block (2 columns × 10 rows) to each cross-treatment. Unreplicated blocks were used to avoid contamination among treatments (Kurahashi, 1995). In total, 440 F$_1$ seedlings were transplanted. Thus, the F$_1$ populations had varying genetic backgrounds but were exposed to the same environmental conditions. Further information about the traits measured on the F$_1$ trees (survival, height, diameter at breast height, leaf nitrogen content, and leaf area per weight) were provided by Goto et al. (2011).

Seed Collection from the F$_1$ Generation

In 2009, 23 years after planting, 353 F$_1$ trees were alive and some of them were in a reproductive stage. Our field surveys during the flowering and seed-maturing seasons showed that 22 trees flowered that year at the F$_1$ planting site (**Table 1**). We collected open-pollinated seeds from the flowering trees on September 6th and 9th, 2009. We selected 13 F$_1$ mother trees that produced a sufficient number of progeny (>20 cones) to use in this study (**Table 1**; see the Supplemental Table for ID-information of the mother trees). The progeny (F$_2$ population) were grouped into four types according to the maternal cross category: L × L, L × H, H × L, and H × H (**Figure 1B**). Collected seeds were stored at −3°C until use.

F$_2$ Common Garden Trial

We conducted a common garden trial of the F$_2$ progeny. Before the trial, viable seeds were selected from all collected seeds using soft X-ray photography (Softex, Kanagawa, Japan). In the spring of 2010, the viable seeds were subjected to a seed stratification treatment and were sown into the UTHF nursery (230 m a.s.l.; **Figure 1A**). One seed was placed in each 4-cm grid of a 1.0 m × 1.2 m block to maintain the identity of all germinated seedlings. The four blocks were separated by buffer spaces. To avoid edge effects for the F$_2$ progeny, control seeds were also sown at the edge of the rows and columns. In total, 2112 seeds were sown in this trial (four blocks × 22 rows × 24 columns). The seedlings were raised until the end of the second growing season. As shown in **Table 2**, the cross-type differences of the mother of the F$_2$ progeny were reflected in some functional traits, such as the germination rates and 2-year heights.

TABLE 1 | Number of F$_1$ trees of *Abies sachalinensis* that were living, flowering, or used to collect open-pollinated seeds in 2009.

Cross-type	Living	Flowering	Seed collection
L × L	144	7	3
L × H	62	4	3
H × L	59	4	3
H × H	88	7	4
Total	353	22	13

Four types of F$_1$ trees were planted at 220 m a.s.l. These trees were originated from controlled reciprocal crosses between low-elevation (L) and high-elevation (H) populations.

TABLE 2 | Germination rates and 2-year heights of F$_2$ progeny in a common garden trial at 230 m a.s.l.

Cross-type of mother	Germination rate (%)	2-year height (mm)
L × L	82.7 (16.1)	80.1 (21.6)
L × H	76.3 (12.8)	72.6 (20.0)
H × L	81.3 (13.2)	77.5 (17.6)
H × H	73.8 (7.8)	63.6 (18.7)
Total	79.3 (14.0)	74.7 (20.9)

The F$_2$ progeny were categorized into four cross-types, according to the cross-types of the F$_1$ mother trees. Standard deviations are shown in parentheses.

Genotyping using SSR Markers

To estimate the genome composition of the F$_2$ progeny, we performed molecular genetic analysis using nuclear microsatellite (nSSR) and chloroplast microsatellite (cpSSR) markers. In the present study, we used four nSSRs: As08, As16, As32 (Lian et al., 2007), and NFH 15 (Hansen et al., 2005), and three cpSSRs: pt30204, pt71936 (Vendramin et al., 1996), and pt30249 (Liepelt et al., 2001).

Samples selected for SSR genotyping consisted of the 22 flowering F$_1$ trees described in **Table 1** plus 80 of their F$_2$ progeny. These 80 F$_2$ progeny were derived from seeds collected in 2009, and were composed of 16 progeny from each of five F$_1$ trees selected across the cross-types (**Table 3**). These seedlings were germinated in an indoor growth chamber.

For the DNA extractions, foliage was collected from the 22 candidate F$_1$ trees and the 80 F$_2$ progeny. DNA was extracted using the DNeasy Plant Mini Kit (Qiagen, Ltd., Crawley, UK), and the PCR reactions were performed using the Multiplex PCR Kit (Qiagen, Ltd., Crawley, UK) following the protocol of Lian et al. (2007). The PCR products were sequenced using an Applied Biosystems 3130xl Genetic Analyzer (Life Technologies, Carlsbad, CA, USA). Genotyping of all samples was performed based on the length of each sequenced fragment using Peak Scanner ver. 1.0 software from Applied Biosystems. Genetic diversity statistics were calculated by GenAlEx ver. 6.5 (Peakall and Smouse, 2012).

Paternity Analysis

We considered the 22 flowering F$_1$ trees as candidate pollen parents of the F$_2$ progeny because no other trees were flowering at the F$_1$ planting site. However, trees outside of the F$_1$ population

TABLE 3 | Genetic diversity statistics of the F$_1$ trees that flowered in 2009 and their open-pollinated F$_2$ progeny.

Population	Mother	N	A	Ho	H$_E$	F$_{IS}$
F$_1$ (flowered in 2009)		22	10.3	0.864	0.838	−0.033
F$_2$	As-F$_1$-C034	16	9.0	0.875	0.735	−0.199
	As-F$_1$-C042	16	7.5	0.906	0.750	−0.216
	As-F$_1$-C148	16	8.8	0.906	0.757	−0.198
	As-F$_1$-C168	16	7.8	0.906	0.749	−0.212
	As-F$_1$-C061	16	8.8	0.906	0.771	−0.175
	Total	80	8.4	0.900	0.752	−0.200

N, number of individuals sampled; A, average number of alleles; H$_O$, observed heterozygosity; H$_E$, expected heterozygosity; F$_{IS}$, fixation index.

may have also served as pollen parents because the F$_1$ planting site was surrounded by mature artificial forests. The origin of these artificial forests was a local lowland forest (200–300 m) that was different from the forests used to create the F$_1$ population.

For paternity analysis, we first inferred paternity of the F$_2$ progeny using four nSSRs. Using CERVUS 3.0 (Kalinowski et al., 2007), we compared the nSSRs genotypes of the 80 F$_2$ progeny to their known seed parents (mothers) and 22 candidate F$_1$ pollen parents (fathers). The most likely father was estimated using maximum-likelihood assignment from the possible allele combinations (genotypes) of the parents. When none of the potential F$_1$ donors were assigned as fathers, we considered pollen contamination and genotyping errors as possible causes. Genotyping errors can lead to inflated estimates of outside pollen flow (Slavov et al., 2005). We set the genotyping error to 1% with an 80% confidence level for the 10000 cycle-simulations for the paternity assignment.

After assigning fathers using CERVUS, we performed a simple exclusion procedure using three cpSSRs. The F$_2$ progeny and the assumed F$_1$ pollen parents were required to have matching cpSSR haplotypes (*Abies* cpSSRs are paternally inherited; Vendramin and Ziegenhagen, 1997). We assumed that CERVUS assigned the correct pollen parent when the genotypes of the F$_2$ progeny and assigned pollen parent matched for all three cpSSR markers (considering genotyping errors). When a complete match was not found, we assumed the pollen came from outside of the F$_1$ planting site.

Combining the results from CERVUS and the exclusion procedure, we estimated the proportion of selfed seeds, seeds sired by the F$_1$ population, and seeds sired by fathers outside the studied F$_1$ population. We also estimated the composition of the F$_1$ trees among four cross-types, based on the genotyping results.

Autumn Freezing Test

We conducted freezing tests on the F$_2$ progeny three times during the autumn season using the method of Eiga and Sakai (1984). The potted seedlings were placed outdoors just before the test. Three freezing tests were conducted at approximately 2-week intervals: on 12 and 26 October, and 11 November (referred to as Tests 1, 2, and 3). Each freezing test was performed at the Institute of Low Temperature Science, Hokkaido University. The potted seedlings were placed in a dark chamber kept at 5°C. After an overnight incubation, the temperature was lowered at a rate of 2°C per hour until the target temperature was reached. The target temperature was maintained for 4 h, and then increased at a rate of 2°C per hour to 5°C. For Test 1, the target temperature was −15°C. For Tests 2 and 3, the target temperatures were −15 and −30°C. The seedlings were kept in the growth chamber for 14 days with a 12-h photoperiod (the photosynthetic photon flux density was 100 μmol m^{-2} s^{-1} during the day). We measured freezing damage using the visual scoring method of Lindgren and Hallgren (1993). Freezing damage was scored using six classes of needle discoloration (brown needles) as follows: 0 (no damage), 1 (1–20% of needles were discolored), 2 (21–40% discolored), 3 (41–60% discolored), 4 (61–80% discolored), and 5 (81–100% discolored). For each temperature at each time point, we used 10 F$_2$ seedlings from each F$_1$ mother tree. For some maternal trees with fewer seedlings, we used seven seedlings. In total, 605 seedlings were used in the freezing test (121 seedlings × 5 test conditions).

Data Analysis

Because no freezing damage was observed in the −15°C treatment in Tests 2 and 3, we excluded these data from subsequent analysis. Therefore, three test conditions were defined: T$_1$ involved freezing at −15°C in Test 1, T$_2$ involved freezing at −30°C in Test 2, and T$_3$ involved freezing at −30°C in Test 3. For each condition, we used a nested ANOVA to study the effect of maternal cross-type and the effect of mother trees on freezing damage in the F$_2$ progeny, using the following model:

$$Y_{ijk} = \mu + C_i + M_j(C_i) + E_{ijk}, \qquad \text{(Model 1)}$$

where Y_{ijk} is the freezing damage score of the k-th progeny (k; 1-10) of the j-th mother tree (j; 1-4) in the i-th cross-type (i; 1-4), μ is the general mean, C_i is the effect of the i-th cross-type, $M_j(C_i)$ is the effect of the j-th mother tree nested in the i-th cross-type, and E_{ijk} is the residual error.

We also evaluated the effect of low-elevation and high-elevation plant genomes on freezing damage using the following statistical model that includes all test conditions. We assumed that freezing damage in the F$_2$ progeny is affected by the genetic inheritance, test conditions, and mother tree. Therefore, we constructed the full model with all assumed effects, as follows:

$$Y_{ijkl} = \mu + C_i + T_l + C_{ij} \times T_l + M_j(C_i) +$$

$$M_j(C_i) \times T_l + E_{ijkl}, \qquad \text{(Models 2–5)}$$

where Y_{ijkl} is the freezing damage score of the k-th progeny of the j-th mother tree in the i-th cross-type under the l-th test condition (l; 1-3), μ, C_i and $M_j(C_i)$ are the same as in Model 1, and T_l is the effect of the l-th test condition. In this full model, we regard the model element C as representing the effect of the plant genome (genetic effect). Then, in the analysis, C, T, and their interaction ($C \times T$) are treated as fixed effects, whereas $M(C)$ and its interaction with T are treated as random effects. Because damage is an ordinal variable (Ishizuka et al., 2015), we used an ordered probit mixed model (Lee, 1992). Specifically, we used the cumulative link mixed model (CLMM) function in

the "ordinal" package of R 3.1.2 (R Development Core Team, 2014). CLMM uses a maximum likelihood approach with the Laplace approximation and adaptive Gauss-Hermite quadrature (see Christensen, 2014).

We then assessed which types of variables were appropriate for the effect of the plant genome (C) to describe the freezing damage score. Therefore, we used the following four types of the genetic effects as C in Models 2-5:

Model 2: Linear effect (0 for L × L; 0.25 for L × H and H × L; 0.5 for H × H)

Model 3: Categorical effect (L × L; L × H; H × L; H × H)

Model 4: Categorical interaction effect ("NoHyb" for L × L and H × H; "Hyb$_{LH}$" for L × H; "Hyb$_{HL}$" for H × L)

Model 5: Linear effect + interaction effect (0 for L × L; 0.25 + α for L × H; 0.25 − α for H × L; 0.5 for H × H).

In Model 2, the proportion of high-elevation genome inherited from the maternal parent was used. Numeric values indicating the linear effect of the maternal genome origin were assigned to C. The highest value was assigned to the H × H cross-type, whereas the lowest value, 0, was assigned to the L × L cross-type. Model 2 should show a good fit if freezing damage was closely related to the amount of high-elevation genome inherited from the maternal trees (i.e., quantitative trait). In Model 3, C was set to one of four categorical values indicating the cross-type of the maternal trees. Model 3 should fit well if the maternal origin is important, but the proportion of high-elevation genome is not. Model 4 included a genome interaction effect. As described above, C was set to one of three categorical values: NoHyb, Hyb$_{LH}$, or Hyb$_{HL}$. In Model 5, we included both the cross-type linear effect and genome interaction effect. C was partitioned into the combination of the variables used in Models 2 and 4.

We used a model selection procedure to exclude the variables that did not improve the model fit. The Akaike Information Criterion (AIC) was used to consider goodness-of-fit and number of parameters (Johnson and Omland, 2004). Within each model (Models 2-5), a backward stepwise procedure from the full model was performed to determine the most effective variable sets. Then, we compared the AIC values among all candidate models and selected the model with the lowest AIC value as the best model for explaining freezing damage in the F$_2$ progeny. All statistical analyses in this study were conducted using R 3.1.2 (R Development Core Team, 2014).

RESULTS

Paternity Analysis

All seven SSR markers were polymorphic. For the nSSR markers, the average number of alleles per locus (N) was 15.0, ranging from 7 for As08 to 21 for As16 and NFH15. For the cpSSR markers, the numbers of haplotypes of Pt30141, Pt30204, and Pt71936 were 15, 13, and 7, respectively. The genotypes of the 22 F$_1$ candidate pollen parents differed from each other (see Supplemental Table). Genetic diversity statistics for the four nSSR markers are summarized for the F$_1$ and F$_2$ materials in **Table 3**.

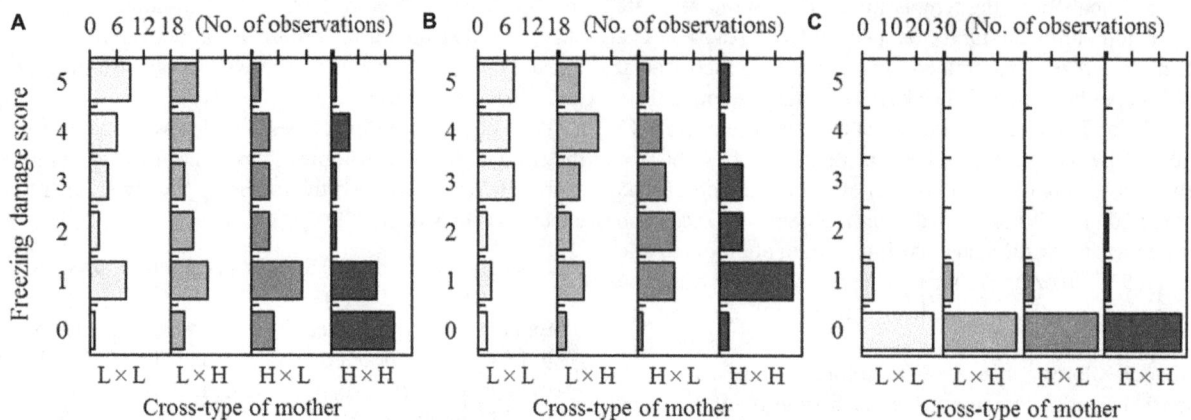

FIGURE 2 | Freezing damage of F$_2$ progeny after a freezing test at −15°C in Test 1 (A), −30°C in Test 2 (B), or −30°C in Test 3 (C). Three freezing tests were conducted in autumn at a 2-week interval. The freezing damage scores range from 5 (complete needle discoloration) to 0 (no discoloration). Observed frequencies were shown by arraying the cross-type of each mother (on the lower x-axis). The maximum scale of the number of observations (on the top x-axis) was set to 18 for each of the four cross-types in Tests 1 and 2, and set to 30 for each of the four cross-types in Test 3.

The genetic characteristics of the F$_2$ progeny were consistent among their maternal groups. The average number of alleles in the F$_2$ progeny was smaller than that in the F$_1$ trees, which was caused by maternal sharing (13 out of 353 F$_1$ trees; **Table 1**). In contrast, their observed heterozygosity (H_O) was higher in the F$_2$ progeny than in the F$_1$ trees, resulting in a negative fixation index (F_{IS}).

Paternity analysis using four nSSR markers showed that 12 F$_2$ progeny had a father from the candidate F$_1$ pollen parent group. By adding the results of cpSSR analysis, seven F$_2$ progeny out of 12 were not excluded. For these seven F$_2$ progeny, the cross-type of all of the assigned F$_1$ fathers was L × L. There were no progeny with genotypes composed solely of maternal alleles, excluding the possibility of selfing. Moreover, most of the F$_2$ progeny had novel alleles in comparison with the candidate F$_1$ pollen donors. Allowing for genotyping errors, the proportions of selfing, mating between F$_1$ trees, and mating with outside trees were 0, 8.7 (7 samples), and 91.3% (73 samples), respectively.

Autumn Freezing Test and Model Selection

Freezing damage scores in the F$_2$ progeny ranged from 0 to 5 for treatment T$_1$ (−15°C in Test 1) and treatment T$_2$ (−30°C in Test 2; **Figure 2**). In T$_1$, the most frequent damage scores for the progeny belonging to the L × L and H × H cross-types were 5 (30.0%) and 0 (45.2%), respectively. For the hybrid cross-types (L × H and H × L), a score of 1 was most common (26.7% for L × H and 36.7% for H × L). In T$_2$, the most frequent damage score for the H × H cross-type was 1 (51.6%). In contrast, high damage scores (4 and 5) were often observed for the L × L, L × H, and H × L cross-types. In T$_3$ (−30°C in Test 3), most progeny showed no damage, regardless of the cross-type (**Figure 2C**).

The ANOVA indicated a significant difference among mother trees within cross-types [M (C)] for treatment T$_1$ (**Table 4**). For treatment T$_2$, a significant difference was detected among maternal cross-types (C), but not among mother trees [M (C)].

TABLE 4 | ANOVA of freezing damage in the F$_2$ progeny for each test condition, including degrees of freedom (DF), mean square (MS) and F-value (F).

Source	DF	T$_1$ MS	T$_1$ F	T$_2$ MS	T$_2$ F	T$_3$ MS	T$_3$ F
C	3	21.47	3.30	13.46	**4.54***	0	–
M (C)	9	6.50	**2.78****	2.97	1.48	0	–
Error	108	2.34		2.00		0	

C, cross-type; M (C), mother tree nested in the cross-type; T_1, −15°C in Test 1; T_2, −30°C in Test 2; T_3, −30°C in Test 3. *P < 0.05, **P < 0.01.

For treatment T$_3$, however, statistical analysis was not possible due to limited variation in freezing damage score, as shown in **Figure 2**.

Model selection for each of the four candidate models (Models 2–5) indicated that the interaction terms $C × T$ and M (C) $× T$ should be excluded from Models 2, 3, and 5 (**Table 5**). In Model 4, the C effect was also excluded. This indicates the effect of the mother trees assumed in Model 4 (i.e., non-hybrid vs. hybrid) was not useful for predicting freezing damage in the F$_2$ progeny. A model comparison based on AIC values revealed that Model 2, which included the cross-type linear effect, was the best model (**Table 5**). The effects included in this model were C, T, and M (C). The estimated coefficients of these variables are shown in **Table 6**. A significant negative effect was observed for C, indicating that the freezing damage tended to be lower in the progeny that inherited a higher proportion of the high-elevation genome.

DISCUSSION

Freezing damage differed among seedlings of *A. sachalinensis* early in the study period, but was barely observed in the final test conducted in November (**Figure 2**). This demonstrates

TABLE 5 | Candidate model components and Akaike Information Criterion (AIC) values after model selection.

Candidate models	Model components					AIC
	C	T	C × T	M (C)	M (C) × T	
Model 2	+ (Linear)	+	−	+	−	**813.23**
Model 3	+ (Categorical)	+	−	+	−	815.10
Model 4	− (Categorical interaction)	+	−	+	−	822.90
Model 5	+ (Linear with categorical interaction)	+	−	+	−	814.99

For each model, variables were retained (represented by +) or excluded (−) by a stepwise procedure. The lowest AIC value, shown in bold, indicates the best model. In Models 2–5, different types of effects were used for model component C.
C, T, C × T: fixed effect of plant genome, the test condition, and their interaction, respectively.
M (C), M (C) × T: random effect of the mother tree nested in the cross-type and its interaction with test condition.

TABLE 6 | Estimated parameters for Model 2, the best model among all candidate models (Models 2–5) for predicting freezing damage in the F_2 progeny (standard errors are in parentheses).

Model 2		Coefficient	p-value
Fixed effect	C (Linear)	−2.440 (0.580)	<0.001
	T (T1)	0	−
	(T2)	0.336 (0.136)	0.014
	(T3)	−8.277 (101.7)	0.935
Random effect	M (C) (13 categories)	≈0 (0.206)	

The negative effect of plant genome (C) indicates that the progeny with a higher proportion of the high-elevation genome tend to show less freezing damage.

the development of freezing tolerance over time. In addition, freezing damage differed among maternal cross-types of the F_2 progeny, even though the mother trees were grown in the same environment (**Figure 2**). Genetic variation in autumn phenology was relatively clear. Seedlings that inherited a greater proportion of the high-elevation genome displayed earlier cold acclimation. We used visual scoring, but alternative quantitative evaluations (e.g., by chlorophyll fluorescence or electrolytic leakage) are also used as indicators of cold hardiness (Lindgren and Hallgren, 1993; Ehlert and Hincha, 2008). Although quantitative evaluation is a powerful tool, the accuracy and efficiency of these alternative measures may be compromised if the plants have small leaves. In the present study, most seedlings developed thin needles, making a quantitative measurement difficult. In contrast, discoloration occurred evenly, making differences in freezing damage clear and comparable among seedlings. In addition, there is a strong correlation ($r = 0.762$) between our scoring method and chlorophyll fluorescence (evaluated by the F_v/F_m value) in *A. sachalinensis* (Ishizuka et al., 2015). Thus, our evaluation method is sufficient for detecting phenotypic variation associated with cold acclimation in this species.

In studies of other boreal conifers, genetic variation in autumn phenology and associations with the climate of origin indicated adaptation to the local climate (Rehfeldt, 1989; Skrøppa and Magnussen, 1993; Oleksyn et al., 1998; Savolainen et al., 2004; Notivol et al., 2007; Mimura and Aitken, 2010). In our previous study using reciprocally transplanted materials of *A. sachalinensis*, we found clinal variation in the timing of autumn cold acclimation along an elevational gradient (Ishizuka et al., 2015). Here, we further evaluated the effects of genetic

background of the trees using molecular markers and statistical models to improve our understanding of the local adaptation and the evolution of autumn cold acclimation.

Paternity Analysis

The combined use of nSSR and cpSSR markers made it possible to assign pollen parents to *A. sachalinensis* seedlings. We assigned 22 candidate F_1 pollen parents to the F_2 progeny with consideration for genotyping error and assignment confidence (Slavov et al., 2005). We assumed that cryptic gene flow had little effect because the genetic origin of the F_1 trees was different from that of the artificial forests surrounding the F_1 planting site. Therefore, when the paternity analysis found no possible fathers among the candidates, we assumed that the pollen parent was outside of the F_1 planting site, and presumably derived from the surrounding mature plantations.

The small paternal contribution from the F_1 trees (8.7%) indicated a high level of pollen flow from outside pollen sources (91.3%). High levels of pollen contamination have been reported to be a severe problem in seed orchard management (Wheeler and Jech, 1992). When a seed orchard is young, the proportion of outside pollen flow tends to be extremely high. For example, Ozawa et al. (2009) performed paternity analysis of collected seeds in a young *Pinus densiflora* seed orchard, and detected 82% outside pollen flow. The results of that study suggested that the amount of pollen brought to the female strobili from the male strobili within the seed orchard was markedly lower than that of immigrant pollen brought from surrounding mature forests. In the present study, only 6.2% of the F_1 parents were flowering due to their young age (29 years old) and small size (**Table 1**). As suggested by Ozawa et al. (2009), the amount of pollen from the F_1 planting site must be lower than that from the surrounding mature plantations. Therefore, the results obtained by paternity analysis are reasonable.

If the timing of pollen shed from surrounding plantations and the receptivity of female strobili of the F_1 trees are mismatched, outside pollination will occur rarely. Indeed, Sasaki (1983) reported that the flowering phenology of *A. sachalinensis* at the plantation forest at 1100 m was delayed by approximately 2 weeks compared with that at 530 m. If the difference in flowering phenology is genetically controlled, the timing of female flowering of F_1 trees derived from a crossing among high-elevation parents may differ from that of male flowering of local (low-elevation) trees. However, the flowering phenology of the F_1

trees which were planted at the same environment overlapped, as described by Goto et al. (2011). These results suggest that environmental factors, rather than ontogenetic control, strongly affected the flowering phenology of *A. sachalinensis* (Sasaki, 1983; Goto et al., 2011). Therefore, it is not surprising that outside pollen flow was predominant at our F_1 planting site.

Evidence of frequent outcrossing was also seen in the genetic parameters obtained using nSSRs. Observed heterozygosity (H_O) increases and the fixation index (F_{IS}) decreases when mating occurs between local and non-local populations (Hamrick and Godt, 1996). In our study, H_O of the F_2 was higher than that of natural stands (Lian et al., 2008; **Table 2**). A large decrease in F_{IS} was also detected between the F_1 and F_2 generations, although there were only small differences within the F_2 (**Table 2**). The negative F_{IS} in F_2 indicated that the chances of mating between an F_1 mother and its relatives were low. These results support our conclusions concerning pollen flow from outside the F_1 planting site.

We concluded that only a few F_1 fathers contributed to the F_2 progeny because of the high levels of pollen flow from the surrounding mature plantations. Moreover, the F_1 trees that were assigned as fathers (seven progeny) were all derived from the L × L cross-type. This indicates that the high-elevation plant genome from the F_1 population was not inherited by the F_2 population to any great degree.

Autumn Freezing Test and Model Selection

Our modeling analysis revealed that the regulation of autumn cold acclimation could be well explained by the linear effect of the genome origin. The model using the expected proportion of high-elevation genome as a fixed effect was selected as the best model (**Table 5**). In this model, variation in autumn cold acclimation was explained by variation in the maternal genome composition. The F_2 progeny that inherited a greater proportion of the high-elevation genome showed earlier cold acclimation.

Because of epigenetic phenomena, such as maternal effects and 'after' effects, the prediction of phenotypic values for quantitative traits can be a complex process (Howe et al., 2003; Johnsen et al., 2005; Kvaalen and Johnsen, 2008; Cané-Retamales et al., 2011). In some cases, the performance of the progeny is subjected to a significant epigenetic effect (Johnsen et al., 2005; Kvaalen and Johnsen, 2008). An experiment with *Picea abies* revealed that autumn phenological traits (date of bud set) was related to the daylength and/or thermal conditions during fertilization and seed maturation (Johnsen et al., 2005). Moreover, the performance of the progeny may be influenced by phenomena related to epistasis (Sykes et al., 2006). Sykes et al. (2006) used the crossbred progeny derived from a specific combination of crosses in *P. taeda*, and successfully measured the effects of genomic interactions in the

chemical contents of woody materials. In the present study, however, the model analysis indicated that epigenetic phenomena and the cross-type-dependent maternal effect were not likely to have played a major role (**Table 5**). In comparison with models considering such effects, the model including only the effect of genome origin showed the best fit. This result could be partly explained by the fact that all maternal trees were exposed to the same environmental conditions. Moreover, in the absence of a specific cross-type effect, it would be reasonable to expect that the autumn cold acclimation of *A. sachalinensis* would show a linear pattern based on the proportion of low-elevation and high-elevation plant genomes.

This interpopulational regulation of autumn phenology may be one of the factors driving the local adaptation of conifers (Aitken and Hannerz, 2001; Howe et al., 2003). For *A. sachalinensis*, clinal variation in the timing of autumn cold acclimation along an elevational gradient is one of the elevation-dependent adaptive traits. Each population adapts to the local climate, with a trade-off between the avoidance of freezing damage and extension of the growth period (Ishizuka et al., 2015). The genome-based control of phenological traits may be the mechanism responsible for the elevation-dependent adaptation of this species.

The genetic control of ecologically relevant traits is important in forest management. Potential genetic changes in future generations must be taken into consideration, particularly considering projections of rapid global warming (Jump and Penuelas, 2005; Aitken et al., 2008; Ishizuka and Goto, 2012). If a target species shows strong genetic variation that is larger than the phenotypic plasticity, it may be important to redistribute the genes that confer optimal performance in the changing climate (Aitken et al., 2008). For *A. sachalinensis*, it may be difficult to track rapid climate change because of the mismatch of locally adapted autumn phenology. However, the presence of adaptive genes and their potential for climatic adaptation have not been sufficiently examined. Future studies may be needed to evaluate this subject in detail by applying powerful molecular techniques, such as the analysis of quantitative trait loci and/or genome scanning.

ACKNOWLEDGMENT

The authors are deeply grateful to Glenn Howe for critical comments on the manuscript.

REFERENCES

Aitken, S. N., and Hannerz, M. (2001). "Genecology and gene resource management strategies for conifer cold hardiness," in *Conifer Cold Hardiness*,

eds F. J. Bigras and S. J. Columbo (Dordrecht: Kluwer Academic Publishers), 23–53.

Aitken, S. N., Yeaman, S., Holliday, J. A., Wang, T. L., and Curtis-McLane, S. (2008). Adaptation, migration or extirpation: climate change outcomes for tree populations. *Evol. Appl.* 1, 95–111. doi: 10.1111/j.1752-4571.2007.00 013.x

Cané-Retamales, C., Mora, F., Vargas-Reeve, F., Perret, S., and Contreras-Soto, R. (2011). Bayesian threshold analysis of breeding values, genetic correlation and heritability of flowering intensity in *Eucalyptus cladocalyx* under arid conditions. *Euphytica* 178, 177–183. doi: 10.1007/s10681-010-0292-y

Christensen, R. H. B. (2014). *Ordinal – Regression Models for Ordinal Data. R Package Version 2015.1-21.* Available at: https://www.cran.r-project.org/package=ordinal/ [Accessed February 15, 2015].

Chuine, I. (2010). Why does phenology drive species distribution? *Philos. Trans. R. Soc. B Biol. Sci.* 365, 3149–3160. doi: 10.1098/rstb.2010.0142

Ehlert, B., and Hincha, K. D. (2008). Chlorophyll fluorescence imaging accurately quantifies freezing damage and cold acclimation responses in *Arabidopsis* leaves. *Plant Methods* 4, 12. doi: 10.1186/1746-4811-4-12

Eiga, S., and Sakai, A. (1984). Altitudinal variation in freezing resistance of Saghalien fir (*Abies sachalinensis*). *Can. J. Bot.* 62, 156–160. doi: 10.1139/b84-025

Fujikawa, S., Ukaji, N., Nagao, M., Yamane, K., Takezawa, D., and Arakawa, K. (2006) "Functional role of winter-accumulating proteins from mulberry tree in adaptation to winter induced stresses," in *Cold Hardiness in Plants: Molecular Genetics, Cell Biology and Physiology*, eds T. H. H. Chen, M. Uemura, and S. Fujikawa (Oxfordshire: CABI Publishers), 181–202.

Goto, S., Iijima, H., Ogawa, H., and Ohya, K. (2011). Outbreeding depression caused by intraspecific hybridization between local and nonlocal genotypes in *Abies sachalinensis*. *Restor. Ecol.* 19, 243–250. doi: 10.1111/j.1526-100X.2009.00568.x

Hamrick, J. L., and Godt, M. J. W. (1996). Effects of life history traits on genetic diversity in plant species. *Philos. Trans. R. Soc. Lon. B Biol. Sci.* 351, 1291–1298. doi: 10.1098/rstb.1996.0112

Hansen, O. K., Vendramin, G. G., Sebastiani, F., and Edwards, K. J. (2005). Development of microsatellite markers in *Abies nordmanniana* (Stev.) Spach and cross-species amplification in the *Abies* genus. *Ecol. Notes* 5, 784–787. doi: 10.1111/j.1471-8286.2005.01062.x

Howe, G. T., Aitken, S. N., Neale, D. B., Jermstad, K. D., Wheeler, N. C., and Chen, T. H. H. (2003). From genotype to phenotype: unraveling the complexities of cold adaptation in forest trees. *Can. J. Bot.* 81, 1247–1266. doi: 10.1139/b03-141

Igarashi, Y. (1996). A late glacial climatic reversion in Hokkaido, northeast Asia, inferred from the Larix pollen record. *Quat. Sci. Rev.* 15, 989–995. doi: 10.1016/S0277-3791(96)00005-4

Ishizuka, W., and Goto, S. (2012). Modeling intraspecific adaptation of *Abies sachalinensis* to local altitude and responses to global warming, based on a 36-year reciprocal transplant experiment. *Evol. Appl.* 5, 229–244. doi: 10.1111/j.1752-4571.2011.00216.x

Ishizuka, W., Ono, K., Hara, T., and Goto, S. (2015). Use of intraspecific variation in thermal responses for estimating an elevational cline in the timing of cold hardening in a sub-boreal conifer. *Plant Biol.* 17, 177–185. doi: 10.1111/plb.12214

Johnsen, Ø., Dæhlen, O. G., Østreng, G., and Skrøppa, T. (2005). Daylength and temperature during seed production interactively affect adaptive performance of *Picea abies* progeny. *New Phytol.* 168, 589–596. doi: 10.1111/j.1469-8137.2005.01538.x

Johnson, J. B., and Omland, K. S. (2004). Model selection in ecology and evolution. *Trends Ecol. Evol.* 19, 101–108. doi: 10.1016/j.tree.2003.10.013

Jump, A. S., and Penuelas, J. (2005). Running to stand still: adaptation and the response of plants to rapid climate change. *Ecol. Lett.* 8, 1010–1020. doi: 10.1111/j.1461-0248.2005.00796.x

Kalinowski, S. T., Taper, M. L., and Marshall, T. C. (2007). Revising how the computer program CERVUS accommodates genotyping error increases success in paternity assignment. *Mol. Ecol.* 16, 1099–1106. doi: 10.1111/j.1365-294X.2007.03089.x

Kato, R. (1952). The vegetation of the Tokyo University Forest in Hokkaido (in Japanese with English summary). *Bull. Tokyo Univ. For.* 43, 1–18.

Kozlowski, T. T., and Pallardy, S. G. (2002). Acclimation and adaptive responses of woody plants to environmental stresses. *Bot. Rev.* 68, 270–334. doi: 10.1663/0006-8101(2002)068[0270:AAAROW]2.0.CO;2

Kurahashi, T. (1995). Growth pattern of the F1 trees derived from reciprocal crosses between low and high elevation populations in *Abies sachalinensis*. *Rep. JSPS KAKENHI* 04806020, 44–49.

Kurahashi, A., and Hamaya, T. (1981). Variation of morphological characters and growth response of Sakhalin fir (*Abies sachalinensis*) in different altitude (in Japanese with English summary). *Bull. Tokyo Univ. For.* 71, 101–151.

Kvaalen, H., and Johnsen, Ø (2008). Timing of bud set in *Picea abies* is regulated by a memory of temperature during zygotic and somatic embryogenesis. *New Phytol.* 177, 49–59.

Lee, J. (1992). Cumulative logit modelling for ordinal response variables: applications to biomedical research. *Bioinformatics* 8, 555–562. doi: 10.1093/bioinformatics/8.6.555

Lian, C. L., Goto, S., and Hogetsu, T. (2007). Microsatellite markers for Sachalin fir (*Abies sachalinensis* Masters). *Mol. Ecol. Notes* 7, 896–898. doi: 10.1111/j.1471-8286.2007.01741.x

Lian, C. L., Goto, S., Kubo, T., Takahashi, Y., Nakagawa, M., and Hogetsu, T. (2008). Nuclear and chloroplast microsatellite analysis of *Abies sachalinensis* regeneration on fallen logs in a subboreal forest in Hokkaido. *Jpn. Mol. Ecol.* 17, 2948–2962. doi: 10.1111/j.1365-294X.2008.03 802.x

Liepelt, S., Kuhlenkamp, V., Anzidei, M., Vendramin, G. G., and Ziegenhagen, B. (2001). Pitfalls in determining size homoplasy of microsatellite loci. *Mol. Ecol. Notes* 1, 332–335. doi: 10.1046/j.1471-8278.2001.00085.x

Lindgren, K., and Hallgren, J. E. (1993). Cold acclimation of *Pinus contorta* and *Pinus sylvestris* assessed by chlorophyll fluorescence. *Tree Physiol.* 13, 97–106. doi: 10.1093/treephys/13.1.97

Mimura, M., and Aitken, S. N. (2010). Local adaptation at the range peripheries of *Sitka spruce*. *J. Evol. Biol.* 23, 249–258. doi: 10.1111/j.1420-9101.2009.01 910.x

Nakata, M., Tanaka, H., and Yagi, H. (1994). Altitudinal changes in vegetation and soils on Mt. Dairoku, central Hokkaido, Japan (in Japanese with English summary). *Jpn. J. Ecol.* 44, 33–47.

Notivol, E., Garcia-Gil, M. R., Alia, R., and Savolainen, O. (2007). Genetic variation of growth rhythm traits in the limits of a latitudinal cline in Scots pine. *Can. J. Forest Res.* 37, 540–551. doi: 10.1139/X06-243

Oleksyn, J., Modrzynski, J., Tjoelker, M. G., Zytkowiak, R., Reich, P. B., and Karolewski, P. (1998). Growth and physiology of *Picea abies* populations from elevational transects: common garden evidence for altitudinal ecotypes and cold adaptation. *Funct. Ecol.* 12, 573–590. doi: 10.1046/j.1365-2435.1998.00 236.x

Ozawa, H., Watanabe, J., Chen, H., Isoda, K., and Watanabe, A. (2009). The impact of phonological and artificial factors on seed quality in a nematode-resistant *Pinus densiflora* seed orchard. *Silvae Genetica* 58, 145–152.

Peakall, R., and Smouse, P. E. (2012). GenAlEx 6.5: genetic analysis in Excel. Population genetic software for teaching and research. *Bioinformatics* 28, 2537–2539.

R Development Core Team (2014). *R: A Language and Environment for Statistical Computing.* Vienna: R Foundation for Statistical Computing.

Rehfeldt, G. E. (1989). Ecological adaptations in Douglas-fir (*Pseudotsuga menziesii* var. glauca) – a synthesis. *Forest Ecol. Manage.* 28, 203–215. doi: 10.1016/0378-1127(89)90004-2

Saho, H., Takahashi, I., and Kurahashi, A. (1994). "Relationship between the elevation of seed collecting sites and the susceptibility of *Abies sachalinensis* to *Scleroderris lagerbergii* in Hokkaido, Japan," in *Proceedings of a Joint Meeting of the IUFRO Working Party, Canker and Shoot Blight of Conifers*, Vallombrosa, 224–247.

Sasaki, T. (1983). Phenology of forest trees and temperature in central Hokkaido, Japan (Japanese title is translated in English by authors). *Shinrin Bunka Kenkyu* 4, 77–86.

Savolainen, O., Bokma, F., Garcia-Gil, R., Komulainen, P., and Repo, T. (2004). Genetic variation in cessation of growth and frost hardiness and consequences for adaptation of *Pinus sylvestris* to climatic changes. *Forest Ecol. Manage.* 197, 79–89. doi: 10.1016/j.foreco.2004.05.006

Skrøppa, T., and Magnussen, S. (1993). Provenance variation in shoot growth components of Norway spruce. *Silvae Genet.* 42, 111–120.

Slavov, G. T., Howe, G. T., and Adams, T. (2005). Pollen contamination and mating patterns in a Douglas-fir seed orchard as measured by simple sequence repeat markers. *Can. J. Forest Res.* 35, 1592–1603. doi: 10.1139/x05-082

Sykes, R., Li, B., Isik, F., Kadla, J., and Chang, H. M. (2006). Genetic variation and genotype by environment interactions of juvenile wood chemical properties in *Pinus taeda* L. *Ann. For. Sci.* 63, 897–904. doi: 10.1051/forest:2006073

Toyooka, H., Sato, M., and Ishizuka, S. (1983). *Distribution Map of the Sasa Group in Hokkaido, Explanatory Note (in Japanese).* Sapporo: Forestry and Forest Products Research Institute, Hokkaido Branch, 36.

Tsumura, Y., and Suyama, Y. (1998). Differentiation of mitochondrial DNA polymorphisms in populations of five Japanese *Abies* species. *Evolution* 52, 1031–1042. doi: 10.2307/2411234

Vendramin, G. G., Lelli, L., Rossi, P., and Morgante, M. (1996). A set of primers for the amplification of 20 chloroplast microsatellites in Pinaceae. *Mol. Ecol.* 5, 595–598. doi: 10.1046/j.1365-294X.1996.00111.x

Vendramin, G. G., and Ziegenhagen, B. (1997). Characterization and inheritance of polymorphic plastid microsatellites in *Abies*. *Genome* 40, 857–864. doi: 10.1139/g97-811

Wheeler, N. C., and Jech, J. H. (1992). The use of electrophoretic markers in seed orchard research. *New Forest.* 6, 311–328. doi: 10.1007/BF00120650

Conflict of Interest Statement: The authors declare that the research was conducted in the absence of any commercial or financial relationships that could be construed as a potential conflict of interest.

7

Novel evidence for within-species leaf economics spectrum at multiple spatial scales

Yu-Kun Hu[1,2], Xu Pan[1,3], Guo-Fang Liu[2], Wen-Bing Li[1], Wen-Hong Dai[1], Shuang-Li Tang[2], Ya-Lin Zhang[2], Tao Xiao[1], Ling-Yun Chen[1], Wei Xiong[1], Meng-Yao Zhou[1], Yao-Bin Song[1]* and Ming Dong[1,2]*

[1] Key Laboratory of Hangzhou City for Ecosystem Protection and Restoration, College of Life and Environmental Sciences, Hangzhou Normal University, Hangzhou, China, [2] State Key Laboratory of Vegetation and Environmental Change, Institute of Botany, Chinese Academy of Sciences, Beijing, China, [3] Institute of Wetland Research, Chinese Academy of Forestry, Beijing, China

Edited by:
Boris Rewald,
University of Natural Resources and
Life Sciences, Vienna, Austria

Reviewed by:
Martín R. Aguiar,
University of Buenos Aires, Argentina
Gerhard Zotz,
Carl von Ossietzky Universität
Oldenburg, Germany

*Correspondence:
Yao-Bin Song
ybsong@hznu.edu.cn;
Ming Dong
dongming@hznu.edu.cn

Leaf economics spectrum (LES), characterizing covariation among a suite of leaf traits relevant to carbon and nutrient economics, has been examined largely among species but hardly within species. In addition, very little attempt has been made to examine whether the existence of LES depends on spatial scales. To address these questions, we quantified the variation and covariation of four leaf economic traits (specific leaf area, leaf dry matter content, leaf nitrogen and phosphorus contents) in a cosmopolitan wetland species (*Phragmites australis*) at three spatial (inter-regional, regional, and site) scales across most of the species range in China. The species expressed large intraspecific variation in the leaf economic traits at all of the three spatial scales. It also showed strong covariation among the four leaf economic traits across the species range. The coordination among leaf economic traits resulted in LES at all three scales and the environmental variables determining variation in leaf economic traits were different among the spatial scales. Our results provide novel evidence for within-species LES at multiple spatial scales, indicating that resource trade-off could also constrain intraspecific trait variation mainly driven by climatic and/or edaphic differences.

Keywords: intraspecific variation, leaf economics spectrum (LES), leaf economic traits, spatial scales, trade-offs, trait relationships, wetland plant

INTRODUCTION

Understanding species' trait variation and covariation is critical to explain species' strategies in response to environmental gradients and ecosystem functioning (Westoby et al., 2002; Garnier et al., 2004; Lavorel and Grigulis, 2012). Some important trade-offs, underpinning ecological strategies, have been found among species, e.g., C-S-R triangle (Grime, 1979), leaf-height-seed (LHS) strategy scheme (Westoby, 1998) and leaf economics spectrum (LES; Wright et al., 2004). LES, describing the covariation among leaf economic traits related to resource acquisition and conservation (Wright et al., 2004; Reich, 2014), provides a paradigm or framework for checking species strategies shaped by evolutionary history (Donovan et al., 2011; Reich, 2014). Recently, LES was found to be modulated by climate and biogeographic factors (Wright et al., 2005; Heberling and Fridley, 2012). Moreover, some researchers attempted to extend LES to 'wood economics spectrum' (Chave et al., 2009) and 'plant economics spectrum' (Freschet et al., 2010, 2012).

In addition, researchers have done much work on the origin of LES (Shipley et al., 2006; Blonder et al., 2011; Vasseur et al., 2012).

Although much progress has been made in LES among species, few studies investigated the application of LES within species (but see Blonder et al., 2013; Niinemets, 2015). The main reasons may be related to the expected lower variation and much less concern about variation within species than that among species (Fajardo and Piper, 2011). However, with increasing concern about intraspecific variation, large variability of functional traits was found within species (Albert et al., 2010; Messier et al., 2010), especially for widespread species because of their genetic and plastic variation (Darwin, 1859; Sides et al., 2014). In addition, recent studies in a number of widespread plant species showed that some trait-based trade-offs within species were either consistent (Fajardo and Piper, 2011; Richardson et al., 2013) or inconsistent (De Frenne et al., 2011; Hajek et al., 2013) with that among species. Therefore, there is need to examine whether LES, a trade-off largely reported among species, exists within species. Among the few studies about within-species LES, Blonder et al. (2013) studied the LES within clones of the tree *Populus tremuloides*, while Niinemets (2015) identified the LES in a shrub, *Quercus ilex*, across the Mediterranean region. Jackson et al. (2013) found that within-species LES occurred in 11 of 16 species across a temperate rain forest. Although these studies provided some evidence for within-species LES, the results were not consistent for all species investigated. Apparently, within-species LES was only examined in forest plant species.

Trait variation exists at all temporal, spatial, and organizational scales: individual, population, species, community, local, and regional (Albert et al., 2010; Messier et al., 2010). It has been known that different spatial scales associated with differences in climatic and/or edaphic conditions can be great drivers of variation and covariation in leaf economic traits (Liu et al., 2010; Messier et al., 2010). Therefore, assessing LES across different spatial scales can provide insight into the causes of LES (Blonder et al., 2013). Among-species LES has been extensively studied and identified at local, regional, and global scales (Díaz et al., 2004; Wright et al., 2004; Freschet et al., 2010; Jackson et al., 2013). As for within-species LES, previous studies were conducted at single organizational or spatial scale like within-clone (Blonder et al., 2013) or regional scale (Jackson et al., 2013; Niinemets, 2015). Therefore, it is still unclear how within-species LES varies across different spatial scales, or whether it is scale-dependent.

Phragmites australis, a perennial grass of Poaceae, is a cosmopolitan wetland species. As a dominant species in many wetland ecosystems, *P. australis* provides a number of important ecosystem services, e.g., water purification, paper production, and energy production (Thevs et al., 2007). It occurs along wide climatic gradients, ranging from temperate to tropic regions and from arid to humid regions in China as well as in the world (Editorial Committee of Wetland Vegetation in China, 1999). Large variation of traits in *P. australis* was found at regional scales due to the wide environmental range and genetic variation (Clevering et al., 2001; Lambertini et al., 2008). Therefore, it is an ideal plant species for the study of intraspecific variation and covariation in traits across different spatial scales. A few studies explored the trait variation in *P. australis* along large-scale environmental gradients (Clevering et al., 2001; Drapikowska and Krzakowa, 2009; Li et al., 2014). But none of them examined the covariation in leaf economic traits of *P. australis* along wide climatic gradients or at multiple spatial scales.

In this study, we aimed to investigate the variation and covariation in leaf economic traits of *P. australis* across different spatial scales (inter-regional, regional, and site) in China. Thus, we carried out a 3-years field investigation on 16 natural wetlands covering most of the geographic range of *P. australis* across China, and quantified four leaf economic traits of *P. australis*: specific leaf area (SLA), leaf dry matter content (LDMC), leaf nitrogen (N), and leaf phosphorus (P) concentration. Specifically, we attempted to answer the following questions: (1) How do leaf economic traits vary across different spatial scales? (2) Do the leaf economic traits vary to form within-species LES across the species range? (3) Does the existence of within-species LES depend on spatial scales? (4) How do climate and soil variables influence leaf economic traits across different spatial scales?

MATERIALS AND METHODS

Study Sites and Sampling Methods

Pragmites australis is distributed over large areas of China: from temperate to tropic regions and from arid to humid regions (Editorial Committee of Wetland Vegetation in China, 1999). Previous studies have found a large extent of trait variation in *P. australis* due to genetic and plastic variation across China (Editorial Committee of Wetland Vegetation in China, 1999; Zhang et al., 2003; An et al., 2012). To investigate the intraspecific variation and covariation in *P. australis*, we chose 16 sites in natural wetlands which covered most of the distribution range of *P. australis* in China (**Figure 1**) and which constituted broad climatic gradients with mean annual temperature (MAT) ranging from 1.6 to 17.4°C and mean annual precipitation (MAP) from 40 to 1702 mm (**Table 1**).

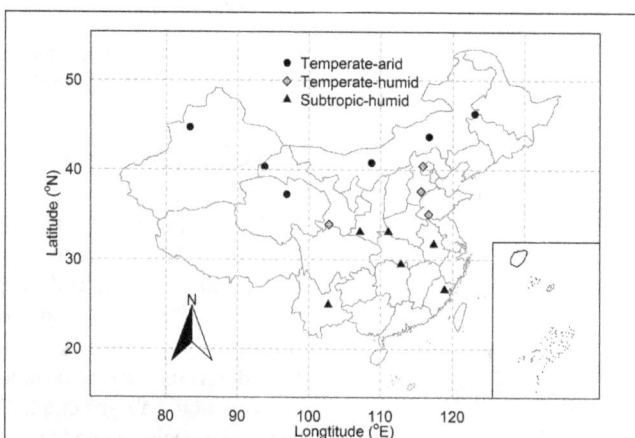

FIGURE 1 | Distribution of sampling sites for *Phragmites australis* across China. Different types of symbols represent sites of different regions.

TABLE 1 | Summary of leaf economic traits of *Phragmites australis* and environmental variables based on plot mean values.

Variables	N	Mean	SD	CV (%)	Minimum	Maximum
Leaf economic trait						
SLA (mm^2 mg^{-1})	53	13.6	3.54	25.9	7.6	24.7
LDMC (mg g^{-1})	53	394	51.9	13.2	287	526
Leaf N (mg g^{-1})	54	25.9	7.42	28.6	10.9	45.4
Leaf P (mg g^{-1})	54	1.5	0.55	36.7	0.7	3.5
Environmental variable						
MAT (°C)	55	10.6	4.66	43.9	1.6	17.4
MAP (mm)	55	543	433	79.9	40	1702
Soil pH	52	8.67	0.70	8.1	6.79	10.61
Soil EC (ms cm^{-1})	53	5.59	9.82	175.6	0.03	42.12
Soil N (mg g^{-1})	52	1.9	1.55	83.8	0.2	7.7
Soil P (mg g^{-1})	51	0.7	0.22	31.5	0.3	1.3
Soil C/N	52	18.9	11.75	62.3	7.4	72.1
Soil organic C (mg g^{-1})	52	18.2	16.20	89.1	2.2	75.4
Soil available N (mg kg^{-1})	52	119	99.9	84.1	12	459
Soil available P (mg kg^{-1})	53	19.2	17.33	90.3	1.5	79.0

SLA, specific leaf area; LDMC, leaf dry matter content. Full names of other variables were given in Section "Materials and Methods."

Three nested spatial scales were identified in this study: inter-regional, regional, and site. At the inter-regional scale, all sampling sites across China were included. For the regional scale, the 16 study sites were distributed in three regions based on temperature and precipitation: temperate-arid, temperate-humid, and subtropic-humid region (Editorial Board of Physical Geography in China Chinese Academy of Sciences, 1985; **Figure 1**). As a result, there were six sites in temperate-arid, four in temperate-humid and six in subtropic-humid region. Within each site, 1–6 plots each about 20 m × 20 m were chosen with a distance of 10–30 km for two adjacent plots. In total, 55 plots across China were chosen. In each plot, several (about 3–10) adult individuals of *P. australis* without obvious symptoms of pathogen or herbivore attack were randomly selected. Then, we picked some mature and fully expanded sun leaves (about 1–5 based on the size of leaves) from each individual. Leaves of *P. australis* from all individuals at each plot were pooled together, divided into three batches and stored in sealed plastic bags within 8 h before being determined for SLA and LDMC (Pérez-Harguindeguy et al., 2013). Leaf sampling and morphological measurements were conducted during the growing seasons from 2012 to 2014. Since sampling was carried out at three adjacent years and wetlands were insensitive to precipitation fluctuation, main climate variation between years, the inter-annual variation in traits was not considered in this study.

Leaf Economic Traits

We selected SLA, LDMC, leaf N and P, which are key traits in LES (Wright et al., 2004; Freschet et al., 2010). Specifically, SLA represents the light intercept capability with per unit of dry-mass investment, and is related to photosynthetic capacity (Pérez-Harguindeguy et al., 2013). LDMC reflects the dry-mass investment in leaves (Freschet et al., 2010). Leaf N, mostly in proteins, is closely related to the mass-based maximum photosynthetic rate, and leaf P, high in nucleic acid, lipid membranes, and bioenergetic molecules such as adenosine triphosphate, is important in metabolic process (Wright et al., 2004). Thus, resource acquisition strategist is generally characterized by high SLA, leaf N and P while resource conservation strategist by high LDMC.

All leaf samples were determined for the four leaf economic traits. The same leaf samples were used during the whole process of measurements. Firstly, one sample from each leaf batch was immersed in water overnight, blotted up water and measured for water-saturated weight. Secondly, the same leaf samples were scanned with a photo scanner (CanoScan LiDE210; Canon, Japan) and weighted after oven-dried at 70°C for 72 h. Thirdly, leaf area of each sample was accessed from scanned photo with ImageJ (http://imagej.nih.gov/ij/). Then, SLA (mm^2 mg^{-1}) of each sample was calculated as the ratio of sample leaf area to oven-dry weight. LDMC (mg g^{-1}) was determined by the ratio of leaf oven-dry weight to water-saturated weight. Leaf N (mg g^{-1}) was determined using an elemental analyzer (vario MICRO cube; Elemental, Germany), while leaf P (mg g^{-1}) was determined using ascorbic acid colorimetric method after H_2SO_4 digestion as described by Bao (2005).

Environmental Variables

Mean annual temperature and MAP were accessed from published studies which were carried out at the same sites. For each plot, three replicates of soil samples from 0 to 15 cm depth were randomly excavated. Soil samples were brought into the laboratory, air dried and passed through a 1 mm-sieve before measurement. For soil pH, 5 g subsample of each soil sample was shaken with 25 mL demineralized water in Eppendorf tube for 30 min at 250 rpm and measured for pH after standing for 30 min. The solution from pH measurement was then used to measure soil electrical conductivity (soil EC, ms cm^{-1}): the solution was centrifuged at 5000 rpm for 5 min, and the supernatant solution was measured for EC. Soil available N (mg kg^{-1}) was measured with the alkaline hydrolysis diffusion method, while soil available P (mg kg^{-1}) was determined by Olsen method (Bao, 2005).

For soil total carbon, nitrogen, and phosphorus content, a small amount of each soil sample which passed through a 0.15 mm-sieve was used. Soil total carbon (soil C, mg g^{-1}), nitrogen (soil N, mg g^{-1}), and phosphorus contents (soil P, mg g^{-1}) were determined following the same methods as for leaf N and P (see above). In addition, we measured soil organic C (mg g^{-1}) by subtracting soil inorganic content from soil total carbon, which were measured using a TOC analyzer (SSM 5000A; Schimadzu, Japan). All air-dry soil samples were oven-dried at 105°C for 6 h and measured for water content, and soil nutrient content of samples were expressed on an oven-dry mass basis.

Statistical Analyses

We used plot-level means of traits for all analyses. Descriptive statistics including mean, standard deviation (SD) and coefficient of variation (CV) were calculated for each trait. To quantify the extent of variation across three spatial scales, we used nested ANOVA with restricted maximum likelihood (REML) method to estimate the variance component across scales (Messier et al., 2010). Because our aim was only to calculate the variance of traits at each scale, the data were not transformed before nested ANOVA (Quinn and Keough, 2002). Instead, SLA, leaf P, MAP, soil N, soil P, soil C/N, soil organic C, available N, available P and EC were $\log_{10}$-transformed before the following data analysis to meet the assumption of approximate normality and residuals homogeneity, while data sets of LDMC, leaf N, MAT and soil pH were not transformed because they met approximately normal distribution. Simple linear regression with ordinary least-squares method was used to explore bivariate trait-trait relationships at three spatial scales (inter-regional, regional, and site). Notably, at site scale, ordinary least-squares regression was only performed for sites with more than three plots, i.e., 12 of total 16 sites. Pearson's correlation analysis was carried out to quantify pairwise relationship between leaf traits and environmental variables at inter-regional and regional scale other than at site scale owing to limited data. Statistical significance of each correlation was assessed after Holm–Bonferroni correction for multiple comparisons. Principal component analysis (PCA) was also used to explore the correlations among traits and account for the majority of variation at inter-regional and regional scales. All analyses were performed in R 3.1.0 (R Core Team, 2014).

RESULTS

Trait Variation in *P. australis* across Different Spatial Scales

The intraspecific variation in four leaf economic traits of *P. australis* was different (**Table 1**). LDMC varied the least, ranging from 287 to 526 mg g^{-1} (CV = 13.2%), while leaf P varied the most, from 0.7 to 3.5 mg g^{-1} (CV = 36.7%). The extent of variation in SLA and leaf N was intermediate with 25.9–28.6% of CV. Moreover, variance partitioning in nested ANOVA showed that variability in the leaf economic traits was different across spatial scales (**Figure 2**). Variability between plots (within sites) accounted for the largest proportion of total variance in SLA, LDMC, and leaf P (34.5, 44.3, and 46.6%, respectively) and the second largest in leaf N (40.0%). Variability between sites differed across traits, 10% of total variance for leaf P and 25–40% for other traits (**Figure 2**). The smallest variance in leaf N (17.0%) was due to difference between regions, which caused the largest variability in leaf P (42.8%). Variance components of SLA and LDMC were 31.3 and 30.8% of total variance at regional scale.

Intraspecific Trait Covariation across Different Scales

At inter-regional scale, strong relationships were found among leaf economic traits of *P. australis* (**Figure 3**). Specifically, SLA was positively correlated with leaf N and P ($r^2 = 0.12$, $p = 0.012$; $r^2 = 0.32$, $p < 0.001$), but was negatively correlated with LDMC ($r^2 = 0.42$, $p < 0.001$). LDMC was significantly negatively correlated with leaf N and P ($r^2 = 0.10$, $p = 0.021$; $r^2 = 0.30$,

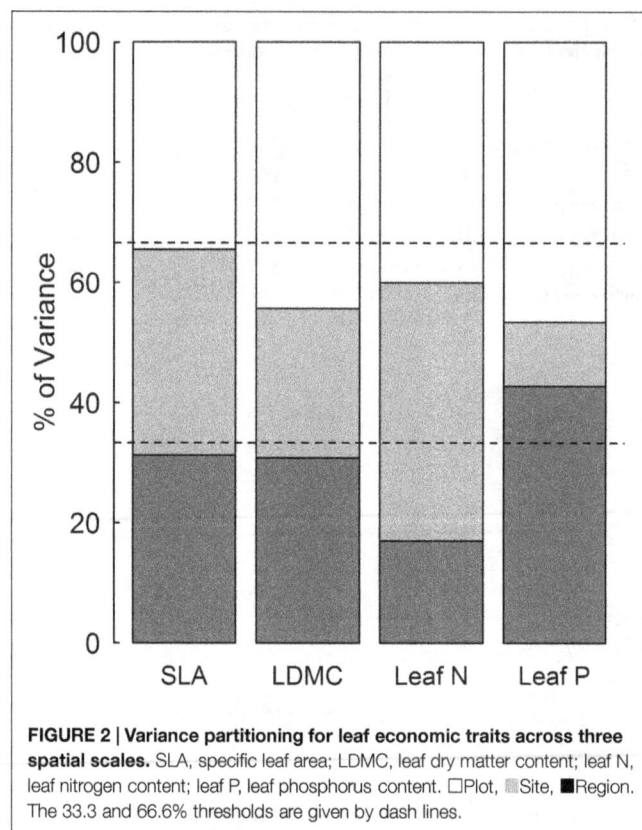

FIGURE 2 | Variance partitioning for leaf economic traits across three spatial scales. SLA, specific leaf area; LDMC, leaf dry matter content; leaf N, leaf nitrogen content; leaf P, leaf phosphorus content. □Plot, ▒Site, ■Region. The 33.3 and 66.6% thresholds are given by dash lines.

$p < 0.001$), while leaf P was positively correlated with leaf N ($r^2 = 0.42$, $p < 0.001$).

At regional scale, significant correlations between SLA and LDMC, between LDMC and leaf P and between leaf N and P remained for three different regions: temperate-arid, temperate-humid, and subtropic-humid regions (**Figure 3**). There were weak relationships between leaf structure traits and leaf N at all three regions (**Figure 3**). Significant correlation between SLA and leaf P was only detected at the temperate-humid and subtropic-humid regions (**Figure 3**). At site scale, significant correlations among leaf economic traits were detected only in a small number of sites which are distributed in different regions (**Table 2**; Supplementary Figure S1).

Environmental Correlates of Leaf Economic Traits across Different Spatial Scales

The correlations between leaf economic traits of *P. australis* and environmental variables (climate and soil factors) were different across different spatial scales (**Table 3**). Specifically, at inter-regional scale, SLA and leaf P were all strongly related to MAT, MAP, soil pH, soil EC and soil C/N, except that the correlations between SLA, MAT, and soil pH were not significant ($p > 0.05$, **Table 3**). Leaf N was significantly related to MAP and soil pH ($p < 0.05$, **Table 3**), and LDMC decreased with increasing soil P and available P (**Table 3**). At regional scale, in temperate-arid region, SLA was significantly correlated with soil EC, while leaf

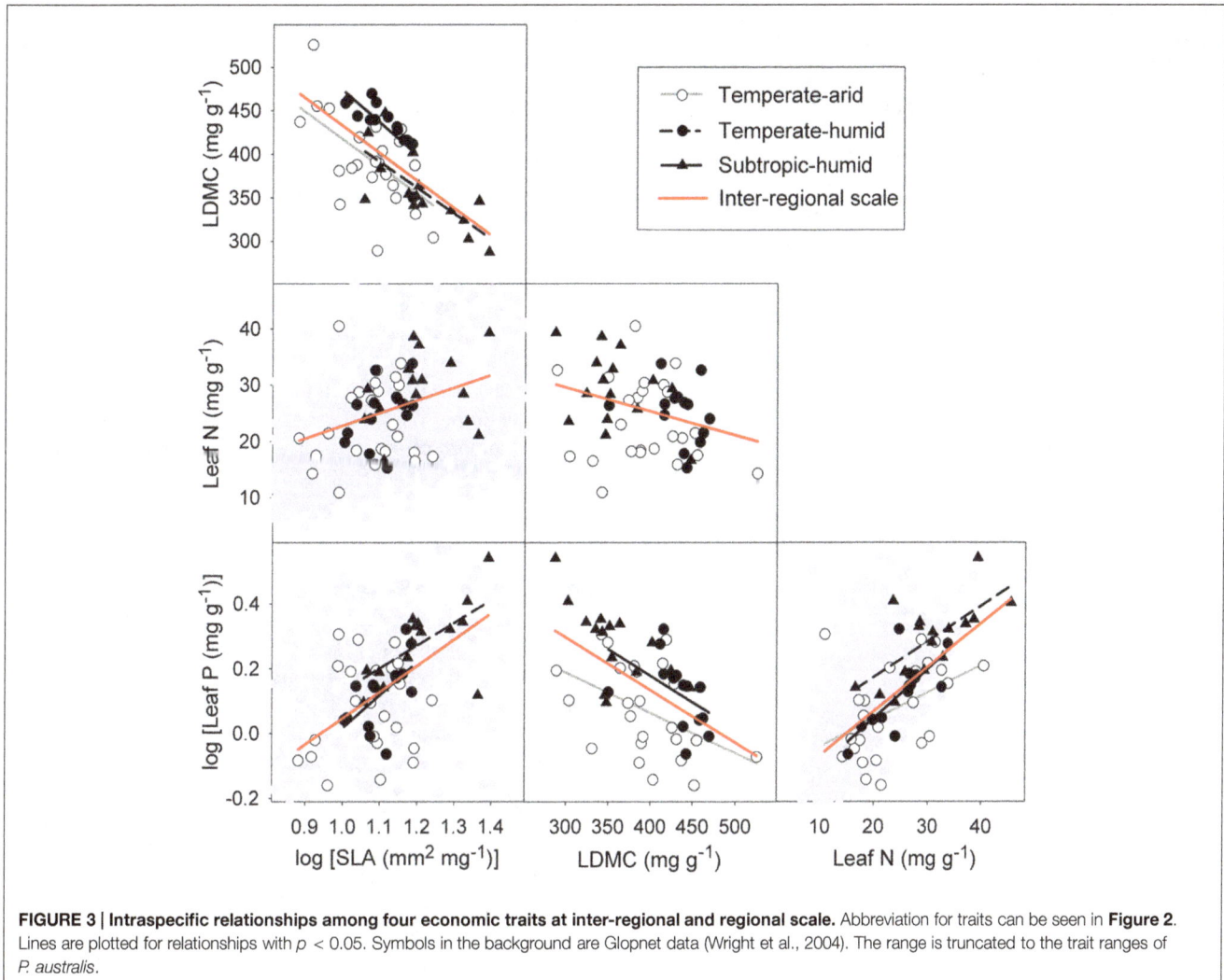

FIGURE 3 | Intraspecific relationships among four economic traits at inter-regional and regional scale. Abbreviation for traits can be seen in **Figure 2**. Lines are plotted for relationships with $p < 0.05$. Symbols in the background are Glopnet data (Wright et al., 2004). The range is truncated to the trait ranges of *P. australis*.

N was significantly related to MAT, and leaf P related to MAP (**Table 3**). In contrast, LDMC was only significantly positively correlated with MAT in temperate-humid region (**Table 3**). In subtropic-humid region, SLA and leaf P were positively related to soil EC, and SLA decreased with increasing MAT (**Table 3**).

DISCUSSION

Multi-scale Intraspecific Variation in Leaf Economic Traits

We found a substantial amount of variation in leaf economic traits for the cosmopolitan wetland species *P. australis* across most of its geographic range in China (**Table 1**, Supplementary Figure S2). Although intraspecific trait variation was less than interspecific variation across China and the globe, it covered a large proportion of interspecific variation at scales either the same as or much larger than that in this study (Supplementary Figure S2). The large trait variation in *P. australis* was consistent with results of Albert et al. (2010) and Fajardo and Piper (2011) in other species along large environmental gradients. Although

the extent of trait variation may depend on species and traits, an increasing number of studies indicate the important role of intraspecific trait variation at community level (Albert et al.,

TABLE 2 | Summary of intraspecific relationships among four leaf economic traits of *P. australis* at site scale.

	Number of sites ($n = 16$)			Direction	Region
	$p < 0.05$	$p < 0.10$	$p < 0.15$		
LDMC-SLA	3	4	5	−	ta, th, sh
Leaf N-SLA	1	1	2	+	th
Leaf P-SLA	0	2	2	+	th, sh
Leaf N-LDMC	1	1	4	−	sh
Leaf P-LDMC	1	2	4	−	th, sh
Leaf P-leaf N	0	3	5	+	th, sh

SLA, specific leaf area; LDMC, leaf dry matter content. Values in the first three columns is number of sites within which trait relationships are significant at $p < 0.05$, $p < 0.10$, and $p < 0.15$, respectively. Direction of trait relationships: +, positive; −, negative. Significant trait relationships ($p < 0.10$) in three regions: ta, temperate-arid; th, temperate-humid; sh, subtropic-humid.

TABLE 3 | Pearson's correlation between leaf economic traits of *P. australis*, climate and soil variables at the inter-regional and regional (temperate-arid, temperate-humid, and subtropical-humid) scales.

Scale	Variables	SLA	LDMC	Leaf N	Leaf P
Inter-regional	MAT	0.28	−0.04	0.20	**0.38**
	MAP	**0.49**	−0.16	**0.49**	**0.60**
	Soil pH	−0.30	0.12	**−0.51**	**−0.42**
	Soil EC	**−0.52**	0.06	−0.22	**−0.39**
	Soil N	0.11	0.05	0.12	−0.09
	Soil P	0.33	**−0.54**	0.01	0.21
	Soil C/N	**−0.54**	0.31	−0.36	**−0.43**
	Soil organic C	0.01	0.12	0.03	−0.17
	Soil available N	0.10	0.04	0.18	−0.02
	Soil available P	0.14	**−0.45**	0.32	0.17
Temperate-arid region	MAT	−0.41	0.17	**−0.57**	−0.43
	MAP	0.44	−0.27	0.55	**0.56**
	Soil pH	−0.06	−0.08	−0.25	−0.03
	Soil EC	**−0.64**	0.26	0.01	−0.08
	Soil N	0.44	0.06	0.13	−0.17
	Soil P	0.45	−0.38	−0.20	0.26
	Soil C/N	−0.47	0.26	−0.26	−0.10
	Soil organic C	0.48	0.02	0.03	−0.20
	Soil available N	0.41	0.08	0.08	−0.16
	Soil available P	0.31	−0.37	0.30	−0.03
Temperate-humid region	MAT	−0.41	**0.80**	0.06	0.03
	MAP	−0.16	−0.05	−0.02	−0.01
	Soil pH	−0.41	0.49	−0.49	−0.28
	Soil EC	0.00	−0.09	0.45	0.28
	Soil N	−0.23	0.39	0.00	−0.17
	Soil P	0.47	−0.50	0.27	0.15
	Soil C/N	0.20	0.03	0.29	0.49
	Soil organic C	−0.26	0.36	0.08	−0.14
	Soil available N	−0.25	0.22	0.22	0.07
	Soil available P	0.52	−0.70	0.56	0.46
Subtropic-humid region	MAT	**−0.75**	0.32	0.15	−0.37
	MAP	−0.61	0.12	0.13	−0.34
	Soil pH	0.62	−0.43	−0.43	0.08
	Soil EC	**0.73**	−0.61	0.55	**0.76**
	Soil N	−0.03	−0.14	0.41	0.47
	Soil P	0.50	−0.54	0.06	0.41
	Soil C/N	0.36	−0.27	0.23	0.22
	Soil organic C	−0.16	0.00	0.36	0.35
	Soil available N	−0.24	−0.03	0.51	0.33
	Soil available P	−0.06	0.02	0.46	0.45

Values are Pearson's correlation coefficients. Values in bold indicate significance at p < 0.05 after Holm-Bonferroni correction for the number of tests. SLA, leaf P, MAP, soil N, soil P, soil C/N, soil organic C, soil available N, soil available P, soil EC were log₁₀-transformed, other variables were not. Full name of variables were given in "Materials and Methods."

2010; Messier et al., 2010; Jackson et al., 2013). These facts imply that intraspecific variation is an important component of trait variation, which should not be neglected in investigating the responses of communities and ecosystems to environmental changes (Fajardo and Piper, 2011; Jackson et al., 2013).

On the other hand, the distribution of variance in SLA and LDMC was relatively uniform among three spatial scales

(**Figure 2**). Since different spatial scales are associated with differences in ecological processes, such as genetic variation, edaphic and climatic conditions (Messier et al., 2010), our results indicate that processes driving variation at different spatial scales were of similar importance for the leaf structural traits considered. In addition, small variance for leaf P at site scale and leaf N at regional scale suggests that climatic condition at regional scale may play a critical role in driving leaf P rather than leaf N. Overall, the variation in all four leaf economic traits at site and regional scale were comparable to that at inter-regional scale (**Figure 2**). The large intraspecific trait variability at local (site) scale was consistent with previous studies within species (Albert et al., 2010; Fu et al., 2013) and among species (Wright et al., 2004). It suggests that environmental heterogeneity (mainly in terms of soil nutrients and water) at local scale plays an important role in driving leaf economic traits of species.

Existence of Within-species LES across the Species Range

We found strong correlations among the four leaf economic traits (SLA, LDMC, leaf N and P) of *P. australis* related to resource acquisition and conservation across the species range in China (**Figure 3**; Supplementary Figure S3A), which indicates that within-species LES occurs in the widespread wetland grass across a large scale. Consistent with results from Blonder et al. (2013) in a clonal tree (*Populus tremuloides*) and Niinemets (2015) in a Mediterranean shrub (*Quercus ilex*), our results provide evidence for within-species LES. This observed LES within species was similar to LES found among species (**Figure 3**; Wright et al., 2004; Freschet et al., 2010). It suggests that the trade-off between resource acquisition and conservation, which has been extensively found among species, is likely to operate within species. Additionally, previous work showed that trait variation in within-species LES could result from environmental difference (Blonder et al., 2013), genetic variation (Vasseur et al., 2012), or both (Jackson et al., 2013; Niinemets, 2015). This study confirmed those, which meant resource trade-off could constrain intraspecific trait variation driven by genetic and environmental variation.

Although some studies, including this one, found within-species LES in different species, it is still far from a general conclusion. It is necessary to examine more species of different growth forms and in different habitats. As within-species LES exists widely, a quantitative comparison is needed to explore the potential differences between intraspecific and interspecific LES. We found some qualitative differences between LES in *P. australis* and among species, e.g., the relationship between leaf N and P (**Figure 3**; Wright et al., 2004). Considering the large extent of trait variation within species and the potential contrasts in LES between within and among species, we argue that intraspecific trait variation needs to be incorporated into worldwide LES and global dynamic vegetation models, which will provide a better understanding and prediction of global change responses (Moran et al., 2015; Niinemets, 2015).

Within-species LES across Different Spatial Scales

The relationships between leaf structural traits (SLA and LDMC) and leaf N of *P. australis* were significant at inter-regional scale, but not at regional scale (**Figure 3**). Meanwhile, there existed some differences in trait relationships between site and regional scales, although the data at site scale was limited. These were consistent with previous studies which already provided some evidence for scale-dependency of trait relationships (Burns and Beaumont, 2009; Liu et al., 2010). Different environmental gradients and biotic factors occurring among different spatial scales were likely to influence the scale-dependency of trait relationships.

Although there were some differences in trait relationships among the three spatial scales, coordination among the leaf economic traits resulted in LES at both inter-regional and regional scales (Supplementary Figure S3). This indicates that trade-off between resource acquisition and conservation, forming within-species LES, constrains the trait variation at large scale (Niinemets, 2015). At local scale, strong correlations among leaf economic traits were found in several sites, for example, in site HS (temperate-humid region), we found significant relationships among traits except between LDMC and leaf N (Supplementary Figure S1). This showed that within-species LES occurred at a local scale, which was consistent with the study by Blonder et al. (2013) on within-clone LES within each site. It suggests that trade-off generating the LES can operate within species at local scale as well (Blonder et al., 2013). In summary, our results demonstrated that within-species LES emerged at multiple spatial scales (inter-regional, regional, and local), although there were a few differences between relationships among leaf economic traits across different scales.

Multi-scale Effects of Climate and Soil Variables on Leaf Economic Traits

At inter-regional scale, leaf economic traits were closely correlated with several climate and soil variables. At regional scale, the few predictors of leaf economic traits were MAT, MAP, and soil EC in temperate-arid region, MAT in temperate-humid region, and MAT and soil EC in subtropic-humid region. These facts indicate that the environmental variables determining leaf economic traits were different among spatial scales, although all these climate and soil factors have been found to influence leaf economic traits (Ordoñez et al., 2009; Fujita et al., 2013; Richardson et al., 2013; Maire et al., 2015).

Contrary to our expectation, soil N, soil organic C and available N, which were closely related to each other, were weak predictors of the four leaf economic traits at inter-regional and regional scales. It may be due to the phosphorus limitation of *P. australis* across wetlands in China, supported by the fact that leaf N/P of *P. australis* in 43/55 plots was higher than 16 (Koerselman and Meuleman, 1996).

Overall, relationships between leaf economic traits and environmental variables were variable among spatial scales. It suggests that distinct environmental variation at different spatial scales shapes the different functional responses of *P. australis*. Therefore, the various environmental variations resulting from multiple spatial scales should be considered in investigating the influence of climate and soil properties on species' leaf functional traits (Ordoñez et al., 2009; De Frenne et al., 2011).

CONCLUSION

A substantial amount of variation in leaf economic traits was found for the cosmopolitan wetland species *P. australis* at three spatial scales across most of its geographic range in China, indicating that variation within species is not negligible. Leaf economic traits were coordinated to form within-species LES in the wetland species *P. australis* across China, which provide novel evidence for existence of LES within species. Our results demonstrated that within-species LES occurred at multiple spatial scales (inter-regional, regional, and site). This improved our understanding on the scale-dependent aspects of within-species LES. Finally, the environmental variables determining variation in leaf economic traits were different among spatial scales. Further studies are needed to investigate the role of scale in trait variation and covariation both within and among species.

ACKNOWLEDGMENTS

We are very grateful to Can Jiang, Hong-Ke Xu, Yan-Fang Fang, Ze-Ning Jin, and Xu-Yan Liu for laboratory assistance and to Johannes H. C. Cornelissen for helpful suggestions. This work was supported by the NSFC (Grants: 31261120580; 31400346; 41401556) and the Innovative R & D grant (201203) from Hangzhou Normal University.

REFERENCES

Albert, C. H., Thuiller, W., Yoccoz, N. G., Soudant, A., Boucher, F., Saccone, P., et al. (2010). Intraspecific functional variability: extent, structure and sources of variation. *J. Ecol.* 98, 604–613. doi: 10.1111/j.1365-2745.2010.01651.x

An, J. X., Wang, Q., Yang, J., and Liu, J. Q. (2012). Phylogeographic analyses of *Phragmites australis* in China: native distribution and habitat preference of the haplotype that invaded North America. *J. Syst. Evol.* 50, 334–340. doi: 10.1111/j.1759-6831.2012.00192.x

Bao, S. D. (2005). *Soil Agricultural Chemistry Analysis.* Beijing: China Agriculture Press.

Blonder, B., Violle, C., Bentley, L. P., and Enquist, B. J. (2011). Venation networks and the origin of the leaf economics spectrum. *Ecol. Lett.* 14, 91–100. doi: 10.1111/j.1461-0248.2010.01554.x

Blonder, B., Violle, C., and Enquist, B. J. (2013). Assessing the causes and scales of the leaf economics spectrum using venation networks in *Populus tremuloides*. *J. Ecol.* 101, 981–989. doi: 10.1111/1365-2745.12102

Burns, K. C., and Beaumont, S. (2009). Scale-dependent trait correlations in a temperate tree community. *Austr. Ecol.* 34, 670–677. doi: 10.1111/j.1442-9993.2009.01973.x

Chave, J., Coomes, D., Jansen, S., Lewis, S. L., Swenson, N. G., and Zanne, A. E. (2009). Towards a worldwide wood economics spectrum. *Ecol. Lett.* 12, 351–366. doi: 10.1111/j.1461-0248.2009.01285.x

Clevering, O. A., Brix, H., and Lukavská, J. (2001). Geographic variation in growth responses in *Phragmites australis*. *Aquat. Bot.* 69, 89–108. doi: 10.1111/j.1461-0248.2009.01285.x

Darwin, C. (1859). *On the Origin of Species by Means of Natural Selection*. London: Murray.

De Frenne, P., Graae, B. J., Kolb, A., Shevtsova, A., Baeten, L., Brunet, J., et al. (2011). An intraspecific application of the leaf-height-seed ecology strategy scheme to forest herbs along a latitudinal gradient. *Ecography* 34, 132–140. doi: 10.1111/j.1600-0587.2010.06399.x

Díaz, S., Hodgson, J. G., Thompson, K., Cabido, M., Cornelissen, J. H. C., Jalili, A., et al. (2004). The plant traits that drive ecosystems: evidence from three continents. *J. Veg. Sci.* 15, 295–304. doi: 10.1111/j.1654-1103.2004.tb02266.x

Donovan, L. A., Maherali, H., Caruso, C. M., Huber, H., and de Kroon, H. (2011). The evolution of the worldwide leaf economics spectrum. *Trends Ecol. Evol.* 26, 88–95. doi: 10.1016/j.tree.2010.11.011

Drapikowska, M., and Krzakowa, M. (2009). Morphological and biochemical variation among common reed (*Phragmites australis*) populations in northwest Poland. *Oceanol. Hydrobiol. Stud.* 38, 29–38. doi: 10.2478/v10009-009-0019-3

Editorial Board of Physical Geography in China Chinese Academy of Sciences (1985). *Physical Geography in China: Climatology*. Beijing: Science Press.

Editorial Committee of Wetland Vegetation in China (1999). *Wetland Vegetation in China*. Beijing: Science Press.

Fajardo, A., and Piper, F. I. (2011). Intraspecific trait variation and covariation in a widespread tree species (*Nothofagus pumilio*) in southern Chile. *New Phytol.* 189, 259–271. doi: 10.1111/j.1469-8137.2010.03468.x

Fu, H., Yuan, G., Zhong, J., Cao, T., Ni, L., and Xie, P. (2013). Environmental and ontogenetic effects on intraspecific trait variation of a macrophyte species across five ecological scales. *PLoS ONE* 8:e62794. doi: 10.1371/journal.pone.0062794

Fujita, Y., van Bodegom, P. M., and Witte, J. P. M. (2013). Relationships between nutrient-related plant traits and combinations of soil N and P fertility measures. *PLoS ONE* 8:e83735. doi: 10.1371/journal.pone.0083735

Freschet, G. T., Aerts, R., and Cornelissen, J. H. C. (2012). A plant economics spectrum of litter decomposability. *Funct. Ecol.* 26, 56–65. doi: 10.1111/j.1365-2435.2011.01913.x

Freschet, G. T., Cornelissen, J. H. C., Van Logtestijn, R. S. P., and Aerts, R. (2010). Evidence of the 'plant economics spectrum' in a subarctic flora. *J. Ecol.* 98, 362–373. doi: 10.1111/j.1365-2745.2009.01615.x

Garnier, E., Cortez, J., Billès, G., Navas, M. L., Roumet, C., Debussche, M., et al. (2004). Plant functional markers capture ecosystem properties during secondary succession. *Ecology* 85, 2630–2637. doi: 10.1890/03-0799

Grime, J. P. (1979). *Plant Strategies and Vegetation Processes*. New York: John Wiley and Sons.

Hajek, P., Hertel, D., and Leuschner, C. (2013). Intraspecific variation in root and leaf traits and leaf-root trait linkages in eight aspen demes (*Populus tremula* and *P. tremuloides*). *Front. Plant Sci.* 4:415. doi: 10.3389/fpls.2013.00415

Heberling, J. M., and Fridley, J. D. (2012). Biogeographic constraints on the world-wide leaf economics spectrum. *Global Ecol. Biogeogr.* 21, 1137–1146. doi: 10.1111/j.1466-8238.2012.00761.x

Jackson, B. G., Peltzer, D. A., and Wardle, D. A. (2013). The within-species leaf economic spectrum does not predict leaf litter decomposability at either the within-species or whole community levels. *J. Ecol.* 101, 1409–1419. doi: 10.1111/1365-2745.12155

Koerselman, W., and Meuleman, A. F. M. (1996). The vegetation N: P ratio: a new tool to detect the nature of nutrient limitation. *J. Appl. Ecol.* 33, 1441–1450. doi: 10.2307/2404783

Lambertini, C., Gustafsson, M. H. G., Frydenberg, J., Speranza, M., and Brix, H. (2008). Genetic diversity patterns in *Phragmites australis* at the

population, regional and continental scales. *Aquat. Bot.* 88, 160–170. doi: 10.1016/j.aquabot.2007.10.002

Lavorel, S., and Grigulis, K. (2012). How fundamental plant functional trait relationships scale-up to trade-offs and synergies in ecosystem services. *J. Ecol.* 100, 128–140. doi: 10.1111/j.1365-2745.2011.01914.x

Li, L. P., Zerbe, S., Han, W. X., Thevs, N., Li, W. P., He, P., et al. (2014). Nitrogen and phosphorus stoichiometry of common reed (*Phragmites australis*) and its relationship to nutrient availability in northern China. *Aquat. Bot.* 112, 84–90. doi: 10.1016/j.aquabot.2013.08.002

Liu, G. F., Freschet, G. T., Pan, X., Cornelissen, J. H. C., Li, Y., and Dong, M. (2010). Coordinated variation in leaf and root traits across multiple spatial scales in Chinese semi-arid and arid ecosystems. *New Phytol.* 188, 543–553. doi: 10.1111/j.1469-8137.2010.03388.x

Maire, V., Wright, I. J., Prentice, I. C., Batjes, N. H., Bhaskar, R., van Bodegom, P. M., et al. (2015). Global effects of soil and climate on leaf photosynthetic traits and rates. *Global Ecol. Biogeogr.* 24, 706–717. doi: 10.1111/geb.12296

Messier, J., McGill, B. J., and Lechowicz, M. J. (2010). How do traits vary across ecological scales? A case for trait-based ecology. *Ecol. Lett.* 13, 838–848. doi: 10.1111/j.1461-0248.2010.01476.x

Moran, E. V., Hartig, F., and Bell, D. M. (2015). Intraspecific trait variation across scales: implications for understanding global change responses. *Glob. Chang. Biol.* doi: 10.1111/gcb.13000 [Epub ahead of print].

Niinemets, Ü. (2015). Is there a species spectrum within the world-wide leaf economics spectrum? Major variations in leaf functional traits in the Mediterranean sclerophyll *Quercus ilex*. *New Phytol.* 205, 79–96. doi: 10.1111/nph.13001

Ordoñez, J. C., Van Bodegom, P. M., Witte, J. P. M., Wright, I. J., Reich, P. B., and Aerts, R. (2009). A global study of relationships between leaf traits, climate and soil measures of nutrient fertility. *Global Ecol. Biogeogr.* 18, 137–149. doi: 10.1111/j.1466-8238.2008.00441.x

Pérez-Harguindeguy, N., Díaz, S., Garnier, E., Lavorel, S., Poorter, H., Jaureguiberry, P., et al. (2013). New handbook for standardised measurement of plant functional traits worldwide. *Aust. J. Bot.* 61, 167–234. doi: 10.1071/BT12225

Quinn, G. P., and Keough, M. J. (2002). *Experimental Design and Data Analysis for Biologists*. London: Cambridge University Press. doi: 10.1017/CBO9780511806384

R Core Team (2014). *R: A Language and Environment for Statistical Computing*. Vienna: R Foundation for Statistical Computing.

Reich, P. B. (2014). The world-wide 'fast–slow' plant economics spectrum: a traits manifesto. *J. Ecol.* 102, 275–301. doi: 10.1111/1365-2745.12211

Richardson, S. J., Allen, R. B., Buxton, R. P., Easdale, T. A., Hurst, J. M., Morse, C. W., et al. (2013). Intraspecific relationships among wood density, leaf structural traits and environment in four co-occurring species of Nothofagus in New Zealand. *PLoS ONE* 8:e58878. doi: 10.1371/journal.pone.0058878

Shipley, B., Lechowicz, M. J., Wright, I., and Reich, P. B. (2006). Fundamental trade-offs generating the worldwide leaf economics spectrum. *Ecology* 87, 535–541. doi: 10.1890/05-1051

Sides, C. B., Enquist, B. J., Ebersole, J. J., Smith, M. N., Henderson, A. N., and Sloat, L. L. (2014). Revisiting Darwin's hypothesis: does greater intraspecific variability increase species' ecological breadth? *Am. J. Bot.* 101, 56–62. doi: 10.3732/ajb.1300284

Thevs, N., Zerbe, S., Gahlert, F., Mijit, M., and Succow, M. (2007). Productivity of reed (*Phragmites australis* Trin. ex Steud.) in continental-arid NW China in relation to soil, groundwater, and land-use. *J. Appl. Bot. Food Qual.* 81, 62–68.

Vasseur, F., Violle, C., Enquist, B., Granier, C., and Vile, D. (2012). A common genetic basis to the origin of the leaf economics spectrum and metabolic scaling allometry. *Ecol. Lett.* 15, 1149–1157. doi: 10.1111/j.1461-0248.2012.01839.x

Westoby, M. (1998). A leaf-height-seed (LHS) plant ecology strategy scheme. *Plant Soil* 199, 213–227. doi: 10.1023/A:1004327224729

Westoby, M., Falster, D. S., Moles, A. T., Vesk, P. A., and Wright, I. J. (2002). Plant ecological strategies: some leading dimensions of variation between species. *Ann. Rev. Ecol. Syst.* 33, 125–159. doi: 10.1146/annurev.ecolsys.33.010802.150452

Wright, I. J., Reich, P. B., Cornelissen, J. H. C., Falster, D. S., Groom, P. K., Hikosaka, K., et al. (2005). Modulation of leaf economic traits and trait relationships by climate. *Global Ecol. Biogeogr.* 14, 411–421. doi: 10.1111/j.1466-822x.2005.00172.x

Wright, I. J., Reich, P. B., Westoby, M., Ackerly, D. D., Baruch, Z., Bongers, F., et al. (2004). The worldwide leaf economics spectrum. *Nature* 428, 821–827. doi: 10.1038/nature02403

Zhang, S. P., Wang, R. Q., Zhang, Z. G., Guo, W. H., Liu, J., and Song, B. M. (2003). Study on morphological variation of *Phragmites australis* in the Yellow River downstream wetland. *Chin. J. Plant Ecol.* 27, 78–85.

Conflict of Interest Statement: The authors declare that the research was conducted in the absence of any commercial or financial relationships that could be construed as a potential conflict of interest.

8

The key factor limiting plant growth in cold and humid alpine areas also plays a dominant role in plant carbon isotope discrimination

Meng Xu[1], Guoan Wang[1]*, Xiaoliang Li[1], Xiaobu Cai[2], Xiaolin Li[1], Peter Christie[1] and Junling Zhang[1]

[1] College of Resources and Environmental Sciences, China Agricultural University, Beijing, China, [2] Tibet Agricultural and Animal Husbandry College, Tibet University, Linzhi, China

Many environmental factors affect carbon isotope discrimination in plants, yet the predominant factor influencing this process is generally assumed to be the key growth-limiting factor. However, to our knowledge this hypothesis has not been confirmed. We therefore determined the carbon isotope composition (δ^{13}C) of plants growing in two cold and humid mountain regions where temperature is considered to be the key growth-limiting factor. Mean annual temperature (MAT) showed a significant impact on variation in carbon isotope discrimination value (Δ) irrespective of study area or plant functional type with either partial correlation or regression analysis, but the correlation between Δ and soil water content (SWC) was usually not significant. In multiple stepwise regression analysis, MAT was either the first or the only variable selected into the prediction model of Δ against MAT and SWC, indicating that the effect of temperature on carbon isotope discrimination was predominant. The results therefore provide evidence that the key growth-limiting factor is also crucial for plant carbon isotope discrimination. Changes in leaf morphology, water viscosity and carboxylation efficiency with temperature may be responsible for the observed positive correlation between Δ and temperature.

Keywords: alpine plants, carbon isotope discrimination, temperature, water availability, key growth-limiting factor

Edited by:
Sebastian Leuzinger,
Auckland University of Technology,
New Zealand

Reviewed by:
Martin Karl-Friedrich Bader,
New Zealand Forest Research
Institute, New Zealand
Zhenzhu Xu,
Institute of Botany – Chinese
Academy of Sciences, China

*Correspondence:
Guoan Wang
gawang@cau.edu.cn

INTRODUCTION

Carbon isotope discrimination in plants reflects a range of physiological responses including stomatal conductance, assimilation rate, altered C:N allocation to carboxylation, and leaf structure (Seibt et al., 2008). Water-use efficiency (WUE), which controls the balance between water use and carbon assimilation within plants, is linked to plant carbon isotope discrimination through the substomatal cavities (Farquhar and Richards, 1984). This relationship has thus resulted in numerous studies on plant isotope discrimination in physiological ecology and the global carbon cycle (e.g., Duquesnay et al., 1998; Wang and Feng, 2012; Cernusak et al., 2013).

As is well acknowledged, plant carbon isotope discrimination may be affected by many environmental factors such as temperature, moisture, altitude, latitude, longitude, solar radiation, air pressure, and atmospheric CO_2 concentration. The fundamental mechanism of how these factors affect plant carbon isotope discrimination is that they can control directly or indirectly the ratio of the intercellular CO_2 concentration (c_i) to the ambient CO_2 concentration (c_a). Previous studies have reported that precipitation has a positive and altitude a negative influence on plant carbon isotope discrimination value (Δ), but the effect of temperature varied (e.g., Körner et al., 1988, 1991; Morecroft and Woodward, 1996; Wang et al., 2008, 2013; Diefendorf et al., 2010; Kohn, 2010). Temperature and water availability are considered to be two of the fundamental influential factors. This is based on the observation that variations in altitude, longitude, and latitude can lead to changes in temperature and/or precipitation. Their effects on plant carbon isotope discrimination will therefore be expressed mainly in the effects of temperature and precipitation. Solar radiation and air pressure also vary with altitude, yet their role in the altitudinal trends in plant carbon isotope discrimination are believed to be rather small compared to temperature and/or precipitation (Körner et al., 1988, 1991; Sparks and Ehleringer, 1997; Wang et al., 2008), with the exceptions of Kelly and Woodward (1995) and Zhu et al. (2010) who demonstrated that decreasing Δ with increasing altitude was primarily attributable to decreasing air pressure rather than air temperature. Their conclusion, however, may not always be reliable because the study areas with different elevations that were used to compare air pressure and temperature effects in their study also experience different precipitation inputs. Although the authors claimed no water stress in these study areas, the contribution of precipitation to Δ cannot be ruled out since Diefendorf et al. (2010) found that plant Δ still shows an increasing trend with precipitation when rainfall amount is more than 1000 mm. Marshall and Linder (2013) showed that mineral nutrition may also have a strong effect on plant carbon isotope discrimination. However, Yao et al. (2011) did not observe any

change in the Δ of a number of species in response to application of N.

Since carbon isotope discrimination in plants is closely related to plant performance and the key growth-limiting factors play a significant role in plant performance, it has been suggested that the key growth-limiting factor is also the predominant factor affecting plant carbon isotope discrimination (Winter et al., 1982; McCarroll and Loader, 2004). However, as far as we know, this hypothesis has not been confirmed because it is difficult to find a site or region where we know with confidence which environmental factor is the key growth-limiting factor influencing the local plants.

In the present study we investigated plant carbon isotope composition ($\delta^{13}C$) in two cold and humid montane regions, Mount Gongga and Mount Segrila, both of which are located on the Qinghai-Tibet Plateau (**Figure 1**). As precipitation is abundant in both regions, water availability can be ruled out as a limiting factor, thereby leaving temperature to be the predominant control for growth of local plants. Our objective was to assess whether or not temperature can exert a dominant impact on carbon isotope discrimination of plants growing in cold and humid montane regions.

MATERIALS AND METHODS

Study Area
Mount Gongga is located in the southeast of the Qinghai-Tibet Plateau in Sichuan Province, southwest China (29°20′–30°00′ N, 101°30′–102°10′ E) with considerable differences in terrain and climate between its east and west slopes. The altitude of the

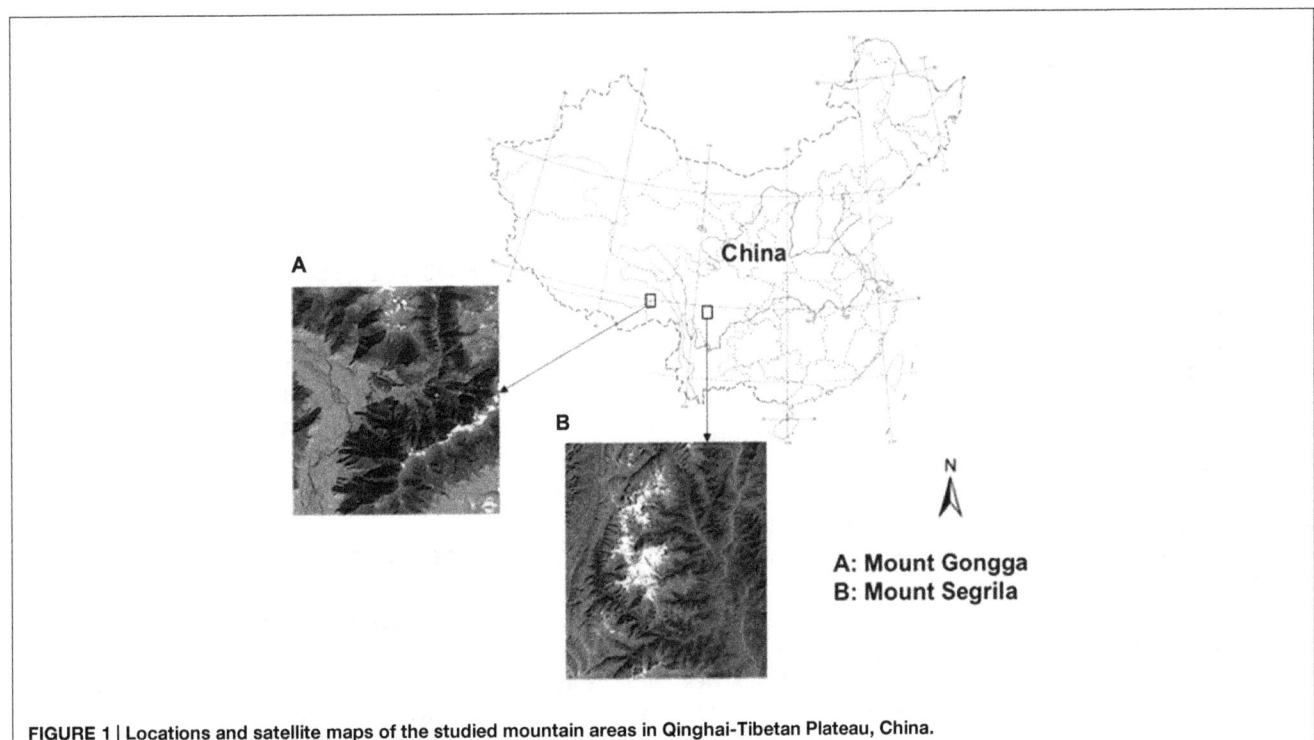

FIGURE 1 | Locations and satellite maps of the studied mountain areas in Qinghai-Tibetan Plateau, China.

east slope ranges from 1100 m (Dadu River valley) to 7600 m above sea level. A continuous vertical vegetation spectrum occurs on this slope, consisting of subtropical evergreen broadleaved vegetation (1100–2200 m, including semi-arid valley with shrubs and grasses below 1500 m, and evergreen broadleaved forests and deciduous broad-leaved forests), temperate coniferous and broad-leaved mixed forests (2200–2800 m), frigid dark coniferous forests (2800–3600 m), alpine subfrigid shrub and meadow vegetation (3600–4200 m), alpine frigid meadow vegetation (4200–4600 m), alpine frigid sparse grass and desert zone (4600–4800 m), and higher altitude alpine ice-and-snow zone (above 4900 m) in sequence. The vertical distribution of soil on the east slope of Mount Gongga is tightly associated with vegetation distribution, and a continuous soil sequence can be observed from 1100 to 4900 m. This consists of yellow-red soil (luvisols; <1500 m), yellow-brown soil (luvisols; 1500–1800 m), brown soil (luvisols; 1800–2200 m), dark-brown soil (luvisols; 2200–2800 m), dark-brown forest soil (luvisols; 2800–3600 m), black mattic soil (cambisols; 3600–4200 m), mattic soil (cambisols; 4200–4600 m), and chilly desert soil (cryosols; >4600 m; Liu and Wang, 2010; Shi et al., 2012). Temperature is certainly the key growth-limiting factor for the plants growing at elevations above 2800 m on Mount Gongga but moisture is definitely not because the climate there is very cold but humid. There are two meteorological observatories (Moxi meteorological observatory, 1640 m above sea level; Hailuogou ecological observatory, 3000 m above sea level) located in the sampling area of Mount Gongga. The mean annual temperature (MAT) and mean summer temperature (MST) above 2800 m are less than 5.2 and 11°C, respectively. However, rainfall is very abundant with a mean annual precipitation (MAP) of 1940 mm at 3000 m, and continues to rise with increasing altitude (Zhong et al., 1997).

Mount Segrila is located on the convergence of the east Nyainqentanglha range and the east Himalaya range, southeast Tibet (29°21′–29°50′ N, 94°28′–94°51′ E). The continuous vertical vegetation and soil spectra of Mount Segrila are described as temperate coniferous and broad-leaved mixed forests (3000–3500 m) with brown soil (luvisols), frigid dark coniferous forests (3500–4200 m) with dark-brown forest soil (luvisols), alpine subfrigid shrub meadow (4200–4500 m) with black mattic soil (cambisols), and alpine frigid meadow and desert zone (4500–5300 m) with mattic soil (cambisols) and desert soil (cryosols). The altitudinal changes in climatic conditions and vegetation spectra were described comprehensively by Xu et al. (2014). There is one meteorological observatories (at 3900 m) located in the sampling area of Mount Segrila; additionally, seven sites with simple meteorological facilities were set in the study area. The MAP and MAT at the elevations above 3100 m are more than 1000 mm and less than 4.2°C, respectively, suggesting that temperature rather than water availability is the predominant growth-limiting factor for local plants (Du et al., 2009).

Plant Sampling

We collected 457 plant samples in total (444 spermatophytes and 13 pteridophytes) in late August 2004 from the east slope of Mount Gongga (from 1200 to 4500 m). Of these, 181 plant samples (97 plant species in total) were collected from elevations above 2800 m and they are all C_3 plants. The influence of human activities, sunshine regime, and location within the canopy, was minimized by restricting the sampling to non-shaded sites far from human habitation. Almost all species that we can find at each sampling altitude were collected. At each site 5–7 plants of each species of interest were identified and the same numbers of leaves were collected from each individual. The leaves from each species at each elevation were pooled to give one composite sample. For herbs and shrubs, the uppermost leaves of each individual were taken. For trees, eight leaves were collected from each individual tree and two leaves at each of the four cardinal directions from positions of full-irradiance 8–10 m above the ground surface. Detailed descriptions of the plant sampling on Mount Gongga have been presented previously (Liu and Wang, 2010; Shi et al., 2012).

Leaves were collected at intervals of about 100 m along an elevational transect from 3000 to 4600 m on the west slope of Mount Segrila. The sampling was conducted in late June 2012 when all plants were actively growing at the higher temperatures of the rainy season. Three sampling quadrats, each 50 m × 50 m at each altitude, were set for the plant sampling. Almost all species that we found at each sampling altitude were collected. The uppermost leaves of herbs and shrubs were collected; the leaves of trees were taken from positions 8–10 m above the ground. The leaves from each sampling quadrat at each altitude were pooled to give one composite sample, giving a total of 45 samples from Mount Segrila.

Soil Sampling and Soil Water Measurement

The samples for soil water content (SWC) measurement were collected in parallel with the plant sampling. The soil sampling on Mount Gongga was conducted in late August 2004 at the end of the rainy season. There was no rain in this study area for at least 1 week based on the meteorological records at the two meteorological observatories. The soil sampling on Mount Segrila was performed in late June 2012 in the rainy season. On Mount Gongga, surface soil samples (0–5, 5–10, and 10–20 cm depth) were obtained for each assigned site (after removing the litter layer) with a soil auger. The soil samples at each locality represented the result of mixing four subsamples randomly taken within a radius of 10 m. On Mount Segrila, three soil cores (2.5 cm diameter, 20 cm depth) were taken randomly from sampling quadrates at each altitude. The soil samples were oven-dried at 105°C to constant weight; the SWC of each sample was the difference between its wet weight and its dry weight divided by its dry weight.

Carbon Isotope Measurement

All plant samples were oven-dried at 65°C and ground to 60 μm mesh using a steel ball mixer mill MM200 (Retsch GmbH, Haan, Germany). The carbon isotope composition ($\delta^{13}C$) of the whole leaf tissue were determined at the Stable Isotope Laboratory of the College of Resources and Environmental Sciences, China Agricultural University, Beijing, China, using a Delta^{Plus}XP mass spectrometer (Thermo Scientific, Bremen, Germany) coupled with an elemental analyzer (FlashEA 1112; CE Instruments, Wigan, UK) in continuous flow mode. The elemental analyzer combustion temperature was 1020°C.

The carbon isotopic composition is reported in the delta notation relative to the V-PDB standard. The standard deviation for the $\delta^{13}C$ measurements is less than 0.15‰. Plant Δ is obtained by the following formula based on Farquhar et al. (1989)

$$\Delta = \frac{\delta^{13}C_{air} - \delta^{13}C_{plant}}{1 + \delta^{13}C_{plant}/1000} \approx \delta^{13}C_{air} - \delta^{13}C_{plant}$$

in which $\delta^{13}C_{air}$ is the carbon isotope ratio of atmospheric CO_2 (−7.8 ‰) and $\delta^{13}C_{plant}$ is the measured $\delta^{13}C$ value of leaf material.

Data Selection

Previous studies have reported that photosynthesis and plant enzyme activity can be strongly inhibited when grown at temperatures below 5°C (Graham and Patterson, 1982; Ramalho et al., 2003). Some woody plants cannot withstand much below −6°C for any length of time and cease growth if the maximum daily temperature is below 9°C (Parker, 1963). Based on these studies, we confidently assume that temperature is the key growth-limiting factor for plants grown in cold and humid alpine areas where MAT is below 5°C. Because the MAT below 2800 m is greater than 5.2°C on Mount Gongga (Zhong et al., 1997), the $\delta^{13}C$ data of the plants grown below 2800 m were excluded in this study. As for Mount Segrila, all $\delta^{13}C$ data obtained were included in the present study because the MAT values at all sampling sites are less than 4.2°C (Du et al., 2009). Note that the direct measurement of temperature at each elevation is not available, MAT data was obtained by linear interpolation with original data from the meteorological observatories and the simple meteorological facilities in the study areas (Zhong et al., 1997; Du et al., 2009).

Statistical Analysis

Bivariate correlation analysis was first performed to examine the links between plant Δ and MAT and SWC. Considering the existence of potential interactions between MAT and SWC, partial correlation analysis, in which MAT and SWC were separately controlled, was applied to describe the actual links between plant Δ and MAT and SWC. Regression analysis was used to constrain the influences of MAT and SWC on plant Δ. Since the plant sampling was conducted in two mountain regions and this might introduce a 'random effect' (Bolker et al., 2008) into the analysis, a linear mixed model was applied to constrain the influence of MAT and SWC on Δ between the regions, in which MAT, SWC and their interaction (MAT × SWC) are defined as the fixed factor, while the study sites (Mount Gongga or Mount Segrila) are defined as the random factor. Multiple stepwise regression was used to eliminate the influence of potential collinearity existing between MAT and SWC. Variables were selected into the model with P-value < 0.05 and excluded with P-value > 0.1. The variable with the largest partial correlation coefficient will be first selected into the predicting

model. All statistical analysis was performed using IBM SPSS Statistics 22.0 (IBM Corporation, New York, NY, USA).

RESULTS

Correlations between Δ and Temperature and Water Availability

The MAT and SWC data and the site-averaged carbon isotope discrimination values (Δ) of plants collected at each sampling site are shown in **Table 1**. In both mountain regions the climate is cold with MAT varying between −5 and 5°C. The surface SWC on Mount Gongga varied from 10.7 to 48.3% with a mean value of 26.7%, while SWC at 0–20 cm depth on Mount Segrila ranged from 25.1 to 76.3% with an average value of 33.1%. There were significant correlations between MAT and SWC on both Mount Gongga ($r = 0.212$, $p = 0.013$) and Mount Segrila ($r = -0.301$, $p = 0.044$).

There was a significantly positive correlation between MAT and Δ of the plants growing on Mount Gongga ($r = 0.565$, $p < 0.001$) and Mount Segrila ($r = 0.456$, $p = 0.02$) in bivariate correlation analysis (**Table 2**), and this impact of MAT on Δ was further expanded after controlling for SWC in partial correlation analysis ($r = 0.602$, $p < 0.001$ for Mount Gongga; $r = 0.553$, $p < 0.001$ for Mount Segrila). The correlation between SWC and Δ, however, was not significant in either mountain region (**Table 2**) and remained non-significant after controlling for MAT, except at Mount Segrila ($r = 0.400$, $p = 0.007$).

Considering that the response of carbon isotope discrimation to environmental factors may be dependent on plant functional type (PFT), the influences of MAT and SWC on Δ were analyzed separately based on PFTs. MAT was positively correlated with Δ for all PFTs as well as *Rhododendron* sp., which is a widely distributed evergreen shrub at elevations of 2800–4200 m, as suggested by both bivariate correlation and partial correlation analyses (**Table 3**). By contrast, the correlation between SWC and Δ was not significant in either type of correlation except for shrubs in the bivariate correlation ($r = 0.289$, $p = 0.044$).

Regression Analysis and Linear Mixed Model of the Relationship between Δ and MAT and SWC

Results of regression analysis reveal that carbon isotope discrimination was significantly influenced by MAT on Mount Gongga ($R^2 = 0.319$, $p < 0.001$, **Figure 2A**), whereas the relationship between MAT and Δ is shown as a unimodal pattern with a turning point at MAT = −1°C on Mount Segrila ($R^2 = 0.365$, $p < 0.001$, **Figure 2B**). Variation in Δ with SWC, however, presents a unimodal pattern on both mountains ($\Delta = 23.5 - 0.307SWC + 0.006SWC^2$, $R^2 = 0.092$, $p = 0.002$ for Mount Gongga; $\Delta = 23.9 - 0.099SWC + 0.001SWC^2$, $R^2 = 0.182$, $p = 0.015$ for Mount Segrila). When analyzed with the whole dataset, variation in Δ was significantly influenced by either MAT ($R^2 = 0.134$, $p < 0.001$, **Figure 2C**) or SWC ($\Delta = 18.9 + 0.053SWC$, $R^2 = 0.148$, $p < 0.001$).

Multiple linear regression analysis shows that MAT and SWC in total accounted for 39.5 and 33.4% of the variance in Δ at Mount Gongga and Mount Segrila, respectively (**Table 4**). Further inclusion of PFTs into the regression model did not increase the estimated R^2 on Mount Gongga. When calculated with the whole dataset, MAT and SWC altogether accounted for 37.9% of the variance in Δ. In view of the significant correlations between MAT and SWC, multiple stepwise regression analysis was applied to eliminate the influence of collinearity existing between the two variables. The results reveal that MAT was the only variable entered in the stepwise regression model of Δ for Mount Gongga ($R^2 = 0.386$, $p < 0.001$, **Table 4**), and the first variable selected into the model of Δ for Mount Segrila ($R^2 = 0.208$, $p = 0.002$) and for the whole dataset ($R^2 = 0.169$, $p < 0.001$). Both MAT and SWC were finally entered in the model of Δ for Mount Segrila ($R^2 = 0.334$, $p < 0.001$) and for the whole dataset ($R^2 = 0.379$, $p < 0.001$).

TABLE 1 | Descriptions of climatic condition, dominant vegetation type, and site-averaged plant carbon isotope discrimination value (Δ) of different sampling sites on Mount Gongga and Mount Segrila.

Sampling mountain	Site no.	Altitude (m a.s.l.)	MAT[a] (°C)	SWC (%)	Site-averaged Δ (‰)	Replicate
Mount Gongga	1	2800	5.2	22.8	21.10	14
	2	2850	4.9	32.5	21.23	4
	3	2860	4.8	32.5	18.26	1
	4	2900	4.6	42.2	21.61	12
	5	3000	4.0	16.7	21.24	12
	6	3100	3.4	34.6	22.36	13
	7	3200	2.8	—	21.21	12
	8	3250	2.5	—	19.71	7
	9	3300	2.2	—	19.31	12
	10	3430	1.4	—	19.45	10
	11	3510	0.94	25.3	20.32	18
	12	3550	0.70	23.7	21.26	7
	13	3600	0.67	22.0	19.62	2
	14	3650	0.10	18.8	18.55	7
	15	3700	−0.20	15.4	20.20	5
	16	3750	−0.50	19.7	18.96	10
	17	3800	−0.80	23.8	17.84	8
	18	3930	−1.58	33.9	18.18	3
	19	3950	−1.7	33.9	18.67	2
	20	4000	−2.0	29.3	17.93	4
	21	4050	−2.3	31.1	18.28	1
	22	4100	−2.6	32.8	18.46	4
	23	4200	−3.2	—	19.04	3
	24	4400	−4.4	30.9	18.75	5
	25	4500	−5.0	18.2	18.74	5
Mount Segrila	1	3135	4.2	33.5	22.25	3
	2	3271	3.3	25.1	21.84	3
	3	3365	2.7	32.1	22.59	3
	4	3456	2.1	45.7	21.93	3
	5	3565	1.4	33.5	22.04	3
	6	3689	0.65	48.7	21.98	3
	7	3770	0.13	76.3	23.06	3
	8	3893	−0.65	56.0	22.37	3
	9	3960	−1.1	36.3	22.59	3
	10	4080	−1.8	71.8	23.23	3
	11	4170	−2.4	45.0	19.94	3
	12	4284	−3.2	50.9	20.86	3
	13	4371	−3.7	48.1	22.18	3
	14	4485	−4.4	36.9	21.61	3
	15	4590	−5.1	46.8	19.49	3

MAT, mean annual temperature; SWC, soil water content. [a]*MAT data was calculated by linear interpolation with original data from Zhong et al. (1997) for Mount Gongga and Du et al. (2009) for Mount Segrila.*

TABLE 2 | Pearson correlations (r) between plant Δ and mean annual temperature (MAT) and soil water content (SWC) on Mount Gongga and Mount Segrila.

	Mount Gongga		Mount Segrila	
	r	p	r	p
Bivariate correlation				
MAT	**0.565**	**<0.001**	**0.456**	**0.02**
SWC	0.225	0.08	0.202	0.183
Partial correlation				
MAT[a]	**0.602**	**<0.001**	**0.553**	**<0.001**
SWC[b]	0.122	0.158	**0.400**	**0.007**

Values presented in bold indicated significant correlations (p < 0.05). [a]Indicates controlling for SWC. [b]Indicates controlling for MAT.

TABLE 3 | Pearson correlations (r) of MAT and SWC with Δ of different plant functional types (PFTs; herbs, shrubs, and trees) as well as Rhododendron sp. growing on Mount Gongga.

	Herbs		Shrubs		Trees		Rhododendron sp.	
	r	p	r	p	r	p	r	p
Bivariate correlation								
MAT	**0.512**	**<0.001**	**0.575**	**<0.001**	**0.527**	**0.007**	**0.690**	**0.003**
SWC	0.157	0.231	**0.289**	**0.044**	0.251	0.300	0.140	0.682
Partial correlation								
MAT[a]	**0.552**	**<0.001**	**0.603**	**<0.001**	**0.521**	**0.027**	**0.724**	**0.018**
SWC[b]	0.031	0.814	0.170	0.248	0.155	0.540	0.068	0.851

Values presented in bold indicated significant correlations (p < 0.05). [a]Indicates controlling for SWC. [b]Indicates controlling for MAT.

In the linear mixed model MAT, SWC and their interaction (MAT × SWC) were defined as the fixed factor and the study sites (Mount Gongga or Mount Segrila) were defined as the random factor. Results show that the sampling mountains did not have any significant effect on the estimated relationship between Δ and MAT or SWC, as the test for the estimated intercept of covariance parameter was not significant ($p = 0.495$, **Table 5**). MAT ($p < 0.001$) and SWC ($p = 0.002$) both significantly affected Δ but their interaction did not ($p = 0.674$).

DISCUSSION

Effect of Soil Water Availability on Plant Carbon Isotope Discrimination

The mechanism of water availability on plant carbon isotope discrimination is that the plants would close their stomata to reduce water loss when moisture decreases, resulting in a lower c_i/c_a ratio and thus less negative $\delta^{13}C$ values. Numerous studies have reported the influence of water availability on plant carbon isotope discrimination (e.g., Wang et al., 2005, 2008; Diefendorf et al., 2010; Kohn, 2010), and a positive correlation between Δ and water availability has been observed on most occasions. In the present study, however, the correlation between Δ and SWC on

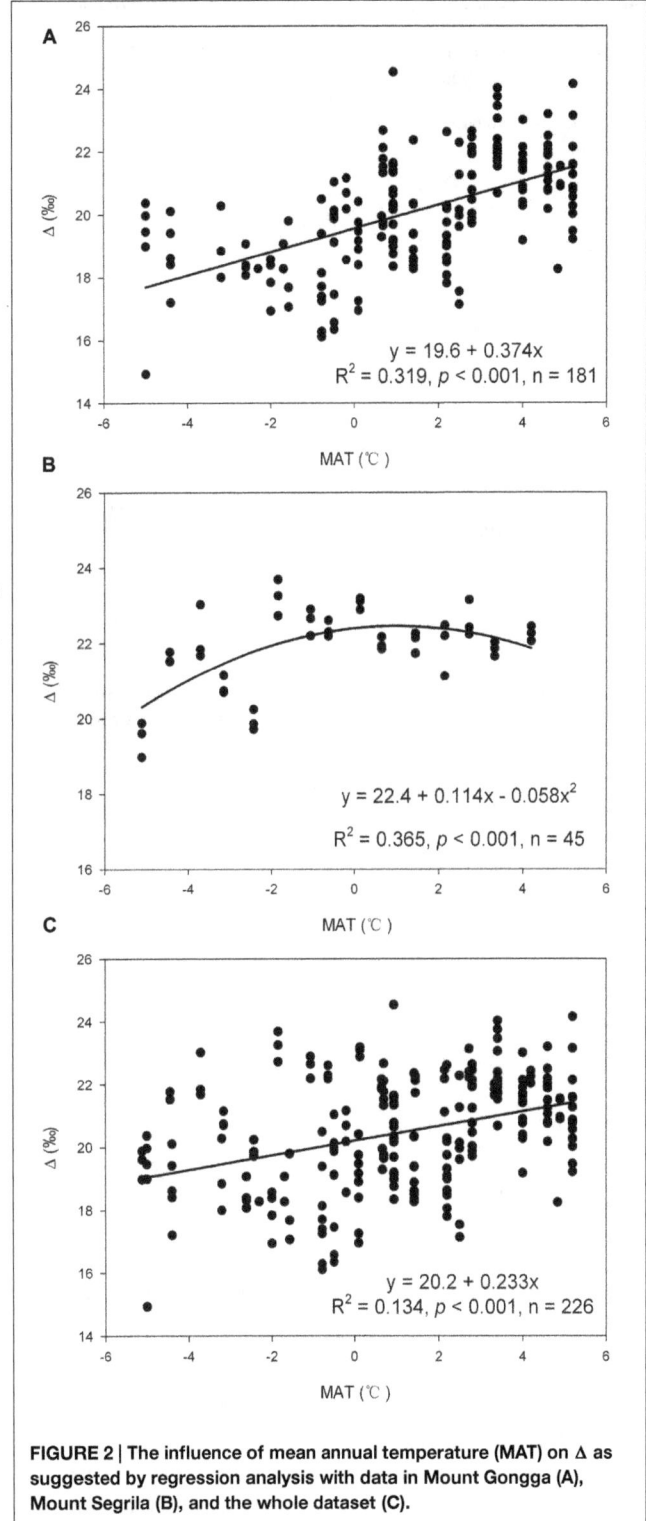

FIGURE 2 | The influence of mean annual temperature (MAT) on Δ as suggested by regression analysis with data in Mount Gongga (A), Mount Segrila (B), and the whole dataset (C).

Mount Gongga was not significant (**Table 2**). It has been observed that carbon isotope discrimination responds differentially over the range of MAP and often becomes nearly constant in wet environments (Kohn, 2010). In the study area of Mount Gongga the water supply is so abundant (with an MAP over 1800 mm)

TABLE 4 | Multiple linear regression of plant Δ against MAT, SWC, and PFT.

Model	Variables entered	R^2	Adjusted R^2	F	p
Variable selection method: Enter					
Mount Gongga					
1	MAT+SWC	0.395	0.386	43.721	<0.001
2	MAT+SWC+PFT	0.382	0.362	19.040	<0.001
Mount Segrila					
1	MAT+SWC	0.334	0.303	10.542	<0.001
Whole dataset					
1	MAT+SWC	0.379	0.372	54.635	<0.001
Variable selection method: Stepwise					
Mount Gongga					
1	MAT	0.386	0.381	84.789	<0.001
Mount Segrila					
1	MAT	0.208	0.189	11.282	0.002
2	MAT+SWC	0.334	0.303	10.542	<0.001
Whole dataset					
1	MAT	0.169	0.165	36.705	<0.001
2	MAT+SWC	0.379	0.372	54.635	<0.001

TABLE 5 | Summary of the linear mixed model of Δ with MAT, SWC and their interaction (MAT × SWC) as the fixed variables while the sampling mountains as the random variables.

Linear mixed model results		
Information criteria		
Akaike information criterion (AIC)		651.679
Bayesian information criterion (BIC)		658.042
Estimation of fixed effect		
Source	*F*	*Significance*
MAT	13.475	<0.001
SWC	10.346	0.002
MAT × SWC	0.177	0.674
Intercept	464.366	0.012
Estimation of covariance parameter		
Parameter	*Wald Z*	*Significance*
Residual	9.407	<0.001
Intercept	0.682	0.495

that water availability is thereby no longer a factor limiting plant growth. Our finding is consistent with the results of Diefendorf et al. (2010) that the correlation of Δ with precipitation is not significant when MAP is over 1800 mm.

In contrast to Mount Gongga, no significant correlation was indicated by the bivariate correlation analysis in Mount Segrila. However, we found a significant relationship between SWC and Δ after controlling for MAT in the partial correlation analysis (**Table 2**). The differentiated results from the two sites may derive from the difference in MAP because the MAP on Mount Segrila varies from 980 to 1300 mm, much less than that on Mount Gongga. Nonetheless, this result indicates a partial influence of water availability on carbon isotope discrimination on Mount Segrila. Moreover, the fact that SWC finally entered the model for Mount Segrila as well as for the whole dataset in multiple

stepwise regression model of variation in Δ (**Table 4**) also suggests that soil water availability has had an effect on carbon isotope discrimination to some extent. As the present study was conducted in areas with high precipitation where plants obtain sufficient water for their growth, our results suggest that even in humid areas, water availability may still be one of the major determining factors shaping the variation in Δ.

Temperature as the Key Factor Influencing Plant Carbon Isotope Discrimination

Variation in Δ with changing temperature has been studied extensively (e.g., Körner et al., 1988, 1991; Morecroft and Woodward, 1990, 1996; Hultine and Marshall, 2000; Treydte et al., 2007; Wang et al., 2013). In the present study we also observed a strong impact of MAT on Δ in two mountain regions (**Figure 2**). MAT together with SWC in total accounted for a large proportion of the variation in Δ of the two montane regions (**Table 4**). Although soil water availability is expected to have certain impact on carbon isotope discrimination in the study areas, this impact is more limited than that of temperature as suggested by the results of partial correlation analysis and stepwise regression (**Tables 2** and **4**). Taking these results together, we believe that temperature, rather than soil water availability, has exerted the key influence on carbon isotope discrimination of the plants growing in these two mountain regions. Since temperature is considered to be the key growth-limiting factor in these two frigid alpine areas, the present study supports the hypothesis that the key growth-limiting factor is also the key factor influencing plant carbon isotope discrimination.

Temperature is one of the most important factors that control plant growth and certain physiological processes related to plant gas exchange activity. A decline in temperature usually results in a reduction in enzyme activity and photosynthetic rate (Beerling, 1994), thus leading to decreased CO_2 assimilation and a lower growth rate as a consequence. Under such circumstances the intercellular CO_2 concentration (c_i) is likely to increase if the ambient CO_2 concentration (c_a), stomatal conductance (g_s) and mesophyll conductance (g_m) all hold constant. An increase in Δ with decreasing temperature is therefore expected and has been observed in most of the studies on the influence of temperature on carbon isotope discrimination (e.g., Pearman et al., 1976; McCarroll and Loader, 2004; Treydte et al., 2007; Wang et al., 2013). Our results, however, are inconsistent with these studies and suggest a decline in Δ with decreasing temperature on Mount Gongga (**Figure 2A**). Similarly, several studies have also reported a positive correlation between Δ and temperature along an elevational gradient (Körner et al., 1988; Hultine and Marshall, 2000) and a latitudal gradient (Körner et al., 1991). Moreover, a decrease in Δ with low temperature was also observed in experiments with controlled environment (Morecroft and Woodward, 1990, 1996). There is still no conclusive explanation for this positive correlation but several possible mechanisms have been proposed. One of these is related to the changes in leaf morphology in response to temperature

(Körner and Diemer, 1987; Körner et al., 1991). Increased leaf thickness has been observed in alpine plants as an adaptation to low temperatures (Körner et al., 1988; Hultine and Marshall, 2000) and this may cause a longer CO_2 diffusion distance from the ambient to the intercellular air space and result in a decline in Δ (Körner and Diemer, 1987; Zhu et al., 2010). Kogami et al. (2001) also reported that the plants growing in highlands had lower CO_2 transfer conductance inside the leaf (g_i) due to greater leaf thickness, thicker mesophyll cell walls and higher mesophyll cell density, resulting in decreased Δ with decreasing temperature. Another explanation is that low temperatures may increase water viscosity (Cernusak et al., 2013). Smith et al. (1984) suggested that lower temperatures might inhibit sapwood water movement and thereby decrease plant water potential, resulting in partial stomatal closure and decreased Δ as a consequence. Finally, the lower Δ with decreasing temperature may also be due to an increase in the efficiency of carbon uptake or carboxylation efficiency at low temperatures, which depends on the amount of active ribulose bisphosphate carboxylase-oxygenase (Rubisco) per unit leaf area (Morecroft and Woodward, 1990).

Changes in temperature and water availability are usually correlated; for example, high temperature can lead to water stress because of high evaporation. These two variables can have synergistic effects on plant growth; co-occurrence of high temperature and water stress was found to constrain plant productivity worldwide (O'Connor et al., 2001). Since carbon isotope discrimination value is an integrated parameter reflecting carbon and water relation, temperature, and water availability are also expected to have interactions on carbon isotope discrimination. Craufurd et al. (1999) found that plant Δ was significantly affected by the interaction of high temperature and water deficit. Xu and Zhou (2005) reported that a perennial grass species decreased its carbon isotope composition in the condition of high nocturnal temperature and water stress. However, the present study showed no significant interaction of temperature and soil water availability on plant Δ ($p = 0.674$, **Table 5**). This result likely suggests that the disturbance of plant water relations by low temperature in the study areas is not significant as reported in previous studies (Norisada et al., 2005; López-Bernal et al., 2015).

CONCLUSION

In the present study we conducted an intensive investigation of plant Δ in two cold and humid mountain regions and analyzed the influence of temperature and soil water availability on the variation in Δ. Temperature, the key growth-limiting factor for the local plants, was found to have a significant influence on carbon isotope discrimination irrespective of study area or PFT but the influence of SWC was relatively weak. Future study should consider temporal dynamics on water availability in relation to plant carbon isotope discrimination. Furthermore, a close relationship between plant carbon isotope discrimination and WUE implies that temperature might potentially affect plant WUE. Ecosystems at high altitudes in Tibetan Plateau are fragile and sensitive to climate change. Elevated temperature may reduce WUE in plants, which may have enormous impacts on productivity and stability of ecosystems in the future.

ACKNOWLEDGMENTS

This research was financially supported by the National Natural Science Foundation of China (Grant No. 41272193, 31272251, and 31421092) and the Ph.D. Program Foundation of the Ministry of Education of China (20100008110003). We would like to thank Ma Yan for analyzing stable carbon isotope ratios in the Isotope Laboratory at the College of Resources and Environment, China Agricultural University.

REFERENCES

Beerling, D. J. (1994). Predicting leaf gas-exchange and C response to the past 30,000 years of global environmental change. *New Phytol.* 128, 425–433. doi: 10.1111/j.1469-8137.1994.tb02988.x

Bolker, B. M., Brooks, M. E., Clark, C. J., Geange, S. W., Poulsen, J. R., Stevens, M. H. H., et al. (2008). Generalized linear mixed models: apractical guide for ecology and evolution. *Trends Ecol. Evol.* 24, 127–135. doi: 10.1016/j.tree.2008.10.008

Cernusak, L. A., Ubierna, N., Winter, K., Holtum, J. A. M., Marshall, J. D., and Farquhar, G. D. (2013). Environmental and physiological determinants of carbon isotope discrimination in terrestrial plants. *New Phytol.* 200, 950–965. doi: 10.1111/nph.12423

Craufurd, P. Q., Wheeler, T. R., Ellis, R. H., Summerfield, R. J., and Williams, J. H. (1999). Effect of temperature and water deficit on water-use efficiency, carbon isotope discrimination, and specific leaf area in peanut. *Crop Sci.* 39, 136–142. doi: 10.2135/cropsci1999.0011183X003900010022x

Diefendorf, A. F., Mueller, K. E., Wing, S. L., Koch, P. L., and Freeman, K. H. (2010). Global patterns in leaf 13C discrimination and implications for studies of past and future climate. *Proc. Natl. Acad. Sci. U.S.A.* 107, 5738–5743. doi: 10.1073/pnas.0910513107

Du, J., Gao, R., Ma, P., Liu, Y., and Zhou, K. (2009). Analysis of stereoscopic climate features on Mt. Seqilha, Tibet. *Plateau Mountain Meteor. Res.* 29, 14–18.

Duquesnay, A., Breda, N., Stievenard, M., and Dupouey, J. L. (1998). Changes of tree-ring (13C and water-use efficiency of beech (*Fagus sylvatica* L.) in northeastern France during the past century. *Plant Cell Environ.* 21, 565–572. doi: 10.1046/j.1365-3040.1998.00304.x

Farquhar, G. D., Hubick, K. T., Condon, A. G., and Richards, R. A. (1989). "Carbon isotope fractionation and plant water-use efficiency," in *Stable Isotopes in Ecological Research*, eds P. W. Rundel, J. R. Ehleringer and K. A. Nagy (New York, NY: Springer Press), 21–40.

Farquhar, G. D., and Richards, P. A. (1984). Isotopic composition of plant carbon correlates with water-use efficiency of wheat genotypes. *Aust. J. Plant Physiol.* 11, 539–552. doi: 10.1071/PP9840539

Graham, D., and Patterson, B. D. (1982). Responses of plants to low, nonfreezing temperatures: proteins, metabolism, and acclimation. *Ann. Rev. Plant Physiol.* 33, 347–372. doi: 10.1146/annurev.pp.33.060182.002023

Hultine, K. R., and Marshall, J. D. (2000). Altitude trends in conifer leaf morphology and stable carbon isotope composition. *Oecologia* 123, 32–40. doi: 10.1007/s004420050986

Kelly, C. K., and Woodward, F. I. (1995). Ecological correlates of carbon isotope composition of leaves: a comparative analysis testing for the effects of temperature, CO_2 and O_2 partial pressures and taxonomic relatedness on 13C. *J. Ecol.* 83, 509–515. doi: 10.2307/2261603

Kogami, H., Hanba, Y. T., Kibe, T., Terashima, I., and Masuzawa, T. (2001). CO_2 transfer conductance, leaf structure and carbon isotope composition of

Polygonum cuspidatum leaves from low and high altitudes. *Plant Cell Environ.* 24, 529–538. doi: 10.1046/j.1365-3040.2001.00696.x

Kohn, M. J. (2010). Carbon isotope compositions of terrestrial C_3 plants as indicators of (paleo) ecology and (paleo) climate. *Proc. Natl. Acad. Sci. U.S.A.* 107, 19691–19695. doi: 10.1073/pnas.1004933107

Körner, C., and Diemer, M. (1987). In situ photosynthetic responses to light, temperature and carbon dioxide in herbaceous plants from low and high altitude. *Funct. Ecol.* 1, 179–194. doi: 10.2307/2389420

Körner, C., Farquhar, G. D., and Roksandic, Z. (1988). A global survey of carbon isotope discrimination in plants from high altitude. *Oecologia* 74, 623–632. doi: 10.1007/BF00380063

Körner, C., Farquhar, G. D., and Wong, S. C. (1991). Carbon isotope discrimination by plants follows latitudinal and altitudinal trends. *Oecologia* 88, 30–40. doi: 10.1007/BF00328400

Liu, X., and Wang, G. (2010). Measurements of nitrogen isotope composition of plants and surface soils along the altitudinal transect of the eastern slope of Mount Gongga in southwest China. *Rapid Commun. Mass Spectrom.* 24, 3063–3071. doi: 10.1002/rcm.4735

López-Bernal, A., García-Tejera, O., Testi, L., Orgaz, F., and Villalobos, F. J. (2015). Low winter temperatures induce a disturbance of water relations in field olive trees. *Trees* 29, 1247–1257. doi: 10.1007/s00468-015-1204-5

Marshall, L. D., and Linder, S. (2013). Mineral nutrition and elevated [CO_2] interact to modify δ13C, an index of gas exchange, in Norway spruce. *Tree Physiol.* 33, 1132–1144. doi: 10.1093/treephys/tpt004

McCarroll, D., and Loader, N. (2004). Stable isotopes in tree rings. *Quat. Sci. Rev.* 23, 771–801. doi: 10.1016/j.quascirev.2003.06.017

Morecroft, M. D., and Woodward, F. I. (1990). Experimental investigations on the environmental determination of δ13C at different altitudes. *J. Exp. Bot.* 41, 1303–1308. doi: 10.1093/jxb/41.10.1303

Morecroft, M. D., and Woodward, F. I. (1996). Experiments on the causes of altitudinal differences in leaf nutrient contents, age and 13C of *Alchemilla alpine. New Phytol.* 134, 471–479. doi: 10.1111/j.1469-8137.1996.tb04364.x

Norisada, M., Hara, M., Yagi, H., and Tange, T. (2005). Root temperature drives winter acclimation of shoot water relations in *Cryptomeria japonica* seedlings. *Tree Physiol.* 25, 1447–1455. doi: 10.1093/treephys/25.11.1447

O'Connor, T. G., Haines, L. M., and Snyman, H. A. (2001). Influence of precipitation and species composition on phytomass of a semi-arid African grassland. *J. Ecol.* 89, 850–860. doi: 10.1046/j.0022-0477.2001.00605.x

Parker, J. (1963). Cold resistance in woody plants. *Bot. Rev.* 29, 124–201. doi: 10.1007/BF02860820

Pearman, G. I., Francey, R. J., and Farser, P. J. B. (1976). Climatic implications of stable isotopes in tree rings. *Nature* 260, 771–772. doi: 10.1038/260771a0

Ramalho, J. C., Quartin, V. L., Leitão, E., Campos, P. S., Carelli, M. L. C., Fahl, J. I., et al. (2003). Cold acclimation ability and photosynthesis among species of the tropical Coffea genus. *Plant Biol.* 5, 631–641. doi: 10.1016/j.jplph.2010.11.013

Seibt, U., Rajabi, A., Griffiths, H., and Berry, J. A. (2008). Carbon isotopes and water use efficiency: sense and sensitivity. *Oecologia* 155, 441–454. doi: 10.1007/s00442-007-0932-7

Shi, W., Wang, G., and Han, W. (2012). Altitudinal variation in leaf nitrogen concentration on the eastern slope of Mount Gongga on the Tibetan Plateau. China. *PLoS ONE* 7:e44628. doi: 10.1371/journal.pone.0044628

Smith, W. K., Young, D. R., Carter, G. A., Hadley, J. L., and McNaughton, G. M. (1984). Autumn stomatal closure in six conifer species of the Central Rocky Mountains. *Oecologia* 63, 237–242. doi: 10.1007/BF00379883

Sparks, J. P., and Ehleringer, J. R. (1997). Leaf carbon isotope discrimination and nitrogen content for riparian trees along elevational transects. *Oecologia* 109, 362–367. doi: 10.1007/s004420050094

Treydte, K., Frank, D., Esper, J., Andreu, L., Bednarz, Z., Berninger, F., et al. (2007). Signal strength and climate calibration of a European tree-ring isotope network. *Geophys. Res. Lett.* 34, 1–6. doi: 10.1029/2007GL031106

Wang, G., and Feng, X. (2012). Response of plant's water use efficiency to increasing atmospheric CO_2 concentration. *Environ. Sci. Technol.* 46, 8610–8620. doi: 10.1021/es301323m

Wang, G., Feng, X., Han, J., Zhou, L., Tan, W., and Su, F. (2008). Paleovegetation reconstruction using (13C of soil organic matter. *Biogeoscis.* 5, 1325–1337. doi: 10.5194/bg-5-1325-2008

Wang, G., Han, J., Zhou, L., Xiong, X., and Wu, Z. (2005). Carbon isotope ratios of plants and occurrences of C4 species under different soil moisture regimes in arid region of Northwest China. *Physiol. Plant.* 125, 74–81. doi: 10.1111/j.1399-3054.2005.00549.x

Wang, G., Li, J., Liu, X., and Li, X. (2013). Variations in carbon isotope ratios of plants across a temperature gradient along the 400 mm isoline of mean annual precipitation in north China and their relevance to paleovegetation reconstruction. *Quat. Sci. Rev.* 63, 83–90. doi: 10.1016/j.quascirev.2012.12.004

Winter, K., Holtum, J. A. M., Edwards, G. E., and O'Leary, M. H. (1982). Effect of low relative humidity on (13C value in two C_3 grasses and in Panicum milioides, a C_3-C_4 intermediate species. *J. Exp. Bot.* 33, 88–91. doi: 10.1093/jxb/33.1.88

Xu, M., Li, X., Cai, X., Gai, J., Li, X., Christie, P., et al. (2014). Soil microbial community structure and activity along a montane elevational gradient on the Tibetan Plateau. *Eur. J. Soil Biol.* 64, 6–14. doi: 10.1016/j.ejsobi.2014.06.002

Xu, Z.-Z., and Zhou, G.-S. (2005). Effects of water stress and nocturnal temperature on carbon allocation in the perennial grass, *Leymus* chinensis. *Physiol. Plant.* 123, 272–280. doi: 10.1111/j.1399-3054.2005.00455.x

Yao, F. Y., Wang, G. A., Liu, X. J., and Song, L. (2011). Assessment of effects of the rising atmospheric nitrogen deposition on nitrogen uptake and long-term water-use efficiency of plants using nitrogen and carbon stable isotopes. *Rapid Commun. Mass Spectrom.* 25, 1827–1836. doi: 10.1002/rcm.5048

Zhong, X. H., Wu, N., Luo, J., Yin, K. P., Tang, Y., and Pan, Z. F. (1997). *Systematic Research of Forest Ecology in Gongga Mountiain (in Chinese).* Chengdu: Chengdu University of Science and Technology Press.

Zhu, Y., Siegwolf, R. T. W., Durka, W., and Körner, C. (2010). Phylogenetically balanced evidence for structural and carbon isotope responses in plants along elevational gradients. *Oecologia* 162, 853–863. doi: 10.1007/s00442-009-1515-6

Conflict of Interest Statement: The authors declare that the research was conducted in the absence of any commercial or financial relationships that could be construed as a potential conflict of interest.

9

Extensive Transcriptome Changes During Natural Onset and Release of Vegetative Bud Dormancy in *Populus*

Glenn T. Howe[1†], David P. Horvath[2†], Palitha Dharmawardhana[1,3], Henry D. Priest[4,5], Todd C. Mockler[3,4] and Steven H. Strauss[1]*

[1] Department of Forest Ecosystems and Society, Oregon State University, Corvallis, OR, USA, [2] Biosciences Research Laboratory, United States Department of Agriculture-Agricultural Research Service, Fargo, ND, USA, [3] Department of Botany and Plant Pathology, Oregon State University, Corvallis, OR, USA, [4] Donald Danforth Plant Science Center, Saint Louis, MO, USA, [5] Division of Biology and Biomedical Sciences, Washington University in Saint Louis, Saint Louis, MO, USA

Edited by:
Karina Vera Rosa Schafer,
Rutgers, The State University
of New Jersey, USA

Reviewed by:
John J. Gaynor,
Montclair State University, USA
Miguel Cervantes-Cervantes,
Rutgers, The State University
of New Jersey, USA

*Correspondence:
Steven H. Strauss
steve.strauss@oregonstate.edu

[†] These authors have contributed
equally to this work.

To survive winter, many perennial plants become endodormant, a state of suspended growth maintained even in favorable growing environments. To understand vegetative bud endodormancy, we collected paradormant, endodormant, and ecodormant axillary buds from *Populus* trees growing under natural conditions. Of 44,441 *Populus* gene models analyzed using NimbleGen microarrays, we found that 1,362 (3.1%) were differentially expressed among the three dormancy states, and 429 (1.0%) were differentially expressed during only one of the two dormancy transitions (FDR p-value < 0.05). Of all differentially expressed genes, 69% were down-regulated from paradormancy to endodormancy, which was expected given the lower metabolic activity associated with endodormancy. Dormancy transitions were accompanied by changes in genes associated with DNA methylation (via RNA-directed DNA methylation) and histone modifications (via Polycomb Repressive Complex 2), confirming and extending knowledge of chromatin modifications as major features of dormancy transitions. Among the chromatin-associated genes, two genes similar to *SPT (SUPPRESSOR OF TY)* were strongly up-regulated during endodormancy. Transcription factor genes and gene sets that were atypically up-regulated during endodormancy include a gene that seems to encode a trihelix transcription factor and genes associated with proteins involved in responses to ethylene, cold, and other abiotic stresses. These latter transcription factors include ETHYLENE INSENSITIVE 3 (EIN3), ETHYLENE-RESPONSIVE ELEMENT BINDING PROTEIN (EBP), ETHYLENE RESPONSE FACTOR (ERF), ZINC FINGER PROTEIN 10 (ZAT10), ZAT12, and WRKY DNA-binding domain proteins. Analyses of phytohormone-associated genes suggest important changes in responses to ethylene, auxin, and brassinosteroids occur during endodormancy. We found weaker evidence for changes in genes associated with salicylic acid and jasmonic acid, and little evidence for important changes in genes associated with gibberellins, abscisic acid, and cytokinin. We identified 315 upstream sequence motifs associated with eight patterns of gene expression, including novel motifs and motifs associated with the circadian clock and responses to photoperiod, cold, dehydration, and ABA. Analogies between flowering and endodormancy suggest important roles for genes similar to *SQUAMOSA-PROMOTER BINDING PROTEIN-LIKE (SPL)*, *DORMANCY ASSOCIATED MADS-BOX (DAM)*, and *SUPPRESSOR OF OVEREXPRESSION OF CONSTANS 1 (SOC1)*.

Keywords: chromatin, ecodormancy, endodormancy, gene expression, paradormancy, phytohormone, transcription factor, QTL

INTRODUCTION

Dormancy, the temporary suspension of growth (Lang et al., 1987), is a regulated process that controls plant growth, development, and architecture. Lang et al. (1987) subdivided dormancy processes into three types: paradormancy, endodormancy, and ecodormancy. Paradormancy denotes the state in which meristem growth (e.g., in buds or seeds) is inhibited by signals from other plant organs. For example, the shoot apex can inhibit the outgrowth of axillary buds by exerting apical dominance—but this state of paradormancy is released and outgrowth of the axillary buds occurs if the apex is removed. Endodormancy denotes the state in which meristem growth is inhibited by signals within the meristem itself. In plants adapted to cold climates, vegetative buds typically become endodormant in the fall and early winter, and prolonged periods of chilling (i.e., temperatures slightly above freezing) are needed before growth can resume, even under favorable environmental conditions. Even after the release of endodormancy, plants may remain ecodormant because of harsh environmental conditions such as cold or drought that are not conducive to cell division and elongation.

The regulated induction and release of bud endodormancy is critical for the survival and long-term growth of perennial plants in temperate, arid, and semiarid climates. Adaptation to local climatic conditions is generally achieved by natural selection in native populations and by artificial selection in forestry and agricultural populations. However, the matching of plant populations and local climatic cycles may become decoupled with rapid climate change. The induction and release of endodormancy are temporally associated with other changes that confer tolerance to cold and other abiotic stresses. Improved understanding of dormancy-associated gene expression may allow us to manipulate plant populations to speed climatic adaptation, and thus mitigate the adverse effects of climate change on forest and agricultural ecosystems.

Environmental and hormonal signals, including short days (SD), cold, ethylene, gibberellin (GA), and abscisic acid (ABA) play direct roles in growth cessation and bud set (Li et al., 2003; Mølmann et al., 2005; Ruonala et al., 2006). In many trees and other perennial plants, SD and low night temperatures in the fall act synergistically to induce growth cessation, vegetative bud set or shoot-tip abscission, and the first stage of cold acclimation (Howe et al., 1999; Mølmann et al., 2005; Ruttink et al., 2007; Rohde et al., 2011a). In some species (e.g., apple pear, *Populus* sp.), cold temperatures alone can induce growth cessation and endodormancy (Mølmann et al., 2005; Heide, 2008; Rohde et al., 2011a). In the model herbaceous perennial plant, leafy spurge (*Euphorbia esula*), cold night temperatures and long days appear to be most effective for inducing endodormancy (Horvath et al., 2010). Although cold can induce endodormancy in some species, extended chilling temperatures release vegetative bud endodormancy in nearly every temperate perennial species examined (Arora et al., 2003). After the release of endodormancy via chilling, warm temperatures in the spring promote cold deacclimation, vegetative bud flush, and the resumption of elongation growth. The quantitative genetics of bud set and bud flush have been well studied, and many quantitative trait loci (QTL) have been identified (Frewen et al., 2000; reviewed in Howe et al., 2003; Jermstad et al., 2003; Scotti-Saintagne et al., 2004; Pelgas et al., 2011; Rohde et al., 2011b).

The phytochrome photoreceptors and components of the circadian clock regulate short-day-induced dormancy in *Populus* and other perennial plants (Howe et al., 1996; Olsen et al., 1997; Ibanez et al., 2010; Kozarewa et al., 2010). In *Populus*, short-day signals induce growth cessation via a regulatory module consisting of poplar homologs of *CONSTANS (CO)* and *FLOWERING LOCUS T (FT)* in *Arabidopsis* (Bohlenius et al., 2006). Ultimately, SD signals lead to changes in poplar cell proliferation via the *Like-APETALA 1 (LAP1)* gene product, which acts on the AINTEGUMENTA-like 1 transcription factor, which is related to a regulator of cell proliferation in *Arabidopsis* (Azeez et al., 2014). In *Populus*, FT2 was also induced by chilling, which subsequently led to the induction of 1,3-β-glucanases, reopening of signal conduits, and release of endodormancy (Rinne et al., 2011). The authors hypothesized that the reopened conduits enabled movement of FT2 and CENTRORADIALIS 1 (CENL1) to locations where they promoted bud flush and shoot elongation (Rinne et al., 2011). The expression of other genes that regulate cold acclimation and other endodormancy-associated processes are induced by SD. Transcription factors such as C-REPEAT/DRE BINDING FACTOR 2/DEHYDRATION RESPONSE ELEMENT-BINDING PROTEIN (CBF/DREB) have been implicated in cold acclimation and endodormancy (Doğramaci et al., 2010). For example, overexpression of a *CBF* gene in apple resulted in the ability to induce endodormancy with SDs (Wisniewski et al., 2011).

Many of the same environmental and hormonal signals that regulate dormancy also regulate cold acclimation and flowering. Thus, it is not surprising that the flowering genes *FT2* and *CENL1* also seem to regulate endodormancy (Bohlenius et al., 2006; Ruonala et al., 2008; Hsu et al., 2011; Rinne et al., 2011). Likewise, proteins suspected of regulating *FT2*, such as those encoded by *DORMANCY ASSOCIATED MADS-BOX (DAM)* genes, have also been implicated in endodormancy regulation (Bielenberg et al., 2008; Horvath et al., 2010; Sasaki et al., 2011; Yamane et al., 2011). Chromatin remodeling processes associated with vernalization may also regulate bud endodormancy in perennials (Horvath et al., 2003), perhaps by modifying the promoters of *DAM* genes (Horvath et al., 2010; Leida et al., 2012). Indeed, chromatin remodeling seems to accompany changes in *Populus* dormancy states (Vining et al., 2012).

Microarray analysis in *Populus* and several other species have identified common signaling processes associated with endodormancy induction and release (Mazzitelli et al., 2007; Ruttink et al., 2007; Halaly et al., 2008; Horvath et al., 2008; Mathiason et al., 2009; Walton et al., 2009; Campbell et al., 2010; Doğramaci et al., 2010; Karlberg et al., 2010). In addition to flowering genes, genes involved in environmental and phytohormone signaling [e.g., photoperiod, cold, oxidative stress, ethylene, auxin, ABA, and jasmonic acid (JA)], chromatin remodeling, and circadian responses are often differentially expressed during the induction and release of endodormancy. However, only a modest number of genes (<15,000) have been assayed in most previous studies, making it difficult to compare differential expression among gene family members. Furthermore, there are few reports in which endodormancy induction *and* release were compared under natural conditions in the same study.

We used analyses of gene expression to infer physiological processes and *cis*-acting motifs associated with the induction and release of endodormancy in *Populus*. We collected vegetative axillary buds from the end of summer through early spring, and then used a NimbleGen genome-scale microarray to measure global changes in gene expression among dormancy states. Our primary objectives were to: (1) identify which individual genes, biological processes, molecular functions, and regulatory pathways were differentially expressed among dormancy states, (2) classify the differentially expressed genes into contrasting gene expression patterns, and (3) identify *cis*-acting elements associated with each gene expression group. We used this approach and previous observations on dormancy physiology and genomics to better understand the processes regulating endodormancy induction and release in *Populus* trees. We report extensive transcriptome remodeling that both confirm and contradict physiological pathway expectations from the published literature.

MATERIALS AND METHODS

Plant Material

We collected axillary buds from the main stem of two, rapidly growing, 3-year-old *Populus trichocarpa* trees (clone Nisqually-1) growing on a field site in Corvallis, OR, USA on five dates between August 2005 and March 2006 (Step 1, **Figure 1**). Average temperatures and precipitation over the collection period are shown in **Supplementary Figure S1**. Separate RNA isolations were performed on a pooled sample of five buds from each of two trees on each date, resulting in two biological replicates that were used for array hybridizations. The buds were dissected in the field using sterile scalpel blades, immediately frozen in liquid nitrogen, and then subsequently stored at −80°C until they were used for RNA isolation. A few buds collected at the same time were fixed in FAA, dehydrated, and then embedded in wax for sectioning (WAX-IT Histology Services, Vancouver, BC, Canada). De-waxed stem sections were stained with Toluidine Blue-O (Jensen, 1962) and photographed.

RNA Isolation

RNA was isolated using a Qiagen RNeasy kit according to the manufacturer's protocol, including a DNase I treatment to remove contaminating genomic DNA (Qiagen, Valencia, CA, USA). The A260/A280 ratios of RNA samples used for hybridizations ranged from 1.8 to 2.0. The absence of contaminating genomic DNA and the integrity of RNA samples were examined by an Agilent 2100 Bioanalyzer (Agilent Technologies, Palo Alto, CA, USA). The RNA Integrity Numbers (RIN; Mueller et al., 2004) of the RNA samples ranged from

8.5 to 10.0, indicating that high-quality RNA was used for the microarray hybridizations.

Microarray Analysis

We measured gene expression using a microarray designed to target all predicted genes in the *P. trichocarpa* nuclear and organellar genomes plus a set of divergent aspen transcripts (Step 2, **Figure 1**). The microarray, which was manufactured by Roche NimbleGen[1], was originally designed to target 55,794 nuclear, 59 mitochondrial, 71 chloroplast, and 49 miRNA gene models based on version 1.1 of the *P. trichocarpa* genome sequence (v1.1; Tuskan et al., 2006), plus 9,995 unigenes derived from aspen ESTs (Sterky et al., 2004). With a few exceptions, each original gene model was represented by two copies of three different 60-mer isothermal probes. Original microarray information is archived in the NCBI Gene Expression Omnibus (GEO) database as accession numbers GPL2699 and GPL7424. To analyze the microarray data using the latest gene models, we used BLASTN (Altschul et al., 1990) to reassign the 194,260 *Populus* probe sequences from the original microarray to the transcript sequences from the *Populus* v3.0 genome assembly (file Ptrichocarpa_210_transcript.fa released as part of Phytozome v9.0[2]). We assigned each array probe to one v3.0 gene model, omitting the filtering of low-complexity probes and using a 75% nucleotide identity cutoff. By reassigning the microarray probes to *Populus* v3.0 transcripts, we were able to measure the expression of 35,048 out of 41,335 v3.0 primary transcripts. For probes that could not be assigned to v3.0 transcripts, we retained the original gene model assignment. These 9,393 v1.1 transcripts were also included in the analyses described below.

Biotin-labeled cRNA was produced using the Ambion MessageAmp[TM] II aRNA amplification kit according to the manufacturer's instructions (Van Gelder et al., 1990; Life Technologies, 2011), and then sent to NimbleGen for fragmentation, hybridization, and detection. Briefly, total RNA (∼1500 ng) from each sample was reversed transcribed using an oligo(dT) primer with a T7 promoter. After second-strand synthesis, the cDNA was used as a template for synthesizing biotin-labeled antisense RNA (cRNA) using *in vitro* transcription with T7 RNA polymerase. For each sample, ∼20 μg cRNA was sent to NimbleGen for fragmentation, hybridization, and detection as described by Kaushik et al. (2005). After hybridization and washing, the arrays were stained with a streptavidin-Cy3 conjugate, and then scanned with a GenePix 4000B microarray scanner.

The complete microarray dataset was deposited in the GEO database[3] as accession GPL20616. We created new NimbleGen design files (ndf and ngd files) based on the probe reassignments described above, and then normalized the data using NimbleScan v2.6. Microarray data were log2 transformed, background corrected, and then normalized across all arrays using the Robust Multiple-array Average (RMA; Irizarry et al., 2003).

---

[1]http://www.nimblegen.com/

[2]http://www.phytozome.net/

[3]http://ncbi.nlm.nih.gov/geo/

Step 1. Collected axillary buds on 5 dates from August to March

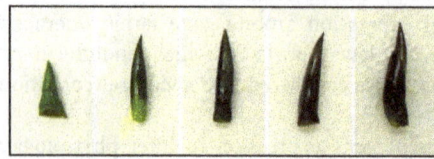

Step 2. Measured changes in steady-state mRNA using Nimblegen whole genome *Populus* microarrays

Step 3. Identified genes differentially expressed among collection dates using ANOVA, and then clustered dates into 3 dormancy states (Para, Endo, Eco)

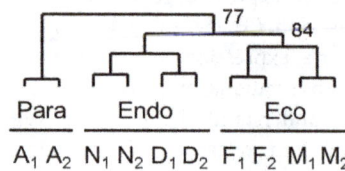

Step 4. Identified genes differentially expressed among dormancy states (DS genes) using ANOVA

SEQ_ID	PARA	ENDO	ECO	DORM_FDR_P
Potri.008G210900	9.5	13.6	11.9	0.0016
Potri.002G107800	10.9	8.4	10.3	0.0017

Step 5. Classified DS genes into 8 gene expression patterns using ANOVA contrasts

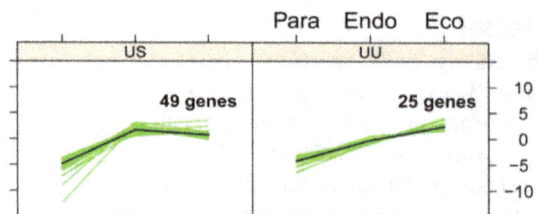

Step 6. Used the ELEMENT program to identify over-represented sequence motifs in subsets of DS genes with specific expression patterns

Step 7. Used Pathway Studio and gene set enrichment analysis (GSEA) to identify over-represented gene ontologies (gene sets) among differentially expressed genes

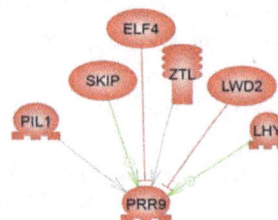

FIGURE 1 | Flow diagram showing the steps used to analyze dormancy related gene expression in *Populus*. Step 1 shows representative axillary buds collected in August, November, December, February, and March (left to right). Step 2 shows the NimbleGen gene expression microarray used to measure relative gene expression. Step 3 shows results of clustering five collection time-points into three dormancy states based on the expression of differentially expressed genes. The dormancy states are paradormant (Para), endodormant (Endo), and ecodormant (Eco). Step 4 shows a section of Supplementary Data File 1, which includes results of analyses of variance (ANOVA). Step 5 shows genes that were classified into two of eight gene expression patterns. Step 6 shows a transcription factor binding to an upstream DNA sequence motif (Evening Element). Step 7 shows a representative regulatory network generated by the Pathway Studio program.

Characterization of Bud Dormancy States and Tests of Differential Expression

We assigned a dormancy state to each collection date using ANOVA and cluster analysis in SAS v9.3 (Statistical Analysis System, Cary, NC, USA). First, we used ANOVA and a false discovery rate (FDR) p-value < 0.05 to identify genes that were differentially expressed among the collection dates. We then used UPGMA and Neighbor-Joining hierarchical clustering to group the collection dates into dormancy states. The UPGMA analysis clustered the collection dates into three distinct clusters: August, November/December, and February/March, which we refer to as paradormant (Para), endodormant (Endo), and ecodormant (Eco) (Step 3, **Figure 1**; see Results), respectively. In the Neighbor-Joining analysis, the February samples clustered with November and December (rather than with March), but this was only weakly supported (i.e., as compared to the UPGMA alternative). Because the UPGMA assignments were judged to be more biologically accurate (i.e., based on morphological observations and past research on *Populus* dormancy), we used the UPGMA groupings for further analyses. We regrouped the samples based on the UPGMA cluster analysis, and then conducted a second ANOVA on the entire dataset to determine which genes were differentially expressed among the assigned dormancy states (treatments; Step 4, **Figure 1**). All subsequent references to 'regulated' or 'differentially expressed' genes refer to the set of 1,362 genes that were differentially expressed among dormancy states at an FDR p-value < 0.05.

Gene Expression Patterns and Sequence Motifs

A priori, we defined eight potential patterns of gene expression that could occur during two dormancy transitions: Para/Endo and Endo/Eco. For each transition, gene expression may either be up-regulated (U), stay the same (S), or be down-regulated (D), which results in eight possible patterns for two transitions when only the differentially expressed genes are tested (i.e., S/S patterns are not possible). That is, 8 patterns = (3 possible changes for the Para/Endo transition × 3 possible changes for the Endo/Eco transition) – 1 pattern (S/S). For each gene, we determined the p-value for each of the eight models using the CONTRAST option of SAS Proc ANOVA, and then used the treatment means and the model p-values to assign the gene expression pattern (Step 5, **Figure 1**). Because we were interested in clustering the genes based on directional changes in gene expression (not differences in mean expression), we first normalized the data using the ANOVA mean square error for each gene.

We used the ELEMENT program (Mockler et al., 2007) to identify sequence motifs that were overrepresented in each gene expression group (Step 6, **Figure 1**). These analyses were conducted using 2 kb of upstream sequence relative to the *Populus* v3.0 transcription start site. Motifs were associated with a particular gene expression pattern when the average number of motifs per sequence (MOTIF_MN_HITS) was significant at an FDR p-value < 0.05 for only one of the gene expression groups. We inferred potential functions of the motifs using the

SIGNALSCAN program and the database of Plant *Cis*-acting Regulatory DNA Elements (PLACE[4]; Higo et al., 1999), and by comparing the motifs to motifs in the PlantCare database[5]) and published literature. We then ranked the motifs based on the number of sequences that contained one or more copies of the motif (SEQ_HIT_P), and identified the top 50 motifs.

Identification of Key Differentially Expressed Genes

We focused attention on genes encoding transcription factors, and genes associated with chromatin, phytohormone responses, or dormancy-related QTL. For each analysis (subset of genes), we classified the genes into four groups: up- or down-regulated from paradormancy to endodormancy, and up- or down-regulated from endodormancy to ecodormancy. Within each group, we ranked genes by FDR p-value, and then focused on the top 15 genes in each group if they had an FDR p-value < 0.05.

Chromatin-associated genes were identified using the *Arabidopsis thaliana* chromatin database (ChromDB; Gendler et al., 2008) and by including genes that had "chromatin" or "histone" in the TAIR10 functional annotation (defline) of the *Populus* v3.0 annotation file or in any of the "full name" aliases listed in the TAIR10 gene aliases text file[6]. Transcription factor genes were identified using the *P. trichocarpa* and *A. thaliana* Plant Transcription Factor Databases v3.0 (TFDB; Jin et al., 2014[7]). Because phytohormones are involved in very large signaling networks with substantial cross-talk, we defined phytohormone-associated genes as those having *direct* roles affecting hormone responses via their influence on hormone metabolism (biosynthesis or inactivation), transport, or signaling. This definition encompassed genes that link hormone receptors to primary downstream transcription factors, but excluded genes that regulate hormone metabolic genes, secondary transcription factors, and other downstream response genes. Genes located within dormancy-related QTL were identified by mapping the gene models shown in Supplementary Table S4 of Rohde et al. (2011b) to the *Populus* v3.0 gene models using the gene model aliases shown in Supplementary Data File 1. Some gene models noted by Rohde et al. (2011b) were excluded if they no longer mapped to the same general region of the *Populus* v3.0 genome, and some new v3.0 gene models were added if they were contiguous to the genes previously described by Rohde et al. (2011b). The genes belonging to each of these subsets are identified in Supplementary Data File 1.

Identification of Differentially Expressed Gene Sets using Gene Set Enrichment Analysis (GSEA)

We used gene set enrichment analysis (GSEA) to identify gene sets that were overrepresented among the differentially expressed genes (Step 7, **Figure 1**). GSEA is a statistical

---

[4]http://www.dna.affrc.go.jp/PLACE/

[5]http://bioinformatics.psb.ugent.be/webtools/plantcare/html/

[6]ftp://ftp.arabidopsis.org/Genes/gene_aliases_20130831.txt

[7]http://planttfdb.cbi.pku.edu.cn/

approach for determining whether sets of genes defined a priori (e.g., genes with a common Gene Ontology term) preferentially occur toward the top or bottom of a ranked list of genes (Subramanian et al., 2005). Using the FDR p-values, we ranked all genes according to their changes in expression between (1) paradormancy versus endodormancy and (2) endodormancy vs. ecodormancy. For each dormancy transition, we conducted two analyses. First, we analyzed the data considering whether the genes were up-regulated or down-regulated. That is, we subtracted the FDR p-value from 1, and then multiplied the result by -1 if the gene was down-regulated. GSEA was then used to identify gene sets that were preferentially located near the top or bottom of this list. Second, we analyzed the data ignoring the direction of change (i.e., no -1 multiplier was used), and then used GSEA to identify genes that were preferentially located near the top of this list. This second analysis was used to identify gene sets whose members contribute to the same biological response via opposite changes in gene expression.

We implemented GSEA using the java application GSEA v2.0.13 (Broad Institute, Cambridge, MA, USA[8]) and 1000 bootstrap replications. We used default parameters, except for setting the minimum and maximum gene set sizes to 5 and 500. Gene sets were considered statistically significant at an FDR p-value of 0.10. We analyzed three datasets: two Gene Ontology (GO) datasets (biological process and molecular function GO categories[9]; Gene Ontology Consortium, 2012) and one Pathway Studio (PS) dataset (Pathway Studio v8, Elsevier). For the GO gene sets, *Populus* genes were assigned using the GO terms or *Arabidopsis* gene assignments downloaded from the Phytozome web site (Supplementary Data File 1; Phytozome v9.0[10]). For the Pathway Studio dataset, we combined seven types of pathways. The combined dataset included sets of genes that encode (1) expression targets, (2) miRNA targets, (3) protein modification targets, (4) proteins regulating disease, (5) proteins regulating cell processes, and (6) binding partners and (7) neighbors of key proteins and biological processes.

RESULTS

Assignment of Bud Dormancy States

We collected axillary buds on five dates between August and March, and then assigned these samples to three dormancy states or treatments based on cluster analysis of gene expression data. Of a total of 44,441 gene models represented on the microarray, 1,206 gene models (36 v1.1 and 1170 v3.0 transcripts) were differentially expressed among months (FDR p-value < 0.05; Supplementary Data File 1). Clustering of these differentially expressed genes produced three well-supported groups consisting of samples collected in (1) August, shortly after terminal bud set, (2) November and December, and (3) February and March (**Figure 2**). We classified the first group (August samples) as paradormant (Para), the November and December

[8]http://www.broadinstitute.org/gsea/index.jsp

[9]http://geneontology.org/

[10]http://www.phytozome.net/

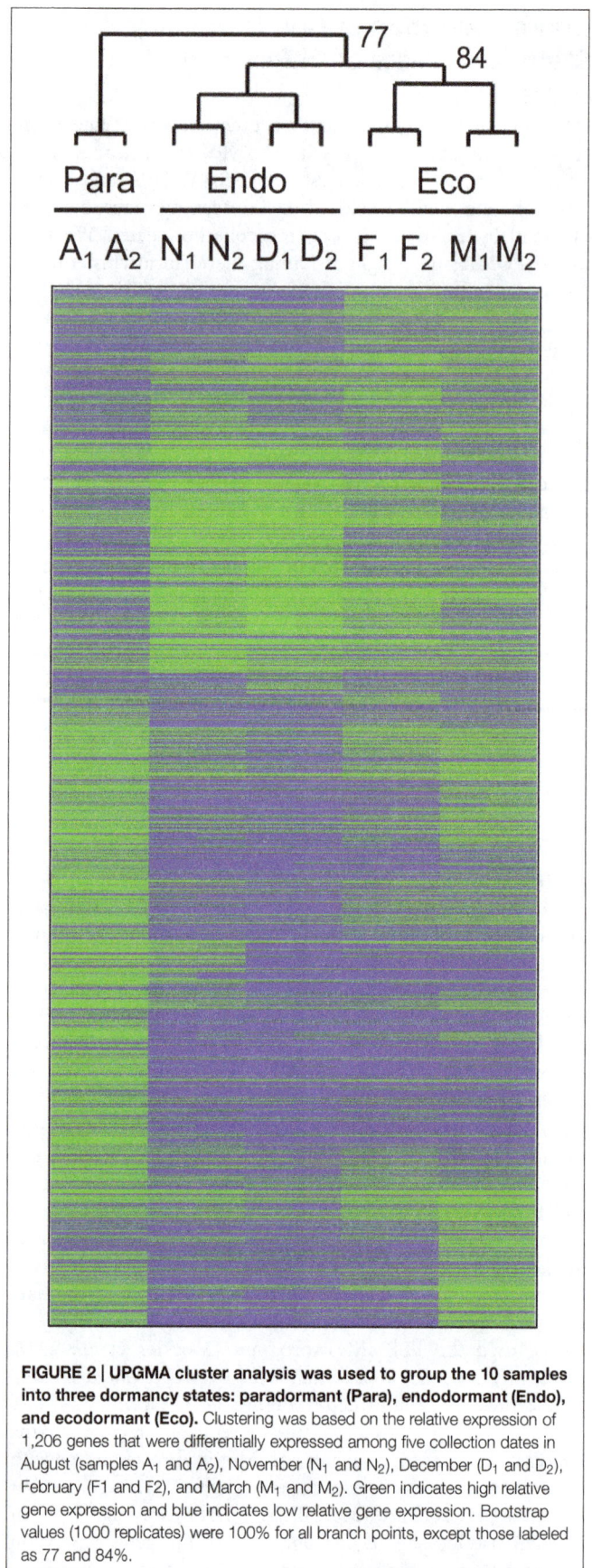

FIGURE 2 | UPGMA cluster analysis was used to group the 10 samples into three dormancy states: paradormant (Para), endodormant (Endo), and ecodormant (Eco). Clustering was based on the relative expression of 1,206 genes that were differentially expressed among five collection dates in August (samples A_1 and A_2), November (N_1 and N_2), December (D_1 and D_2), February (F1 and F2), and March (M_1 and M_2). Green indicates high relative gene expression and blue indicates low relative gene expression. Bootstrap values (1000 replicates) were 100% for all branch points, except those labeled as 77 and 84%.

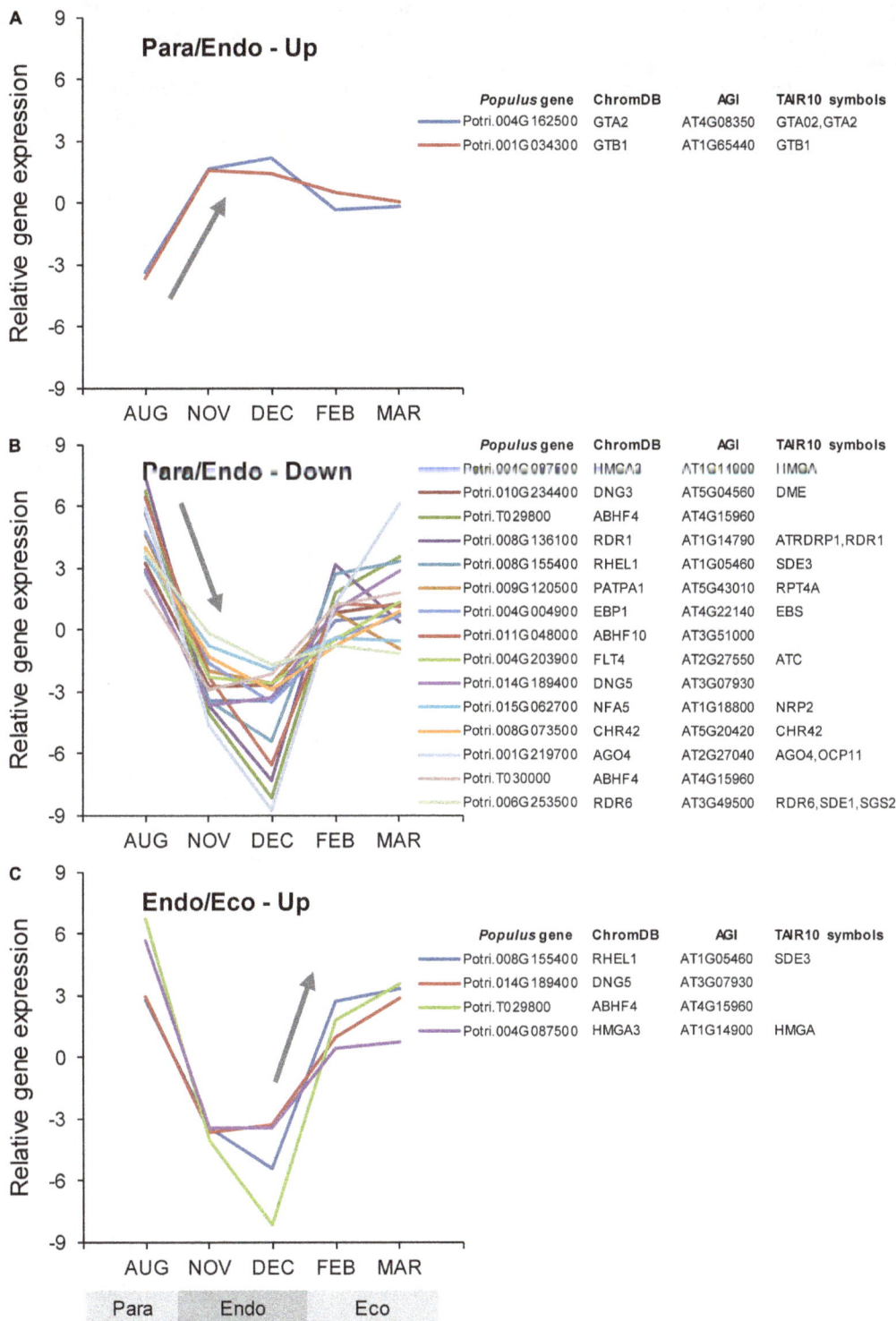

FIGURE 3 | Relative expression of differentially expressed chromatin-associated genes. Chromatin-associated genes were identified using the *Arabidopsis thaliana* chromatin database (ChromDB; Gendler et al., 2008). Genes were classified into four groups: up-regulated or down-regulated from paradormancy to endodormancy **(A,B)**, or up-regulated or down-regulated from endodormancy to ecodormancy **(C)**. Within each group, we ranked genes by FDR *p*-value, and then displayed the top 15 genes for each group if they had a FDR *p*-value < 0.05. Only four genes were differentially expressed between endodormancy and ecodormancy, and all of these were up-regulated **(C)**. Gene expression values are the means for each month normalized to a mean of zero and a standard deviation equal to the ANOVA RMSE from the analysis of gene expression differences among months. ChromDB is the ChromDB identifier, *Populus* gene is the *P. trichocarpa* gene-locus name, and AGI and TAIR10 symbols are the *Arabidopsis* gene identifiers and gene symbols from the *P. trichocarpa* v3.0 annotations (http://www.phytozome.net/).

samples as endodormant (Endo) and the February and March samples as ecodormant (Eco). Morphological observations of bud development (**Supplementary Figure S2**), and past dormancy research on *P. trichocarpa* and other cottonwoods (see Discussion) support the assignment of these dormancy states.

Major Transcriptome Changes Occur during Endodormancy Induction and Release

Of the 44,441 gene models analyzed, 1,362 genes (v1.1 = 43; v3.0 = 1,319) were differentially expressed among dormancy states (FDR *p*-value < 0.05; Supplementary Data File 1). Based on analyses of individual dormancy transitions, however, 1,523 genes (v1.1 = 25; v3.0 = 1,498) were differentially expressed between paradormancy and ecodormancy, and the majority of these (*n* = 1,168) were down-regulated. Only 293 transcripts (v1.1 = 13; v3.0 = 280) were differentially expressed between endodormancy and ecodormancy, and the majority of these (*n* = 193) were up-regulated. A total of 190 genes (v1.1 = 4; v3.0 = 186) were differentially expressed during both transitions. Of all the v3.0 differentially expressed genes discussed above, only 3.2% were novel—i.e., had no *Arabidopsis* match (Supplementary Data File 1). We found no evidence that transcripts associated with the mitochondrial, chloroplast, or miRNA gene models were differentially expressed. However, results for the organelle gene models are difficult to interpret because total RNA was reversed transcribed using an oligo(dT) primer.

Gene Set Enrichment Analysis (GSEA)

In addition to single-gene analyses, we used GSEA to help identify differentially expressed genes sets, which are groups of genes that share a common biological function, chromosomal location, or regulation (Subramanian et al., 2005). Compared to single-gene analyses, GSEA has the potential to identify biologically relevant genes, even when the results of single gene analyses are not statistically significant, and may yield insights that are not obvious from reviewing long lists of statistically significant genes. In particular, GSEA can be valuable for identifying important regulatory pathways. Significant gene sets identified using the GO-term, Pathway Studio, and phytohormone analyses are presented in Supplementary Tables S1–S8, and overrepresented ontologies of particular interest are described below.

Differential Expression of Chromatin-associated Genes
Overview

Among the *Populus* v3.0 genes on the array, 727 were identified as being chromatin-associated, 21 of which were differentially expressed between adjacent dormancy states (Supplementary Data File 1). During the transition from paradormancy to endodormancy, 19 genes were down-regulated and 2 were up-regulated. During the transition from endodormancy to ecodormancy, no genes were down-regulated, and 4 were up-regulated. Four genes were differentially expressed during both

dormancy transitions. Changes in expression for the top genes (ranked on FDR *p*-value) are shown in **Figure 3**.

Chromatin-associated Gene Sets

Three chromatin-associated Pathway Studio gene sets were expressed at lower levels during endodormancy—with significant changes in expression at each of the two dormancy transitions. These gene sets were 'Neighbors of RDR6' (RNA-DEPENDENT RNA POLYMERASE 6), 'Regulators of cytosine methylation,' and 'Regulators of maintenance of DNA methylation' (Supplementary Tables S5 and S6). Sixteen other gene sets were differentially expressed during one of the two dormancy transitions, 13 of which were down-regulated from paradormancy to endodormancy. These Pathway Studio gene sets were 'Neighbors of SWN' (SWINGER), 'Expression targets of RDR6,' 'Neighbors of DCL1' (DICER-LIKE 1), 'Neighbors of CMT3' (CHROMOMETHYLASE3), 'Regulators of DNA methylation,' 'Neighbors of HDA6' (HISTONE DEACETYLASE A6), 'Neighbors of histone,' 'Binding partners of FIE' (FERTILIZATION INDEPENDENT ENDOSPERM), 'Expression targets of DET1' (DEETIOLATED 1), 'Regulators of chromatin remodeling,' 'Neighbors of PRMT11' (PROTEIN ARGININE METHYLTRANSFERASE 11), 'Neighbors of AGO7' (ARGONAUT7), 'Neighbors of FIE,' 'Neighbors of HD1' (HISTONE DEACETYLASE 1), 'Neighbors of polycomb complex,' and the GO molecular function gene set, 'Histone-lysine *n*-methyltransferase activity.' Finally, five additional Pathway Studio gene sets were differentially expressed, from paradormancy to endodormancy, but with no common pattern of expression among the gene set members ('Up- or down-regulated' gene sets in Supplementary Tables S5 and S6). These gene sets were 'Binding partners of DDB1A' (DNA DAMAGE-BINDING PROTEIN 1A), 'Neighbors of DCL2' (DICER-LIKE 2), 'Regulators of histone methylation,' 'Binding partners of JAZ10' (JASMONATE ZIM-DOMAIN PROTEIN 10), and 'Neighbors of PKL' (PICKLE).

Chromatin-associated Genes

Of the genes shown in **Figure 3**, four genes were down-regulated from paradormancy to endodormancy and then up-regulated from endodormancy to ecodormancy. One of these genes (Potri.004G087500; HMGA3) is similar to a gene that encodes a HIGH MOBILITY GROUP A (HMGA) protein in *Arabidopsis*. The second gene (Potri.008G155400; RHEL1) is similar to *Arabidopsis SILENCING DEFECTIVE 3* (*SDE3*). The third gene (Potri.014G189400; DNG5) is similar to genes that encode DNA glycosylases involved in gene silencing and chromatin remodeling. The last gene (Potri.T029800; ABHF4) is similar to a gene that encodes an alpha/beta-hydrolase in *Arabidopsis*. Two genes had atypical patterns of expression—being strongly up-regulated from paradormancy to endodormancy, and then down-regulated from endodormancy to ecodormancy. The first gene (Potri.004G162500; GTA2), which is listed as encoding a GLOBAL TRANSCRIPTION FACTOR GROUP A2 (GTA2) protein in **Figure 3**, is similar to *Arabidopsis SPT5-2*, an ortholog of yeast *SUPPRESSOR OF TY 5* (Durr et al., 2014). The second gene (Potri.001G034300; GTB1), which is listed as encoding a GLOBAL TRANSCRIPTION FACTOR GROUP B1 (GTB1) protein, is similar to the *SPT6*-like (*SPT6L*) gene from *Arabidopsis* (Li et al., 2010; Gu et al., 2012). Other distinct chromatin-associated genes were also differentially expressed (see ChromDB classifications PATPA1, EPB1, FLT4, NFA5, and CHR42 in **Figure 3**). These genes encode proteins that may be involved in histone ubiquitination and methylation, chromatin assembly or disassembly, histone and DNA binding, and chromatin remodeling.

Differential Expression of Transcription Factor Genes
Overview
Among the *Populus* v3.0 genes on the array, 2,469 were identified as transcription factors, 117 of which were differentially expressed between adjacent dormancy states (Supplementary Data File 1). During the transition from paradormancy to endodormancy, 89 genes were down-regulated and 19 were up-regulated. During the transition from endodormancy to ecodormancy, 5 genes were down-regulated and 19 were up-regulated. Fifteen genes were differentially expressed during both dormancy transitions. Changes in expression for the top genes (ranked on FDR *p*-value) are shown in **Figure 4**.

Transcription Factor Gene Sets
Nine transcription factor gene sets were differentially expressed during both dormancy transitions. Four were expressed at higher levels during endodormancy: 'Neighbors of EIN3' (ETHYLENE INSENSITIVE 3), 'Expression targets of EIN3,' 'Neighbors of RHL41' (RESPONSIVE TO HIGH LIGHT 41), and 'Expression targets of WRKY'' (Supplementary Tables S5 and S6). The other five gene sets were expressed at lower levels during endodormancy: 'Neighbors of JLO' (JAGGED LATERAL ORGAN), 'Neighbors of SEU' (SEUSS), 'Neighbors of RPL' (REPLUMLESS), 'Neighbors of ARF2' (AUXIN RESPONSE FACTOR 2), and 'Neighbors of BASIC-HELIX-LOOP-HELIX PROTEIN.'

Transcription Factor Genes
Of the genes shown in **Figure 4**, five were differentially expressed during both dormancy transitions. Two genes were down-regulated from paradormancy to endodormancy and then up-regulated from endodormancy to ecodormancy. One of these genes (Potri.009G134000) is similar to a gene that encodes MYB DOMAIN PROTEIN 4 (MYB4) in *Arabidopsis*. The second gene (Potri.004G230100) is similar to *Arabidopsis VERNALIZATION1* (*VRN1*). Three other genes had atypical patterns of expression—being strongly up-regulated from paradormancy to endodormancy, and then down-regulated from endodormancy to ecodormancy. Potri.008G210900 is similar to a gene that encodes an ETHYLENE-RESPONSIVE ELEMENT BINDING PROTEIN (EBP), Potri.014G017300 is similar to the *SALT TOLERANCE ZINC FINGER* (*STZ*) gene, and Potri.001G066900 is similar to an *Arabidopsis* gene that encodes a trihelix transcription factor.

Differential Expression of Phytohormone-associated Genes
Overview
Among the *Populus* v3.0 genes on the array, 718 genes were identified as being putatively involved in hormone synthesis, catabolism, signaling, or transport. Of these, 41 were differentially expressed between adjacent dormancy states (Supplementary Data File 1). During the transition from paradormancy to endodormancy, 38 genes were down-regulated and only 2 were up-regulated. No genes were down-regulated from endodormancy to ecodormancy, but 8 genes were up-regulated. Seven genes were differentially expressed during both dormancy transitions. Changes in expression for the top genes (ranked on FDR *p*-value) are shown in **Figure 5**.

Auxin-associated Gene Expression
The auxin-associated gene set was mostly down-regulated from paradormancy to endodormancy, and then up-regulated thereafter (Supplementary Tables S7 and S8). Changes in Pathway Studio gene sets provide additional support for the importance of auxin associated genes. One auxin-associated gene set ('Neighbors of ARF6,' AUXIN RESPONSE FACTOR 6) was up-regulated from paradormancy to endodormancy, but three other key gene sets, 'Binding partners of ARF1,' 'Neighbors of ARF2,' and 'Binding partners of TIR1' (TRANSPORT INHIBITOR RESPONSE 1), were down-regulated (Supplementary Tables S5 and S6). Two gene sets associated with ARF2 were subsequently up-regulated from endodormancy to ecodormancy.

Ethylene-associated Gene Expression
The ethylene gene set was up-regulated from paradormancy to endodormancy, and then down-regulated from endodormancy to ecodormancy (Supplementary Tables S7 and S8). More specifically, one of only two phytohormone genes that were significantly up-regulated from paradormancy to endodormancy is similar to a gene that encodes the CTR1 (CONSTITUTIVE TRIPLE RESPONSE 1) protein, which is a negative regulator of the ethylene response pathway in *Arabidopsis*. Changes in other genes that participate in ethylene responses were described above (see Differential Expression of Transcription Factor Genes).

GA-associated Gene Expression
Gibberellin-associated genes were generally up-regulated from paradormancy to endodormancy, but did not change from endodormancy to ecodormancy (Supplementary Tables S7 and S8). We then focused attention on genes encoding GA-20 oxidases and GA-2-oxidases because of their potential involvement in endodormancy. We identified genes encoding GA-20-oxidases and GA-2-oxidases based on similarities to *Arabidopsis* genes and the information presented in Gou et al. (2011), but none was differentially expressed. In fact, no individual GA-related genes were differentially expressed.

ABA-associated Gene Expression
The ABA-associated gene set did not change among dormancy states (Supplementary Tables S7 and S8), but our analyses of individual ABA genes identified four genes that were

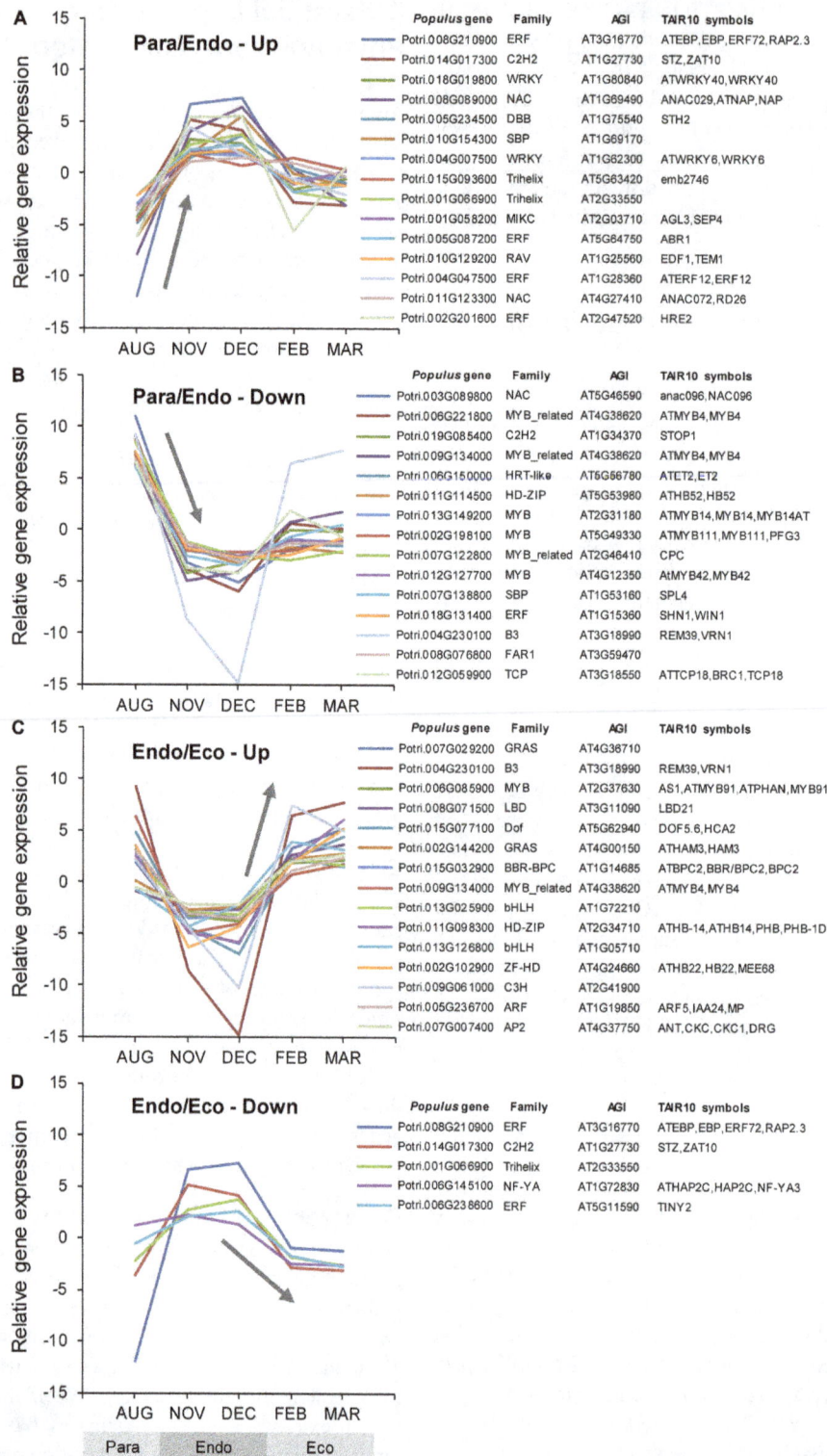

FIGURE 4 | Relative expression of differentially expressed transcription factor genes. Transcription factor genes were identified using the *P. trichocarpa* and *A. thaliana* Plant Transcription Factor Databases v3.0 (TFDB; Jin et al., 2014; http://planttfdb.cbi.pku.edu.cn/). Genes were classified into four groups: up-regulated or down-regulated from paradormancy to endodormancy **(A,B)**, or up-regulated or down-regulated from endodormancy to ecodormancy **(C,D)**. Within each group, we ranked genes by FDR *p*-value, and then displayed the top 15 genes for each group if they had a FDR *p*-value < 0.05. Gene expression values are the means for each month normalized to a mean of zero and a standard deviation equal to the ANOVA RMSE from the analysis of gene expression differences among months. Family is the TFDB family designation for the corresponding *Arabidopsis* gene, *Populus* gene is the *P. trichocarpa* gene-locus name, and AGI and TAIR10 symbols are the *Arabidopsis* gene identifiers and gene symbols from the *P. trichocarpa* v3.0 annotations (http://www.phytozome.net/).

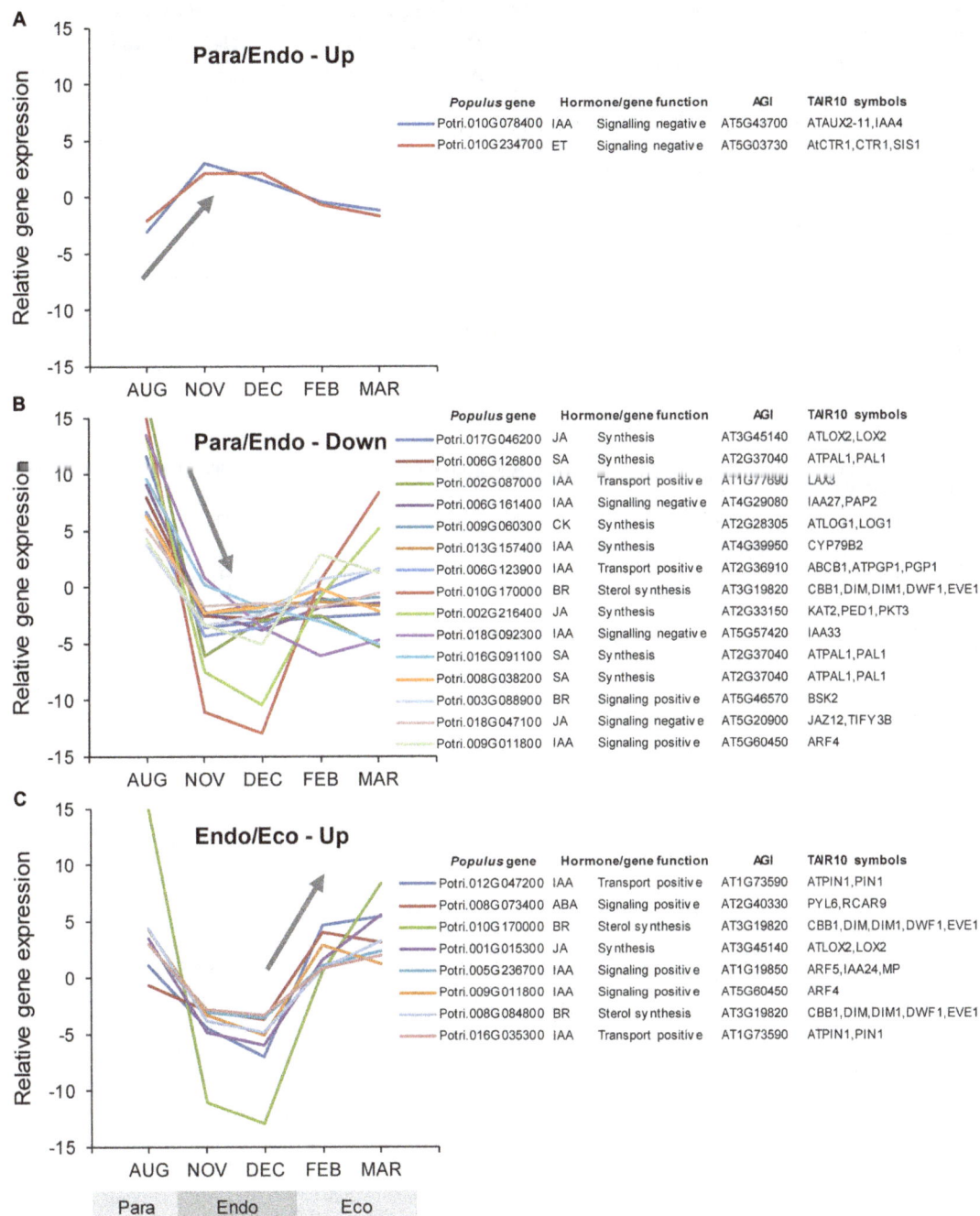

FIGURE 5 | Relative expression of differentially expressed phytohormone-associated genes. Genes were classified into four groups: up-regulated or down-regulated from paradormancy to endodormancy **(A,B)**, or up-regulated or down-regulated from endodormancy to ecodormancy **(C)**. Within each group, we ranked genes by FDR p-value, and then displayed the top 15 genes for each group if they had a FDR p-value < 0.05. All genes differentially expressed between endodormancy and ecodormancy were up-regulated **(C)**. Gene expression values are the means for each month normalized to a mean of zero and a standard deviation equal to the ANOVA RMSE from the analysis of gene expression differences among months. Abbreviations for phytohormones: ABA, abscisic acid; BR, brassinosteroid; CK, cytokinin; ET, ethylene; GA, gibberellin; IAA, indole-3-butyric acid; JA, jasmonic acid, and SA, salicylic acid. *Populus* gene is the *P. trichocarpa* gene-locus name and AGI and TAIR10 symbols are the *Arabidopsis* gene identifiers and gene symbols from the *P. trichocarpa* v3.0 annotations (http://www.phytozome.net/).

differentially expressed between one or more dormancy states. One gene (Potri.008G0734000), similar to a gene that encodes a PYL (PYRABACTIN RESISTANCE-LIKE) ABA receptor, was

down-regulated slightly from paradormancy to endodormancy, and then significantly up-regulated from endodormancy to ecodormancy (**Figure 5**). Three other genes were significantly down-regulated from paradormancy to endodormancy, including genes that likely encode proteins involved in ABA transport (Potri.001G175700), ABA biosynthesis (Potri.003G176300), and positive regulation of ABA signaling (Potri.013G112500). The Pathway Studio gene set, 'Neighbors of ABF2' (ABA RESPONSIVE ELEMENTS-BINDING FACTOR 2), was significantly down-regulated from paradormancy to endodormancy. In contrast, the gene set, 'Neighbors of ABF3,' was down-regulated from endodormancy to ecodormancy.

Brassinosteroid-associated Gene Expression

The brassinosteroid (BR) gene set showed nearly the same expression pattern as the auxin gene set—general down-regulation from paradormancy to endodormancy, and then up-regulation from endodormancy to ecodormancy (but at a FDR p-value of 0.115). In addition, the Pathway Studio gene set, 'Binding partners of BES1' (BRI1-EMS SUPRESSOR 1), was down-regulated from paradormancy to endodormancy. Analyses of individual genes identified three BR-related genes (Potri.010G170000, Potri.008G084800, and Potri.003G088900) that showed the same general pattern of gene expression (**Figure 5**). The first two genes are similar to *Arabidopsis CABBAGE1 (CBB1)* and the third is similar to *BRASSINOSTEROID-SIGNALING KINASE 2 (BSK2)*.

Salicylic-acid-associated Gene Expression

Among the phytohormone gene sets, the SA-associated gene set had the strongest evidence for differential expression (Supplementary Tables S7 and S8). Four of the seven SA-associated genes that were differentially expressed were similar to *Arabidopsis PAL1 (PHENYLALANINE AMMONIA-LYASE 1)*, and all of these were down-regulated from paradormancy to endodormancy.

Jasmonic-acid-associated Gene Expression

Although the JA-associated gene set was not differentially expressed, the Pathway Studio gene sets, 'Neighbors of JA,' 'Neighbors of MEJA,' 'Expression targets of COI1' (CORONATINE INSENSITIVE 1), and 'Neighbors of COI1' were all down-regulated from paradormancy to endodormancy. Furthermore, we saw this same pattern of expression for all six JA-associated genes that were differentially expressed, the top three of which are shown in **Figure 5**. Five of these six genes are associated with JA synthesis (Supplementary Data File 1) and one is associated with negative JA signaling. This latter gene (Potri.018G047100) is similar to *Arabidopsis JASMONATE ZIM-DOMAIN 12 (JAZ12)*. Another Pathway Studio gene set, 'Binding partners of JAZ10,' was differentially regulated, but with no consistent pattern among the gene set members.

Cytokinin-associated Gene Expression

Although the hormone gene set was not differentially expressed, the Pathway Studio gene set, 'Neighbors of cytokinin,' and three individual genes (Potri.009G060300, Potri.010G027100, and Potri.016G044100) were all down-regulated from paradormancy to endodormancy.

Differential Expression of Genes Associated with Bud Set QTL

Among the *Populus* v3.0 genes on the array, 2,181 were identified as being associated with bud set QTL. These genes covered genomic regions ranging from 1.9 Mbp for QTL LG13, to 7.3 Mbp for QTL LG3. A total of 103 genes were differentially expressed using two different criteria. Seventy of these genes were differentially expressed among the three dormancy states (*F*-test), whereas 94 were differentially expressed between either of the two adjacent dormancy states (Supplementary Data File 1). During the transition from paradormancy to endodormancy, 67 genes were down-regulated and 19 were up-regulated. Six genes were down-regulated from endodormancy to ecodormancy, and 15 genes were up-regulated. Fourteen genes were differentially expressed during both dormancy transitions. Changes in expression for the top genes (ranked on FDR p-value) are shown in **Figure 6**. Differentially expressed genes were well distributed among the six QTL (described below).

Quantitative trait loci LG3 and LG5 each mapped near 16 differentially expressed genes (i.e., genes differentially expressed among the three dormancy states *or* between adjacent dormancy states), but none had any obvious regulatory role in bud set. In contrast, several potential regulatory genes were found among the 19 differentially expressed genes located near QTL LG6. These include genes that seem to encode a chromatin-associated DCL protein (Potri.006G188800) and a JAZ protein involved in JA signaling (Potri.006G217200; discussed above). Among the 19 differentially expressed genes located near QTL LG8a, are genes that seem to encode proteins involved in chromatin remodeling (Potri.008G073500), positive ABA signaling (Potri.008G073400), BR synthesis (Potri.008G084800), responses to far-red light (Potri.008G076800), organization of lateral organ boundaries (Potri.008G071500), and a NAC-domain transcription factor associated with leaf senescence (Potri.008G089000). QTL LG8b maps near 24 differentially expressed genes, including two chromatin-associated genes (Potri.008G155400 and Potri.008G136100), an Aux/IAA gene (Potri.008G172400), and a gene similar to *MITOTIC ARREST-DEFICIENT 2* (Potri.008G179600). Nine differentially expressed genes mapped near QTL LG13, including a gene (Potri.013G011400) that seems to encode a plant homeodomain (PHD) finger family protein that was up-regulated during endodormancy, and a gene (Potri.013G025900) that may encode a bHLH transcription factor that was strongly down-regulated and mapped near the center of the QTL.

Patterns of Gene Expression and Regulatory Motifs

With three dormancy states, eight different patterns of gene expression are possible (**Figure 7**). We used ANOVA to identify genes that were differentially expressed among dormancy states (FDR p-value < 0.05), and then used contrast statements to assign each gene to a specific pattern of gene expression. The numbers of genes assigned to each expression pattern ranged from 25 to 470 (**Figure 7**); the gene expression pattern assigned to each gene is available in Supplementary Data File 1.

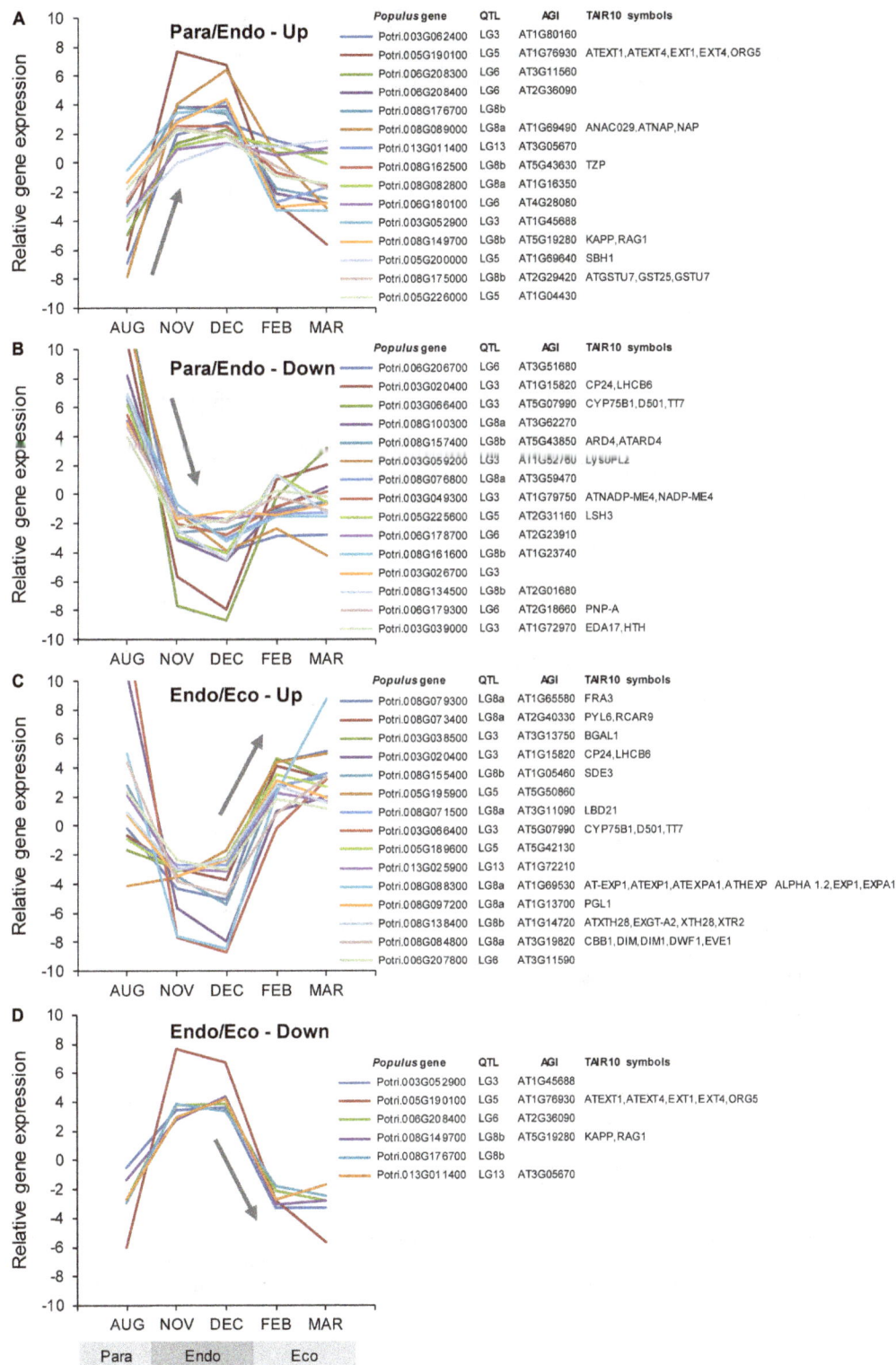

FIGURE 6 | Relative expression of differentially expressed genes located near dormancy-related QTL. Genes were classified into four groups: up-regulated or down-regulated from paradormancy to endodormancy **(A,B)**, or up-regulated or down-regulated from endodormancy to ecodormancy **(C,D)**. Within each group, we ranked genes by FDR p-value, and then displayed the top 15 genes for each group if they had a FDR p-value < 0.05. Gene expression values are the means for each month normalized to a mean of zero and a standard deviation equal to the ANOVA RMSE from the analysis of gene expression differences among months. QTL is the QTL designation from Rohde et al. (2011b), *Populus* gene is the *P. trichocarpa* gene-locus name, and AGI and TAIR10 symbols are the *Arabidopsis* gene identifiers and gene symbols from the *P. trichocarpa* v3.0 annotations (http://www.phytozome.net/).

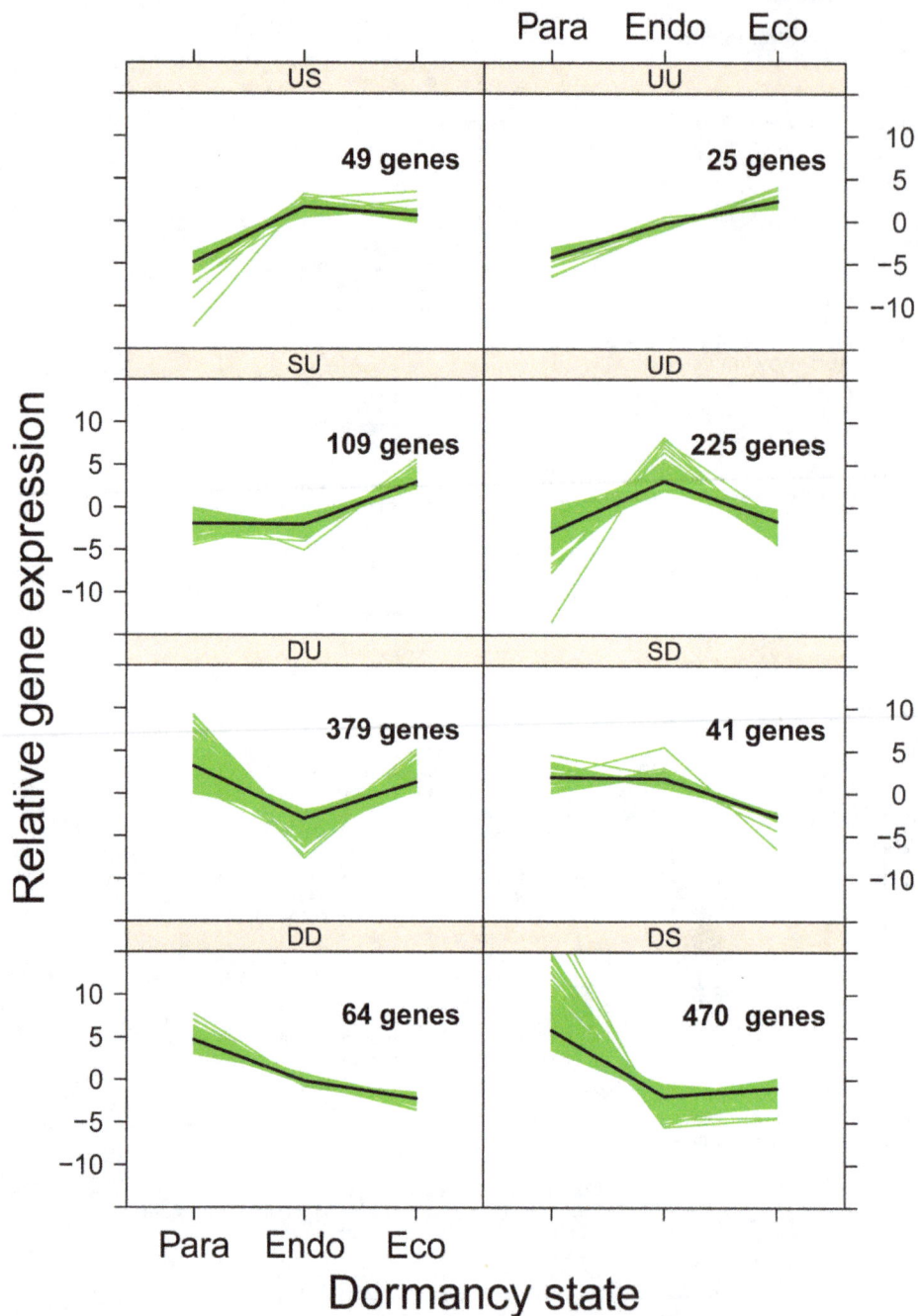

FIGURE 7 | Expression patterns of 1,362 genes that were differentially expressed among dormancy states. Genes were classified into eight gene expression patterns associated with two dormancy transitions: paradormancy to endodormancy and endodormancy to ecodormancy. For each transition, gene expression was either up-regulated (U), the same (S), or down-regulated (D). Solid green lines represent the mean normalized expression for each gene, and the solid black lines represent the mean expression of all genes shown in the panel (i.e., gene expression group).

We subsequently tested whether genes sharing a common pattern of gene expression have common upstream regulatory motifs. For each gene expression group, we analyzed 2 kb of upstream sequence, identifying 315 unique, overrepresented sequence motifs (Supplementary Data File 2). We found only a few unique motifs for four expression patterns (SU and US = 2; SD and UU = 3). Larger numbers of motifs were found for two expression patterns (DD = 16 and DU = 64). The largest numbers of motifs were found for patterns UD ($n = 103$) and DS ($n = 122$). The putative functions of the top 50 overrepresented motifs are shown in **Table 1**.

TABLE 1 | Top 50 upstream sequence motifs overrepresented in eight gene expression pattern-groups.

Motif no.	Pattern	Sequence motif	Seq p-value	Motif p-value	Place sites and other motifs	Proposed functions
208	UD	AAATATCT	5.82E-09	1.63E-14	GATABOX, ROOTMOTIFTAPOX1, EVENINGELEMENTLIKE*	Cold, light, and circadian responses
214	UD	GCCGAC	1.51E-08	4.92E-08	LTRECOREATCOR15, DRECRTCOREAT, CBFHV	Cold, dehydration responses
143	DU	AAAAATCA	2.26E-08	1.50E-07	ARR1AT	Cytokinin response
211	UD	ACGTGTCC	7.87E-08	1.59E-08	ACGTABREMOTIFA2OSEM, ABRELATERD1, ACGTATERD1, GADOWNAT	ABA, GA, and light responses
221	UD	ATGTCGG	2.88E-07	8.44E-07	LTRECOREATCOR15	Cold, dehydration responses
213	UD	CCGACA	3.45E-07	1.73E-08	LTRECOREATCOR15	Cold, dehydration responses
223	UD	GCCGACA	2.08E-06	1.13E-06	LTRECOREATCOR15, DRECRTCOREAT, CBFHV	Cold, dehydration responses
219	UD	GTCGGCA	2.20E-06	5.39E-07	LTRECOREATCOR15, DRECRTCOREAT, CBFHV	Cold, dehydration responses
41	DS	GAAAAATA	2.26E-06	1.91E-06	GT1CONSENSUS, GT1GMSCAM4	Stress and light responses
212	UD	CCGAC	2.52E-06	1.62E-08	LTRECOREATCOR15	Cold, dehydration responses
147	DU	GTTTTTA	3.61E-06	2.20E-06		
61	DS	TATAATAA	4.18E-06	1.51E-05		
229	UD	ATGTCGGC	5.88E-06	8.31E-06	LTRECOREATCOR15, DRECRTCOREAT, CBFHV	Cold, dehydration responses
42	DS	TATAATA	6.52E-06	1.95E-06		
224	UD	AGCCGCC	6.66E-06	2.77E-06	AGCBOXNPGLB, GCCCORE	Ethylene and other responses
311	US	GGTGAAC	1.05E-05	1.58E-06	GTGANTG10	Pollen expression
55	DS	AATTATTA	1.69E-05	9.17E-06	POLASIG3	Polyadenylation-like motif
210	UD	AATATCT	2.13E-05	1.12E-09	GATABOX, ROOTMOTIFTAPOX1, EVENINGELEMENTLIKE*	Cold, light, and circadian responses
264	UD	GCCGCC	2.24E-05	1.03E-04	GCCCORE	Ethylene and other responses
297	UD	CCGTC	2.45E-05	1.12E-03		
142	DU	AAAATAAC	2.52E-05	8.32E-08	TATABOX5	TATA-box motif
69	DS	AAGTTTAT	2.74E-05	2.56E-05		
44	DS	AATTATAT	3.01E-05	3.37E-06		
161	DU	AGTAAAAA	3.13E-05	1.97E-05	CACTFTPPCA1	Widely distributed cis-acting element
245	UD	CATGTCGG	3.70E-05	5.14E-05	LTRECOREATCOR15	Cold, dehydration responses
118	DS	GGTAAAA	3.96E-05	3.21E-04	GT1CONSENSUS	Stress and light responses
53	DS	GTATTTTA	4.51E-05	7.81E-06		
231	UD	GTCGGCAA	4.55E-05	1.84E-05	LTRECOREATCOR15, DRECRTCOREAT, CBFHV	Cold, dehydration responses
230	UD	AAGCCGCC	4.56E-05	1.26E-05	AGCBOXNPGLB, GCCCORE	Ethylene and other responses
244	UD	CCGACAC	5.00E-05	5.06E-05	LTRECOREATCOR15	Cold, dehydration responses
153	DU	AATCATGG	5.62E-05	7.96E-06	ARR1AT	Cytokinin response
261	UD	CGAGGATA	5.69E-05	9.10E-05	GATABOX, MYBST1	Light and MYB responses
63	DS	CTAGTCGC	6.79E-05	1.95E-05		
265	UD	ACCGT	7.03E-05	1.13E-04		
258	UD	CACGCCA	7.81E-05	7.29E-05		
21	DS	ATATAAT	8.23E-05	9.13E-10		
207	SU	CGTAC	8.24E-05	8.29E-04	CURECORECR	SBP response
144	DU	AAATATTT	8.25E-05	1.61E-07	ROOTMOTIFTAPOX1	Starch degradation gene expression
64	DS	AAATAATA	8.63E-05	1.98E-05	POLASIG3, TATABOX5	Polyadenylation-like and TATA motifs
12	DD	AACGAC	8.76E-05	1.96E-04		Auxin response
105	DS	GTTAAAAA	9.57E-05	1.58E-04		
263	UD	GCCGCCC	9.64E-05	9.74E-05	GCCCORE	Ethylene and other responses
96	DS	ACCGCACG	1.03E-04	1.33E-04		
108	DS	AAACTTTA	1.16E-04	1.68E-04	DOFCOREZM, NTBBF1ARROLB, TAAAGSTKST1	Auxin and Dof responses

(Continued)

TABLE 1 | Continued

Motif no.	Pattern	Sequence motif	Seq *p*-value	Motif *p*-value	Place sites and other motifs	Proposed functions
243	UD	AGGACACG	1.18E-04	5.01E-05		
155	DU	TAATAAAA	1.20E-04	1.01E-05	POLASIG1	Polyadenylation-like motif
250	UD	ACGTGTC	1.23E-04	6.25E-05	ACGTABREMOTIFA2OSEM, ABRELATERD1, ACGTATERD1, GADOWNAT	ABA, GA, and light responses
59	DS	ATAAAAAT	1.29E-04	1.12E-05	SEF4MOTIFGM7S	Widely distributed *cis*-acting element
114	DS	CCGCACG	1.30E-04	2.33E-04		
181	DU	ACGTGAT	1.34E-04	1.79E-04	GTGANTG10, ABRELATERD1, ACGTATERD1, RHERPATEXPA7	ABA, cytokinin, and light responses

The motif number can be used to find additional motif information in Supplementary Data File 1. 'Pattern' is the gene expression pattern, where the first letter indicates the direction of change from paradormancy to endodormancy (U = up-regulated, D = down-regulated, S = same or no change), and the second letter indicates the direction of change from endodormancy to ecodormancy. For example, a pattern of 'DS' indicates that the gene was down-regulated from paradormancy to endodormancy, and then did not change from endodormancy to ecodormancy. 'Seq p-value' is the random probability of finding the observed number of genes with at least one sequence motif. 'Motif p-value' is the random probability of finding the observed mean number of motifs per sequence. PLACE sites are the names of related motifs based on a search of the Database of Plant Cis-acting Regulatory DNA Elements (http://www.dna.affrc.go.jp/PLACE/index.html). Non-PLACE site names are designated with an asterisk.

DISCUSSION

Rationale for the Classification of Dormancy Treatments

We classified the monthly time points into three dormancy treatments based on previous research on *P. trichocarpa* and other cottonwoods, and by grouping the monthly samples based on patterns of gene expression. The first sample was classified as 'paradormant' because black cottonwood shoots were still elongating on August 1. Furthermore, in eastern cottonwood, endodormancy is not evident until 2–3 weeks after SD-induced bud set, and does not peak until about 7 weeks after bud set (Howe et al., 1999). The induction of endodormancy seems to progress in a similar fashion in black cottonwood (Frewen et al., 2000; Chen et al., 2002). The November and December samples were classified as 'endodormant.' As noted above, SD-induced endodormancy peaked about 7 weeks after SD-induced bud set in eastern cottonwood, and was readily measurable for the next 4 weeks (Howe et al., 1999). This would place peak endodormancy somewhere between our November 1 and December 1 collection dates. Information on the release of endodormancy is also available from experiments on an F_2 population of hybrids between black cottonwood and eastern cottonwood (Chen et al., 2002). Based on these data and other research on balsam poplar (Farmer and Reinholt, 1986), we classified the February and March samples as 'ecodormant.'

Differential Expression among Dormancy States

During the transition from paradormancy to endodormancy, most of the differentially expressed genes ($n = 913$; 67%) were down-regulated. In contrast, during the transition from endodormancy to ecodormancy, the two largest groups of genes ($n = 513$ and 519; 38% each) were those that were either up-regulated or did not change. These patterns are consistent with the lower cell division and metabolic activity that occurs during

endodormancy. Below, we focus on genes that had opposite changes in expression between the two dormancy transitions ($n = 604$; 44%). That is, genes that were clearly expressed at either higher or lower levels during endodormancy compared to the other two dormancy states. Finally, the smaller number of genes whose expression was atypically higher during endodormancy were of particular interest ($n = 225$; 17%).

Differential Expression of Chromatin-associated Genes
Overview

Large-scale changes in chromatin are associated with plant developmental changes and responses to the environment. These include (1) covalent modifications to histones or DNA and (2) non-covalent remodeling of chromatin, including changes in nucleosome position or stability, and substitution of one histone type for another (Gentry and Hennig, 2014). Several transcriptomic, physiological, and genetic studies have implicated chromatin modifications and remodeling in the regulation of endodormancy (Ruttink et al., 2007; Horvath et al., 2008, 2010; Karlberg et al., 2010). We found additional support for this link, identifying differentially expressed gene sets associated with 'histone,' 'histone methylation,' 'chromatin remodeling,' 'DNA methylation,' 'cytosine methylation,' and 'maintenance of DNA methylation.' The expression of other gene sets and individual genes are discussed in more detail below.

RNA-directed DNA Methylation (RdDM)

We observed many changes in genes and gene sets associated with transcriptional gene silencing (TGS) via RNA-directed DNA methylation (RdDM). RdDM is a gene silencing process that is regulated by the methylation and demethylation of DNA at target loci. In general, RNA-DEPENDENT RNA POLYMERASE (RDR) copies single-stranded transcripts into double-stranded RNAs (dsRNAs) that are then processed by DICER-like (DCL) proteins into short interfering RNAs (siRNAs). These siRNAs subsequently associate with ARGONAUT (AGO) proteins to form RNA-induced silencing complexes (RISCs) that mediate

DNA methylation and TGS (Matzke and Mosher, 2014). In *Arabidopsis*, the 'canonical' RdDM pathway involves RDR2, DCL3, and AGO4 (Matzke and Mosher, 2014). Gene silencing is complicated because post-transcriptional gene silencing (PTGS) involves similar machinery, and the two processes (TGS and PTGS) interact (Matzke and Mosher, 2014). The genes typically associated with PTGS are RDR6, DCL2, DCL4, and AGO1 (Matzke and Mosher, 2014). After TGS is established, DNA methylation and gene silencing can be maintained by CMT3 (Matzke and Mosher, 2014). Differentially expressed genes and gene sets included genes that seem to encode RDR (RDR1, RDR6), DCL (DCL1, DCL2), ARGONAUT (AGO4, AGO7), and CHROMOMETHYLASE 3 (CMT3). Because of the complexity and similarities of the TGS and PTGS pathways, we did not attempt to link particular *Populus* genes and gene sets to each of these two gene silencing pathways (i.e., TGS versus PTGS). Endodormancy related changes in RdDM pathways have been observed in other perennial species. For example, a gene similar to *Arabidopsis AGO4* and a gene that may be functionally similar to *DCL4* were also down-regulated during endodormancy in leafy spurge (Horvath et al., 2008).

Histone Modifications

An assortment of gene sets involved in histone modifications were differentially expressed, including neighbors of 'histone' and 'polycomb complex.' In *Arabidopsis*, the Polycomb Repressive Complex 2 (PRC2) participates in stable gene silencing. PRC2 methylates histone H3, resulting in the repression of gene expression (Kim et al., 2012). For example, PRC2 controls flowering via histone methylation of *FT* chromatin (Jiang et al., 2008) and represses *FLOWERING LOCUS C (FLC)* during vernalization (De Lucia et al., 2008). Other differentially expressed gene sets included neighbors of 'SWN,' 'FIE,' and 'PKL.' In *Arabidopsis*, *SWN* and *FIE* encode core components of PRC2 (Deng et al., 2013), whereas *PKL* encodes a DNA-binding helicase which seems to associate with PRC2 target loci to enhance histone modification (Zhang et al., 2012). Interestingly, aspen genes similar to *FIE* and *PKL* were up-regulated during SD-induced bud set (Ruttink et al., 2007). Although genes that presumably encode 'neighbors' of FIE and PKL were differentially expressed in our study, the genes themselves were not. That is, we did not see differential expression of two *PKL*-like genes (Potri.006G262200 and Potri.018G021100) and one *FIE*-like gene (Potri.001G417300) in our study.

'Binding partners of DDB1A' was another differentially expressed gene set that seems to be associated with gene silencing via PRC2. In *Arabidopsis,* DDB1A is a component of the CULLIN 4 (CUL4)/DDB1 ubiquitin ligase complex that functions in a wide array of plant processes, including flowering, photomorphogenesis, and parental imprinting (Hou et al., 2014). The CUL4/DDB1 complex seems to interact with histone tails to repress the transcription of genes involved in photomorphogenesis (Benvenuto et al., 2002), and an association between CUL4/DDB1A and PRC2 seems to regulate flowering time in *Arabidopsis* (Dumbliauskas et al., 2011; Pazhouhandeh et al., 2011). The differential expression of a 'DET1' gene set provides a specific link to endodormancy. The DET1 protein interacts with CONSTITUTIVE PHOTOMORPHOGENIC 10 (COP10) and the CUL4/DDB1 complex to regulate responses to light and temperature (Delker et al., 2014).

Gene activation and silencing also involve histone acetylation and deacetylation. In general, histone acetylation is associated with gene activation, whereas deacetylation is associated with gene silencing. Two differentially expressed gene sets were associated with histone deacetylases—HDA6 and HDA19 (also known as HD1). HDA6 is a histone deacetylase that has been identified as a component of the *Arabidopsis* RdDM machinery (To et al., 2011a). In particular, deacetylation of histone H3 seems to be important for the subsequent methylation of histone H3 described above (To et al., 2011a). In *Arabidopsis*, HDA6 is involved in the regulation of flowering, senescence, leaf development, the circadian clock, and responses to salt stress, ABA, and JA (Wu et al., 2008b; Chen et al., 2010; To et al., 2011b; Liu et al., 2014), whereas HDA19 regulates seed maturation and flower development (Liu et al., 2014). Other differentially expressed genes and gene sets are connected to these histone deacetylases. For example, JAZ proteins recruit HDA6 to inhibit JA signaling (Zhu et al., 2011).

One of the strongly differentially expressed genes (Potri.014G189400; DNG5) is a putative homolog of the vertebrate gene *MBD4 (METHYL-CPG-BINDING DOMAIN 4;* Ramiro-Merina et al., 2013). MBD proteins may recruit histone deacetylases such as HDA6—thereby acting as the 'bridges' between DNA methylation and histone deacetylation (Liu et al., 2012).

Finally, two *SPT*-like genes had atypical patterns of expression—being strongly up-regulated from paradormancy to endodormancy, and then down-regulated from endodormancy to ecodormancy. *Arabidopsis* SPT5-2 is part of the SPT4/SPT5 transcript elongation factor that seems to link transcription elongation, histone modification, and chromatin remodeling in yeast and *Arabidopsis* (Hartzog and Fu, 2013; Durr et al., 2014). Furthermore, an *Arabidopsis SPT5* homolog (*KTF1/RDM3/SPT5-like*) has been linked to AGO4-mediated gene silencing (Karlowski et al., 2010; Hartzog and Fu, 2013). A *Populus* gene similar to *Arabidopsis SPT6L* was also atypically expressed at higher levels during endodormancy—and *Arabidopsis SPT6L* seems to interact with AGO proteins to regulate embryo development (Gu et al., 2012). Thus, it is curious that a gene similar to *AGO4* (Potri.001G219700) was down-regulated during endodormancy in our study and in other plants (Horvath et al., 2008).

Other Chromatin-associated Genes

Three other genes were clearly expressed at lower levels during endodormancy. The first gene, Potri.004G087500, is similar to *Arabidopsis HMGA*. HMGA proteins interact with A/T-rich DNA, altering the chromatin structure and transcription of their target genes (Reeves, 2010). The second gene, Potri.008G155400, is similar to *Arabidopsis SILENCING DEFECTIVE 3 (SDE3)*, which is more clearly associated with PTGS. The third gene, Potri.T029800, is similar to an *Arabidopsis* gene (*ALPHA/BETA*

HYDROLASE F4; *ABHF4*) that is associated with the alpha/beta-hydrolase superfamily of proteins with unknown function.

Roles of Chromatin-associated Genes in Endodormancy Induction and Release

Dormancy transitions were accompanied by changes in multiple genes associated with DNA methylation (e.g., via RdDM) and histone modifications (e.g., via PRC2). However, although we expected to see increased expression of TGS components during endodormancy—the opposite was true—most chromatin-associated genes and gene sets were expressed at lower levels during endodormancy. One explanation is that reduced gene silencing may activate genes that positively induce and maintain endodormancy. *DAM* genes, for example, which seem to participate in the induction and maintenance of endodormancy, show reduced histone methylation (Horvath et al., 2008; Leida et al., 2012). Second, reduced expression of chromatin-associated genes may simply reflect the lower cell division and metabolic activity that occurs during endodormancy. Most of the other non-chromatin-associated genes were also expressed at lower levels during endodormancy. Third, subtle changes in dormancy-specific TGS may have been swamped by other processes. We compared paradormant buds (not actively growing meristems) to endodormant buds, and only subtle differences in the complement of active and silenced genes may exist between these two dormant states. Finally, the transition from paradormancy to endodormancy may involve a transient increase in gene silencing activity—an increase that we missed with our 1–2 month sampling interval.

Differential Expression of Transcription Factor Genes

Ethylene-associated Transcription Factors

We found substantial evidence for the differential expression of genes associated with ethylene responses, including transcription factors. Gene sets associated with EIN3 and EIL1 (EIN3-LIKE 1) were expressed at higher levels during endodormancy. The EIN3/EIL1 transcription factors act downstream of the signaling protein EIN2 to positively regulate the ethylene response pathway, including leaf senescence (Kim et al., 2014). The induction of ethylene responses during endodormancy is also supported by the differential expression of gene sets associated with the ETHYLENE-RESPONSIVE ELEMENT BINDING PROTEIN (EREBP), EIN2 membrane protein, EIN4 ethylene receptor—as well as individual genes that encode ETHYLENE RESPONSE FACTOR (ERF) proteins, which belong to the APETALA 2 (AP2)/EREBP family of transcription factors. Five of the six ERFs described in **Figure 4** seemed to be expressed at higher levels during endodormancy. Although the details differ, ERF genes have also been implicated in bud dormancy in hybrid aspen and Japanese apricot (Rohde et al., 2007; Zhong et al., 2013). Kim et al. (2014) proposed that AtNAP and other 'senescence-associated' NAC transcription factors act downstream of EIN2, and genes that seem to encode AtNAP and other NAC proteins were differentially expressed (**Figure 4**). Gene sets associated with EIN2, an ethylene signaling component, and EIN4, an ethylene receptor, were also expressed

at higher levels during endodormancy. Changes in ethylene-associated transcription factors and other genes support broader physiological evidence that ethylene has an important functional role in bud dormancy (Ruonala et al., 2006; Rohde et al., 2007). Finally, our results suggest that JA interacts with ethylene to regulate bud dormancy. This is in agreement with indications of JA signaling during dormancy transitions observed in leafy spurge and Japanese apricot (Horvath et al., 2008; Zhong et al., 2013). In *Arabidopsis*, EIN3/EIL is a 'key integration node' that integrates signaling by ET and JA (Zhu et al., 2011), and gene sets associated with CORONATINE INSENSITIVE 1 (COI1) and JASMONATE ZIM-DOMAIN 10 (JAZ10) were differentially expressed. COI1 is a JA receptor and JAZ10 is a transcriptional repressor (Zhu et al., 2011). Furthermore, HDA6, a histone deacetylase involved in gene silencing (discussed above), interacts with JAZ proteins and COI1 to repress EIN3/EIL1-mediated transcription and JA signaling (Zhu et al., 2011).

WRKY DNA-binding Domain Transcription Factors

WRKY transcription factors, which contain the WRKY DNA-binding domain, have been described as 'major hubs' in abiotic stress signaling (Tripathi et al., 2014). Several genes associated with WRKY transcription factors were differentially expressed. These include gene sets associated with *Arabidopsis* WRKY, WRKY33, and WRKY70 (Supplementary Tables S5 and S6), and individual genes similar to *Arabidopsis WRKY5*, *WRKY6*, *WRKY27*, *WRKY40*, and *WRKY33* (**Figure 4**; Supplementary Data File 1). There seem to be about 100 WRKY genes in *P. trichocarpa*, many of which are induced by SA, JA, cold, drought, salinity, or wounding (He et al., 2012; Jiang et al., 2014). Thus, the WRKY transcription factors provide one potential link between dormancy-associated gene expression and the phytohormones JA and SA (discussed below).

Cold-responsive Transcription Factors

Another gene set that was strongly associated with endodormancy was 'Neighbors of RHL41' (ZINC FINGER PROTEIN 12, ZAT12). ZAT12 is one of the transcription factors induced very quickly after exposure to cold temperatures (Vogel et al., 2005; Park et al., 2015). Other genes encoding 'first wave' transcription factors were differentially expressed in our study as well (e.g., *WRKY33* and *ZAT10/STZ*; **Figure 4**), but others were not, including genes encoding the C-REPEAT BINDING FACTORS (CBFs). Induction of *ZAT12* leads to the induction of some cold-responsive (COR) genes, and the repression of others; and overexpression of *ZAT12* leads to enhanced freezing tolerance (Vogel et al., 2005). A gene that seems to encode a ZAT10/STZ transcription factor was expressed at higher levels during endodormancy (**Figure 4**). In *Arabidopsis*, the expression of *ZAT10/STZ* is regulated by cold, drought, and salt; and overexpression of *ZAT10/STZ* enhances drought tolerance (Sakamoto et al., 2004). Although *CBF* genes are clearly involved in acclimation to cold and drought (Thomashow, 2010), *CBF*-like genes were not differentially expressed in our study. Nonetheless, CBF binding motifs were significantly overrepresented in the promoters of genes that were up-regulated during endodormancy (discussed below). In a similar study of aspen, none of the four

CBF-like genes was up-regulated during 5 weeks of SDs (Karlberg et al., 2010). Although another CBF-like gene was up-regulated for a short time, it was then down-regulated to almost the original level during the endodormant period. In our study, we may have missed transient increases in CBF gene expression because of the 1–2-month sampling interval we used.

Dormancy-associated Transcription Factors

Genes encoding other transcription factors were expected to be differentially expressed. For example, DORMANCY ASSOCIATED MADS-BOX (DAM) genes are putative transcription factors found in perennial plants that have been directly linked to vegetative endodormancy. They are similar to two genes, SHORT VEGETATIVE PHASE (SVP) and AGAMOUS-LIKE 24 (AGL24), that encode transcription factors regulating flowering time in Arabidopsis. In peach (Prunus persica), a deletion of DAM genes resulted in trees that were unable to become endodormant, and DAM expression is enhanced during endodormancy in several perennial species (Horvath et al., 2010; Jiménez et al., 2010). In our study, several DAM-like (SVP-like) genes were differentially expressed (Potri.005G155700, Potri.017G044500, and Potri.002G105600), but they were down-regulated during endodormancy, unlike the DAM genes in leafy spurge and peach (Horvath et al., 2010; Jiménez et al., 2010). A different DAM-like gene (Potri.007G010800) was up-regulated during the induction of endodormancy in hybrid aspen (Ruttink et al., 2007), and strongly down-regulated in early flushing trees that were overexpressing EARLY BUD-BREAK 1 (EBB1; Yordanov et al., 2014). However, this gene was not differentially expressed in our study. Finally, one gene (Potri.001G328400) was highly up-regulated in our December and February samples—but these differences were not significant among months or dormancy states (FDR p-value = 0.07 to 0.11). This gene encodes an unusual truncated transcript, which is reminiscent of the truncated splice variant of the endodormancy-induced DAM transcript in leafy spurge (Horvath et al., 2013). Overall, these disparate results provide limited insight into the connections between endodormancy and DAM-like genes in Populus.

Other Transcription Factors

Other gene sets associated with transcription factors were strongly down-regulated during endodormancy. JLO, SEU, RPL, and ARF2 seem to have various roles in auxin signaling, including organization of the shoot apical meristem and organ development (Franks et al., 2006; Sluis and Hake, 2015). This suggests they could be involved in the formation or development of new leaf primordia. If so, their patterns of expression (i.e., lower expression during endodormancy) are consistent with the cessation of primordia initiation and development that occurs during dormancy induction.

Genes that seem to encode MYB transcription factors were also common among the genes that were down-regulated from paradormancy to endodormancy (Figure 4). Given that the MYB family is very large, and endodormancy is associated with a general reduction in metabolic activity, the significance of these changes is uncertain. Nonetheless, MYB14 represses the CBF

regulon in Arabidopsis, (Chen et al., 2013), but we did not see differential expression of the Populus CBF/DREB genes in our study (discussed above).

Finally, genes encoding other flowering-associated transcription factors were also differentially expressed, including SQUAMOSA PROMOTER BINDING PROTEIN-LIKE (SPL) and SUPPRESSOR OF OVEREXPRESSION OF CONSTANS 1 (SOC1). These and other flowering-associated genes are discussed in more detail below.

Differential Expression of Phytohormone-associated Genes
Auxin-associated Gene Expression

Auxin-associated genes were generally down-regulated during endodormancy, with most changes expected to lead to a reduction in auxin signaling. This, and other more specific changes in gene expression, suggests that auxin signaling undergoes important changes during dormancy transitions. The first step in auxin signaling involves an interaction between auxin, an auxin receptor such as TIR1, and AUXIN/INDOLE 3-ACETIC ACID (Aux/IAA) proteins. This interaction ultimately leads to the degradation of the Aux/IAA proteins, which normally repress ARF transcription factors. The reduction in Aux/IAA leads to enhanced transcription of auxin-inducible genes by ARF and downstream auxin responses (Korasick et al., 2014). In our study, genes that encode binding partners of TIR1 were down-regulated from paradormancy to endodormancy, and a gene that was strongly up-regulated (Potri.010G078400) is similar to the Arabidopsis gene that encodes IAA4. This latter change, in particular, is consistent with a reduction in auxin signaling and auxin responses during endodormancy. Furthermore, gene sets associated with three ARF transcription factors were differentially expressed between dormancy states. In Arabidopsis, ARFs are also negatively regulated by miRNAs, including miR160 (Rhoades et al., 2002; Paponov et al., 2009). In our study, the gene set associated with miR160A was significantly down-regulated from paradormancy to endodormancy (Supplementary Table S5). Likewise, genes that encode neighbors and targets of miR393A and miR393B were differentially expressed between paradormancy to endodormancy. This miRNA seems to negatively regulate the gene encoding TIR1 (Liu et al., 2009). Corresponding changes in auxin and auxin-associated gene expression have been found in other species. In silver birch, auxin declined during SD-induced endodormancy (Li et al., 2003), and auxin-associated genes were down-regulated during endodormancy in the cambial meristem of Populus (Baba et al., 2011) and the buds of leafy spurge (Horvath et al., 2008). Our results concur, and because of its atypical pattern of expression, point to a particularly important role for the IAA4-like gene (Potri.010G078400).

Down-regulation of auxin transport also seems to occur during endodormancy. For example, genes similar to two Arabidopsis genes involved in auxin transport were down-regulated from paradormancy to endodormancy. The first gene (Potri.002G087000) is similar to a gene that encodes the auxin influx carrier, LAX3 (LIKE AUX3). The second (Potri.006G123900) is similar to a gene that encodes an ATP-BINDING CASSETTE (ABC) transporter that regulates basipetal auxin transport. Consistent with these changes, two other genes (Potri.012G047200 and Potri.016G035300) were up-regulated from endodormancy to ecodormancy, both of which are similar to the *Arabidopsis PIN1* (*PIN-FORMED 1*), which encodes a putative auxin efflux transporter (Sluis and Hake, 2015).

Finally, changes in the expression of genes associated with the synthesis of phenylpropanoids and flavonoids may affect auxin responses. As discussed below, these genes were mostly down-regulated during endodormancy, which could *enhance* auxin transport, but also destabilize auxin levels by increasing auxin oxidation (Brown et al., 2001; Buer and Muday, 2004; Peer et al., 2013).

Ethylene-associated Gene Expression

The ethylene-associated gene set was up-regulated during endodormancy, which was opposite from what we observed for auxin and BR. However, this gene set included a mix of genes with positive and negative effects on ethylene signaling. For example, genes that seem to encode negative (CTR1) and positive (ERF5) regulators of ethylene signaling were both up-regulated during endodormancy. Although this seems counterintuitive, genes similar to *CRT1* and *ERF* are induced by exogenous ethylene, demonstrating some degree of coordinate regulation (Vahala et al., 2013; Zou et al., 2014). Other related gene sets were up-regulated during endodormancy, including those associated with the ethylene receptor EIN4 and the transcription factors EIN2 and EIN3. Up-regulation was also observed for multiple genes that seem to encode ERF transcription factors, at least some of which have been specifically associated with responses to ethylene. Ethylene has been implicated in bud endodormancy, perhaps in concert with ABA. Ruttink et al. (2007), for example, concluded that ethylene and ABA act sequentially during SD-induced bud dormancy in *Populus*. Although they emphasized the transient nature of increases in ethylene-associated gene expression, we observed up-regulation of multiple ethylene-associated genes over a two-month period under natural conditions. Interacting roles for ethylene and ABA during endodormancy induction and release have also been reported in birch and grape, which is consistent with our results (Ruonala et al., 2006; Ophir et al., 2009; Zheng et al., 2015).

GA- and ABA-associated Gene Expression

Perhaps the most consistent associations between bud endodormancy and phytohormones are the opposing changes in GA and ABA—GA levels tend to decline and ABA levels increase early in the induction of endodormancy (reviewed in Olsen, 2010). For example, SD-induced down-regulation of GA-20-oxidase reduces the levels of active GAs in *Salix* and *Populus* (reviewed in Olsen, 2010), and up-regulation of GA-2-oxidases may lead to GA inactivation (Zawaski and Busov, 2014). However, we did not observe differential expression of any genes that seem to encode GA-20-oxidase or GA-2-oxidase, and the changes in other genes associated with GA and ABA were contrary to expectations. It is possible that

transient changes in GA- and ABA-related gene expression were missed due to our monthly sampling scheme. Most previous analyses focused on the early stages of dormancy induction, often focusing on changes in response to SDs. Our results suggest that GA and ABA have (at most) modest roles in the maintenance of endodormancy *per se*. For example, *Populus* trees genetically engineered to underexpress or overexpress the *ABA INSENSITIVE 3* (*ABI3*) transcription factor still became endodormant under SDs (Ruttink et al., 2007), and other studies suggest that ABA is primarily involved in the formation of the dormant bud, not in the maintenance of endodormancy (reviewed in Olsen, 2010).

Brassinosteroid-associated Gene Expression

Brassinosteroid has non-redundant roles as a "potent" growth-promoting phytohormone (Schröder et al., 2014). Although clear involvement in bud endodormancy has not been previously established, changes in the expression BR-related genes point in this direction. First, the BR hormone-associated gene set, 'Binding partners of BES1,' was down-regulated during endodormancy. BES1 is a BR-regulated transcription factor, and low BES1 signaling helps ensure stem cell quiescence in plants (Vilarrasa-Blasi et al., 2014). Second, three genes expected to enhance BR signaling were all down-regulated during endodormancy (i.e., compared to paradormancy or ecodormancy). Two of these genes are similar to an *Arabidopsis* gene (*CBB1*) that encodes a sterol reductase involved in the early steps of BR synthesis (Choe et al., 1999). The third gene is similar to *BRASSINOSTEROID SIGNALING KINASE 2* (*BSK2*), which belongs to a family of genes that encode positive regulators of BR signaling (Sreeramulu et al., 2013). Thus, it appears that down-regulation of BR biosynthetic and signaling genes may help maintain the reduced cell division associated with endodormancy. Because of crosstalk between BR and SA signaling (Divi et al., 2010), there may be a link between the changes in BR-associated genes described above, and the changes in SA associated genes described below.

Salicylic-acid-associated Gene Expression

During the transition from paradormancy to endodormancy, we found clear evidence for changes in the expression SA-associated genes. Genes with the strongest patterns of differential expression include three genes similar to *Arabidopsis PAL1*, which encodes a key enzyme in SA biosynthesis. All three genes were down-regulated from paradormancy to endodormancy. Although connections between SA and bud endodormancy have not been widely reported, SA promoted endodormancy release in grape (i.e., as measured by H_2O_2 production; Or et al., 2000). Nonetheless, changes in SA-associated genes may be more closely related to general changes in the phenylpropanoid pathway, rather than endodormancy *per se*—such as changes in gene sets associated with anthocyanin and flavonoid biosynthesis (Nugroho et al., 2002). Although genes associated with proanthocyanin production have been associated with seed dormancy (Debeaujon et al., 2000), corresponding gene sets were mostly down-regulated in our study, including gene sets associated with GLABRA 3, ENHANCER OF GLABRA 3,

TRANSPARENT TESTA 2 (TT2), TT8, TRANSPARENT TESTA GLABRA 1 (TTG1), and CAPRICE (Supplementary Tables S5 and S6). Although SA may contribute to endodormancy, the changes in phenylpropanoid and SA gene expression may have no direct role in endodormancy.

Jasmonic-acid-associated Gene Expression

We found consistent evidence for down-regulation of many JA-associated genes and gene sets from paradormancy to endodormancy, including genes involved in JA synthesis and signaling. Because enhanced JA signaling is associated with responses to stress and decreased growth, many of these changes contradict the expected roles that JA might have during endodormancy. Given that these changes coincide with broader reductions in gene expression during endodormancy, their relevance is uncertain. Nonetheless, because JAZ proteins repress JA signaling, the down-regulation of a gene similar to *JAZ12* (Potri.018G047100) could lead to increased JA-repressed growth during endodormancy. Finally, because JA regulates anthocyanin accumulation through COI, down-regulation of JA signaling could contribute to the reduction in in phenylpropanoid and SA-associated gene expression described above. Additional efforts are needed to identify the mechanisms underlying the seemingly contradictory responses in JA signaling and synthesis.

Cytokinin-associated Gene Expression

We found limited evidence to suggest that changes in cytokinin (CK) gene expression are important during endodormancy induction and release. Although one gene set, 'Neighbors of cytokinin,' and three individual genes were all down-regulated from paradormancy to endodormancy, these changes mirror the broader reductions in gene expression that occurred during endodormancy. Therefore, their relevance is uncertain. On balance, however, these changes are consistent with the hypothesis that cytokinin acts as an antagonist of auxin-mediated apical dominance by promoting the outgrowth of paradormant buds (Mueller and Leyser, 2011).

Flowering Genes and Processes are Associated with Vegetative Bud Dormancy

We previously presented a general model for endodormancy involving *FT*-based regulatory networks analogous to the networks that regulate flowering (reviewed by Horvath, 2009). Consistent with this model, our current results suggest an important role for genes similar to *SPL* in *Populus*. In *Populus*, *FT2* inhibits growth cessation and bud set. Thus, down-regulation of *FT2* seems to be an important early step in the induction of endodormancy (Bohlenius et al., 2006; Ruonala et al., 2008; Hsu et al., 2011). In *Arabidopsis*, photoperiodic regulation of *FT* involves a network of *miR156*, SPL proteins, *miR172*, and a set of AP2-like transcription factors, including AP2, TOE1-3 (TARGET OF EARLY ACTIVATION TAGGED 1-3), SCHLAFMÜTZE (SMZ), and SCHNARCHZAPFEN (SNZ; Aukerman and Sakai, 2003; Schmid et al., 2003; Jung et al., 2007). A model has emerged in which SPLs positively regulate *miR172*, which normally represses AP2, TOEs, SMZ, and SNZ (Wellmer

and Riechmann, 2010). Thus, up-regulation of SPLs represses these AP2-like transcription factors, leading to an increase in *FT* expression and the promotion of flowering. In our study, eight *SPL*-like genes were down-regulated from paradormancy to endodormancy. By analogy to the flowering pathway described above, down-regulation of *SPL* genes should lead to down-regulation of *miR172*, up-regulation of genes similar to *AP2*, *TOE*, *SMZ*, and *SNZ*, repression of *FT2*, and ultimately, growth cessation, bud set, and endodormancy. However, despite the consistent down-regulation of *SPL*-like genes in our study, two other observations contradict this simple model. First, three *DAM*-like (*SVP*-like) genes were unexpectedly down-regulated in our study, which is opposite of what has been seen in other perennial plants (discussed above). Furthermore, because SVP negatively regulates *miR172* in *Arabidopsis* (Cho et al., 2012), down-regulation of *DAM/SVP* genes in *Populus* is ultimately expected to increase, rather than decrease, the expression of *FT2*. However, SVP is not uniformly up-regulated during flower development. In *Arabidopsis*, *SVP* is repressed in early flower development to prevent flower reversion, and in late flower development to allow the activation of *SEPALATA3* (Lee and Lee, 2010). Therefore, our results may indicate that the timing of *DAM/SVP* expression is important in *Populus* as well. Second, *FT2* itself was not differentially expressed. However, this was not surprising because *FT2* expression decreases dramatically after only a few SD (Resman et al., 2010). Therefore, the longer-term changes in *SPL* gene expression that we observed may help keep the expression of *FT2* and other flowering-related genes at already low levels, rather than being the direct, early cause of *FT2* down-regulation.

Like *FT*, *SOC1* (also known as *AGL20*) is considered a major integrator of flowering signals in *Arabidopsis*. In our study, two *SOC1*-like genes and the gene set "Binding partners of AGL20" were down-regulated from paradormancy to endodormancy, and similar results were observed in other studies of *Populus* (Ruttink et al., 2007) and leafy spurge (Horvath et al., 2008; data not shown). *SOC1* expression generally promotes flowering in *Arabidopsis*. More specifically, studies in *Arabidopsis* (Tao et al., 2012) show that SOC1 directly down-regulates *AP2*, *TOE1*, and *SMZ*, which are downstream targets of *SPLs* and *miR172* (described above). Finally, allelic variation in a *SOC1* homolog has been linked to dormancy in apricot (Trainin et al., 2013). Given the functional and regulatory similarities between *DAM* and *FLC*, a repressor of *SOC1*, a similar mechanism involving *SOC1*-like genes might regulate vegetative bud endodormancy.

Despite the potential connections between endodormancy and flowering-like genes, the genes described above have been implicated in diverse processes, including endodormancy induction (e.g., photoperiodism), endodormancy release (e.g., chilling), cold acclimation, flowering, and fruit development (Preston and Sandve, 2013). For example, in *Arabidopsis*, there seems to be a feedback loop between cold responses and flowering-time that involves interactions between CBF, SOC1, and FLC (Seo et al., 2009). Therefore, it will be challenging to dissect the functional significance of dormancy-associated changes in expression, particularly for genes that serve as key regulatory hubs.

Differential Expression of Genes Associated with Bud Set QTL

Rohde et al. (2011b) identified six robust QTL associated with various components of bud set, and then compared the QTL locations to the locations of differentially expressed genes (Ruttink et al., 2007). We conducted a similar analysis using a larger number of genes and updated gene models (*Populus* v3.0). Although hundreds of genes were associated with each QTL, the number of differentially expressed genes ranged from 9 to 24. We then examined relationships between the QTL and the various classes of genes described above (chromatin-associated, transcription factor, and hormone-associated) to identify promising QTL candidates.

We found 14 differentially expressed genes that are reasonable QTL candidates based on their differential expression and putative functions. These include four genes that seem to be involved in chromatin remodeling (e.g., *DCL4*). Other differentially expressed genes seem to be involved in BR synthesis (*CBB1*) or phytohormone signaling via ABA (*PYL*), JA (*JAZ12*), or auxin (*IAA13*). Additional genes are more generally associated with responses to far-red light (*FAR-RED IMPAIRED RESPONSIVE 1*), organization of lateral organ boundaries (*LOB DOMAIN-CONTAINING PROTEIN 21*), and mitotic arrest (*MAD2*). Finally, we also found genes that seem to encode a NAC domain transcription factor associated with leaf senescence, and other PHD and bHLH transcription factors that are not as well characterized. In addition to these candidates, a few other differentially expressed genes with unknown functions should be considered.

Overall, these results suggest that differential expression can be used to reduce a large number of positional candidate genes (>2,000) to a much smaller set of plausible QTL candidate genes. Nonetheless, it would still be challenging to investigate all of these candidates in detail using functional genomics. For example, the genes underlying these QTL may not be differentially expressed, or the changes in expression may be transient. Furthermore, transgenic functional approaches often cause major, poorly timed perturbations that may lead to responses that do not accurately reflect natural gene functions. A combination of approaches including fine-scale mapping, association genetics, analyses of gene expression, and subtle gene perturbations will be needed to understand the roles of the numerous genes that appear to regulate dormancy transitions.

Upstream Sequence Motifs are Associated with Specific Patterns of Gene Expression

Motifs associated with photoperiodic responses and circadian patterns of gene expression were highly enriched in some gene expression pattern groups. For example, two of the top-ranked motifs match binding sites for two central transcription factors that regulate the circadian clock, LATE ELONGATED HYPOCOTYL (LHY) and CIRCADIAN CLOCK ASSOCIATED 1 (CCA1). The first motif matches the EVENING ELEMENT-LIKE (EEL) motif (AATATCT). The EEL and the EVENING ELEMENT itself (EE, AAAATATCT) are important regulators

of circadian clock and cold-responsive genes (Mikkelsen and Thomashow, 2009). The EEL motif was our top-ranked motif, being significantly overrepresented in genes that were up-regulated during endodormancy. As summarized by Hsu and Harmer (2014), most clock components regulate the transcription of genes that contain EE, or are regulated by other clock components through EE in their own promoters. The second motif matches a binding site (AAAAATCA) that is found in the target genes of CCA1 (Maxwell et al., 2003), being enriched in genes that were down-regulated during endodormancy.

Other motifs previously associated with circadian patterns of gene expression were also found, including G-box (CACGTG), I-box core (GATAA), PIF-binding E-box (PBE-box; CACATG), and the CAB2 DET1-ASSOCIATED FACTOR 1 binding site motif (CDA-1; CAAAA). Both the G-box and PBE-box elements are binding sites for PIF3, a transcription factor that interacts with phytochromes A and B (Zhang et al., 2013). Furthermore, the HORMONE UP AT DAWN (HUD) element, which has the same sequence as the PBE-box, is overrepresented in the promoters of phytohormone genes (Michael et al., 2008). We also found many motifs that matched the circadian elements described in Table 1 of Smieszek et al. (2014), only some of which were identified using the PLACE database. Photoperiodic and circadian regulation of endodormancy has been well documented (Howe et al., 1996; Rohde and Bhalerao, 2007; Horvath et al., 2010), and was also highlighted by our GSEA. Gene sets associated with two components of the circadian clock, ZEITLUPE (ZTL) and LIGHT-REGULATED WD 2 (LWD2), were up-regulated during endodormancy. ZTL is a blue light photoreceptor that regulates photoperiodic responses (Kim et al., 2007), and LWD2 encodes a WD (tryptophan and aspartate) protein that contributes to clock function (Wu et al., 2008a).

A second class of overrepresented motifs was broadly associated with responses to cold, dehydration, and ABA. The most prominent of these is the RCCGAC motif, which is the core of the C-repeat (CRT) element, also known as the dehydration responsive element (DRE). The CRT/DRE element is the binding site for CBFs (C-REPEAT BINDING FACTORS), some of the most important transcription factors involved in cold-induced gene expression (Benedict et al., 2006). CBFs regulate gene expression in response to cold and dehydration. Other transcription factors, such as the ABA RESPONSIVE ELEMENT BINDING PROTEIN (AREB), regulate gene expression in response to cold and dehydration in an ABA-dependent manner. These transcription factors bind to ABA responsive elements (ABRE; ACGT core), ABRE-like elements (ABREL; AGCTG), and G-box elements (Mikkelsen and Thomashow, 2009). We found many enriched motifs that contain these core sequences.

We found a number of dehydrin-like genes that were up-regulated from paradormancy to endodormancy, and these genes provide good models for how different combinations of *cis*-regulatory elements can lead to different patterns of gene expression. For example, Zolotarov and Stromvik (2015) identified 14 conserved motifs in 350 dehydrin promoters from 51 plant genomes, many of which were similar to motifs that were overrepresented in our differentially expressed genes, as well as in genes that have been associated with responses to cold, dehydration, ABA, and light.

We also found motifs associated with other phytohormones, including ethylene, auxin, SA, and JA. Among the top-ranked motifs, we found four with the GCC-box core motif (GCCGCC), which serves as a binding site for ethylene-responsive genes (Ohme-Takagi and Shinshi, 1995). These motifs were highly enriched in genes that were up-regulated from paradormancy to endodormancy. This coincides with our GSEA results indicating that ethylene is an important regulator of endodormancy. We also found five motifs that matched two auxin responsive elements (NTBBF1ARROLB and TGA-element; **Table 1**), and all five motifs were enriched in genes that were down-regulated from paradormancy to endodormancy. Again, this supports our GSEA results indicating that auxin related genes were mostly down-regulated from paradormancy to endodormancy. We found a large number of motifs that are associated with responses to cytokinin, but their significance is unclear; we saw no strong trends in cytokinin-related gene expression. The longest matching PLACE motif was CPBCSPOR (TATTAG), which exhibits cytokinin-enhanced protein binding, but the other two PLACE motifs, ARR1AT (NGATT) and RHERPATEXPA7 (KCACGW), are much less specific. We also found a small number of motifs that have been non-specifically associated with JA and SA.

In sum, our analyses of promoter motifs showed clear associations between patterns of endodormancy-related gene expression and two broad classes of genes—those associated the circadian clock and photoperiodic responses, and those associated with phytohormone-mediated responses to cold and dehydration. An understanding of the finer details of gene regulation are complicated by the fact that many of the consensus motifs are short and widely distributed among plant promoters involved in responses to light, biotic and abiotic stresses, and phytohormones. Furthermore, about 45% of the enriched motifs had no assigned functions, suggesting that more work is needed to understand the functions of these motifs and their potential roles in the regulation endodormancy-associated processes. Further insights could be gained by analyses that focus on understanding how the numbers, distributions, and combinations of motifs are associated with genes known to have specific patterns of gene expression across *Populus* species.

CONCLUSION

Our work highlights both the conserved nature and the extraordinary complexity of transcriptome changes associated with vegetative dormancy. For example, we confirmed and elaborated upon earlier evidence from studies of chromatin remodeling. We found multiple genes associated with DNA methylation (e.g., via RdDM) and histone modifications (e.g., via PRC2) that were differentially expressed during the induction and release of endodormancy. We identified 19 chromatin-associated genes that were down-regulated during endodormancy, and two genes that were strongly and atypically up-regulated. These

latter two genes are similar to *Arabidopsis SPT5-2* and *SPT6L*, which encode proteins described as 'global' transcription factors. We also identified links to genes that regulate the onset of flowering, pointing to potentially important roles for genes similar to *SPL*, *DAM/SVP*, and *SOC1*. Differential expression of *SPL* genes corroborates earlier observations and implicates miRNA-associated regulatory pathways in the repression of *FT2* during endodormancy.

A number of surprises emerged from our analyses of phytohormone-related genes. Changes in genes encoding GA-20-oxidase and GA-2-oxidase were not observed, and changes in genes associated with ABA were contrary to expectations. Although we may have missed transient changes associated with short-day-induced bud set, these results suggest that these phytohormones have relatively narrow windows of action. In contrast, we saw clearer evidence for changes in the expression of genes associated with ethylene, auxin, BR, SA, and JA. For example, genes and gene sets that were atypically up-regulated during endodormancy included those associated with responses to ethylene (*EIN3, EBP, ERFs*), and a gene similar to *Arabidopsis IAA4*. However, genes associated with auxin, BR, SA, and JA were mostly down-regulated during endodormancy. Because of the general down-regulation of metabolic activity and gene expression during endodormancy, the biological significance of these changes warrants further study.

Other genes that were atypically up-regulated during endodormancy included those encoding transcription factors associated with responses to cold and other abiotic stresses (*ZAT10/STZ, ZAT12/RHL41, WRKY*), and a gene that seems to encode a trihelix transcription factor. The down-regulation of other transcription factor genes was consistent with changes known to accompany endodormancy. These include genes with various roles in auxin signaling, organization of the shoot apical meristem, and organ development.

We identified many novel and previously identified promoter motifs that appear to regulate these dormancy-associated changes in gene expression. The most common motifs were those associated with the circadian clock and others associated with responses to photoperiod, cold, dehydration, and ABA. Among the most common motifs were the EVENING ELEMENT-LIKE motif, a binding site found in genes targeted by CCA1, CBF-binding sites, and various ABA responsive elements.

Finally, we found many differentially expressed genes that were located near bud set QTL, some of which are clear candidates for having functional roles in the induction of endodormancy. These latter genes are potential targets for basic research and for manipulating dormancy-associated processes using molecular breeding and transgenic approaches. Additional gene expression, fine-scale mapping, functional, and population genetic studies should help elucidate the roles of the many genes and biological processes we identified.

ACKNOWLEDGMENT

We thank Scott Kolpak and Kori Ault for help with tables and figures.

REFERENCES

Altschul, S. F., Gish, W., Miller, W., Myers, E. W., and Lipman, D. J. (1990). Basic local alignment search tool. *J. Mol. Biol.* 215, 403–410. doi: 10.1016/s0022-2836(05)80360-2

Arora, R., Rowland, L. J., and Tanino, K. (2003). Induction and release of bud dormancy in woody perennials: a science comes of age. *HortScience* 38, 911–921.

Aukerman, M. J., and Sakai, H. (2003). Regulation of flowering time and floral organ identity by a microRNA and its APETALA2-like target genes. *Plant Cell* 15, 2730–2741. doi: 10.1105/Tpc.016238

Azeez, A., Miskolczi, P., Tylewicz, S., and Bhalerao, R. P. (2014). A tree ortholog of APETALA1 mediates photoperiodic control of seasonal growth. *Curr. Biol.* 24, 717–724. doi: 10.1016/j.cub.2014.02.037

Baba, K., Karlberg, A., Schmidt, J., Schrader, J., Hvidsten, T. R., Bako, L., et al. (2011). Activity-dormancy transition in the cambial meristem involves stage-specific modulation of auxin response in hybrid aspen. *Proc. Natl. Acad. Sci. U.S.A.* 108, 3418–3423. doi: 10.1073/pnas.1011506108

Benedict, C., Skinner, J. S., Meng, R., Chang, Y. J., Bhalerao, R., Huner, N. P. A., et al. (2006). The CBF1-dependent low temperature signalling pathway, regulon and increase in freeze tolerance are conserved in *Populus* spp. *Plant Cell Environ.* 29, 1259–1272. doi: 10.1111/j.1365-3040.2006.01505.x

Benvenuto, G., Formiggini, F., Laflamme, P., Malakhov, M., and Bowler, C. (2002). The photomorphogenesis regulator DET1 binds the amino-terminal tail of histone H2B in a nucleosome context. *Curr. Biol.* 12, 1529–1534. doi: 10.1016/S0960-9822(02)01105-3

Bielenberg, D. G., Wang, Y., Li, Z. G., Zhebentyayeva, T., Fan, S. H., Reighard, G. L., et al. (2008). Sequencing and annotation of the evergrowing locus in peach [*Prunus persica* (L.) Batsch] reveals a cluster of six MADS-box transcription factors as candidate genes for regulation of terminal bud formation. *Tree Genet. Genomes* 4, 495–507. doi: 10.1007/s11295-007-0126-9

Bohlenius, H., Huang, T., Charbonnel-Campaa, L., Brunner, A. M., Jansson, S., Strauss, S. H., et al. (2006). CO/FT regulatory module controls timing of flowering and seasonal growth cessation in trees. *Science* 312, 1040–1043. doi: 10.1126/science.1126038

Brown, D. E., Rashotte, A. M., Murphy, A. S., Normanly, J., Tague, B. W., Peer, W. A., et al. (2001). Flavonoids act as negative regulators of auxin transport *in vivo* in *Arabidopsis*. *Plant Physiol.* 126, 524–535. doi: 10.1104/Pp.126.2.524

Buer, C. S., and Muday, G. K. (2004). The *transparent testa4* mutation prevents flavonoid synthesis and alters auxin transport and the response of *Arabidopsis* roots to gravity and light. *Plant Cell* 16, 1191–1205. doi: 10.1105/Tpc.020313

Campbell, M. A., Gleichsner, A., Alsbury, R., Horvath, D., and Suttle, J. (2010). The sprout inhibitors chlorpropham and 1,4-dimethylnaphthalene elicit different transcriptional profiles and do not suppress growth through a prolongation of the dormant state. *Plant Mol. Biol.* 73, 181–189. doi: 10.1007/s11103-010-9607-6

Chen, L. T., Luo, M., Wang, Y. Y., and Wu, K. (2010). Involvement of *Arabidopsis* histone deacetylase HDA6 in ABA and salt stress response. *J. Exp. Bot.* 61, 3345–3353. doi: 10.1093/jxb/erq154

Chen, T. H. H., Howe, G. T., and Bradshaw, H. D. Jr. (2002). Molecular genetic analysis of dormancy-related traits in poplars. *Weed Sci.* 50, 232–240. doi: 10.2307/4046369

Chen, Y., Chen, Z. L., Kang, J. Q., Kang, D. M., Gu, H. Y., and Qin, G. J. (2013). AtMYB14 regulates cold tolerance in *Arabidopsis*. *Plant Mol. Biol. Rep.* 31, 87–97. doi: 10.1007/s11105-012-0481-z

Cho, H. J., Kim, J. J., Lee, J. H., Kim, W., Jung, J. H., Park, C. M., et al. (2012). SHORT VEGETATIVE PHASE (SVP) protein negatively regulates miR172 transcription via direct binding to the pri-miR172a promoter in *Arabidopsis*. *FEBS Lett.* 586, 2332–2337. doi: 10.1016/j.febslet.2012.05.035

Choe, S., Dilkes, B. P., Gregory, B. D., Ross, A. S., Yuan, H., Noguchi, T., et al. (1999). The *Arabidopsis dwarf1* mutant is defective in the conversion of 24-methylenecholesterol to campesterol in brassinosteroid biosynthesis. *Plant Physiol.* 119, 897–907. doi: 10.1104/pp.119.3.897

Debeaujon, I., Leon-Kloosterziel, K. M., and Koornneef, M. (2000). Influence of the testa on seed dormancy, germination, and longevity in *Arabidopsis*. *Plant Physiol.* 122, 403–413. doi: 10.1104/Pp.122.2.403

Delker, C., Sonntag, L., James, G. V., Janitza, P., Ibanez, C., Ziermann, H., et al. (2014). The DET1-COP1-HY5 pathway constitutes a multipurpose signaling module regulating plant photomorphogenesis and thermomorphogenesis. *Cell Rep.* 9, 1983–1989. doi: 10.1016/j.celrep.2014.11.043

De Lucia, F., Crevillen, P., Jones, A. M. E., Greb, T., and Dean, C. (2008). A PHD-polycomb repressive complex 2 triggers the epigenetic silencing of *FLC* during vernalization. *Proc. Natl. Acad. Sci. U.S.A.* 105, 16831–16836. doi: 10.1073/pnas.0808687105

Deng, W. W., Buzas, D. M., Ying, H., Robertson, M., Taylor, J., Peacock, W. J., et al. (2013). *Arabidopsis* polycomb repressive complex 2 binding sites contain putative GAGA factor binding motifs within coding regions of genes. *BMC Genomics* 14:593. doi: 10.1186/1471-2164-14-593

Divi, U. K., Rahman, T., and Krishna, P. (2010). Brassinosteroid-mediated stress tolerance in *Arabidopsis* shows interactions with abscisic acid, ethylene and salicylic acid pathways. *BMC Plant Biol.* 10:151. doi: 10.1186/1471-2229-10-151

Doğramacı, M., Horvath, D., Chao, W., Foley, M., Christoffers, M., and Anderson, J. (2010). Low temperatures impact dormancy status, flowering competence, and transcript profiles in crown buds of leafy spurge. *Plant Mol. Biol.* 73, 207–226. doi: 10.1007/s11103-010-9621-8

Dumbliauskas, E., Lechner, E., Jaciubek, M., Berr, A., Pazhouhandeh, M., Alioua, M., et al. (2011). The *Arabidopsis* CUL4-DDB1 complex interacts with MSI1 and is required to maintain MEDEA parental imprinting. *EMBO J.* 30, 731–743. doi: 10.1038/emboj.2010.359

Durr, J., Lolas, I. B., Sorensen, B. B., Schubert, V., Houben, A., Melzer, M., et al. (2014). The transcript elongation factor SPT4/SPT5 is involved in auxin-related gene expression in *Arabidopsis*. *Nucleic Acids Res.* 42, 4332–4347. doi: 10.1093/Nar/Gku096

Farmer, R. E., and Reinholt, R. W. (1986). Genetic-variation in dormancy relations of balsam poplar along a latitudinal transect in Northwestern Ontario. *Silvae Genet.* 35, 38–42.

Franks, R. G., Liu, Z. C., and Fischer, R. L. (2006). SEUSS and LEUNIG regulate cell proliferation, vascular development and organ polarity in *Arabidopsis* petals. *Planta* 224, 801–811. doi: 10.1007/s00425-006-0264-6

Frewen, B. E., Chen, T. H. H., Howe, G. T., Davis, J., Rohde, A., Boerjan, W., et al. (2000). Quantitative trait loci and candidate gene mapping of bud set and bud flush in *Populus*. *Genetics* 154, 837–845.

Gendler, K., Paulsen, T., and Napoli, C. (2008). ChromDB: the chromatin database. *Nucleic Acids Res.* 36, D298–D302. doi: 10.1093/nar/gkm768

Gene Ontology Consortium (2012). The Gene Ontology: enhancements for 2011. *Nucleic Acids Res.* 40, D559–D564. doi: 10.1093/nar/gkr1028

Gentry, M., and Hennig, L. (2014). Remodelling chromatin to shape development of plants. *Exp. Cell Res.* 321, 40–46. doi: 10.1016/j.yexcr.2013.11.010

Gou, J. Q., Ma, C., Kadmiel, M., Gai, Y., Strauss, S., Jiang, X. N., et al. (2011). Tissue-specific expression of *Populus* C-19 GA 2-oxidases differentially regulate above– and below-ground biomass growth through control of bioactive GA concentrations. *New Phytol.* 192, 626–639. doi: 10.1111/j.1469-8137.2011.03837.x

Gu, X. L., Wang, H., Huang, H., and Cui, X. F. (2012). SPT6L encoding a putative WG/GW-repeat protein regulates apical-basal polarity of embryo in *Arabidopsis*. *Mol. Plant* 5, 249–259. doi: 10.1093/mp/ssr073

Halaly, T., Pang, X. Q., Batikoff, T., Crane, O., Keren, A., Venkateswari, J., et al. (2008). Similar mechanisms might be triggered by alternative external stimuli that induce dormancy release in grape buds. *Planta* 228, 79–88. doi: 10.1007/s00425-008-0720-6

Hartzog, G. A., and Fu, J. H. (2013). The Spt4-Spt5 complex: a multi-faceted regulator of transcription elongation. *Biochim. Biophys. Acta* 1829, 105–115. doi: 10.1016/j.bbagrm.2012.08.007

He, H. S., Dong, Q., Shao, Y. H., Jiang, H. Y., Zhu, S. W., Cheng, B. J., et al. (2012). Genome-wide survey and characterization of the *WRKY* gene family in *Populus trichocarpa*. *Plant Cell Rep.* 31, 1199–1217. doi: 10.1007/s00299-012-1241-0

Heide, O. M. (2008). Interaction of photoperiod and temperature in the control of growth and dormancy of *Prunus* species. *Sci. Hortic.* 115, 309–314. doi: 10.1016/j.scienta.2007.10.005

Higo, K., Ugawa, Y., Iwamoto, M., and Korenaga, T. (1999). Plant cis-acting regulatory DNA elements (PLACE) database: 1999. *Nucleic Acids Res.* 27, 297–300. doi: 10.1093/nar/27.1.297

Horvath, D. (2009). Common mechanisms regulate flowering and dormancy. *Plant Sci.* 177, 523–531. doi: 10.1016/j.plantsci.2009.09.002

Horvath, D. P., Anderson, J. V., Chao, W. S., and Foley, M. E. (2003). Knowing when to grow: signals regulating bud dormancy. *Trends Plant Sci.* 8, 534–540. doi: 10.1016/j.tplants.2003.09.013

Horvath, D. P., Chao, W. S., Suttle, J. C., Thimmapuram, J., and Anderson, J. V. (2008). Transcriptome analysis identifies novel responses and potential regulatory genes involved in seasonal dormancy transitions of leafy spurge (*Euphorbia esula* L.). *BMC Genomics* 9:536. doi: 10.1186/1471-2164-9-536

Horvath, D. P., Kudrna, D., Talag, J., Anderson, J. V., Chao, W. S., Wing, R., et al. (2013). BAC library development and clone characterization for dormancy-responsive *DREB4A*, *DAM*, and *FT* from leafy spurge (*Euphorbia esula*) identifies differential splicing and conserved promoter motifs. *Weed Sci.* 61, 303–309. doi: 10.1614/Ws-D-12-00175.1

Horvath, D. P., Sung, S., Kim, D., Chao, W., and Anderson, J. (2010). Characterization, expression and function of *DORMANCY ASSOCIATED MADS-BOX* genes from leafy spurge. *Plant Mol. Biol.* 73, 169–179. doi: 10.1007/s11103-009-9596-5

Hou, P., Ren, P. F., Zeng, D. E., Yu, G. R., Tang, W., Zheng, X., et al. (2014). Differential expression patterns and a novel interaction factor of Damaged DNA Binding Protein 1A (DDB1A) and DDB1B in *Arabidopsis thaliana*. *J. Plant Biol.* 57, 239–244. doi: 10.1007/s12374-014-0221-z

Howe, G. T., Aitken, S. N., Neale, D. B., Jermstad, K. D., Wheeler, N. C., and Chen, T. H. H. (2003). From genotype to phenotype: unraveling the complexities of cold adaptation in forest trees. *Can. J. Bot.* 81, 1247–1266. doi: 10.1139/B03-141

Howe, G. T., Davis, J., Jeknic, Z., Chen, T. H. H., Frewen, B., Bradshaw, H. D., et al. (1999). Physiological and genetic approaches to studying endodormancy-related traits in *Populus*. *HortScience* 34, 1174–1184.

Howe, G. T., Gardner, G., Hackett, W. P., and Furnier, G. R. (1996). Phytochrome control of short-day-induced bud set in black cottonwood. *Physiol. Plant.* 97, 95–103. doi: 10.1034/j.1399-3054.1996.970115.x

Hsu, C. Y., Adams, J. P., Kim, H. J., No, K., Ma, C. P., Strauss, S. H., et al. (2011). *FLOWERING LOCUS T* duplication coordinates reproductive and vegetative growth in perennial poplar. *Proc. Natl. Acad. Sci. U.S.A.* 108, 10756–10761. doi: 10.1073/pnas.1104713108

Hsu, P. Y., and Harmer, S. L. (2014). Wheels within wheels: the plant circadian system. *Trends Plant Sci.* 19, 240–249. doi: 10.1016/j.tplants.2013.11.007

Ibanez, C., Kozarewa, I., Johansson, M., Ogren, E., Rohde, A., and Eriksson, M. E. (2010). Circadian clock components regulate entry and affect exit of seasonal dormancy as well as winter hardiness in *Populus* trees. *Plant Physiol.* 153, 1823–1833. doi: 10.1104/pp.110.158220

Irizarry, R. A., Hobbs, B., Collin, F., Beazer-Barclay, Y. D., Antonellis, K. J., Scherf, U., et al. (2003). Exploration, normalization, and summaries of high density oligonucleotide array probe level data. *Biostatistics* 4, 249–264. doi: 10.1093/biostatistics/4.2.249

Jensen, W. A. (1962). *Botanical Histochemistry: Principles and Practice*. San Francisco, CA: W. H. Freeman.

Jermstad, K. D., Bassoni, D. L., Jech, K. S., Ritchie, G. A., Wheeler, N. C., and Neale, D. B. (2003). Mapping of quantitative trait loci controlling adaptive traits in coastal Douglas fir. III. Quantitative trait loci-by-environment interactions. *Genetics* 165, 1489–1506.

Jiang, D., Wang, Y., Wang, Y., and He, Y. (2008). Repression of *FLOWERING LOCUS C* and *FLOWERING LOCUS T* by the *Arabidopsis* polycomb repressive complex 2 components. *PLoS ONE* 3:e3404. doi: 10.1371/journal.pone.0003404

Jiang, Y. Z., Duan, Y. J., Yin, J., Ye, S. L., Zhu, J. R., Zhang, F. Q., et al. (2014). Genome-wide identification and characterization of the *Populus* WRKY transcription factor family and analysis of their expression in response to biotic and abiotic stresses. *J. Exp. Bot.* 65, 6629–6644. doi: 10.1093/Jxb/Eru381

Jiménez, S., Li, Z. G., Reighard, G. L., and Bielenberg, D. G. (2010). Identification of genes associated with growth cessation and bud dormancy entrance using a dormancy-incapable tree mutant. *BMC Plant Biol.* 10:25. doi: 10.1186/1471-2229-10-25

Jin, J., Zhang, H., Kong, L., Gao, G., and Luo, J. (2014). PlantTFDB 3.0: a portal for the functional and evolutionary study of plant transcription factors. *Nucleic Acids Res.* 42, D1182–D1187. doi: 10.1093/nar/gkt1016

Jung, J. H., Seo, Y. H., Seo, P. J., Reyes, J. L., Yun, J., Chua, N. H., et al. (2007). The *GIGANTEA*-regulated MicroRNA172 mediates photoperiodic flowering independent of *CONSTANS* in *Arabidopsis*. *Plant Cell* 19, 2736–2748. doi: 10.1105/tpc.107.054528

Karlberg, A., Englund, M., Petterle, A., Molnar, G., Sjodin, A., Bako, L., et al. (2010). Analysis of global changes in gene expression during activity-dormancy cycle in hybrid aspen apex. *Plant Biotechnol.* 27, 1–16. doi: 10.5511/plantbiotechnology.27.1

Karlowski, W. M., Zielezinski, A., Carrere, J., Pontier, D., Lagrange, T., and Cooke, R. (2010). Genome-wide computational identification of WG/GW Argonaute-binding proteins in *Arabidopsis*. *Nucleic Acids Res.* 38, 4231–4245. doi: 10.1093/Nar/Gkq162

Kaushik, N., Fear, D., Richards, S. C. M., McDermott, C. R., Nuwaysir, E. F., Kellam, P., et al. (2005). Gene expression in peripheral blood mononuclear cells from patients with chronic fatigue syndrome. *J. Clin. Pathol.* 58, 826–832. doi: 10.1136/jcp.2005.025718

Kim, H. J., Hong, S. H., Kim, Y. W., Lee, I. H., Jun, J. H., Phee, B. K., et al. (2014). Gene regulatory cascade of senescence-associated NAC transcription factors activated by ETHYLENE-INSENSITIVE2-mediated leaf senescence signalling in *Arabidopsis*. *J. Exp. Bot.* 65, 4023–4036. doi: 10.1093/Jxb/Eru112

Kim, S. Y., Lee, J., Eshed-Williams, L., Zilberman, D., and Sung, Z. R. (2012). EMF1 and PRC2 cooperate to repress key regulators of *Arabidopsis* development. *PLoS Genet.* 8:e1002512. doi: 10.1371/journal.pgen.1002512

Kim, W. Y., Fujiwara, S., Suh, S. S., Kim, J., Kim, Y., Han, L. Q., et al. (2007). ZEITLUPE is a circadian photoreceptor stabilized by GIGANTEA in blue light. *Nature* 449, 356–360. doi: 10.1038/Nature06132

Korasick, D. A., Westfall, C. S., Lee, S. G., Nanao, M. H., Dumas, R., Hagen, G., et al. (2014). Molecular basis for AUXIN RESPONSE FACTOR protein interaction and the control of auxin response repression. *Proc. Natl. Acad. Sci. U.S.A.* 111, 5427–5432. doi: 10.1073/pnas.1400074111

Kozarewa, I., Ibanez, C., Johansson, M., Ogren, E., Mozley, D., Nylander, E., et al. (2010). Alteration of *PHYA* expression change circadian rhythms and timing of bud set in *Populus*. *Plant Mol. Biol.* 73, 143–156. doi: 10.1007/s11103-010-9619-2

Lang, G. A., Early, J. D., Martin, G. C., and Darnell, R. L. (1987). Endo-, para-, and ecodormancy: physiological terminology and classification for dormancy research. *HortScience* 22, 371–377.

Lee, J., and Lee, I. (2010). Regulation and function of SOC1, a flowering pathway integrator. *J. Exp. Bot.* 61, 2247–2254. doi: 10.1093/jxb/erq098

Leida, C., Conesa, A., Llácer, G., Badenes, M. L., and Ríos, G. (2012). Histone modifications and expression of *DAM6* gene in peach are modulated during bud dormancy release in a cultivar-dependent manner. *New Phytol.* 193, 67–80. doi: 10.1111/j.1469-8137.2011.03863.x

Li, C., Junttila, O., Ernstsen, A., Heino, P., and Palva, E. T. (2003). Photoperiodic control of growth, cold acclimation and dormancy development in silver birch (*Betula pendula*) ecotypes. *Physiol. Plant.* 117, 206–212. doi: 10.1034/j.1399-3054.2003.00002.x

Li, L., Ye, H. X., Guo, H. Q., and Yin, Y. H. (2010). *Arabidopsis* IWS1 interacts with transcription factor BES1 and is involved in plant steroid hormone brassinosteroid regulated gene expression. *Proc. Natl. Acad. Sci. U.S.A.* 107, 3918–3923. doi: 10.1073/pnas.0909198107

Life Technologies (2011). *MessageAmp II aRNA Amplification Kit, Part Number 1751M*. Carlsbad, CA: Life Technologies Corporation, 38.

Liu, D. M., Song, Y., Chen, Z. X., and Yu, D. Q. (2009). Ectopic expression of miR396 suppresses *GRF* target gene expression and alters leaf growth in *Arabidopsis*. *Physiol. Plant.* 136, 223–236. doi: 10.1111/j.1399-3054.2009.01229.x

Liu, X., Yang, S., Zhao, M., Luo, M., Yu, C. W., Chen, C. Y., et al. (2014). Transcriptional repression by histone deacetylases in plants. *Mol. Plant* 7, 764–772. doi: 10.1093/mp/ssu033

Liu, X., Yu, C. W., Duan, J., Luo, M., Wang, K., Tian, G., et al. (2012). HDA6 directly interacts with DNA methyltransferase MET1 and maintains transposable element silencing in *Arabidopsis*. *Plant Physiol.* 158, 119–129. doi: 10.1104/pp.111.184275

Mathiason, K., He, D., Grimplet, J., Venkateswari, J., Galbraith, D., Or, E., et al. (2009). Transcript profiling in *Vitis riparia* during chilling requirement fulfillment reveals coordination of gene expression patterns with optimized bud break. *Funct. Integr. Genomics* 9, 81–96. doi: 10.1007/s10142-008-0090-y

Matzke, M. A., and Mosher, R. A. (2014). RNA-directed DNA methylation: an epigenetic pathway of increasing complexity. *Nat. Rev. Genet.* 15, 394–408. doi: 10.1038/Nrg3683

Maxwell, B. B., Andersson, C. R., Poole, D. S., Kay, S. A., and Chory, J. (2003). HY5, Circadian Clock-Associated 1, and a cis-element, DET1 dark response element, mediate DET1 regulation of *Chlorophyll a/b-Binding Protein 2* expression. *Plant Physiol.* 133, 1565–1577. doi: 10.1104/pp.103.025114

Mazzitelli, L., Hancock, R. D., Haupt, S., Walker, P. G., Pont, S. D. A., McNicol, J., et al. (2007). Co-ordinated gene expression during phases of dormancy release in raspberry (*Rubus idaeus* L.) buds. *J. Exp. Bot.* 58, 1035–1045. doi: 10.1093/Jxb/Erl266

Michael, T. P., Breton, G., Hazen, S. P., Priest, H., Mockler, T. C., Kay, S. A., et al. (2008). A morning-specific phytohormone gene expression program underlying rhythmic plant growth. *PLoS Biol.* 6:e225. doi: 10.1371/journal.pbio.0060225

Mikkelsen, M. D., and Thomashow, M. F. (2009). A role for circadian evening elements in cold-regulated gene expression in *Arabidopsis*. *Plant J.* 60, 328–339. doi: 10.1111/j.1365-313X.2009.03957.x

Mockler, T. C., Michael, T. P., Priest, H. D., Shen, R., Sullivan, C. M., Givan, S. A., et al. (2007). The DIURNAL project: DIURNAL and circadian expression profiling, model-based pattern matching, and promoter analysis. *Cold Spring Harb. Symp. Quant. Biol.* 72, 353–363. doi: 10.1101/sqb.2007.72.006

Mølmann, J. A., Asante, D. K. A., Jensen, J. B., Krane, M. N., Ernstsen, A., Junttila, O., et al. (2005). Low night temperature and inhibition of gibberellin biosynthesis override phytochrome action and induce bud set and cold acclimation, but not dormancy in *PHYA* overexpressors and wild-type of hybrid aspen. *Plant Cell Environ.* 28, 1579–1588. doi: 10.1111/j.1365-3040.2005.01395.x

Mueller, D., and Leyser, O. (2011). Auxin, cytokinin and the control of shoot branching. *Ann. Bot.* 107, 1203–1212. doi: 10.1093/Aob/Mcr069

Mueller, O., Lightfoot, S., and Schröder, A. (2004). *RNA Integrity Number (RIN) Standardization of RNA quality control. Application Note [Online]*. Available at: http://www.chem.agilent.com/Library/applications/5989-1165EN.pdf

Nugroho, L. H., Verberne, M. C., and Verpoorte, R. (2002). Activities of enzymes involved in the phenylpropanoid pathway in constitutively salicylic acid-producing tobacco plants. *Plant Physiol. Biochem.* 40, 755–760. doi: 10.1016/S0981-9428(02)01437-7

Ohme-Takagi, M., and Shinshi, H. (1995). Ethylene-inducible DNA-binding proteins that interact with an ethylene-responsive element. *Plant Cell* 7, 173–182. doi: 10.2307/3869993

Olsen, J. E. (2010). Light and temperature sensing and signaling in induction of bud dormancy in woody plants. *Plant Mol. Biol.* 73, 37–47. doi: 10.1007/s11103-010-9620-9

Olsen, J. E., Junttila, O., Nilsen, J., Eriksson, M. E., Martinussen, I., Olsson, O., et al. (1997). Ectopic expression of oat phytochrome A in hybrid aspen changes critical daylength for growth and prevents cold acclimatization. *Plant J.* 12, 1339–1350. doi: 10.1046/j.1365-313x.1997.12061339.x

Ophir, R., Pang, X. Q., Halaly, T., Venkateswari, J., Lavee, S., Galbraith, D., et al. (2009). Gene-expression profiling of grape bud response to two alternative dormancy-release stimuli expose possible links between impaired mitochondrial activity, hypoxia, ethylene-ABA interplay and cell enlargement. *Plant Mol. Biol.* 71, 403–423. doi: 10.1007/s11103-009-9531-9

Or, E., Vilozny, I., Eyal, Y., and Ogrodovitch, A. (2000). The transduction of the signal for grape bud dormancy breaking induced by hydrogen cyanamide may involve the SNF-like protein kinase GDBRPK. *Plant Mol. Biol.* 43, 483–494. doi: 10.1023/A:1006450516982

Paponov, I. A., Teale, W., Lang, D., Paponov, M., Reski, R., Rensing, S. A., et al. (2009). The evolution of nuclear auxin signalling. *BMC Evol. Biol.* 9:126. doi: 10.1186/1471-2148-9-126

Park, S., Lee, C. M., Doherty, C. J., Gilmour, S. J., Kim, Y., and Thomashow, M. F. (2015). Regulation of the *Arabidopsis* CBF regulon by a complex low-temperature regulatory network. *Plant J.* 82, 193–207. doi: 10.1111/Tpj.12796

Pazhouhandeh, M., Molinier, J., Berr, A., and Genschik, P. (2011). MSI4/FVE interacts with CUL4-DDB1 and a PRC2-like complex to control epigenetic regulation of flowering time in *Arabidopsis*. *Proc. Natl. Acad. Sci. U.S.A.* 108, 3430–3435. doi: 10.1073/pnas.1018242108

Peer, W. A., Cheng, Y., and Murphy, A. S. (2013). Evidence of oxidative attenuation of auxin signalling. *J. Exp. Bot.* 64, 2629–2639. doi: 10.1093/Jxb/Ert152

Pelgas, B., Bousquet, J., Meirmans, P. G., Ritland, K., and Isabel, N. (2011). QTL mapping in white spruce: gene maps and genomic regions underlying adaptive traits across pedigrees, years and environments. *BMC Genomics* 12:145. doi: 10.1186/1471-2164-12-145

Preston, J. C., and Sandve, S. R. (2013). Adaptation to seasonality and the winter freeze. *Front. Plant Sci.* 4:167. doi: 10.3389/Fpls.2013.00167

Ramiro-Merina, A., Ariza, R. R., and Roldan-Arjona, T. (2013). Molecular characterization of a putative plant homolog of MBD4 DNA glycosylase. *DNA Repair* 12, 890–898. doi: 10.1016/j.dnarep.2013.08.002

Reeves, R. (2010). Nuclear functions of the HMG proteins. *Biochim. Biophys. Acta* 1799, 3–14. doi: 10.1016/j.bbagrm.2009.09.001

Resman, L., Howe, G., Jonsen, D., Englund, M., Druart, N., Schrader, J., et al. (2010). Components acting downstream of short day perception regulate differential cessation of cambial activity and associated responses in early and late clones of hybrid poplar. *Plant Physiol.* 154, 1294–1303. doi: 10.1104/pp.110.163907

Rhoades, M. W., Reinhart, B. J., Lim, L. P., Burge, C. B., Bartel, B., and Bartel, D. P. (2002). Prediction of plant microRNA targets. *Cell* 110, 513–520. doi: 10.1016/S0092-8674(02)00863-2

Rinne, P. L. H., Welling, A., Vahala, J., Ripel, L., Ruonala, R., Kangasjärvi, J., et al. (2011). Chilling of dormant buds hyperinduces *FLOWERING LOCUS T* and recruits GA-inducible 1,3-β-glucanases to reopen signal conduits and release dormancy in *Populus*. *Plant Cell* 23, 130–146. doi: 10.1105/tpc.110.081307

Rohde, A., Bastien, C., and Boerjan, W. (2011a). Temperature signals contribute to the timing of photoperiodic growth cessation and bud set in poplar. *Tree Physiol.* 31, 472–482. doi: 10.1093/treephys/tpr038

Rohde, A., Storme, V., Jorge, V., Gaudet, M., Vitacolonna, N., Fabbrini, F., et al. (2011b). Bud set in poplar – genetic dissection of a complex trait in natural and hybrid populations. *New Phytol.* 189, 106–121. doi: 10.1111/j.1469-8137.2010.03469.x

Rohde, A., and Bhalerao, R. P. (2007). Plant dormancy in the perennial context. *Trends Plant Sci.* 12, 217–223. doi: 10.1016/j.tplants.2007.03.012

Rohde, A., Ruttink, T., Hostyn, V., Sterck, L., Van Driessche, K., and Boerjan, W. (2007). Gene expression during the induction, maintenance, and release of dormancy in apical buds of poplar. *J. Exp. Bot.* 58, 4047–4060. doi: 10.1093/Jxb/Erm261

Ruonala, R., Rinne, P. L. H., Baghour, M., Moritz, T., Tuominen, H., and Kangasjarvi, J. (2006). Transitions in the functioning of the shoot apical meristem in birch (*Betula pendula*) involve ethylene. *Plant J.* 46, 628–640. doi: 10.1111/j.1365-313X.2006.02722.x

Ruonala, R., Rinne, P. L. H., Kangasjarvi, J., and van der Schoot, C. (2008). *CENL1* expression in the rib meristem affects stem elongation and the transition to dormancy in *Populus*. *Plant Cell* 20, 59–74. doi: 10.1105/tpc.107.056721

Ruttink, T., Arend, M., Morreel, K., Storme, V., Rombauts, S., Fromm, J., et al. (2007). A molecular timetable for apical bud formation and dormancy induction in poplar. *Plant Cell* 19, 2370–2390. doi: 10.1105/tpc.107.052811

Sakamoto, H., Maruyama, K., Sakuma, Y., Meshi, T., Iwabuchi, M., Shinozaki, K., et al. (2004). *Arabidopsis* Cys2/His2-type zinc-finger proteins function as transcription repressors under drought, cold, and high-salinity stress conditions. *Plant Physiol.* 136, 2734–2746. doi: 10.1104/pp.104.046599

Sasaki, R., Yamane, H., Ooka, T., Jotatsu, H., Kitamura, Y., Akagi, T., et al. (2011). Functional and expressional analyses of *PmDAM* genes associated with endodormancy in Japanese apricot. *Plant Physiol.* 157, 485–497. doi: 10.1104/pp.111.181982

Schmid, M., Uhlenhaut, N. H., Godard, F., Demar, M., Bressan, R., Weigel, D., et al. (2003). Dissection of floral induction pathways using global expression analysis. *Development* 130, 6001–6012. doi: 10.1242/dev.00842

Schröder, F., Lisso, J., Obata, T., Erban, A., Maximova, E., Giavalisco, P., et al. (2014). Consequences of induced brassinosteroid deficiency in *Arabidopsis* leaves. *BMC Plant Biol.* 14:309. doi: 10.1186/S12870-014-0309-0

Scotti-Saintagne, C., Bodenes, C., Barreneche, T., Bertocchi, E., Plomion, C., and Kremer, A. (2004). Detection of quantitative trait loci controlling bud burst and height growth in *Quercus robur* L. *Theor. Appl. Genet.* 109, 1648–1659. doi: 10.1007/s00122-004-1789-3

Seo, E., Lee, H., Jeon, J., Park, H., Kim, J., Noh, Y. S., et al. (2009). Crosstalk between cold response and flowering in *Arabidopsis* is mediated through the flowering-time gene *SOC1* and its upstream negative regulator *FLC*. *Plant Cell* 21, 3185–3197. doi: 10.1105/tpc.108.063883

Sluis, A., and Hake, S. (2015). Organogenesis in plants: initiation and elaboration of leaves. *Trends Genet.* 31, 300–306. doi: 10.1016/j.tig.2015.04.004

Smieszek, S. P., Yang, H. X., Paccanaro, A., and Devlin, P. F. (2014). Progressive promoter element combinations classify conserved orthogonal plant circadian gene expression modules. *J. R. Soc. Interface* 11, 1–11. doi: 10.1098/Rsif.2014.0535

Sreeramulu, S., Mostizky, Y., Sunitha, S., Shani, E., Nahum, H., Salomon, D., et al. (2013). BSKs are partially redundant positive regulators of brassinosteroid signaling in *Arabidopsis*. *Plant J.* 74, 905–919. doi: 10.1111/Tpj.12175

Sterky, F., Bhalerao, R. R., Unneberg, P., Segerman, B., Nilsson, P., Brunner, A. M., et al. (2004). A *Populus* EST resource for plant functional genomics. *Proc. Natl. Acad. Sci. U.S.A.* 101, 13951–13956. doi: 10.1073/pnas.0401641101

Subramanian, A., Tamayo, P., Mootha, V. K., Mukherjee, S., Ebert, B. L., Gillette, M. A., et al. (2005). Gene set enrichment analysis: a knowledge-based approach for interpreting genome-wide expression profiles. *Proc. Natl. Acad. Sci. U.S.A.* 102, 15545–15550. doi: 10.1073/pnas.0506580102

Tao, Z., Shen, L., Liu, C., Liu, L., Yan, Y., and Yu, H. (2012). Genome-wide identification of SOC1 and SVP targets during the floral transition in *Arabidopsis*. *Plant J.* 70, 549–561. doi: 10.1111/j.1365-313X.2012.04919.x

Thomashow, M. F. (2010). Molecular basis of plant cold acclimation: insights gained from studying the CBF cold response pathway. *Plant Physiol.* 154, 571–577. doi: 10.1104/pp.110.161794

To, T. K., Kim, J. M., Matsui, A., Kurihara, Y., Morosawa, T., Ishida, J., et al. (2011a). *Arabidopsis* HDA6 regulates locus-directed heterochromatin silencing in cooperation with MET1. *PLoS Genet.* 7:e1002055. doi: 10.1371/journal.pgen.1002055

To, T. K., Nakaminami, K., Kim, J. M., Morosawa, T., Ishida, J., Tanaka, M., et al. (2011b). *Arabidopsis* HDA6 is required for freezing tolerance. *Biochem. Biophys. Res. Commun.* 406, 414–419. doi: 10.1016/j.bbrc.2011.02.058

Trainin, T., Bar-Ya'akov, I., and Holland, D. (2013). *ParSOC1*, a MADS-box gene closely related to *Arabidopsis AGL20/SOC1*, is expressed in apricot leaves in a diurnal manner and is linked with chilling requirements for dormancy break. *Tree Genet. Genomes* 9, 753–766. doi: 10.1007/s11295-012-0590-8

Tripathi, P., Rabara, R. C., and Rushton, P. J. (2014). A systems biology perspective on the role of WRKY transcription factors in drought responses in plants. *Planta* 239, 255–266. doi: 10.1007/s00425-013-1985-y

Tuskan, G. A., DiFazio, S., Jansson, S., Bohlmann, J., Grigoriev, I., Hellsten, U., et al. (2006). The genome of black cottonwood, *Populus trichocarpa* (Torr. & Gray). *Science* 313, 1596–1604. doi: 10.1126/science.1128691

Vahala, J., Felten, J., Love, J., Gorzsas, A., Gerber, L., Lamminmaki, A., et al. (2013). A genome-wide screen for ethylene-induced ethylene response factors (ERFs) in hybrid aspen stem identifies *ERF* genes that modify stem growth and wood properties. *New Phytol.* 200, 511–522. doi: 10.1111/Nph.12386

Van Gelder, R. N., von Zastrow, M. E., Yool, A., Dement, W. C., Barchas, J. D., and Eberwine, J. H. (1990). Amplified RNA synthesized from limited quantities of heterogeneous cDNA. *Proc. Natl. Acad. Sci. U.S.A.* 87, 1663–1667. doi: 10.1073/pnas.87.5.1663

Vilarrasa-Blasi, J., Gonzalez-Garcia, M. P., Frigola, D., Fabregas, N., Alexiou, K. G., Lopez-Bigas, N., et al. (2014). Regulation of plant stem cell quiescence by a brassinosteroid signaling module. *Dev. Cell* 30, 36–47. doi: 10.1016/j.devcel.2014.05.020

Vining, K. J., Pomraning, K. R., Wilhelm, L. J., Priest, H. D., Pellegrini, M., Mockler, T. C., et al. (2012). Dynamic DNA cytosine methylation in the *Populus trichocarpa* genome: tissue-level variation and relationship to gene expression. *BMC Genomics* 13:27. doi: 10.1186/1471-2164-13-27

Vogel, J. T., Zarka, D. G., Van Buskirk, H. A., Fowler, S. G., and Thomashow, M. F. (2005). Roles of the CBF2 and ZAT12 transcription factors in configuring the low temperature transcriptome of *Arabidopsis*. *Plant J.* 41, 195–211. doi: 10.1111/j.1365-313X.2004.02288.x

Walton, E. F., Wu, R.-M., Richardson, A. C., Davy, M., Hellens, R. P., Thodey, K., et al. (2009). A rapid transcriptional activation is induced by the dormancy-breaking chemical hydrogen cyanamide in kiwifruit (*Actinidia deliciosa*) buds. *J. Exp. Bot.* 60, 3835–3848. doi: 10.1093/jxb/erp231

Wellmer, F., and Riechmann, J. L. (2010). Gene networks controlling the initiation of flower development. *Trends Genet.* 26, 519–527. doi: 10.1016/j.tig.2010.09.001

Wisniewski, M., Norelli, J., Bassett, C., Artlip, T., and Macarisin, D. (2011). Ectopic expression of a novel peach (*Prunus persica*) CBF transcription factor in apple (*Malus x domestica*) results in short-day induced dormancy and increased cold hardiness. *Planta* 233, 971–983. doi: 10.1007/s00425-011-1358-3

Wu, J. F., Wang, Y., and Wu, S. H. (2008a). Two new clock proteins, LWD1 and LWD2, regulate *Arabidopsis* photoperiodic flowering. *Plant Physiol.* 148, 948–959. doi: 10.1104/pp.108.124917

Wu, K., Zhang, L., Zhou, C., Yu, C. W., and Chaikam, V. (2008b). HDA6 is required for jasmonate response, senescence and flowering in *Arabidopsis*. *J. Exp. Bot.* 59, 225–234. doi: 10.1093/jxb/erm300

Yamane, H., Ooka, T., Jotatsu, H., Hosaka, Y., Sasaki, R., and Tao, R. (2011). Expressional regulation of *PpDAM5* and *PpDAM6*, peach (*Prunus persica*) dormancy-associated MADS-box genes, by low temperature and dormancy-breaking reagent treatment. *J. Exp. Bot.* 62, 3481–3488. doi: 10.1093/jxb/err028

Yordanov, Y. S., Ma, C., Strauss, S. H., and Busov, V. B. (2014). *EARLY BUD-BREAK 1 (EBB1)* is a regulator of release from seasonal dormancy in poplar trees. *Proc. Natl. Acad. Sci. U.S.A.* 111, 10001–10006. doi: 10.1073/pnas.1405621111

Zawaski, C., and Busov, V. B. (2014). Roles of gibberellin catabolism and signaling in growth and physiological response to drought and short-day photoperiods in *Populus* trees. *PLoS ONE* 9:e86217. doi: 10.1371/journal.pone.0086217

Zhang, H., Bishop, B., Ringenberg, W., Muir, W. M., and Ogas, J. (2012). The CHD3 remodeler PICKLE associates with genes enriched for trimethylation of histone H3 lysine 27. *Plant Physiol.* 159, 418–432. doi: 10.1104/pp.112.194878

Zhang, Y., Mayba, O., Pfeiffer, A., Shi, H., Tepperman, J. M., Speed, T. P., et al. (2013). A quartet of PIF bHLH factors provides a transcriptionally centered signaling hub that regulates seedling morphogenesis through differential expression-patterning of shared target genes in *Arabidopsis*. *PLoS Genet.* 9:e1003244. doi: 10.1371/journal.pgen.1003244

Zheng, C. L., Halaly, T., Acheampong, A. K., Takebayashi, Y., Jikumaru, Y., Kamiya, Y., et al. (2015). Abscisic acid (ABA) regulates grape bud dormancy, and dormancy release stimuli may act through modification of ABA metabolism. *J. Exp. Bot.* 66, 1527–1542. doi: 10.1093/Jxb/Eru519

Zhong, W. J., Gao, Z. H., Zhuang, W. B., Shi, T., Zhang, Z., and Ni, Z. J. (2013). Genome-wide expression profiles of seasonal bud dormancy at four critical stages in Japanese apricot. *Plant Mol. Biol.* 83, 247–264. doi: 10.1007/s11103-013-0086-4

Zhu, Z., An, F., Feng, Y., Li, P., Xue, L., Mu, A., et al. (2011). Derepression of ethylene-stabilized transcription factors (EIN3/EIL1) mediates jasmonate and ethylene signaling synergy in *Arabidopsis*. *Proc. Natl. Acad. Sci. U.S.A.* 108, 12539–12544. doi: 10.1073/pnas.1103959108

Zolotarov, Y., and Stromvik, M. (2015). *De novo* regulatory motif discovery identifies significant motifs in promoters of five classes of plant dehydrin genes. *PLoS ONE* 10:e0129016. doi: 10.1371/journal.pone.0129016

Zou, Y., Zhang, L., Rao, S., Zhu, X. Y., Ye, L. L., Chen, W. X., et al. (2014). The relationship between the expression of ethylene-related genes and papaya fruit ripening disorder caused by chilling injury. *PLoS ONE* 9:e116002. doi: 10.1371/journal.pone.0116002

Conflict of Interest Statement: The authors declare that the research was conducted in the absence of any commercial or financial relationships that could be construed as a potential conflict of interest.

10

Qualitative Distinction of Autotrophic and Heterotrophic Processes at the Leaf Level by Means of Triple Stable Isotope (C–O–H) Patterns

Adam Kimak[1,2], Zoltan Kern[3]* and Markus Leuenberger[1,2]

[1] *Climate and Environmental Physics, Physics Institute, University of Bern, Bern, Switzerland, [2] Oeschger Centre for Climate Change Research, University of Bern, Bern, Switzerland, [3] Institute for Geological and Geochemical Research, Research Centre for Astronomy and Earth Sciences, Hungarian Academy of Sciences (MTA), Budapest, Hungary*

Foliar samples were harvested from two oaks, a beech, and a yew at the same site in order to trace the development of the leaves over an entire vegetation season. Cellulose yield and stable isotopic compositions ($\delta^{13}C$, $\delta^{18}O$, and δD) were analyzed on leaf cellulose. All parameters unequivocally define a juvenile and a mature period in the foliar expansion of each species. The accompanying shifts of the $\delta^{13}C$-values are in agreement with the transition from remobilized carbohydrates (juvenile period), to current photosynthates (mature phase). While the opponent seasonal trends of $\delta^{18}O$ of blade and vein cellulose are in perfect agreement with the state-of-art mechanistic understanding, the lack of this discrepancy for δD, documented for the first time, is unexpected. For example, the offset range of 18 permil (oak veins) to 57 permil (oak blades) in δD may represent a process driven shift from autotrophic to heterotrophic processes. The shared pattern between blade and vein found for both oak and beech suggests an overwhelming metabolic isotope effect on δD that might be accompanied by proton transfer linked to the Calvin-cycle. These results provide strong evidence that hydrogen and oxygen are under different biochemical controls even at the leaf level.

Keywords: leaf cellulose, hydrogen, oxygen, carbon, isotope, seasonal, intra-annual, juvenile-mature foliage

Edited by:
Jian-Guo Huang,
South China Botanical Garden, CAS,
China

Reviewed by:
Bao Yang,
Cold and Arid Regions Environmental
and Engineering Research Institute,
CAS, China
Ze-Xin Fan,
Xishuangbanna Tropical Botanical
Garden, CAS, China

*Correspondence:
Zoltan Kern
kern@geochem.hu

INTRODUCTION

Stable isotopes in plant physiology are used across a broad scale. They have also been applied to paleoclimate studies since tree-ring carbohydrates act as fingerprints of current climatic conditions (Gessler et al., 2014; Saurer et al., 2014). In previous decades, most of the studies were performed using only a single isotopic composition rather than a multi-isotope approach (Scheidegger et al., 2000). The following can be summarized for each element and its isotopic composition.

Carbon

Carbon isotopes are assimilated from atmospheric carbon-dioxide, carrying intra-annual variations of $\delta^{13}C$ of carbon-dioxide (CO_2) (Keeling et al., 1995). According to Farquhar et al. (1982), photosynthesis discrimination against ^{13}C by Rubisco is dependent on the ratio of intercellular to atmospheric CO_2 concentrations (c_i/c_a), which are themselves controlled by stomatal conductance and driven by the CO_2 assimilation rate (Farquhar et al., 1989). However, annual tree ring formation is partly supported by stored carbohydrates, especially at the beginning of the growing season, thus inducing higher degrees of coherence with previous years (Helle and Schleser, 2004; Kagawa et al., 2006; Cernusak et al., 2009). Furthermore, similar seasonal variability of $\delta^{13}C$ has been also found in ryegrass. However, an opposite trend has been documented for $\delta^{15}N$ (Wang and Schjoerring, 2012). Since, carbon isotopes originate from atmospheric CO_2 driven by climate associated variables, $\delta^{13}C$ measurements of tree rings are frequently used for the determination of past weather extremes (Raffalli-Delerce et al., 2004; Kress et al., 2010), and changes in past water use efficiency (Saurer et al., 2014) among others. However, these regional signals stored in tree rings can be altered or concealed by local conditions (Saurer et al., 1997).

Oxygen

Oxygen (and hydrogen) isotopes originate from source water, taken up mainly as groundwater, although in some special cases fog, dew, or water vapor in general can also act as alternative water sources (Roden et al., 2009; Studer et al., 2015).

According to the depths of water uptake, the source water can be fractionated by evaporative gradients in the soil, in particular for shallow-rooting trees, providing possible seasonal variations of $\delta^{18}O$ as well as significant phase-shifts (White, 1989; Treydte et al., 2014) in source water (Ehleringer and Dawson, 1992; Gessler et al., 2014). In addition, no fractionation has been observed from roots to leaves when the water is transported from soil into the xylem of roots, trunk and twigs (White et al., 1985; Dawson and Ehleringer, 1993; Treydte et al., 2014).

Further processes do influence leaf water oxygen isotopes, such as evaporative enrichment during transpiration (Farquhar and Lloyd, 1993) and the Péclet-effect, which represents the modified Craig-Gordon model by including the correction of the diffusion of evaporative ^{18}O-enriched water back to the leaf lamina by advective transport of unenriched xylem water (Farquhar and Lloyd, 1993; Gessler et al., 2014). Moreover, almost every metabolic step is followed by a certain degree of isotope fractionation, resulting in an overall enrichment of plant organic matter by $+27‰$ compared to lamina leaf water (Deniro and Epstein, 1981; Yakir and Deniro, 1990).

Hydrogen

At the leaf level water molecules are split in photosynthesis. Hydrogen from leaf water is transferred to nicotinamide adenine dinucleotide ($NADP^+$), which is further transformed to NADPH which in turn strongly prefers light hydrogen (^1H), resulting in a very depleted isotopic composition for hydrogen (Yakir and Deniro, 1990; Luo et al., 1991). However, further metabolic steps, such as the Calvin-cycle (Yakir, 1992; Hayes, 2001), the exchange of C-bound H with cell water (Luo and Sternberg, 1992; Yakir, 1992) and the biosynthesis of plant organic material (Yakir and Deniro, 1990; Luo and Sternberg, 1992) initiate isotopic enrichments in hydrogen isotopic composition.

Although, the isotopic fractionations of metabolic processes, including carbon- (Gessler et al., 2008), oxygen- (Barbour et al., 2004), and hydrogen-isotopes (Roden and Ehleringer, 1999), are mostly well-understood (Gessler et al., 2014), a complete seasonal variability of isotopic composition (C–O–H) in leaf cellulose,

containing all the major constructive elements has yet to be presented.

The development of leaves, similar to the case with tree rings, is influenced by previous years' stored carbohydrates and current climate conditions, and is driven by species-specific turnover from heterotrophic- to autotrophic-based metabolism (Keel et al., 2007; Keel and Schädel, 2010). At the beginning of the growing season, the major impact on cellulose synthesis is generated by remobilized carbohydrates traced by tree ring (Helle and Schleser, 2004; Kagawa et al., 2006) and leaf cellulose analyses (Leavitt and Long, 1985; Stokes et al., 2009). However, the inter- and intra-annual distribution of carbohydrate storage/usage can vary, depending on local climate conditions (Kimak and Leuenberger, 2015). In addition, the remaining soluble carbohydrates that support the tree against frost damage in winter, described by the increasing soluble carbohydrate concentration in the xylem of roots, stem, and branches (Loescher et al., 1990; Barbaroux and Bréda, 2002; Barbaroux et al., 2003; Helle and Schleser, 2004), can be also used for plant organic matter formation in spring. According to Eglin et al. (2009), the aforementioned effect of remaining carbohydrates was also documented at leaf level, including leaf cellulose, starch, and water-soluble fractions. After the first leaf cells are developed and are capable of photosynthesis, the usage of reserves decreases and concurrently the usage of newly assimilated carbohydrates increases until it becomes the major source of cellulose synthesis. The period during which the tree uses carbohydrate from reserves in parallel with direct assimilates for cellulose formation is called earlywood (EW) (Huang et al., 2014). In deciduous species, the earlywood cells contain typically large vessels (in particular in oaks) for the increased nutrient and water transport at the beginning of growing season. The period of the growing season dominated by autotrophic metabolism is called latewood (LW), representing smaller cells besides the higher cell density and excluding the aforementioned large vessels (Feuillat et al., 1997). Consequently, newly assimilated carbohydrates used for LW cells record the current climate conditions. This is why LW is generally used for climate research in preference to EW (McCarroll and Loader, 2004; Kern et al., 2013). Consequently, the biochemical pathways supporting cellulose synthesis change during the growing season, where principally heterotrophic metabolism becomes autotrophic, thereby having an impact on the concentration of non-structural carbohydrate (Gilson et al., 2014), cellulose and lipid synthesis (Sessions, 2006; Tipple et al., 2013) among others.

In this study, we investigated the leaves and needles of long living deciduous- and evergreen species, including two oak trees (*Quercus robur* L.), a beech (*Fagus sylvatica* L.) and a yew tree (*Taxus baccata* L.). We assumed that carbon, oxygen, and hydrogen follow a similar isotopic seasonality in the foliage as have been documented in the xylem. The leaves were collected sequentially, through a complete growing season between April and October in 2012. Besides the alpha-cellulose yield, the stable isotopic composition of carbon, oxygen, and hydrogen was simultaneously measured in extracted alpha-cellulose (Loader et al., 2014) for the characterization of leaf development and intra-annual changes of cellulose synthesis at the leaf level.

MATERIALS AND METHODS

Site Description

The sampling site is located in Bern, Switzerland and lies within the temperate climate zone, having a humid continental climate with warm summer ("Cfb" in Köppen-Geiger climate classification; Kottek et al., 2006). Climate norms for the 1981–2010 period, measured at the nearest meteorological station (Bern-Zollikofen; 46.9286°N, 7.4313°E, 553 m a.s.l.),were as follows: an average annual air temperature of +8.8°C, the mean temperature of the hottest month (July), +18.3°C, and for the coldest (January), −0.4°C. Due to the effect of its being located Alpine region, the annual precipitation is 1059 mm and August is the wettest month (116 mm), while February is the driest (55 mm) (Seiz and Foppa, 2011).

The sampling site is situated at the margin of the rural area (46.9283°N, 7.4317°E, 609 m) at the NE slope of Gurten Hill, on the outskirt of the City of Bern (**Figure 1**). The trees are mature specimens, growing next to each other and lacking any dominance effects. According to the Swiss regional phenological phases, the unfolding occurs at end of April (Rutishauser et al., 2007), similar to in central Europe (Michelot et al., 2012). Our observed unfolding date for the oak was a couple of days before, i.e., 116 Day Of Year (DOY), while for the beech and yew it happened between 116 and 122 DOY in 2012.

Sampling of Plant Material

Two deciduous and one evergreen species, namely two pedunculate oaks (*Quercus robur* L.), a European beech (*Fagus*

sylvatica L.), and a European yew (*Taxus baccata* L.) were sampled. The current foliar samples were harvested weekly in the first 3 weeks of the leaf development and with 11–21 day intervals through the rest of the growing season, depending on site accessibility, until leaf fall in October. The sampling covers the time period 116–281 DOY for the oaks, and 122–281 DOY for the beech and yew. The leaves were harvested from the same branches (at the same height and from the same side of the canopy) consistently to minimize sampling noise due to within-canopy isotopic variability. Two oak trees were sampled regularly to provide a verification test for the representativeness and reproducibility of the isotopic signature yielded by the sampling strategy. Regarding the sequentially harvested branch of the oaks, the right marginal oak tree was situated in a much more shaded sector of the canopy, while the branch of the middle tree bowed over a more open area, and hence represented a relatively more sun-exposed canopy sector. The samples were dried at room temperature, subsequently the broadleaves were separated for vein and blade. The "vein" phase contained the petiole, mid rib, and major veins.

Cellulose Extraction

Foliar samples were finely cut, and this was followed by measurements of the dry weight (DW) of the vein and blade samples before any chemical preparations were used, and afterwards alpha-cellulose was extracted and dried by the modified Jayme-Wise method (Boettger et al., 2007).

The alpha-cellulose yield results represent the ratio of the mass of DW to the mass of extracted alpha-cellulose yield, and are given in %.

$$\text{alpha cellulose yield [\%]} = \frac{m_{\text{alpha cellulose yield}}}{m_{\text{dry weight}}} \times 100 \quad (1)$$

We should note that, depending on the grain size of fine cut leaf/needle material, some part of it will stick in the pores of the filter during extraction. Therefore, we use the expression "yield" rather than "content" of cellulose.

Sample Preparation and Isotope Analysis of Leaf Alpha-cellulose

Overall, separated leaves of two deciduous and complete needles of an evergreen tree species were analyzed. The samples were homogenized to decrease possible uncertainties (Laumer et al., 2009), subsequently oven-dried (50°C), weighed, and placed in silver capsules.

For triple-isotope analysis (Loader et al., 2014), we used conventional Isotope Ratio Mass Spectrometry coupled to a Pyrolyser (HEKAtech GmbH, Germany), which is similar to the previously used TC/EA the technical details are described by Leuenberger (2007). However, this approach is extended to measurements of non-exchangeable hydrogen of alpha-cellulose using the on-line equilibration method (Filot et al., 2006). Results are reported in permil (‰) relative to the Vienna Pee Dee Belemnite (VPDB) for carbon and to Vienna Standard Mean

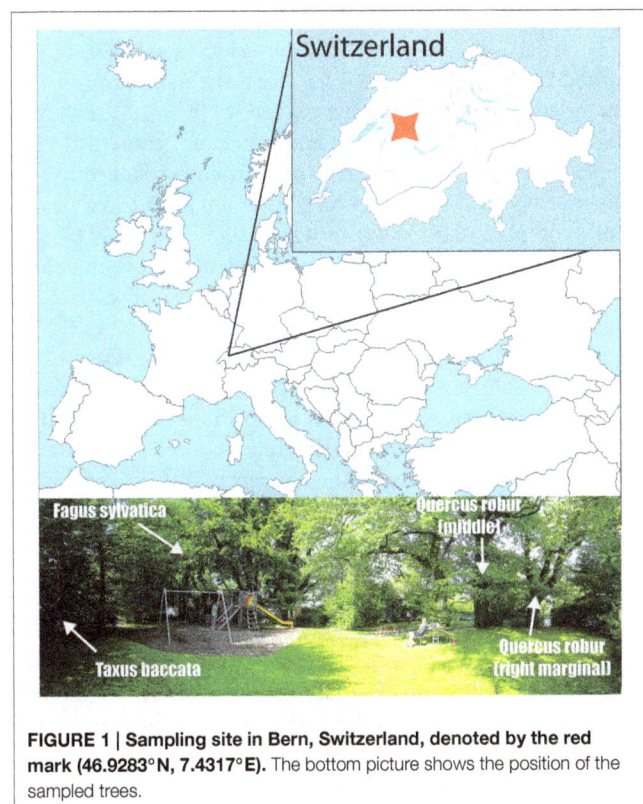

FIGURE 1 | Sampling site in Bern, Switzerland, denoted by the red mark (46.9283°N, 7.4317°E). The bottom picture shows the position of the sampled trees.

Ocean Water (VSMOW) for hydrogen and oxygen (Coplen, 1994), using the traditional "δ" notation

$$\delta X[‰] = \left(\frac{R_{Sample} - R_{Standard}}{R_{Standard}} \right) \cdot 1000 \qquad (2)$$

where X stands for D, ^{13}C, ^{18}O, R stands for ^{2}H/^{1}H, ^{13}C/^{12}C, or ^{18}O/^{16}O in a sample and standard, correspondingly.

Samples were measured in triplicates and their standard deviations were 90% to within 0.2‰ for carbon, 93% to within 3‰ for hydrogen and 90% to within 0.3‰ for oxygen isotopes with overall mean standard deviations of 0.12, 1.83, and 0.28‰ for carbon, hydrogen and oxygen, respectively.

Meteorological Data

Meteorological datasets are supported by the STARTWAVE atmospheric water monitoring system (Morland et al., 2006), providing a homogenized high resolution (10 min) air temperature and relative humidity dataset in Bern. However, we measured daily precipitation events and their hydrogen and oxygen isotopic composition in 2012. The δD and δ^{18}O of precipitation are expressed in permil (‰) relative to the Vienna Standard Mean Ocean Water (VSMOW) for hydrogen and oxygen (**Figure 2**).

Mechanistic Modeling

For a better understanding of the seasonal leaf expansion (leaf cellulose), we used the Péclet-modified Craig-Gordon Model (PMCG) (Kahmen et al., 2011) to calculate hydrogen and oxygen isotopic composition of leaf water taking into consideration results from related other studies (Dongmann et al., 1974; Farquhar and Lloyd, 1993; Roden and Ehleringer, 2000).

$$\delta D_{LC} = p_{ex} \times p_x \times (\delta D_{SW} + \varepsilon_{HH})$$
$$+ (1 - p_{ex}p_x) \times (\delta D_{LW} + \varepsilon_{HA}) \qquad (3)$$
$$\delta^{18}O_{LC} = p_{ex} \times p_x \times (\delta^{18}O_{SW} + \varepsilon_{OH})$$
$$+ (1 - p_{ex}p_x) \times (\delta^{18}O_{LW} + \varepsilon_{OA}) \qquad (4)$$

where δD_{LC} and $\delta^{18}O_{LC}$ represent the hydrogen and oxygen isotopic composition of leaf cellulose, p_{ex} represents the proportion of exchangeable oxygen in cellulose formed from soluble carbohydrates, p_x represents the proportion of source water, δD_{SW} and $\delta^{18}O_{SW}$ represent the isotopic composition of source water, $\varepsilon_{HH}/\varepsilon_{OH}$ and $\varepsilon_{HA}/\varepsilon_{OA}$ represent the overall heterotrophic and autotrophic fractionation between hydrogen and oxygen atoms of source water and photosynthate, respectively. Finally, δD_{LW} and $\delta^{18}O_{LW}$ represent the isotopic composition of leaf water.

Consequently, the hydrogen and oxygen isotopic composition of leaf water (δD_{LW}, $\delta^{18}O_{LW}$) can be determined as

$$\delta D_{LW} = \frac{\delta D_{LC} - p_{ex}p_x (\delta D_{SW} + \varepsilon_{HH})}{(1 - p_{ex}p_x)} - \varepsilon_{HA} \quad (5)$$

$$\delta^{18}O_{LW} = \frac{\delta^{18}O_{LC} - p_{ex}p_x (\delta^{18}O_{SW} + \varepsilon_{OH})}{(1 - p_{ex}p_x)} - \varepsilon_{OA} \quad (6)$$

The leaf water isotope values were calculated using current measurements of leaf cellulose δD_{LC} and $\delta^{18}O_{LC}$, 0.4 for p_{ex}, 0.65 for p_x after Sternberg and Cernusak (Cernusak et al., 2005; Sternberg, 2009), 21.7‰ for ε_{OH} (Guy et al., 1993), and 27‰ for ε_{OA} (Epstein et al., 1977; Deniro and Epstein, 1981; Sternberg and Deniro, 1983). In addition, during water uptake no isotope fractionation occurs (White et al., 1985), therefore we defined p_x as 1 regarding the isotopic composition of vein water. This assumption is equivalent to the vein leaf water value being equal to the source water in the case of the absence of a back diffusion effect of enriched leaf water from blades into veins. ε_{HH} and ε_{HA} are calculated as +158 and −171‰, respectively (Yakir and Deniro, 1990). For the isotopic composition of source water, we assumed homogenized δD and δ^{18}O of site precipitation as indicated by the observed low fractionation in the soil at a different site (Treydte et al., 2014). The aforementioned mechanistic models were applied to all species and leaf compartments to determine the δD_{LW} and $\delta^{18}O_{LW}$.

Statistics

The *t*-tests of independent-samples were conducted on unrelated groups of tree species or leaf compartments (vein vs. blade) using cellulose yield and isotope values as parameters to compare their group means objectively. The difference was taken to be significant for $p < 0.05$. STATISTICA (StatSoft Inc., 2011) was used for calculations.

RESULTS

Alpha-cellulose Yield of DW

Similar foliar cellulose yields were observed in the case of the oaks and beech (**Figure 3**). Yew needles had a significantly lower alpha-cellulose yield than deciduous species ($p < 10^{-8}$ and $p < 10^{-7}$ in any comparison for full and extraction failure excluded datasets, respectively), which may be plausibly explained by the distinct anatomical structures (i.e., lower amounts of xylem) of needles compared to broadleaves (Spjut, 2007).

Anomalous cellulose yields indicated some extraction failure in three cases (see **Figure 3**). A vein sample for beech yielded an unusually low amount of cellulose, indicating sample loss, probably due to the opening of the filter bag. However, all of the derived isotope ratios fitted nicely into the seasonal trends, and therefore the isotopic values were kept. Two yew needles (the fourth and the last) yielded a higher amount of cellulose than the rest of the samples. Although the corresponding δ^{13}C-values fit nicely into the seasonal trend, the water isotopes clearly depart from the inter-annual trend. It may be supposed that further depleted residuum remained in the filter bag besides the alpha-cellulose with regard to its oxygen and hydrogen composition. These values were omitted from the discussion.

As may be logically expected, the lowest amount of alpha cellulose was found in the leaf samples taken at the initiation of the leaf development. Moreover, at this time the blades and veins of oak and beech similarly started from 5 to 7%. In the course of the following days, the characteristics of the blades and veins evolved, and simultaneously the alpha-cellulose yield increased,

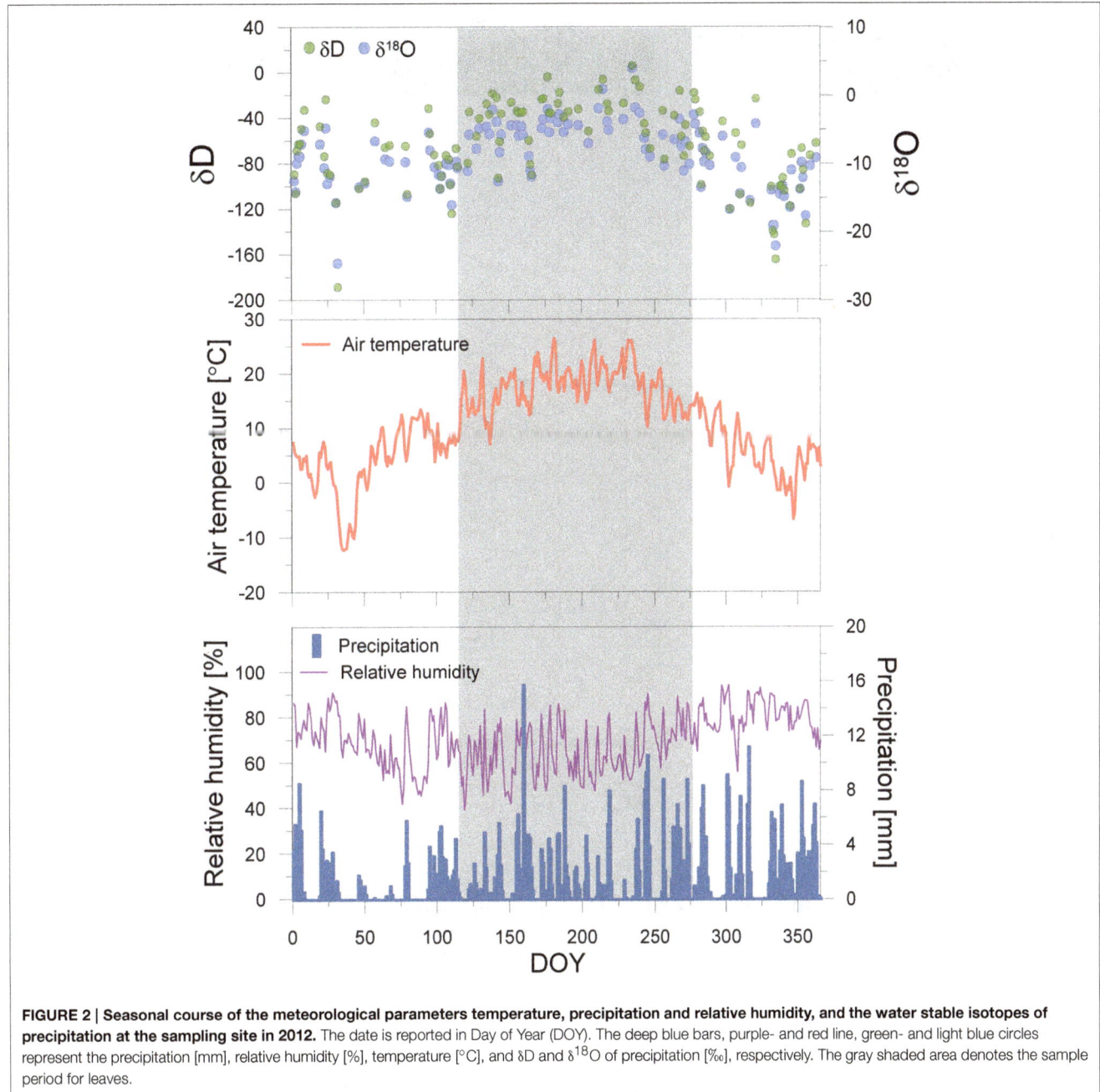

FIGURE 2 | Seasonal course of the meteorological parameters temperature, precipitation and relative humidity, and the water stable isotopes of precipitation at the sampling site in 2012. The date is reported in Day of Year (DOY). The deep blue bars, purple- and red line, green- and light blue circles represent the precipitation [mm], relative humidity [%], temperature [°C], and δD and δ18O of precipitation [‰], respectively. The gray shaded area denotes the sample period for leaves.

until the signals became continuously consistent for blades (10–12%), and also for veins (16%). Based on these characteristically distinct patterns, seasonal leaf development can be divided into a juvenile and a mature phenophase. Increasing cellulose yields indicate that foliage was still in the expansion phase (juvenile period) while the stabilized cellulose yields suggest that the foliage is complete (mature phase). The juvenile phase was shortest (<16 days) for beech, evidencing more dynamic blade expansion than it is for oak (Gond et al., 1999; Bequet et al., 2011), while a twice-as-long juvenile phase was observed for oak and yew (>30 days).

More specifically, for oaks, after the mature phase had begun, the cellulose content did not change, inducing a steady offset with a ~7% higher cellulose yield for veins. On the contrary, for beech, the alpha-cellulose yield of blades and veins frequently fluctuated between 1.5 and 5%, with an average offset of ~4% during the mature phase. For yew, the cellulose yield signal is weaker during the juvenile phase period. It starts from ~3% and became steady at around the 5% level representing a balanced leaf expansion.

In addition, the juvenile and mature phases of foliar development were accompanied by characteristic changes in the stable isotope composition of leaf cellulose.

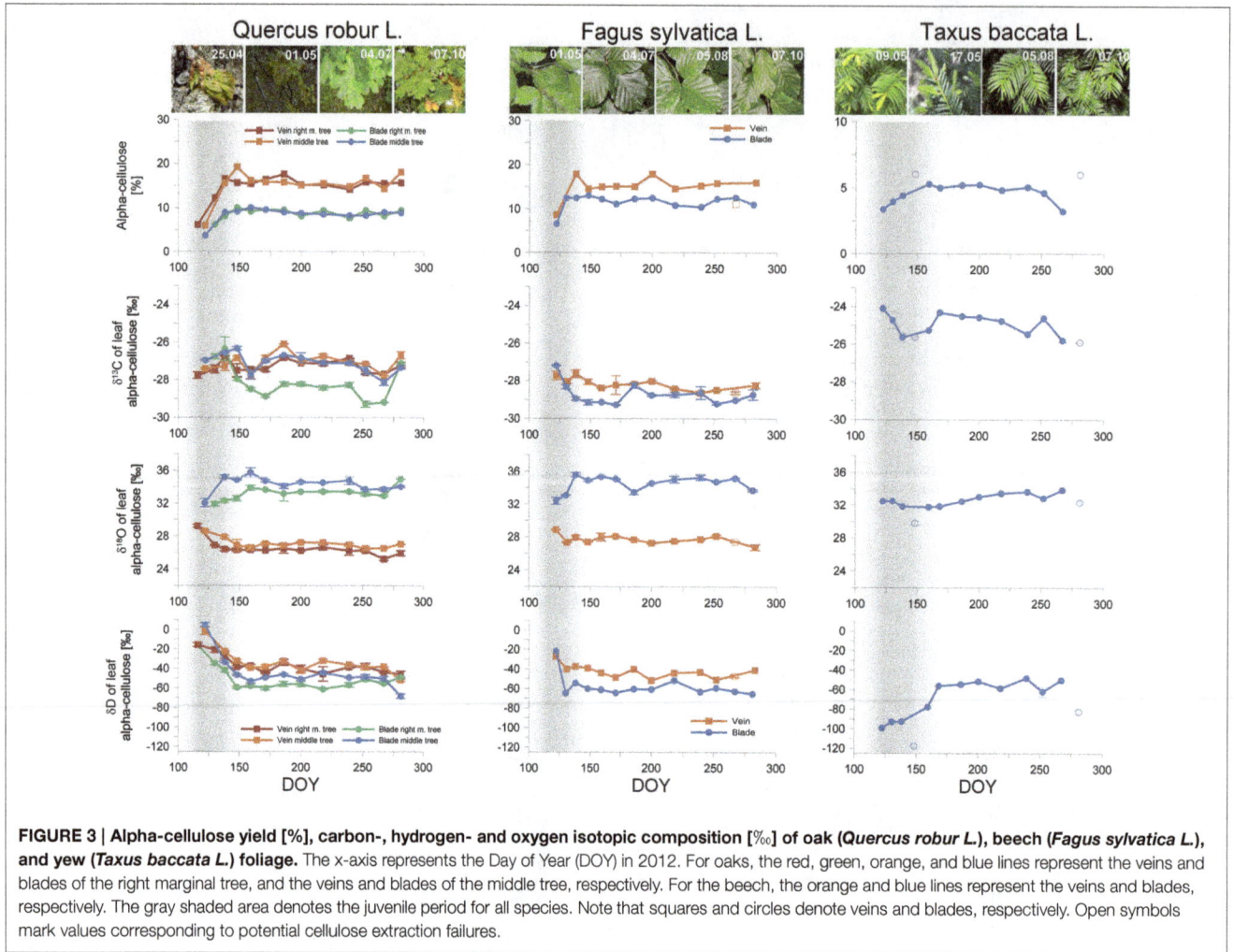

FIGURE 3 | Alpha-cellulose yield [%], carbon-, hydrogen- and oxygen isotopic composition [‰] of oak (*Quercus robur L.*), beech (*Fagus sylvatica L.*), and yew (*Taxus baccata L.*) foliage. The x-axis represents the Day of Year (DOY) in 2012. For oaks, the red, green, orange, and blue lines represent the veins and blades of the right marginal tree, and the veins and blades of the middle tree, respectively. For the beech, the orange and blue lines represent the veins and blades, respectively. The gray shaded area denotes the juvenile period for all species. Note that squares and circles denote veins and blades, respectively. Open symbols mark values corresponding to potential cellulose extraction failures.

The Carbon Isotopic Composition of Leaf Alpha-cellulose

We documented the expected seasonality only for blades and needles, starting with more enriched values and decreasing afterwards, reflecting the previously documented seasonal carbon isotope pattern in tree rings (Damesin et al., 1998; Helle and Schleser, 2004). The veins of oaks and beech show no clear difference in $\delta^{13}C$ during the growing season (**Figure 3**). However, the carbon isotopic composition of oak leaves had a slightly more positive mean during the mature phase ($\delta^{13}C_{oak-blade} = -27.84 \pm 0.86‰$, $\delta^{13}C_{oak-vein} = -27.10 \pm 0.42‰$) compared to beech leaves ($\delta^{13}C_{beech-blade} = -28.91 \pm 0.34‰$, $\delta^{13}C_{beech-vein} = -28.33 \pm 0.22‰$). Nevertheless, yew needles presented more positive values than either deciduous species, with an annual mean of $\delta^{13}C_{yew} = -24.96 \pm 0.54‰$ (**Figure 3**).

The Oxygen Isotopic Composition of Leaf Alpha-cellulose

Oxygen isotope measurements show a distinct individual characterization of the foliage type (broadleaf vs. needle), and for structural parts (vein vs. blade) as well. Veins are less enriched compared to blades, starting from 29‰ and decreasing to 27‰ for oaks and starting from 29‰ and decreasing to 28‰ for beech during the juvenile phase (**Figure 3**). In the mature phase, the $\delta^{18}O$ veins of both tree species were practically constant. Blade cellulose started from 32‰ in both cases, increasing during the juvenile phase up to 36‰, remaining constant during the mature phase. The yew represented an annual mean value of $\delta^{18}O_{yew} = 33.21 \pm 0.52‰$, with no visible sign of a juvenile-mature difference (**Figure 3**). However, the mean values of the mature phase are $\delta^{18}O_{oak-vein} = 26.57 \pm 0.57$ ‰, $\delta^{18}O_{oak-blade} = 33.89 \pm 0.65‰$ for the oaks, and $\delta^{18}O_{beech-vein} = 27.61 \pm 0.42‰$, $\delta^{18}O_{beech-blade} = 34.85 \pm 0.57‰$ for the beech.

The Hydrogen Isotopic Composition of Leaf Alpha-cellulose

The seasonal variability found for hydrogen isotopes is not similar to that of carbon and oxygen isotopes.

Since the blades and veins follow a distinct signal in the cases of carbon and oxygen, in the case of hydrogen the only difference observed was the higher enrichment of veins compared to blades for deciduous species (**Figure 3**). The first oak leaf samples started with highly enriched values and decreased during the juvenile phase. The mean values are $\delta D_{oak-blade} = -53.34 \pm 6.26‰$, $\delta D_{oak-vein} = -39.94 \pm 5.24‰$ during the mature phase. However, the offset between the first sampling and the mean mature phase values varies between 18 and 57‰. For beech, a very similar seasonal pattern occurred, but due to the shorter juvenile phase the offset was less resolved. The annual dataset begins from $-20‰$ for vein and blade and decreases to $-40‰$ for vein and $-60‰$ for blade, in a similar way to the case of the oaks, described above. The mean values during the mature phase are $\delta D_{beech-blade} = -60.31 \pm 3.69‰$, $\delta D_{beech-vein} = -38.75 \pm 4.43‰$, resulting in an offset of 20–40‰.

The yew presented a different intra-annual hydrogen isotopic composition change (**Figure 3**). During the juvenile phase, the lowest δD-values were observed between -100 and $-90‰$ which subsequently increased during the mature phase to $\delta D_{yew} = -54.38 \pm 4.86‰$, resulting in an offset of 44.65‰.

The Results of Leaf Water Simulation

Applying the Péclet-modified Craig-Gordon Model (PMCG) we estimated hydrogen and oxygen isotopic composition of leaf water regarding veins, blades and needles (**Figure 4**). Since, we defined p_x as 1 in order not to perform fractionation during water transport from the roots to the leaf level, the δD and $\delta^{18}O$ of veins became less positive than blades. $\delta^{18}O_{vein}$ of deciduous species varies between 5.8 and 8.37‰ ($\delta^{18}O_{vein-oakmid} = 7.10‰$, $\delta^{18}O_{vein-oakright} = 5.84‰$ and $\delta^{18}O_{vein-beech} = 8.37‰$) while the

$\delta^{18}O_{blade}$ varies in the range from 12.61 to 14.29‰. Yew, similarly as oak and beech blades, represents 12.05‰ for $\delta^{18}O$ of needle.

We found an offset between veins and blades but similar trends for hydrogen isotopes. δD of oak and beech veins are less positive ($\delta D_{vein-oakmid} = 27.84‰$, $\delta D_{vein-oakright} = 24.53‰$ and $\delta D_{vein-beech} = 20.45‰$) than blades ($\delta D_{blade-oakmid} = 60.52‰$, $\delta D_{blade-oakright} = 55.27‰$ and $\delta D_{blade-beech} = 48.39‰$). As it was found for oxygen isotopes, yew fits to deciduous species, namely $\delta D_{yew} = 51.09‰$.

DISCUSSION

The stable carbon, oxygen, and hydrogen isotope composition of the alpha-cellulose of foliage of the three species was monitored throughout the growing season in 2012 (from 115 DOY until 281 DOY). The investigated trees are growing next to each other, and therefore the atmospheric (weather, ambient CO_2, etc.) and soil moisture conditions were identical for all trees. Consequently, we can compare the leaf cellulose formation of the studied species by investigating the isotopic composition of all elements (C–O–H) of cellulose molecules. Besides, tree ring anatomy and the pathways of metabolic processes also change during growing season (Hayes, 2001). At the very beginning of the growing season, when the tree is not yet ready for CO_2 fixation, remobilized carbohydrates dominate the sources. These reserves are converted into metabolic intermediates and applied to proteins, carbohydrates, lipids, etc. through biosynthesis, providing indirect sources of carbohydrates from previous years. During the juvenile phase, similarly to the EW phase in tree rings,

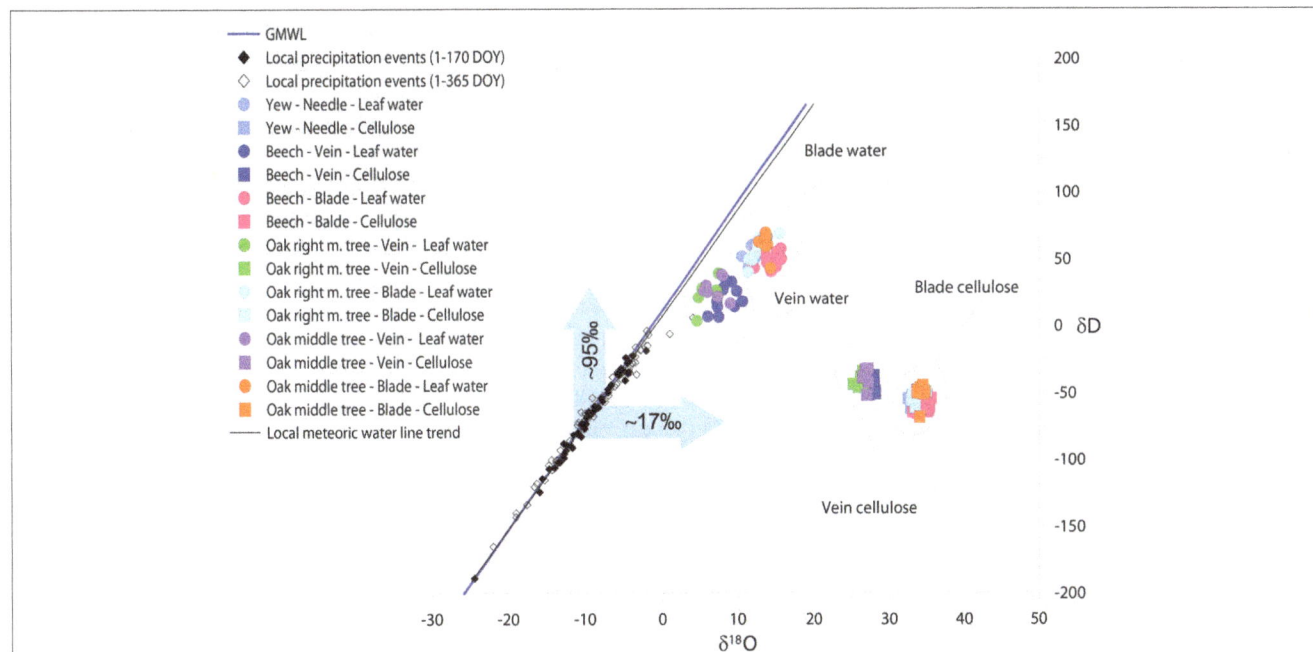

FIGURE 4 | The relationship between δD and δ¹⁸O of local precipitation, species specific calculated leaf water, and measured leaf cellulose in mature leaves (DOY 150–365). Different colors denote different trees. Circles, squares, open, and closed diamonds denote the isotopic composition of leaf water, cellulose, precipitation events during 1–365 DOY and precipitation events during 1–170 DOY, respectively. The blue arrows denote the isotopic shift of δD and δ¹⁸O of vein leaf water compared to the annual mean δD and δ¹⁸O of precipitation. The blue and black lines represent the global and local meteoric water line, respectively.

the dominance of indirect carbohydrate sources decreases besides the increase of products formed by photosynthesis. Despite depending on climate conditions during summer, carbohydrates can be formed by either heterotrophic metabolism under non-optimal or by autotrophic metabolism under optimal conditions (Kimak and Leuenberger, 2015), new assimilates control the isotopic signal in tree-ring cellulose. In addition, depending on whether the tree utilizes carbohydrates directly or indirectly, the product serves as an input of downstream processes fingerprinted in the stable isotopic composition of tree ring cellulose (Kagawa et al., 2006).

Coherency between Individual Trees

The sampled oak trees presented a strikingly similar seasonal course regarding cellulose yield and water isotope signals. The carbon isotope fluctuations presented are also similar, although the mature blade of the right marginal oak tree (**Figure 3** green line) had an offset (1–1.5‰) compared to other blades or veins. This offset can plausibly be explained by a shaded canopy effect with decreased light conditions (see Section Sampling of the Plant Material) since decreasing irradiance in the canopy generates more negative $\delta^{13}C$-values in the foliage (Hanba et al., 1997) due to the decreased assimilation rate and consequently the increased c_i/c_a ratio relative to the stomatal conductance (Jones, 1998). Note, that the offset is not presented for the first and the last leaf samples therefore we cannot exclude other influences that could explain this variable offset.

The strong coherency found between the sampled trees verifies the sampling strategy and confirms that representative isotopic signals can be observed allowing a physiological interpretation of the plants.

Quercus Robur

The alpha-cellulose yield of the leaf dry weight shows well-expressed differences between phenological phases, clarifying the switch from the juvenile to mature phases of leaves (**Figure 3**). The juvenile phase started with a lower alpha-cellulose content and increased until the leaf area maximizes along with the maturing of the leaves. The juvenile-mature turning point came at ~150 DOY, consequently the juvenile phase lasted ~35 days. In addition, the alpha-cellulose yield of veins and blades indicates that veins are more readily comparable with the alpha-cellulose yield of tree rings than blades, although the observed veins contain less alpha-cellulose than tree rings (Rowell, 2012). However, the steady offset between veins and blades manifest in progressive leaf development after budburst.

Although the expected seasonality of $\delta^{13}C$ is not clearly represented in the case of oaks, the values of $\delta^{13}C$ show a good agreement with previous investigation documenting $\delta^{13}C$-values in the range from 25.5 to 28‰ regarding oak bulk leaf material (Eglin et al., 2009; Maunoury-Danger et al., 2012) having an inter-annual variability of 2.5‰ that was also documented for other species living under different environmental condition (Ehleringer et al., 1992; Wang and Schjoerring, 2012).

Veins for both oak trees indicate positive peaks during the juvenile phase, and a decreasing trend until the mature phase. However, on 159 DOY, notably less enriched values occurred

that might be an influence of stomatal conductance, consequently on the ratio of c_i/c_a and discrimination against heavy isotopes by Rubisco, indicated by considerable amounts of precipitation between 150 DOY and 175 DOY (**Figures 2, 3**). In addition, this observation was also made in the case of the beach leaves and the yew too. On the contrary, the oxygen isotopic composition of veins and blades showed remarkable differences over the growing season, indicating either (i) alternative metabolic processes, or (ii) alternative sources for oxygen isotopes. Since the source water is directly transported from the soil through the roots and stem to the branches without any fractionation (White et al., 1985), the responsible fractionation are probably the biochemical processes acting at leaf level. The backward projected juvenile trends of vein and blade $\delta^{18}O$ cellulose point to a shared source water with a ~30.5‰ oxygen isotope composition. Over time the $\delta^{18}O_{blade}$ presented an increasing trend due to leaf formation related to higher evaporative enrichment, until it is maximized by the end of the juvenile phase, and remained consistent during mature phase. Consequently, the difference between the usage of indirectly and directly synthetized carbohydrates is approximately 3‰. Contrary, $\delta^{18}O_{vein}$ showed a moderately decreasing trend during the juvenile phase that we assign to the diminishing role of more enriched soluble carbohydrates and sap water stored over winter that is mixed and is subsequently dominated by source water. However, if we assume the first point was formed from soluble carbohydrates and old sap water and $\delta^{18}O_{LW}$ is formed from source water we can establish the difference (at ~1–2‰).

The seasonality of the hydrogen isotopic composition of oak leaf cellulose showed individual trends compared to the previously discussed isotopes. In particular, the δD of veins and blades represented more enriched values induced by indirectly produced carbohydrates and decreased until the end of the juvenile period. These patterns provide clear evidence for the transition between different metabolic pathways. However, due to the higher exchange rate of hydrogen within biochemical processes (Augusti et al., 2006), no significant difference was found for the δD of veins and blades of the particular trees, namely ~0‰ for the middle tree and approximately −20‰ for the right marginal tree. Despite this, we found strong correlations between the veins and blades of the right marginal tree ($r^2 = 0.82$) and the middle tree ($r^2 = 0.96$), the blades are slightly shifted toward more depleted compositions during mature phase. The offset between blades and veins is 16.75‰ for the right marginal tree, 11.81‰ for the middle tree.

However, the difference between indirectly and directly formed carbohydrates is 22, 24, 53, and 21‰ for the blades and veins of the right marginal tree, blades, and veins of the middle tree, respectively. These differences are supported by δD of leaf wax n-alkanes (nC-27 and nC-29) measurements (Newberry et al., 2015).

Fagus Sylvatica

Although, oak and beech are both deciduous, many differences occur in their tree ring anatomy. The beech xylem ring includes many vessels scattered over the entire year, therefore the visual determination of juvenile and mature phases is more ambiguous

(Čufar et al., 2008). However, our observations verify the existing characteristics of intra-annual differences in phenological phases. Alpha-cellulose yield results document faster leaf expansion and consequently, a shorter leaf growing period (<16 days) for beech than for oaks and yew. The veins, similar to those of oaks, contain a higher proportion of alpha-cellulose than blades, but the alpha-cellulose content of the blades is closer to that of veins, and dissimilar to that of oak, resulting in a mean offset of ~4% during mature phase (**Figure 3**).

Despite beech leaf cells being formed faster than those of oaks, we found a clear seasonality in intra-annual isotope measurements. The carbon isotopic composition of leaf alpha-cellulose confirms the aforementioned faster leaf cellulose development. Thus, a decreasing trend occurred during the juvenile phase, starting with more enriched $\delta^{13}C$ until the leaf area and therefore, the assimilation rate is maximized. In addition, veins did not mirror the blades' variability. We assign this to the higher usage of reserved carbohydrates.

The $\delta^{18}O$ of leaf compartments showed similar results to those of the oaks (**Figure 3**). Since blades are supposed to follow the gradual increase of evaporative enrichment at leaf level, it displays a positive trend through the juvenile period. However, we found lower seasonal variation for veins, which might indicate a faster consumption of the remaining soluble carbohydrates and sap water from winter and the earlier usage of soil water.

The hydrogen isotopic composition of veins and blades demonstrates practically the same seasonal variability as found in oaks. The observed fractionation between indirect and direct cellulose synthesis can be estimated as 17.4‰ for veins and 38.6‰ for blades.

Taxus Baccata

As the single evergreen in our study, the yew has a species-specific annual xylem ring anatomy (Vaganov et al., 2009) and isotopic composition (Barbour et al., 2002). Despite, intra-annual changes in $\delta^{13}C$ of the tree ring cellulose, typically reflecting environmental conditions (Leavitt and Long, 1991), the influence of previous years' storage on current year xylem formation is also well-documented (Schubert and Jahren, 2011). The needles of Taxus baccata include more mesophyll than vascular tissue (Schirone et al., 2010), therefore these results should be regarded as a blade-dominated mixture of blade and vein signals. This has to be taken into consideration when a comparison is to be made with the broadleaf patterns.

We found a lower percentage of alpha-cellulose yield for yew than for deciduous species, and the seasonal variability also decreased. In particular, needle formation was accompanied by a moderated increase from 3 to 5% during the juvenile phase, and remained constant up to the end of the mature phase.

Surprisingly, the $\delta^{13}C$ of needles shows similar activity for the yew as compared to the oaks and beech, namely the impact of previous years' carbohydrates fingerprinted in juvenile phase tissues, indicating a negative tendency at the start of growing period. Moreover, the less enriched $\delta^{13}C$-values after 150 DOY, documented for all species, might be the aforementioned collective impact of heavy precipitation events between 150 and 175 DOY (**Figure 3**). In addition, yew needles have significantly

($p < 10^{-14}$ in any combination) more enriched values (3–5‰) than deciduous species.

Due to the lack of leaf tissue separation for yew, the oxygen isotopic composition represents also a bulk of veins and blades, including values closer to that of the $\delta^{18}O$ of deciduous species. It might be reasoned that the domination of blade tissues within needles is responsible for this finding.

In contrast to the oaks and beech, the yew has a particular seasonality in the case of hydrogen isotopes (**Figure 3**). Due to the presence of needles from previous years, the variability of indirect and direct cellulose synthesis disappears and the direct assimilates became the major driver for cellulose formation during the juvenile phase, too. Therefore, the δD-values of leaf cellulose represent the current water conditions, consequently the δD of precipitation. These results are in agreement with the findings of Treydte et al. (2014) namely, that the oxygen isotopic composition of tree ring cellulose predominantly mirrors the seasonal trends of the source water.

The Characterization of Leaf Compartments Based on δD and $\delta^{18}O$ Measurements and Model Results

Since leaf cellulose synthesis, in common with tree-ring cellulose synthesis, is supported by the reserves from previous years during the leaf expansion stage, the carbon isotopic composition of leaf cellulose is also influenced by stored carbohydrate, affecting or even masking the expected seasonal variations at the leaf level. On the contrary, the stored carbohydrates have much less influence on hydrogen and oxygen isotopes, due to their exchangeability. There is no exchange during cellulose synthesis in the positions of certain oxygen atoms (Waterhouse et al., 2013), thus decreasing the impact which might be expected of carbohydrate storage on cellulose formation during the juvenile phase.

The exchangeability of hydrogen isotopes in cellulose varies widely. Generally, the higher the isotope exchange rate for specific molecule positions is, the more representative the isotope composition of that particular position for physiological processes is. While in the opposite case, specific molecule positions are influenced primarily by source water, and so the actual climate conditions (Augusti et al., 2008). Because in alpha-cellulose the majority of hydrogen positions with strong isotope exchange rates are represented, our results of bulk veins and blades, we expect that they should reflect mainly the seasonal changes in metabolic processes.

The water isotopes in vein and blade cellulose clearly indicate temporal and spatial patterns, including (i) the seasonal changes of the autotrophic and heterotrophic metabolisms, and (ii) distinction of cellulose synthesis between leaf tissues. For species where the separation of blades and veins was available, the oxygen and hydrogen signal is clearly different at the beginning of leaf formation driven by autotrophic metabolism compared to the mature phase, where both oxygen and hydrogen fractionations are dominated by heterotrophic metabolism (**Figure 3**). During the leaf expansion phase, the evaporative enrichment and the Péclet-effect exercise an increasing effect on oxygen isotopes, until the leaves are fully developed, and the aforementioned processes are maximized for blades, while these impacts are less effective for veins.

As with oxygen, hydrogen isotopes also represent the heterotrophic metabolism at the beginning of leaf formation and become less enriched due to the accelerated exchange of $NADP^+$-NADPH reactions during photosynthesis. However, the hydrogen results of blades and veins share the same seasonality but with more enriched isotopic composition characterizes the veins (15–20‰).

As may be seen in **Figure 4**, a discrepancy was found between the isotopic signal of vein water and the measured isotopic ratio of site precipitation (source water), where hydrogen and oxygen isotopic compositions are more enriched. Due to the fact that water uptake is not accompanied by isotope fractionation, we attribute this to a mixing effect, particularly as vein cellulose is formed not only by the source water but by a mixture of non-enriched soil water and evaporative enriched laminar water originating from the photosynthetic active part of the leaf. These effects result in an approximately +95‰ enrichment for hydrogen and +17‰ for oxygen for modeled isotopic composition of vein water compared to local precipitation events.

CONCLUSIONS

We investigated the stable carbon, oxygen, and hydrogen isotope signatures of the leaf samples of two deciduous and one evergreen species during the growing season in 2012. Although, the deciduous species have good coherences, the characterization by the stable isotope approach revealed species-specific variations between phenology phases and activities. Yew has the lowest and beech the highest degree of leaf expansion activity. However, after the leaves are complete (mature phase), the values represent

the foliage variability of the sampling site, exhibiting very good coherences.

Separated leaf tissues combined with the triple isotope approach allow us to document seasonality in leaf development driven by carbohydrate sources from previous years on the one hand, and by the alternation of heterotrophic and autotrophic metabolism on the other. In addition, due to the different utilization of carbon, oxygen, and hydrogen isotopes, alternative metabolic processes produce individual seasonality for each isotope for the studied species.

AUTHOR CONTRIBUTIONS

AK performed the stable isotope analyses and drafted the first version of the manuscript. ZK conceived the study and collected the samples. ML contributed to the evaluation of the results. All authors provided comments to improve the manuscript.

FUNDING

AK and ML acknowledge the financial support for the Swiss National Science Foundation (SNF-iTREE, project number: 136295). ZK and ML thanks the Scientific Exchange Programme between Switzerland and Hungary (project code: 10.255) for the financial support. ZK acknowledges the support for the "Lendület" program of the Hungarian Academy of Sciences (LP2012-27/2012).

ACKNOWLEDGMENTS

The authors are thankful to Peter Nyfeler for the laboratory assistance to Sonja Keel, Ansghar Kahmen, and Marc Andre Cormier for the helpful comments on the study.

REFERENCES

Augusti, A., Betson, T. R., and Schleucher, J. (2006). Hydrogen exchange during cellulose synthesis distinguishes climatic and biochemical isotope fractionations in tree rings. *New Phytol.* 172, 490–499. doi: 10.1111/j.1469-8137.2006.01843.x

Augusti, A., Betson, T. R., and Schleucher, J. (2008). Deriving correlated climate and physiological signals from deuterium isotopomers in tree rings. *Chem. Geol.* 252, 1–8. doi: 10.1016/j.chemgeo.2008.01.011

Barbaroux, C., and Bréda, N. (2002). Contrasting distribution and seasonal dynamics of carbohydrate reserves in stem wood of adult ring-porous sessile oak and diffuse-porous beech trees. *Tree Physiol.* 22, 1201–1210. doi: 10.1093/treephys/22.17.1201

Barbaroux, C., Bréda, N., and Dufrêne, E. (2003). Distribution of above-ground and below-ground carbohydrate reserves in adult trees of two contrasting broad-leaved species (*Quercus petraea* and *Fagus sylvatica*). *New Phytol.* 157, 605–615. doi: 10.1046/j.1469-8137.2003.00681.x

Barbour, M. M., Roden, J. S., Farquhar, G. D., and Ehleringer, J. R. (2004). Expressing leaf water and cellulose oxygen isotope ratios as enrichment above source water reveals evidence of a Peclet effect. *Oecologia* 138, 426–435. doi: 10.1007/s00442-003-1449-3

Barbour, M., Walcroft, A. S., and Farquhar, G. (2002). Seasonal vartiaton in $\delta^{13}C$ and $\delta^{18}O$ of cellulose from growth rings of *Pinus radiata*. *Plant Cell Environ.* 25, 1483–1499. doi: 10.1046/j.0016-8025.2002.00931.x

Bequet, R., Campioli, M., Kint, V., Vansteenkiste, D., Muys, B., and Ceulemans, R. (2011). Leaf area index development in temperate oak and beech forests is driven by stand characteristics and weather conditions. *Trees* 25, 935–946. doi: 10.1007/s00468-011-0568-4

Boettger, T., Haupt, M., Knöller, K., Weise, S. M., Waterhouse, J. S., Rinne, K. T., et al. (2007). Wood cellulose preparation methods and mass spectrometric analyses of $\delta^{13}C$, $\delta^{18}O$, and nonexchangeable δ^2H values in cellulose, sugar, and starch: an interlaboratory comparison. *Anal. Chem.* 79, 4603–4612. doi: 10.1021/ac0700023

Cernusak, L. A., Farquhar, G. D., and Pate, J. S. (2005). Environmental and physiological controls over oxygen and carbon isotope composition of Tasmanian blue gum, *Eucalyptus globulus*. *Tree Physiol.* 25, 129–146. doi: 10.1093/treephys/25.2.129

Cernusak, L. A., Tcherkez, G., Keitel, C., Cornwell, W. K., Santiago, L. S., Knohl, A., et al. (2009). Why are non-photosynthetic tissues generally ^{13}C enriched compared with leaves in C3 plants? Review and synthesis of current hypotheses. *Funct. Plant Biol.* 36, 199–213. doi: 10.1071/FP08216

Coplen, T. B. (1994). Reporting of stable hydrogen, carbon, and oxygen isotopic abundances (Technical Report). *Pure Appl. Chem.* 66, 273–276. doi: 10.1351/pac199466020273

Čufar, K., Prislan, P., De Luis, M., and Gričar, J. (2008). Tree-ring variation, wood formation and phenology of beech (Fagus sylvatica) from a representative site in Slovenia, SE Central Europe. *Trees* 22, 749–758. doi: 10.1007/s00468-008-0235-6

Damesin, C., Rambal, S., and Joffre, R. (1998). Seasonal and annual changes in leaf δ¹³C in two cooccurring Mediterranean oaks: relations to leaf growth and drought progression. *Funct. Ecol.* 12, 778–785. doi: 10.1046/j.1365-2435.1998.00259.x

Dawson, T. E., and Ehleringer, J. R. (1993). Isotopic enrichment of water in the "woody" tissues of plants: implications for plant water source, waer uptake, and other studies which use the stable isotopic composition of cellulose. *Geochim. Cosmochim. Acta* 57, 3487–3492. doi: 10.1016/0016-7037(93)90554-A

Deniro, M. J., and Epstein, S. (1981). Isotopic composition of cellulose from aquatic organisms. *Geochim. Cosmochim. Acta* 45, 1885–1894. doi: 10.1016/0016-7037(81)90018-1

Dongmann, G., Nürnberg, H. W., Förstel, H., and Wagener, K. (1974). On the enrichment of $H_2^{18}O$ in the leaves of transpiring plants. *Radiat. Environ. Biophys.* 11, 41–52. doi: 10.1007/BF01323099

Eglin, T., Fresneau, C., Lelarge-Trouverie, C., Francois, C., and Damesin, C. (2009). Leaf and twig δ¹³C during growth in relation to biochemical composition and respired CO_2. *Tree Physiol.* 29, 777–788. doi: 10.1093/treephys/tpp013

Ehleringer, J. R., and Dawson, T. E. (1992). Water uptake by plants: perspectives from stable isotope composition. *Plant Cell Environ.* 15, 1073–1082. doi: 10.1111/j.1365-3040.1992.tb01657.x

Ehleringer, J. R., Phillips, S. L., and Comstock, J. P. (1992). Seasonal variation in the carbon isotopic composition of desert plants. *Funct. Ecol.* 6, 396–404. doi: 10.2307/2389277

Epstein, S., Thompson, P., and Yapp, C. J. (1977). Oxygen and hydrogen isotopic ratios in plant cellulose. *Science* 198, 1209–1215. doi: 10.1126/science.198.4323.1209

Farquhar, G. D., Ehleringer, J. R., and Hubick, K. T. (1989). Carbon isotope discrimination and photosynthesis. *Annu. Rev. Plant Physiol. Plant Mol. Biol.* 40, 503–537. doi: 10.1146/annurev.pp.40.060189.002443

Farquhar, G. D., and Lloyd, J. (1993). "Carbon and oxygen isotope effects in the exchange of carbon dioxide between terrestrial plants and the atmosphere," in *Stable Isotopes and Plant Carbon-water Relations*, eds B. Saugier, J. R. Ehleringer, A. E. Hall, and G. D. Farquhar (San Diego,CA: Academic Press), 47–70.

Farquhar, G., Oleary, M., and Berry, J. (1982). On the relationship between carbon isotope discrimination and the intercellular carbon dioxide concentration in leaves. *Funct. Plant Biol.* 9, 121–137. doi: 10.1071/pp9820121

Feuillat, F., Dupouey, J. L., Sciama, D., and Keller, R. (1997). A new attempt at discrimination between *Quercus petraea* and *Quercus robur* based on wood anatomy. *Can. J. For. Res.* 27, 343–351. doi: 10.1139/x96-174

Filot, M. S., Leuenberger, M., Pazdur, A., and Boettger, T. (2006). Rapid online equilibration method to determine the D/H ratios of non-exchangeable hydrogen in cellulose. *Rapid Commun. Mass Spectrom.* 20, 3337–3344. doi: 10.1002/rcm.2743

Gessler, A., Ferrio, J. P., Hommel, R., Treydte, K., Werner, R. A., and Monson, R. K. (2014). Stable isotopes in tree rings: towards a mechanistic understanding of isotope fractionation and mixing processes from the leaves to the wood. *Tree Physiol.* 34, 796–818. doi: 10.1093/treephys/tpu040

Gessler, A., Tcherkez, G., Peuke, A. D., Ghashghaie, J., and Farquhar, G. D. (2008). Experimental evidence for diel variations of the carbon isotope composition in leaf, stem and phloem sap organic matter in *Ricinus communis*. *Plant Cell Environ.* 31, 941–953. doi: 10.1111/j.1365-3040.2008.01806.x

Gilson, A., Barthes, L., Delpierre, N., Dufrêne, É., Fresneau, C., and Bazot, S. (2014). Seasonal changes in carbon and nitrogen compound concentrations in a *Quercus petraea* chronosequence. *Tree Physiol.* 34, 716–729. doi: 10.1093/treephys/tpu060

Gond, V., De Pury, D. G. G., Veroustraete, F., and Ceulemans, R. (1999). Seasonal variations in leaf area index, leaf chlorophyll, and water content; scaling-up to estimate fAPAR and carbon balance in a multilayer, multispecies temperate forest. *Tree Physiol.* 19, 673–679. doi: 10.1093/treephys/19.10.673

Guy, R. D., Fogel, M. L., and Berry, J. A. (1993). Photosynthetic fractionation of the stable isotopes of oxygen and carbon. *Plant Physiol.* 101, 37–47.

Hanba, Y. T., Mori, S., Lei, T. T., Koike, T., and Wada, E. (1997). Variations in leaf d13C along a vertical profile of irradiance in a temperate Japanese forest. *Oecologia* 110, 253–261. doi: 10.1007/s004420050158

Hayes, J. M. (2001). Fractionation of carbon and hydrogen isotopes in biosynthetic processes. *Rev. Mineral. Geochem.* 43, 225–277. doi: 10.2138/gsrmg.43.1.225

Helle, G., and Schleser, G. H. (2004). Beyond CO_2 fixation by Rubisco—an interpretation of ¹³C/¹²C variations in tree rings from novel intra-seasonal studies on broad-leaf trees. *Plant Cell Environ.* 27, 367–380. doi: 10.1111/j.0016-8025.2003.01159.x

Huang, J. G., Deslauriers, A., and Rossi, S. (2014). Xylem formation can be modeled statistically as a function of primary growth and cambium activity. *New Phytol.* 203, 831–841. doi: 10.1111/nph.12859

Jones, H. G. (1998). Stomatal control of photosynthesis and transpiration. *J. Exp. Bot.* 49, 387–398. doi: 10.1093/jxb/49.Special_Issue.387

Kagawa, A., Sugimoto, A., and Maximov, T. C. (2006). Seasonal course of translocation, storage and remobilization of ¹³C pulse-labeled photoassimilate in naturally growing Larix gmelinii saplings. *New Phytol.* 171, 793–803. doi: 10.1111/j.1469-8137.2006.01780.x

Kahmen, A., Sachse, D., Arndt, S. K., Tu, K. P., Farrington, H., Vitousek, P. M., et al. (2011). Cellulose d¹⁸O is an index of leaf-to-air vapor pressure difference (VPD) in tropical plants. *Proc. Natl. Acad. Sci.* 108, 1981–1986. doi: 10.1073/pnas.1018906108

Keel, S. G., and Schädel, C. (2010). Expanding leaves of mature deciduous forest trees rapidly become autotrophic. *Tree Physiol.* 30, 1253–1259. doi: 10.1093/treephys/tpq071

Keel, S. G., Siegwolf, R. T., Jäggi, M., and Korner, C. (2007). Rapid mixing between old and new C pools in the canopy of mature forest trees. *Plant Cell Environ.* 30, 963–972. doi: 10.1111/j.1365-3040.2007.01688.x

Keeling, C. D., Whorf, T. P., Wahlen, M., and van der Plichtt, J. (1995). Interannual extremes in the rate of rise of atmospheric carbon dioxide since 1980. *Nature* 375, 666–670. doi: 10.1038/375666a0

Kern, Z., Patkó, M., Kázmér, M., Fekete, J., Kele, S., and Pályi, Z. (2013). Multiple tree-ring proxies (earlywood width, latewood width and δ¹³C) from pedunculate oak (*Quercus robur L.*), Hungary. *Quaternary Int.* 293, 257–267. doi: 10.1016/j.quaint.2012.05.037

Kimak, A., and Leuenberger, M. (2015). Are carbohydrate storage strategies of trees traceable by early–latewood carbon isotope differences? *Trees* 29, 859–870. doi: 10.1007/s00468-015-1167-6

Kottek, M., Grieser, J., Beck, C., Rudolf, B., and Rubel, F. (2006). World Map of the Köppen-Geiger climate classification updated. *Meteorol. Z.* 15, 259–263. doi: 10.1127/0941-2948/2006/0130

Kress, A., Saurer, M., Siegwolf, R. T. W., Frank, D. C., Esper, J., and Bugmann, H. (2010). A 350 year drought reconstruction from Alpine tree ring stable isotopes. *Glob. Biogeochem. Cycles* 24:GB2011. doi: 10.1029/2009GB003613

Laumer, W., Andreu, L., Helle, G., Schleser, G. H., Wieloch, T., and Wissel, H. (2009). A novel approach for the homogenization of cellulose to use micro-amounts for stable isotope analyses. *Rapid Commun. Mass Spectrom.* 23, 1934–1940. doi: 10.1002/rcm.4105

Leavitt, S. W., and Long, A. (1985). Stable-carbon isotopic composition of maple sap and foliage. *Plant Physiol.* 78, 427–429. doi: 10.1104/pp.78.2.427

Leavitt, S. W., and Long, A. (1991). Seasonal stable-carbon isotope variability in tree rings: possible paleoenvironmental signals. *Chem. Geol.* 87, 59–70. doi: 10.1016/0168-9622(91)90033-S

Leuenberger, M. (2007). "To what extent can ice core data contribute to the understanding of plant ecological developments of the past?," in *Stable Isotopes as Indicators of Ecological Change*, eds T. E. Dawson and R. T. W (Siegwolf. Elsevier Inc), 211–233.

Loader, N. J., Street-Perrott, F. A., Daley, T. J., Hughes, P. D. M., Kimak, A., Levaniè, T., et al. (2014). Simultaneous determination of stable carbon, oxygen, and hydrogen isotopes in cellulose. *Anal. Chem.* 87, 376–380. doi: 10.1021/ac502557x

Loescher, W. H., McCamant, T., and Keller, J. D. (1990). Carbohydrate reserves, translocaation and storage in woody plant roots. *Hortscience* 25, 274–281.

Luo, Y.-H., and Sternberg, L. (1992). Spatial D/H heterogeneity of leaf water. *Plant Physiol.* 99, 348–350. doi: 10.1104/pp.99.1.348

Luo, Y.-H., Steinberg, L., Suda, S., Kumazawa, S., and Mitsui, A. (1991). Extremely low D/H Ratios of photoproduced hydrogen by cyanobacteria. *Plant Cell Physiol.* 32, 897–900.

Maunoury-Danger, F., Chemidlin Prevost Boure, N., Ngao, J., Berveiller, D., Brechet, C., Dufrene, E., et al. (2012). Carbon isotopic signature of CO_2 emitted by plant compartments and soil in two temperate deciduous forests. *Ann. For. Sci.* 70, 173–183. doi: 10.1007/s13595-012-0249-5

McCarroll, D., and Loader, N. J. (2004). Stable isotopes in tree rings. *Quat. Sci. Rev.* 23, 771–801. doi: 10.1016/j.quascirev.2003.06.017

Michelot, A., Simard, S., Rathgeber, C., Dufrêne, E., and Damesin, C. (2012). Comparing the intra-annual wood formation of three European species (*Fagus sylvatica*, *Quercus petraea* and *Pinus sylvestris*) as related to leaf phenology and non-structural carbohydrate dynamics. *Tree Physiol.* 32, 1033–1045. doi: 10.1093/treephys/tps052

Morland, J., Deuber, B., Feist, D. G., Martin, J., Nyeki, S., Kämpfer, N., et al. (2006). The STARTWAVE atmospheric water database. *Atmosp. Chem. Phys.* 6, 2039–2056. doi: 10.5194/acp-6-2039-2006

Newberry, S. L., Kahmen, A., Dennis, P., and Grant, A. (2015). n-Alkane biosynthetic hydrogen isotope fractionation is not constant throughout the growing season in the riparian tree *Salix viminalis*. *Geochim. Cosmochim. Acta* 165, 75–85. doi: 10.1016/j.gca.2015.05.001

Raffalli-Delerce, G., Masson-Delmotte, V., Dupouey, J. L., Stievenard, M., Breda, N., and Moisselin, J. M. (2004). Reconstruction of summer droughts using tree-ring cellulose isotopes: a calibration study with living oaks from Brittany (western France). *Tellus B* 56, 160–174. doi: 10.1111/j.1600-0889.2004.00086.x

Roden, J. S., and Ehleringer, J. R. (1999). Observations of hydrogen and oxygen isotopes in leaf water confirm the Craig-Gordon model under wide-ranging environmental conditions. *Plant Physiol.* 120, 1165–1174. doi: 10.1104/pp.120.4.1165

Roden, J. S., and Ehleringer, J. R. (2000). There is no temperature dependence of net biochemical fractionation of hydrogen and oxygen isotopes in tree-ring cellulose. *Isotopes Environ. Health Stud.* 36, 303–317. doi: 10.1080/10256010008036389

Roden, J. S., Johnstone, J. A., and Dawson, T. E. (2009). Intra-annual variation in the stable oxygen and carbon isotope ratios of cellulose in tree rings of coast redwood (*Sequoia sempervirens*). *Holocene* 19, 189–197. doi: 10.1177/0959683608098959

Rowell, R. M. (2012). *Handbook of Wood Chemistry and Wood Composites.* Boca Raton, FL: CRC Press.

Rutishauser, T., Luterbacher, J., Jeanneret, F., Pfister, C., and Wanner, H. (2007). A phenology-based reconstruction of interannual changes in past spring seasons. *J. Geophys. Res.* 112:G04016. doi: 10.1029/2006JG000382

Saurer, M., Borella, S., Schweingruber, F., and Siegwolf, R. (1997). Stable carbon isotopes in tree rings of beech: climatic versus site-related influences. *Trees* 11, 291–297. doi: 10.1007/s004680050087

Saurer, M., Spahni, R., Frank, D. C., Joos, F., Leuenberger, M., Loader, N. J., et al. (2014). Spatial variability and temporal trends in water-use efficiency of European forests. *Glob. Change Biol.* 20, 3700–3712. doi: 10.1111/gcb.12717

Scheidegger, Y., Saurer, M., Bahn, M., and Siegwolf, R. (2000). Linking stable oxygen and carbon isotopes with stomatal conductance and photosynthetic capacity: a conceptual model. *Oecologia* 125, 350–357. doi: 10.1007/s004420000466

Schirone, B., Ferreira, R. C., Vessella, F., Schirone, A., Piredda, R., and Simeone, M. C. (2010). *Taxus baccata* in the Azores: a relict form at risk of imminent extinction. *Biodivers. Conserv.* 19, 1547–1565. doi: 10.1007/s10531-010-9786-0

Schubert, B. A., and Jahren, A. H. (2011). Quantifying seasonal precipitation using high-resolution carbon isotope analyses in evergreen wood. *Geochim. Cosmochim. Acta* 75, 7291–7303. doi: 10.1016/j.gca.2011.08.002

Seiz, G., and Foppa, N. (2011). National Climate Observing System of Switzerland (GCOS Switzerland). *Adv. Sci. Res.* 6, 95–102. doi: 10.5194/asr-6-95-2011

Sessions, A. L. (2006). Seasonal changes in D/H fractionation accompanying lipid biosynthesis in Spartina alterniflora. *Geochim. Cosmochim. Acta* 70, 2153–2162. doi: 10.1016/j.gca.2006.02.003

Spjut, R. W. (2007). A phytogeographical analysis of Taxus (Taxaceae) based on leaf anatomical characters. *J. Bot. Res. Inst. Tex.* 1, 291–332. Available online at: http://www.jstor.org/stable/41971412?seq=1#page_scan_tab_contents

StatSoft Inc. (2011). *STATISTICA (Data Analysis Software System), Version 10.* Available online at: www.statsoft.com.

Sternberg, L. (2009). Oxygen stable isotope ratios of tree-ring cellulose: the next phase of understanding. *New Phytol.* 181, 553–562. doi: 10.1111/j.1469-8137.2008.02661.x

Sternberg, L., and Deniro, M. J. D. (1983). Biogeochemical implications of the isotopic equilibrium fractionation factor between the oxygen atoms of acetone and water. *Geochim. Cosmochim. Acta* 47, 2271–2274. doi: 10.1016/0016-7037(83)90049-2

Stokes, V. J., Morecroft, M. D., and Morison, J. I. L. (2009). Comparison of leaf water use efficiency of oak and sycamore in the canopy over two growing seasons. *Trees* 24, 297–306. doi: 10.1007/s00468-009-0399-8

Studer, M. S., Siegwolf, R. T. W., Leuenberger, M., and Abiven, S. (2015). Multi-isotope labelling of organic matter by diffusion of ^{2}H/^{18}O-H$_2$O vapour and ^{13}C-CO$_2$ into the leaves and its distribution within the plant. *Biogeosciences* 12, 1865–1879. doi: 10.5194/bg-12-1865-2015

Tipple, B. J., Berke, M. A., Doman, C. E., Khachaturyan, S., and Ehleringer, J. R. (2013). Leaf-wax n-alkanes record the plant-water environment at leaf flush. *Proc. Natl. Acad. Sci. U.S.A.* 110, 2659–2664. doi: 10.1073/pnas.1213875110

Treydte, K., Boda, S., Graf Pannatier, E., Fonti, P., Frank, D., Ullrich, B., et al. (2014). Seasonal transfer of oxygen isotopes from precipitation and soil to the tree ring: source water versus needle water enrichment. *New Phytol.* 202, 772–783. doi: 10.1111/nph.12741

Vaganov, E. A., Schulze, E. D., Skomarkova, M. V., Knohl, A., Brand, W. A., and Roscher, C. (2009). Intra-annual variability of anatomical structure and d^{13}C values within tree rings of spruce and pine in alpine, temperate and boreal Europe. *Oecologia* 161, 729–745. doi: 10.1007/s00442-009-1421-y

Wang, L., and Schjoerring, J. K. (2012). Seasonal variation in nitrogen pools and ^{15}N/^{13}C natural abundances in different tissues of grassland plants. *Biogeosciences* 9, 1583–1595. doi: 10.5194/bg-9-1583-2012

Waterhouse, J. S., Cheng, S., Juchelka, D., Loader, N. J., McCarroll, D., Switsur, V. R., et al. (2013). Position-specific measurement of oxygen isotope ratios in cellulose: isotopic exchange during heterotrophic cellulose synthesis. *Geochim. Cosmochim. Acta* 112, 178–191. doi: 10.1016/j.gca.2013.02.021

White, J. W. C. (1989). "Stable hydrogen isotope ratios in plants: a review of current theory and some potential applications," in *Stable Isotopes in Ecological Research*, eds P. W. Rundel, J. R. Ehleringer and K. A. Nagy (New York, NY: Springer), 142–162.

White, J. W. C., Cook, E. R., Lawrence, J. R., and Wallace, S., B. (1985). The DH ratios of sap in trees: implications for water sources and tree ring DH ratios. *Geochim. Cosmochim. Acta* 49, 237–246. doi: 10.1016/0016-7037(85)90207-8

Yakir, D. (1992). Variations in the natural abundance of oxygen-18 and deuterium in plant carbohydrates. *Plant Cell Environ.* 15, 1005–1020. doi: 10.1111/j.1365-3040.1992.tb01652.x

Yakir, D., and Deniro, M. J. (1990). Oxygen and hydrogen isotope fractionation during cellulose metabolism in *Lemna gibba* L. *Plant Physiol.* 93, 325–332. doi: 10.1104/pp.93.1.325

Conflict of Interest Statement: The authors declare that the research was conducted in the absence of any commercial or financial relationships that could be construed as a potential conflict of interest.

11

Assessing Conifer Ray Parenchyma for Ecological Studies: Pitfalls and Guidelines

Georg von Arx [1]*, Alberto Arzac [2], José M. Olano [3] and Patrick Fonti [1]

[1] Landscape Dynamics Research Unit, Swiss Federal Institute for Forest, Snow and Landscape Research WSL, Birmensdorf, Switzerland, [2] Departamento de Biología Vegetal y Ecología, Facultad de Ciencia y Tecnología, Universidad del País Vasco, Leioa, Spain, [3] Departamento de Ciencias Agroforestales, Escuela Universitaria de Ingenierías Agrarias, Instituto Universitario de Investigación en Gestión Forestal Sostenible-Universidad de Valladolid, Soria, Spain

Edited by:
Achim Braeuning,
University of Erlangen-Nuremberg,
Germany

Reviewed by:
Eryuan Liang,
Chinese Academy of Sciences, China
Francisco Artigas,
Meadowlands Environmental
Research Institute, USA

*Correspondence:
Georg von Arx
georg.vonarx@wsl.ch

Ray parenchyma is an essential tissue for tree functioning and survival. This living tissue plays a major role for storage and transport of water, nutrients, and non-structural carbohydrates (NSC), thus regulating xylem hydraulics and growth. However, despite the importance of rays for tree carbon and water relations, methodological challenges hamper knowledge about ray intra- and inter-tree variability and its ecological meaning. In this study we provide a methodological toolbox for soundly quantifying spatial and temporal variability of different ray features. Anatomical ray features were surveyed in different cutting planes (cross-sectional, tangential, and radial) using quantitative image analysis on stem-wood micro-sections sampled from 41 mature Scots pines (*Pinus sylvestris*). The percentage of ray surface (PERPAR), a proxy for ray volume, was compared among cutting planes and between early- and latewood to assess measurement-induced variability. Different tangential ray metrics were correlated to assess their similarities. The accuracy of cross-sectional and tangential measurements for PERPAR estimates as a function of number of samples and the measured wood surface was assessed using bootstrapping statistical technique. Tangential sections offered the best 3D insight of ray integration into the xylem and provided the most accurate estimates of PERPAR, with 10 samples of $4\,mm^2$ showing an estimate within ±6.0% of the true mean PERPAR (relative 95% confidence interval, CI95), and 20 samples of $4\,mm^2$ showing a CI95 of ±4.3%. Cross-sections were most efficient for establishment of time series, and facilitated comparisons with other widely used xylem anatomical features. Earlywood had significantly lower PERPAR (5.77 vs. 6.18%) and marginally fewer initiating rays than latewood. In comparison to tangential sections, PERPAR was systematically overestimated (6.50 vs. 4.92%) and required approximately twice the sample area for similar accuracy. Radial cuttings provided the least accurate PERPAR estimates. This evaluation of ray parenchyma in conifers and the presented guidelines regarding data accuracy as a function of measured wood surface and number of samples represent an important methodological reference for ray quantification, which will ultimately improve the understanding of the fundamental role of ray parenchyma tissue for the performance and survival of trees growing in stressed environments.

Keywords: cutting plane, measured wood surface, measurement accuracy, non-structural carbohydrates (NSC), number of samples, ray density, ray dimensions, ray volume

INTRODUCTION

Parenchyma in the xylem is a neglected living tissue in ecological and eco-physiological research, despite its essential role for tree carbon and water relations and thus tree survival in response to environmental stochasticity. More specifically, it is important for buffering temporal imbalances between carbon uptake and consumption (McDowell and Sevanto, 2010), and has been principally documented to be involved in the storage and transport of water and nutrients (Witt and Sauter, 1994; Gartner et al., 2000). Additionally, parenchyma provides carbohydrates and water for refilling embolized conduits (Salleo et al., 2009; Brodersen and Mcelrone, 2013; Spicer, 2014), is involved in the formation of heartwood (Bamber, 1976), the defense against pathogens (Hudgins et al., 2006), in wounding processes (Arbellay et al., 2010, 2012), and contributes to the mechanical strength of the wood (Burgert and Eckstein, 2001; Fonti and Frey, 2002).

Due to this important role for tree functioning, pioneer studies on parenchyma mainly focused on its anatomical characterization and quantification among different species (e.g., Myer, 1922; Bannan, 1937; Brown et al., 1949). It has been observed that ray parenchyma represents 3–12% and 5–42% of the overall stem xylem tissue in conifers and angiosperms, respectively (Myer, 1922; Panshin and De Zeeuw, 1980; Brandes and Barros, 2008), making ray parenchyma by far the most important living tissue in the xylem despite substantial contribution of axial parenchyma in some angiosperm species (Spicer, 2014). Parenchyma in the stem sapwood stores a large proportion (25–40%) of the overall non-structural carbohydrates (NSC) reserves of a tree, which makes it a more important NSC reservoir than the phloem or the leaves (Jacquet et al., 2014).

Recently, triggered by the increasing awareness of global change impact on forest ecosystems and tree mortality (Allen et al., 2010; Choat et al., 2012; Lloret et al., 2013; Hereş et al., 2014), there is a renewed and increased need for a better ecological understanding of the role of parenchyma (mainly rays) for tree performance and survival in response to environmental variability (e.g., Olano et al., 2013; Fonti et al., 2015), and particularly with respect to carbon and water balance in stressed trees (Pruyn et al., 2005; Salleo et al., 2006; Esteban et al., 2012; Barnard et al., 2013; Gruber et al., 2013; Rosell et al., 2014). However, the understanding of the ecological role of this tissue has been hampered, mainly due to the scarce data about how parenchyma varies along ecological gradients and/or during tree life, and due to the diversity of methods used for quantification, making comparisons difficult. A review of the measured parameters used for ecological investigations is summarized in **Table 1**. In general, ray properties have been quantified either in terms of "amount" as proxy for investigating variability in tree vigor, growth conditions and storage capacity; and in terms of "spatial distribution" to explore the integration of rays in the 3D xylem network, which is key for its function as storage and transport tissue connecting the xylem and phloem (Spicer, 2014). These observations seem to reveal that within-tree variability is relatively small compared to the variability observed among individuals (Wimmer and Grabner, 2000; Olano et al.,

2013; Fonti et al., 2015). Moreover, it has been observed that ray features co-vary at within-tree level related to organ and age, thus reflecting functional needs and/or allometric scaling in relation to variation in conduit size (Carlquist, 1982; Lev-Yadun and Aloni, 1995; Fonti et al., 2015). An open question in this regard is whether ray characteristics might also show some intra-annual variability. Overall, the observed variations suggest a rather strong intrinsic control of the ray characteristics (Aloni, 2013), leaving less room for identifying plastic responses to external environmental factors. Nevertheless, higher percentage of ray tissue appears to be related to tree performance (**Table 1**); and annual time series of anatomical features such as the number of initiating rays and the percentage of ray tissue to respond to environmental conditions (Eckstein, 2013; Olano et al., 2013).

Disentangling the comparably weak environmental signal from the strong intrinsic and allometric component of ray variability requires both a mechanistic understanding of the processes triggering ray formation, and an appropriate methodological toolbox for a sufficiently accurate quantification. However, the relatively few studies performed so far were using different metrics (e.g., ray area, density, size) quantified on different wood cutting planes (i.e., tangential, cross-sectional, or radial planes; see **Figure 1**) of varying measured wood surface, thus hampering comparisons among studies. Consequently, many questions as to the best practice remain unanswered. The studies involving parenchyma quantification have mostly used tangential sections, in which typically a surface of just about $1\,mm^2$ per sample was measured (e.g., Pruyn et al., 2005; Esteban et al., 2010, 2012), but sometimes surfaces up to $4\,mm^2$ were also considered (Fonti et al., 2015). Few ecological studies— for practical reasons when building up time-series along tree rings—have also quantified rays in cross-sections from 5-mm increment cores (Olano et al., 2013; Arzac, 2014; Fonti et al., 2015), whereas radial sections have hardly been, if at all, used for ray quantification.

In this study we aimed at clarifying some methodological issues related to the quantification of ray tissue in conifer wood and giving some guidelines for ecological studies. In particular we focused on the influence of the methodological approach on the quantification of ray features by comparing several ray metrics in (i) different cutting planes, (ii) within the annual ring (earlywood vs. latewood), and (iii) as a function of the measured wood surface. These aspects were addressed using *Pinus sylvestris* L. as a model conifer species.

MATERIALS AND METHODS

Study Material

Pinus sylvestris (Scots Pine) is a sub-boreal evergreen conifer with one of the largest distribution ranging between Scotland and northeast Asia (Nikolov and Helmisaari, 1992) at altitudes between 0 to 2700 m asl. As most conifers, it contains uniseriate rays characterized by a single layer of parenchyma cells (Lev-Yadun and Aloni, 1995; **Figure 1**) facultatively embraced by tracheid cells on the upper and lower extremities, and sometimes enclosing resin ducts (Brown et al., 1949).

TABLE 1 | Literature review of measured anatomical ray parameters and the inferred ecological interpretation of their variability.

Ray parameter	Plane*	Documented and hypothesized variability and potential functional meaning
AMOUNT AND DIMENSIONS: RELATED TO TREE VIGOR, GROWTH CONDITIONS, MAXIMUM STORAGE CAPACITY		
Percentage of ray surface (PERPAR)	C, T, R	-Increase in individuals with higher growth rate (Harlow, 1927; Bannan, 1954, 1965; Gregory, 1977; Fonti et al., 2015), dominant individuals (Myer, 1922), lowland individuals (Myer, 1922), fully illuminated individuals (Hartig, 1894), individuals with larger leaf area (Gartner et al., 2000), with increasing aridity (Esteban et al., 2010, 2012; von Arx et al., 2012), after wounding (Lev-Yadun and Aloni, 1992; Arbellay et al., 2010, 2012) -Responses to short-term climate variability (Olano et al., 2013)
Number of continuing rays (in time series)	C, R	-Responses to short-term climate variability (Olano et al., 2013)
Number of initiating rays (in time series)	C, R	-Responses to short-term climate variability (Olano et al., 2013)
Ray height in metric units	T, R	- Increase with conduit size (Carlquist, 1982; Lev-Yadun and Aloni, 1995; Fonti et al., 2015), distance from pith (DeSmidt, 1922; Lev-Yadun and Aloni, 1995; Falcon-Lang, 2005; Aloni, 2013), with increasing growth rate (Bannan, 1965; Lev-Yadun, 1998; but see White and Robards, 1966), also at same distance from pith (Bannan, 1937) - Decreased close to wounding (Lev-Yadun and Aloni, 1992) - Responses to short-term climate variability (Wimmer and Grabner, 2000), environmental stress (Wimmer, 2002)
Ray height in cell counts	T, R	- Increased from pith to bark (Weinstein, 1926; Gregory, 1977; Lev-Yadun and Aloni, 1992), with growth rate (Bannan, 1965; Gregory and Romberger, 1975)
Ray width in metric units	T	- Increase with growth rate (White and Robards, 1966; Wimmer and Grabner, 2000) - Decreased after wounding (Lev-Yadun and Aloni, 1992)
Ray cell size	T, R	- Increased after wounding (Lev-Yadun and Aloni, 1992)
SPATIAL DISTRIBUTION: FUNCTIONAL RAY INTEGRATION IN THE 3D XYLEM NETWORK, SEASONAL CHANGES IN STORAGE SPACE REQUIREMENTS		
Ray density (No·mm^{-2})	C, T, R	- Increase from bark to pith (Lev-Yadun, 1998; Falcon-Lang, 2005), in branches compared to stems (Jaccard, 1915), with higher growth rate [(White and Robards, 1966; Fonti et al., 2015); but see (Lev-Yadun, 1998)], after wounding (Lev-Yadun and Aloni, 1992), with increasing aridity (Esteban et al., 2012) - Reduced due to air pollution (Von Schneider and Halbwachs, 1989)
Ray cell density (No·mm^{-2})	T	- Responses to short-term climate variability (Wimmer and Grabner, 2000) - Increase with aridity (Esteban et al., 2012)
Position of initiating rays (in time series)	C, R	- Preferential initiation of rays in latewood to meet seasonal changes in storage space requirements?

Parameter accessible in: C, cross-section; T, tangential section; R, radial section.

Wood samples for this study were obtained in May 2013 in the xeric Pfynwald forest located in the Swiss Rhone Valley (Valais). The climate at this site is continental with 600–700 mm of annual precipitation and a mean annual temperature of 10.1°C (data from 1981 to 2010 of the weather station of Sion, at 20 km distance from the site, MeteoSwiss). Cores with a diameter of 10 mm were taken at stem breast height from 40 mature Scots pines of 10–13 m height, 12–30 cm dbh, and aged between 45 and 135 years. Ring widths in the cores were measured using a LinTab device and time series were cross-dated using COFECHA (Grissino-Mayer, 2001) to assign each ring to the correct calendar year. The 20 outmost annual rings from all the 40 cores were considered for cross-sectional analyses, whereas a subset of six cores was used for additional tangential and radial analyses at three locations each separated by five to seven annual rings. In addition, a single stem disc of 14 cm diameter (58 years old) was collected from a felled individual, split in three radial bars of 20 mm width and 20 mm height at offsets of 120°, and used for tangential and radial analyses in the sapwood rings 1989, 2002, and 2012.

Sample Processing and Measurement of Ray Features

Wood sample where processed according to Schweingruber and Poschlod (2005) for the anatomical characterization and quantification of the rays. Therefore, 10–15-μm permanent thin sections were produced with a sledge microtome (Gärtner et al., 2014), placed on a slide and stained with Alcian blue (1% solution in acetic acid) and safranin (1% solution in ethanol) to differentiate between unlignified (blue) and lignified (red) cells. Afterwards, sections were dehydrated using a series of ethanol solutions of increasing concentrations, washed with xylol, and finally permanently preserved by embedding them into Eukitt glue (Gärtner and Schweingruber, 2013). Overlapping images covering the entire samples were captured with a Nikon D90 digital camera mounted on a Nikon Eclipse 50i optical microscope with 40× magnification and merged using PTGUI

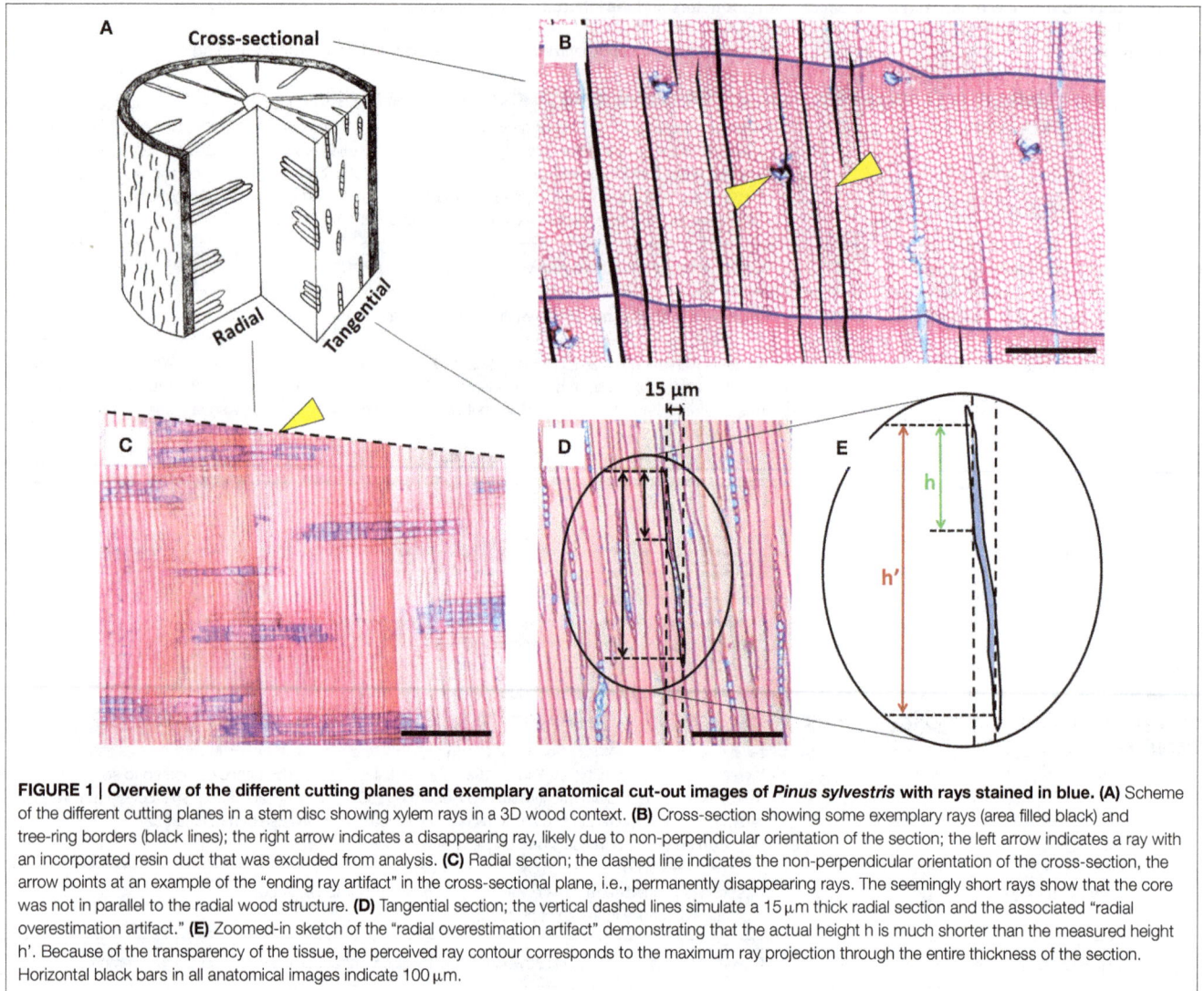

FIGURE 1 | Overview of the different cutting planes and exemplary anatomical cut-out images of *Pinus sylvestris* with rays stained in blue. (A) Scheme of the different cutting planes in a stem disc showing xylem rays in a 3D wood context. **(B)** Cross-section showing some exemplary rays (area filled black) and tree-ring borders (black lines); the right arrow indicates a disappearing ray, likely due to non-perpendicular orientation of the section; the left arrow indicates a ray with an incorporated resin duct that was excluded from analysis. **(C)** Radial section; the dashed line indicates the non-perpendicular orientation of the cross-section, the arrow points at an example of the "ending ray artifact" in the cross-sectional plane, i.e., permanently disappearing rays. The seemingly short rays show that the core was not in parallel to the radial wood structure. **(D)** Tangential section; the vertical dashed lines simulate a 15 μm thick radial section and the associated "radial overestimation artifact." **(E)** Zoomed-in sketch of the "radial overestimation artifact" demonstrating that the actual height h is much shorter than the measured height h'. Because of the transparency of the tissue, the perceived ray contour corresponds to the maximum ray projection through the entire thickness of the section. Horizontal black bars in all anatomical images indicate 100 μm.

v8.3.10 Pro (New House Internet Services B.V., Rotterdam, the Netherlands). In the (subset) of the 40 cores, the average measured width and surface in cross-sectional, tangential and radial sections were 6.69 mm, 4.87 and 9.39 mm², respectively. In the stem disc the average measured width in cross-sections was 19 mm, while the average measured surface in tangential sections was 28 mm².

The outlines of all rays and tree ring borders were vectorized manually in the images using a tailored version of ROXAS 1.6 (von Arx and Dietz, 2005; von Arx and Carrer, 2014), a specialized image analysis tool for wood cell anatomical measurements based on Image-Pro Plus (Media Cybernetics, Silver Spring, Maryland, USA). As a result ROXAS provided several statistics in a global and annual resolution such as individual ray area (all planes), individual ray length (cross- and radial sections) and ray height (tangential sections), the number and position of initiating (NEWRAY) and disappearing rays within the ring (cross-sections), and the percentage of ray surface in the xylem (PERPAR; all cutting planes).

Study Design

Several trials were used to identify reliable quantification methods and assess distinctness of ray metrics based on the subset of the six cores. First, the consistency of the measured percentage of ray surface in the xylem (PERPAR) in the three cutting planes (cross-, tangential, radial plane) was evaluated by comparing values from the same annual rings. Second, to explore the potential of more efficient ray quantification, the relationship of individual ray area and ray length (cross-sections) and ray height (tangential sections) was investigated by linear regression analyses. Third, the similarity of different ray parameters in the tangential sections such as PERPAR, ray density, ray area, ray height, and ray width was assessed by correlation analyses. Ray width was estimated from ray area through division by ray height.

The variability of ray parameters was assessed in two different ways. First, the inter-annual variability in cross-sectional PERPAR was determined in the full set of 40 cores and time series of 20 rings each by calculating the coefficient of variation (CV) and the mean sensitivity (Cook and Pederson,

2011), i.e., the average percentage change from one yearly value to the next. The same sample set was used to assess the intra-annual variability of cross-sectional PERPAR and the proportion of initiating rays (NEWRAY) between early- and latewood using t-test. Second, the variability of the obtained PERPAR values was assessed as a function of the measured wood surface by bootstrapping. Using tangential sections from each of the three locations in the three radial bars of the stem disc, one thousand 1-mm^2 measurement windows were randomly placed and the PERPAR values were extracted and pooled to a single data pool ($n = 9000$). Bootstrapped values based on 1000 replications for PERPAR mean and coefficient of variation of its estimation (CV) were obtained by randomly combining an increasing number of individual 1-mm^2 values from the data pool, thus simulating the measurement of increasing wood surface (from 1 to 15 mm^2). The same procedure was repeated for the cross-sectional images from the same annual rings as the tangential sections for steps of 1-mm width (from 1 to 15 mm). Since mean ring width was around 1 mm (1.047 mm), a measured width of 1 mm in cross-sections corresponded to 1 mm^2 in tangential sections, thus allowing direct comparisons of the obtained variability between the cutting planes. The calculated CVs were used to calculate the relationship between measured wood width/surface, number of samples (n), and estimated accuracy. These calculations were based on Equation (1):

$$CI95 = 2 \cdot CV \cdot n^{-0.5} \tag{1}$$

which expresses the relative 95% confidence interval (CI95; given as a percentage of the mean) as a function of CV and n. In fact

CI95 approximately corresponds to twice the standard error (SE), SE equals to SD divided by the square root of the number of samples (n), and SD is CV multiplied by the mean.

RESULTS

Results from the six cores indicated that the percentage of ray surface (PERPAR) varied substantially depending on the cutting plane. However, PERPAR values within a cutting plane and core were rather similar (mean CV = 0.12, 0.16, and 0.17 for cross-, tangential and radial sections, respectively; **Figure 2**). Mean PERPAR value for cross-sections was 6.50%, which was significantly higher than for the tangential sections (4.92%; $t = 5.606$; $P < 0.001$). PERPAR was largest in the radial sections (16.84%), deviating significantly from the values of both other cutting planes ($t = -6.287$; $P < 0.001$ and $t = -7.641$; $P < 0.001$ for cross-sectional and tangential planes, respectively). Despite the differences, PERPAR values in cross- and tangential sections were significantly correlated ($r = 0.502$; $P = 0.034$).

The analysis of the subset of six cores showed that individual ray area was highly correlated with ray length in the cross-sectional ($n = 2017$; $R^2 = 0.926$; $P < 0.001$) plane and ray height in the tangential plane ($n = 13.218$; $R^2 = 0.847$; $P < 0.001$). In tangential sections correlation analyses revealed that PERPAR is significantly correlated with all other parameters except for mean ray width (*Pearsons's* $r = 0.196$, $P = 0.435$; **Figure 3**). At the section level we also observed that mean ray

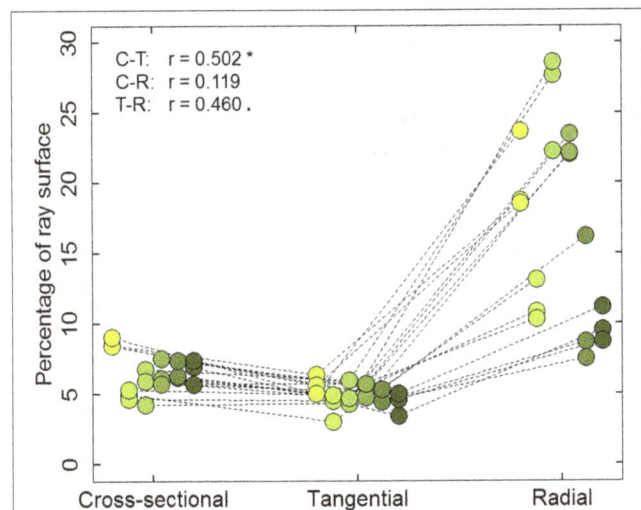

FIGURE 2 | Percentage of ray surface (PERPAR) measured on the cross-sectional, tangential and radial planes at three tree-rings (1989, 2002, and 2012) along the radial cores of six mature Scots pine trees. Symbols of the same color represent the PERPAR values of a specific cutting. Dotted lines connect the measurements from the same tree and ring. Symbols of the same cutting plane are jittered along the x-axis for easier interpretation. Left upper data indicates the *Pearson* correlation coefficient (r) between cross- and tangential sections (C-T), cross- and radial sections (C-R), radial and tangential sections (R-T), respectively. *P ≤ 0.05; · P ≤ 0.1.

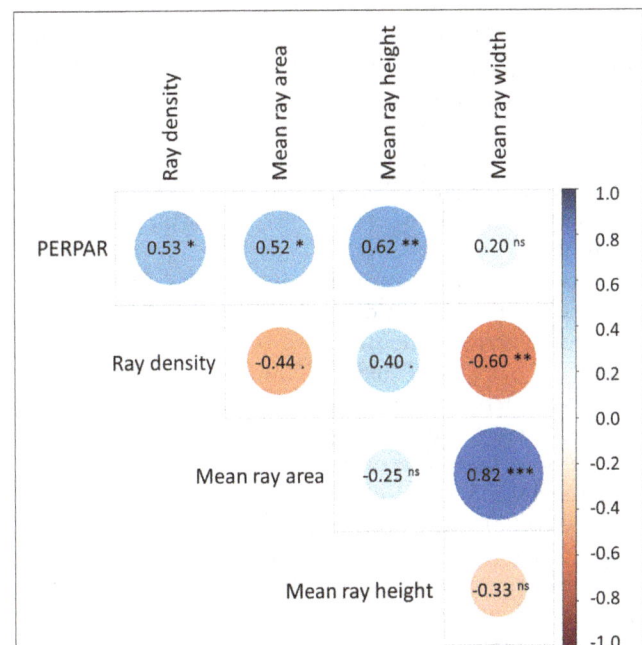

FIGURE 3 | *Pearson's* correlation matrix between the ray parameters measured in tangential sections (n = 18). The parameters include percentage of ray surface (PERPAR), ray density, mean ray area, mean ray height and mean ray width based on measurement from six mature Scots pine trees. For each tree, three locations separated by five to seven annual rings were analyzed. ***P ≤ 0.001; **P ≤ 0.01; *P ≤ 0.05; · P ≤ 0.1; ns, not significant.

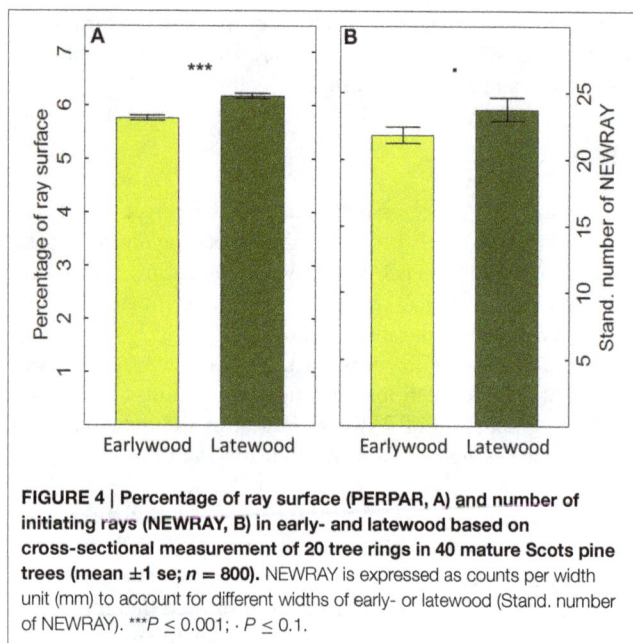

FIGURE 4 | Percentage of ray surface (PERPAR, A) and number of initiating rays (NEWRAY, B) in early- and latewood based on cross-sectional measurement of 20 tree rings in 40 mature Scots pine trees (mean ±1 se; n = 800). NEWRAY is expressed as counts per width unit (mm) to account for different widths of early- or latewood (Stand. number of NEWRAY). ***$P \leq 0.001$; · $P \leq 0.1$.

DISCUSSION

Estimates of Percentage of Ray Surface Depend on the Cutting Plane

In this study we evaluated several methodologies for quantifying ray features in conifers. Tangential and cross-sections proved to be suitable, but with strengths and limitations in terms of accuracy and investment, and with differences in the information they register. In contrast radial sections were generally unsuitable for ray quantification. The large differences in percentage of ray surface (PERPAR) estimation (up to three times) among cutting planes likely reflects sampling artifacts linked to the sheet-like orientation and fusiform shape of uniseriate rays (see **Figures 1D,E**). In fact, in a radial section, the position and contour of the ray outline may change substantially from the lower to the upper side of the section. Because of the transparency of the tissue, the perceived ray outline is the maximum ray projection through the entire thickness of the section, which leads to a systematic overestimation ("radial overestimation artifact"). This effect–although much weaker—also occurs in cross-sections due to the tapering toward the upper and lower ray extremities, supposedly explaining the overestimated values (**Figure 2**). In contrast, the tangential sections are robust in this respect and therefore likely produced most accurate estimates of PERPAR, and therefore relative ray volume (Myer, 1922). The significant correlation between tangential and cross-sectional data confirms the systematic nature of the larger cross-sectional values. With respect to the quantification of the area of individual rays, the strong allometric relationship allows accurate estimation of ray area as a function of ray length in cross-sections (Olano et al., 2013) and ray height in tangential sections. Using such relationships could significantly increase the efficiency of the measurement procedure.

PERPAR and Ray Width are the Most Complementary Tangential Ray Parameters

Among the different tangential ray parameters, PERPAR positively correlated with all other ray metrics but mean ray width, which suggests PERPAR incorporates the information from most other metrics (**Figure 4**). Notably, the significant correlations with mean ray area and height are in line with previously observed consistent patterns of these metrics in relation to tree vigor and growth conditions (e.g., Bannan, 1954, 1965; White and Robards, 1966; Fonti et al., 2015). Moreover, the missing relationship between mean ray height and width indicates that their ratio mainly varied among the analyzed images (i.e., trees and/or annual rings), since within the images individual ray height well scaled to the width (mean *Pearson's* $r = 0.475$; $P < 0.001$). This variability is mainly due to the variability in ray width: supplementary analyses revealed that the width of individual rays is less strongly correlated with ray area (mean *Pearson's* $r = 0.750$; $P < 0.001$) than ray height (mean *Pearson's* $r = 0.899$; $P < 0.001$). Mean ray width also showed a larger variability among trees and/or annual rings than mean ray height (CV $= 0.169$ vs. 0.094). Together with its independence from

width was highly correlated to mean ray area ($r = 0.824$; $P < 0.001$), but was unrelated to mean ray height ($r = -0.326$, $P = 0.186$).

The 40 cores showed an overall mean CV in the annual PERPAR of 0.121 (ranging from 0.069 to 0.171 among individuals) and a mean sensitivity (MS) of 10.8% (ranging from 6.0 to 17.2%). Moreover, PERPAR in the cross-sectional plane was significantly smaller in the earlywood (5.77%) than in the latewood (6.18%; $t = 6.139$; $P < 0.001$; **Figure 4A**). Similarly, the number of initiating rays (NEWRAY) after correction for unequal tissue contributions to the overall ring area differed marginally between earlywood (21.82 rays/mm) and latewood (23.75 rays/mm; $t = -1.810$, $P = 0.070$; **Figure 4B**).

Bootstrapping analyses indicated that the variability of PERPAR mean estimation decreased with increasing measured wood width (cross-sections)/surface (tangential sections) and number of samples (**Figure 5**). However, the relative 95% confidence interval (CI95) was about twice as large in cross- than in tangential sections for a given measured wood surface. In cross-sections, CI95 curves flattened and got almost linear after a measured wood width of 8 mm, while this already occurred with 4–5 mm² in tangential sections. Measurements from 10 cross-sections on 5-mm increment cores resulted in a CI95 of ±10.1% of the true mean PERPAR, doubling the number of samples to 20 reduced CI95 to ±7.0%, while increasing the measured width to 8-mm wide strips provided an accuracy of CI95 ±7.7% with 10 samples. In tangential sections, 10 of the 1-mm² surfaces frequently used in previous studies (see Introduction) only provided a CI95 of ±12.7% of the true mean PERPAR. In contrast, measuring 2 × 2 mm areas in 10 samples provided a CI95 of ±6.0%, while it was reduced to even ±4.3% with twenty 2 × 2 mm samples.

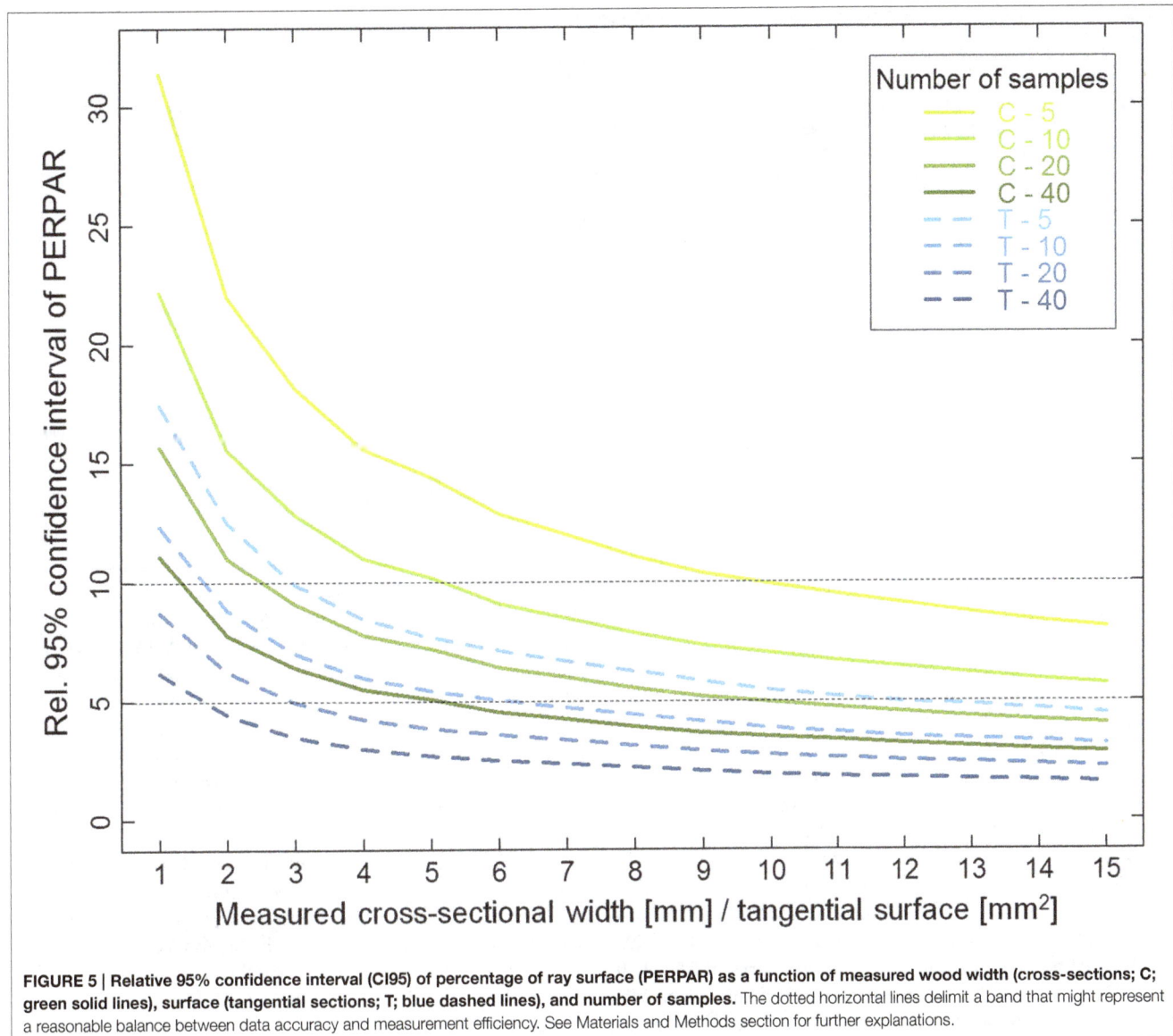

FIGURE 5 | Relative 95% confidence interval (CI95) of percentage of ray surface (PERPAR) as a function of measured wood width (cross-sections; C; green solid lines), surface (tangential sections; T; blue dashed lines), and number of samples. The dotted horizontal lines delimit a band that might represent a reasonable balance between data accuracy and measurement efficiency. See Materials and Methods section for further explanations.

PERPAR, mean ray width might therefore be a more promising parameter for ecological studies than mean ray height.

Spatial Variability of Ray Features within Tree Rings

The higher PERPAR and marginally larger number of initiating rays (NEWRAY) observed in the latewood compared to earlywood was unexpected when considering the radial orientation of the rays. In fact, once initiated, rays grow and extend to keep the connection with the cambium and phloem (Fischer and Höll, 1992; Spicer, 2014). Such small intra-annual differences could indicate slightly larger parenchyma cells (DeSmidt, 1922) and/or a higher ray initiation rate in latewood than in earlywood. They potentially evidence an advantage of having a larger storage and transport capacity close to the cambium to support the onset of cambial activity in the following growing season. However, the small intra-annual difference in

PERPAR values also indicates that it should suffice to only roughly balance early- and latewood in a proportional way when analyzing PERPAR in tangential sections. Additionally, if the section was not taken fully parallel to the orientation of the rays, the number of NEWRAY in cross-sections may be overestimated (see **Figure 1B**). The extent of overestimation in ray initiation is probably directly related to the number of disappearing rays ("ending ray artifact"; Eckstein, 2013), because they are both directly linked to the orientation of the section (see **Figure 1C**). Nevertheless, time series of the number of initiating rays from samples with an orientation problem still represent valid data if standardized before statistical analysis.

Accuracy of PERPAR Values Depending on Measurement Width and Area

The accuracy of PERPAR greatly changed depending on the measured wood width (cross-sections), surface (tangential

TABLE 2 | Potential and pitfalls of anatomical ray quantification in conifers for different cutting planes.

	Cross-section	Tangential section	Radial section
Potential	- Efficient creation of annual time series - Clear assignment to early- vs. latewood - Comparison with other anatomical variables on same image (ring width, tracheid size, cell wall thickness, etc.)	- Most accurate estimate of relative ray volume - Good estimation of ray density in 3D and connectivity with conduits - Robust against deviations in cutting orientation - Commonly used 5-mm cores provide accurate results based on 4-mm^2 measured areas and $n = 10$	- Inspection of cutting orientation in cross-sections to assess the quality of the number of initiating rays
Pitfalls	- Systematic overestimation of relative ray volume - Non-perpendicular cuttings: overestimation of number of initiating rays; permanently disappearing rays "ending ray artifact" - Transiently disappearing rays - Large sample size ($n \geq 20$) required to obtain accurate results with commonly used 5-mm wide cores	- Time consuming when creating annual time series	- Substantial overestimation of ray surface ("radial overestimation artifact") - Very unreliable for most of the parameters

sections), and the number of samples (**Figure 5**). Yet, our results suggest that common measurement strategies—10 cross-sections of 5-mm increment cores or 10 tangential planes of 1 mm^2—provide rather inaccurate estimates of PERPAR. They may thus be often inappropriate to extract an ecological signal considering that year-to-year variability (MS) in PERPAR is about 20% (Olano et al., 2013; Fonti et al., 2015), or even only about 11% as observed in the 40 cores analyzed here. This study estimates how an increased number of samples [since, CI95 decreases with the square root of the number of samples (cf. Equation 1)], or expansion of the measured wood width/area will increase the accuracy. Although our assessment was based on a very detailed analysis of a single stem disc, it cannot be excluded that the CI95 curves may be shifted for Scots pines from other populations or different species. However, we speculate here, that a CI95 in the range between ±5 and ±10% (dotted horizontal lines in **Figure 5**), corresponding to approximately half the CV or MS, could represent a reasonable balance between data accuracy and measurement efficiency.

CONCLUSIONS

Our results suggest that both cross- and tangential sections are suitable for quantitative approaches, whereas radial sections are generally unsuitable due to strong sampling artifacts. **Table 2** summarizes the major potential and pitfalls recognized in this study. Our main conclusions are that the quantification of rays in cross-sections is generally very efficient and allows establishing annual time series of (consistently overestimated) ray volume and number of initiating rays that can be compared to other anatomical traits such as tracheid dimensions, cell wall thickness and resin ducts, and related to time series of environmental conditions. Tangential sections seem more suitable to accurately estimate ray volume and investigate the spatial integration of rays in the xylem such as the connectivity of conduits to rays, or, more generally, research into structure-function relationships. In addition, tangential ray width registers information independent

from PERPAR, which makes it a promising complementary parameter for ecological studies. The choice between tangential and cross-sections will therefore depend on the specific study question and on the available lab capacities. In this context, this study presents for the first time a very concrete guidelines (**Figure 5**) for estimating data accuracy depending on the size of the measured wood width (cross-sections), surface (tangential sections), and number of samples, helping to define a suitable sampling strategy, although the latter also depends on the known or expected responsiveness and variability of the target ray features.

Since, most conifer species display a similar ray architecture and a relatively narrow range of ray volume, we are confident that most of our results are representative for other conifer species as well. A similar assessment might be applied to identify the best methodology in angiosperm species. We are convinced that the methodological guidelines presented here are necessary to foster the establishment of robust quantifications, which will ultimately improve the understanding of the fundamental role of ray parenchyma tissue for the performance and survival of trees growing in stressed environments.

ACKNOWLEDGMENTS

Evaluations were based on data from the long-term irrigation experiment Pfynwald, which is part of the Swiss Long-term Forest Ecosystem Research Programme LWF (www.lwf.ch). We are indebted to L. Matěju, L. Schneider, and G. Juste for the preparation of anatomical samples, and to S. Meier and J. Helfenstein for assistance in the field. This study was supported by a grant from the Swiss State Secretariat for Education, Research and Innovation SERI (SBFI C12.0100) and from the Spanish Ministry of Economy and Competitivity with UE FEDER funds (CGL2012-34209). A. Arzac was supported by a FPI-EHU predoctoral grant. The project profited from discussions within the framework of the COST Action STReESS (COST-FP1106).

REFERENCES

Allen, C. D., Macalady, A. K., Chenchouni, H., Bachelet, D., Mcdowell, N., Vennetier, M., et al. (2010). A global overview of drought and heat-induced tree mortality reveals emerging climate change risks for forests. *For. Ecol. Manag.* 259, 660–684. doi: 10.1016/j.foreco.2009. 09.001

Aloni, R. (2013). "The role of hormones in controlling vascular differentiation," in *Cellular Aspects of Wood Formation,* ed J. Fromm (Berlin Heidelberg: Springer), 99–139.

Arbellay, E., Fonti, P., and Stoffel, M. (2012). Duration and extension of anatomical changes in wood structure after cambial injury. *J. Exp. Bot.* 63, 3271–3277. doi: 10.1093/jxb/ers050

Arbellay, E., Stoffel, M., and Bollschweiler, M. (2010). Wood anatomical analysis of Alnus incana and Betula pendula injured by a debris-flow event. *Tree Physiol.* 30, 1290–1298. doi: 10.1093/treephys/tpq065

Arzac, A. (2014). *Exploring the Potential of Parenchyma Ray as a Proxy for Climate and Plant Resource Levels.* Ph.D. Universidad del Pais Vasco, Leioa.

Bamber, R. K. (1976). Heartwood, its function and formation. *Wood Sci. Technol.* 10, 1–8. doi: 10.1007/BF00376379

Bannan, M. W. (1937). Observations on the distribution of xylem-ray tissue in conifers. *Ann. Bot.* 1, 717–726.

Bannan, M. W. (1954). Ring width, tracheid size, and ray volume in stem wood of *Thuja occidentalis* L. *Can. J. Bot. Rev.* 32, 466–479.

Bannan, M. W. (1965). Ray contacts and rate of anticlinal division in fusiform cambial ceils of some Pinaceae. *Can. J. Bot.* 43, 487–508. doi: 10.1139/b65-055

Barnard, D. M., Lachenbruch, B., McCulloh, K. A., Kitin, P., and Meinzer, F. C. (2013). Do ray cells provide a pathway for radial water movement in the stems of conifer trees? *Am. J. Bot.* 100, 322–331. doi: 10.3732/ajb.1200333

Brandes, A. F. D. N., and Barros, C. F. (2008). Wood anatomy of eight liana species of Leguminosae family from Atlantic Rain Forest. *Acta Bot. Brasilica* 22, 465–480. doi: 10.1590/S0102-33062008000200015

Brodersen, C. R., and McElrone, A. J. (2013). Maintenance of xylem network transport capacity: a review of embolism repair in vascular plants. *Front. Plant Sci.* 4:108. doi: 10.3389/fpls.2013.00108

Brown, H. P., Panshin, A. J., and Forsaith, C. C. (1949). *Structure, Identification, Defects and Uses of the Commercial Woods of the United States.* New York, NY: McGraw-Hill Book Company, Inc.

Burgert, I., and Eckstein, D. (2001). The tensile strength of isolated wood rays of beech (Fagus sylvatica L.) and its significance for the biomechanics of living trees. *Trees Struct. Funct.* 15, 168–170. doi: 10.1007/s004680000086

Carlquist, S. (1982). Wood anatomy of Illicium (Illiciaceae): phylogenetic, ecological, and functional interpretations. *Am. J. Bot.* 69, 1587–1598. doi: 10.2307/2442914

Choat, B., Jansen, S., Brodribb, T. J., Cochard, H., Delzon, S., Bhaskar, R., et al. (2012). Global convergence in the vulnerability of forests to drought. *Nature* 491, 752–755. doi: 10.1038/nature11688

Cook, E., and Pederson, N. (2011). "Uncertainty, emergence, and statistics in dendrochronology," in *Dendroclimatology,* eds M. K. Hughes, T. W. Swetnam, and H. F. Diaz (Dordrecht; Heidelberg; London; New York, NY: Springer), 77–112.

DeSmidt, W. J. (1922). Studies of the distribution and volume of the wood rays in slippery elm (Ulmus Fulva Michx.). *J. For.* 20, 352–362.

Eckstein, D. (2013). 'A new star' – but why just parenchyma for dendroclimatology? *New Phytol.* 198, 328–330. doi: 10.1111/nph.12229

Esteban, L., Martín, J., De Palacios, P., and Fernández, F. (2012). Influence of region of provenance and climate factors on wood anatomical traits of *Pinus nigra* Arn. subsp. *salzmannii. Eur. J. For. Res.* 131, 633–645. doi: 10.1007/s10342-011-0537-x

Esteban, L., Martín, J., De Palacios, P., García Fernández, F., and López, R. (2010). Adaptive anatomy of *Pinus halepensis* trees from different Mediterranean environments in Spain. *Trees Struct. Func.* 24, 19–30. doi: 10.1007/s00468-009-0375-3

Falcon-Lang, H. J. (2005). Intra-tree variability in wood anatomy and its implications for fossil wood systematics and palaeoclimatic studies. *Palaeontology* 48, 171–183. doi: 10.1111/j.1475-4983.2004.00429.x

Fischer, C., and Höll, W. (1992). Food reserves of scots pine (*Pinus sylvestris* L.). *Trees* 6, 147–155. doi: 10.1007/BF00202430

Fonti, P., and Frey, B. (2002). Is the ray volume a possible factor influencing ring shake occurrence in chestnut wood? *Trees Struct. Funct.* 16, 519–522. doi: 10.1007/s00468-002-0193-3

Fonti, P., Tabakova, M., Kirdyanov, A., Bryukhanova, M., and von Arx, G. (2015). Variability of ray anatomy of Larix gmelinii along a forest productivity gradient in Siberia. *Trees Struct. Funct.* 29, 1165–1175. doi: 10.1007/s00468-015-1197-0

Gartner, B. L., Baker, D. C., and Spicer, R. (2000). Distribution and vitality of xylem rays in relation to tree leaf area in Douglas-fir. *IAWA J.* 21, 389–401. doi: 10.1163/22941932-90000255

Gärtner, H., Lucchinetti, S., and Schweingruber, F. H. (2014). New perspectives for wood anatomical analysis in dendrosciences: the GSL1-microtome. *Dendrochronologia* 32, 47–51. doi: 10.1016/j.dendro.2013.07.002

Gärtner, H., and Schweingruber, F. H. (2013). *Microscopic Preparation Techniques for Plant Stem Analysis.* Remagen: Kessel Publishing House.

Gregory, R. A., and Romberger, J. A. (1975). Cambial activity and height of uniseriate vascular rays in conifers. *Bot. Gaz.* 136, 246–253. doi: 10.1086/336810

Gregory, R. A. (1977). Cambial activity and ray cell abundance in Acer saccharum. *Can. J. Bot.* 55, 2559–2564. doi: 10.1139/b77-293

Grissino-Mayer, H. (2001). Evaluating crossdating acuracy: a manual and tutorial for the computer program COFECHA. *Tree Ring Res.* 57, 205–221.

Gruber, A., Pirkebner, D., and Oberhuber, W. (2013). Seasonal dynamics of mobile carbohydrate pools in phloem and xylem of two alpine timberline conifers. *Tree Physiol.* 33, 1076–1083. doi: 10.1093/treephys/tpt088

Harlow, W. M. (1927). The effect of site on the structure and growth of white cedar Thuya occidentalis L. *Ecology* 8, 453–470. doi: 10.2307/1930153

Hartig, R. (1894). Untersuchungen ueber die entstehung und die eigenschaften des Eichenholzes. *Forstl. Naturwiss. Z.* 4, 1–13, 49–68, 172–203.

Hereş, A.-M., Camarero, J., López, B., and Martínez-Vilalta, J. (2014). Declining hydraulic performances and low carbon investments in tree rings predate Scots pine drought-induced mortality. *Trees Struct. Funct.* 28, 1737–1750. doi: 10.1007/s00468-014-1081-3

Hudgins, J. W., Ralph, S., Franceschi, V. R., and Bohlmann, J. (2006). Ethylene in induced conifer defense: cDNA cloning, protein expression, and cellular and subcellular localization of 1-aminocyclopropane-1-carboxylate oxidase in resin duct and phenolic parenchyma cells. *Planta* 224, 865–877. doi: 10.1007/s00425-006-0274-4

Jaccard, P. (1915). Über die Verteilung der Markstrahlen bei den Coniferen. *Ber. Dtsch. Bot. Ges.* 33, 492–498.

Jacquet, J.-S., Bosc, A., O'Grady, A., and Jactel, H. (2014). Combined effects of defoliation and water stress on pine growth and non-structural carbohydrates. *Tree Physiol.* 34, 367–376. doi: 10.1093/treephys/tpu018

Lev-Yadun, S., and Aloni, R. (1992). The role of wounding and partial girdling in differentiation of vascular rays. *Int. J. Plant Sci.* 153, 348–357. doi: 10.1086/297039

Lev-Yadun, S., and Aloni, R. (1995). Differentiation of the ray system in woody plants. *Bot. Rev.* 61, 45–84. doi: 10.1007/BF02897151

Lev-Yadun, S. (1998). The relationship between growth-ring width and ray density and ray height in cell number in the earlywood of *Pinus halepensis* and Pinus pinea. *IAWA J.* 19, 131–139. doi: 10.1163/22941932-90001515

Lloret, F., Martinez-Vilalta, J., Serra-Diaz, J. M., and Ninyerola, M. (2013). Relationship between projected changes in future climatic suitability and demographic and functional traits of forest tree species in Spain. *Clim. Change* 120, 449–462. doi: 10.1007/s10584-013-0820-6

McDowell, N. G., and Sevanto, S. (2010). The mechanisms of carbon starvation: how, when, or does it even occur at all? *New Phytol.* 186, 264–266. doi: 10.1111/j.1469-8137.2010.03232.x

Myer, J. E. (1922). Ray volumes of the commercial woods of the United States and their significance. *J. For.* 20, 337–351.

Nikolov, R., and Helmisaari, N. (1992). "Silvics of the circumpolar boreal forest tree species," in *A Systems Analysis of the Global Boreal Forest,* eds H. Shugart, R. Leeman, and G. Bonan (New York, NY: Cambridge University Press), 13–84.

Olano, J. M., Arzac, A., García-Cervigón, A. I., von Arx, G., and Rozas, V. (2013). New star on the stage: amount of ray parenchyma in tree rings shows a link to climate. *New Phytol.* 198, 486–495. doi: 10.1111/nph.12113

Panshin, A., and De Zeeuw, C. (1980). *Textbook of Wood Technology. Part 1. Formation, Anatomy, and Properties of Wood.* New York, NY: McGraw-Hill.

Pruyn, M. L., Gartner, B. L., and Harmon, M. E. (2005). Storage versus substrate limitation to bole respiratory potential in two coniferous tree species of contrasting sapwood width. *J. Exp. Bot.* 56, 2637–2649. doi: 10.1093/jxb/eri257

Rosell, J. A., Gleason, S., Méndez-Alonzo, R., Chang, Y., and Westoby, M. (2014). Bark functional ecology: evidence for tradeoffs, functional coordination, and environment producing bark diversity. *New Phytol.* 201, 486–497. doi: 10.1111/nph.12541

Salleo, S., Trifilo, P., Esposito, S., Nardini, A., and Lo Gullo, M. A. (2009). Starch-to-sugar conversion in wood parenchyma of field-growing *Laurus nobilis* plants: a component of the signal pathway for embolism repair? *Funct. Plant Biol.* 36, 815–825. doi: 10.1071/FP09103

Salleo, S., Trifilo, P., and Lo Gullo, M. A. (2006). Phloem as a possible major determinant of rapid cavitation reversal in stems of *Laurus nobilis* (laurel). *Funct. Plant Biol.* 33, 1063–1074. doi: 10.1071/FP06149

Schweingruber, F. H., and Poschlod, P. (2005). *Growth Rings in Herbs and Shrubs: Life Span, Age Determination and Stem Anatomy.* Bern: Haupt.

Spicer, R. (2014). Symplasmic networks in secondary vascular tissues: parenchyma distribution and activity supporting long-distance transport. *J. Exp. Bot.* 65, 1829–1848. doi: 10.1093/jxb/ert459

von Arx, G., Archer, S. R., and Hughes, M. K. (2012). Long-term functional plasticity in plant hydraulic architecture in response to supplemental moisture. *Ann. Bot.* 109, 1091–1100. doi: 10.1093/aob/mcs030

von Arx, G., and Carrer, M. (2014). ROXAS - a new tool to build centuries-long tracheid-lumen chronologies in conifers. *Dendrochronologia* 32, 290–293. doi: 10.1016/j.dendro.2013.12.001

von Arx, G., and Dietz, H. (2005). Automated image analysis of annual rings in the roots of perennial forbs. *Int. J. Plant Sci.* 166, 723–732. doi: 10.1086/431230

Von Schneider, M., and Halbwachs, G. (1989). Anatomische und morphologische Untersuchungen zur Regenerationsfähigkeit einer durch Fluorimmissionen geschädigten Fichte. *Eur. J. For.Pathol.* 19, 29–46. doi: 10.1111/j.1439-0329.1989.tb00767.x

Weinstein, A. I. (1926). Summary of literature relating to the volume, distribution, and effects of medullary rays in wood. *J. Forest.* 24, 915–925.

White, D. J. B., and Robards, A. W. (1966). Some effects of radial growth rate upon the rays of certain ring-porous hardwoods. *J. Inst. Wood Sci.* 17, 45–52.

Wimmer, R., and Grabner, M. (2000). A comparison of tree-ring features in Picea abies as correlated with climate. *IAWA J.* 21, 403–416. doi: 10.1163/22941932-90000256

Wimmer, R. (2002). Wood anatomical features in tree-rings as indicators of environmental change. *Dendrochronologia* 20, 21–36. doi: 10.1078/1125-7865-00005

Witt, W., and Sauter, J. J. (1994). Starch metabolism in poplar wood ray cells during spring mobilization and summer deposition. *Physiol. Plant.* 92, 9–16. doi: 10.1111/j.1399-3054.1994.tb06648.x

Conflict of Interest Statement: The authors declare that the research was conducted in the absence of any commercial or financial relationships that could be construed as a potential conflict of interest.

12

Seasonal Patterns of Fine Root Production and Turnover in a Mature Rubber Tree (*Hevea brasiliensis* Müll. Arg.) Stand- Differentiation with Soil Depth and Implications for Soil Carbon Stocks

Jean-Luc Maeght[1,2]*, Santimaitree Gonkhamdee[3], Corentin Clément[4], Supat Isarangkool Na Ayutthaya[3], Alexia Stokes[2] and Alain Pierret[5]

[1] Institut de Recherche pour le Développement, UMR 242/iEES – Paris (IRD-UPMC-CNRS-UPEC-UDD-INRA), Bondy, France, [2] INRA, UMR-AMAP, Montpellier, France, [3] Khon Kaen University, Faculty of Agriculture, Khon Kaen, Thailand, [4] International Water Management Institute, Vientiane, Laos, [5] Institut de Recherche Pour le Développement, UMR IEES-Paris – Department of Agricultural Land Management (DALaM), Vientiane, Laos

Edited by:
Sergio Rossi,
Université du Québec à Chicoutimi,
Canada

Reviewed by:
Petra Fransson,
Swedish University of Agricultural
Sciences, Sweden
Aidan M. Keith,
Centre for Ecology and Hydrology, UK

*Correspondence:
Jean-Luc Maeght
jean-luc.maeght@ird.fr

Fine root dynamics is a main driver of soil carbon stocks, particularly in tropical forests, yet major uncertainties still surround estimates of fine root production and turnover. This lack of knowledge is largely due to the fact that studying root dynamics *in situ*, particularly deep in the soil, remains highly challenging. We explored the interactions between fine root dynamics, soil depth, and rainfall in mature rubber trees (*Hevea brasiliensis* Müll. Arg.) exposed to sub-optimal edaphic and climatic conditions. A root observation access well was installed in northern Thailand to monitor root dynamics along a 4.5 m deep soil profile. Image-based measurements of root elongation and lifespan of individual roots were carried out at monthly intervals over 3 years. Soil depth was found to have a significant effect on root turnover. Surprisingly, root turnover increased with soil depth and root half-life was 16, 6–8, and only 4 months at 0.5, 1.0, 2.5, and 3.0 m deep, respectively (with the exception of roots at 4.5 m which had a half-life similar to that found between depths of 1.0 and 2.5 m). Within the first two meters of the soil profile, the highest rates of root emergence occurred about 3 months after the onset of the rainy season, while deeper in the soil, root emergence was not linked to the rainfall pattern. Root emergence was limited during leaf flushing (between March and May), particularly within the first two meters of the profile. Between soil depths of 0.5 and 2.0 m, root mortality appeared independent of variations in root emergence, but below 2.0 m, peaks in root emergence and death were synchronized. Shallow parts of the root system were more responsive to rainfall than their deeper counterparts. Increased root emergence in deep soil toward the onset of the dry season could correspond to a drought acclimation mechanism, with the relative importance of deep water capture increasing once rainfall ceased. The considerable soil depth regularly explored by fine roots, even though significantly less than in surface layers in terms of root length density and biomass, will impact strongly the evaluation of soil carbon stocks.

Keywords: deep roots, root phenology, root turnover, soil carbon, root access well, drought

INTRODUCTION

Fine root production and turnover represent 22% of terrestrial net primary production globally (McCormack et al., 2015). Yet there are still major uncertainties about the mechanisms that control fine root production and turnover. With the growing global demand for food and plant-derived commodities, unraveling such mechanisms is becoming increasingly important, particularly with the concomitantly pressing need to develop more sustainable agro-ecosystems. In recent years, rubber tree (*Hevea brasiliensis* Müll. Arg.) plantations have rapidly expanded, especially in marginal regions where the climate is much drier than in the species' natural range and where seasonal drought occurs (Carr, 2011). Soaring prices of natural rubber in the late 2010s influenced governmental policies regarding the expansion of *H. brasiliensis* cultivation. In Thailand, the world's leading latex producer, where the surface area planted with *H. brasiliensis* was multiplied by a factor of 75 between 1980 and 2008, from 24,000 ha^{-1} to 1.8 million ha^{-1} (Carr, 2011). Given the stress that tapping, i.e., the process by which the latex is collected, already imposes on *H. brasiliensis*, the sustainability and profitability of latex production in such areas could greatly benefit from adapting tapping modalities by taking into account the physiological response of trees to water availability (Boithias et al., 2011; Junjittakarn, 2012).

As roots are conduits for nutrients and water from the soil to plants, they have a determining role with regard to tree resilience to a range of environmental constraints, especially water stress (Boyce, 2005; Lobet et al., 2013). Fine roots are also an integrative indicator of plant response to environmental factors (Edwards et al., 2004) and we assume that root production or elongation of *H. brasiliensis* is synchronized with rainfall patterns (Green et al., 2005), although there exists evidence of endogenous controls of root growth (Abramoff and Finzi, 2015). Therefore, we expect that fine root growth is arrested during the dry season, but no data exist to support or refute this hypothesis. Fine roots of trees contribute to soil water extraction (Danjon et al., 2013), while a variable (and most often poorly quantified) share of the water demand is supplied by deep roots (Maeght et al., 2013). The measurement of root growth and survival *in situ* along a deep soil profile is an approach that can bring essential information to understanding how trees cope with water-limiting conditions and tree resilience to such constraints.

Quantifying fine root phenology and mortality down a soil profile, and particularly in deep soils, will also impact the evaluation of belowground carbon stocks, an area where data are scarce (Wauters et al., 2008). This proportion of soil carbon stocks could well contribute to the balance between the release and accumulation of carbon fluxes, currently described as the "missing sink" (Esser et al., 2011). However, to observe and analyze roots non-destructively within the soil matrix is still a major scientific challenge (Virginia and Jarrell, 1987), especially in deep soil layers and most studies have focused on the superficial layers of soil (Stone and Kalisz, 1991; Maeght et al., 2013). Our knowledge of fine root lifespan is also limited, particularly at depth (Eissenstat and Yanai, 1997).

We hypothesize that: (i) rooting in general is deeper than commonly assumed, (ii) fine root phenology and mortality are synchronized with annual patterns of precipitation, (iii) fine roots growing deep in the soil contribute to the resilience of *H. brasiliensis* to frequently occurring drought conditions in N. E. Thailand. Therefore, we measured seasonal patterns of fine root production and turnover in a mature stand of *H. brasiliensis*, down a 4.5 m soil profile during a 3-year observation period. We examined the interactions between fine root dynamics, rainfall, and soil depth and estimated the relative contribution of fine and deep roots to soil carbon.

MATERIALS AND METHODS

Study Site and Climate

The field site was located at Baan Sila Khu-Muang village, Buriram province in North East Thailand (N 15°16'23'', E 103°04'51.3'', 150 m a.s.l.). This region is part of the non-traditional areas for *H. brasiliensis* cultivation established since the 1990s. The experiment was set up in 2006 in a monoclonal plot of 14 years old trees (RRIM 600 clone developed by the Rubber Research Institute of Malaysia), planted at 2.5 m × 7.0 m spacing (~570 trees ha^{-1}) that had already been tapped for over 4 years to produce latex. Tapping was performed using a semi-spiral cut 2 days out of 3 and is largely adapted to the local climatic conditions. Tapping is usually discontinued for 3–4 months during the dry season. The maximum leaf area index (LAI), measured using 91 m^1 litter traps during the defoliation period (December–February; Isarangkool Na Ayutthaya et al., 2010), was estimated to be 3.9 ± 0.7 (mean ± standard deviation).

This marginal area for rubber tree cultivation is subjected to the Southeast Asian monsoon, with heavy rainfall between April and October. Local microclimate was monitored automatically with a Minimet weather station (Skye Instruments Ltd, UK) attached to a data logger recording air temperature, relative humidity, incident short wave radiation and rainfall at 30-min intervals. Reference evapotranspiration (ET0) was calculated according to Allen et al. (1998) using the data collected from the weather station.

Soil Properties

The soil at the study site was a deep loamy sand with limited water retention capacity, developed on fine sand or coarse silt deposits with a homogeneous sandy loam texture throughout the profile. The Ap horizon was a 0.25 m thick remnant from previous sugar cane (*Sacharum officinarum L.*) cultivation (Hartmann et al., 2006). Clay, silt, and sand contents were 100, 100, and 800 $g\ kg^{-1}$, respectively. The clay content increased with depth: from 150 $g\ kg^{-1}$ in the Bt1 horizon (0.25–0.50 m) to 200 $g\ kg^{-1}$ in the Bt2 (0.50–1.0 m). Silt content was similar in all soil layers throughout the soil profile (100 $g\ kg^{-1}$) while sand decreased to 700 $g\ kg^{-1}$ at a depth of 1.0 m. From 100 to about 4.0 m, these properties remained fairly stable. The laterite layer was found at a depth of 6.0–7.0 m, as previously observed

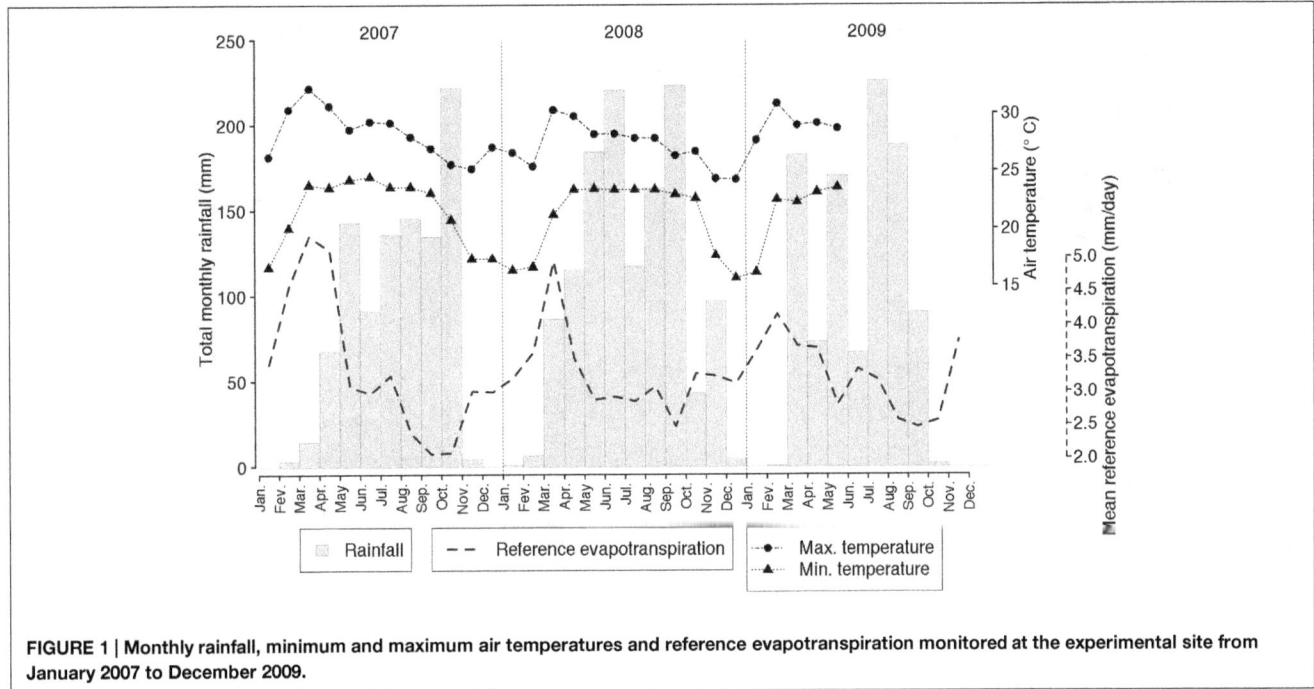

FIGURE 1 | Monthly rainfall, minimum and maximum air temperatures and reference evapotranspiration monitored at the experimental site from January 2007 to December 2009.

in this region (Cawte and Boyd, 2010). The water table was not found within the first 7 m of the profile, even during the rainy season. The soil was acidic with a pH ranging from 5.0 to 5.3. Soil organic carbon content was lower than 10 g kg^{-1} in the topsoil (Isarangkool Na Ayutthaya et al., 2011). Typical bulk soil density was 1.55 g cm^{-3} to a depth of 3.0 m (Gonkhamdee et al., 2009). Additional soil properties can be found in Hartmann et al. (2006) and Isarangkool Na Ayutthaya et al. (2011).

Root Growth Monitoring and Rooting Profiles

Soil Coring

To quantify carbon stocks associated with fine roots, root samples were collected at depths corresponding to the depths covered by root windows. We extracted undisturbed soil cores using standard soil sample steel rings (diameter 53 mm, height 50 mm and 100 cm^3 internal volume, Eijkelkamp Giesbeek, The Netherlands), in the vicinity of the root access well ($n = 12$ at soil depths 0.25, 0.50, 0.75, and 1.0 m; $n = 5$ at soil depths 1.6, 2.8, and 4.0 m). Root samples were analyzed according to Pierret et al. (2007b). Roots were first washed free of soil from the undisturbed soil cores and then imaged using a flatbed scanner (Epson Perfection V700 Photo scanner; Seiko Epson Corp., Japan) in light transmission mode, at a spatial resolution of 600 dpi (pixel size of 0.0423 mm). Special care was taken to separate every root from each other as much as possible, since overlapping roots are known to impair accurate length recovery. Specific root length (SRL) values, i.e., the length of root per unit dry root biomass, obtained from fine root samples collected within the first meter of the soil profile, were used to estimate the root dry biomass (RDB) distribution along the 4.5 m

profile observed in the well, based on the following equation:

$$RDB = RLD \times [Z]/SRL \qquad (1)$$

where RDB (in Mg ha^{-1}) is the RDB in a soil depth layer of thickness [Z] (m), RLD is the root length density (m m^{-3}) calculated from soil cores, in this soil layer and SRL the specific root length (m g^{-1}).

Root Access Well

Root growth was studied using the access well technique described in Maeght et al. (2013). An access well (0.9 m in diameter and 4.5 m deep) was installed in July 2006 at a distance of 1.35 m from two trees and 0.5 m aside from a tree row. The access well observation technique is an evolution of basic techniques for root observation at transparent interfaces with soil (Smit et al., 2000). A total of nine observation windows were cut through the concrete walls of the well in staggered rows (with 1.0 and 0.5 m horizontal and vertical spacing, respectively). Each root window included a specifically designed metallic frame supporting, on its upper side, a piece of 8 mm thick glass (0.25 m × 0.30 m) pressed against the soil at a 45° angle by means of two threaded rod actuators. On the frame's lower side, two guide rails allow the insertion of a standard flatbed scanner. Overall, given the geometrical arrangement of windows, the soil depth increments that were accessible were 0.4–0.6, 0.9–1.1, 1.4–1.6, 1.9–2.1, 2.4–2.6, 2.9–3.1, 3.4–3.6, 3.9–4.1, and 4.4–4.6 m. For simplicity, these are referred to as 0.5, 1.0, 1.5, 2.0, 2.5, 3.0, 3.5, 4.0, and 4.5 m hereafter. Due to time and financial constraints, it was not possible to build and monitor replicate root access wells within the framework of this field experiment. More details about the set up of the root access well set be found in Gonkhamdee et al. (2009) and Maeght et al. (2013).

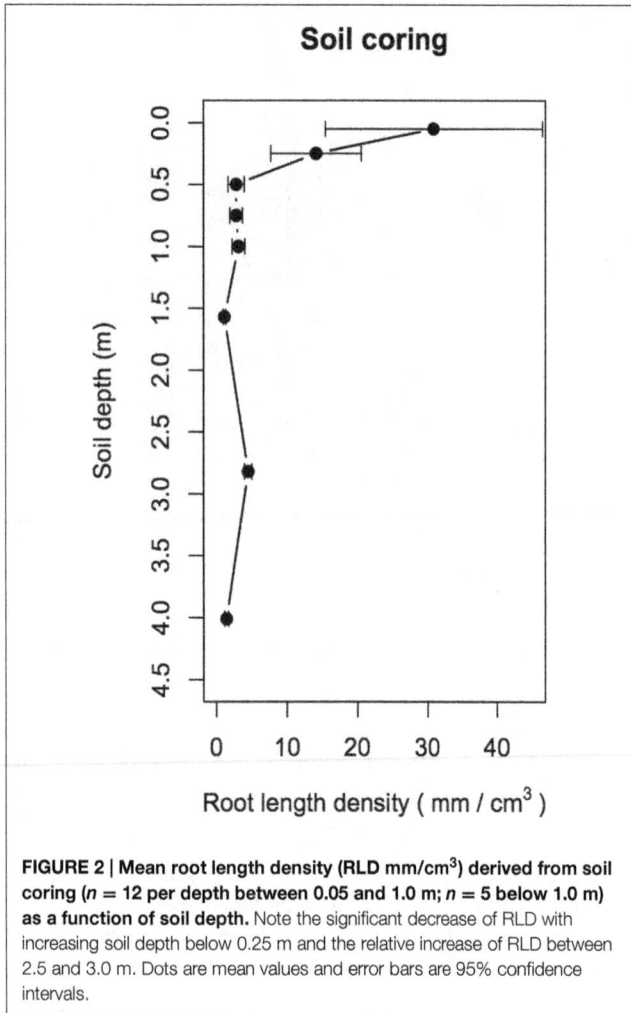

FIGURE 2 | Mean root length density (RLD mm/cm³) derived from soil coring (*n* = 12 per depth between 0.05 and 1.0 m; *n* = 5 below 1.0 m) as a function of soil depth. Note the significant decrease of RLD with increasing soil depth below 0.25 m and the relative increase of RLD between 2.5 and 3.0 m. Dots are mean values and error bars are 95% confidence intervals.

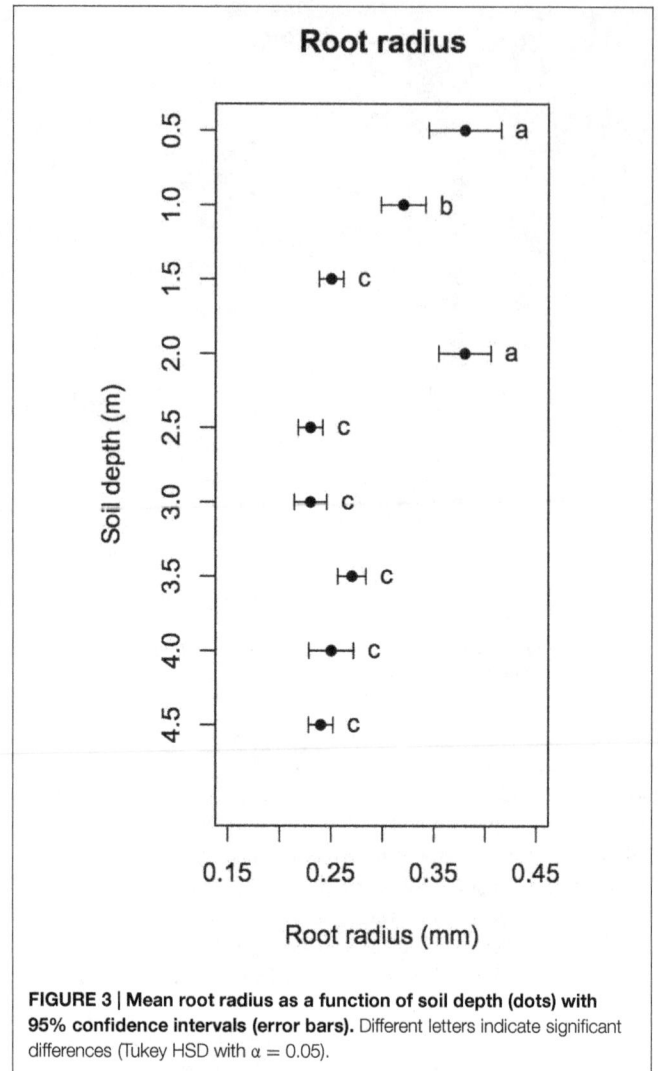

FIGURE 3 | Mean root radius as a function of soil depth (dots) with 95% confidence intervals (error bars). Different letters indicate significant differences (Tukey HSD with $\alpha = 0.05$).

Images of roots were taken using a flatbed scanner (HP Scanjet 4370 Photo scanner at 200 DPI – Hewlett-Packard Development Company, California) and custom software which offers a convenient, faster and more accurate record than manual techniques (Zoon and Tienderen, 1990; Kaspar and Ewing, 1997). Root windows were scanned at monthly intervals during 3 years starting in January 2007. This scanning was started 6 months after setting up the access well to avoid recording overproduction of roots at the onset, as often occurs in mini-rhizotron experiments (Yuan and Chen, 2012). Long-term observations are also highly recommended to avoid the risk of overestimating fine root turnover (Strand et al., 2008). Root growth monitoring was conducted following a procedure described in Maeght et al. (2007). Images of the soil and roots in direct contact with each window were used to estimate root length, radius, and time of root appearance/disappearance (from which root turnover was inferred). We used the Gimp freeware package[1] to digitize roots and classify them as live or senescent. Senescent roots are often difficult to identify with certainty (Majdi et al., 2005). We considered roots as senescent when they exhibited no elongation

and/or radius expansion for at least two successive observation dates and when their color turned from white to brown. Senescent roots were considered dead when their color changed to dark brown/black or when they completely disappeared from one observation to the next. A total of more than 1500 roots distributed in 300 images were processed.

Root emergence was quantified as the number of roots appearing between two monthly observations divided by the number of root windows from which this number of roots was derived (number of new roots per cm^2 and per month). Likewise, root mortality was quantified as the number of roots that disappeared between two monthly observations divided by the number of root windows from which this number of roots was derived (number of senescent/dead roots per root window and per month). 95% confidence intervals were computed as an indicator of the variability of root emergence/mortality across windows.

Assuming that all roots had emerged at the same time, the half-life represents the time after which half of all the roots would have died. Half-life values simplify comparisons between the survival

---

[1] www.gimp.org

of roots that emerged at the onset of the observation period and for which the time of disappearance could actually be recorded and that of roots that emerged much later and which died after the end of the observation period (right-censored data).

Analysis of Root Sample Images

Root length measurements were obtained using IJ_Rhizo's implementation of the method developed by Kimura et al. (1999). IJ_Rhizo (Pierret et al., 2013) is a software designed to measure roots washed from soil samples and developed in the ImageJ[2] macro language. The approach developed by Kimura et al. (1999) is based on discriminating each pixel of the medial axis (or skeleton) of each digitized root according to its number of orthogonal and diagonal (vertical or horizontal) neighbors. We also used IJ_Rhizo to compute frequency distributions of root radius classes (i.e., the cumulated root length par root radius class).

Statistical Analyses

All numerical data processing and statistical analyses were performed within the R environment (version 3.0.2; R Development Core Team, 2013). We first explored our dataset using a principal component analysis (PCA; "ade4" package, version 1.6-2). RLD and root radius values are reported as mean ± 95% confidence interval. We applied analysis of variance with Tukey's HSD *post hoc* tests to determine differences in root radius at different soil depths. We assessed fine root survival at the depth of each root using a Kaplan–Meier survival analysis (Kaplan and Meier, 1958) implemented in the "*survival*" package (version 2.37-4) of the R environment. For each individual root, we recorded the time of emergence and the time to either an event (death) or the end of the study (i.e., roots that were still alive at the time of the last observation were right-censored). Differences in survivorship of roots that emerged at different soil depths (regardless of their actual time of emergence) were assessed by *post hoc* pairwise comparisons using the Mann–Whitney test with a Bonferroni correction. Additionally, a Cox proportional hazards regression model was used to test whether root radius had an influence on root survival. Fine root emergence and mortality determined for each root observation window at monthly intervals are reported as pseudo-medians derived from the Wilcoxon test ± 95% confidence intervals.

A PCA (Supplementary Figure S1) indicated that root emergence was partly explained by rainfall and evapotranspiration, but the two first axes accounted for less than 50% of the variance. Therefore, we resorted here to a more descriptive analysis of fine root dynamics. As (i) roots with lifespans of 30 months and more only occurred between the surface and a depth of 2.0 m, (ii) roots were thicker above 2.0 m (with the exception of the 1.5 m depth increment) and (iii) at depths below 2.0 m, roots emerged at least 12 months later than at depths above 2.0 m, we chose to analyze separately root dynamics above and below the soil depth of 2.0 m.

[2]http://rsbweb.nih.gov/ij/

RESULTS

Climate Measurements

Total annual rainfall during the 3-year period over which the experiment was conducted, was 965, 1265, and 1002 mm, in 2007, 2008, and 2009, respectively (average: 1077 mm; **Figure 1**). Air temperatures increased during the dry season and decreased following the end of the rainy season (with a range from +8.3 to +40.3°C). Reference evapotranspiration (ET0) was found to roughly follow the monsoon regime, with a peak toward the end of the dry season and a subsequent decrease throughout the rainy season (**Figure 1**).

Rooting Profiles

Mean fine RLD derived from soil coring decreased by about one order of magnitude from a depth of 0.05 to 0.5 m, then declined slightly from 0.5 to 1.5 m before further increasing at 2.82 m ($F_{7,55} = 7.49$, $p < 0.001$; **Figure 2**). A *post hoc* Tukey test showed that mean fine RLD was significantly higher at 0.05 m than at all other depths ($p < 0.05$) and that fine root RLD between 0.25 and 4.0 m were not significantly different from each other.

Mean root radius measured in root windows significantly varied with soil depth ($F_{8,89} = 34.15$, $p < 0.001$), reaching 0.38 ± 0.03 mm at 0.5, 0.38 ± 0.02 mm at 2.0 m and 0.32 ± 0.02 mm at 1.0 m. At all other soil depths, mean root radius was fairly constant at 0.23–0.27 mm (**Figure 3**). A *post hoc* Tukey HSD test showed that mean root radii at 0.5 and 2.0 m were significantly higher than those at 1.0 m at $p < 0.05$ and that mean root radius at 1.0 m was itself significantly higher than those at all other soil depths ($p < 0.05$).

Root Emergence and Age Distributions

Roots near the soil surface (0.0–0.5 m) had a significantly longer life span compared to that in deeper soil layers (**Figure 4A**). There was a significant effect of soil depth on root age ($c^2 = 94.93$, $p < 0.001$). Roots with life spans of 30 months and more were only observed between the surface and a depth of 2.0 m.

Roots first emerged in layers close to the soil surface (0.5 and 1.0 m) and at 3.5 m, i.e., within the first 6 months of the observation period. However, at depths of 2.5, 3.0, and 4.0 m, roots did not emerge until 11–17 months after the beginning of the observation period (**Figure 4B**). Root emergence differed significantly depending on soil depth ($F_{8,89} = 34.15$, $p < 0.001$). Root emergence occurred significantly earlier ($p < 0.05$) in the three top windows (means were: 11, 14, and 17 months at 0.5, 1.0, and 1.5 m, respectively) than in the deeper windows (means were: 21–25 months).

Root Survival

Kaplan and Meier (1958) curves showed that, overall, root half-life decreased with soil depth, with the half-life of roots at 0.5 m being in the order of 500 days (>16 months; turnover of 0.73 yr^{-1}). The half-life of roots between 1.0 and 2.5 m was about 180–250 days (6–8 months; turnover 1.46–2.03 yr^{-1}) and that of roots at 3.0 m and below dropped to less than 120 days

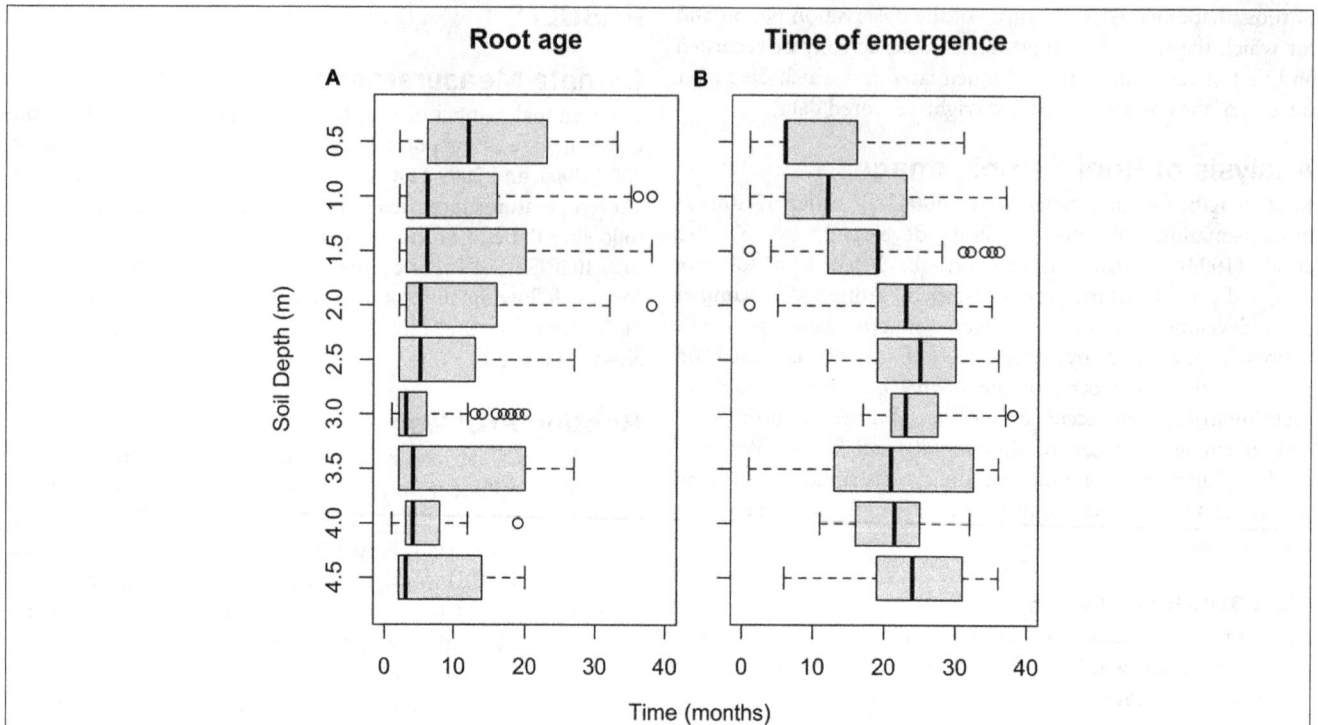

FIGURE 4 | (A) Box-whisker plot of root age distributions in each root window (for live and dead roots combined). **(B)** Distribution of the times of emergence of roots for each root window. The central vertical line indicates the median value, and the left and right edges of boxes (hinges) correspond to the 25th and 75th percentile values, while the whiskers extend 1.5× beyond the spread of the hinges. Data points outside this range (outliers) are indicated with circles.

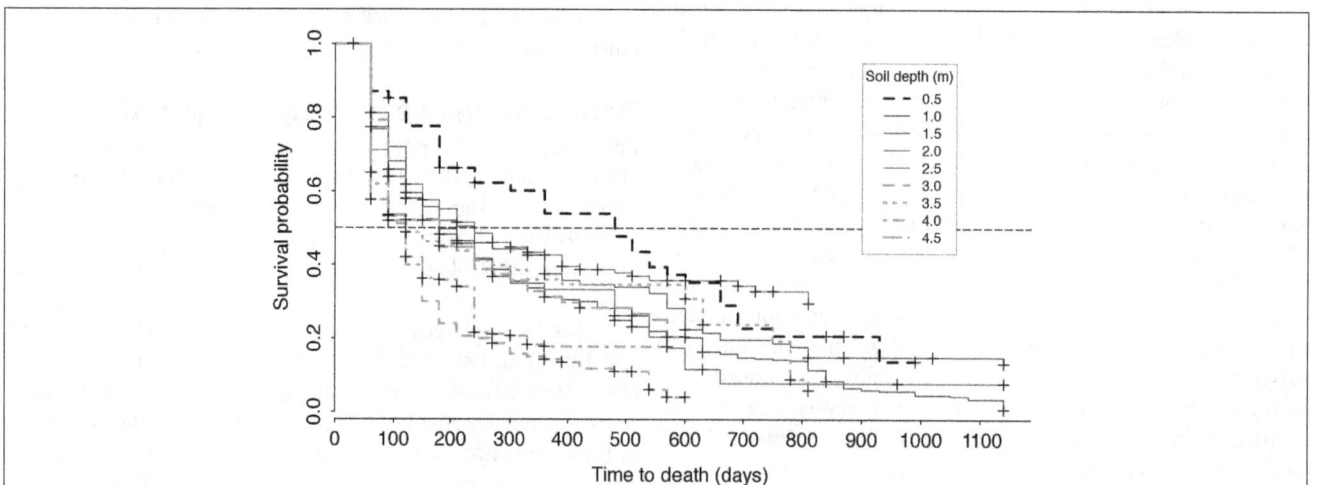

FIGURE 5 | Kaplan–Meier survival probability of roots depending on soil depth (see legend). The dotted horizontal line indicates a survival probability of 50%: the abscissa at which a given Kaplan–Meier curves intersects this line represents the half-life of roots corresponding to this survival curve, i.e., the estimated time required for half of the roots observed at any given point in time to have died.

(~4 months; turnover 3.04 yr^{-1}), with the exception of roots at 4.5 m; the latter had a half-life similar to that found between 1.0 and 2.5 m (**Figure 5**). Soil depth significantly affected root half-life values ($\chi^2 = 89.9, p < 0.001$) and survival at 0.5 m was longer than that at all other depths ($p < 0.05$) except for 2.5 and 1.5 m, while there was no difference in root survival between depths of 3.5–4.5 m.

Differences in root survival might be related in part, to root branching order, with higher branching order roots (Pregitzer et al., 2002), living longer, i.e., thicker roots observed at 0.5 m (**Figure 3**). This could be putatively associated with slower turnover compared to lower order roots (Yao et al., 2009; Sun et al., 2012). However, a Cox proportional hazards model including root radius as a

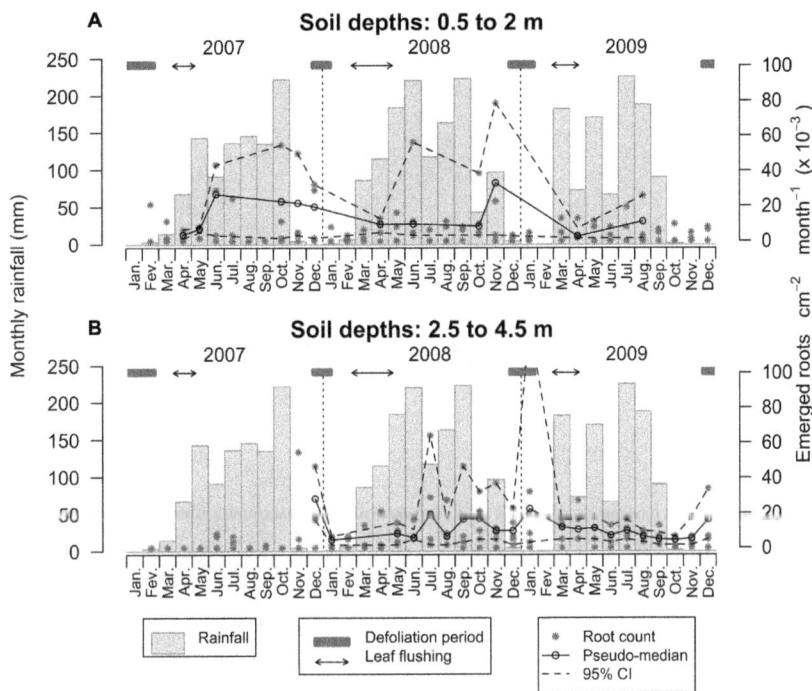

FIGURE 6 | Dynamics of root emergence over 3 years within the top 0.5–2.0 m of the soil profile (A) and between 2.5 and 4.5 m (B). Asterisks are emerging root counts determined in every root window at monthly intervals. Open circles represent the pseudo-median of emerging root counts in root windows ($n = 4$ and $n = 5$ for the upper and lower plots, respectively) and dashed lines are 95% confidence intervals estimated using the Wilcoxon signed rank test.

covariate of soil depth showed that root survival was clearly influenced by soil depth ($p < 0.001$) and not by root radius ($p = 0.25$).

Root Dynamics as a Function of Soil Depth and Rainfall

Root emergence between soil depths of 0.5 and 2.0 m ranged from 1.60 to 107.42 $\times$ 10^{-3} emerging roots cm^{-2} $month^{-1}$, with an average of 7.33 $\times$ 10^{-3} emerging roots cm^{-2} $month^{-1}$ (**Figure 6A**). Despite much variability, root emergence tended to be lowest in the first 3–4 months of each observed year, followed by an increase that lasted at least until the month of November (although there was much variability between depth increments and observation years, **Figure 6A**). During the first 2 years, root emergence increased approximately 3 months after the first rainfall, usually in the month of June. Root emergence could occur at relatively high rates during defoliation but was generally low during leaf flushing (**Figure 6A**). The dynamics of root emergence observed below 2.0 m was radically different with root emergence ranging from 1.60 to 128.27 $\times$ 10^{-3} emerging roots cm^{-2} $month^{-1}$, with a mean of 7.11 $\times$ 10^{-3} emerging roots cm^{-2} $month^{-1}$ (**Figure 6B**). There was very limited root growth until the 11th month of the observation period – or 14 months after root windows were installed – (i.e., until the onset of the first dry season and during leaf fall). Beyond that point in time, root emergence subsided until February 2008 and increased again and remained relatively high for 1 year (with a mean of 11.81 $\times$ 10^{-3} emerging roots cm^{-2} $month^{-1}$). Subsequently,

root emergence slowed down and became more stable over time with a mean of 6.57 $\times$ 10^{-3} emerging roots cm^{-2} $month^{-1}$ (**Figure 6B**).

The range of root mortality between soil depths of 0.5 and 2.0 m was 1.60 to 40.08 $\times$ 10^{-3} dead roots cm^2 month, with an average of 6.33 $\times$ 10^{-3} emerging roots cm^{-2} $month^{-1}$ (**Figure 7A**). Root mortality at these soil depths was relatively stable over time with the highest mortality rates observed from August to January. Below a depth of 2.0 m, the range of root mortality was 1.60–78.56 $\times$ 10^{-3} dead roots cm^{-2} $month^{-1}$, with an average of 5.28 $\times$ 10^{-3} emerging roots cm^{-2} $month^{-1}$ (**Figure 7B**). Following the initial period of root emergence at these soil depths, root mortality tended to remain at relatively high levels from June 2008 until June 2009 (a mean of 10.04 $\times$ 10^{-3} emerging roots cm^{-2} $month^{-1}$), beyond which it stabilized at a lower level of 6.65 $\times$ 10^{-3} emerging roots cm^{-2} $month^{-1}$.

Bivariate plots of monthly root length variations as a function of (1) average monthly rainfall, (2) monthly average of minimum daily soil temperature, and (3) average reference evapotranspiration are given in Supplementary Figure S2. There was a weak yet significant (as indicated by the low R-squared and p-values of the regressions) positive relationship between, on the one hand, average monthly rainfall and monthly root length variations (Supplementary Figure S2A) and on the other hand, monthly average of minimum daily soil temperature and monthly root length variations (Supplementary Figure S2C).

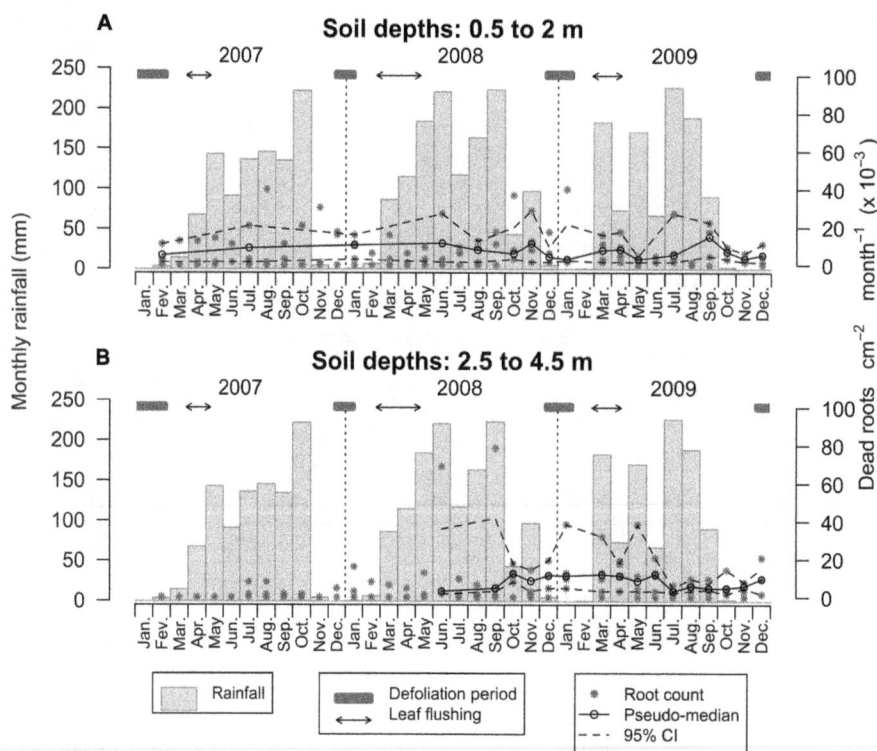

FIGURE 7 | Dynamics of root mortality over 3 years within the top 0.5–2.0m of the soil profile (A) and between 2.5 and 4.5 m (B). Asterisks are emerging root counts determined in every root window at monthly intervals. Open circles represent the pseudo-median of dead root counts in the deeper root windows ($n = 4$ and $n = 5$ for the upper and lower plots, respectively) and dashed lines are 95% confidence intervals estimated using the Wilcoxon signed rank test.

DISCUSSION

Fine Root Emergence

We showed that root phenology differed along the soil profile and was not synchronized with rainfall patterns as we had hypothesized, particularly below a depth of 2.0 m. Within the first 2 m of the soil profile, the highest rates of root emergence occurred about 3 months after the onset of the rainy season, while deeper in the soil, root emergence remained low until the 11th month of the observation period and was not correlated with the rainfall pattern. Therefore, the shallow parts of the root system were more responsive to rainfall, as roots near the soil surface capture water from rainfall more readily than deeper roots. Deep roots only emerged once rainfall became scarcer and may reflect the need for trees to use increasingly deeper water resources during the dry season.

Below 2.0 m, the first peak of root emergence rates occurred in November and December 2007, followed by a period of high root emergence that spanned from July 2008 to January 2009 (with the maximum peak in January 2009). Surprisingly, the highest emergence rates below 2.0 m occurred during the period of aerial dormancy, i.e., with no leaves supplying resources for root growth through photosynthesis. Similar results, whereby broadleaf tree root growth occurs significantly during a period of aerial dormancy, were also found in a

Mediterranean climate in *Juglans regia* L. (Germon et al., under revision). Abramoff and Finzi (2015) suggest that endogenous cuing (i.e., any factor that affects growth other than climate), and subsequent allocation of stored non-structural carbohydrates (NSC) are dominant drivers of root growth in subtropical and Mediterranean trees. Although the climate in our study was tropical, distinct rainy seasons are present, but soil and air temperatures remain warm, therefore water supply is likely the main limiting climatic factor, particularly in the upper soil horizons where less buffering exists against soil drying. In deeper soils, thermic and hydric buffering should thus allow for more constant rates of root growth throughout the year if endogenous cuing is not the main driver of growth. However, we found that the peak of deep root growth was delayed until late into the dry season. As tree root and stem NSC usually decline during the growing season and re-accumulate during aerial dormancy (Richardson et al., 2013), NSC re-accumulation rates may differ between shallow and deep roots, with a time lag resulting in delayed deep root growth. Nevertheless, as root emergence rates were so different between shallow and deep roots, it may be that the drivers between the two compartments are separate and distinct, with rainfall driving shallow root growth and endogenous cuing driving deep root growth. However, further studies using isotopes would be needed to support this hypothesis (Trumbore et al., 2006).

Between soil depths of 0.5 and 2.0 m, root mortality was relatively disconnected from variations in root emergence, although higher mortality values occurred toward the end of the rainy season, as did the highest emergence values. Below 2.0 m, from 2008 onward, peaks in root emergence and death were largely synchronized, e.g., in June and September 2008 as well as April–May 2009, suggesting the existence of a mechanism for the replacement of recently senesced roots. It is also possible that deep roots that first grew after the installation of the well, began to die, either because they could not be maintained by the tree (too costly in terms of resources) or because the relatively high rates of emergence during the second year were a response to the disturbance caused by the well installation, as often occurs in rhizotron experiments (Strand et al., 2008). Roots growing in the direction of the well may have had reduced access to resources (since the volume occupied by the well was inaccessible) thus suppressing root emergence (feedback response). Observed differences in root emergence could also be related to the presence/absence of roots in the immediate vicinity of observation windows, prior to their installation which increased the probability for early root emergence in windows near pre-existing roots. The total higher and lower rates of root emergence observed in 2008 and 2009, respectively, may also have been influenced by the total annual precipitation, which in 2009 (1002 mm) was only 79% of that in 2008 (1265 mm). The year 2007 was the driest during the observation period with only 965 mm total annual precipitation. However, the second semester of year 2009, was the driest second semester of the monitoring period with only 78% of the rainfall that had been monitored for the same period in the two preceding years. In the first 3 months of the rainy season 2009 (April–June) it rained less than 60% of that for the same period in 2008. However, the reduced emergence within the first 2 m, putatively related to drier conditions during the rainy season, was not counterbalanced by increased emergence at depth. Therefore, while our data support the hypothesis that deep root emergence might correspond to a safety net against water stress during the dry season, they do not point at the existence of a similar mechanism against dry periods occurring during the wet season.

The peak of evapotranspiration that occurs every year around March–April, was highest in 2007, intermediate in 2008 and lowest in 2009 (it did not occur in 2009 as high rainfall occurred as soon as March 2009). We hypothesize that high evaporative demand may be a signal that triggers root growth at the onset of the rainy season, particularly near the soil surface and the low evapotranspiration observed in 2009 may have resulted in a weaker pattern of root emergence in that year.

At all depths, in 2008, root emergence reached its lowest level throughout the period during which trees had already shed old leaves but not yet started to grow new leaves (i.e., February). Although this pattern could not be observed in the previous and following years, it might correspond to a dormancy state prior to the resumption of physiological activity with the first rains of the season.

Fine Root Turnover

As generally reported in the literature (Chen and Brassard, 2013), we found that soil depth had a significant effect on root turnover, but rather unexpectedly, root turnover increased with soil depth from 0.73 yr^{-1} at 0.5 m to 2-3 yr^{-1} at greater depths, in sharp contrast with what is generally reported (Wu et al., 2013). However, most studies have been concerned with soil depth ranges that were much shallower than those in our study (Baddeley and Watson, 2005; Chen and Brassard, 2013). Similarly, we did not find any evidence that root radius had an influence on longevity, although this has also been reported in the literature (Baddeley and Watson, 2005; Chen and Brassard, 2013). Furthermore, we did not find a linear increase in root turnover with soil depth over the whole 4.5 m range investigated, which is consistent with the theory that several factors, both intrinsic and extrinsic, control fine root life span (Chen and Brassard, 2013). It is known that environmental parameters (e.g., temperature, water content, N availability, CO_2 partial pressure) influence fine root turnover to variable degrees (Vogt et al., 1996). Therefore, we hypothesize that during dry periods, deeper distal roots underwent a physiological pruning process, whereby peripheral organs died, as also observed in shoots of *H. brasiliensis* (Chen and Cao, 2015).

Fine Root Biomass and Carbon

Assuming that the RLD values that we obtained from soil coring are homogeneous over large volumes of soil, it can be inferred that fine root biomass below a depth of 1.0 m could account for more than half of the overall fine root biomass of the rubber trees measured [4.8 t ha^{-1} between 0.0 and 1.0 m compared to 5.8 t ha^{-1} between 1.0 and 4.0 m, with a mean SRL of $\sim$14 m/g^{-1} for roots $\leq$1 mm in diameter (Pierret et al., 2007a)]. As roots may also be present below a depth of 4 m (the bedrock was found at 7–8 m), total root biomass may be underestimated. Assuming that rubber tree root tissues have a mean organic carbon content of approximately 47% (Wauters et al., 2008), our results show that rubber tree roots $\leq$0.5 mm in diameter, on average account for about 5 t C ha^{-1}. This value is similar, although slightly higher, than the 1.91–3.72 5 t C ha^{-1} range reported by Wauters et al. (2008) for coarser roots (2.5–25 mm in diameter) for a range of rubber tree clones from Western Ghana and Brazil. Similarly Cheng et al. (2007) estimated carbon stocks of 16.50 t C ha^{-1} for roots of all sizes, in rubber tree plantations at Hainan Island, China and Yuen et al. (2013) calculated total carbon stocks of the order of 4–32 t C ha^{-1} for rubber trees at six locations in Southeast Asia. Hence, the presence of fine roots at fairly low length densities over considerable soil depths can have important implications with regard to soil carbon accounting. As recently pointed out by Yuen et al. (2013), more attention should be given to sampling roots at appropriate depths if we are to improve baseline data on belowground carbon stocks. In addition, it should be acknowledged that there are still major uncertainties regarding (1) the reliability of coring versus imaging techniques for quantifying fine root biomass and turnover (Yuan and Chen, 2012) and (2) the way different fine root definitions might influence such quantifications (McCormack et al., 2015).

CONCLUSION

We explored the interactions between fine root dynamics, the rainfall regime and soil along a 4.5 m profile using a root access-well. Our results reveal that root growth dynamics in the upper 2 m of soil surface were related mainly to precipitation patterns (Chairungsee et al., 2013). Deeper in the soil, root growth was more independent of rainfall and was likely driven by internal tree carbon allocation. We show that fine root production will impact soil carbon stocks and was higher than commonly reported (e.g., Brunner and Godbold, 2007), particularly at depth. Such an input of fine root related carbon in soils could be all the more significant considering the slow breakdown of fine roots in some sub-tropical tree species (Xiong et al., 2012). One major limitation of this work is that observations are taken from a single location, which means that inference and conclusions cannot be generalized. The results of this study thus advocate in favor of more field studies aimed at assessing precisely the production and fate of fine roots, not only near the soil surface but also very deep in the soil.

AUTHOR CONTRIBUTIONS

JLM and AP designed the experimental setup, implemented it in the field, analysed the data and wrote the paper.

SG and SINA performed data collection in the field. CC, AS and SG have contributed significantly to the data analysis, discussing the results and writing of the paper.

ACKNOWLEDGMENTS

This research was funded by the French Institute of Research for Development (IRD), France, the French Institute for Natural Rubber (IFC), France, and Michelin/Socfinco/SIPH Rubber Tree Plantations Companies. We would also like to thank all our Thai counterparts from Khon Kaen University (KKU), Land Development Department (LDD), and the owner of the plantation (Mr. Chaipat Sirichaiboonwat) who kindly welcomed us. The authors also wish to thanks Drs. D. Nandris and F. Do (IRD) for their interest in and support of this research.

REFERENCES

Abramoff, R. Z., and Finzi, A. C. (2015). Are above- and below-ground phenology in sync? *New Phytol.* 205, 1054–1061. doi: 10.1111/nph.13111

Allen, R. G., Pereira, L. S., Raes, D., Smith, M., and Ab, W. (1998). *Crop Evapotranspiration - Guidelines for Computing Crop Water Requirements - FAO Irrigation and Drainage Paper 56.* Rome: FAO, 1–15.

Baddeley, J. A., and Watson, C. A. (2005). Influences of root diameter, tree age, soil depth and season on fine root survivorship in *Prunus avium. Plant Soil* 276, 15–22. doi: 10.1007/s11104-005-0263-6

Boithias, L., Do, F. C., Isarangkool Na Ayutthaya, S., Junjittakarn, J., Siltecho, S., and Hammecker, C. (2011). Transpiration, growth and latex production of a *Hevea brasiliensis* stand facing drought in northeast Thailand: the use of the WaNuLCAS model as an exploratory tool. *Exp. Agric.* 48, 49–63. doi: 10.1017/S001447971100086X

Boyce, K. C. (2005). "The evolutionary history of roots and leaves," in *Vascular Transport in Plants,* eds N. M. Holbrook and M. A. Zwieniecki (San Diego, CA: Elsevier), 479–500.

Brunner, I., and Godbold, D. L. (2007). Tree roots in a changing world. *J. For. Res.* 12, 78–82. doi: 10.1007/s10310-006-0261-4

Carr, M. K. V. (2011). The water relations of rubber (*Hevea brasiliensis*): a review. *Exp. Agric.* 48, 176–193. doi: 10.1017/S0014479711000901

Cawte, H. J., and Boyd, W. E. (2010). Laterite nodules: a credible source of iron ore in iron age northeast Thailand? *Geoarchaeology* 25, 626–644. doi: 10.1002/gea.20326

Chairungsee, N., Gay, F., Thaler, P., Kasemsap, P., Thanisawanyangkura, S., Chantuma, A., et al. (2013). Impact of tapping and soil water status on fine root dynamics in a rubber tree plantation in Thailand. *Front. Plant Sci.* 4:538. doi: 10.3389/fpls.2013.00538

Chen, H. Y. H., and Brassard, B. W. (2013). Intrinsic and extrinsic controls of fine root life span. *Crit. Rev. Plant Sci.* 32, 151–161. doi: 10.1080/07352689.2012.734742

Chen, J.-W., and Cao, K.-F. (2015). A possible link between hydraulic properties and leaf habits in *Hevea brasiliensis. Funct. Plant Biol.* 42, 718–726. doi: 10.1071/FP14294

Cheng, C.-M., Wang, R.-S., and Jiang, J.-S. (2007). Variation of soil fertility and carbon sequestration by planting *Hevea brasiliensis* in Hainan Island, China. *J. Environ. Sci.* 19, 348–352. doi: 10.1016/S1001-0742(07)60057-6

Danjon, F., Stokes, A., and Bakker, M. R. (2013). "Root systems of woody plants," in *Plant Roots Hidden Half,* 4th Edn, eds A. Eshel and T. Beeckman (Boca Raton, FL: CRC Press), 29.1–29.21.

Edwards, E. J., Benham, D. G., Marland, L. A., and Fitter, A. H. (2004). Root production is determined by radiation flux in a temperate grassland community. *Glob. Chang. Biol.* 10, 209–227. doi: 10.1111/j.1365-2486.2004.00729.x

Eissenstat, D. M., and Yanai, R. D. (1997). The ecology of root lifespan advances in ecological research. *Adv. Ecol. Res.* 27, 1–60. doi: 10.1016/S0065-2504(08)60005-7

Esser, G., Kattge, J., and Sakalli, A. (2011). Feedback of carbon and nitrogen cycles enhances carbon sequestration in the terrestrial biosphere. *Glob. Chang. Biol.* 17, 819–842. doi: 10.1111/j.1365-2486.2010.02261.x

Gonkhamdee, S., Maeght, J. L., Do, F., and Pierret, A. (2009). Growth dynamics of fine *Hevea brasiliensis* roots along a 4.5-m soil profile. *Khon Kaen Agric. J.* 37, 265–276.

Green, J. J., Dawson, L. A., Proctor, J., Duff, E. I., and Elston, D. A. (2005). Fine root dynamics in a tropical rain forest is influenced by rainfall. *Plant Soil* 276, 23–32. doi: 10.1007/s11104-004-0331-3

Hartmann, C., Lesturgez, G., Do, F. C., Maeght, J. L., Isarangkool, S., Ayutthaya, N., et al. (2006). "Rubber tree trunk phloem necrosis (TPN) in northeast thailand: 1. Investigations on soil heterogeneities linked to disease outbreak," in *Proceedings of the International Natural Rubber Conference - IRRDB Annual Meetinge,* Ho Chi Minh City, 139–156.

Isarangkool Na Ayutthaya, S., Do, F. C., Pannengpetch, K., Junjittakarn, J., Maeght, J.-L., Rocheteau, A., et al. (2010). Transient thermal dissipation method of xylem sap flow measurement: multi-species calibration and field evaluation. *Tree Physiol.* 30, 139–148. doi: 10.1093/treephys/tpp092

Isarangkool Na Ayutthaya, S., Do, F. C., Pannangpetch, K., Junjittakarn, J, Maeght, J.-L., Rocheteau, A., et al. (2011). Water loss regulation in mature *Hevea brasiliensis:* effects of intermittent drought in the rainy season and hydraulic regulation. *Tree Physiol.* 31, 751–762. doi: 10.1093/treephys/tpr058

Junjittakarn, J. (2012). Short term effects of latex tapping on micro-changes of trunk girth in *Hevea brasiliensis*. *Aust. J. Crop Sci.* 6, 65–72.

Kaplan, E. L., and Meier, P. (1958). Nonparametric estimation from incomplete observations. *J. Am. Stat. Assoc.* 53, 457–481. doi: 10.1080/01621459.1958.10501452

Kaspar, T. C., and Ewing, R. P. (1997). ROOTEDGE: software for measuring root length from desktop scanner images. *Agron. J.* 89, 932. doi: 10.2134/agronj1997.00021962008900060014x

Kimura, K., Kikuchi, S., and Yamasaki, S. (1999). Accurate root length measurement by image analysis. *Plant Soil* 216, 117–127. doi: 10.1023/A:1004778925316

Lobet, G., Hachez, C., and Draye, X. (2013). "Root water uptake and water flow in the soil–root domain," in *Plant Roots: The Hidden Half*, eds Y. Waisel, A. Eshel, and U. Kafkafi (Boca Raton, FL: CRC Press), 24–21.

Maeght, J.-L., Pierret, A., Sanwangsri, M., and Hammecker, C. (2007). "Field monitoring of rice rhizosphere dynamics in saline soils of NE Thailand," in *Proceedings of the International Conference Rhizosphere 2, 26-31 August 2007*, Montpellier, 26–31.

Maeght, J.-L., Rewald, B., and Pierret, A. (2013). How to study deep roots—and why it matters. *Front. Plant Sci.* 4:299. doi: 10.3389/fpls.2013.00299

Majdi, H., Pregitzer, K., Morén, A.-S., Nylund, J.-E., and Ågren, G. I. (2005). Measuring fine root turnover in forest ecosystems. *Plant Soil* 276, 1–8. doi: 10.1007/s11104-005-3104-8

McCormack, M. L., Dickie, I. A., Eissenstat, D. M., Fahey, T. J., Fernandez, C. W., Guo, D., et al. (2015). Tansley review: redefining fine roots improves understanding of below- ground contributions to terrestrial biosphere processes. *New Phytol.* 207, 505–518. doi: 10.1111/nph.13363

Pierret, A., Doussan, C., Pages, L., Do, F. C., Gonkhamdee, S., Maeght, J. L., et al. (2007a). Is impeded root growth related to the occurrence of rubber tree trunk phloem necrosis (TPN)? Preliminary results from NE Thailand. *Paper Presented at IRRDB Annual Meeting*, Siem Reap, 489–498.

Pierret, A., Latchackak, K., Sengtaheuanghoung, O., and Valentin, C. (2007b). Interactions between root growth, slope and soil detachment depending on land use: a case study in a small mountain catchment of Northern Laos. *Plant Soil* 301, 51–64. doi: 10.1007/s11104-007-9413-3

Pierret, A., Gonkhamdee, S., Jourdan, C., and Maeght, J.-L. (2013). IJ_Rhizo: an open-source software to measure scanned images of root samples. *Plant Soil* 373, 531–539. doi: 10.1007/s11104-013-1795-9

Pregitzer, K. S., DeForest, J. L., Burton, A. J., Allen, M. F., Ruess, R. W., and Hendrick, R. L. (2002). Fine root architecture of nine North American trees. *Ecol. Monogr.* 72, 293–309.

R Development Core Team (2013). *R: A Language and Environment for Statistical Computing v.3.0.0.* Vienna: R Foundation for Statistical Computing. Available at: http://www.R-project.org/ [accessed April 2013].

Richardson, S. J., Allen, R. B., Buxton, R. P., Easdale, T. A., Hurst, J. M., Morse, C. W., et al. (2013). Intraspecific relationships among wood density, leaf structural traits and environment in four co-occurring species of Nothofagus in New Zealand. *PLoS ONE* 8:e58878. doi: 10.1371/journal.pone.0058878

Smit, A. L., Bengough, A. G., Engels, C., van Noordwijk, M., Pellerin, S., and van de Geijn, S. C. (2000). *Root Methods: A Handbook.* Berlin: Springer-Verlag, 587.

Stone, E. L., and Kalisz, P. J. (1991). On the maximum extent of tree roots. *For. Ecol. Manage.* 46, 59–102. doi: 10.1016/0378-1127(91)90245-Q

Strand, A. E., Pritchard, S. G., McCormack, M. L., Davis, M. A., and Oren, R. (2008). Irreconcilable differences: fine-root life spans and soil carbon persistence. *Science* 319, 456–458. doi: 10.1126/science.1151382

Sun, J., Gu, J., and Wang, Z. (2012). Discrepancy in fine root turnover estimates between diameter-based and branch-order-based approaches: a case study in two temperate tree species. *J. For. Res.* 23, 575–581. doi: 10.1007/s11676-012-0297-6

Trumbore, S., Da Costa, E. S., Nepstad, D. C., Barbosa De Camargo, P., Martinelli, L. A., Ray, D., et al. (2006). Dynamics of fine root carbon in Amazonian tropical ecosystems and the contribution of roots to soil respiration. *Glob. Change Biol.* 12, 217–229. doi: 10.1111/j.1365-2486.2005.001063.x

Virginia, R. A., and Jarrell, W. M. (1987). "Approaches for studying the function of deep root systems," in *Plant Response to Stress: Functional Analysis in Mediterranean Ecosystems*, Vol. 15, eds J. D. Tenhunen, F. M. Catarino, O. L. Lange, and W. C. Oechel (Berlin: Springer), 107–127.

Vogt, K. A., Vogt, D. J., Palmiotto, P. A., Boon, P., O'Hara, J., and Asbjornsen, H. (1996). Review of root dynamics in forest ecosystems grouped by climate, climatic forest type and species. *Plant Soil* 187, 159–219. doi: 10.1007/BF00017088

Wauters, J. B., Coudert, S., Grallien, E., Jonard, M., and Ponette, Q. (2008). Carbon stock in rubber tree plantations in Western Ghana and Mato Grosso (Brazil). *For. Ecol. Manage.* 255, 2347–2361. doi: 10.1016/j.foreco.2007.12.038

Wu, Y., Deng, Y., Zhang, J., Wu, J., Tang, Y., Cao, G., et al. (2013). Root size and soil environments determine root lifespan: evidence from an alpine meadow on the Tibetan Plateau. *Ecol. Res.* 28, 493–501. doi: 10.1007/s11284-013-1038-9

Xiong, Y., Fan, P., Fu, S., Zeng, H., and Guo, D. (2012). Slow decomposition and limited nitrogen release by lower order roots in eight Chinese temperate and subtropical trees. *Plant Soil* 363, 19–31. doi: 10.1007/s11104-012-1290-8

Yao, S., Merwin, I. A., and Brown, M. G. (2009). Apple root growth, turnover, and distribution under different orchard groundcover management systems. *HortScience* 44, 168–175.

Yuan, Z. Y., and Chen, H. Y. H. (2012). Indirect methods produce higher estimates of fine root production and turnover rates than direct methods. *PLoS ONE* 7:e48989. doi: 10.1371/journal.pone.0048989

Yuen, J. Q., Ziegler, A. D., Webb, E. L., and Ryan, C. M. (2013). Uncertainty in below-ground carbon biomass for major land covers in Southeast Asia. *For. Ecol. Manage.* 310, 915–926. doi: 10.1016/j.foreco.2013.09.042

Zoon, F. C., and Tienderen, P. H. (1990). A rapid quantitative measurement of root length and root branching by microcomputer image analysis. *Plant Soil* 126, 301–308. doi: 10.1007/BF00012833

Conflict of Interest Statement: The authors declare that the research was conducted in the absence of any commercial or financial relationships that could be construed as a potential conflict of interest.

13

Zinc Oxide Nanoparticles Affect Biomass Accumulation and Photosynthesis in *Arabidopsis*

Xiaoping Wang, Xiyu Yang, Siyu Chen, Qianqian Li, Wei Wang, Chunjiang Hou, Xiao Gao, Li Wang and Shucai Wang*

Key Laboratory of Molecular Epigenetics of Ministry of Education, Northeast Normal University, Changchun, China

Edited by:
Nelson Marmiroli,
University of Parma, Italy

Reviewed by:
Zhenzhu Xu,
Chinese Academy of Sciences, China
Jane Geisler-Lee,
Southern Illinois University
Carbondale, USA

*Correspondence:
Shucai Wang
wangsc550@nenu.edu.cn

Dramatic increase in the use of nanoparticles (NPs) in a variety of applications greatly increased the likelihood of the release of NPs into the environment. Zinc oxide nanoparticles (ZnO NPs) are among the most commonly used NPs, and it has been shown that ZnO NPs were harmful to several different plants. We report here the effects of ZnO NPs exposure on biomass accumulation and photosynthesis in *Arabidopsis*. We found that 200 and 300 mg/L ZnO NPs treatments reduced *Arabidopsis* growth by ~20 and 80%, respectively, in comparison to the control. Pigments measurement showed that Chlorophyll a and b contents were reduced more than 50%, whereas carotenoid contents remain largely unaffected in 300 mg/L ZnO NPs treated *Arabidopsis* plants. Consistent with this, net rate of photosynthesis, leaf stomatal conductance, intercellular CO_2 concentration and transpiration rate were all reduced more than 50% in 300 mg/L ZnO NPs treated plants. Quantitative RT-PCR results showed that expression levels of chlorophyll synthesis genes including *CHLOROPHYLL A OXYGENASE* (CAO), *CHLOROPHYLL SYNTHASE* (CHLG), *COPPER RESPONSE DEFECT 1* (CRD1), *MAGNESIUM-PROTOPORPHYRIN IX METHYLTRANSFERASE* (CHLM) and *MG-CHELATASE SUBUNIT D* (CHLD), and photosystem structure gene *PHOTOSYSTEM I SUBUNIT D-2* (PSAD2), *PHOTOSYSTEM I SUBUNIT E-2* (PSAE2), *PHOTOSYSTEM I SUBUNIT K* (PSAK) and *PHOTOSYSTEM I SUBUNIT K* (PSAN) were reduced about five folds in 300 mg/L ZnO NPs treated plants. On the other hand, elevated expression, though to different degrees, of several carotenoids synthesis genes including *GERANYLGERANYL PYROPHOSPHATE SYNTHASE 6* (GGPS6), *PHYTOENE SYNTHASE* (PSY) *PHYTOENE DESATURASE* (PDS), and *ZETA-CAROTENE DESATURASE* (ZDS) were observed in ZnO NPs treated plants. Taken together, these results suggest that toxicity effects of ZnO NPs observed in *Arabidopsis* was likely due to the inhibition of the expression of chlorophyll synthesis genes and photosystem structure genes, which results in the inhibition of chlorophylls biosynthesis, leading to the reduce in photosynthesis efficiency in the plants.

Keywords: nanoparticles, ZnO, biomass, chlorophylls, carotenoid, gene expression, *Arabidopsis*

INTRODUCTION

Nanoparticles (NPs), also known as particulate nanomaterials (NMs), are particle materials with at least one dimension in the nanoscale (1–100 nm). Because of their particular properties, such as small size, high surface-to-volume ratio and unique physical and chemical properties, the use of NPs in industries and a wide range of consumer products are increasing greatly (Stampoulis et al., 2009). These positive commercial advances have stimulated a rapidly increasing production of engineered NPs, which made nanotechnologies a rapidly developing field with an expectation that the annual value of nanotechnology-related products is going to reach one trillion dollars in 2015 (Roco, 2005). However, the increasing usage of NPs greatly increased the likelihood of their release into the environment, and has raised concerns about the impacts of NMs on health and the environment (Xia et al., 2009; Shvedova et al., 2010).

Studies in plants have demonstrated that at least some NPs can be uptaken (Etxeberria et al., 2006; Chen et al., 2010; Kurepa et al., 2010; Schwab et al., 2015a), transported (Liu et al., 2009; Cifuentes et al., 2010; Zhao et al., 2012; Wang et al., 2012b), and accumulated in specific subcellular locations such as cell vacuoles, nuclei and plasmodesmata (Kurepa et al., 2010; Schwab et al., 2015b), and NPs can alter plant physiological processes, and influence plant growth and development (Stampoulis et al., 2009; Ma et al., 2010; Burklew et al., 2012; García-Sánchez et al., 2015). For example, ultra small anatase TiO_2 NPs have been shown to be able to enter into plant cells, accumulate in subcellular locations such as cell vacuoles and nuclei of root cells, and cause reorganization and elimination of microtubules, resulting in inhibition of root elongation in Arabidopsis (Kurepa et al., 2010; Wang et al., 2011). CuO NPs have been shown to be able to transport in maize via xylem and phloem (Wang et al., 2012b). Whereas silver NPs (Ag NPs) and zinc oxide NPs (ZnO NPs) treatment lead to increase in contents of free radicals, including reactive oxygen and nitrogen species (ROS/RNS) and hydrogen peroxide (H_2O_2) in duckweed (Thwala et al., 2013).

Effects of several NPs, including Ag NPs (Qian et al., 2013; Thwala et al., 2013; Geisler-Lee et al., 2014), aluminum oxide NPs (Al_2O_3 NPs) (Burklew et al., 2012; Riahi-Madvar et al., 2012), silicon dioxide NPs (SiO_2 NPs) (Zhang et al., 2015), and ZnO NPs (Ma et al., 2013) have been studied in several different plant species. These experiments indicated that NPs have negative effects on several different aspects of plant growth and development including seed germination (Ma et al., 2013), root elongation (Wang et al., 2012a), biomass accumulation (Zhao et al., 2013). NPs have also been shown to be able to induce oxidative stress and alter gene expression in plants (Wang et al., 2013a).

ZnO NPs are among the most commonly used NPs in a variety of applications such as personal care, ceramics, paints, pigments, foods, batteries, and semiconductors (Ju-Nam and Lead, 2008), which increased the potential of their direct release to the environment. Thus the ecological risk of ZnO NPs is an important topic. As a matter of fact, the toxicity effects of ZnO NPs have been observed in several different plants species including Arabidopsis (Lee et al.,

2010), buckweed (Lee et al., 2013), wheat (Du et al., 2011), dotted duckmeat (Thwala et al., 2013), cucumber (Zhao et al., 2013), rapeseed (Kouhi et al., 2014), alfalfa (Bandyopadhyay et al., 2015), and cowpea (Wang et al., 2013b). Most of the experiments show that the inhibition effects of ZnO NPs on plant growth and development are likely due to the induction of oxidative stress (Hernandez-Viezcas et al., 2011; Thwala et al., 2013; Bandyopadhyay et al., 2015). In consistent with this, transcriptome analysis in Arabidopsis have shown that most of the genes induced by ZnO NPs are ontology groups annotated as stress responsive, including both abiotic and biotic stimuli (Landa et al., 2012).

To further investigate the mechanisms of the toxicity effects of ZnO NPs on plant growth and development, we examined the effects of ZnO NPs on biomass accumulation and photosynthesis in Arabidopsis. We show that expose to ZnO NPs led to decrease in biomass accumulation in both shoots and roots, and that chlorophylls, but not carotenoid contents were decreased in plants treated ZnO NPs. Consistent with this observation, quantitative RT-PCR results showed that the expression levels of some chlorophyll synthesis genes and photosystem genes examined were decreased in response to ZnO NPs treatment.

MATERIALS AND METHODS

Plant Materials and Growth Conditions

Arabidopsis (Arabidopsis thaliana) ecotype Columbia (Col-0) was used for the experiments. Seeds were sown directly into soil pots and kept in a growth room at 22°C with a light density (photosynthetic active radiation) of approximately 120 μmol $m^{-2}s^{-1}$, and a light/dark photoperiod of 16 h/8 h.

ZnO NPs Treatment

Soil pots were prepared by filling the 5.5 cm × 5.5 cm pots with TS-1 white peat bedding substrate (Epagma) moistened thoroughly with suspensions containing 0, 50, 100, 200, 250, and 300 mg/L ZnO NPs with particle size <50 nm (Sigma), respectively. Substrate moistened with supernatants from 300 mg/L ZnO NPs suspensions were used as controls. The supernatants were obtained by centrifuging 300 mg/L ZnO NPs suspensions for 10 min at 4000 rpm, and filtered through 0.22 um-diameter filters. Arabidopsis seeds were then germinated and grown in the soil pots. After germination, extra seedlings were removed to ensure that every pot contains 12 plants. The plants were watered regularly till 4-week-old, then watered once every 2 days for two times with 50 ml suspensions containing ZnO NPs at the same concentrations, or supernatants from 300 mg/L ZnO NPs suspensions as described above.

Measurement of Net Rate of Photosynthesis, Leaf Stomatal Conductance, Intercellular CO_2 Concentration, and Transpiration Rate

Net rate of photosynthesis (P_N), leaf stomatal conductance (g_s), intercellular CO_2 concentration (Ci) and transpiration rate (E) of the fully expanded fifth rosette leaves of 6-week-old plants were measured by using a portable open-flow, gas-exchange system (LI-6400; LICOR Biosciences, Lincoln, NE, USA) on the morning before the plants were harvested. The ambient CO_2 concentration was 360 ± 10 μmol mol^{-1}, the air temperature was 22°C, and humidity was about 50%. A total of four leaves per pot were measured, and measurements were repeated four times for each leaf.

The net rate of photosynthesis was calculated as CO_2 uptake in micromole per square meter leaf area per second (μmol CO_2 m^{-2}s^{-1}), leaf stomatal conductance as water vapor in mole per square meter leaf area per second (mol H_2O m^{-2}s^{-1}), intercellular CO_2 concentration as CO_2 in micromole per mole intercellular air (μmol CO_2 mol^{-1}), and the transpiration rate as water loss in millimole per square meter leaf area per second (mmol H_2O m^{-2}s^{-1}).

Chlorophylls and Carotenoids Measurement

Chlorophylls and carotenoids contents were measured as described by Lichtenhaler and Wellburn (1983). Briefly, Chlorophylls and Carotenoids were extracted from rosette leaves from 6-week-old plants with 100% alcohol. Absorption of the extracts was measured using a spectrophotometer (721 TYPE, Shanghai analysis instrument co., LTD). Contents of Chlorophyll a (Chl a), Chlorophyll b (Chl b), total chlorophylls (C_T), and Carotenoids (Car) were calculated using the formula described (Lichtenhaler and Wellburn, 1983).

Biomass and Water Content Measurement

Six-week-old plants from each pot were harvested, washed thoroughly with tap water, and then distilled water. After removing excess water with paper towels, the roots and shoots were separated, and fresh weight (FW) was recorded. The samples were then oven-dried at 80°C for 15 min, followed by vacuum-dry at 40°C to a constant mass before dry weight (DW) was recorded. FW and DW per plant were calculated respectively, by divided the FW or DW per pot with 12. Water contents were calculated by using the formula: (FW-DW)/DW.

RNA Isolation and Quantitative RT-PCR (qRT-PCR)

Total RNA were isolated from rosette leaves of 6-week-old *Arabidopsis* plants by using EasyPure™ Plant RNA Kit (Transgene) and following the manufacturer's protocols.

First strand cDNA was synthesized using 2 μg total RNA by Oligo(dT)-primed reverse transcription using the EazyScript First-Strand DNA Synthesis Super Mix (TransGen Biotech) by following the manufacturer's instructions. Quantitative RT-PCR (qRT-PCR) was used to examine the expression of chlorophylls synthesis genes, carotenoids synthesis genes, and photosystem structure genes. *Arabidopsis* gene *ACTIN2* (*ACT2*) were used as a inner control for qRT-PCR. The primers used for qRT-PCR examination of *ACT2, MAGNESIUM-PROTOPORPHYRIN IX*

METHYLTRANSFERASE (*CHLM*), *Mg-chelatase subunit D* (*CHLD*), *GERANYLGERANYL PYROPHOSPHATE SYNTHASE 6* (*GGPS6*), *PHYTOENE SYNTHASE* (*PSY*), *PHYTOENE DESATURASE* (*PDS*), *ZETA-CAROTENE DESATURASE* (*ZDS*), *PHOTOSYSTEM I SUBUNIT K* (*PSAK*), and *PHOTOSYSTEM I SUBUNIT K* (*PSAN*) have been described previously (Qin et al., 2007; Stephenson and Terry, 2008; Liu et al., 2015).

The primer pairs used for qRT-PCR examination of *CHLOROPHYLL A OXYGENASE* (*CAO*), *CHLOROPHYLL SYNTHASE* (*CHLG*), *COPPER RESPONSE DEFECT 1* (*CRD1*), *PHOTOSYSTEM I SUBUNIT D-2* (*PSAD2*) and *PHOTOSYSTEM I SUBUNIT E-2* (*PSAE2*) were listed in **Table1**.

Statistical Analysis

Statistical analysis was performed as described previously (Wang et al., 2015). Briefly, data were analyzed by one-way analysis of variance (ANOVA) using the statistical software SPSS 22.0 (IBM Inc., New York, USA), and compared by student-Newman–Keuls (*q*-test).

RESULTS

ZnO NPs Affect Growth and Biomass Accumulation in *Arabidopsis*

After 1 month of growth, ZnO NPs treated plants showed an obvious decrease in the rosette size, and the decrease in the rosette sizes were positively related with the concentrations of ZnO NPs applied (**Figure 1A**), suggesting that ZnO NPs inhibited *Arabidopsis* growth. Quantitative analysis showed that ZnO NPs at 100 mg/L had little, if any effects on the FW and DW of the plants (**Figures 1B,C**), however, an about 20% of decrease on both fresh and dry weight was observed for 200 mg/L ZnO NPs treated plants, and that for 300 mg/L ZnO NPs treated plants was about 80% (**Figures 1B,C**). Water contents, on the other hand, were reduced only in 300 mg/L ZnO NPs treated plants (**Figure 1D**). Detailed analysis showed that ZnO NPs at high concentration has more severe inhibition effects on root growth than shoot growth (**Figures 1E,F**). As a result, root/shoot ratio reduced about 40% in 300 mg/L ZnO NPs treated plants, whereas ZnO NPs at relative lower concentration has little, if any effects on root/shoot ratio of the plants (**Figure 1G**).

TABLE 1 | Primers used in this study.

Primers	Sequences
CAO-qF	5'-AGTCCTTCTGCTTTATCTCTC-3'
CAO-qR	5'-TTCTCAACTAATCCACTCTCA-3'
CHLG-qF	5'-GAGATTTGTTGTGCGTGCGG-3'
CHLG-qR	5'-CCAGTGGAGGCCAAGTGACT-3'
CRD1-qF	5'-AAGAGGAAACTGGATAGAA-3'
CRD1-qR	5'-AAAGAAGTAACCAAAGGAA-3'
PSAD2-qF	5'-CAAACACACCATCACCAATC-3'
PSAD2-qR	5'-ACCTCGTACCTAAAGCCAAA-3'
PSAE-qF	5'-CACCACCATTGTGTCTTTCT-3'
PSAE-qR	5'-TTGACCTTGGATCCTCTCTT-3'

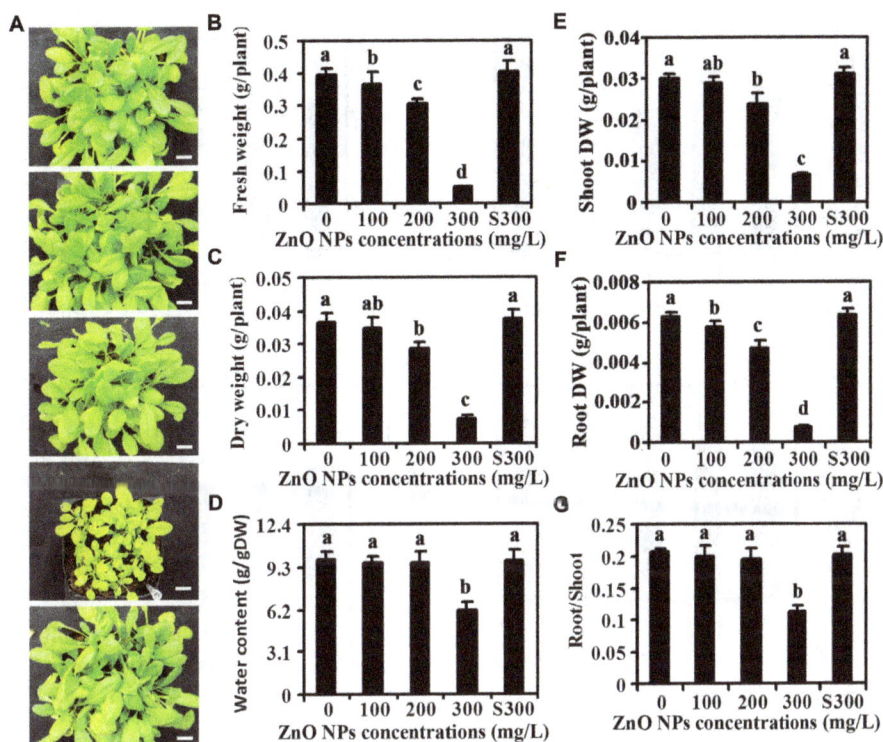

FIGURE 1 | Effects of ZnO NPs on growth of *Arabidopsis* plants. (A) Photographs of 1-month-old *Arabidopsis* plants treated with 0, 100, 200, 300 mg/L ZnO NPs, or supernatant from 300 mg/L ZnO NPs suspensions (from top to bottom). Bar, 9 mm. **(B)** Fresh weight of 6-week-old *Arabidopsis* plants treated with 0, 100, 200, 300 mg/L ZnO NPs, or supernatant from 300 mg/L ZnO NPs suspensions. **(C)** Dry weight (DW) of 6-week-old *Arabidopsis* plants treated with 0, 100, 200, 300 mg/L ZnO NPs, or supernatant from 300 mg/L ZnO NPs suspensions. **(D)** Water content in 6-week-old *Arabidopsis* plants treated with 0, 100, 200, 300 mg/L ZnO NPs, or supernatant from 300 mg/L ZnO NPs suspensions. **(E)** Shoot DW of 6-week-old *Arabidopsis* plants treated with 0, 100, 200, 300 mg/L ZnO NPs, or supernatant from 300 mg/L ZnO NPs suspensions. **(F)** Root DW of 6-week-old *Arabidopsis* plants treated with 0, 100, 200, 300 mg/L ZnO NPs, or supernatant from 300 mg/L ZnO NPs suspensions. **(G)** Root/Shoot DW ratio of 6-week-old *Arabidopsis* plants treated with 0, 100, 200, 300 mg/L ZnO NPs, or supernatant from 300 mg/L ZnO NPs suspensions. Data in **(B–G)** represent the mean ± standard deviation (SD) of four replicates. Different letters indicate significantly different ($p < 0.05$).

Chlorophylls, but not Carotenoids Contents were Affected by ZnO NPs Treatment

Yellow leaf color observed in ZnO NPs treated plants (**Figure 1A**) indicates that ZnO NPs may affect chlorophylls contents in the plants. To test this, we measured contents of chlorophylls including Chl a and Chl b in the plants. As shown in **Figure 2A**, ZnO NPs at lower concentrations has little, if any effects on Chl a contents, however, an about 50% reduce in Chl a contents was observed in 300 mg/L ZnO NPs treated plants. On the other hand, even at lower concentrations, i.e., 100 mg/L ZnO NPs treatment decreased Chl b contents in plants, and the degrees of decrease increased following the increase in the ZnO NPs concentrations (**Figure 2B**). Consistent with this, Chl a/b ratio increased, though not to a high degree in ZnO NPs treated plants (**Figure 2C**), and total chlorophylls contents also reduced in ZnO NPs treated plants (**Figure 2D**). On the other hand, ZnO NPs treatment increased carotenoids contents (**Figure 2E**). However, a positive relation between the increases of carotenoids contents and the concentrations of ZnO NPs was not observed (**Figure 2E**). As a consequence, carotenoids/chlorophylls ratio

increased in ZnO NPs treated plants, and an about three-fold increase was observed in 300 mg/L ZnO NPs treated plants (**Figure 2F**).

Leaf Photosynthesis in ZnO NPs Treated Plants was Reduced

Because ZnO NPs treatment resulted in reduction of chlorophylls contents (**Figure 2**), we suspected that photosynthesis may be affected in ZnO NPs treated plants. To test this, we examined net rate of photosynthesis, leaf stomatal conductance, intercellular CO_2 concentration and transpiration rate of the fully expanded fifth rosette leaves. As shown in **Figure 3A**, ZnO NPs treatment at all the concentrations tested decreased the net rate of photosynthesis, with an about 60% decrease observed in 300 mg/L ZnO NPs treated plants. An even higher degree of decrease was observed for the leaf stomatal conductance, in the plants treated with 300 mg/L ZnO NPs, the leaf stomatal conductance was only about 30% of that in the control plants treated with water only or with supernatants from 300 mg/L ZnO NPs suspensions (**Figure 3B**). The intercellular CO_2 concentration was also greatly reduced in ZnO NPs treated

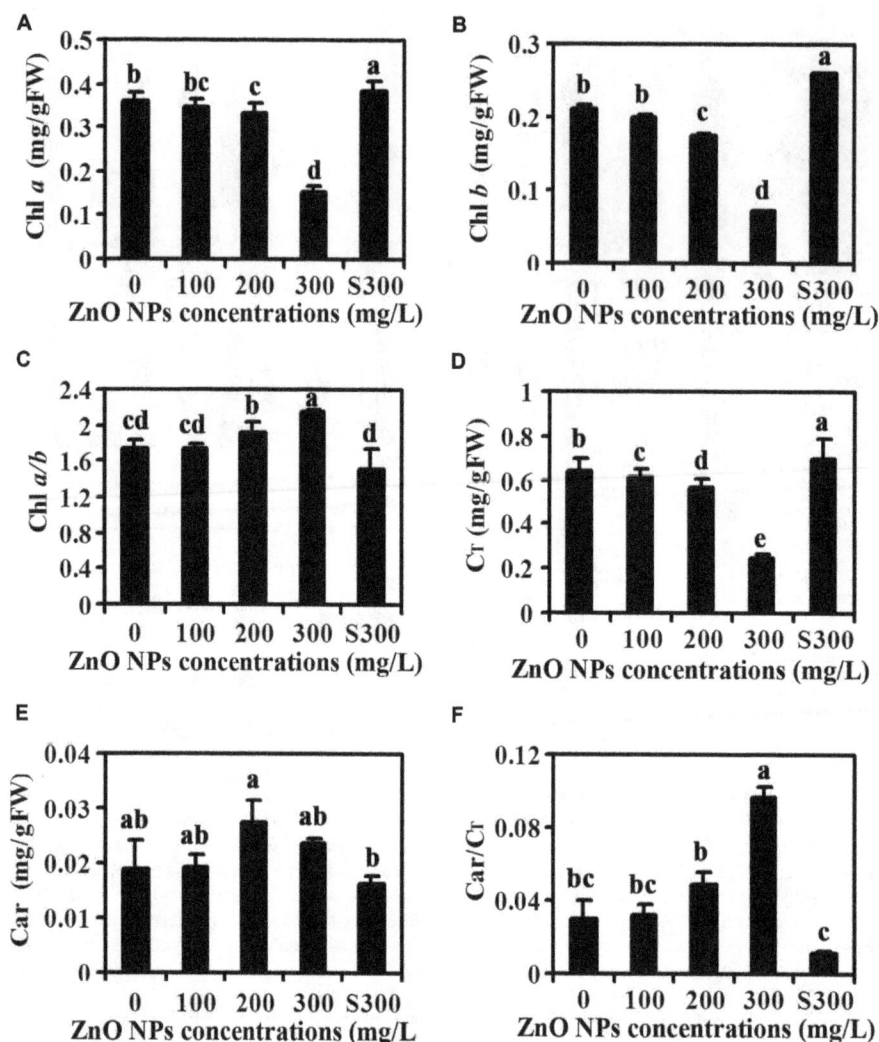

FIGURE 2 | Effects of ZnO NPs on chlorophylls and carotenoids contents in *Arabidopsis* rosette leaves. Chlorophyll a **(A)**, Chlorophyll b **(B)**, Chlorophyll a/b ratio **(C)**, total chlorophylls **(D)**, carotenoids **(E)** and carotenoids/total chlorophylls ratio **(F)** in rosette leaves of 6-week-old *Arabidopsis* plants treated with 0, 100, 200, 300 mg/L ZnO NPs, or supernatant from 300 mg/L ZnO NPs suspensions. Chlorophylls and carotenoids were extracted from rosette leaves from 6-week-old *Arabidopsis* plants using alcohol, and measured using a spectrophotometer. Data represent the mean ± SD of four replicates. Different letters indicate significantly different ($p < 0.05$).

plants, with an about 20% decrease in 200 mg/L ZnO NPs treated plants, and a more than 60% decrease in 300mg/L ZnO NPs treated plants (**Figure 3C**). A similar trend of decrease on transpiration rate was also observed in plants watered with ZnO NPs, i.e., an about 10, 30, and 60% decrease, respectively in plants treated with 100, 200, and 300 mg/L ZnO NPs (**Figure 3D**).

Expression Levels of Chlorophyll Synthesis Genes in ZnO NPs Treated Plants were Reduced

Several genes including *CHLOROPHYLL A OXYGENASE* (*CAO*), *CHLOROPHYLL SYNTHASE* (*CHLG*), *COPPER RESPONSE DEFECT 1* (*CRD1*), *MAGNESIUM-PROTOPORPHYRIN*

IX METHYLTRANSFERASE (*CHLM*), and *MG-CHELATASE SUBUNIT D* (*CHLD*) have been shown to be involved chlorophyll synthesis (Tanaka et al., 1998; Lange and Ghassemian, 2003; Rzeznicka et al., 2005; Pontier et al., 2007; Kobayashi et al., 2008). Having shown that ZnO NPs treatments reduced chlorophylls contents in *Arabidopsis* (**Figure 2**), we wanted to further examine if expression of the genes involved in chlorophyll synthesis were affected. Total RNA was isolated from rosette leaves, and quantitative RT-PCR was used to examine the expression of the chlorophyll synthesis genes. As shown in **Figure 4**, expression of all the genes examined including *CAO*, *CHLG*, *CRD1*, *CHLM* and *CHLD* was reduced in the plants treated with ZnO NPs at all concentrations. In 300 mg/L ZnO NPs treated plants, an ~5–7-fold decrease of the genes examined was observed (**Figure 4**). It should be noted that a slight, i.e., ~1.3-fold,

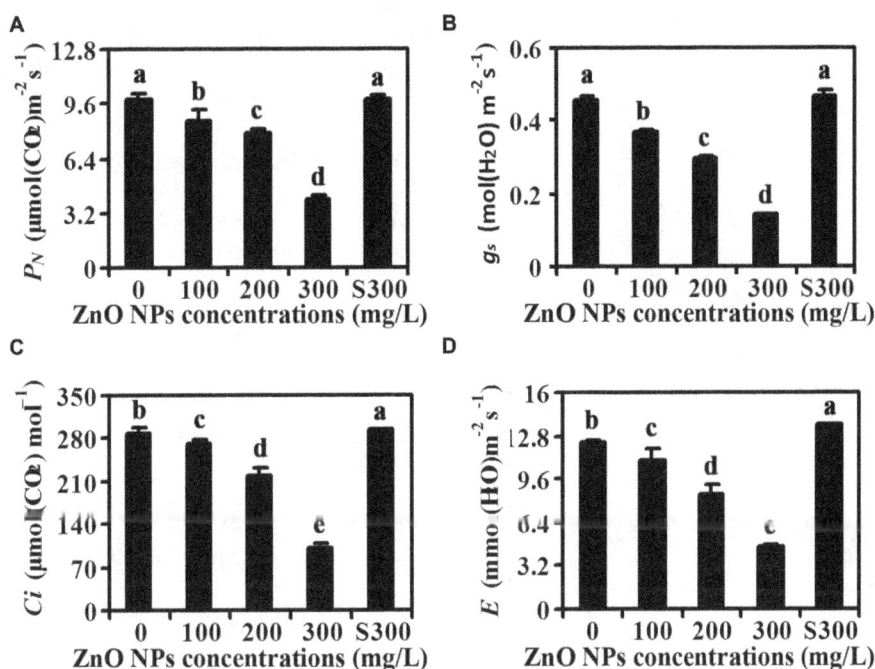

FIGURE 3 | Effects of ZnO NPs on photosynthesis of *Arabidopsis* plants. (A) Net photosynthetic rate (P_N), **(B)** stomatal conductance (g_s), **(C)** intercellular CO_2 concentration (C_i), and **(D)** transpiration rate (E) in rosette leaves of 6-week-old *Arabidopsis* plants treated with 0, 100, 200, 300 mg/L ZnO NPs, or supernatant from 300 mg/L ZnO NPs suspensions. The fifth rosette leaves were used for the measurement by using a portable open-flow, gas-exchange system. Data represent the mean ± SD of four replicates. Different letters indicate significantly different ($p < 0.05$).

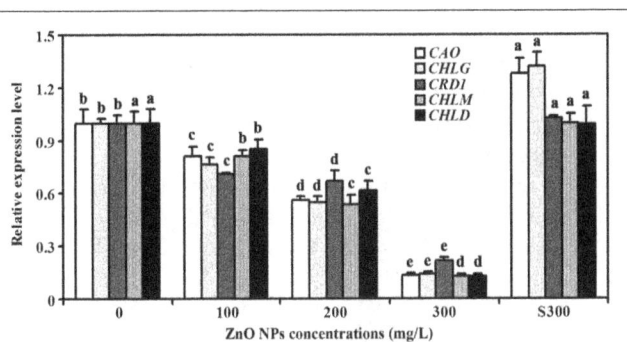

FIGURE 4 | Effects of ZnO NPs on the expression of chlorophyll synthesis genes in *Arabidopsis*. RNA was isolated from rosette leaves of *Arabidopsis* plants treated with 0, 100, 200, 300 mg/L ZnO NPs, or supernatant from 300 mg/L ZnO NPs suspensions, and qRT-PCR was used to examine the expression of genes involved chlorophyll synthesis. Expression of *ACT2* was used as a reference gene, and expression of corresponding genes in *Arabidopsis* in the absence of ZnO NPs was set as 1. Data represent the mean ± SD of three replicates. Different letters indicate significantly different ($p < 0.05$).

Expression Levels of Carotenoids Synthesis Genes were Increased in ZnO NPs Treated Plants

Because carotenoids contents were increased slightly in ZnO NPs treated plants, we also examined the expression of several genes that have been reported to be involved in carotenoids synthesis, including *GERANYLGERANYL PYROPHOSPHATE SYNTHASE 6 (GGPS6)*, *PHYTOENE SYNTHASE (PSY)*, *PHYTOENE DESATURASE (PDS)*, and *ZETA-CAROTENE DESATURASE (ZDS)* (Lange and Ghassemian, 2003; Dong et al., 2007; Qin et al., 2007; Rodríguez-Villalón et al., 2009). We found that the expression of these genes was increased in response to ZnO NPs treatments, but to different degrees. In 300 mg/L ZnO NPs treated plants, an about 2-, 4-, 2- and 10- fold increase for *GGPS6*, *PSY*, *PDS* and *ZDS*, respectively was observed (**Figure 5**).

Expression Levels of Photosystem Structure Genes in ZnO NPs Treated Plants were Reduced

In addition to the expression of the chlorophylls and carotenoids synthesis genes, we also examined the expression of photosystem structure genes including *PHOTOSYSTEM I SUBUNIT D-2 (PSAD2)*, *PHOTOSYSTEM I SUBUNIT E-2 (PSAE2)*, *PHOTOSYSTEM I SUBUNIT K (PSAK)*, and *PHOTOSYSTEM I SUBUNIT K (PSAN)* (Varotto et al., 2002; Friso et al., 2004; Ihnatowicz et al., 2004; Peltier et al., 2004). We found that the expression of these genes was also reduced in ZnO NPs treated

increase for the expression of *CAO* and *CHLG* was observed in plants treated with supernatant of 300 mg/L ZnO NPs, whereas the expression of *CRD1*, *CHLM* and *CHLD* in plants treated with supernatant of 300 mg/L ZnO NPs remained largely unchanged when compared with that in water watered control plants (**Figure 4**).

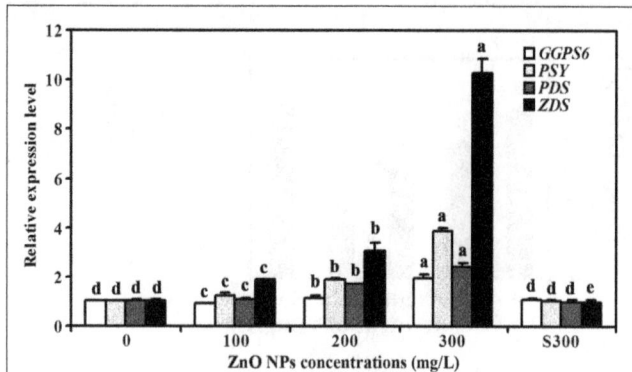

FIGURE 5 | Effects of ZnO NPs on the expression of carotenoid synthesis genes in Arabidopsis. RNA was isolated from rosette leaves of *Arabidopsis* plants treated with 0, 100, 200, 300 mg/L ZnO NPs, or supernatant from 300 mg/L ZnO NPs suspensions, and qRT-PCR was used to examine the expression of genes involved carotenoid synthesis. Expression of *ACT2* was used as a reference gene, and expression of corresponding genes in *Arabidopsis* in the absence of ZnO NPs was set as 1. Data represent the mean ± SD of three replicates. Different letters indicate significantly different ($p < 0.05$).

plants (**Figure 6**). As a control, a slight increased expression of all the four genes was observed in plants treated with supernatant of 300 mg/L ZnO NPs (**Figure 6**).

DISCUSSION

Previous experiments have shown that toxic effects of ZnO NPs on plants are likely caused by the induction of oxidative stress in plants (Hernandez-Viezcas et al., 2011; Thwala et al., 2013; Bandyopadhyay et al., 2015). Consistent with this, most of the genes induced by ZnO NPs in *Arabidopsis* are relative

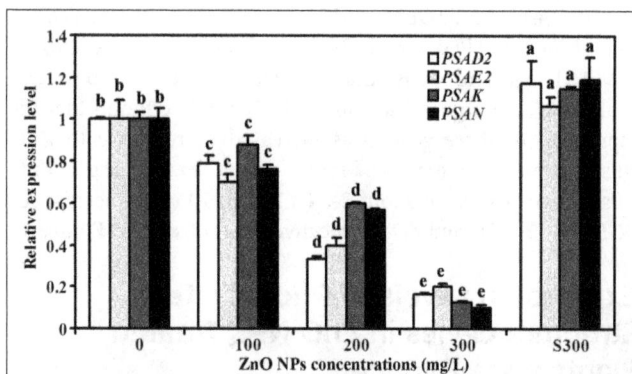

FIGURE 6 | Effects of ZnO NPs on the expression of photosystem genes in Arabidopsis. RNA was isolated from rosette leaves of *Arabidopsis* plants treated with 0, 100, 200, 300 mg/L ZnO NPs, or supernatant from 300 mg/L ZnO NPs suspensions, and qRT-PCR was used to examine the expression of photosystem genes. Expression of *ACT2* was used as a reference gene, and expression of corresponding genes in *Arabidopsis* in the absence of ZnO NPs was set as 1. Data represent the mean ± SD of three replicates. Different letters indicate significantly different ($p < 0.05$).

to stress responses (Landa et al., 2012). We provide evidence here in this report that ZnO NPs are toxic to *Arabidopsis* plants, and their toxic effects may due to the effects of ZnO NPs on the expression of chlorophyll synthesis and photosystem genes.

First, similar to the results obtained in other plants such as cucumber (Zhao et al., 2013), ZnO NPs treated *Arabidopsis* plants have reduced rosette size and biomass accumulation, and the reduction is positively related to the concentrations of ZnO NPs used (**Figure 1**). Our data also show that the ZnO NPs have similar inhibition effects on biomass accumulation in shoots and roots (**Figure 1**). Whereas plants treated with supernatant of 300 mg/L ZnO NPs suspensions are similar to control plants in both rosette size and biomass accumulation (**Figure 1**). These results suggest that ZnO NPs are toxic to *Arabidopsis*, though we could not rule out the possibility that the toxic effects may be partially due to the release of Zn^{2+} by ZnO NPs into the soil (Cornelis et al., 2014), it is unlikely that the toxic effects observed was due to the possible release of Zn^{2+} or other ions by ZnO NPs to the supernatants.

Second, contents of chlorophyll a and chlorophyll b were reduced in rosette leaves of ZnO NPs treated plants, whereas those in plants treated with supernatants of 300 mg/L ZnO NPs suspensions remain largely unaffected (**Figure 2**). Net rate of photosynthesis, leaf stomatal conductance, intercellular CO_2 concentration and transpiration rate were all reduced in ZnO NPs treated plants (**Figure 3**). Carotenoids contents, on the other hand, showed slightly increase in ZnO NPs treated plants (**Figure 2**), suggesting that reduction in photosynthesis is caused by reduced chlorophylls contents. Nevertheless, these results suggest that toxic effects of ZnO NPs on *Arabidopsis* are likely caused by reduced chlorophylls contents in the plants, which in turn limited photosynthesis in the plants, leading to the reduce in biomass accumulation.

Third, consistent with the observation that chlorophylls contents were reduced in ZnO NPs treated plants, expression of chlorophyll synthesis genes was reduced in ZnO NPs treated plants (**Figure 4**). We also found the expression of photosystem genes was reduced in ZnO NPs treated plants (**Figure 6**). On the other hand, the expression of some chlorophyll synthesis and photosystem genes was slightly increased in plants treated with supernatant from 300 mg/L ZnO NPs suspensions (**Figures 4 and 6**), further confirmed that the toxic effects observed in ZnO NPs treated plants were not due to the Zn^{2+} or other ions released by ZnO NPs in to the supernatants. Though carotenoids contents only slightly increased in ZnO NPs treated plants (**Figure 2**), we found the expression of some carotenoids synthesis genes, especially *ZDS*, was dramatically increased in ZnO NPs treated plants (**Figure 5**). Considering that carotenoids is also involved in photosynthesis, and ZnO NPs are toxic to plants, it will be of great interest to find out why ZnO NPs treatment increased the expression of carotenoids synthesis genes in *Arabidopsis*.

It should be noted that when compared with a single plant or plants grown at low density, plants growth at high density showed intraspecific competition (Geisler et al., 2012; Masclaux et al., 2012; Pierik and de Wit, 2014), which led to the upregulation of photosynthesis genes (Geisler et al., 2012). However, because all the pots in our experiments contained the same number of plants, the difference of the gene expression observed was unlikely due to intraspecific competition.

Nevertheless, our results show that ZnO NPs inhibited plant growth and biomass accumulation in *Arabidopsis*, and that in addition to the induction of oxidative stresses, regulation of chlorophyll synthesis and photosystem genes expression may also contribute to the toxic effects of ZnO NPs in *Arabidopsis*.

REFERENCES

Bandyopadhyay, S., Plascencia-Villa, G., Mukherjee, A., Rico, C. M., José-Yacamán, M., Peralta-Videa, J. R., et al. (2015). Comparative phytotoxicity of ZnO NPs, bulk ZnO, and ionic zinc onto the alfalfa plants symbiotically associated with *Sinorhizobium meliloti* in soil. *Sci. Total Environ.* 515–516, 60–69. doi: 10.1016/j.scitotenv.2015.02.014

Burklew, C. E., Ashlock, J., Winfrey, W. B., and Zhang, B. (2012). Effects of aluminum oxide nanoparticles on the growth, development, and microRNA expression of tobacco (*Nicotiana tabacum*). *PLoS ONE* 7:e34783. doi: 10.1371/journal.pone.0034783

Chen, R., Ratnikova, T. A., Stone, M. B., Lin, S., Lard, M., Huang, G., et al. (2010). Differential uptake of carbon nanoparticles by plant and mammalian cells. *Small* 6, 612–617. doi: 10.1002/smll.200901911

Cifuentes, Z., Custardoy, L., de la Fuente, J. M., Marquina, C., Ibarra, M. R., Rubiales, D., et al. (2010). Absorption and translocation to the aerial part of magnetic carbon-coated nanoparticles through the root of different crop plants. *J. Nanobiotecg* 8:26. doi: 10.1186/1477-3155-8-26

Cornelis, G., Hund-Rinke, K., Kuhlbusch, T., van den Brink, N., and Nickel, C. (2014). Fate and bioavailability of engineered nanoparticles in soils: a review. *Crit. Rev. Env. Sci. Technol.* 44, 2720–2764. doi: 10.1080/10643389.2013.829767

Dong, H., Deng, Y., Mu, J., Lu, Q., Wang, Y., Xu, Y., et al. (2007). The *Arabidopsis* spontaneous cell death1 gene, encoding a zeta-carotene desaturase essential for carotenoid biosynthesis, is involved in chloroplast development, photoprotection and retrograde signalling. *Cell Res.* 17, 458–470. doi: 10.1038/cr.2007.37

Du, W., Sun, Y., Ji, R., Zhu, J., Wu, J., and Guo, H. (2011). TiO2 and ZnO nanoparticles negatively affect wheat growth and soil enzyme activities in agricultural soil. *J. Environ. Monit.* 13, 822–828. doi: 10.1039/c0em00611d

Etxeberria, E., Gonzalez, P., Baroja-Fernandez, E., and Romero, J. P. (2006). Fluid phase endocytic uptake of artificial nano-spheres and fluorescent quantum dots by sycamore cultured cells. *Plant Signal. Behav.* 1, 196–200. doi: 10.4161/psb.1.4.3142

Friso, G., Giacomelli, L., Ytterberg, A. J., Peltier, J. B., Rudella, A., Sun, Q., et al. (2004). In-depth analysis of the thylakoid membrane proteome of *Arabidopsis thaliana* chloroplasts: new proteins, new functions, and a plastid proteome database. *Plant Cell* 16, 478–499. doi: 10.1105/tpc.017814

García-Sánchez, S., Bernales, I., and Cristobal, S. (2015). Early response to nanoparticles in the *Arabidopsis* transcriptome compromises plant defence and root-hair development through salicylic acid signalling. *BMC Genomics* 16:341. doi: 10.1186/s12864-015-1530-4

Geisler, M., Gibson, D. J., Lindsey, K. J., Millar, K., and Wood, A. J. (2012). Upregulation of photosynthesis genes, and down-regulation of stress defense genes, is the response of *Arabidopsis thaliana* shoots to intraspecific competition. *Bot. Stud.* 53, 85–96.

Geisler-Lee, J., Brooks, M., Gerfen, J. R., Wang, Q., Fotis, C., Sparer, A., et al. (2014). Reproductive toxicity and life history study of silver nanoparticle effect, uptake and transport in *Arabidopsis thaliana*. *Nanomaterials* 4, 301–318. doi: 10.3390/nano4020301

Hernandez-Viezcas, J. A., Castillo-Michel, H., Servin, A. D., Peralta-Videa, J. R., and Gardea-Torresdey, J. L. (2011). Spectroscopic verification of zinc absorption and distribution in the desert plant *Prosopis juliflora*-velutina (velvet mesquite) treated with ZnO nanoparticles. *Chem. Eng. J.* 170, 346–352. doi: 10.1016/j.cej.2010.12.021

Ihnatowicz, A., Pesaresi, P., Varotto, C., Richly, E., Schneider, A., Jahns, P., et al. (2004). Mutants for photosystem I subunit D of *Arabidopsis thaliana*: effects on photosynthesis, photosystem I stability and expression of nuclear genes for chloroplast functions. *Plant J.* 37, 839–852. doi: 10.1111/j.1365-313X.2004.02011.x

Ju-Nam, Y., and Lead, J. R. (2008). Manufactured nanoparticles: an overview of their chemistry, interactions and potential environmental implications. *Sci. Total Environ.* 400, 396–414. doi: 10.1016/j.scitotenv.2008.06.042

Kobayashi, K., Mochizuki, N., Yoshimura, N., Motohashi, K., Hisabori, T., and Masuda, T. (2008). Functional analysis of *Arabidopsis thaliana* isoforms of the Mg-chelatase CHLI subunit. *Photochnol. Photobiol. Sci.* 7, 1188–1195. doi: 10.1039/b802604c

Kouhi, S. M. M., Lahouti, M., Ganjeali, A., and Entezar, M. H. (2014). Comparative phytotoxicity of ZnO nanoparticles, ZnO microparticles, and Zn^{2+} on rapeseed (*Brassica napus* L.): investigating a wide range of concentrations. *Toxico. Environ. Chem.* 96, 861–868. doi: 10.1080/02772248.2014.994517

Kurepa, J., Paunesku, T., Vogt, S., Arora, H., Rabatic, B. M., Lu, J., et al. (2010). Uptake and distribution of ultrasmall anatase TiO_2 Alizarin red S nanoconjugates in *Arabidopsis thaliana*. *Nano Lett.* 10, 2296–2302. doi: 10.1021/nl903518f

Landa, P., Vankova, R., Andrlova, J., Hodek, J., Marsik, P., Storchova, H., et al. (2012). Nanoparticle-specific changes in *Arabidopsis thaliana* gene expression after exposure to ZnO, TiO_2, and fullerene soot. *J. Hazard. Mater.* 241–242, 55–62. doi: 10.1016/j.jhazmat.2012.08.059

Lange, B. M., and Ghassemian, M. (2003). Genome organization in *Arabidopsis thaliana*: a survey for genes involved in isoprenoid and chlorophyll metabolism. *Plant Mol. Biol.* 51, 925–948. doi: 10.1023/A:1023005504702

Lee, C. W., Mahendra, S., Zodrow, K., Li, D., Tsai, Y. C., Braam, J., et al. (2010). Developmental phytotoxicity of metal oxide nanoparticles to *Arabidopsis thaliana*. *Environ. Toxicol. Chem.* 29, 669–675. doi: 10.1002/etc.58

Lee, S., Kim, S., Kim, S., and Lee, I. (2013). Assessment of phytotoxicity of ZnO NPs on a medicinal plant, Fagopyrum *esculentum*. *Environ. Sci. Pollut. Res. Int.* 20, 848–854. doi: 10.1007/s11356-012-1069-8

Lichtenhaler, H. K., and Wellburn, A. R. (1983). Determination of total carotenoid and chlorophyll a and b of leaf extract in different solvent. *Biochem. Soc. Trans.* 603, 591–592. doi: 10.1042/bst0110591

Liu, Q., Chen, B., Wang, Q., Shi, X., Xiao, Z., Lin, J., et al. (2009). Carbon nanotubes as molecular transporters for walled plant cells. *Nano Lett.* 9, 1007–1010. doi: 10.1021/nl803083u

Liu, S., Hu, Q., Luo, S., Li, Q., Yang, X., Wang, X., et al. (2015). Expression of wild-type PtrIAA14.1, a poplar Aux/IAA gene causes morphological changes in *Arabidopsis*. *Front. Plant Sci.* 6:388. doi: 10.3389/fpls.2015.00388

Ma, H., Williams, P. L., and Diamond, S. A. (2013). Ecotoxicity of manufactured AnO nanoparticles – a review. *Environ. Pollut.* 172, 76–85. doi: 10.1016/j.envpol.2012.08.011

Ma, X., Geiser-Lee, J., Deng, Y., and Kolmakov, A. (2010). Interactions between engineered nanoparticles (ENPs) and plants: phytotoxicity, uptake and accumulation. *Sci. Total. Environ.* 408, 3053–3061. doi: 10.1016/j.scitotenv.2010.03.031

Masclaux, F., Bruessow, F., Schweizer, F., Gouhier-Darimont, C., Keller, L., and Reymond, P. (2012). Transcriptome analysis of intraspecific competition in *Arabidopsis thaliana* reveals organ-specific signatures related to nutrient acquisition and general stress response pathways. *BMC Plant Biol.* 12:227. doi: 10.1186/1471-2229-12-227

ACKNOWLEDGMENTS

This work was supported by the International Science & Technology Cooperation Program of China (2014DFA31740), the Key Laboratory of Molecular Epigenetics of MOE (130014542), a startup fund from Northeast Normal University (www.nenu.edu.cn), and the Programme for Introducing Talents to Universities (B07017). The funders had no role in study design, data collection and analysis, decision to publish, or preparation of the manuscript.

Peltier, J. B., Ytterberg, A. J., Sun, Q., and van Wijk, K. J. (2004). New functions of the thylakoid membrane proteome of *Arabidopsis thaliana* revealed by a simple, fast, and versatile fractionation strategy. *J. Biol. Chem.* 279, 49367–49383. doi: 10.1074/jbc.M406763200

Pierik, R., and de Wit, M. (2014). Shade avoidance: phytochrome signalling and other aboveground neighbour detection cues. *J. Exp. Bot.* 65, 2815–2824. doi: 10.1093/jxb/ert389

Pontier, D., Albrieux, C., Joyard, J., Lagrange, T., and Block, M. A. (2007). Knock-out of the magnesium protoporphyrin IX methyltransferase gene in *Arabidopsis*. Effects on chloroplast development and on chloroplast-to-nucleus signaling. *J. Biol. Chem.* 282, 2297–2304. doi: 10.1074/jbc.M610286200

Qian, H., Peng, X., Han, X., Ren, J., Sun, L., and Fu, Z. (2013). Comparison of the toxicity of silver nanoparticles and silver ions on the growth of terrestrial plant model *Arabidopsis thaliana*. *J. Environ. Sci.* 25, 1947–1955. doi: 10.1016/S1001-0742(12)60301-5

Qin, G., Gu, H., Ma, L., Peng, Y., Deng, X. W., Chen, Z., et al. (2007). Disruption of phytoene desaturase gene results in albino and dwarf phenotypes in *Arabidopsis* by impairing chlorophyll, carotenoid, and gibberellin biosynthesis. *Cell Res.* 17, 471–482. doi: 10.1038/cr.2007.40

Riahi-Madvar, A., Rezaee, F., and Jalali, V. (2012). Effects of alumina nanoparticles on morphological properties and antioxidant system of *Triticum aestivum*. *Ira. J. Plant Physiol.* 3, 595–603.

Roco, M. C. (2005). Environmentally responsible development of nanotechnology. *Environ. Sci. Technol* 39, 106A–112A. doi: 10.1021/es053199u

Rodríguez-Villalón, A., Gas, E., and Rodríguez-Concepción, M. (2009). Colors in the dark: a model for the regulation of carotenoid biosynthesis in etioplasts. *Plant Signal. Behav.* 4, 965–967. doi: 10.4161/psb.4.10.9672

Rzeznicka, K., Walker, C. J., Westergren, T., Kannangara, C. G., von Wettstein, D., Merchant, S., et al. (2005). Xantha-l encodes a membrane subunit of the aerobic Mg-protoporphyrin IX monomethyl ester cyclase involved in chlorophyll biosynthesis. *Proc. Natl. Acad. Sci. U.S.A.* 102, 5886–5891. doi: 10.1073/pnas.0501784102

Schwab, F., Tanner, S., Schulin, R., Rotzetter, A., Stark, W., von Quadt, A., et al. (2015a). Dissolved cerium contributes to uptake of Ce in the presence of differently sized CeO_2-nanoparticles by three crop plants. *Metallomics* 7, 466–477. doi: 10.1039/c4mt00343h

Schwab, F., Zhai, G., Kern, M., Turner, A., Schnoor, J. L., and Wiesner, M. R. (2015b). Barriers, pathways and processes for uptake, translocation and accumulation of nanomaterials in plants – critical review. *Nanotoxicology* doi: 10.3109/17435390.2015.1048326 [Epub ahead of print].

Shvedova, A. A., Kagan, V. E., and Fadeel, B. (2010). Close encounters of the small kind: adverse effects of man-made materials interfacing with the nano-cosmos of biological systems. *Annu. Rev. Pharmacol.* 50, 63–88. doi: 10.1146/annurev.pharmtox.010909.105819

Stampoulis, D., Sinha, S. K., and White, J. C. (2009). Assay-dependent phytotoxicity of nanoparticles to plants. *Environ. Sci. Technol.* 43, 9473–9479. doi: 10.1021/es901695c

Stephenson, P. G., and Terry, M. J. (2008). Light signalling pathways regulating the Mg-chelatase branchpoint of chlorophyll synthesis during de-etiolation in *Arabidopsis thaliana*. *Photochem. Photobiol. Sci.* 7, 1243–1252. doi: 10.1039/b802596g

Tanaka, A., Ito, H., Tanaka, R., Tanaka, N. K., Yoshida, K., and Okada, K. (1998). Chlorophyll a oxygenase (CAO) is involved in chlorophyll b formation from chlorophyll a. *Proc. Natl. Acad. Sci. U.S.A.* 95, 12719–12723. doi: 10.1073/pnas.95.21.12719

Thwala, M., Musee, N., Sikhwivhilu, L., and Wepener, V. (2013). The oxidative toxicity of Ag and ZnO nanoparticles towards the aquatic plant Spirodela punctuta and the role of testing media parameters. *Environ. Sci. Proc. Impacts* 15, 1830–1843. doi: 10.1039/c3em00235g

Varotto, C., Pesaresi, P., Jahns, P., Lessnick, A., Tizzano, M., Schiavon, F., et al. (2002). Single and double knockouts of the genes for photosystem I subunits G, K, and H of *Arabidopsis*. Effects on photosystem I composition, photosynthetic electron flow, and state transitions. *Plant Physiol.* 129, 616–624. doi: 10.1104/pp.002089

Wang, H., Wu, F., Meng, W., White, J. C., Holden, P. A., and Xing, B. (2013a). Engineered nanoparticles may induce genotoxicity. *Environ. Sci. Technol.* 47, 13212–13214. doi: 10.1021/es404527d

Wang, P., Menzies, N. W., Lombi, E., McKenna, B. A., Johannessen, B., Glover, C. J., et al. (2013b). Fate of ZnO nanoparticles in soils and cowpea (*Vigna unguiculata*). *Environ. Sci. Technol.* 47, 13822–13830. doi: 10.1021/es403466p

Wang, S., Kurepa, J., and Smalle, J. A. (2011). Ultra-small TiO(2) nanoparticles disrupt microtubular networks in *Arabidopsis thaliana*. *Plant Cell Environ.* 34, 811–820. doi: 10.1111/j.1365-3040.2011.02284.x

Wang, X., Han, H., Liu, X., Gu, X., Chen, K., and Lu, D. (2012a). Multi-walled carbon nanotubes can enhance root elongation of wheat (*Triticum aestivum*) plants. *J. Nanopart. Res.* 14:841. doi: 10.1007/s11051-012-0841-5

Wang, Z., Xie, X., Zhao, J., Liu, X., Feng, W., White, J. C., et al. (2012b). Xylem- and phloem-based transport of CuO nanoparticles in maize (*Zea mays L.*). *Environ. Sci. Technol.* 46, 4434–4441. doi: 10.1021/es204212z

Wang, X., Jiang, P., Ma, Y., Geng, S., Wang, S., and Shi, D. (2015). Physiological strategies of sunflower exposed to salt or alkali stresses: restriction of ion transport in the cotyledon node zone and solute accumulation. *Agron. J.* 107, 2181–2192. doi: 10.2134/agronj15.0012

Xia, T., Li, N,. and Nel, A. E. (2009). Potential health impact of nanoparticles. *Annu. Rev. Publ. Health* 30, 137–150. doi: 10.1146/annurev.publhealth.031308.100155

Zhang, W., Ebbs, S. D., Musante, C., White, J. C., Gao, C., and Ma, X. (2015). Uptake and accumulation of bulk and nanosized cerium oxide particles and ionic cerium by radish (*Raphanus sativus L.*). *J. Agric. Food Chem.* 63, 382–390. doi: 10.1021/jf5052442

Zhao, L., Sun, Y., Hernandez-Viezcas, J. A., Servin, A. D., Hong, J., Niu, G., et al. (2013). Influence of CeO2 and ZnO nanoparticles on cucumber physiological markers and bioaccumulation of Ce and Zn: a life cycle study. *J. Agric. Food Chem.* 61, 11945–11951. doi: 10.1021/jf404328e

Zhao, L. J., Peralta-Videa, J. R., Ren, M. H., Varela-Ramirez, A., Li, C. Q., Hernandez-Viezcas, J. A., et al. (2012). Transport of Zn in a sandy loam soil treated with ZnO NPs and uptake by corn plants: electron microprobe and confocal microscopy studies. *Chem. Eng. J.* 184, 1–8. doi: 10.1016/j.cej.2012.01.041

Conflict of Interest Statement: The authors declare that the research was conducted in the absence of any commercial or financial relationships that could be construed as a potential conflict of interest.

14

Detection of Photosynthetic Performance of *Stipa bungeana* Seedlings under Climatic Change using Chlorophyll Fluorescence Imaging

Xiliang Song[1,2], Guangsheng Zhou[1,3]*, Zhenzhu Xu[1,2], Xiaomin Lv[1,2] and Yuhui Wang[1,2]*

[1] State Key Laboratory of Vegetation and Environmental Change, Institute of Botany, Chinese Academy of Science, Beijing, China, [2] University of Chinese Academy of Sciences, Beijing, China, [3] Chinese Academy of Meteorological Sciences, China Meteorological Administration, Beijing, China

Edited by:
Boris Rewald,
University of Natural Resources and
Life Sciences, Vienna, Austria

Reviewed by:
Hazem M. Kalaji,
Warsaw University of Life Sciences,
Poland
Dusan Lazar,
Palacky University, Czech Republic

*Correspondence:
Guangsheng Zhou
gszhou@ibcas.ac.cn;
Yuhui Wang
yuhuiwang@ibcas.ac.cn

In this study, the impact of future climate change on photosynthetic efficiency as well as energy partitioning in the *Stipa bungeana* was investigated by using chlorophyll fluorescence imaging (CFI) technique. Two thermal regimes (room temperature, T_0: 23.0/17.0°C; High temperature, T_6: 29.0/23.0°C) and three water conditions (Control, W_0; Water deficit, W_{-30}; excess precipitation, W_{+30}) were set up in artificial control chambers. The results showed that excess precipitation had no significant effect on chlorophyll fluorescence parameters, while water deficit decreased the maximal quantum yield of photosystem II (PSII) photochemistry for the dark-adapted state (F_v/F_m) by 16.7%, with no large change in maximal quantum yield of PSII photochemistry for the light-adapted state (F_V'/F_M') and coefficient of the photochemical quenching (q_P) at T_0 condition. Under T_6 condition, high temperature offset the negative effect of water deficit on F_v/F_m and enhanced the positive effect of excess precipitation on F_v/F_m, F_v'/F_m', and q_P, the values of which all increased. This indicates that the temperature higher by 6°C will be beneficial to the photosynthetic performance of *S. bungeana*. Spatial changes of photosynthetic performance were monitored in three areas of interest (AOIs) located on the bottom, middle and upper position of leaf. Chlorophyll fluorescence images (F_v/F_m, actual quantum yield of PSII photochemistry for the light-adapted state (Φ_{PSII}), quantum yield of non-regulated energy dissipation for the light-adapted state (Φ_{NO}) at T_0 condition, and Φ_{PSII} at T_6 condition) showed a large spatial variation, with greater value of Φ_{NO} and lower values of F_v/F_m and Φ_{PSII} in the upper position of leaves. Moreover, there was a closer relationship between Φ_{PSII} and Φ_{NO}, suggesting that the energy dissipation by non-regulated quenching mechanisms played a dominant role in the yield of PSII photochemistry. It was also found that, among all measured fluorescence parameters, the F_v/F_m ratio was most sensitive to precipitation change at T_0, while Φ_{PSII} was the most sensitive indicator at T_6.

Keywords: *Stipa bungeana*, chlorophyll fluorescence imaging, photosynthetic efficiency, energy partitioning, high temperature, precipitation change

INTRODUCTION

High temperature and water stress as abiotic stress factors will limit plant growth and reduce crop productivity (Boyer, 1982; Wahid et al., 2007), and they always occur simultaneously in that high temperature increases both evaporation and potential evapotranspiration and exacerbates the negative influence of water deficit (Machado and Paulsen, 2001; Osório et al., 2011). Models of global climate change have predicted that the globally averaged surface temperature will be 1.5–4.0°C higher till 2100 and the extreme precipitation events will occur more frequently than before (IPCC, 2013). According to the study by Xu et al. (2009), temperature and precipitation change determine the physiological response of perennial grass to new environmental conditions to a large extent. Among all plant physiological functions, photosynthesis plays a pivotal role in plant carbon uptake, plant growth and biomass accumulation. It is commonly considered that stomatal limitation which influences the substomatal CO_2 concentration is the main reason for the reduction of photosynthesis under moderate water deficit (Cornic, 2000). The limitation on CO_2 assimilation may damage the balance between photochemical activity in photosystem II (PSII) and electron requirement for photosynthesis, resulting in the photodamage of PSII centers. Although plant photosynthetic apparatus appears to be highly resistant to water deficit (Giardi et al., 1996; Petsas and Grammatikopoulos, 2009; Zivcak et al., 2014), temperature rising can change the response of photosynthesis to water stress (Chaves et al., 2002). Among all plant physiological activities, photosynthesis has been proved to be most sensitive to high temperature and can be inhibited entirely by heat stress before other plant physiological symptoms occur (Berry and Bjorkman, 1980). High temperature damages several photosynthetic functions, such as Calvin cycle, photosystem I (PSI) and PSII. Many studies have reported that the cooperative effect of water stress and high temperature is more drastic than their single effect (Albert et al., 2011; Thomey et al., 2011; Bauweraerts et al., 2013). When water and heat stress occur simultaneously, water stress may impose a certain effect on the photosynthesis together with temperature through oxidative damage (Chaves et al., 2002). On this basis, the inhibitory effect and damage on photosynthesis can be studied when the two stresses coexist, even at a low light intensity.

For the quantitative detection of the changes in the photosynthetic apparatus and photosynthetic activity under various environmental stresses, chlorophyll fluorescence measurement has been demonstrated to be a fast, non-destructive, sensitive and reliable method (Berry and Bjorkman, 1980; Havaux, 1992; Martínez-Carrasco et al., 2002; Mielke

et al., 2003; Xu et al., 2004; Xu and Zhou, 2006; Swoczyna et al., 2010; Tuba et al., 2010; Ogaya et al., 2011; Brestic et al., 2014; Kalaji et al., 2014; Lazár, 2015). However, the conventional chlorophyll fluorescence measurement approach is based on point measurements and cannot exhibit the physiological status of a whole plant (Lichtenthaler and Miehé, 1997; Ehlert and Hincha, 2008). Furthermore, habitual heterogeneity of photosynthetic activity over the leaf surface makes this approach highly error-prone (Ehlert and Hincha, 2008). To overcome these problems, a more advanced technique, chlorophyll fluorescence imaging (CFI), was developed to take a powerful role in identifying spatial heterogeneity of leaf photosynthetic performance (Omasa et al., 1987; Baker and Rosenqvist, 2004; Ivanov and Bernards, 2015). This provides new possibilities to understand the regulation mechanism of photosynthesis, and to assess the properties of the photosynthetic apparatus and the extent to which the plants are affected by different stresses (Gorbe and Calatayud, 2012; Shaw et al., 2014; Humplík et al., 2015; Ivanov and Bernards, 2015). One of the first works on experiments with CFI were carried out by Omasa et al. (1987). In their work, the analysis of CFI was proved to be a useful method in early warning diagnosis, functional analysis of disorders during environment stress and plant's ability to recover. CFI can also be used to study plant response to dynamic climate control as image information is the most intuitive, easily comprehensible, and provides useful information on plant status (Omasa, 1990; Calatayud et al., 2006; Gorbe and Calatayud, 2012). Because CFI detects fluorescence signal pixel-by-pixel, it also provides huge amount of data which can be used for a sophisticated statistical treatments which can lead to an early detection of plant stress (Lazár et al., 2006).

As a key vegetation type dominating the typical steppe in Loess Plateau, *Stipa bungeana* is a useful plant species which can control water loss and soil erosion and improve the ecological environment effectively for its developed root system. It is also a type of appetizing forage with high nutritive value for livestocks. Hence, the research concerning the photosynthetic physiological responses of *S. bungeana* to the major stresses becomes increasingly important in the context of the predicted future climatic changes. In the present study, it was confirmed that CFI is a useful and convenient method for detecting the physiological mechanism of response to higher temperature and precipitation change. Moreover, the spatial variations of chlorophyll a (Chl a) fluorescence parameters of *S. bungeana* under different environmental stresses were analyzed. This work aims to evaluate the impact of high temperature and precipitation change on the photosynthetic performance and the utilization of excess excitation energy in photosynthetic apparatus of *S. bungeana*. Specifically, the following questions were addressed: (1) Are there any negative or positive impact of high temperature and precipitation change on the photosynthetic apparatus of *S. bungeana*? (2) What are the mechanisms of PSII photoprotection for *S. bungeana*? (3) Which is the most sensitive fluorescence parameter in predicting the impact of future climate change on *S. bungeana*?

Abbreviations: AOI, areas of interest; Chl, chlorophyll; CFI, chlorophyll fluorescence imaging; PSII, photosystem II; F_v/F_m, maximal quantum yield of PSII photochemistry for the dark-adapted state; qP, coefficient of the photochemical quenching; Φ_{PSII}, actual quantum yield of PSII photochemistry for the light-adapted state; Φ_{NPQ}, quantum yield of regulated energy dissipation of PSII for the light-adapted state; Φ_{NO}, quantum yield of non-regulated energy dissipation of PSII for the light-adapted state; F_v'/F_m', maximal quantum yield of PSII photochemistry for the light-adapted state.

MATERIALS

Plant Material and Growth Conditions

To understand the effects of high temperature and precipitation change on *S. bungeana*'s photosynthesis characteristics, the changes of water and heat conditions were controlled for the seedlings germinated from seeds. The experiment was carried out at the Institute of Botany, Chinese Academy of Sciences. The seeds of *S. bungeana* were obtained from the grassland in Dongsheng (39°82′N, 110°00′E), Inner Mongolia. They were sterilized by soaking in 0.7% potassium permanganate solution for 8 min and rinsed. Then, these seeds were sown in plastic pots wrapped with plastic film. Each plastic pot was filled with 4.08 kg of dry soil and planted with four plants. In the chestnut soil, the organic carbon content was $12.3\,g\cdot kg^{-1}$ and the total nitrogen content was $1.45\,g\cdot kg^{-1}$. Polyethylene pots were used as the experimental containers, which were lined with plastic bags to prevent water leakage.

Different temperature and precipitation treatments were set according to the monthly average temperature and precipitation during *S. bungeana* 's blooming stage in the past 30 years (1978-2007). Considering the diurnal temperature variations, two temperature treatments 23.0/17.0°C (T_0) and 29.0/23.0°C (T_6) were selected for experiment. Three precipitation regimes were set: average monthly precipitation over 30 years (W_0: 82.3 mm); the average increased by 30% (W_{+30}); the average decreased by 30% (W_{-30}). All the plants were grown in a naturally illuminated glasshouse (the CO_2 concentration was maintained at 390 ppm with a photosynthetic photon flux density of 1000 μmol photons $m^{-2}\cdot s^{-1}$) and the timing used for day/night regime was 16 h light/8 h dark.

Imaging of Chlorophyll Fluorescence Measurement

In order to investigate the spatial heterogeneity of Chl fluorescence parameters, Chl fluorescence imaging of leaves was performed by using an imaging-PAM fluorometer (Walz, Effeltrich, Germany). Chl a fluorescence parameter was measured in the healthy and fully expanded leaves of three plants from each treatment. To evaluate spatial heterogeneity, three areas of interest (AOI, AOI type: Rectangle) in the same leaf were selected, the first one in the bottom position of the leaf (AOI1), the second one in the middle position (AOI2) and the third one in the upper position (AOI3). Three replicates of each plant were used for AOI determination. All plants were placed in dark for 10 min before measurement. Images of maximum fluorescence in the dark-adapted state, F_m, was determined by applying a blue saturation pulse. The saturation pulse intensity was 8000 μmol photons $m^{-2}\cdot s^{-1}$ for 0.8 s. Minimum Chl fluorescence yield F_0 was determined using low frequency light pulses (0.5 μmol photons $m^{-2}\cdot s^{-1}$). Then the images of maximal quantum yield of PSII photochemistry for the dark-adapted state F_v/F_m were captured, and F_v/F_m ratio were obtained as $\frac{F_m-F_0}{F_m}$. To determine the maximum fluorescence yield in the light-adapted state (F_m') and Chl fluorescence during actinic illumination (F_s), actinic illumination (336 μmol photons

$m^{-2}\cdot s^{-1}$) was switched on and saturating pulses were applied at 20 s intervals for 5 min. All the fluorescence levels for the light-adapted state of the sample were determined at the end of 5 min. The maximal quantum yield of PSII photochemistry for the light-adapted state was estimated by the F_v'/F_m' and was calculated by measuring the above same parameters (F_0' and F_m') on light-adapted leaves. In light-adapted state, the F_0' level was estimated using the approximation of Oxborough and Baker (1997): $F_0' = \frac{F_0}{(F_v/F_m)+(F_0/F_m)}$. The actual quantum yield of PSII photochemistry for the light-adapted state (Φ_{PSII}) could be calculated by the formula: $\Phi_{PSII} = \frac{F_m'-F_s}{F_m'} = \frac{\Delta F}{F_m'}$ (Genty et al., 1989). The coefficient of the photochemical quenching (q_P), which was used for the estimation of the fraction of open PSII centers, was calculated as: $q_P = 1 - \left(\frac{F_s-F_0'}{F_m'-F_0'}\right)$ (Bilger and Schreiber, 1987). The quantum yields of PSII photochemical energy dissipation (Φ_{PSII}), non-regulated (Ψ_{NO}), and regulated (Φ_{NPQ}) thermal energy dissipation for the light-adapted state could be used to reflect the utilization of photons which are absorbed by the PSII antennae (Lazár, 2015). It has been proved that $\Phi_{PSII} + \Phi_{NPQ} + \Phi_{NO} = 1$ (Hendrickson et al., 2004; Kramer et al., 2004; Lazár, 2015). Φ_{NO} in PSII was calculated by the equation $\Phi_{NO} = \frac{1}{\left[(NPQ+1+q_L)\left(\frac{F_m}{F_0}-1\right)\right]}$ and ΦNPQ was calculated by $\Phi_{NPQ} = 1 - \Phi_{PSII} - \frac{1}{\left[(NPQ+1+q_L)\left(\frac{F_m}{F_0}-1\right)\right]}$, separately. At last, it should be noted that all the chlorophyll fluorescence parameters were calculated by the Imaging Win v2.32 software.

Statistical Analysis

All statistical analysis was performed using SPSS 18.0 (SPSS, Chicago, Illinois, USA). The mean with standard deviation ($\pm$SD) of each treatment was shown. The parameters were analyzed by One-/Two-way analysis of variance (ANOVA) followed by Duncan's multiple range test (Duncan, 1955). The graphing were performed using Origin 9.0 software (Origin Lab, USA).

RESULTS AND DISCUSSION

In this study, fluorescence imaging technique was used to provide real-time information of photosynthetic performance of *S. bungeana* under different heat and water conditions. The change of images revealed the spatial variation of photosynthetic efficiency in the leaves of *S. bungeana* under different climate environments.

Chlorophyll Fluorescence Parameters in Temperature Warming and Precipitation Change Leaves

Different chlorophyll fluorescence parameters were measured for the leaves of *S. bungeana* to determine the impact of high temperature and precipitation change on the photosynthesis. Maximal quantum yield of PSII photochemistry for the dark-adapted state (F_v/F_m) has been widely used as an indicator of environmental stress. It can reveal the potential electron transport of maximal PSII quantum yield in the dark-adapted state. The imaging of q_P and F_v'/F_m' facilitates the evaluation of their variations (Oxborough and Baker, 1997). The fraction of the open PSII can be quantified by the parameter q_P (Lazár, 2015). The light-induced non-photochemical quenching is a process that regulates energy conversion in PSII to protect plants from photoinhibition. It represents the plant's ability to dissipate excess light energy that cannot be utilized in CO_2 assimilation (Müller et al., 2001). Our results showed that under T_0 condition, F_v/F_m in water-deficient (W_{-30}) plants significantly decreased by 16.7% compared with the normally-watered (W_0) plants. There is no significant difference between the over-watered (W_{+30}) plants and W_0 plants. Moreover, there is a great change in F_m but not F_0 under W_{-30} condition, suggesting that the decrease in F_v/F_m was due to the decrease in F_m. Except for F_v/F_m, there were no significant changes in the other chlorophyll fluorescence parameters such as F_v'/F_m' and q_P in both W_{-30} and W_{+30} plants. The results indicate that (1) excess precipitation had no effect on *S. bungeana* at room temperature; (2) *S. bungeana* suffered from water deficit (decrease in F_v/F_m), and water stress inhibited plant's ability in thermal energy dissipation (Zivcak et al., 2014). This can be explained by the fact that an extreme decrease in trans-thylakoid pH gradient was not generated owing to cooperative consumption of light energy by CO_2 fixation and photorespiration (Müller et al., 2001).

At the T_6 condition, high temperature offset the negative effect of water deficit on F_v/F_m, and enhanced the positive effect of excess precipitation on F_v/F_m, F_v'/F_m' and q_P, leading to the increase in value. This indicates that the temperature higher by 6°C will be beneficial to the photosynthetic performance of *S. bungeana*. However, in the study by Xu and Zhou (2006), the combination of severe water stress and high temperature exhibited adverse effects on the PSII function of *Leymus chinensis*, which is similar to Petsas and Grammatikopoulos (2009)'s conclusion that PSII function of *Phlomis fruticosa* was progressively suppressed under long-term water deficit. This obvious difference may be explained by that *S. bungeana* can well adapt to stress environment for its well-developed root system (Cheng et al., 2012).

Furthermore, in the leaves of plants under optimum temperature and water condition, the mean value of F_v/F_m was 0.678 (**Table 1**), which was lower than the typical value of 0.83 for non-photoinhibited leaves (Björkman and Demmig, 1987). There are two possible reasons to explain the difference. One is the usage of a different intensity of illumination during plants growing (Brestic et al., 2014) and the timing used for day/night regime was also different when compared to natural conditions. The other one is *S. bungeana* grown in the plastic pots with a small size under weak illumination may limit *S. bungeana*'s normal growth (Xu and Zhou, 2006). Therefore, the use of pots inside the greenhouse still requires further investigation.

Utilization of Excess Excitation Energy under High Temperature and Abnormal Water Conditions

F_v/F_m is known to be a sensitive indicator of plant photosynthetic performance (Björkman and Demmig, 1987). It reflects the maximum efficiency of photosynthetic apparatus converting the absorbed light energy into chemical energy, and has been widely used for the detection of photoinhibition (Dickmann et al., 1992; Herppich and Peckmann, 2000). The plants under stress have a lower value of F_v/F_m than those under normal environment (Papageorgiou and Govindjee, 2004; Tuba et al., 2010; Shaw et al., 2014). Calatayud et al. (2013) proposed several reasons for why the F_v/F_m ratio is preferable for the research of environmental stress. Firstly, it can be measured rapidly for the dark-adapted samples. Secondly, it is very useful for quick screening of stress-suffered plants in large quantities. Lastly, unlike the other parameters such as Φ_{PSII} or q_P, it does not need an extended period of illumination as a single saturation pulse is enough. However, the decrease of F_v/F_m can only reflect the degree of environmental stress, while the utilization of excess excitation energy is still unknown. To solve the problem, three fluorescence parameters which divides the allocation of absorbed light energy into three fractions: (1) utilized by PSII photochemistry (Φ_{PSII}); (2) thermally dissipated via ΔpH and xanthophyll-dependent energy quenching (Φ_{NPQ}); (3) non-regulated energy dissipation (Φ_{NO}) (Demmig-Adams et al., 1996; Lazár, 2015). Among the

---

TABLE 1 | Effects of precipitation treatments on maximum and minimum fluorescence yield in dark (F_m and F_0, respectively), maximal quantum yield of PSII photochemistry for the dark-adapted state (F_v/F_m), coefficient of the photochemical quenching (q_P) and maximal quantum yield of PSII photochemistry for the light-adapted state (F_v'/F_m') of *Stipa bungeana* leaf under ambient temperature (T_0) and high temperature (T_6) conditions.

Temperature treatments	Water treatments	Chl fluorescence parameters				
		F_m	F_0	F_v/F_m	F_v'/F_m'	q_P
T_0	W_{+30}	0.557 ± 0.130 ab	0.170 ± 0.048 a	0.691 ± 0.076 a	0.368 ± 0.109 a	0.627 ± 0.158 a
	W_0	0.577 ± 0.121 a	0.181 ± 0.031 a	0.678 ± 0.078 a	0.426 ± 0.129 a	0.690 ± 0.150 a
	W_{-30}	0.417 ± 0.181 b	0.172 ± 0.054 a	0.565 ± 0.096 b	0.448 ± 0.066 a	0.560 ± 0.195 a
T_6	W_{+30}	0.555 ± 0.090 a	0.164 ± 0.036 a	0.702 ± 0.058 a	0.524 ± 0.100 a	0.600 ± 0.088 a
	W_0	0.548 ± 0.122 a	0.193 ± 0.069 a	0.652 ± 0.081 ab	0.461 ± 0.069 ab	0.461 ± 0.113 b
	W_{-30}	0.522 ± 0.136 a	0.233 ± 0.112 a	0.569 ± 0.111 b	0.409 ± 0.129 b	0.400 ± 0.121 b

Different letters indicate significant differences (p < 0.05) between water treatments according Duncan test. Values shown are means ± standard deviation (SD) of nine to twelve replicates.

latter three fluorescence parameters, Φ_{PSII} reflects light-induced protective mechanism while Φ_{NO} reflects a basal quenching which is not regulated by light. They shed a light on the study of plant's capacity to cope with excess excitation energy, and have been widely used to determine QA redox state and excitation energy fluxes in order to gain a better understanding of the stress response mechanisms (Calatayud et al., 2006; Osório et al., 2011).

As seen in **Figure 1**, the precipitation change had no significant effect ($p > 0.05$) on Φ_{PSII}, Φ_{NPQ}, and Φ_{NO} at T_0 condition, though Φ_{PSII} decreased by 20.8 and 16.7% under W_{+30} and W_{-30} conditions and Φ_{NO} increased by 21.0 and 23.9%, respectively. At the T_6 condition, high temperature slightly decreased the value of Φ_{PSII} at W_0 and W_{-30} conditions. This suggests that heat dissipation of the excess light energy was activated to protect the photosynthetic apparatus from photoinhibitory damage (Ort and Baker, 2002; Omasa and Takayama, 2003). Whereas excess precipitation under T_6 condition significantly increased Φ_{PSII}, indicating that the increased precipitation can enhance the protective mechanism of PSII. In addition, Φ_{NPQ} showed no obvious change under all environment conditions and the change of Φ_{NO} was opposite to Φ_{PSII}. This means that there was much overlap between Φ_{PSII} and Φ_{NO}, indicating that energy dissipation by non-regulated quenching mechanisms tends to dominate the yield of PSII photochemistry under drought and heat stress, with the xanthophyll cycle-mediated thermal dissipation playing possibly

a less important role. Similar results were also reported by Osório et al. (2011).

Spatial Heterogeneity of Chlorophyll Fluorescence Parameters under Various Temperature and Water Conditions

In **Table 2**, the value of Chl a fluorescence was obtained from three leaves in each treatment. CFI reveals spatial changes in three areas of interest (AOIs) of the same leaves of *S. bungeana*, i.e., the bottom position of leaf, middle position of leaf and upper position of leaf as shown in **Figure 2**. For each AOI, the values of fluorescence parameter of all pixels within this area were averaged. In **Figure 2**, the images of a single leaf are used to show the heterogeneous distribution of light utilization (changes in Φ_{PSII}, Φ_{NPQ}, and Φ_{NO}) and photosynthetic activity (change in F_v/F_m) over the surface of the whole leaf. The observation of changed image color is an intuitive process. Pixel-value images of F_v/F_m, Φ_{PSII}, Φ_{NPQ}, and Φ_{NO} were displayed with the help of a false color code, ranging from black (0.000) to pink (ending 1.000).

According to **Table 2**, it was found that CFI exhibited the spatial changes in different AOIs of the leaf of *S. bungeana*. Three different AOI were considered for each leaf. Each datum in the table is the mean value of the corresponding AOI from all leaves. At T_0 condition, excess precipitation did not alter the

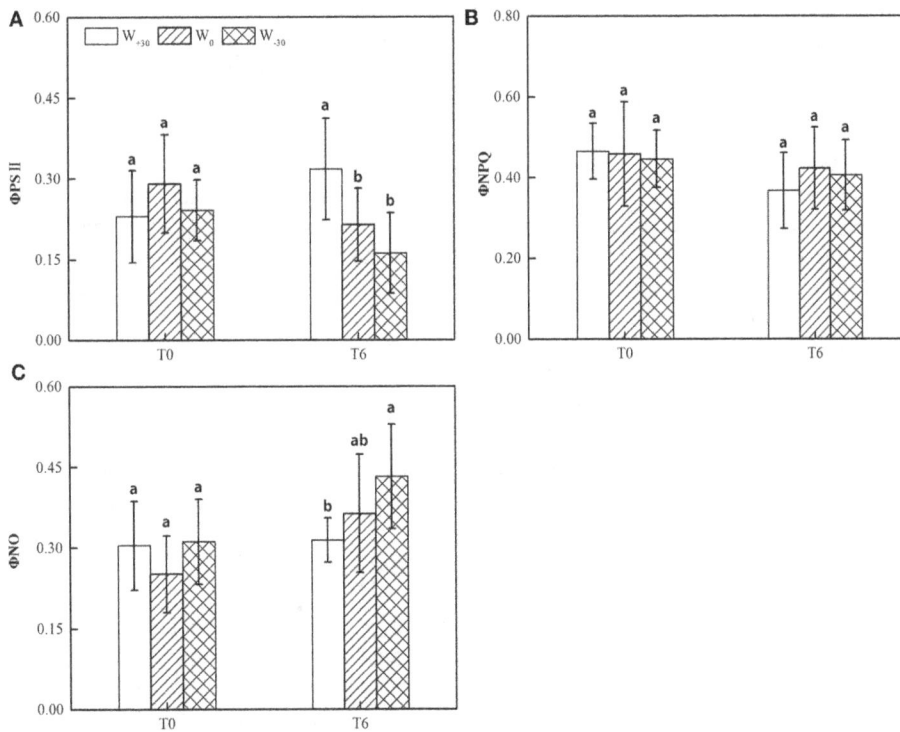

FIGURE 1 | (A) Actual quantum yield of PSII photochemistry for the light-adapted state (Φ_{PSII}), **(B)** quantum yield of regulated energy dissipation of PSII for the light-adapted state (Φ_{NPQ}), **(C)** quantum yield of non-regulated energy dissipation of PSII for the light-adapted state (Φ_{NO}) in the leaves of *S. bungeana* under two temperature treatments and three water treatments. Different letters indicate significant difference ($p < 0.05$) between water treatments in Duncan test. The values are expressed as mean ± standard deviation (SD), which was calculated with nine to twelve replicates.

TABLE 2 | Effect of high temperature and precipitation change on maximal quantum yield of PSII photochemistry for the dark-adapted state (F_v/F_m), actual quantum yield of PSII photochemistry for the light-adapted state (Φ_{PSII}), quantum yield of regulated energy dissipation of PSII for the light-adapted state (Φ_{NPQ}), and quantum yield of non-regulated energy dissipation of PSII for the light-adapted state (Φ_{NO}) of *Stipa bungeana* in different AOIs.

	Precipitation change	T_0				T_6			
		F_v/F_m	Φ_{PSII}	Φ_{NPQ}	Φ_{NO}	F_v/F_m	Φ_{PSII}	Φ_{NPQ}	Φ_{NO}
AOI1	W_{+30}	0.73 ± 0.02 a	0.31 ± 0.05 a	0.44 ± 0.04 a	0.25 ± 0.01 a	0.72 ± 0.05 a	0.37 ± 0.12 a	0.32 ± 0.07 a	0.31 ± 0.06 ab
	W_0	0.76 ± 0.00 a	0.35 ± 0.05 a	0.42 ± 0.14 a	0.23 ± 0.08 a	0.71 ± 0.02 a	0.28 ± 0.06 a	0.44 ± 0.10 a	0.28 ± 0.03 b
	W_{-30}	0.66 ± 0.02 b	0.30 ± 0.02 a	0.46 ± 0.06 a	0.24 ± 0.06 a	0.64 ± 0.10 a	0.23 ± 0.10 a	0.39 ± 0.09 a	0.38 ± 0.05 a
AOI2	W_{+30}	0.54 ± 0.10 b	0.24 ± 0.05 ab	0.45 ± 0.08 a	0.31 ± 0.10 a	0.70 ± 0.04 a	0.31 ± 0.03 a	0.36 ± 0.06 a	0.33 ± 0.04 a
	W_0	0.68 ± 0.02 a	0.31 ± 0.04 a	0.42 ± 0.11 a	0.27 ± 0.08 a	0.63 ± 0.11 a	0.19 ± 0.05 b	0.45 ± 0.02 a	0.36 ± 0.05 a
	W_{-30}	0.73 ± 0.02 a	0.22 ± 0.03 b	0.47 ± 0.08 a	0.31 ± 0.05 a	0.53 ± 0.11 a	0.13 ± 0.03 b	0.43 ± 0.11 a	0.44 ± 0.12 a
AOI3	W_{+30}	0.49 ± 0.04 a	0.14 ± 0.04 a	0.51 ± 0.08 a	0.35 ± 0.09 a	0.68 ± 0.09 a	0.27 ± 0.11 a	0.42 ± 0.13 a	0.31 ± 0.04 a
	W_0	0.59 ± 0.03 a	0.21 ± 0.11 a	0.54 ± 0.15 a	0.26 ± 0.08 a	0.62 ± 0.09 a	0.17 ± 0.05 a	0.37 ± 0.16 a	0.46 ± 0.15 a
	W_{-30}	0.61 ± 0.09 a	0.20 ± 0.06 a	0.41 ± 0.09 a	0.39 ± 0.04 a	0.53 ± 0.12 a	0.13 ± 0.03 a	0.40 ± 0.10 a	0.48 ± 0.12 a

Different letters indicate significant differences between water treatments at the same part of leaf ($p < 0.05$) according Duncan test. Values shown are means ± standard deviation (SD) of three replicates.

F_v/F_m in AOI1, but reduced the F_v/F_m by 13.0 and 9.2% in AOI2 and AOI3, respectively. Water deficit significantly decreased the F_v/F_m by 15.0% in AOI1, but there was no significant change in AOI2 and AOI3. At T_6 condition, excess precipitation increased the F_v/F_m in AOI1, AOI2, and AOI3 by 2.8, 12.6, and 9.0%, respectively. This means that a 6°C higher temperature is beneficial for F_v/F_m under abundant water condition. In contrast, water deficit decreased the F_v/F_m in AOI1, AOI2, and AOI3 by 9.2, 14.9, and 16.8%, respectively. This suggests a reduction in light energy utilization by chloroplasts in the photosynthesis. According to the changes of F_v/F_m and the results of One-way ANOVA in **Table 2**, it can be concluded as follows. The middle position of leaf (AOI2) is most sensitive to excess precipitation under both T_0 and T_6 condition, while the bottom position (AOI1) and upper position (AOI3) are most sensitive to water deficit under both temperature conditions.

The contribution of different pathways to energy partitioning in PSII complexes is shown in **Table 2**. In AOI1 and AOI3, the actual quantum yield of PSII photochemistry for the light-adapted state (Φ_{PSII}) which can indirectly reflect linear electron transport was not affected by precipitation change at both T_0 and T_6 treatment. This confirms that photoinhibition is not induced under these conditions. In AOI2, high temperature (T_6) improved the effect of W_{+30} on F_v/F_m by 63.6%. This change of Φ_{PSII} resulted from changes in the total non-photochemical quenching capacity ($\Phi_{NPQ}+ \Phi_{NO}$). The quantum yield of regulated energy dissipation (Φ_{NPQ}) was quite similar in all AOIs under different environment conditions, indicating that no excess light energy was produced by precipitation change and high temperature. The values of quantum yield of non-regulated energy dissipation (Φ_{NO}) were low and similar in all AOIs at T_0. This means that there were sufficient photochemical conversion and protective regulatory mechanisms in the whole leaf. At T_6 condition, water deficit increased Φ_{NO} in whole leaf, indicating that high temperature exacerbated the negative effect of water deficit on energy dissipation. Both photochemical

energy conversion and protective regulatory mechanism were not enough.

The Two-way ANOVA (**Table 3**) indicated that, F_v/F_m was significantly influenced by precipitation change at T_0 condition and varied greatly at different AOIs ($p<0.01$), exhibiting significant interaction of AOI and precipitation change ($p < 0.01$). The other chlorophyll fluorescence parameters such as Φ_{PSII} and Φ_{NO} exhibited significant difference across AOIs, but were not affected by precipitation change. Under condition, only Φ_{PSII} was significantly affected by both AOI and precipitation change, but the interaction was not significant ($p>0.05$). To conclude, F_v/F_m is most sensitive to precipitation change at T_0 condition, while Φ_{PSII} is the most sensitive indicator at T_6 condition. However, in the study by Lazár et al. (2006), even if there are no changes in the mean value of a fluorescence parameter, there can be the changes in shapes of statistical distributions of fluorescence parameter which is an early indication of a plant stress. Base on this, we should not only find the most sensitive parameter (Kalaji et al., 2014) and the most sensitive species (Swoczyna et al., 2010) but also find the best (statistical) method for detection of the stresses is more important. The use of fluorescence imaging and the detection of photosynthetic performance of *Stipa bungeana* response to climatic change still requires further investigation.

CONCLUSION

Chlorophyll fluorescence imaging provided detailed intuitive information on the spatial heterogeneity of chlorophyll fluorescence parameters of *S. bungeana* and facilitated the investigation of plant photosynthetic performance under various temperature and water conditions. Our results showed that *S. bungeana* has strong ability in protecting photosynthetic apparatus against the photoinhibitory damage from drought, and a 6°C higher temperature could offset the negative effect of water deficit to a certain extent. On the other hand, excess precipitation

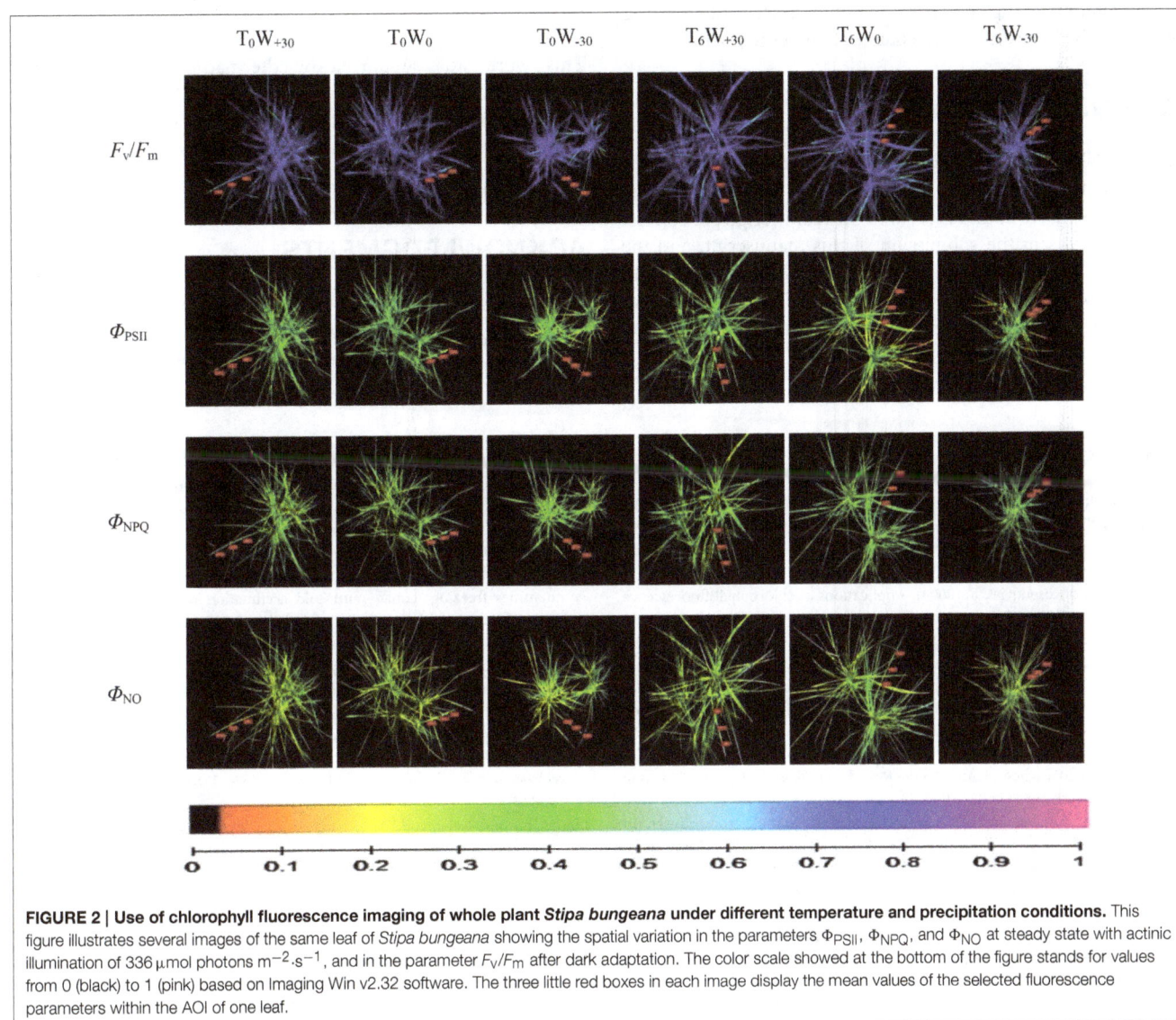

FIGURE 2 | Use of chlorophyll fluorescence imaging of whole plant *Stipa bungeana* under different temperature and precipitation conditions. This figure illustrates several images of the same leaf of *Stipa bungeana* showing the spatial variation in the parameters Φ_{PSII}, Φ_{NPQ}, and Φ_{NO} at steady state with actinic illumination of $336 \, \mu\text{mol}$ photons $\text{m}^{-2}\cdot\text{s}^{-1}$, and in the parameter F_v/F_m after dark adaptation. The color scale showed at the bottom of the figure stands for values from 0 (black) to 1 (pink) based on Imaging Win v2.32 software. The three little red boxes in each image display the mean values of the selected fluorescence parameters within the AOI of one leaf.

TABLE 3 | Multiple range test among effects of areas of interest (AOI) and watering treatments on *Stipa bungeana* leaf Chl fluorescence parameters under ambient temperature (T_0) and high temperature (T_6) conditions based on the Two-way ANOVA.

		F_v/F_m			Φ_{PSII}			Φ_{NPQ}			Φ_{NO}		
		df	F	P	df	F	P	df	F	P	df	F	P
T_0	AOI	2	20.918	0.000	2	14.455	0.000	2	0.587	0.566	2	3.815	0.042
	Precipitation	2	8.041	0.003	2	3.051	0.072	2	0.090	0.914	2	1.879	0.182
	AOI × Precipitation	4	5.160	0.006	4	0.692	0.607	4	0.847	0.514	4	0.648	0.635
T_6	AOI	2	2.166	0.144	2	4.713	0.023	2	0.223	0.802	2	2.661	0.097
	Precipitation	2	5.272	0.016	2	10.366	0.001	2	0.717	0.502	2	4.622	0.024
	AOI × Precipitation	4	0.198	0.936	4	0.075	0.989	4	0.661	0.627	4	0.926	0.471

df, Degree of freedom; F, F-value, used for Homogeneity of variance test; P, Significant level.

had no significant effect on PSII at room temperature. But high temperature had a positive effect on PSII and significantly enhanced the photosynthesis of *S. bungeana*. We also found that it is energy dissipation by non-regulated quenching mechanisms rather than the xanthophyll cycle-mediated thermal dissipation that plays an important role in dominating the yield of PSII photochemistry under climate change. This study also found that F_v/F_m measured in AOIs was the most sensitive indicator to precipitation change under room temperature, while Φ_{PSII} is more sensitive to precipitation change at higher temperature.

AUTHOR CONTRIBUTIONS

GZ and YW conceived the experiment, YW, XL, and ZX conducted the experiment, XS analyzed the results and wrote the manuscript. All authors reviewed the manuscript. No conflict of interest exits in the submission of this manuscript, and the manuscript is approved by all authors for publication. We would like to declare that the work described was original research that has not been published previously, and not under consideration for publication elsewhere, in whole or in part.

FUNDING

This work was supported by the national Basic research Program of China [grant number 2010CB951301] and Chinese Academy of Sciences "Strategic Priority Research Program-Climate Change: Carbon Budget and Relevant Issues" [grant number XDA-05050408].

ACKNOWLEDGMENTS

We thank Jun Chen, Liping Tan, Tao Liu, Bingrui Jia, Yanling Jiang, Feng Zhang, Jian Song, Zhixiang Yang, Hui Wang, Yaohui Shi, Yanhui Hou for their help during the experiment.

REFERENCES

Albert, K. R., Ro-Poulsen, H., Mikkelsen, T. N., Michelsen, A., van der Linden, L., and Beier, C. (2011). Interactive effects of elevated CO_2, warming, and drought on photosynthesis of *Deschampsia flexuosa* in a temperate heath ecosystem. *J. Exp. Bot.* 62, 4253–4266. doi: 10.1093/jxb/err133

Baker, N. R., and Rosenqvist, E. (2004). Applications of chlorophyll fluorescence can improve crop production strategies: an examination of future possibilities. *J. Exp. Bot.* 55, 1607–1621. doi: 10.1093/jxb/erh196

Bauweraerts, I., Wertin, T. M., Ameye, M., McGuire, M. A., Teskey, R. O., and Steppe, K. (2013). The effect of heat waves, elevated CO_2 and low soil water availability on northern red oak (*Quercus rubra* L.) seedlings. *Glob. Chang. Biol.* 19, 517–528. doi: 10.1111/gcb.12044

Berry, J., and Bjorkman, O. (1980). Photosynthetic response and adaptation to temperature in higher plants. *Annu. Rev. Plant Physiol.* 31, 491–543. doi: 10.1146/annurev.pp.31.060180.002423

Bilger, W., and Schreiber, U. (1987). "Excitation energy and electron transfer," in *Photosynthesis*, eds Govindjee, J. Barber, W. A. Cramer, J. H. C. Goedheer, J. Lavorel, R. Marcelle, and B. Zilinskas (Dordrecht: Springer), 157–162.

Björkman, O., and Demmig, B. (1987). Photon yield of O_2 evolution and chlorophyll fluorescence characteristics at 77 K among vascular plants of diverse origins. *Planta* 170, 489–504. doi: 10.1007/BF00402983

Boyer, J. S. (1982). Plant productivity and environment. *Science* 218, 443–448. doi: 10.1126/science.218.4571.443

Brestic, M., Zivcak, M., Olsovska, K., Shao, H. B., Kalaji, H. M., and Allakhverdiev, S. I. (2014). Reduced glutamine synthetase activity plays a role in control of photosynthetic responses to high light in barley leaves. *Plant Physiol. Biochem.* 81, 74–83. doi: 10.1016/j.plaphy.2014.01.002

Calatayud, A., Roca, D., and Martínez, P. F. (2006). Spatial-temporal variations in rose leaves under water stress conditions studied by chlorophyll fluorescence imaging. *Plant Physiol. Biochem.* 44, 564–573. doi: 10.1016/j.plaphy.2006.09.015

Calatayud, Á., San Bautista, A., Pascual, B., Maroto, J. V., and López-Galarza, S. (2013). Use of chlorophyll fluorescence imaging as diagnostic technique to predict compatibility in melon graft. *Sci. Hortic.* 149, 13–18. doi: 10.1016/j.scienta.2012.04.019

Chaves, M. M., Pereira, J. S., Maroco, J., Rodrigues, M. L., Ricardo, C. P., Osorio, M. L., et al. (2002). How plants cope with water stress in the field? Photosynthesis and growth. *Ann. Bot.* 89, 907–916. doi: 10.1093/aob/mcf105

Cheng, J., Cheng, J., Hu, T., Shao, H., and Zhang, J. (2012). Dynamic changes of *Stipa bungeana* steppe species diversity as better indicators for soil quality and sustainable utilization mode in yunwu mountain nature reserve, Ningxia, China. *CLEAN Soil Air Water* 40, 127–133. doi: 10.1002/clen.201000438

Cornic, G. (2000). Drought stress inhibits photosynthesis by decreasing stomatal aperture – not by affecting ATP synthesis. *Trends Plant Sci.* 5, 187–188. doi: 10.1016/S1360-1385(00)01625-3

Demmig-Adams, B., et al. (1996). Using chlorophyll fluorescence to assess the fraction of absorbed light allocated to thermal dissipation of excess excitation. *Physiol. Plant.* 98, 253–264. doi: 10.1034/j.1399-3054.1996.980206.x

Dickmann, D. I., Liu, Z., Nguyen, P. V., and Pregitzer, K. S. (1992). Photosynthesis, water relations, and growth of two hybrid populus genotypes during a severe drought. *Can. J. Forest Res.* 22, 1094–1106. doi: 10.1139/x92-145

Duncan, D. B. (1955). Multiple range and multiple F tests. *Biometrics* 11, 1–42. doi: 10.2307/3001478

Ehlert, B., and Hincha, D. (2008). Chlorophyll fluorescence imaging accurately quantifies freezing damage and cold acclimation responses in Arabidopsis leaves. *Plant Methods* 4, 1–7. doi: 10.1186/1746-4811-4-12

Genty, B., Briantais, J.-M., and Baker, N. R. (1989). The relationship between the quantum yield of photosynthetic electron transport and quenching of chlorophyll fluorescence. *Biochim. Biophys. Acta* 990, 87–92. doi: 10.1016/S0304-4165(89)80016-9

Giardi, M. T., Cona, A., Geiken, B., Kučera, T., Masojídek, J., and Mattoo, A. K. (1996). Long-term drought stress induces structural and functional reorganization of photosystem II. *Planta* 199, 118–125. doi: 10.1007/BF00196888

Gorbe, E., and Calatayud, A. (2012). Applications of chlorophyll fluorescence imaging technique in horticultural research: a review. *Sci. Hortic.* 138, 24–35. doi: 10.1016/j.scienta.2012.02.002

Havaux, M. (1992). Stress Tolerance of Photosystem II *in vivo*: antagonistic effects of water, heat, and photoinhibition stresses. *Plant Physiol.* 100, 424–432. doi: 10.2307/4274644

Hendrickson, L., Furbank, R. T., and Chow, W. S. (2004). A simple alternative approach to assessing the fate of absorbed light energy using chlorophyll fluorescence. *Photosynth. Res.* 82, 73–81. doi: 10.1023/B:PRES.0000040446.87305.f4

Herppich, W. B., and Peckmann, K. (2000). Influence of drought on mitochondrial activity, photosynthesis, nocturnal acid accumulation and water relations in the CAM plants prenia sladeniana(ME-type) and crassula lycopodioides(PEPCK-type). *Ann. Bot.* 86, 611–620. doi: 10.1006/anbo.2000.1229

Humplík, J., Lazár, D., Husièková, A., and Spíchal, L. (2015). Automated phenotyping of plant shoots using imaging methods for analysis of plant stress responses a review. *Plant Methods* 11, 1–10. doi: 10.1186/s13007-015-0072-8

IPCC (2013). *Climatic Change. The Physical Science Basis.* Stockholm: Intergovernmental Panel on Climate Change.

Ivanov, D. A., and Bernards, M. A. (2015). Chlorophyll fluorescence imaging as a tool to monitor the progress of a root pathogen in a perennial plant. *Planta* 243, 263–279. doi: 10.1007/s00425-015-2427-9

Kalaji, H. M., Oukarroum, A., Alexandrov, V., Kouzmanova, M., Brestic, M., Zivcak, M., et al. (2014). Identification of nutrient deficiency in maize and tomato plants by *in vivo* chlorophyll a fluorescence measurements. *Plant Physiol. Biochem.* 81, 16–25. doi: 10.1016/j.plaphy.2014.03.029

Kramer, D., Johnson, G., Kiirats, O., and Edwards, G. (2004). New fluorescence parameters for the determination of QA redox state and excitation energy fluxes. *Photosynth. Res.* 79, 209–218. doi: 10.1023/B:PRES.0000015391.99477.0d

Lazár, D. (2015). Parameters of photosynthetic energy partitioning. *J. Plant Physiol.* 175, 131–147. doi: 10.1016/j.jplph.2014.10.021

Lazár, D., Sušila, P., and Nauš, J. (2006). Early detection of plant stress from changes in distributions of chlorophyll a fluorescence parameters measured with fluorescence imaging. *J. Fluoresc.* 16, 173–176. doi: 10.1007/s10895-005-0032-1

Lichtenthaler, H. K., and Miehé, J. A. (1997). Fluorescence imaging as a diagnostic tool for plant stress. *Trends Plant Sci.* 2, 316–320. doi: 10.1016/S1360-1385(97)89954-2

Machado, S., and Paulsen, G. (2001). Combined effects of drought and high temperature on water relations of wheat and sorghum. *Plant Soil* 233, 179–187. doi: 10.1023/A:1010346601643

Martínez-Carrasco, R., Sánchez-Rodriguez, J., and Pérez, P. (2002). Changes in chlorophyll fluorescence during the course of photoperiod and in response to drought in casuarina equisetifolia forst and Forst. *Photosynthetica* 40, 363–368. doi: 10.1023/A:1022618823538

Mielke, M. S., de Almeida, A.-A. F., Gomesa, F. P., Aguilarb, M. A. G., and Mangabeira, P. A. O. (2003). Leaf gas exchange, chlorophyll fluorescence and growth responses of *Genipa americana* seedlings to soil flooding. *Environ. Exp. Bot.* 50, 221–231. doi: 10.1016/S0098-8472(03)00036-4

Müller, P., Li, X.-P., and Niyogi, K. K. (2001). Non-Photochemical quenching. A response to excess light energy. *Plant Physiol.* 125, 1558–1566. doi: 10.2307/4279788

Ogaya, R., Peñuelas, J., Asensio, D., and Llusià, J. (2011). Chlorophyll fluorescence responses to temperature and water availability in two co-dominant Mediterranean shrub and tree species in a long-term field experiment simulating climate change. *Environ. Exp. Bot.* 73, 89–93. doi: 10.1016/j.envexpbot.2011.08.004

Omasa, K. (1990). "Physical methods," in *Plant Sciences*, Vol. 11, *Modern Methods of Plant Analysis*, eds H.-F. Linskens and J. F. Jackson (Berlin; Heidelberg: Springer), Ch. 8, 203–243.

Omasa, K., and Takayama, K. (2003). Simultaneous measurement of stomatal conductance, non-photochemical quenching, and photochemical yield of photosystem ii in intact leaves by thermal and chlorophyll fluorescence imaging. *Plant Cell Physiol.* 44, 1290–1300. doi: 10.1093/pcp/pcg165

Omasa, K., Shimazaki, K.-I., Aiga, I., Larcher, W., and Onoe, M. (1987). Image analysis of chlorophyll fluorescence transients for diagnosing the photosynthetic system of attached leaves. *Plant Physiol.* 84, 748–752.

Ort, D. R., and Baker, N. R. (2002). A photoprotective role for O$_2$ as an alternative electron sink in photosynthesis? *Curr. Opin. Plant Biol.* 5, 193–198. doi: 10.1016/S1369-5266(02)00259-5

Osório, M. L., Osório, J., Vieira, A. C., Gonçalves, S., and Romano, A. (2011). Influence of enhanced temperature on photosynthesis, photooxidative damage, and antioxidant strategies in *Ceratonia siliqua* L. seedlings subjected to water deficit and rewatering. *Photosynthetica* 49, 3–12. doi: 10.1007/s11099-011-0001-7

Oxborough, K., and Baker, N. (1997). Resolving chlorophyll a fluorescence images of photosynthetic efficiency into photochemical and non-photochemical components – calculation of qP and Fv-/Fm-; without measuring Fo. *Photosyn. Res.* 54, 135–142. doi: 10.1023/A:1005936823310

Papageorgiou, G., Govindjee (eds.). (2004). *Chlorophyll Fluorescence: A Signature of Photosynthesis*. Dordrecht: Kluwer Academic Publishers.

Petsas, A., and Grammatikopoulos, G. (2009). Drought resistance and recovery of photosystem II activity in a Mediterranean semi-deciduous shrub at the seedling stage. *Photosynthetica* 47, 284–292. doi: 10.1007/s11099-009-0044-1

Shaw, A. K., Ghosh, S., Kalaji, H. M., Bosa, K., Brestic, M., Zivcak, M., et al. (2014). Nano-CuO stress induced modulation of antioxidative defense and photosynthetic performance of Syrian barley (*Hordeum vulgare* L.). *Environ. Exp. Bot.* 102, 37–47. doi: 10.1016/j.envexpbot.2014.02.016

Swoczyna, T., Kalaji, H. M., Pietkiewicz, S., Borowski, J., and Zaraś-Januszkiewicz, E. (2010). Photosynthetic apparatus efficiency of eight tree taxa as an indicator of their tolerance to urban environments. *Dendrobiology* 63, 65–75.

Thomey, M. L., Collins, S. L., Vargas, R., Johnson, J. E., Brown, R. F., Natvig, D. O., et al. (2011). Effect of precipitation variability on net primary production and soil respiration in a Chihuahuan Desert grassland. *Glob. Chang. Biol.* 17, 1505–1515. doi: 10.1111/j.1365-2486.2010.02363.x

Tuba, T., Saxena, D. K., Srivastava, K., Singh, S., Czobel, S., and Kalaji, H. M. (2010). Chlorophyll a fluorescence measurements for validating the tolerant bryophytes for heavy metal (Pb) biomapping. *Res. Commun.* 98, 1501–1508.

Wahid, A., Gelani, S., Ashraf, M., and Foolad, M. R. (2007). Heat tolerance in plants: an overview. *Environ. Exp. Bot.* 61, 199–223. doi: 10.1016/j.envexpbot.2007.05.011

Xu, Z.-Z., Zhou, G.-S., and Li, H. (2004). Responses of chlorophyll fluorescence and nitrogen level of Leymus chinensis Seedling to changes of soil moisture and temperature. *J. Environ. Sci.* 16, 666–669. doi: 10.3321/j.issn:1001-0742.2004.04.029

Xu, Z. Z., and Zhou, G. S. (2006). Combined effects of water stress and high temperature on photosynthesis, nitrogen metabolism and lipid peroxidation of a perennial grass Leymus chinensis. *Planta* 224, 1080–1090. doi: 10.1007/s00425-006-0281-5

Xu, Z. Z., Zhou, G. S., and Shimizu, H. (2009). Effects of soil drought with nocturnal warming on leaf stomatal traits and mesophyll cell ultrastructure of a perennial grass. *Crop Sci.* 49, 1843–1851. doi: 10.2135/cropsci2008.12.0725

Zivcak, M., Kalaji, H. M., Shao, H.-B., Olsovska, K., and Brestic, M. (2014). Photosynthetic proton and electron transport in wheat leaves under prolonged moderate drought stress. *J. Photochem. Photobiol. B* 137, 107–115. doi: 10.1016/j.jphotobiol.2014.01.007

Conflict of Interest Statement: The authors declare that the research was conducted in the absence of any commercial or financial relationships that could be construed as a potential conflict of interest.

15

Differential Toxicity of Bare and Hybrid ZnO Nanoparticles in Green Pea (*Pisum sativum* L.): A Life Cycle Study

Arnab Mukherjee[1,2], *Youping Sun*[3], *Erving Morelius*[1,2], *Carlos Tamez*[1,2], *Susmita Bandyopadhyay*[1,2], *Genhua Niu*[3], *Jason C. White*[4], *Jose R. Peralta-Videa*[1,2,5] *and Jorge L. Gardea-Torresdey*[1,2,5*]

[1] Environmental Science and Engineering, The University of Texas at El Paso, El Paso, TX, USA, [2] University of California Center for Environmental Implications of Nanotechnology, The University of Texas at El Paso, El Paso, TX, USA, [3] Texas A&M AgriLife Research Center at El Paso, El Paso, TX, USA, [4] Department of Analytical Chemistry, The Connecticut Agricultural Experiment Station, New Haven, CT, USA, [5] Department of Chemistry, The University of Texas at El Paso, El Paso, TX, USA

Edited by:
Nelson Marmiroli,
University of Parma, Italy

Reviewed by:
Stephen Ebbs,
Southern Illinois University, USA
Filip Pošcic,
University of Udine, Italy

*Correspondence:
Jorge L. Gardea-Torresdey
jgardea@utep.edu

The effect of surface or lattice modification of nanoparticles (NPs) on terrestrial plants is poorly understood. We investigated the impact of different zinc oxide (ZnO) NPs on green pea (*Pisum sativum* L.), one of the highest consumed legumes globally. Pea plants were grown for 65 d in soil amended with commercially available bare ZnO NPs (10 nm), 2 wt% alumina doped (Al_2O_3@ZnO NPs, 15 nm), or 1 wt% aminopropyltriethoxysilane coated NPs (KH550@ZnO NP, 20 nm) at 250 and 1000 mg NP/kg soil inside a greenhouse. Bulk (ZnO) and ionic Zn (zinc chloride) were included as controls. Plant fresh and dry biomass, changes in leaf pigment concentrations, elements (Zn, Al, Si), and protein and carbohydrate profile of green pees were quantified upon harvest at 65 days. With the exception of the coated 1000 mg/kg NP treatment, fresh and dry weight were unaffected by Zn exposure. Although, all treated plants showed higher tissue Zn than controls, those exposed to Al_2O_3@ZnO NPs at 1000 mg/kg had greater Zn concentration in roots and seeds, compared to bulk Zn and the other NP treatments, keeping Al and Si uptake largely unaffected. Higher Zn accumulation in green pea seeds were resulted in coated ZnO at 250 mg/kg treatments. In leaves, Al_2O_3@ZnO NP at 250 mg/kg significantly increased Chl-*a* and carotenoid concentrations relative to the bulk, ionic, and the other NP treatments. The protein and carbohydrate profiles remained largely unaltered across all treatments with the exception of Al_2O_3@ZnO NPs at 1000 mg/kg where sucrose concentration of green peas increased significantly, which is likely a biomarker of stress. Importantly, these findings demonstrate that lattice and surface modification can significantly alter the fate and phytotoxic effects of ZnO NPs in food crops and seed nutritional quality. To the authors' knowledge, this is the first report of a life cycle study on comparative toxicity of bare, coated, and doped ZnO NPs on a soil-grown food crop.

Keywords: bare, doped, coated, ZnO nanoparticles, phytotoxicity, dissolution, seed quality

INTRODUCTION

Engineered nanoparticles (ENPs), due to their high surface to volume ratio and greater numbers of atoms at the particle surface, have been widely used in the fields of medicine, agriculture (nano-fertilizers and nano-pesticides), manufacturing, electronics, and energy production (Ghormade et al., 2011; Roco, 2011; Bandyopadhyay et al., 2013; Gardea-Torresdey et al., 2014). It has been estimated that the global nanotechnology market will exceed to $3 trillion by the year 2020 (Venkatesan et al., 2004). In recent years, hybrid ENPs, e.g., doped and coated nanomaterials (NMs), have received increased attention due to their potential applications in microelectronics, semiconductors, optical device fabrication, and optics (Venkatesan et al., 2004; Ozgur et al., 2005; Dhiman et al., 2012). Commercially, available silane coupling agent (KH550) coated ZnO NPs and alumina doped (Al_2O_3) ZnO NPs are two of the important hybrid NPs and are being used in the fabrication of detectors and optoelectronic devices (Zhang et al., 2010; Thandavan et al., 2015), preparation of novel polymer-inorganic nanocomposites, among others (Abdolmaleki et al., 2012). Unique properties, such as, high reactivity and bio-compatibility are two reasons for concern related to potential toxicity to biota. The rapidly increasing production and use have elevated the likelihood of ENP exposure in the environment (Mukherjee et al., 2014a,b). However, very little is known about the environmental health and safety of these newer hybridized materials.

The literature has shown mixed effects of NP exposure on various animals, plants, and microorganisms; depending upon their species, growth conditions, NP type and exposure concentrations, among others. For example, Montalvo et al. (2015) reported improved phosphorus bioavailability through the application of hydroxyapatite nanoparticles to wheat (*Triticum aestivum).* Application of nanomaterials toward nano-fertilizer development and plant disease suppression is described elsewhere (Liu and Lal, 2015; Servin et al., 2015). Conversely, ample evident of negative effects could also be found in the literature. For example, growth can be negatively affected by ENPs exposure (Lin and Xing, 2007; Sinha et al., 2011; Bandyopadhyay et al., 2012a,b; Gaiser et al., 2012; Hawthorne et al., 2012; Mukherjee et al., 2014b; Rico et al., 2014). There are several reports on the toxicity of different ENPs on food crops (Lin and Xing, 2007; Lee et al., 2008; Navarro et al., 2008; Sinha et al., 2011; Bandyopadhyay et al., 2012a,b; Gaiser et al., 2012; Hawthorne et al., 2012; Zhao et al., 2013a, 2014a; Rico et al., 2014; Mukherjee et al., 2014a,b). However, a mechanistic understanding of the impact of ENPs on edible/crop plants is needed for accurate exposure and risk assessment, but this knowledge remains elusive. "The Nanotechnology Consumers Products Inventory" identifies zinc oxide (ZnO) NP as the fifth most widely used material in terms of use in the consumer products (Maynard and Evan, 2006). ZnO NPs are commonly used in personal care products, anti-microbial agents, paints, and photovoltaics (Szabo et al., 2003; Hernandez-Viezcas et al., 2013). However, ZnO NPs have been shown to be potentially toxic in the environment (Kahru and Dubourguier, 2010). For instance,

a 5-day exposure study with ZnO NP-DI water suspension in petri dishes showed root growth inhibition in ryegrass (*Lolium perenne*), radish (*Raphanus sativus*), and rape (*Brassica napus*) (Lin and Xing, 2007). NPs can also exert phytotoxicity by disrupting the water and nutrient pathways in plants (Szabo et al., 2003; Lin and Xing, 2008; Kahru and Dubourguier, 2010; Lopez-Moreno et al., 2010; De La Rosa et al., 2011). Lopez-Moreno et al. (2010) reported on the genotoxicity of ZnO NPs to soybean (*Glycine max*). A reduction in wheat (*Triticum aestivum*) biomass upon ZnO exposure, along with elevated reactive oxygen species (ROS) level, was reported by Dimkpa et al. (2012). Zhao et al. (2013a) observed reduction in chlorophyll production in corn (*Zea mays*) grown in soil amended with ZnO NPs at 800 mg/kg. Importantly, the toxicity of ZnO NPs may often be due to its greater dissolution or release of Zn^{2+} ions into the growth media as a function of small particle size, opposed to the induction of oxidative stress by the parent ENPs (Hendry and Jones, 1980; Nel et al., 2006; Xia et al., 2006; Du et al., 2011; Kim et al., 2011; Priester et al., 2012). For example, released Zn^{2+} ions from the dissolution of ZnO NPs can displace the central Mg^{2+} of chlorophyll, effectively disabling the photosynthetic core, causing phytotoxicity (Rebeiz and Castelfranco, 1973; Hendry and Jones, 1980; Kupper et al., 1996; Oberdorster et al., 2005). There are very few reports on the effects of NPs on seed quantity, quality, or nutritional content. For instance, CeO_2 NPs change the nutritional quality of wheat (*Triticum aestivum* L.) (Rico et al., 2014). The fruit quality of soybean was impacted by ZnO and CeO_2 NPs (Priester et al., 2012). However, there appears to be no information available on the comparative toxicity of bare, doped, and coated ZnO NPs on green pea (*Pisum sativum* L.).

The aim of this work was to evaluate the effect of surface coating and lattice doping on ZnO NP-plant interactions. Green pea was chosen because of its high global consumption and nutritional value (Iqbal et al., 2006). Pea plants were exposed to different concentrations of ENPs and bulk ZnO and zinc chloride. The accumulation/uptake of Zn, Al (present in doped NP), and Si (present in KH550 coating) in different plant tissues, as well as the mineral, carbohydrate, and protein content in seeds were also determined.

MATERIALS AND METHODS

Soil Sampling
The soil was collected from the field at Texas AgriLife Research Center, El Paso, TX (31°41′44.98″N; 106°17′ 01.36″ W, top 20 cm) and is a sandy loam with 3.73% clay, 12.15% silt, and 84.1% sand (Zhao et al., 2013a). The experiment was conducted in a 1:1 mixture of the native soil with high organic matter potting soil [Miracle-Gro Garden Soil for Flowers & Vegetables; N-P-K = 0.09-0.05-0.07] so as to improve the soil quality in terms of soil porosity, and water retention capacity, among others.

Pot Preparation
The bare ZnO NPs (10 nm commercial spheroid, Meliorum Technologies, New York) were obtained from the University

of California Center for Environmental Implications of Nanotechnology (UC CEIN). Two percent wt Al_2O_3@ZnO (15 nm), and one percent wt KH550 coated ZnO NPs (20 nm) were obtained from US Research Nanomaterials (http://www.us-nano.com). ENPs and bulk ZnO were added as dry powder at 0 (control), 250 and 1000 mg NPs/kg of soil in black plastic containers (Ns-400; diameter: 20 cm; tall: 12.5 cm; volume: 3.925 L; Nursery Supplies). To achieve 1 wt% dissolved Zn, equivalent amount of 5 and 20 mg/kg zinc chloride was dissolved in 50 mL Millipore water (MPW) and added to the soil for ionic treatments. The soil was vigorously mixed with spatulas to maximize particle/ion homogeneity. Early rise variety of green pea (Seeds of Change, USDA organic, Home Depot, life cycle 65 days) were immersed in 4% bleach solution and rinsed three times with tap water. Seeds were soaked overnight in regular tap water and were sown in the test pots for a 65-day growth period. Two hundred milliliters of nutrient solution per day [0.72 g/L 15 N– 2.2 P– 12.5 K (Peters 15-5-15); EC = 1.80 dS/m; pH = 6.62] was added to each pot and the pots were maintained for 24 h in the green house for stabilization. The daily light integral (photosynthetically active radiation) was 15.3 ± 3.1 mol/m^2/d. The greenhouse temperature was maintained at 26.9 ± 8.6°C during the day and 13.7 ± 4.3°C at night. The relative humidity was 41.6 ± 19.1%.

Zeta Potential, Size, and pH of the NP Suspensions

Particles were dispersed in 10 mL Millipore water (MPW) to achieve 250 and 1000 mg/L concentrations, sonicated for 10 min, and kept undisturbed for 1 h and the zeta potential and size were measured using a Zetasizer Nano-ZS 90, (Malvern Instruments Ltd., UK). The pH of the supernatants was measured. Each analysis was performed in triplicate.

Dissolution of Different NPs in Soil Solution

Release of Zn^{2+} was measured by dispersing all the NPs and bulk ZnO in soil solution containing 5 g of soil mixture (1:1) in 20 mL DI water at a concentration of 1000 mg/kg soil. Zinc chloride was excluded due to its complete solubility in water. Each measurement was done in three replicates at three sampling intervals of 15, 30, and 45 days. These samples were used for the time-dependent dissolution study. Multiple serial centrifugations were used to remove suspended particles from the solution and to isolate the dissolved Zn ions. At the predetermined time (15, 30, and 45 days) intervals, samples were taken and centrifuged at 5000 rpm (Eppendorf AG bench centrifuge 5417R, Hamburg, Germany), and 2 mL aliquot of the supernatant was collected and centrifuged at 14000 rpm for 30 min. Subsequently, this supernatant was transferred and centrifuged again at 14000 rpm for 45 min. This process was repeated three times to remove particulate matters (Bandyopadhyay et al., 2015). The final supernatant was diluted to 15 mL with 2% HNO_3 and elemental concentrations were measured using ICP-OES/MS as described below.

Elemental Analysis of Soil, Plant Tissues, and Seeds

For each replicate, 1 g of native and 1:1 soil were collected separately from the stock pile and grounded in mortar-pestle. Approximately, 200 mg of soil portions were digested in a microwave acceleration reaction system (CEM MARSx, Mathews, NC) with 1:4 plasma pure HNO_3 (trace metals ≤ 1 ppb) and H_2O_2 at 195°C for 30 min (ramp 5 min; hold 25 min) in triplicate (Packer et al., 2007). Sixty five-day old pea plants were harvested and roots were washed with 0.01 M HNO_3, with subsequent rinsing in DI water. The tissues were then oven dried at 70°C for 2 days (Fisher Scientific Isotemp., Pittsburgh, PA; USA). The seeds were dried at room temperature for a week. Different tissues were weighed and digested similar to that described above. The digested samples were analyzed for elemental content using a Perkin Elmer optima 4300 DV inductively coupled plasma optical emission spectrometer (ICP-OES) or ICP-MS (ELAN DRC II; Perkin-Elmer) as required.

Chlorophyll and Carotenoid Estimation in Leaf

Approximately 0.5 gram fresh, razor blade chopped leaves were placed into 15 mL tubes. Five mL acetone was added and the samples were shaken overnight on a horizontal shaker (Revco Scientific DS1473AVA, 115 volts, 60 Hz, 7 amps). The supernatants were collected and absorbance was measured at 470, 645, and 662 nm using a Perkin Elmer Lambda 14 UV/Vis spectrometer (single-beam mode, Perkin-Elmer, Uberlinger, Germany). Concentrations of Chl-a, b, and total carotenoids were measured according to a previously described method (Wellburn, 1994).

Determination of Starch, Total Soluble Sugars, and Reducing Sugars in Seeds

The total soluble sugar extraction was performed following the method of Verma and Dubey (2001) with little modification. A sample of 100 mg of dried pea seed was ground in 2 mL of 80% ethanol and then heated (80°C) in a water bath for 30 min. After cooling to room temperature, the extracts were centrifuged at 14000 rpm for 30 min (Thermo Scientific, Soruall T1, U.S.); a process that was repeated twice. All supernatants were combined and the total soluble sugar content was determined spectrophotometrically (λ = 490 nm, single-beam mode, Perkin-Elmer, Uberlinger, Germany) following the method of Dubois et al. (1956). The reducing sugar content was measured spectrophotometrically (λ = 620 nm) by the procedure of Somogyi (1952). In both cases, sugar content was determined against a standard calibration curve of glucose. The amount of non-reducing sugar was determined by subtracting the value of reducing sugar from total sugar.

Seed starch was also determined following the method of Verma and Dubey; the residue from total sugar extraction was used to measure the starch content (Verma and Dubey, 2001). The precipitate was dried at 70°C for 24 h, 2 mL of MPW was added, and the mixture was heated in a water bath at

95°C for 15 min. After cooling to ambient temperature, 1 mL of concentrated sulfuric acid was added. The suspension was stirred for 15 min, and the final volume was adjusted to 5 mL using MPW. The supernatant was centrifuged at 3000 rpm for 20 min, and the extraction was repeated once using 50% sulfuric acid. The supernatants were combined and diluted to 10 mL. The starch content was quantified following the method of Dubois et al. (1956) and expressed in mg/100 g dry weight.

Protein Fractionation in Seeds

Protein fractionation was performed according to Chen and Bushuk (1970). Dried pea seeds (100 mg) were extracted sequentially with 2 mL each of water, 0.5 mol/L NaCl, 70% ethanol, and 0.05 M acetic acid for 2 h. The extracted protein in each step was labeled as albumin (water soluble), globulin (salt-soluble), prolamin (alcohol-soluble), and glutelin (acid-soluble), respectively. Each fraction was centrifuged at 14,000 rpm; the supernatants were collected and analyzed using the methods of Bradford (1976).

Statistical Analysis

Unless otherwise noted, all the treatments were replicated four times. Data (means $\pm$ SE) were reported as averages of four replicates. We have run the Two-way ANOVA considering treatment type and concentration as variables with 6 and 2 levels respectively, followed by Tukey-HSD multiple comparison for the means to check the individual effects and their interactions at $p \leq 0.05$ (R version 3.1.3). Pairwise comparison tests (with adjusted p-values) between concentration and nanoparticle type revealed high significance in some cases. However, we plotted only those interactions which resonate with the research goals, promote clarity and ease of comparison.

RESULTS AND DISCUSSION

Size, Zeta Potential, and pH of Different Particles in MPW

In MPW, doped NPs had lower hydrodynamic diameter values than the other particles (Table S1). As expected, at 250 and 1000 mg/L, bulk ZnO possessed significantly higher diameter (1627 $\pm$ 198.9 and 9324 $\pm$ 236.8 nm) followed by coated

(526.6 $\pm$ 14.2 and 608.5 $\pm$ 11.9 nm), bare-ZnO (397.5 $\pm$ 25.3 and 290.9 $\pm$ 20.2 nm), and doped NPs (362.2 $\pm$ 20.7 and 244.1 $\pm$ 25.6 nm). Interestingly, as the concentration increased, the size of aggregates of bare and doped nanoparticles decreased but that of the coated NP increased. This may be due to higher rate of aggregation and co-precipitation of bare and doped NPs with other suspended NPs and/or soil particles at the higher concentration, leaving behind smaller aggregates in the suspension. Conversely, the coated NP, due to its high negative surface charge (-26.4 ± 7.09 mV, Table S1), can form hydrogen bonding in MPW, leading to greater stability of the dispersion with larger NPs in diameter compared to bare and doped forms.

With the exception of coated nanoparticles, all particles showed positive zeta potential. The order of magnitude was: coated < bare-ZnO < doped < bulk. The higher zeta potential for doped NPs compared to bare-ZnO NPs can be attributed to the fact that in doped NPs, Al^{3+} replaced Zn^{2+} in the ZnO lattice, which increases the surface potential. The negative zeta potential of the coated nanoparticles is understandable, given the nature of the surface coating; The ethoxy groups present in aminopropyltriethoxy-silane (KH550) can hydrolyze readily in water and generate hydroxyl silane (Wang et al., 2012). Thus, the attachment of KH550 onto the surface of ZnO NPs and corresponding hydrolysis could create a negative surface charge through the activity of the oxygen atoms, yielding a negative zeta potential. Although, there are differences in the numerical values of soil pH (7.7–8.5) among the treatments, these differences are not statistically significant (Table S1). This might be due to fewer number of replicates (three) and/or short exposure period.

Particle Dissolution

Dissolution data (mg/kg soil) of all the treatment is shown in **Figure 1**. We found no significant differences in dissolution across the three types of NPs in soil suspension (**Figure 1**). Similarly, the amount of released Si and Al remained unaffected at a given time, most probably, due to: (i) the very low amount of Al and Si in doped (2 wt%) and coated (1 wt%) NPs, respectively, with regard to the mass of ZnO NP and (ii) high background concentrations of Si and Al coming from the soil could make it difficult to quantify the source-specificity (NP vs. soil) of those two elements. On the other side, a variation in Zn dissolution

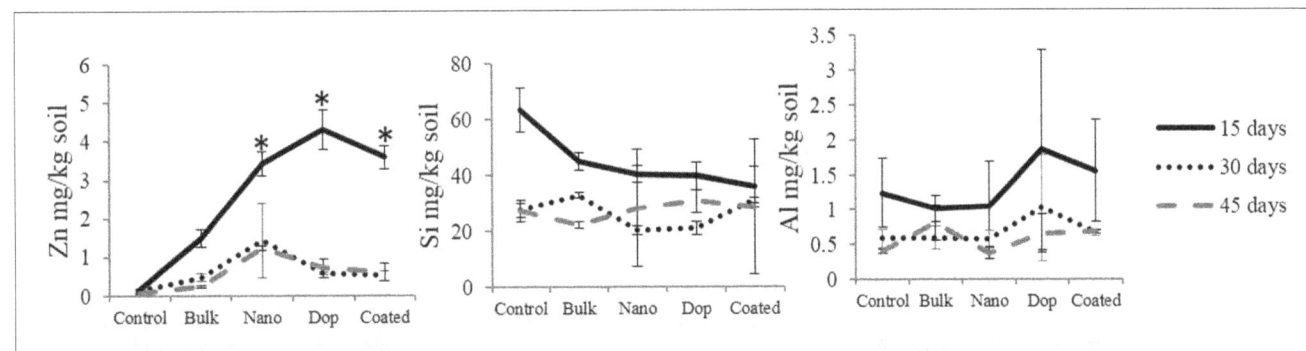

FIGURE 1 | Zinc, silicon, and aluminum dissolution from all the particles after 15, 30, and 45 days at 1000 mg/kg soil concentration. Data points with same/no symbols (*) represent no statistical significance at $p \leq 0.05$.

was observed with time. For instance, at 15th day, the amount of released Zn for all three NPs varied from 3.4 to 4.3 mg/kg soil but this difference was not enough to reach statistical significance. As expected, bulk ZnO particles released 1.5 mg Zn/kg soil, which was significantly less than all nano treatments ($p \leq$ 0.05). This could be attributed to the larger size of the bulk particles, which yields far less surface area and subsequently, less dissolution from the ZnO. The amount of dissolved zinc did not change between 30 and 45 days. Interestingly, the extent of dissolution after 30 and 45 days was lower than that of 15 days. This may be due to the production of zinc hydroxide that precipitates from solution and/or sorption of zinc ions to the different soil components, leaving behind fewer zinc ions upon reaching equilibrium (Wang et al., 2013; Zhao et al., 2013a). It is important to note that the dissolution study was neither intended to guide the toxicological experiments nor to elucidate the dissolution kinetics, but to check the trend of NP-dissolution in this particular soil type. It is noteworthy that we assumed 1 wt% zinc dissolution, although our dissolution study showed 0.06–0.43 wt% dissolution. The reasons for this difference are that the current study was performed under a closed system whereas the actual experiment was conducted under constant irrigation; this could significantly increase zinc dissolution. This is in part why we preferred to consider a "higher" amount reported in the literature, which was 1 wt% of ZnO NP (Bian et al., 2011). To achieve ~1 wt% of zinc ion, we added 2 wt% of $ZnCl_2$ as the molar mass of Zn (65.4) is close to half the molar mass of

$ZnCl_2$ (136.3). As such, 5 and 20 mg $ZnCl_2$/kg soil approximate 1 wt% Zn^{2+} dissolution from 250 and 1000 mg ZnO NPs /kg soil, respectively.

Fresh/Dry Weights and Zinc/Aluminum/Silicon Bioaccumulation in Root/Stem/Leaf

The elemental analysis of the native soil and 1:1 soil mix is shown in Table S2. Significant changes were observed in the elemental composition between the native and 1:1 soil (Table S2). The total amount of Zn, K, Mg, S, Mn, P, and Mo increased with the amendment of organic matter rich potting soil. However, the Fe concentration decreased and the Ca and Cu concentrations were unaltered. There was a slight numerical decrease in the pH values in the native (7.7 ± 0.18) and 1:1 soil (7.2 ± 0.08) but the difference was not of statistical significance.

The biomass of plants exposed to Zn treatments is shown in Figure S1. At 250 mg/kg treatments, regardless of particle type, the fresh and dry weight were unaffected by Zn treatment. Similarly, at 1000 mg/kg, nano, doped and ion-exposed plants had equivalent biomass, compared to the unexposed controls. However, the 1000 mg/kg bulk and coated NP treatments significantly increased the fresh weight, relative to the control plants; although the same trend was seen for dry weight, the differences were not statistically significant (Figure S1).

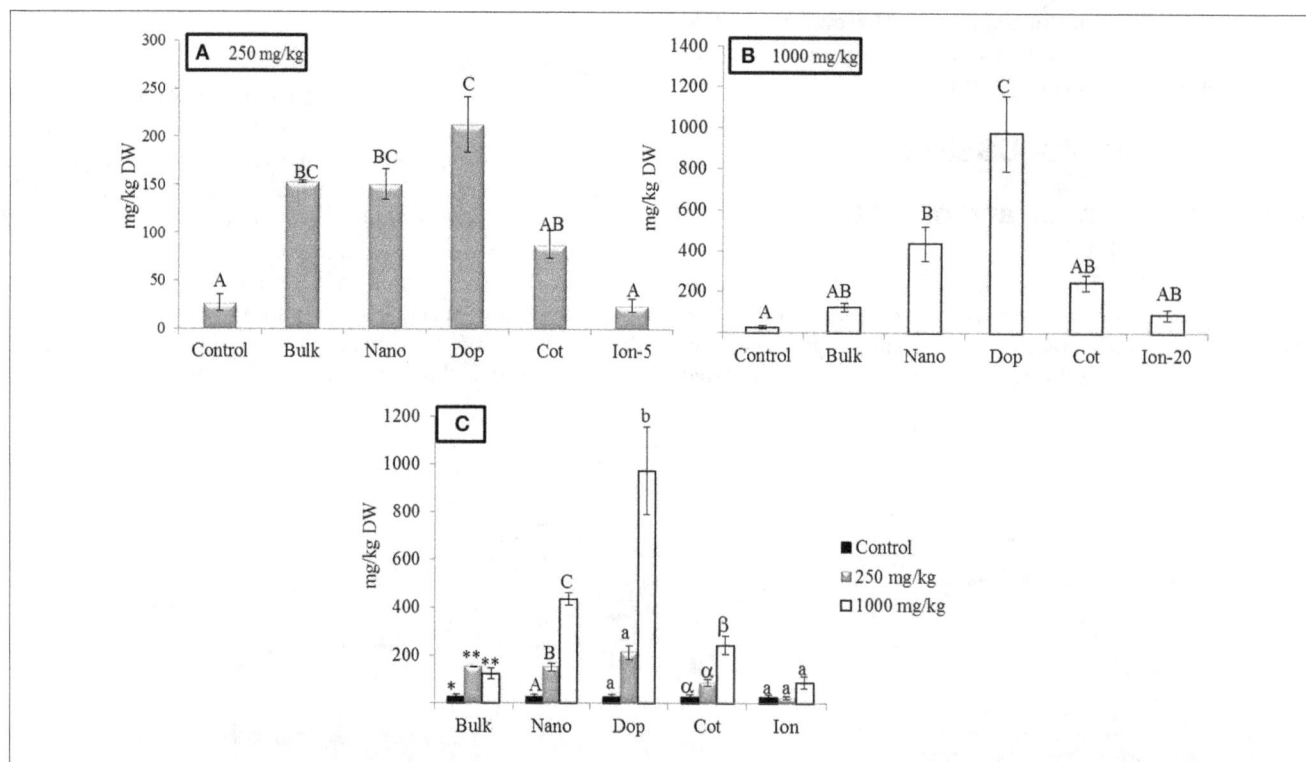

FIGURE 2 | Zinc bioaccumulation in root tissues. (A,B) Show the effects of different treatments at 250 and 1000 mg/kg exposure, respectively. (C) Shows the comparison among control, 250, and 1000 mg/kg treatments for each type of treatment separately. Bars are mean ± SE. Bars with same letters/symbols represent no statistical significance at $p \leq 0.05$. Upper case, lower case, and symbols are mutually exclusive.

As expected, Zn treatments increased root Zn (**Figure 2**). At 250 mg/kg exposure, roots showed 5.8, 5.8, and 3 times more Zn for bulk, bare, and coated NPs, respectively, compared to controls. The doped NP exposure yielded a root Zn concentration 8 times higher than controls. Moreover, increases at 1000 mg/kg bare NP and doped treatments were 16–36 times higher than controls (**Figure 2**). The level of Zn in the 5 mg/kg ion exposed ("Ion-5") plants was equivalent to that of the control. The bulk, coated, and ion exposures did have nominal concentrations that were higher than the controls but large variability among these specific replicates resulted in statistical insignificance in these treatments. Concentration dependent increases in Zn content were evident for the nano, doped, and coated treatments; these trends were less clear for the bulk and ion exposures.

Similar to the roots, green pea stems showed significant increase in Zn accumulation upon exposure, with the exception of the ionic zinc treatment (**Figure 3**). Increased Zn accumulation was in the following order: at 250 mg/kg, with the increases relative to control stems expressed parenthetically: bulk (5x), bare (7x), doped (4.7x), and coated (7x); at 1000 mg/kg, the values were as follows: bulk (9x), bare (11x), doped (20x), and coated (9x) (**Figure 3**). Unlike the roots, at 250 mg/kg there were no significance differences across the nanoparticle treatments. However, at 1000 mg/kg, similar to the roots, the accumulation of Zn from the doped nanoparticle treatment was significantly greater than the other nanoparticles. Similar to the roots, all ZnO treatments exhibited concentration dependent increases in zinc at the two exposure levels; this trend was not evident for the ion exposure. In leaves, all amendments except the ion treatment showed 4.6–5.3 fold increases in zinc uptake with exposure at 250 mg/kg but there were no differences among the particle types. At 1000 mg/kg, only the nano and doped treatments resulted in values significantly above the controls (5.5–11 times; **Figure 4**).

No concentration dependent changes in Al and Si uptake were observed. Al and Si uptake by pea roots, stems, and leaves were largely unaffected across the different treatments. However, a few exceptions to this overall trend were noted. In stems, at the 1000 mg/kg treatments, doped and coated NPs accumulation showed 2.7 to 3.3 fold decreases in Al uptake compared to control (Figure S2). Similarly, silicon uptake into pea roots was decreased significantly at 250 mg/kg bare (2.6x) and doped (2x) treatments compared to control (Figure S3). In roots at 1000 mg/kg, bare nanoparticle exposure resulted in a 2.4 times decrease in Si content but no other differences were of statistical significance compared to control.

Previously, Mukherjee et al. reported the differential effects of bare-ZnO NPs, bulk ZnO, and iron doped ZnO (Fe@ZnO)

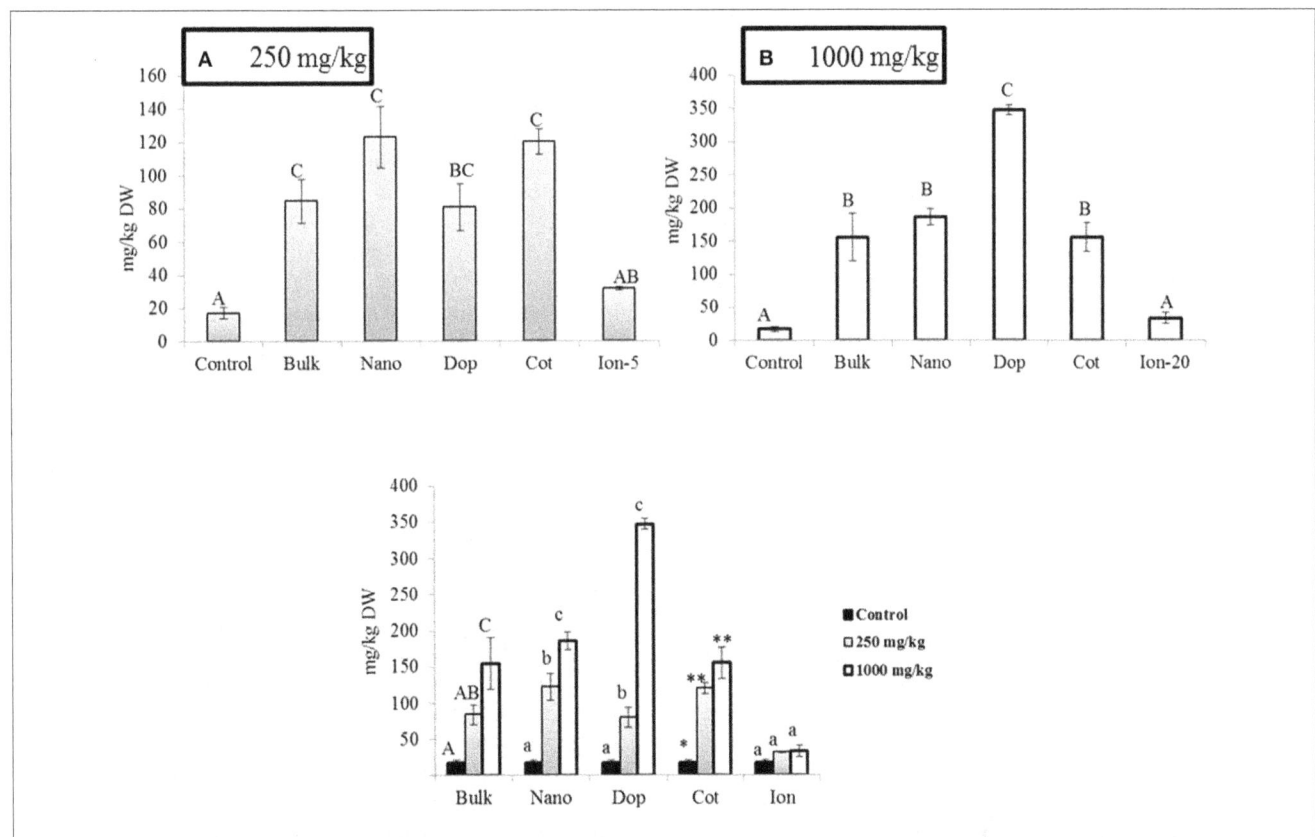

FIGURE 3 | Zinc bioaccumulation in stem tissues. (A,B) show the effects of different NP treatments at 250 and 1000 mg/kg exposure, respectively. The bottom graph shows the comparison among control, 250, and 1000 mg/kg treatments for each type of treatment separately. Bars are mean ± SE. Bars with same letters/symbols represent no statistical significance at $p \leq 0.05$. Upper case, lower case, and symbols are mutually exclusive.

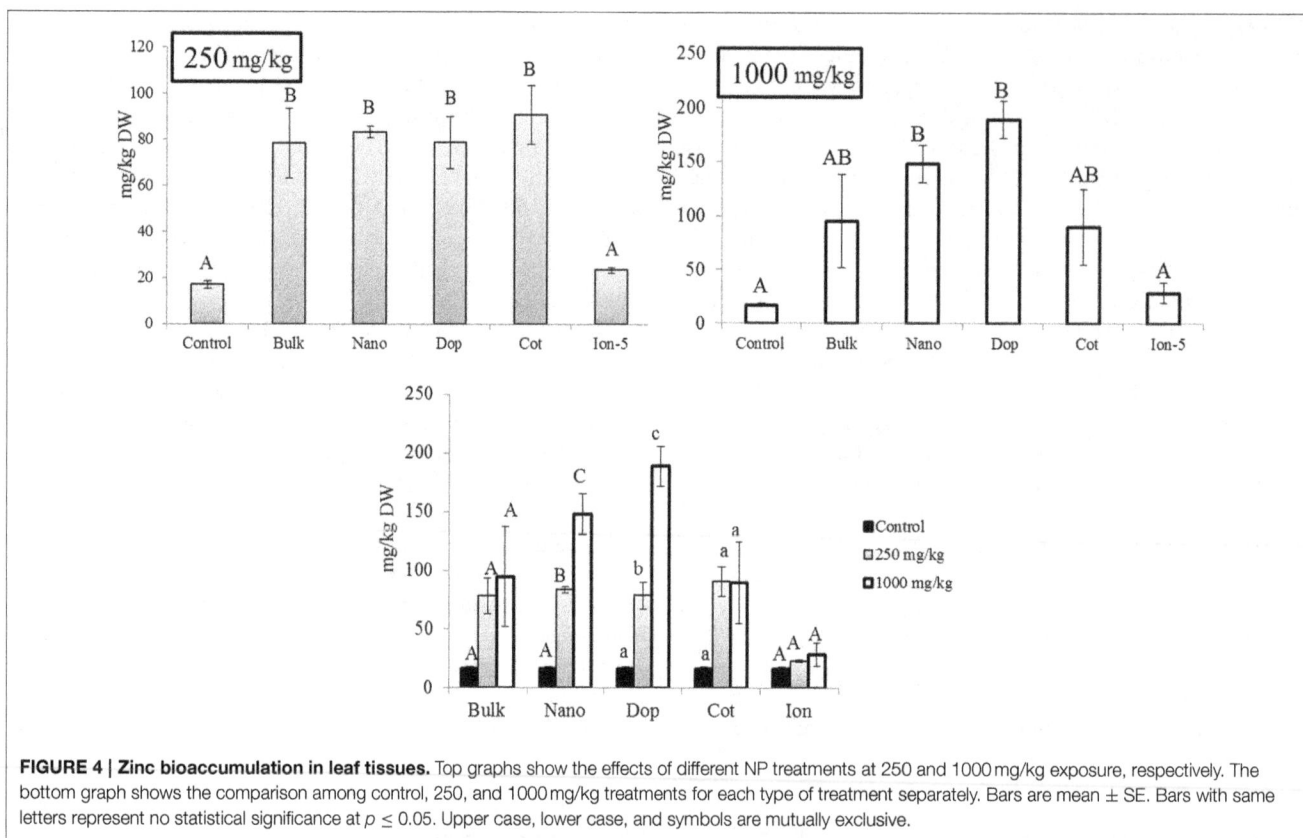

FIGURE 4 | Zinc bioaccumulation in leaf tissues. Top graphs show the effects of different NP treatments at 250 and 1000 mg/kg exposure, respectively. The bottom graph shows the comparison among control, 250, and 1000 mg/kg treatments for each type of treatment separately. Bars are mean ± SE. Bars with same letters represent no statistical significance at $p \leq 0.05$. Upper case, lower case, and symbols are mutually exclusive.

NPs on green peas cultivated in a growth chamber (Mukherjee et al., 2014a,b). At 250 mg/kg, in all the tissues, bulk and bare-ZnO NPs showed similar 3–6 fold increases in zinc uptake compared to control. However, in agreement with our current data, at higher concentrations (500 mg/kg) bare-ZnO NP showed 2.5 to 4 times higher Zn bioaccumulation compared to bulk treatment (Mukherjee et al., 2014a). Conversely, Mukherjee et al. (2014b) reported that roots of green pea exposed to 500 mg/kg Fe@ZnO showed lower Zn uptake (9x) compared to the NP treatment (12x). In addition, our current findings also indicate an opposite trend with a 36 fold increase in Zn uptake at higher concentrations of alumina doped, compared to bare-ZnO NPs. Therefore, changes in the doping agents (i.e., alumina or iron) can clearly change the uptake behavior of Zn in higher plants. Increases in element uptake from Al_2O_3@ZnO treatment compared to Fe@ZnO could be attributed to i) higher (more positive) surface charge due to alumina doping, which ensures greater adhesion/absorption to the root surface and ii) higher ion dissolution. At 1000 mg/kg, ZnO NPs@KH550 showed less (5x–9x) uptake across all tissues compared to all other particles. This might be attributed to larger size in soil and high negative surface change which exerts a repulsive force to the negatively charged root surface. Additionally, silicon has been proven to reduce the bioavailability of zinc ions in plants (Gu et al., 2011). Therefore, silicon released from the dissolution of coated NPs (KH550 or 3-aminopropyltriethoxysilane) may be another cause for reduced Zn uptake compared to bare and doped NPs. In case of bare ZnO NP, intermediate size, and zeta potential could be two of

the most important governing factors for keeping the extent of zinc uptake in-between doped and coated NPs. Although we do not know the reason for different Zn accumulation as a function of particle/exposure type, ion dissolution could be an important determinant too. Our data showed that ion release was greatest for doped, but the levels did not reach statistical significance (**Figure 1**).

It has been reported that Si is not an essential element for plant growth (Epstein, 1999). However, the presence of Si in the coated NP makes it important to quantify the Si uptake in different plant tissues. The soil type was sandy-loam with ∼84% sand. The loading of the coating agent KH550 is only 1 wt% of NP. The Si content in the coated NP is negligibly small compared to that of soil. A similar scenario exists for Al content in soil (>6000 mg/kg soil), which was much higher than that in alumina doped NPs (2 wt% of NP). Consequently, due to very high background values, it is difficult to identify the effects of coating and doping on Si and Al uptake, respectively. There was a numerical decrease in Al and Si (except 1000 mg/kg doped) content in roots. Similar results were reported by Wang et al. (2013) where higher concentration of Zn (500 mg Zn/kg soil), lowered the bioavailability of Al "due to formation of ZnAl-layered double hydroxide (ZnAl-LDH)." Another, reason could be the coexistence of Si and Al with ZnO NPs, followed by adsorption onto the clay minerals (Zhao et al., 2013a). Silicon induced apoplastic binding of Al could also explain the lower translocation of Al through the shoot system of the plant (Wang et al., 2004). Moreover, alumina hydrolysis occurs in the acidic media (Balint et al., 2001) and

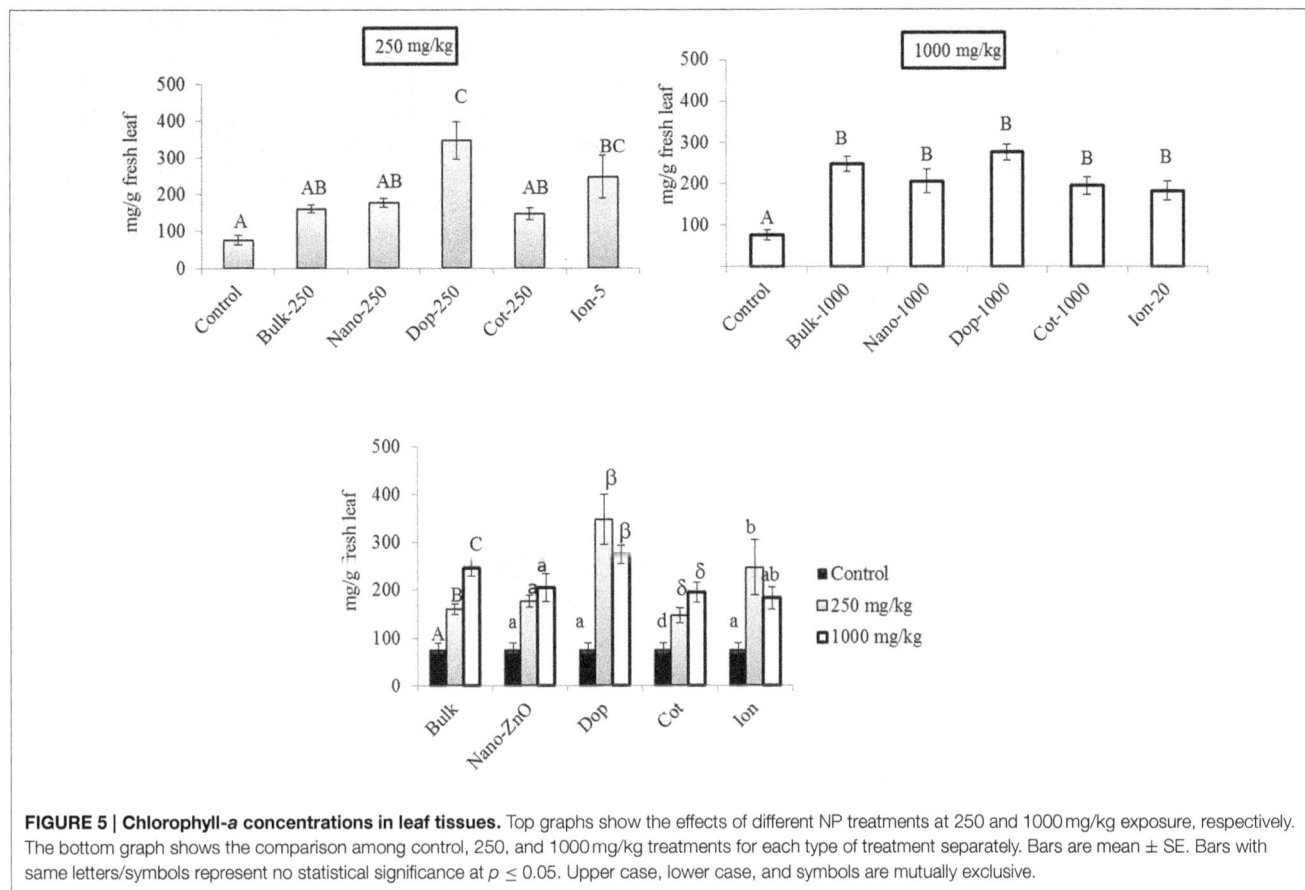

FIGURE 5 | Chlorophyll-*a* concentrations in leaf tissues. Top graphs show the effects of different NP treatments at 250 and 1000 mg/kg exposure, respectively. The bottom graph shows the comparison among control, 250, and 1000 mg/kg treatments for each type of treatment separately. Bars are mean ± SE. Bars with same letters/symbols represent no statistical significance at $p \leq 0.05$. Upper case, lower case, and symbols are mutually exclusive.

the pH of the test media was in the basic range. That could be another reason for little or no dissolution of alumina in the soil. Nonetheless, synchrotron studies are essential to establish the relationship between NP composition and bioavailability. Ongoing speciation studies are focused on identifying the modes of interaction among bare, coated, and doped ZnO NPs with soil particles and higher plants.

From the above results, it is clear that the phyto-toxicological response of green pea from exposure to these particles was very different. At the highest concentration, bare and doped NPs showed the greatest bioaccumulation in all the parts of the plant. However, no observable sign of toxicity was observed. Therefore, it is evident that the amount of zinc present in compound/particles is not the only determining factor for NP toxicity; the form (bare, coated, and doped) of ZnO NPs also plays a crucial role.

Chlorophyll and Carotenoids in Leaf

At 250 mg/kg, the amount of Chl-*a* increased with Zn exposure, although statistically significant increases were observed only with doped and ion treatments (3.2x–4.5x), compared to control (**Figure 5**). At 1000 mg/kg, all treatments resulted in 2.4–3.6 fold significant increases in Chl- *a*, compared to control, although there were no significant differences among the types of Zn amendments (**Figure 5**). Interestingly, there were no differences in the amount of chlorophyll-*b* (Chl-*b*) with Zn exposure (Figure

S4). Similar to the leaves at 250 mg/kg, the total carotenoid content trended upward with Zn exposure but only the doped and ion treatment enhancements (10x and 7x, respectively) were of statistical significance (Supporting Information Figure S5). The same trend was evident at 1000 mg/kg but only the bulk and doped particles resulted in statistically significant increases.

Our findings are in good agreement with previous reports. For example, Prasad et al. (2012) reported higher chlorophyll content in peanut at 1000 mg/kg ZnO NP (25 nm) treatment. No effect on Chl-*b* in corn was observed at 400 mg/kg ZnO (Zhao et al., 2013a). Zhao et al. reported an increasing trend (but statistically insignificant) in total chlorophyll content in cucumber (*Cucumis sativus*) treated with 400 and 800 mg/kg bare-ZnO NP in soil (Zhao et al., 2013b). Zinc is an essential micronutrient in plants (Hansch and Mendel, 2009) but above a "threshold" concentration, the element can generate toxicity in different plant species (Broadley et al., 2007; Zhao et al., 2013b). For instance, Kupper et al. (1996) reported that zinc can substitute the central metal atom magnesium (Mg^{2+}) in chlorophyll, causing a breakdown of the photosynthetic process. It has been reported that above 200 mg/kg (threshold value) in leaf tissues, *Bacopa monniera* and *Lolium perenne* L. *cv* Apollo showed phytotoxicological responses (Ali et al., 2000; Bonnet, 2000). In our study, the maximum Zn concentration in leaf was <300 mg/kg DW. This value is likely less than the threshold Zn tolerance value (not determined here) for green pea leaves under our particular growth condition. Moreover, at 1000 mg/kg, carotenoid concentrations increased up to 9 fold, compared to control. Carotenoids are photo-absorbing pigments which might have protected Chl-a from photooxidation (Lichtenthaler, 1987). In leaf tissues, the unchanged (Chl-b) or increased (Chl-a, carotenoids) pigment content clearly suggests little or no toxicity to photosynthetic pigment production with Zn exposure. However, these findings may not exclude the possibility of damage to other components of the photosynthetic apparatus, e.g., electron transport chains and photosynthetic enzyme activities. Further biochemical investigations are warranted to evaluate the effects of ZnO NP exposure on other complex photosynthetic components.

Effects of NPs on Green Pea Seed Quality

Exposure to Zn, regardless of type, generally had little effect on the green pea pod characteristics. The pod length, pod weight, and number of seeds per pod did not change as a function of treatment, with the exception of doped 250 mg/kg nanoparticles (data not shown). Here, the number of seeds per pod decreased

by 33% compared to that of bare ZnO NP treatment. Unlike bulk treatments, bare, doped, and coated NPs showed increase in Zn uptake at 250 mg/kg treatment, compared to control (**Figure 6**). At 1000 mg/kg, the Zn content increased by 2–2.5 times in all NP and bulk treatments as compared to control. The ionic treatments did not show any significant change in Zn uptake at 5 mg/kg or 20 mg/kg. Concentrations of Cu, Mg, and K in the seed did not change significantly with Zn exposure (data not shown). The Fe level was significantly elevated by the coated (250 mg/kg) and doped (1000 mg/kg) treatments. In addition, at 1000 mg/kg coated treatment, P and Mn were significantly increased (**Figures 6B–D**).

Overall, Zn exposure, regardless of type or concentration, had little impact on the protein or carbohydrate profile of the green pea seeds. The amount of acid-soluble (glutelin), salt-soluble (globulin), water-soluble (albumin), and alcohol-soluble (prolamin) protein fractions remained unaltered in all treatments (Figure S6). There was a decrease in glutelin amount (50%) at 1000 mg/kg doped treatment, compared to control, but due to large variability and modest replicate numbers, the

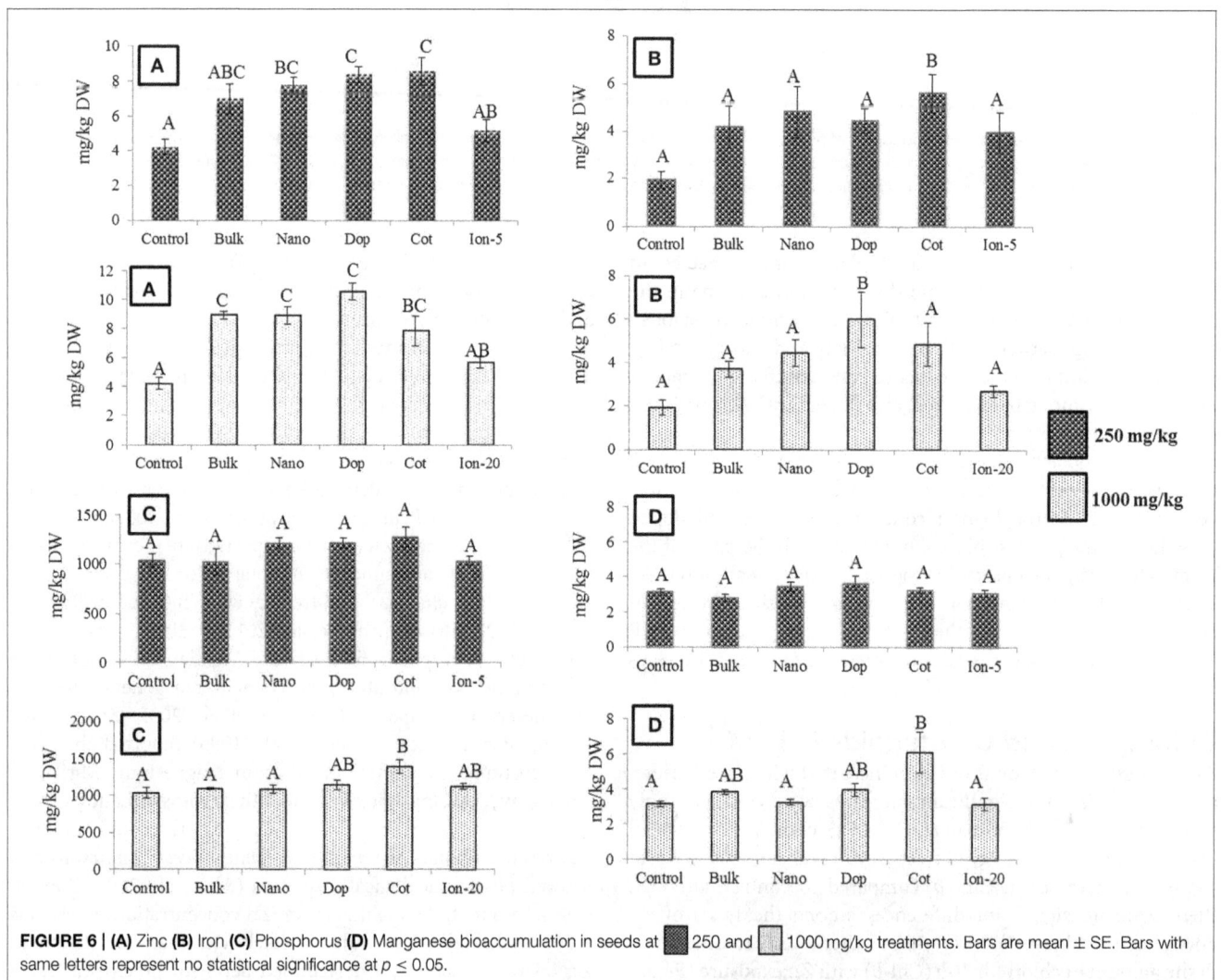

FIGURE 6 | (A) Zinc **(B)** Iron **(C)** Phosphorus **(D)** Manganese bioaccumulation in seeds at ▓ 250 and ▒ 1000 mg/kg treatments. Bars are mean ± SE. Bars with same letters represent no statistical significance at $p \leq 0.05$.

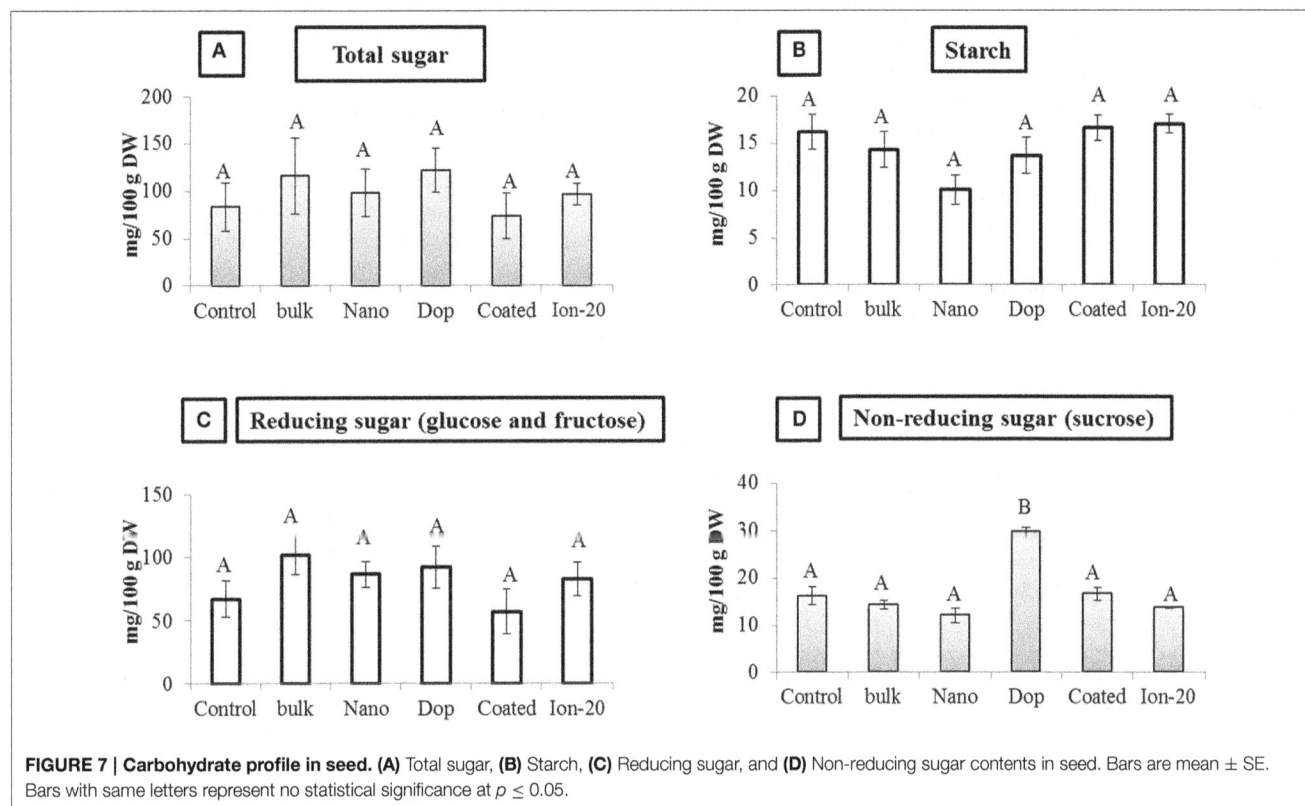

FIGURE 7 | Carbohydrate profile in seed. (A) Total sugar, **(B)** Starch, **(C)** Reducing sugar, and **(D)** Non-reducing sugar contents in seed. Bars are mean ± SE. Bars with same letters represent no statistical significance at $p \leq 0.05$.

decrease was statistically insignificant. The amount of total sugar, starch, reducing sugars (glucose and fructose), and non-reducing sugar (sucrose) also remained largely unaltered. The exception was the 1000 mg/kg doped NP treatment where the sucrose content of pea seeds was significantly increased by 1.8 fold compared to all other treatments (**Figure 7**). Higher sucrose concentration in green pea at 1000 mg/kg doped treatment may be less of concern for seed quality but more problematic as an indicator of plant stress (Koch, 2004; Levitz, 2004; Zhao et al., 2014b). It has been reported that reducing and non-reducing sugars can contribute to the signaling pathways related to stress (Koch, 2004; Levitz, 2004; Zhao et al., 2014b).

As mentioned earlier, green pea plants were chosen to evaluate the effects of NP exposure because of the crop worldwide production and consumption. Green pea seeds are rich in protein, certain minerals, and vitamins and have modest calorific content (Iqbal et al., 2006). Raw green peas are excellent source of vitamin K, C, B1, B9, A, B6, B3, and B2. The crop is also rich in Mn, P, Mg, Cu, Fe, Zn, and K (Iqbal et al., 2006). Among major legumes (i.e., lentil, green peas, and common bean, among others), green pea is the second best protein source (24.9/100 g raw green pea, Iqbal et al., 2006). It has been reported that a cup of raw green peas (=137.75 g) provides 30.3% fiber, 14.7% of protein, and only 6% calories as measured against typical daily nutritional values (Iqbal et al., 2006). There are very few reports available in the literature investigating the effect of nanoparticle exposure in soil under field-like conditions on pea seed quality. Several similar studies have been published focusing on bare-ZnO and CeO_2 NPs exposure. For instance,

Rico et al. (2014) treated wheat plants at 0, 125, 250, and 500 mg/kg soil, and found changes in nutrient content (S and Mn), amino acid, and fatty acid profiles upon exposure to CeO_2. Our findings agree well with Priester et al. (2012) where a 2.5 fold increase in zinc uptake by soybean pods was observed upon exposure to 500 mg/kg bare-ZnO NP as compared to controls. Peralta-Videa et al. (2014) found increased zinc concentration in soybean pods at 50, 100, and 500 mg/kg bare-ZnO treatments. Moreover, at "medium" concentration (100 mg/kg), significant bioaccumulation of Cu and Mn in soybean pods were also observed. Similarly, Zhao et al. (2014b) reported that treatment with 400 and 800 mg ZnO NP/kg soil resulted in changes of micronutrient and carbohydrate content without any alteration in protein profile of cucumber fruit. Elevated levels of Zn in the seeds was likely due to the enhanced mobility of Zn^{2+} ions (Broadley et al., 2007; Wang et al., 2013) generated from the dissolution of NPs in soil. In terms of cellular uptake, there are different transporter genes and pathways present, which regulate the mobility of different metals across the plasma membrane. For example, Mn transport is regulated by natural resistance-associated macrophage protein (*Nramp*) transporters and zinc-regulated transporter/iron-regulated transporter (*ZRT/IRT1*)-related protein (*ZIP*) transporters, among others (Pittman, 2005). Currently, we have no information regarding the interaction among specific metal transporters and different NPs. As such, characterizing potential correlations between macro/macro nutrient uptake in seed with different NP exposure is too speculative with the current knowledge base. However, considering all the above data, it can be said that the mineral/nutrient concentration in the edible tissue was affected differentially by nanoparticle type, with the coated and doped ZnO exerting the greatest effects. Similarly, under high dose exposure (1000 mg/kg), doped NP altered (1.8 times higher) the carbohydrate profile (sucrose) of the seed. The implications of these NP-induced changes in fruit content/quality are currently unknown but are the subject of intense investigation.

In summary, our study investigated the comparative phytotoxicity of bare-ZnO NPs, $Al_2O_2@ZnO$, and ZnO@KH550 NPs on green pea plants in terms of biomass, element bioaccumulation, changes in leaf photosynthetic pigment, along with the changes in seed quality. Our results confirmed that, in spite of possessing larger size in the commercial form, alumina doped ZnO NPs (15 nm) have greater effects on plant and seed quality, compared to bare-ZnO NPs (10 nm). The seed quality was affected most by the doped NPs at 1000 mg/kg where nutrient content and carbohydrate profile (sucrose) changed. It was suggested in the literature that doping (Fe doped ZnO NPs) could decrease the phytotoxicological effects of bare-ZnO NPs to higher plants (Mukherjee et al., 2014b). Nevertheless, our findings clearly demonstrate that $Al_2O_3@ZnO$ NP treatments exerted more negative effects on green pea when compared to bare and coated ZnO NP. Therefore, the doping agents certainly play a crucial role in the phytotoxicological responses of NP exposure to the higher plants. Although, the mechanism is unknown, ion release and coating facilitated uptake of intact NPs are possible pathways of concern. Additional study into the broader implications of NP doping and coating type on food safety and on the fate and disposition of these materials in the environment is warranted.

SUPPORTING INFORMATION

Two tables listing details on particle characterization and elemental composition of native and 1:1 soil. Six figures describe different physiological and biochemical parameters of root, stem, leaf, and seeds.

ACKNOWLEDGMENTS

This work was supported by funding from the National Science Foundation and the Environmental Protection Agency under Cooperative Agreement Number DBI-0830117, as well as USDA-AFRI (#2011-67006-30181). Any opinions, findings, and conclusions or recommendations expressed in this material are those of the author(s) and do not necessarily reflect the views of the National Science Foundation or the Environmental Protection Agency. This work has not been subjected to EPA review and no official endorsement should be inferred. This work was supported by Grant 2G12MD007592 from the National Institutes on Minority Health and Health Disparities (NIMHD), a component of the National Institutes of Health (NIH). The authors also acknowledge the USDA grant number 2011-38422-30835 and the NSF Grant # CHE-0840525 and DBI-1429708. JG acknowledges the Dudley family for the Endowed Research Professorship in Chemistry and the Academy of Applied Science/US Army REAP program at UTEP, grant # W11NF-10-2-0076, sub-grant 13-7. The authors also acknowledge Dr. Chuan (River) Xiao biochemistry research laboratory for spectroscopic measurements. We greatly appreciate generous help from Mr. Prithish Banerjee, Department of Statistics, West Virginia University with statistical analysis.

REFERENCES

Abdolmaleki, A., Mallakpour, S., and Borandeh, S. (2012). Effect of silane-modified ZnO on morphology and properties of bionanocomposites based on poly(ester-amide) containing tyrosine linkages. *Poly. Bull.* 69, 15–28. doi: 10.1007/s00289-011-0685-7

Ali, G., Srivastava, P. S., and Iqbal, M. (2000). Influence of cadmium and zinc on growth and photosynthesis of *Bacopa monniera* cultivated *in vitro*. *Biol. Plant.* 43, 599–601. doi: 10.1023/a:1002852016145

Balint, I., Miyazaki, A., and Aika, K. (2001). Alumina Dissolution during Impregnation with $PdCl_4^{2-}$ in the Acid pH Range. *Chem. Mater.* 13, 932–938. doi: 10.1021/cm000693i

Bandyopadhyay, S., Peralta-Videa, J. R., and Gardea-Torresdey, J. L. (2013). Advanced analytical techniques for the measurement of nanomaterials in food and agricultural samples: a review. *Environ. Eng. Sci.* 30, 118–125. doi: 10.1089/ees.2012.0325

Bandyopadhyay, S., Peralta-Videa, J. R., Hernandez-Viezcas, J. A., Montes, M. O., Keller, A. A., and Gardea-Torresdey, J. L. (2012a). Microscopic and spectroscopic methods applied to the measurements of nanoparticles in the environment. *Appl. Spectros. Rev.* 47, 180–206. doi: 10.1080/05704928.2011.637186

Bandyopadhyay, S., Peralta-Videa, J. R., Plascencia-Villa, G., Jose-Yacaman, M., and Gardea-Torresdey, J. L. (2012b). Comparative toxicity assessment of CeO_2 and ZnO nanoparticles towards Sinorhizobium meliloti, a symbiotic alfalfa associated bacterium: use of advanced microscopic and spectroscopic techniques. *J. Hazard. Mat.* 241, 379–386. doi: 10.1016/j.jhazmat.2012.09.056

Bandyopadhyay, S., Plascencia-Villa, G., Mukherjee, A., Rico, C. M., Jose-Yacaman, M., Peralta-Videa, J. R., et al. (2015). Comparative phytotoxicity of ZnO NPs, bulk ZnO, and ionic zinc onto the alfalfa plants symbiotically associated with Sinorhizobium meliloti in soil. *Sci. Total. Environ.* 515-516, 60–69. doi: 10.1016/j.scitotenv.2015.02.014

Bian, S. W., Mudunkotuwa, I. A., Rupasinghe, T., and Grassian, V. H. (2011). Aggregation and dissolution of 4 nm ZnO nanoparticles in aqueous environments: influence of pH, Ionic Strength, Size, and Adsorption of Humic Acid. *Langmuir* 27, 10, 6059–6068. doi: 10.1021/la200570n

Bonnet, M. (2000). Effects of zinc and influence of Acremonium lolii on growth parameters, chlorophyll a fluorescence and antioxidant enzyme activities of ryegrass (Lolium perenne L. cv Apollo). *J. Exp. Bot.* 51, 945–953. doi: 10.1093/jexbot/51.346.945

Bradford, M. M. (1976). A rapid and sensitive method for the quantitation of microgram quantities of protein utilizing the principle of protein-dye binding. *Anal. Biochem.* 72, 248–254.

Broadley, M. R., White, P. J., Hammond, J. P., Zelko, I., and Lux, A. (2007). Zinc in plants. *New Phytol.* 173, 677–702. doi: 10.1111/j.1469-8137.2007.01996.x

Chen, C. H., and Bushuk, W. (1970). Nature of proteins in triticale and its parental species.1. solubility characteristics and amino acid composition of endosperm proteins. *Can. J. Plant Sci.* 50, 9–14.

De La Rosa, G., Lopez-Moreno, M. L., Hernandez-Viezcas, J., Montes, M. O., Peralta-Videa, J. R., and Gardea-Torresdey, J. L. (2011). Toxicity and biotransformation of ZnO nanoparticles in the desert plants Prosopis juliflora-velutina, Salsola tragus and Parkinsonia florida. *Int. J. Nanotechnol.* 8, 492–506. doi: 10.1504/IJNT.2011.040190

Dhiman, P., Batoo, K. M., Kotnala, R. K., and Singh, M. (2012). Fe-doped ZnO nanoparticles synthesised by solution combustion method. *Micro. Nano. Lett.* 7, 1333–1335. doi: 10.1049/mnl.2012.0862

Dimkpa, C. O., Mclean, J. E., Britt, D. W., and Anderson, A. J. (2012). Bioactivity and biomodification of Ag, ZnO, and CuO nanoparticles with relevance to plant performance in agriculture. *Ind. Biotechnol.* 8, 344–357. doi: 10.1089/ind.2012.0028

Du, W. C., Sun, Y. Y., Ji, R., Zhu, J. G., Wu, J. C., and Guo, H. Y. (2011). TiO$_2$ and ZnO nanoparticles negatively affect wheat growth and soil enzyme activities in agricultural soil. *J. Environ. Mon.* 13, 822–828. doi: 10.1039/C0em00611d

Dubois, M., Gilles, K. A., Hamilton, J. K., Rebers, P. A., and Smith, F. (1956). Colorimetric method for determination of sugars and related substances. *Anal. Chem.* 28, 350–356. doi: 10.1021/Ac60111a017

Epstein, E. (1999). Silicon. *Ann. Rev. Plant Physiol. Plant Mol. Biol* 50, 641–664. doi: 10.1146/annurev.arplant.50.1.641

Gaiser, B. K., Fernandes, T. F., Jepson, M. A., Lead, J. R., Tyler, C. R., Baalousha, M., et al. (2012). Interspecies comparisons on the uptake and toxicity of silver and cerium dioxide nanoparticles. *Environ. Toxicol. Chem.* 31, 144–154. doi: 10.1002/Etc.703

Gardea-Torresdey, J. L., Rico, C. M., and White, J. C. (2014). Trophic transfer, transformation, and impact of engineered nanomaterials in terrestrial environments. *Environ. Sci. Technol.* 48, 2526–2540. doi: 10.1021/Es4050665

Ghormade, V., Deshpande, M. V., and Paknikar, K. M. (2011). Perspectives for nano-biotechnology enabled protection and nutrition of plants. *Biotechnol. Advan.* 29, 792–803. doi: 10.1016/j.biotechadv.2011.06.007

Gu, H.-H., Zhan, S.-S., Wang, S.-Z., Tang, Y.-T., Chaney, R. L., Fang, X.-H., et al. (2011). Silicon-mediated amelioration of zinc toxicity in rice (Oryza sativa L.) seedlings. *Plant Soil* 350, 193–204. doi: 10.1007/s11104-011-0894-8

Hansch, R., and Mendel, R. R. (2009). Physiological functions of mineral micronutrients (Cu, Zn, Mn, Fe, Ni, Mo, B, Cl). *Curr. Opin. Plant Biol.* 12, 259–266. doi: 10.1016/j.pbi.2009.05.006

Hawthorne, J., Musante, C., Sinha, S. K., and White, J. C. (2012). Accumulation and phytotoxicity of engineered nanoparticles to Cucurbita Pepo. *Int. J. Phytoremed.* 14, 429–442. doi: 10.1080/15226514.2011.620903

Hendry, G. A., and Jones, O. T. (1980). Haems and chlorophylls: comparison of function and formation. *J. Med. Genet.* 17, 1–14.

Hernandez-Viezcas, J. A., Castillo-Michel, H., Andrews, J. C., Cotte, M., Rico, C., Peralta-Videa, J. R., et al. (2013). In Situ Synchrotron X-ray fluorescence mapping and speciation of CeO$_2$ and ZnO Nanoparticles in Soil Cultivated Soybean (Glycine max). *Acs Nano* 7, 1415–1423. doi: 10.1021/Nn305196q

Iqbal, A., Khalil, I. A., Ateeq, N., and Khan, M. S. (2006). Nutritional quality of important food legumes. *Food Chem.* 97, 331–335. doi: 10.1016/j.foodchem.2005.05.011

Kahru, A., and Dubourguier, H. C. (2010). From ecotoxicology to nanoecotoxicology. *Toxicology* 269, 105–119. doi: 10.1016/j.tox.2009.08.016

Kim, S., Kim, J., and Lee, I. (2011). Effects of Zn and ZnO nanoparticles and Zn^{2+} on soil enzyme activity and bioaccumulation of Zn in Cucumis sativus. *Chem. Ecol.* 27, 49–55. doi: 10.1080/02757540.2010.529074

Koch, K. (2004). Sucrose metabolism: regulatory mechanisms and pivotal roles in sugar sensing and plant development. *Curr. Opin. Plant Biol.* 7, 235–246. doi: 10.1016/j.pbi.2004.03.014

Kupper, H., Kupper, F. C., and Spiller, M. (1996). Environmental relevance of heavy metal-substituted chlorophylls using the example of water plants. *J. Exp. Bot.* 47, 259–266. doi: 10.1093/Jxb/47.2.259

Lee, W. M., An, Y. J., Yoon, H., and Kweon, H. S. (2008). Toxicity and bioavailability of copper nanoparticles to the terrestrial plants mung bean (Phaseolus radiatus) and wheat (Triticum aestivum): plant agar test for water-insoluble nanoparticles. *Environ. Toxicol. Chem.* 27, 1915–1921. doi: 10.1897/07-481.1

Levitz, S. A. (2004). Interactions of Toll-like receptors with fungi. *Microbes Infect.* 6, 1351–1355. doi: 10.1016/j.micinf.2004.08.014

Lichtenthaler, H. K. (1987). Chlorophylls and carotenoids - pigments of photosynthetic biomembranes. *Methods Enzymol.* 148, 350–382.

Lin, D. H., and Xing, B. S. (2007). Phytotoxicity of nanoparticles: inhibition of seed germination and root growth. *Environ. Pollut.* 150, 243–250. doi: 10.1016/j.envpol.2007.01.016

Lin, D. H., and Xing, B. S. (2008). Root uptake and phytotoxicity of ZnO nanoparticles. *Environ. Sci. Technol.* 42, 5580–5585. doi: 10.1021/Es800422x

Liu, R., and Lal, R. (2015). Potentials of engineered nanoparticles as fertilizers for increasing agronomic productions. *Sci. Total Environ.* 514, 131–139. doi: 10.1016/j.scitotenv.2015.01.104

Lopez-Moreno, M. L., De La Rosa, G., Hernandez-Viezcas, J. A., Castillo-Michel, H., Botez, C. E., Peralta-Videa, J. R., et al. (2010). Evidence of the Differential Biotransformation and Genotoxicity of ZnO and CeO$_2$ Nanoparticles on Soybean (Glycine max) Plants. *Environ. Sci. Technol.* 44, 7315–7320. doi: 10.1021/Es903891g

Maynard, A., and Evan, M. (2006). *The Nanotechnology Consumers Products Inventory* [Online]. Available online at: http://www.nanotechproject.org/process/files/2753/consumer_product_inventory_analysis_handout.pdf [Accessed 06/30/2014]

Montalvo, D., McLaughlin, M. J., and Degryse, F. (2015). Efficacy of Hydroxyapatite Nanoparticles as Phosphorus Fertilizer in Andisols and Oxisols. *Soil Sci. Soc. Am. J.* 79, 2, 551–558. doi: 10.2136/sssaj2014.09.0373

Mukherjee, A., Peralta-Videa, J. R., Bandyopadhyay, S., Rico, C. M., Zhao, L. J., and Gardea-Torresdey, J. L. (2014a). Physiological effects of nanoparticulate ZnO in green peas (Pisum sativum L.) cultivated in soil. *Metallomics* 6, 132–138. doi: 10.1039/C3mt00064h

Mukherjee, A., Pokhrel, S., Bandyopadhyay, S., Madler, L., Peralta-Videa, J. R., and Gardea-Torresdey, J. L. (2014b). A soil mediated phyto-toxicological study of iron doped zinc oxide nanoparticles (Fe@ZnO) in green peas (Pisum sativum L.). *Chem. Eng. J.* 258, 394–401. doi: 10.1016/j.cej.2014.06.112

Navarro, E., Baun, A., Behra, R., Hartmann, N. B., Filser, J., Miao, A. J., et al. (2008). Environmental behavior and ecotoxicity of engineered nanoparticles to algae, plants, and fungi. *Ecotoxicology* 17, 372–386. doi: 10.1007/s10646-008-0214-0

Nel, A., Xia, T., Madler, L., and Li, N. (2006). Toxic potential of materials at the nanolevel. *Science* 311, 622–627. doi: 10.1126/science.1114397

Oberdorster, G., Oberdorster, E., and Oberdorster, J. (2005). Nanotoxicology: an emerging discipline evolving from studies of ultrafine particles. *Environ. Health Perspect.* 113, 823–839. doi: 10.1289/Ehp.7339

Ozgur, U., Alivov, Y. I., Liu, C., Teke, A., Reshchikov, M. A., Dogan, S., et al. (2005). A comprehensive review of ZnO materials and devices. *J. Appl. Physics* 98, 041301. doi: 10.1063/1.1992666

Packer, A. P., Lariviere, D., Li, C. S., Chen, M., Fawcett, A., Nielsen, K., et al. (2007). Validation of an inductively coupled plasma mass spectrometry (ICP-MS) method for the determination of cerium, strontium, and titanium in ceramic materials used in radiological dispersal devices (RDDs). *Anal. Chim. Acta* 588, 166–172. doi: 10.1016/j.aca.2007.02.024

Peralta-Videa, J. R., Hernandez-Viezcas, J. A., Zhao, L., Diaz, B. C., Ge, Y., Priester, J. H., et al. (2014). Cerium dioxide and zinc oxide nanoparticles alter the nutritional value of soil cultivated soybean plants. *Plant Physiol. Biochem.* 80, 128–135. doi: 10.1016/j.plaphy.2014.03.028

Pittman, J. K. (2005). Managing the manganese: molecular mechanisms of manganese transport and homeostasis. *New Phytol.* 167, 733–742. doi: 10.1111/j.1469-8137.2005.01453.x

Prasad, T. N. V. K.V., Sudhakar, P., Sreenivasulu, Y., Latha, P., Munaswamy, V., Reddy, K. R., et al. (2012). Effect of Nanoscale Zinc Oxide Particles on the Germination, Growth and Yield of Peanut. *J. Plant Nutr.* 35, 905–927. doi: 10.1080/01904167.2012.663443

Priester, J. H., Ge, Y., Mielke, R. E., Horst, A. M., Moritz, S. C., Espinosa, K., et al. (2012). Soybean susceptibility to manufactured nanomaterials with evidence for food quality and soil fertility interruption. *Proc. Natl. Acad. Sci. U.S.A.* 109, E2451–E2456. doi: 10.1073/pnas.1205431109

Rebeiz, C. A., and Castelfranco, P. A. (1973). Protochlorophyll and chlorophyll biosynthesis in cell-free systems from higher-plants. *Ann. Rev. Plant Physiol. Plant Mol. Biol.* 24, 129–172. doi: 10.1146/annurev.pp.24.060173.0 01021

Rico, C. M., Lee, S. C., Rubenecia, R., Mukherjee, A., Hong, J., Peralta-Videa, J. R., et al. (2014). Cerium oxide nanoparticles impact yield and modify nutritional parameters in wheat (*Triticum aestivum* L.). *J. Agr. Food Chem.* 62, 9669–9675. doi: 10.1021/Jf503526r

Roco, M. C. (2011). "The long view of nanotechnology development: the national nanotechnology initiative at 10 years," in *Nanotechnology Research Directions for Societal Needs in 2020, Retrospective and Outlook,* Vol. 1, eds M. C. Roco, C. A. Mirkin, and M. C. Hersam (Amsterdam: Springer), 1–28.

Servin, A., Elmer, W., Mukherjee, A., De la Torre-Roche, R., Hamdi, H., White, J. C., et al. (2015). A review of the use of engineered nanomaterials to suppress plant disease and enhance crop yield. *J. Nanoparticle Res.* 17, 92. doi: 10.1007/s11051-015-2907-7

Sinha, R., Karan, R., Sinha, A., and Khare, S. K. (2011). Interaction and nanotoxic effect of ZnO and Ag nanoparticles on mesophilic and halophilic bacterial cells. *Biores. Technol.* 102, 1516–1520. doi: 10.1016/j.biortech.2010.07.117

Somogyi, M. (1952). Notes on Sugar Determination. *J. Biol. Chem.* 195, 19–23.

Szabo, T., Nemeth, J., and Dekany, I. (2003). Zinc oxide nanoparticles incorporated in ultrathin layer silicate films and their photocatalytic properties. *Colloids Surfaces Physicochem. Eng. Aspects* 230, 23–35. doi: 10.1016/j.colsurfa.200 3.09.0

Thandavan, T. M. K., Gani, S. M. A., San Wong, C., and Md Nor, R. (2015). Enhanced Photoluminescence and raman properties of Al-Doped ZnO nanostructures prepared using thermal chemical vapor deposition of methanol assisted with heated brass. *PLoS ONE* 10:e0121756. doi: 10.1371/journal.pone.0121756

Venkatesan, M., Fitzgerald, C. B., Lunney, J. G., and Coey, J. M. D, (2004). Anisotropic ferromagnetism in substituted zinc oxide. *Phys. Rev. Lett.* 93:177206. doi: 10.1103/PhysRevLett.93.177206

Verma, S., and Dubey, R. S. (2001). Effect of cadmium on soluble sugars and enzymes of their metabolism in rice. *Biol. Plant.* 44, 117–123. doi: 10.1023/A:1017938809311

Wang, J., Yu, J., Zhu, X., and Kong, X. Y. (2012). Preparation of hollow TiO_2 nanoparticles through TiO_2 deposition on polystyrene latex particles and characterizations of their structure and photocatalytic activity. *Nanoscale Res Lett.* 7:646. doi: 10.1186/1556-276X-7-646

Wang, P., Menzies, N. W., Lombi, E., Mckenna, B. A., Johannessen, B., Glover, C. J., et al. (2013). Fate of ZnO Nanoparticles in Soils and Cowpea (*Vigna unguiculata*). *Environ. Sci. Technol.* 47, 13822–13830. doi: 10.1021/Es403466p

Wang, Y. X., Stass, A., and Horst, W. J. (2004). Apoplastic binding of aluminum is involved in silicon-induced amelioration of aluminum toxicity in maize. *Plant Physiol.* 136, 3762–3770. doi: 10.1104/pp.104.045005

Wellburn, A. R. (1994). The spectral determination of chlorophylls a and b, as well as total carotenoids, using various solvents with spectrophotometers of different resolution. *J. Plant Physiol.* 144, 307–313. doi: 10.1016/s0176-1617(11)81192-2

Xia, T., Kovochich, M., Brant, J., Hotze, M., Sempf, J., Oberley, T., et al. (2006). Comparison of the abilities of ambient and manufactured nanoparticles to induce cellular toxicity according to an oxidative stress paradigm. *Nano Lett.* 6, 1794–1807. doi: 10.1021/Nl061025k

Zhang, W. Y., He, D. K., Liu, Z. Z., Sun, L. J., and Fu, Z. X. (2010). Preparation of transparent conducting Al-doped ZnO thin films by single source chemical vapor deposition. *Optoelectronics Advan. Materials Rapid Commun.* 4, 1651–1654. doi: 10.1103/PhysRevLett.93.177206

Zhao, L. J., Hernandez-Viezcas, J. A., Peralta-Videa, J. R., Bandyopadhyay, S., Peng, B., Munoz, B., et al. (2013a). ZnO nanoparticle fate in soil and zinc bioaccumulation in corn plants (Zea mays) influenced by alginate. *Environ. Sci. Process. Impacts* 15, 260–266. doi: 10.1039/C2em30610g

Zhao, L. J., Peralta-Videa, J. R., Peng, B., Bandyopadhyay, S., Corral-Diaz, B., Osuna-Avila, P., et al. (2014a). Alginate modifies the physiological impact of CeO_2 nanoparticles in corn seedlings cultivated in soil. *J. Environ. Sci. China* 26, 382–389. doi: 10.1016/S1001-0742(13)60559-8

Zhao, L. J., Peralta-Videa, J. R., Rico, C. M., Hernandez-Viezcas, J. A., Sun, Y. P., Niu, G. H., et al. (2014b). CeO_2 and ZnO nanoparticles change the nutritional qualities of cucumber (*Cucumis sativus*). *J. Agr. Food Chem.* 62, 2752–2759. doi: 10.1021/Jf405476u

Zhao, L. J., Sun, Y. P., Hernandez-Viezcas, J. A., Servin, A. D., Hong, J., Niu, G. H., et al. (2013b). Influence of CeO_2 and ZnO nanoparticles on cucumber physiological markers and bioaccumulation of Ce and Zn: a life cycle study. *J. Agri. Food Chem.* 61, 11945–11951. doi: 10.1021/jf404328e

Conflict of Interest Statement: The authors declare that the research was conducted in the absence of any commercial or financial relationships that could be construed as a potential conflict of interest.

The handling editor Nelson Marmiroli declares that, despite hosting the Research Topic "Nanotoxicology and environmental risk assessment of engineered nanomaterials (ENMs) in plants" together with co-author Jason C. White, the review process was handled objectively.

16

Short day transcriptomic programming during induction of dormancy in grapevine

Anne Y. Fennell[1]*, Karen A. Schlauch[2], Satyanarayana Gouthu[3], Laurent G. Deluc[3], Vedbar Khadka[1], Lekha Sreekantan[1], Jerome Grimplet[4], Grant R. Cramer[2] and Katherine L. Mathiason[1]

[1] Northern Plains BioStress Laboratory, Plant Science Department, South Dakota State University, Brookings, SD, USA, [2] Department of Biochemistry and Molecular Biology, University of Nevada, Reno, Reno, NV, USA, [3] Department of Horticulture, Oregon State University, Corvallis, OR, USA, [4] Instituto de Ciencias de la Vid y del Vino (CSIC, Universidad de La Rioja, Gobierno de La Rioja), Logroño, Spain

Edited by:
Glenn Thomas Howe,
Oregon State University, USA

Reviewed by:
Jorunn Elisabeth Olsen,
Norwegian University of Life Sciences,
Norway
Amy Brunner,
Virginia Polytechnic Institute and State
University, USA
Jason P. Londo,
United States Department of
Agriculture-Agricultural Research
Service, USA

*Correspondence:
Anne Y. Fennell
anne.fennell@sdstate.edu

Bud dormancy in grapevine is an adaptive strategy for the survival of drought, high and low temperatures and freeze dehydration stress that limit the range of cultivar adaptation. Therefore, development of a comprehensive understanding of the biological mechanisms involved in bud dormancy is needed to promote advances in selection and breeding, and to develop improved cultural practices for existing grape cultivars. The seasonally indeterminate grapevine, which continuously develops compound axillary buds during the growing season, provides an excellent system for dissecting dormancy, because the grapevine does not transition through terminal bud development prior to dormancy. This study used gene expression patterns and targeted metabolite analysis of two grapevine genotypes that are short photoperiod responsive (*Vitis riparia*) and non-responsive (*V. hybrid*, Seyval) for dormancy development to determine differences between bud maturation and dormancy commitment. Grapevine gene expression and metabolites were monitored at seven time points under long (LD, 15 h) and short (SD, 13 h) day treatments. The use of age-matched buds and a small (2 h) photoperiod difference minimized developmental differences and allowed us to separate general photoperiod from dormancy specific gene responses. Gene expression profiles indicated three distinct phases (perception, induction and dormancy) in SD-induced dormancy development in *V. riparia*. Different genes from the NAC DOMAIN CONTAINING PROTEIN 19 and WRKY families of transcription factors were differentially expressed in each phase of dormancy. Metabolite and transcriptome analyses indicated ABA, trehalose, raffinose and resveratrol compounds have a potential role in dormancy commitment. Finally, a comparison between *V. riparia* compound axillary bud dormancy and dormancy responses in other species emphasized the relationship between dormancy and the expression of *RESVERATROL SYNTHASE* and genes associated with C3HC4-TYPE RING FINGER and NAC DOMAIN CONTAINING PROTEIN 19 transcription factors.

Keywords: *Vitis riparia*, Seyval, bud, VitisNet, ABA, resveratrol, trehalose, raffinose

INTRODUCTION

Wild grapevine species (*Vitaceae*) are predominately native to the northern hemisphere; however, the production of grape cultivars is widely distributed, and the grapevine is one of the temperate fruit crops most frequently damaged by winter freezing temperatures. Development of dormancy allows the grapevine to better tolerate the stress of unfavorable winter temperatures, but may limit the production range (Fennell, 2004). Indeed, winter sub-zero and chilling temperatures limit the production range of many cultivars, and as climate changes regional microenvironments, they are subjected to increasingly variable dormancy, acclimation and spring bud break conditions. Therefore, it is necessary to have a better understanding of the mechanisms involved in dormancy. This will allow us to match cultivars appropriately with growing sites, improve cultural practices that minimize freezing injury, and aid in breeding and selecting grapevines for sustained winter survival.

In many temperate woody plants, growth cessation and dormancy development are necessary for acclimation to occur, and a decreasing day length promotes the transition to a winter tolerant state. For example, short days (SD) induces growth cessation, terminal bud set and dormancy in birch, chestnut, oak, peach, and poplar trees. Dormancy results from multiple sequential suites of gene expression from early perception, terminal bud set and finally transition into dormancy (Ruttink et al., 2007; Jiménez et al., 2010; Santamaría et al., 2011; Ueno et al., 2013). Over-expression of *ABSCISIC ACID INSENSITIVE3 (ABI3), FLOWERING TIME (FT)* or *PHYTOCHROME A (PHYA)* resulted in delayed bud maturation and dormancy in *Populus* (Olsen et al., 1997; Rhode et al., 2002; Böhlenius et al., 2006). In transgenic poplar over-expressing an *Avena sativa PHYA*, transgenic and wild type plants ceased growth and formed terminal buds under SD; however, only the wild type became dormant (Ruonala et al., 2008). It was suggested that dormancy development was not determined by signals from the leaf, but was dependent on apical and rib meristem properties. In ethylene insensitive transgenic birch, SD did not induce terminal bud set; however, the ethylene-insensitive trees did become dormant. These studies indicate that growth cessation and terminal bud set are distinct developmental events, and that dormancy is a separate developmental process (Rinne et al., 2001; Rhode et al.,

2002; Ruonala et al., 2006). Transgenic and natural mutants that are disrupted in terminal bud set, coupled with natural temporal studies, highlight the role of SD in cessation of stem elongation, terminal bud set and dormancy development. However, the mechanisms specific to the signaling cascade resulting in tissue dormancy still need clarification.

The compound axillary buds of the seasonally indeterminate grapevine provide an alternative model to explore bud dormancy. In contrast to many of the tree model systems commonly used to study dormancy development, grapevines do not set terminal buds, rather, the shoot tip abscises in response to decreasing photoperiod and/or low temperature (Fennell and Mathiason, 2002). The grapevine produces compound axillary buds containing primary, secondary and tertiary meristems throughout the growing season. The grapevine compound axillary buds remain paradormant during the growing season and will break and grow if the apical portion of the shoot is damaged or removed. The primary and secondary meristems typically contain vegetative and inflorescence primordia, whereas the tertiary meristem is predominately composed of vegetative primordia (Mullins et al., 1992). Thus, the compound axillary grapevine buds have already completed developmental processes that are associated with photoperiod induced growth cessation and dormancy in the terminal bud-forming model species (Ruttink et al., 2007; Jiménez et al., 2010; Sreekantan et al., 2010; Santamaría et al., 2011; Ueno et al., 2013). Grapevine genotypes vary in their response to SD, allowing selection of genotypes that are photoperiod responsive and non-responsive for dormancy development for dissecting the dormancy processes (Wake and Fennell, 2000; Fennell and Mathiason, 2002). Previous studies in *V. riparia* and Seyval grapevines show no differences in bud dormancy status in 14 days of long day (LD) or SD treatment, and buds in both genotypes show uncommitted tendril/floral primordia (Fennell and Hoover, 1991; Wake and Fennell, 2000; Sreekantan et al., 2010). In *V. riparia*, dormancy induction is evident after 21 days of SD by a delay in bud break, and dormancy depth increases at 28 days of SD, showing only 40% bud break. After 42 days of SD, *V. riparia* vines are dormant; neither vines nor individual nodes resume growth after 4 weeks of LD forcing conditions (**Figure 1**; Wake and Fennell, 2000). *V. riparia* bud water content decreases and freezing tolerance increases in SD, with a 5°C difference in freezing tolerance between LD and SD vines grown under optimal temperature conditions (Fennell and Mathiason, 2002). In contrast, Seyval is not dormant after 42 days of SD because the buds readily resume growth upon decapitation (Wake and Fennell, 2000). With these striking differences in photoperiod induced dormancy response, the grapevine system provides an excellent model system for dormancy development; therefore, this study compared global gene and key metabolite expression in *V. riparia* and *V. hybrid* Seyval. Analyses of age-matched buds from vines that were grown under LD (paradormant) or SD (dormancy induced) allowed us to dissect the molecular progress of bud dormancy. In this manuscript, dormancy refers to the endodormant phase wherein buds will not break and grow, even under favorable environmental conditions. All key terms and physiological response summary are provided in Supplementary Tables 1A,B.

Abbreviations: ABA, Abscisic acid; ABAGE, ABA glucose ester; CV, coefficient of variation; D, day of treatment; DE, differentially expressed; DEG, differentially expressed gene; DPA, diphaseic acid; FDR, false discovery rate; GSEA, gene set enrichment analysis; LT, lethal bud freezing temperature; LD or SD respectively, long day or short day; PCA, Principal components analysis; VRL or VRS respectively, *Vitis riparia* (VR) long or short photoperiod treatment; SVL or SVS respectively, Seyval (SV) long or short photoperiod treatment. VR_Photo_Time_Probesets, Significant *V. riparia* photoperiod x time probesets; SV_Photo_Time_Probesets, Significant Seyval photoperiod x time interaction probesets; Geno_Photo_Time_Probesets, Signficant genotype x photoperiod x time probesets; VR_Phase_Specific_DEGs, *V. riparia* dormancy phase specific differentially expressed genes; VR_Perception_Phase_DEGs, VR_Induction_Phase_DEGs, or VR_Dormancy_Phase_DEGs, *V. riparia* perception, induction or dormancy phase specific differentially expressed genes.

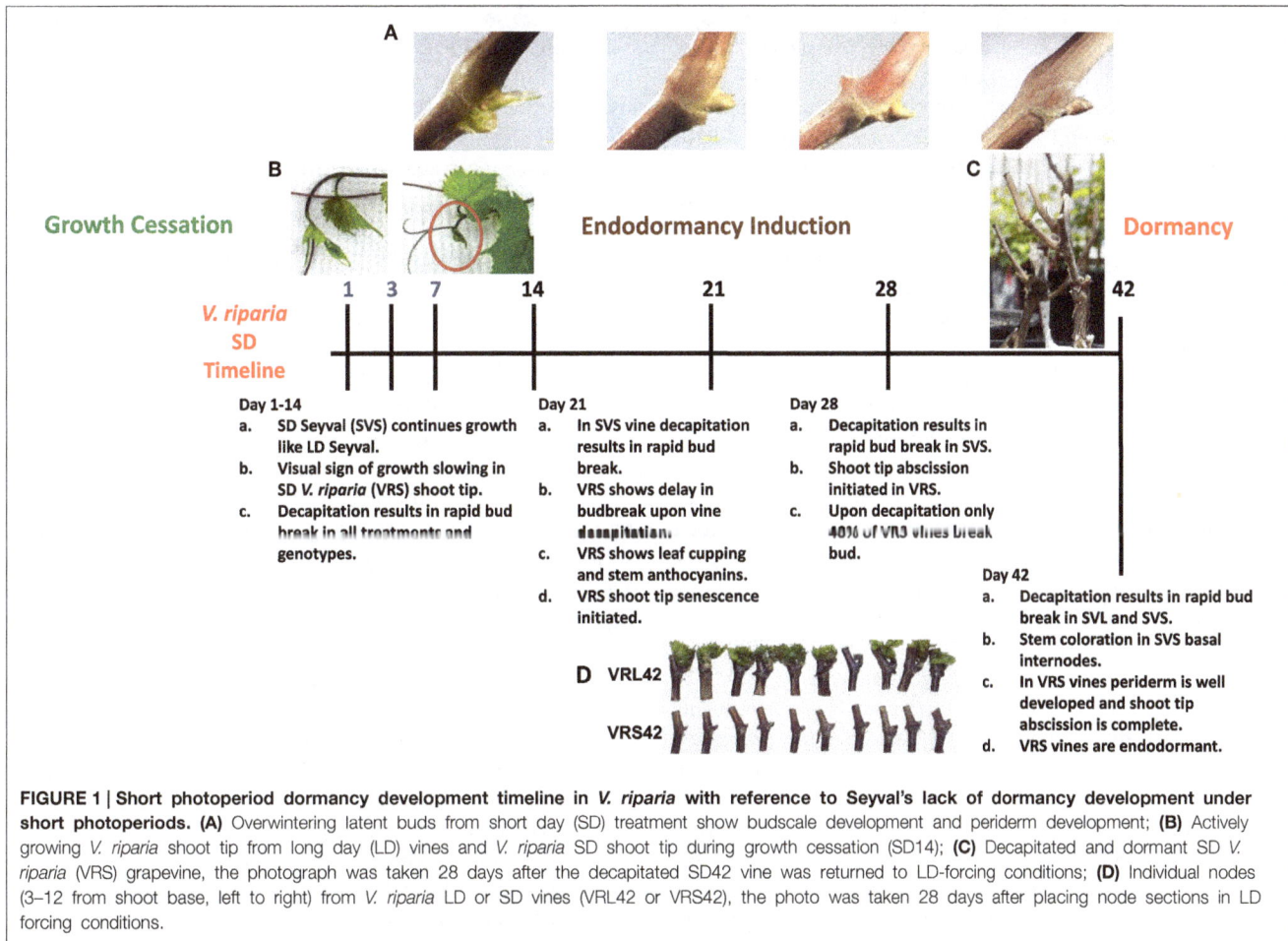

FIGURE 1 | Short photoperiod dormancy development timeline in *V. riparia* with reference to Seyval's lack of dormancy development under short photoperiods. (A) Overwintering latent buds from short day (SD) treatment show budscale development and periderm development; **(B)** Actively growing *V. riparia* shoot tip from long day (LD) vines and *V. riparia* SD shoot tip during growth cessation (SD14); **(C)** Decapitated and dormant SD *V. riparia* (VRS) grapevine, the photograph was taken 28 days after the decapitated SD42 vine was returned to LD-forcing conditions; **(D)** Individual nodes (3–12 from shoot base, left to right) from *V. riparia* LD or SD vines (VRL42 or VRS42), the photo was taken 28 days after placing node sections in LD forcing conditions.

MATERIALS AND METHODS

Plant Materials

Potted, spur-pruned 2- to 6-year-old vines of *V. riparia* and *V. hybrid Seyval* were removed from cold storage and grown in LD (15 h) at 25/20 ± 3°C day/night temperatures with 600–1400 mol m^{-2} s^{-1} photosynthetic photon flux in a climate-controlled unshaded glass greenhouse (En Tech Control Systems Inc., Montrose, Minn.) in Brookings, South Dakota (44.3 N). Vines were grown in 19 L pots with shoots trained vertically. When the grapevines reached 12–15 nodes (30 days post bud break), five-vine experimental units were randomized within each of two photoperiod treatments with the same temperature and light intensity as initial growing conditions.

Photoperiod Treatments

A split plot design was used to provide LD and SD treatments (15 h and 13 h, respectively). Forty two experimental units were randomized into each photoperiod plot to provide replicates for each time point and treatment. The SD treatment was imposed using an automated white-covered black-out system (735 ft^2 × 12 ft ceiling height; Van Rijn Enterprises LTD; Grassie, Ontario).

Each photoperiod treatment (735 ft^2) contained 42 experimental units (two genotypes × seven time points × three replicates) and spare experimental units. Five days after randomization of the experimental units (35 days post bud break) the SD photoperiod treatment was started in the SD plot and LD continued in the LD plot. Buds were harvested from each five-vine experimental unit into liquid nitrogen in separate tubes for each experimental unit. All experimental units were harvested between 8:30 and 11:30 a.m. at 1, 3, 7, 14, 21, 28, and 42 days of the LD or SD treatments. The buds were harvested from nodes 3 to 12, from the shoot base, for the entire experimental unit. Treatments were conducted between May and June with three replicate experimental units for each treatment combination in each of two consecutive years, providing six replicates in total.

RNA Extraction

Total RNA was extracted using a modified method of Chang et al. (1993). DNA was removed by incubation with 1 unit per microgram (μg) RNase-free DNase (Promega, Madison WI) at 37°C for 30 min. RNA was purified using RNeasy plant mini columns (Qiagen, Valencia CA). RNA quality and quantity were verified with an Agilent (Santa Clara, CA) 2100 Bioanalyzer RNA 6000 nano chip.

Microarray Data Acquisition and Analysis

Messenger RNA was converted to cDNA using reverse transcriptase and oligo dT primers containing a T7 RNA polymerase promoter sequence. Biotinylated complementary RNAs (cRNAs) were synthesized *in vitro* using T7 RNA polymerase in the presence of biotin-labeled UTP/CTP, purified, fragmented and hybridized with the GeneChip® *Vitis vinifera* (Grape) Genome Array ver. 1.0 cartridge (Affymetrix®, Santa Clara, CA). The hybridized arrays were washed and stained with streptavidin phycoerythrin and biotinylated anti-streptavidin antibody using an Affymetix Fluidics Station 400. Microarrays were scanned using a Hewlett-Packard GeneArray® Scanner and image data were collected and processed on a GeneChip® workstation using Affymetrix® GCOS software.

Expression data were first subjected to a series of rigorous quality control steps to ensure data reproducibility and overall quality following protocols previously described (Cramer et al., 2007; Tattersall et al., 2007). Average background and noise metrics were examined for consistency across all 168 arrays, as indicated by the Affymetrix GeneChip® Operating Software Users Guide. Probesets with less than 10% present calls across all arrays and Affymetrix control probesets were excluded. Raw intensity values of the remaining probesets were processed by Robust Multi-Array Average (RMA) (Irizarry et al., 2003) using the R package affy (Gautier et al., 2004). After pre-processing and normalization, a batch effect correction was conducted on all 168 arrays: to adjust for the batch effect stemming from two different years of sampling and array processing, an empirical Bayes method robust to outliers in small sample sizes was used (Johnson et al., 2007). This decreased the number of probesets with a significant batch effect from 10,773 to none. Expression data across years were combined after adjusting for batch effect, resulting in an experimental design with six replicates for each of the 28 (genotype × photoperiod × time point) treatment combinations. An additional quality control step was performed on each set of replicates: First, for each probeset, any set of six replicates having a coefficient of variation greater than 0.28 and one outlier more than 1.55 standard deviations from the mean across the six replicates was examined closely. The one outlying data point was deleted in these 3674 sets of replicates (1% of all sets of replicates). Second, any set of replicates with a notably large coefficient of variation (CV > 0.475) was completely excluded from further analyses (1% of all sets of replicates in the experiment). This yielded a dataset of 13,587 probesets with 98.8% of all expression data retained. This additional quality control step reduced the average coefficient of variation across replicates to 0.15. We found that these thresholds allowed us to identify gross outlying individual data points within replicates (Miller et al., 2009; Aw et al., 2010; Kuhn et al., 2010; Altick et al., 2012). Analyses were performed on these quality-controlled expression data of 13,587 probesets. Microarray data have been deposited in PlexDB (http://plexdb.org, VV18).

Principal component analysis (PCA) was applied to quality-controlled expression data using the correlation matrix to visualize any trends in the expression data (Gordon, 1999; Baldi and Hatfield, 2002; Knudsen, 2002; Stekel, 2003).

A Three-way ANOVA and false discovery rate (FDR; Benjamini and Hochberg, 1995) was used to test for significant main and interaction effects. The following standard model was used for this analysis: $y_{ijkl} = G_{il} + P_{jl} + T_{kl} + GP_{ijl} + GT_{ikl} + PT_{jkl} + GPT_{ijkl} + \varepsilon_{ijkl}$, where y_{ijkl} denotes the log2 gene expression value (signal) measured for genotype i, photoperiod j, time point k, and biological replicate l, with $1 \leq I \leq 2$, $1 \leq j \leq 2$, $1 \leq k \leq 7$, and $1 \leq l \leq 6$. The terms G_i, P_j, and T_k represent the main effects of genotype, photoperiod, and time point, respectively; the terms GP_{ij}, GT_{ik}, and PT_{jk} represent the two-way interactions between genotype and photoperiod, genotype and time point, and photoperiod and time point, respectively. The term GPT_{ijk} represents the effect of the three-way interaction between genotype, photoperiod and time point as described in Kerr et al. (2000). There were 6331 probesets with a significant two-way photoperiod × time effect (FDR adjusted p-value for the two-way interaction effect $p < 0.05$). These were examined by a *post-hoc* Tukey's Honest Significant Difference Test (HSD) with an adjustment for multiple comparisons for statistically significant photoperiod effects at each time point. Statistical significance was defined by an FDR adjusted p-value ($p < 0.05$). This analysis allowed examination of the significant expression between the age-matched LD and SD buds, and resulted in 3185 probesets that were differentially expressed (DE) between LD and SD at the same time point. Many probesets were DE at more than one time point. Data for all DE Photo_Time_Probesets are included in Supplementary Table 2A with a column indicating probesets that map to the same gene (Grimplet et al., 2012).

Array results were verified using real-time PCR of three replicates of RNA for five genes exhibiting different expression patterns across all time points as described in (Sreekantan et al., 2010). These genes referred to *EARLY LIGHT INDUCIBLE PROTEIN (ELIP1)*, *HISTONE H3*, *STRESS ENHANCED PROTEIN 2 (SEP2)*, *PHOSPHENOLPYRUVATE CARBOXYKINASE (PEPCK)*, and *INDOEACETIC ACID-INDUCED PROTEIN 6 (IAA6)* (Supplementary Table 3).

All DE probesets were annotated using the *Vitis* manual curation and gene annotation from Grimplet et al. (2012; Additional File 2.2). The functional categories (MIPS) were classified using the MIPS categorization for GeneChip® *Vitis vinifera* (Grape) Genome Array in Plexdb (http://www.plexdb.org/modules/PD_probeset/annotation.php?genechip=Grape, Cramer manual curation). There are multiple probesets for some *V. vinifera* (12X V1 assembly) genes represented on the Affymetrix GeneChip® 16K *Vitis vinifera* (Grape) Genome Array ver. 1.0 (11,249 unique gene identifiers, Grimplet et al., 2012). After annotation of probesets with the *Vitis* gene annotation, multiple probesets associated with a single gene were collapsed to one unique representative (Unique ID) to determine the (1) number of unique DEGs at a time point or (2) percent of genes in 11 major functional categories for each time point (Photo_Time_DEGs, Supplementary Figure 1).

The Three-way ANOVA identified 2365 probesets with a significant three-way genotype × photoperiod × time effect (Geno_Photo_Time_Probesets, Supplementary Table 2B). Statistical significance was defined by an FDR adjusted p-value ($p < 0.05$). These Geno_Photo_Time_Probesets were also examined by a *post-hoc* Tukey's HSD test with adjustment for multiple comparisons for statistically significant photoperiod effects between *V. riparia* LD and SD (VRL and VRS; Supplementary Table 2C: VR_Photo_Time_Probesets) and between Seyval LD and SD (SVL and SVS; Supplementary Table 2D: SV_Photo_Time_Probesets) at each time point. These datasets in Supplementary Table 2 include all significant Geno_Photo_Time_Probesets, with a column indicating multiple probesets that map to same Unique ID. The DE probesets or genes are those that were DE between photoperiods. The DEGs are identified as up-regulated or down-regulated in SD relative to the LD expression level for their respective genotype and time point.

The genes specific to three phases of short day induced dormancy in *V. riparia* were determined by excluding from the VR_Photo_Time_Probesets, those that were also found in the SV_Photo_Time_Probesets (Supplementary Tables 2C,D). Multiple probesets associated with a single gene were collapsed to one unique representative (Unique ID) to determine (1) the number of unique DEGs at a time point (each dormancy phase DEG was counted only once in a specific phase); however, a Unique ID may occur in more than one phase (**Table 1**, VR_Phase_Specific_DEGs; Supplementary Tables 4A–C; VR_Perception_Phase_DEGs, VR_Induction_Phase_DEGs, VR_Dormancy_Phase_DEGs).

Gene Set Enrichment Analysis

Gene Set Enrichment Analysis (GSEA) was conducted using GSEA-P 2.0 (http://www.broad.mit.edu/GSEA) on the quality-controlled expression values of 13,587 probesets and all VitisNet molecular networks including at least 10 genes (Supplementary Table 5A; Subramanian et al., 2005, 2007; Grimplet et al., 2009, 2012). The recommended GSEA-P 2.0 default parameters of 1000 permutations, false discovery q-value ($q < 0.25$) and nominal p-value ($p < 0.05$) were used to discover enriched molecular networks during dormancy development (Subramanian et al., 2007). Significant enrichments in VitisNet molecular networks were determined using pairwise comparisons of VRS and VRL and SVS and SVL at each time point. The significant enriched networks in common to both genotypes at the same time point were excluded since Seyval does not go dormant in response to the SD.

TABLE 1 | Number of differentially expressed genes (DEGs) categorized in perception, induction and dormancy phases of grapevine.

Phase	VR_Photo_Time_Probesets	SV_Photo_Time_Probesets	VR_Phase_Specific_DEGs
Perception	359	344	238
Induction	493	252	461
Dormancy	1317	167	1006

Phases are defined as: perception (D01, D03, and D07), induction (D14 and D21) and dormancy (D28 and D42). Values are VR_Photo_Time_Probesets and SV_Photo_Time_Probesets identified from post-hoc Tukey's HSD test of significant three-way interaction effects. (Supplementary Tables 2C,D). The number of DEGs specific to dormancy phases in V. riparia (VR_Phase_Specific_DEGs) and not in common with Seyval (SV) (Supplementary Tables 4A–C). Individual probesets may occur in more than one phase.

Dormancy Gene Set Comparisons

Comparisons of VR_Phase_Specific_DEGs were made with dormancy related gene lists from other dormancy gene expression studies to discover potential dormancy candidate genes. The dormancy studies included: axillary buds (*V. vinifera*, grapevine and *Euphorbia esula*, leafy spurge), terminal buds (*Populus*, hybrid poplar; *Castanea sativa*, chestnut and *Quercus petraea*, oak), cambium (*Populus*, hybrid poplar) and dormant seeds (*Arabidopsis thaliana*) (Supplementary Tables 6A–G; Cadman et al., 2006; Ruttink et al., 2007; Horvath et al., 2008; Resman et al., 2010; Santamaría et al., 2011; Díaz-Riquelme et al., 2012; Ueno et al., 2013). The best *Arabidopsis* match for the DEGs in each species was used for gene set comparisons. The datasets from these studies were determined using different platforms and they varied in size. Consequently, the number of cross-referenced genes varies widely with each comparison (Supplementary Table 6).

Hormone Analysis

Buds, as described in the plant materials section, were used for the quantification of abscisic acid and other related compounds using LC/MS/MS under Multiple Reaction Monitoring mode following the established method by Owen and Abrams (2009). Acquisition of the mass spectral data was performed according to Deluc et al. (2009).

Metabolite Extraction and Derivatization of the Metabolites

All tissue samples were homogenized in liquid nitrogen and lyophilized while keeping them frozen throughout the freeze-drying procedure. Freeze-dried bud tissue (100 mg) was placed in a standard screw-cap-threaded glass vial. Polar metabolites were extracted with a water/chloroform protocol according to previously established procedures (Broeckling et al., 2005). The aqueous phase, after 1 h of extraction, containing 12.5 mg l^{-1} of ribitol as an internal standard, was evaporated over-night in a vacuum concentrator and the tube was then returned to the $-80°C$ freezer until use. Polar samples were derivatized by adding 120 μl of 15 mg ml^{-1} of methoxyamine HCl in pyridine solution, were sonicated until all crystals disappeared, and then incubated at 50°C for 30 min. One hundred and twenty μl of MSTFA + 1% TMCS (Sigma-Aldrich, Inc., St. Louis, MO, USA) were added, the samples were incubated at 50°C for 30 min, and then immediately taken for analysis with a Thermo Finnigan Polaris Q230 GC-MS (Thermo Electron Corporation, Waltham, MA, USA). Derivatized samples (120 μl) were transferred to a 200-μl silanized vial insert and run at an injection split of 200:1 and 10:1 to bring the large and weak peaks to a concentration within the range of the detector. The inlet and transfer lines were held at 240°C and 320°C, respectively. Separation was achieved using a temperature program of 80°C for 3 min, and then increasing the temperature 5°C min^{-1} to 315°C, where it was held for 17 min. This was accomplished using a 60-m DB-5MS column (J&W Scientific, 0.25 mm ID, 0.25 μm film thickness) and a constant flow of 1.0 ml min^{-1}. All organic acids, sugars and amino acids were verified with standards purchased from Sigma-Aldrich.

Metabolite Data Processing

Metabolites were identified in the chromatograms using the software Xcalibur (1.3; Thermo Electron Corporation). The software matched the mass spectrum in each peak against three different metabolite libraries: NIST (v2.0: http://www.nist.gov/srd/mslist.htm), Golm (T_MSRI_ID: http://gmd.mpimp-golm.mpg.de) and a Cramer Lab, University of Nevada Reno, custom-created library (V1) made from Sigma-Aldrich standards. Quantification of the area of the chromatogram peaks was determined using Xcalibur and normalized as a ratio of the area of the compound peak to the area of the ribitol internal standard. The accumulation of ABA, ABA conjugates and ABA catabolites (ABA, ABA-GE, DPA, PA, 7′OH-ABA, and Neo-PA) were compared in both genotypes (*Vitis riparia* and Seyval) under SD and LD. Metabolite significant differences were determined using Three-way ANOVA and Tukey's HSD test at $p < 0.05$ [$n = 5$, one entire replicate (28 samples) was lost during shipping].

RESULTS

Transcriptome Profiles Showed a distinct Photoperiod by time Response (Photo_Time)

Gene expression was examined in age-matched LD buds (paradormant) and SD induced buds. The PCA showed distinct differences between the two genotypes (**Figure 2**). A dramatic separation of *V. riparia* gene expression under LD and SD was observed after 21 days of the photoperiod treatment. In contrast,

the PCA showed limited differences in LD and SD buds across time in Seyval.

There were 3185 Photo_Time_Probesets DE in SD relative to respective LD, and some were DE at more than one time point (Supplementary Table 2A). After annotation of probesets to unique gene identifiers, the number of unique photoperiod × time differentially expressed genes (Photo_Time_DEGs) showed a bimodal pattern with the greatest differential expression at 1–7 and 28–42 days of photoperiod treatment (D01, D03, D07, D28 and D42, respectively; Supplementary Table 2A; **Figure 3**; Grimplet et al., 2012). The number of Photo_Time_DEGs that were DE at only one time point was most common at days 1–14 of photoperiod treatment (Supplementary Figure 1A). At D28, the majority of the Photo_Time_DEGs were also DE at D42 (Supplementary Figure 1A, Supplementary Table 2A).

The functional categories of metabolism, transcription, cellular transport, cellular communication, cell rescue/defense and unclassified genes showed a greater number of up-regulated than down-regulated genes at day one of photoperiod treatment (D01; Supplementary Figure 1B; Supplementary Table 2A). A greater number of Photo_Time_DEGs related to protein processes were up-regulated than down-regulated at day three (D03). In contrast, a greater number of Photo_Time_DEGs related to metabolism, energy and transport were down-regulated than up-regulated at day three of SD (Supplementary Figure 1C; Supplementary Table 2A). There was a greater number of Photo_Time_DEGs in functional categories of protein processes, cellular transport, and cellular communication up-regulated than down-regulated on day seven of photoperiod

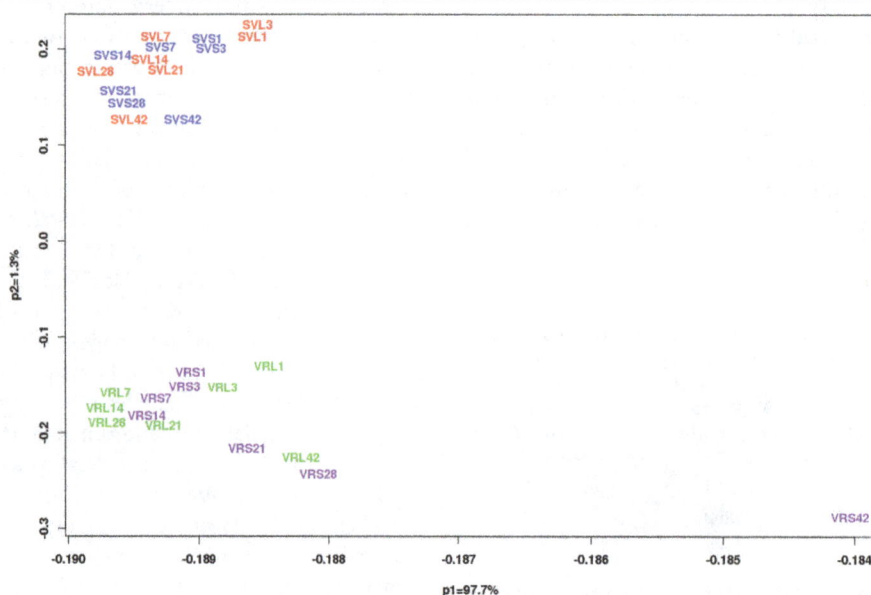

FIGURE 2 | Principal component analysis (PCA) of genotype (*V. riparia* and Seyval) and time (D01, D03, D07, D14, D21, D28, and D42 of treatment).
Analysis was conducted on the quality-controlled data (13,587 probesets in 6 replicates) using a correlation matrix. Photoperiod treatments are color coded for each genotype. *V. riparia* [LD, green (VRL1, 3, 7, 14, 21, 28, and 42); SD, purple (VRS1, 3, 7, 14, 21, 28, and 42)] and Seyval [LD, red (SVL1, 3, 7, 14, 21, 28, and 42) and SD, blue (SVS1, 3, 7, 14, 21, 28, and 42)].

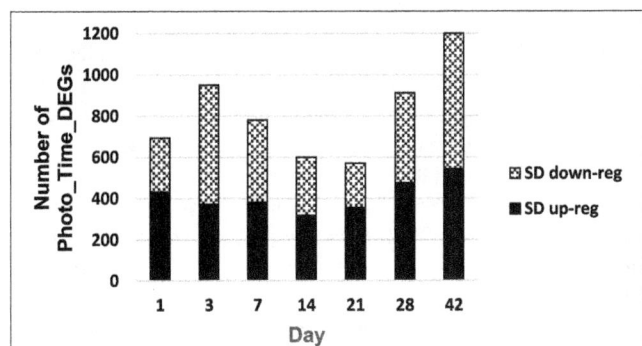

FIGURE 3 | Number of photoperiod x time differentially expressed genes (Photo_Time_DEGs) that (1) had a significant photoperiod x time interaction effect and (2) were differentially expressed between photoperiod treatments on a particular day.

treatment. In contrast, there were more Photo_Time_DEGs in the categories of energy and transcription down-regulated than upregulated on day seven (D07, Supplementary Figure 1D; Supplementary Table 2A).

At 14 and 21 days of the photoperiod treatment (D14, D21), there was a greater number of Photo_Time_DEGS expressed at consecutive time points (consecutive Photo_Time_DEGs) than observed in D01 to D07. This period corresponded to the SD-induced delay in bud break in *V. riparia* (**Figure 1**; Supplementary Figure 1A). This period had the lowest number of Photo_Time_DEGs (**Figure 3**; Supplementary Figure 1A; Supplementary Table 2A). A greater number of metabolism, and protein processes genes were up-regulated than down-regulated on D14 (**Figure 3**; Supplementary Figure 1E). More Photo_Time_DEGs related to metabolism, cell rescue/defense, interaction with environment, developmental processes and unclassified genes were up-regulated than down-regulated on D21 of photoperiod treatment (**Figure 3**; Supplementary Figure 1F).

The majority of the Photo_Time_DEGs on D28 were also DE on D42 (consecutive Photo_Time_DEGs; Supplementary Figure 1A, Supplementary Table 2A). In addition, there were consecutive Photo_Time_DEGs from D21 through D42 (Supplementary Table 2A). Days 3 through 28 had a greater number of Photo_Time_DEGs down-regulated in SD in the functional category of energy (primarily photosynthesis) (Supplementary Figures 1C–H). Days 14, 21, and 28 had a greater number of DEGs up-regulated in SD relative to LD in the metabolism functional category (Supplementary Tables 2E–G). There were two peak periods with greater numbers of Photo_Time_DEGs (**Figure 3**). The number of Photo_Time_DEGS increased to D07, and there were fewer DEGs on D14 and D21 than on D01, D03, and D07. The number of DEGs increased again during D28 and D42. This pattern and the changing prevalence of the functional categories in relation to the physiological responses of *V. riparia*, indicated three phases of development: Perception (SD01–SD07), Induction (SD14–SD21) and Dormancy (SD28–SD42) (**Figure 3**).

Differential Photoperiod Responses Indicated Three Phases of Dormancy Development in *V. riparia*

A *post-hoc* ANOVA Tukey's HSD test identified 2365 Genotype_Photo_Time_Probesets that were DE at one or more time points during the photoperiod treatment (Supplementary Table 2B). There were a similar number of VR_Photo_Time_Probesets and SV_Photo_Time_Probesets DE at one or more time points in the perception phase (**Table 1**). The majority of these were DE at only one time point (**Table 1**; Supplementary Tables 2C,D). There was a two times greater number of VR_Photo_Time_Probesets DE than SV_Photo_Time_Probesets DE during the induction phase (days 14 and 21) (**Table 1**). In addition, over half the VR_Photo_Time_Probesets were DE at both 14 and 21 days, and a greater number were up-regulated than down-regulated (Supplementary Table 2C). In contrast, in Seyval, there was only one consecutive DE probeset in the induction phase. The majority of the DE SV_Photo_Time_Probesets occurred on day 14 (Supplementary Table 2D). The dormancy phase (Days 28 and 42) had the greatest number VR_Photo_Time_Probesets, and one-third of these were consecutively DE (Supplementary Table 2C). The rest of the dormancy phase DE VR_Photo_Time_probesets were DE at only one time point, with 80% of these DE on D42. This increased number occurred when the vines became unable to break bud upon decapitation. In contrast, there were fewer DE SV_Photo_Time_Probesets at these time points, and only four were DE at consecutive time points.

In the VR_Perception_Phase_DEGs, genes associated with the functional categories of response to environment (hormone signaling and cell rescue/defense responses), as well as other unknown/unclassified genes were prominent (Supplementary Table 4A). A large number of metabolism-related genes were expressed during the induction phase. For example, there were 30% metabolism-related genes up-regulated, including fatty acid biosynthesis, polysaccharide biosynthesis and degradation, phenylpropanoid biosynthesis and auxin signaling (Supplementary Table 4B). Down-regulated VR_Phase_Specific_DEGs were involved in photosynthesis and energy-related processes. Typical up- and down-regulation patterns of *V. riparia* in comparison to Seyval during dormancy development are presented in **Figures 4A,B**. In the VR_Dormancy_Phase_DEGs, there were equal numbers of up- and down-regulated genes, with up-regulated DEGs in the functional categories involving amino acid biosynthesis, transcription, protein processes and response to environment (hormone signaling and defense responses) (Supplementary Table 4C). Two major functional categories down-regulated in VR_Dormancy_Phase_DEGs under SD were photosynthesis/energy related and cellular communication. These data indicated a major reprogramming in *V. riparia* metabolism during the induction phase in response to a 13 h SD that continues through dormancy. It further highlights the absence of induction in Seyval in response to the SD.

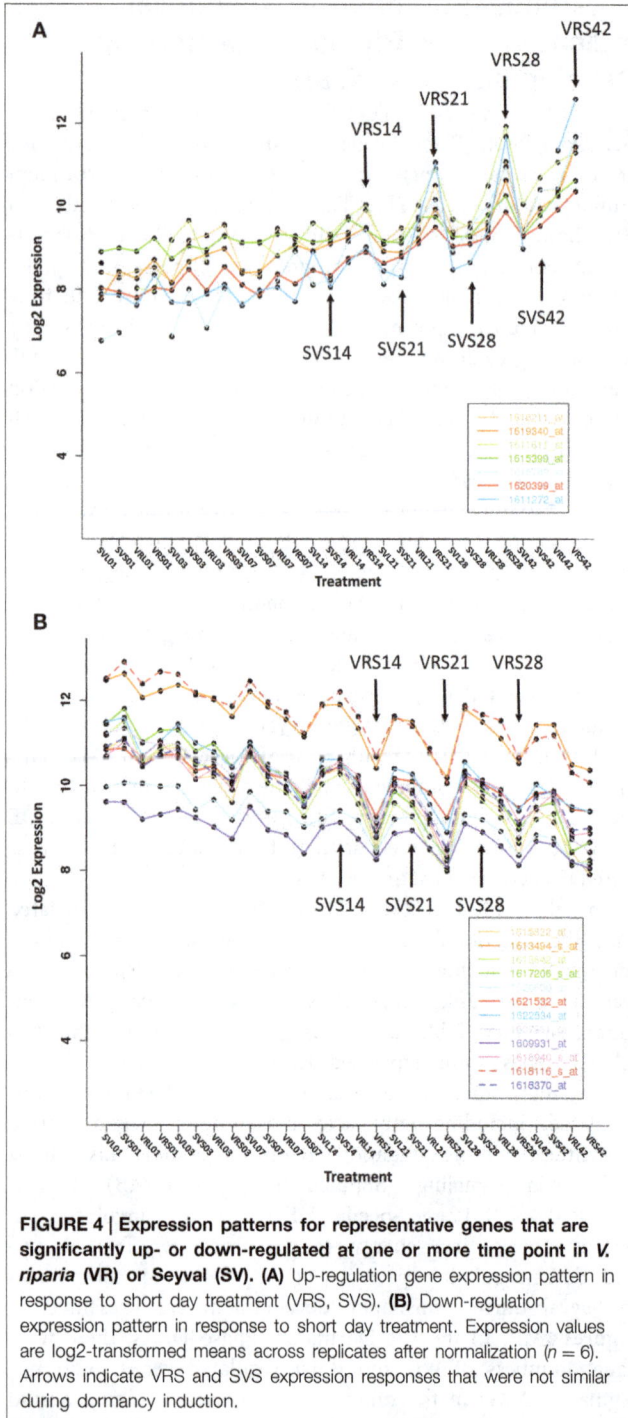

FIGURE 4 | Expression patterns for representative genes that are significantly up- or down-regulated at one or more time point in *V. riparia* (VR) or Seyval (SV). (A) Up-regulation gene expression pattern in response to short day treatment (VRS, SVS). **(B)** Down-regulation expression pattern in response to short day treatment. Expression values are log2-transformed means across replicates after normalization (*n* = 6). Arrows indicate VRS and SVS expression responses that were not similar during dormancy induction.

Gene Set Enrichment Analysis (GSEA-P 2.0) Distinguished Potential Molecular Network Signatures for Dormancy Development

Pairwise comparisons were conducted for VRS vs. VRL and SVS vs. SVL, and VitisNet molecular networks with a statistically significant over-representation of genes up-regulated in SD vs. LD at each time point in *V. riparia* and Seyval were

identified (Supplementary Tables 5B,C, respectively). A total of 128 enriched networks (*SD* = 55 and *LD* = 73) were identified in *V. riparia* and 114 enriched networks (*SD* = 31 and *LD* = 73) were identified in Seyval. Molecular networks enriched in LD represent networks active during paradormancy. Molecular networks with significant over-representation of genes up-regulated in SD relative to LD in both *V. riparia* and Seyval were removed from the VRS list. Therefore, the networks significantly enriched in VRS alone were used to guide exploration of DEGs in the dormancy phases. Briefly, the perception phase showed significant enrichment in ZF-C3HC4, WRKY, and NAC transcription factor family molecular networks. The induction phase was characterized by significant enrichment in molecular networks associated with cell wall, starch and sugar metabolism and phenylpropanoid biosynthesis. Finally, the dormancy phase showed significant enrichment in molecular networks associated with amino acid metabolism and RNA and protein processing.

Dormancy Gene Expression Profiles in Diverse Species and Tissues Pinpointed Potential Dormancy Candidate Genes

Phase specific genes up-regulated in the *V. riparia* were compared with DEGs from Tempranillo compound axillary bud and showed two *C3HC4-TYPE RING FINGER (ZF-C3HC4)*, a *NAC DOMAIN CONTAINING PROTEIN 19* and *RESVERATROL SYNTHASE* were expressed in both *Vitis* species (Supplementary Table 6A, Díaz-Riquelme et al., 2012). *Euphorbia esula* (leafy spurge) and *V. riparia* axillary buds both had auxin signaling genes and an aquaporin gene up-regulated during dormancy (Supplementary Table 6B, Horvath et al., 2008). A comparison of poplar terminal bud and cambium tissue DEGs with *V. riparia* identified the largest number of DEGs in common between species, including genes associated with carbohydrate degradation, secondary product biosynthesis (phenylpropanoids), protein processes (folding, transport) and rescue/defense related categories. Several phenylpropanoid pathway and cell wall related genes were up-regulated in both species. These included several phenylpropanoid biosynthesis, cell wall synthesis and late embryogenesis abundant protein genes (Supplementary Tables 6C,F; Ruttink et al., 2007; Resman et al., 2010). Although, the *RESVERATROL SYNTHASE* gene was not DE in the poplar terminal buds, a *STILBENE SYNTHASE* gene was DE in poplar terminal bud and cambium tissue. Genes that were DE in *Arabidopsis* seed and *V. riparia*, included auxin and ABA signaling, ZF-C3HC4 and *NAC DOMAIN CONTAINING PROTEIN 19* genes (Supplementary Table 6G). The majority of the DEGs found in the *V. riparia* and *Arabidopsis* comparison were up-regulated in *V. riparia*, and included auxin and ABA signaling genes. A three-way comparison of grape bud, hybrid poplar and *Arabidopsis* seed data sets showed ABA, ethylene (*ETHYLENE INSENSITIVE* 3, *EIN3*) and auxin signaling (*DORMANCY/AUXIN ASSOCIATED*, *DRM1*) genes and *Em PROTEIN GEA6* DE in all three species during dormancy. The differential expression of these genes in different tissue types (axillary bud, cambium and seed embryo) indicates a strong potential role in dormancy development.

Abscisic Acid, Raffinose, Trehalose, and Resveratrol Metabolites Increased During Dormancy

ABA significantly accumulated from day 3 until day 21, at which time it peaked in the VRL and VRS treatments (**Figure 5A**). The ABA concentration was significantly greater from SD14 to SD28 relative to LD, with the largest differences observed at day 21. Analyses of the conjugate form of ABA (ABA-GE) did not show any significant trend over time or between treatments, which might indicate a weak implication of conjugation events in changing the ABA levels during dormancy (Supplementary Table 7A).

A clear differential trend in glucose, fructose and sucrose concentrations was not observed (Supplementary Table 7B)

In contrast, there were changes in polysaccharides known to contribute to freezing tolerance and cold acclimation. There was little raffinose accumulation during the perception phase, but there was increased accumulation during induction phase under SD with >2-fold increase from VRS28 to VRS42 (**Figure 5B**, Supplementary Table 7B). A similar trend was found in *V. riparia* with trehalose, with an increasing trend between VRS28 and VRS42 (2-fold, **Figure 5C**). The stilbenoid, resveratrol, increased from day 21 until day 42 in a similar manner as found for raffinose. This accumulation was most pronounced in *V. riparia*, with greater abundance in VRS42 than VRL42 (>3-fold, **Figure 5D**).

In general, there was no difference in the accumulation of the organic acids between the LD and SD in either genotype. Glycerate, malate, succinate, ribonate, and glucarate all decreased in both LD and SD in *V. riparia* and Seyval during bud development (Supplementary Table 7B). The amino acids also

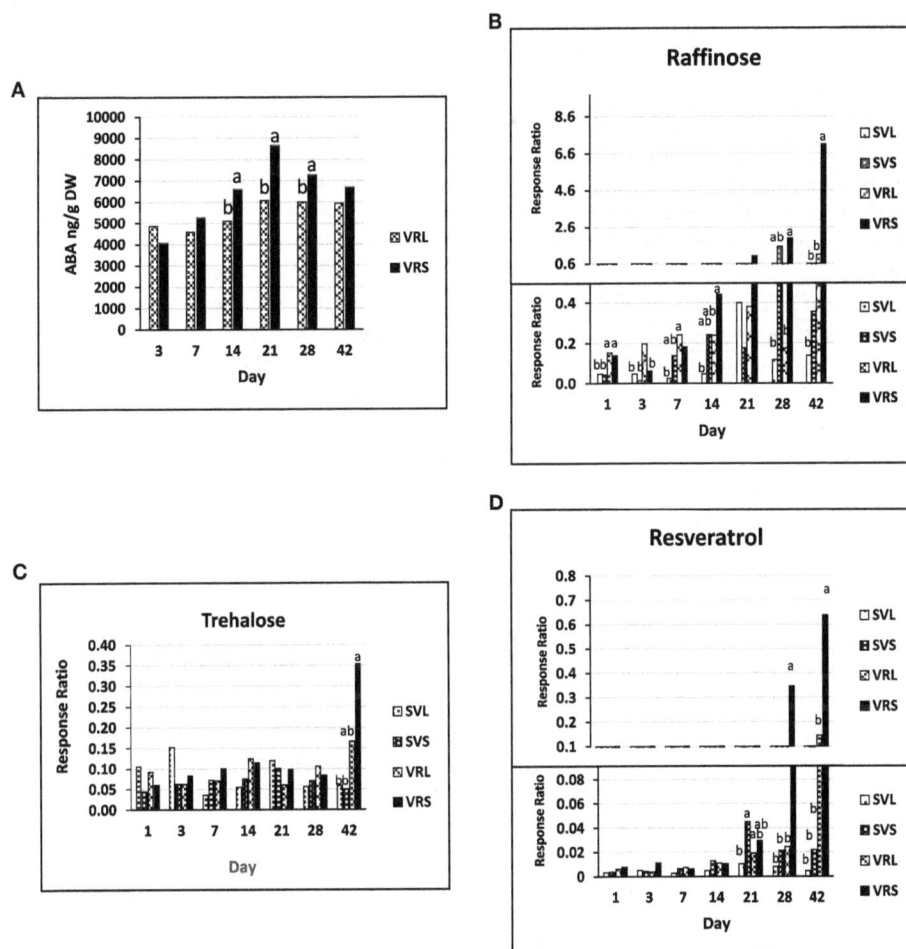

FIGURE 5 | Metabolites in *V. riparia* and Seyval under long or short day conditions (VRL or VRS and SVL or SVS). (A) Abscisic acid concentration in *V. riparia* long or short day buds (VRL or VRS, respectively). Values are ABA concentration per gram dry weight (DW) **(B)** Raffinose accumulation relative to ribitol in *V. riparia* and Seyval long or short day buds (VRL, VRS, SVL, SVS; respectively). **(C)** Trehalose accumulation relative to ribitol (response ratio) in VRL, VRS, SVL and SVS. **(D)** Resveratrol accumulation relative to ribitol (response ratio) in VRL, VRS, SVL, and VRS. Values are mean and standard error of abundance. Tukey's HSD significant differences ($p < 0.05$, $n = 5$) between SVL, SVS, VRL, and VRS at a given time point are noted by different letters, no letter indicates no difference found at that time point.

had similar patterns of decreasing concentrations. However, in VRS, there was an increase in leucine at 21 days, but with increasing age, the leucine levels became equivalent. In contrast, proline showed a bimodal increase, with an increased concentration in the samples VRS21 and VRS42.

DISCUSSION

Sensitivity to environmental cues and the ability to acclimate or become dormant to withstand stressful conditions is a hallmark of perennial plants. In widely distributed woody species, response to the reliably constant annual photoperiod cycle is used to regulate the growth cycle by initiating dormancy and acclimation processes; therefore, contributing to low winter temperature survival (Cooke et al., 2012). While much work in woody plants has focused on growth cessation and terminal bud development in response to SD, axillary buds are also subject to dormancy cycles. In the seasonally indeterminate grapevine, shoot tips abscise rather than set terminal buds, and dormancy development in the compound axillary bud is the critical factor for seasonal cycling and winter survival. The grapevine latent axillary bud develops continuously during the growing season, initiates floral meristems prior to the onset of decreasing photoperiod or low temperature cues, and is maintained in a paradormant state until either shoot apical dominance is disrupted or dormancy is initiated. In *Vitis* species, dormancy is induced by SDs and/or low temperatures (**Figure 1**, Fennell and Hoover, 1991; Schnabel and Wample, 1987). This differential response provides the ability to compare genotypes that are photoperiod responsive and nonresponsive for dormancy development. This removes the confounding effects of temperature, allowing one to characterize the early perception and dormancy commitment cascade resulting in the dormant state. Previous studies have shown that *V. riparia* has a 13 h critical photoperiod and Seyval has an 11 h critical photoperiod for growth cessation (Wake and Fennell, 2000). However, Seyval requires a synergistic low temperature and photoperiod interaction for dormancy induction (Garris et al., 2009). Therefore, only a 2 h photoperiod change was used in this study. *V. riparia* responds to SD (13 h) at normal growing temperatures by decreasing water content, increasing freezing tolerance, abscising the shoot tip and developing dormancy in latent axillary buds prior to leaf abscision (**Figure 1**, Supplementary Table 1B; Fennell and Hoover, 1991; Wake and Fennell, 2000). In contrast, Seyval slows growth and shows only a 2°C increase in freezing tolerance, and the buds remain paradormant under SD. These distinct genotype differences provided the ability to separate metabolite and gene expression changes as a general photoperiod response or a dormancy specific response. Because Seyval does not become dormant under 13 h photoperiod, genes that are DE in both genotypes were considered a general response to change in photoperiod. Thus, the DEGs in common to both genotypes were removed to create the *V. riparia* dormancy phase specific gene sets (**Table 1**). The remainder of the discussion focuses on the SD induced dormancy-specific responses in *V. riparia*.

Short Photoperiod, Dormancy-specific Responses Characterized Three Phases of Dormancy Development
Perception Phase

Differential gene expression in response to the 2 h difference in photoperiod provided evidence for a perception phase (D01, D03, and D07). The majority of differentially expressed probesets were different at only one of the time points during this phase (VR_Photo_Time_Probesets, Supplementary Table 2C). The number of differentially expressed probesets decreased during the 7-day perception phase, and there was a striking increase in the number of differentially expressed probesets during the induction phase (VR_Photo_Time_Probesets, **Table 1**). This suggested a pattern of perception and photoperiod entrainment, followed by a downstream trend of sustained gene activity, resulting in dormancy in *V. riparia*. The GSEA of the perception phase indicated that SD resulted in (1) down-regulation of genes associated with protein processes and fatty acid biosynthesis and (2) up-regulation of genes associated with transcription factors (ZF-C3HC4, WRKY, and NAC), sugars, and amino acid metabolism, and cytokinin signaling networks (Supplementary Table 5A). The perception phase showed multiple fatty acid biosynthesis and transport and protein processing and transport genes down-regulated in SD (Supplementary Table 4A). Sugars have been recognized as signaling molecules in dormancy (Anderson et al., 2005; Horvath et al., 2008). There were two endochitinase and alpha-glucan, alpha-amylase and alpha-galactosidases (polysaccharide degradation) genes, up-regulated in SD during the perception phase (Supplementary Table 4A). Several are also found among the poplar bud DEGs (Supplementary Table 6C). There were multiple genes associated with glutamate, asparagine, methionine, and cysteine that were up-regulated in SD during the perception phase that were not differentially expressed during the other phases (Supplementary Tables 4A–C). This indicates the potential for amino acids to act as signaling molecules in response to photoperiod, which has been shown in abiotic stress responses (Forde and Lea, 2007; Less and Galilli, 2008).

A *WRKY65* transcription factor gene, which responds to water stress and sugar starvation, was up-regulated in VRS01 and again in VRS42 (Supplementary Table 4A; Contento et al., 2004; Wang et al., 2014). This is the first indication of a photoperiod and dormancy response for *WRKY65*. No WRKY transcription factor genes were differentially expressed in leafy spurge, poplar, chestnut or oak buds; but, this may be because those studies used greater time intervals to monitor dormancy development (Ruttink et al., 2007; Horvath et al., 2008; Santamaría et al., 2011; Ueno et al., 2013). There were multiple genes encoding zinc finger binding proteins of the ZF-C3HC4 family differentially expressed throughout dormancy development, and these were both up- and down-regulated. *ZF-C3HC4* genes were up-regulated in both the perception phase and the dormancy phase. These were also DE in poplar and chestnut bud dormancy, indicating a consistent role in SD induced dormancy (Supplementary Tables 6B–F). The ZF-C3HC4 and WRKY families of transcription factors are large families, and functional roles have been linked with biotic and abiotic stress and developmental processes. The *ZF-C3HC4* genes are differentially expressed in response to SD in multiple dormancy systems (grapevine, poplar, and chestnut buds), suggesting they have a specific role in dormancy induction (Supplementary Tables 6A,C,D).

There were four genes related to flowering that were consecutively up-regulated in the perception phase, including a *MADS-BOX PROTEIN (AGL20/SOC1)* gene (VR_Preception_Phase_DEGs, Supplementary Table 4A). Flowering time genes have been implicated in dormancy because perception of photoperiod plays a role in floral initiation and growth cessation in several woody species (Böhlenius et al., 2006). Two additional transcription factor genes *FLOWERING LOCUS C (FLC)*, a negative regulator of *AGL20*, and a *SQUAMOSA PROMOTER-BINDING PROTEIN (SBP)* were up-regulated during the perception phase (Supplementary Table 4A). Both *AGL20* and the MADS-box transcription factor gene *FLC* were up-regulated in *V. riparia* SD on D01. Furthermore, *SBP* was one of the few floral related genes up-regulated on D07. Although these genes have roles in floral transitioning, the transient differential expression of *AGL20* and its negative regulator at the same time suggest an additional role in SD perception that can contribute to dormancy.

Ethylene and auxin signaling related genes were up-regulated in response to SD in *V. riparia*. Two auxin signaling genes, *DORMANCY/AUXIN ASSOCIATED* and *DOMAIN REARRANGED METHYLASE (DRM1)/AUXIN ASSOCIATED* were up-regulated only on day one in SD (Supplementary Tables 2C, 4A). In kiwi axillary buds, *DRM1* is up-regulated during the entire endodormancy period, and down-regulated during bud break (Wood et al., 2013). In contrast, the *DORMANCY/AUXIN ASSOCIATED* genes were up-regulated only in VRS01, apparently serving as SD perception genes. Ethylene signaling-related genes that were up-regulated in SD were an *ETHYLENE RESPONSE FACTOR (AP2/EREBP)* and ethylene biosynthesis inhibiting E8 protein, *ETHYLENE INSENSITIVE3 (EIN3)*. In addition, BTB/POZ DOMAIN-CONTAINING or TRAF transcription factor genes were also up-regulated (Supplementary Table 4A). The BTB/POZ DOMAIN-CONTAINING or TRAF transcription factors can assemble with *AP2/EREBP*. These signaling genes interact with other signaling pathways, as *EIN3* is noted to be involved in sugar-mediated signaling (Chao et al., 1997; Solano et al., 1998). In poplar terminal bud set, ethylene signaling occurred after 2 weeks of SD, before ABA-mediated bud maturation (Ruttink et al., 2007). However, ethylene signaling occurred early and transiently in the grapevine axillary bud perception phase (Supplementary Tables 2C, 4A). Thus, the dormancy perception phase in *V. riparia* was characterized by a rapid and transient up-regulation of several genes that encode transcription factors, ethylene and auxin signaling proteins and floral transition proteins. The temporary nature of expression of the floral transition genes also indicates a potential dual role for these genes, possibly a recruitment of their photoperiod responsiveness to serve as SD signals for the induction phase (Böhlenius et al., 2006; Sreekantan et al., 2010).

Induction Phase

The induction phase (D14 and D21) was characterized by an increased number of probesets differentially expressed at consecutive time points (Supplementary Table 2C). The GSEA showed an enhanced up-regulation of carbohydrate metabolism (starch/sucrose and nucleotide metabolism), protein processing and phenylpropanoid biosynthesis molecular networks. By comparison, there was an enrichment in the set of down-regulated genes related to photosynthesis/energy, floral development, and circadian rhythm.

The enrichment of these molecular networks indicated changes in metabolic programming and an enhancement of secondary metabolism. Increases in bud carbohydrates are common during seasonal progression of vegetative maturation, and changes in photoperiod have been shown to reconfigure carbohydrate metabolism (Anderson et al., 2005; Ruttink et al., 2007). It should be noted that a much smaller photoperiod difference was used in this study than in studies of poplar, peach and birch; therefore the differences in carbohydrate dynamics may be less prominent in this grapevine system (Ruonala et al., 2006; Jiménez et al., 2010; Santamaría et al., 2011). A shift in the Tempranillo grapevine carbohydrate metabolism genes was noted under natural conditions. Díaz-Riquelme et al. (2012) report an increase in expression of starch and sucrose metabolism genes during grapevine bud maturation and dormancy development under field conditions. Metabolic analysis in this study indicated that there were increases in fructose, glucose, glucose-6 phosphate, sucrose, and maltose concentrations. However, carbohydrate accumulation dynamics did not show a direct association with dormancy induction patterns (Supplementary Table 7B).

Many cell wall related genes were up-regulated during the induction phase, particularly genes encoding biosynthesis enzymes (cellulose synthases, polygalacturonases, pectate lyases, invertases, and xyloglucan endotransglucosylases) (VR_Induction_Phase_DEGs, Supplementary Table 4B). Coincidently, we also found a significant number of genes in the phenylpropanoid biosynthesis network, including genes related to lignin biosynthesis. The up- and down-regulation of genes related to transport during the induction phase was not apparent in the GSEA (Supplementary Table 5A). But a distinct suite of genes encoding transporters (potassium, sulfate, nitrate, and mitochondrial and vesicle related) and *TONOPLAST INTRINSIC PROTEIN* genes (*TIP1; TIP3*) were up-regulated in SD (Supplementary Table 4B). An aquaporin located in the vacuole membrane (TIP1) is associated with vesicle targeting (Hachez et al., 2006). In addition, the genes encoding two vesicle associated proteins and a lipid transfer protein were up-regulated, but only during the induction phase.

Several genes encoding transcription factors belonging to families ZF-C3HC4, WRKY, NAC DOMAIN CONTAINING and MYB DOMAIN CONTAINING, as well as hormone signaling genes related to cell cycle and interaction with the environment were differentially expressed during consecutive time points in the induction phase (Supplementary Tables 2C, 4B). Five zinc finger transcription factor genes (*ZF-C3HC4*) were up-regulated including a *HISTONE MONO-UBIQUITINATION 2 (HUB2)* gene which has a role in chromatin remodeling during seed dormancy (Liu et al., 2007). Several WRKY transcription factors are involved in seed dormancy and abiotic stress tolerance (Rushton et al., 2010; Chen et al., 2012). Separate WRKY transcription factor genes were up-regulated in SD relative to LD in each of perception, induction and dormancy phases (Supplementary Table 4), and this is the first indication of their photoperiod regulation. The MYB transcription factor gene family is large and impacts a variety of developmental processes. In this study, two *MYB DOMAIN PROTEIN* genes (*MYB4* and *MYB14*) were up-regulated during the induction phase. *MYB4* increases freezing tolerance in transgenic *Arabidopsis* and apple plants (Pasquali et al., 2008). *MYB14* regulates isoprenoid and flavonoid metabolism in transgenic spruce (Bedon et al., 2010; Bomal et al., 2014). Flavonoid biosynthesis genes were up-regulated in grapevine buds in response to SD, as were many genes related to fatty acid and isoprenoid metabolism (Supplementary Table 4B). The NAC transcription factor has been described in shoot and embryo meristems. More recently, many NAC domain transcription factors have been associated with drought stress, but no information previously existed on their regulation by photoperiod (Aida et al., 1997; Puranik et al., 2012; Jensen et al., 2013; You et al., 2015). Three *NAC* genes were differentially expressed. The *NAC 87* gene was also up-regulated in response to SD-induced dormancy in chestnut terminal buds, further emphasizing their potential role in photoperiod induced dormancy (Santamaría et al., 2011).

In the DE hormone signaling gene set, there are several cell cycle related genes that were up-regulated (*ALPHA-EXPANSIN PRECURSER, ALPHA-EXPANSIN 3* and *BETA-EXPANSIN*). This indicated that bud development continued during the induction phase. Similarly, a suite of hormone signaling genes were up-regulated (*SNAKIN-1, E8 PROTEIN, AUXIN RESPONSIVE SAUR29, AUXIN BINDING PROTEIN ABP19a PRECURSOR, ASSOCIATED WITH PLASMAMEMBRANE-19 (AWPM-19)*). An increase in *AWPM-19* expression was associated with ABA-induced freezing tolerance in wheat suspension cells (Koike et al., 1997). This study showed an increased ABA accumulation during the induction of bud dormancy that peaked in VRS21, and was maintained until VRS42 in comparison with VRL21 and VRL42. Control of free ABA levels through conjugation or degradation may influence the ABA mediation of the dormancy cycle (El Kayal et al., 2011). Analysis of the genes related to ABA homeostasis showed few or no changes for genes associated with the rate limiting steps of ABA biosynthesis (*9-CIS-EPOXYCAROTENOID DIOZYGENASE*, NCED; *ZEAXANTHIN EPOXIDASE ZEP*, ABA1; and *ABA DEFICIENT 2*, ABA2). One NCED gene was up-regulated in VRS28-42, and another was down-regulated in VRS42. No changes were observed in expression for the transcript involved in the conjugation of ABA into ABA-GE, which was consistent with similar levels of ABA-GE in VRS and VRL from D14 to D42 (Supplementary Table 7A). On the other hand, there was one gene involved in ABA degradation (ABA 8′hydroxylase) for which the expression level decreased from D14 until D42 in

VRS (Supplementary Table 7A). We observed a similar trend for the same gene in Seyval, but the measurement of ABA in Seyval was too variable to draw any definitive conclusion. However, it is tempting to hypothesize that ABA levels under SD may be directly related to the inactivation pathway through its degradation, rather than through its synthesis as recently proposed by Zheng et al. (2015). Several factors influence *ABA-8′HYDROXYLASE* gene expression, including the ABA level itself and low temperature (Krochko et al., 1998; Zhou et al., 2007). Further, investigation will be needed to ascertain the influence of photoperiod on the gene regulation of *ABA-8′HYDROXYLASE* during endodormancy. The sustained up-regulation of gibberellin, ABA, auxin and ethylene signaling and biosynthesis genes during the induction phase clearly indicates the potential role of these hormones in the reprogramming of bud development toward induction. However, limited coverage of the hormone biosynthesis and catabolism pathways in this study limits our ability to define distinct relationships with dormancy processes.

Dormancy Phase

The dormancy phase was characterized by a doubling of the number of differentially expressed *V. riparia* probesets in comparison with the induction phase (Table 1). There was no significant molecular network enrichment for VRS28, but VRS42 showed significant enrichment in amino acid, protein and RNA processes networks (Supplementary Table 5B). As was noted for the induction phase, there were more genes related to transport that were up-regulated than were down-regulated (Supplementary Table 4C). Furthermore, we observed a greater number of protein and vesicular transport related genes in the dormancy phase. Protein processing, phosphorylation, folding and stabilization genes were up-regulated in the dormancy phase, and there were two times the number in the dormancy phase in comparison with the induction phase (Supplementary Tables 4B,C). This indicates the potential for increased post-transcriptional processes related to dormancy.

Starch and sucrose metabolism genes were predominantly up-regulated during the dormancy phase. Increases in bud carbohydrates are common during seasonal progression of vegetative maturation, and changes in photoperiod reconfigure carbohydrate metabolism (Anderson et al., 2005; Ruttink et al., 2007). Maltose increased through the induction phase, maintaining a greater level under SD in both *V. riparia* and Seyval (Supplementary Table 7B). Maltose has been found in all organs of dormant grapevines (Stoev et al., 1960). In the dormancy phase, raffinose and trehalose increased in *V. riparia* with two-fold greater abundance in VRS42 buds (**Figures 5B,C**). In poplar, maltose concentrations are low, whereas raffinose and sucrose concentrations are high during dormancy, and it was suggested that enhanced expression of *SUCROSE SYNTHASE* and *GALACTINOL SYNTHASE* could drive carbohydrate accumulation toward oligosaccharides (Ruttink et al., 2007). Similarly, four *GALACTINOL SYNTHASE* genes in this study were up-regulated in VRS relative to SVS during the dormancy phase, or throughout the SD treatment. Raffinose is associated with freezing tolerance in grapevines, and *V. riparia* vines showed 5°C increased bud freezing tolerance in response SD (Fennell and Hoover, 1991; Stushnoff et al., 1993; Hamman et al., 1996; Jones et al., 1999; Fennell and Mathiason, 2002). Trehalose accumulated in a similar pattern as did raffinose. In potatoes, enhanced levels of trehalose-6-phosphate maintain tuber dormancy (Debast et al., 2011). Trehalose accumulates in small quantities in plants, and has been proposed to be a component of a sugar signaling system involving ABA and *SNF1-RELATED KINASE 1* (*SnRK1*) (Paul et al., 2008; Sonnewald and Sonnewald, 2014). It is also possible that raffinose and trehalose serve as storage carbohydrates that are recycled into glucose at bud break.

The phenylpropanoid biosynthesis network was enriched during the dormancy phase, and *RESVERATROL SYNTHASE* and two stilbene biosynthesis genes were up-regulated. Resveratrol increased in abundance, as did raffinose during the induction and dormancy phases (**Figure 5D**). However, a distinct relationship between metabolite levels and genes differentially expressed in the dormancy phase was not noted. Resveratrol increases in berries in response to dehydration and with tissue maturation in native grapevine species, and has been implicated in disease resistance and hypersensitive cell death (Chang et al., 2011; Deluc et al., 2011; Degu et al., 2014). Because resveratrol can act as an antioxidant, it may play a role in oxidative stress during low temperature and other abiotic stresses (Wang et al., 2010). It is also possible that resveratrol increases with maturation and provides disease protection, rather than being specific to dormancy (Langcake and Pryce, 1976).

There were 27 transcription factor families represented in the *V. riparia* phase specific genes (VR_Phase_Specific_DEGs). Twenty of these families had genes up-regulated during the dormancy phase, and seven also had members up-regulated in the induction phase (Supplementary Tables 4B,C). Of note were genes encoding members of the MYB, NAC, and WRKY transcription factor families. The MYB14 transcription factor binds to the promoter of *STILBENE SYNTHASE* and induces expression of the encoded enzyme in resveratrol biosynthesis (Fang et al., 2014). The MYB, NAC, ZINC FINGER, and LIM domain protein transcription factors have also been associated with secondary cell wall biosynthesis, and several cell wall and lignin biosynthesis genes and members of these transcription factor families are up-regulated during the dormancy phase (Oh et al., 2003; Zhong et al., 2010). However, it appears that only the *MYB14* gene is in common with the previously detected transcription factors in other dormancy systems (Supplementary Table 6). Three additional NAC family genes, *NAC19, NAC78,* and *NAC87* were up-regulated (Supplementary Table 4C). *NAC19* is a positive regulator of ABA signaling (Jensen et al., 2013). *NAC78* and *WRKY75* were up-regulated in rice overexpressing *MYB4* and associated with increased stress tolerance (Park et al., 2010). *NAC19* and *NAC87* are also up-regulated in Tempranillo grapevine buds during the transition into dormancy (Díaz-Riquelme et al., 2012). Several ZINC FINGER transcription factor genes were up-regulated during dormancy, the best characterized of these is *DOF1.* *DOF1,* a negative regulator of seed germination, appears to

integrate light and hormone signaling. In *Arabidopsis, DOF1* acts between *PHYTOCHROME INTERACTING FACTOR-LIKE 5 (PIL5)* and *GIBERELLIN 3 OXYDASE,* resulting in lower levels of gibberellin, whereas *PIL5* appears to result in increased ABA levels, thus maintaining seed dormancy (Lau and Deng, 2010). Another transcription factor gene also found in dormant seeds is *BTB/POZ DOMIAN CONTAINING PROTEIN* which encodes a TRAF transcription factor found in dormant tea and rice seed (Chen et al., 2010). A *SCARECROW TRANSCRIPTION FACTOR 14 (SCL14)* gene and *SUPEROXIDE DISMUTASE* gene were up-regulated in the dormancy phase. *SCL14* is a key transcriptional co-activator that acts with *SUPEROXIDE DISMUTASE* and other oxygen and radical detoxification genes in reactive oxygen species detoxification functions (Farmer and Mueller, 2013).

The transcription factor gene expression patterns indicated over 73 transcription factor genes (from 33 transcription factor gene families) as photoperiod regulated. Twenty of these are indicated as potential markers for the dormancy phase. Similarly, there was a large number of transport-related genes, including two vesicular transport genes that were up-regulated specifically during the dormancy phase. There were more transport genes up-regulated than down-regulated during the dormancy phase, and a distinctly different set than were found during the induction phase. Finally, there were several hormone signaling genes (genes associated with ethylene, cytokinin, auxin, and ABA) that were up-regulated during dormancy, indicating continued hormonal regulation of processes during dormancy. A comparison of the *V. riparia* dormancy phase differentially expressed genes with the Tempranillo bud dormancy related genes showed 113 genes in common, with 84% of these showing an up- or down-regulation pattern similar to that in *V. riparia* (Supplementary Table 6). However, most of the transcription factor and hormone signaling genes specific to the *V. riparia* dormancy phase are not present in the Tempranillo data set. It is possible that activation of these occurred for only a short period during early dormancy development and, therefore, they were not detected in the Tempranillo study. Comparisons across several woody dormancy systems highlight the commonality between poplar terminal bud dormancy and grapevine axillary bud dormancy. A three way comparison between poplar, leafy spurge and *V. riparia* dormancy associated genes showed flavonoid and phenylpropanoid biosynthesis genes in common. This analysis also highlighted one lignin biosynthesis gene (*Caffeic acid O-methyltransferase*) that is common to potato, raspberry, poplar, leafy spurge, and grapevine dormancy (Horvath et al., 2008). This gene might be used as a marker gene for dormancy induction.

CONCLUSIONS

Development of dormancy is a temporal and spatially regulated process in perennial species. In this study, the time frame for dormancy development in *V. riparia* was similar to that found in poplar and birch (Ruttink et al., 2007; Ruonala et al., 2008). This is striking because a smaller change in photoperiod (2 h) was used in this study, in comparison with the birch and poplar

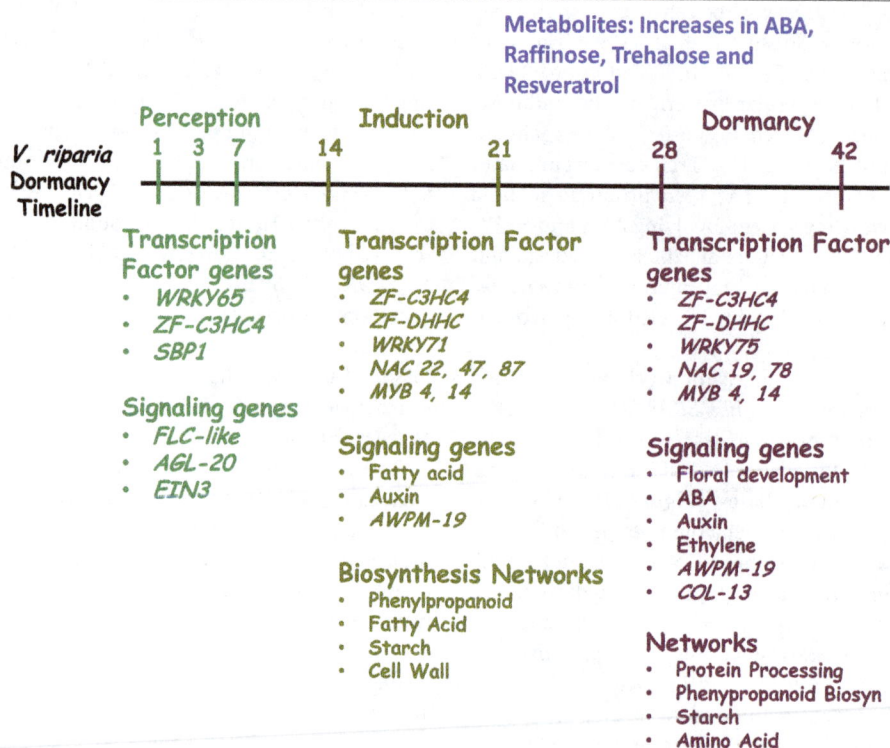

FIGURE 6 | Summary of the differentially expressed transcription factor and signaling genes and the enriched molecular networks during perception, induction or dormancy phase. Perception phase (D01, D03, D07) is in green; induction phase (D14, D21) is in olive and dormancy phase (D28, D42) is in purple. Metabolites differentially abundant during dormancy are in blue.

studies (6- to 8-h difference) (Ruonala et al., 2008; Ruttink et al., 2007). The similarity in time-frame indicates common photoperiod perception mechanisms between different perennial plant species. However, differences in downstream metabolite pools are likely a result of the larger photoperiod differences used in other studies (Ruttink et al., 2007). Furthermore, the 13 h photoperiod, and differential response type of the *V. riparia* and Seyval grapevines used in this study, allowed the separation of general photoperiod responses from those involved in dormancy development. Analysis of the dynamics of gene expression distinguished three phases: perception, induction and dormancy. Each phase was characterized by a distinct set of differentially expressed genes, with the dormancy phase containing the most. The induction and dormancy phases are characterized by different transcription factor and transporter sets of differentially expressed genes (**Figure 6**), including the NAC and WRKY transcription factor families, which have been commonly associated with water and salt stresses. The transitions from SD perception to dormancy indicate that *NAC22*, *NAC47*, *NAC87*, and *WRKY71* are characteristic of induction phase, whereas *NAC19*, *NAC78*, and *WRKY75* are characteristic of the dormancy phase (Supplementary Table 4). During induction, many changes in gene expression can be associated with known changes in cell walls, carbohydrates and hormone signaling, that have also been associated with cold acclimation. Finally, four metabolites (ABA, raffinose, trehalose,

and resveratrol) increased in concert with the transition to dormancy in *V. riparia*. While raffinose plays a role in freezing tolerance, an important characteristic of dormant overwintering buds, the specific role of trehalose and resveratrol remains to be studied. It is interesting that dormancy induction in *Populus* promotes an up-regulation of resveratrol synthase, but whether this is a photoperiodic response or a response to reduced water content in buds is not known. The distinct phases of dormancy development and photoperiod response characteristics distinguished in this study provide a framework for dissecting both early and downstream dormancy processes and interacting components.

AUTHOR CONTRIBUTIONS

AF, GRC, and KAS developed the experimental design. AF, SG, LGD, KM and LS conducted experiments. KAS, AF, JG, GRC, LGD, SG, and VK contributed to data analysis.

ACKNOWLEDGMENTS

This work was supported by National Science Foundation Plant Genome Research Grant DBI0604755, South Dakota State University Experiment Station and the South Dakota State University Functional Genomics Core Facility and the University of Nevada, Reno Proteomics Center. The Nevada Proteomics

Center and the Nevada Center for Bioinformatics are supported by a grant from the National Institute of General Medical Sciences (P20GM103440). We thank Drs. Gary Hatfield and Jixiang Wu for providing an external statistical review of this manuscript.

REFERENCES

Aida, M., Ishida, T., Fukai, H., Fujisawa, H., and Tasaka, M. (1997). Genes involved in organ separation in *Arabidopsis*: an analysis of the cup-shaped cotyledon mutant. *Plant Cell* 9, 841–857. doi: 10.1105/tpc.9.6.841

Altick, A., Feng, C. Y., Johnson, A. L., Schlauch., K., and Von Bartheld, C. S. (2012). Differences in gene expression between strabismic and normal human extraocular muscles. *Invest. Ophthalmol. Vis. Sci.* 53, 5168–5177. doi: 10.1167/iovs.12-9785

Anderson, J. V., Gesch, R. W., Jia, Y., Chao, W S., and Horvath, D. P. (2005). Seasonal shifts in dormancy status, carbohydrate metabolism, and related gene expression in crown buds of leafy spurge. *Plant Cell Envir.* 28, 1567–1578. doi: 10.1111/j.1365-3040.2005.01393.x

Aw, T., Schlauch, K., Keeling, C. I., Young, S., Bearfield, J. C., Blomquist, G. J., et al. (2010). Functional genomics of mountain pine beetle (Dendroctonus ponderosae) midguts and fat bodies. *BMC Genomics* 11:215. doi: 10.1186/1471-2164-11-215

Baldi, P., and Hatfield, G. W. (2002). *DNA Microarrays and Gene Expression*. Cambridge: University Press.

Bedon, F., Bomal, C., Caron, S., Levasseur, C., Boyle, B., Mansfield, S. D., et al. (2010). Subgroup 4 R2R3-MYBs in conifer trees: gene family expansion and contribution to the isoprenoid- and flavonoid-oriented response. *J. Exp. Bot.* 61, 3847–3864. doi: 10.1093/jxb/erq196

Benjamini, Y., and Hochberg, Y. (1995). Controlling the false discovery rate: a practical and powerful approach to multiple testing. *J. R. Statist. Soc. B* 57, 289–300.

Böhlenius, H., Huang, T., Charbonnel-Campaa, L., Brunner, A. M., Jansson, S., Strauss, S. H., et al. (2006). CO/FT regulatory module controls timing of flowering and seasonal growth cessation in trees. *Science* 312, 1040–1043. doi: 10.1126/science.1126038

Bomal, C., Duval, I., Giguere, I., Fortin, É., Caron, S., Stewart, D., et al. (2014). Opposite action of R2R3-MYBs from different subgroups on key genes of the shikimate and monolignol pathways in spruce. *J. Exp. Bot.* 65, 495–408. doi: 10.1093/jxb/ert398

Broeckling, C. D., Hulman, D. V., Farag, M. A., Smith, J. T., May, G. D., Mendes, P., et al. (2005). Metabolic profiling of medicago truncatula cell cultures reveals the effects of biotic and abiotic elicitor on metabolism. *J. Exp. Bot.* 56, 323–336. doi: 10.1093/jxb/eri058

Cadman, C. S. C., Toorop, P. E., Hilhorst, H. W. M., and Finch-Savage, W. E. (2006). Gene expression profiles of *Arabidops* Cvi seeds during dormancy cycling indicate a common underlying dormancy control mechanism. *Plant J.* 46, 805–822. doi: 10.1111/j.1365-313X.2006.02738.x

Chang, S., Puryear, J., and Cairney, J. (1993). A simple efficient method for isolating RNA from pine trees. *Plant Mol. Biol. Rep.* 11, 113–116. doi: 10.1007/BF02670468

Chang, X., Heene, E., Qiao, F., and Nick, P. (2011). The phytoalexin resveratrol regulates the initiation of hypersensitive cell death in *Vitis* cell. *PLoS ONE* 6:e264405. doi: 10.1371/journal.pone.0026405

Chao, Q., Rothenberg, M., Solano, R., Roman, G., Terzaghi, W., and Ecker, J. R. (1997). Activation of the ethylene gas response pathway in Arabidopsis by nuclear protein ETHYLENE-INSENSITIVE3 and related proteins. *Cell* 89, 1133–1144. doi: 10.1016/S0092-8674(00)80300-1

Chen, L., Song, L., Li, S., Zhang, L., Zou, C., and Yu, D. (2012). The role of WRKY transcription factors in plant abiotic stresses. *Biochimica et Biophysica Acta* 1819, 120–128. doi: 10.1016/j.bbagrm.2011.09.002

Chen, Q., Yang, L., Ahmad, P., Wan, X., and Hu, X. (2010). Proteomic profiling and redox status alteration of recalcitrant tea (*Camellia sinensis*) seed in response to desiccation. *Planta* 233, 583–592. doi: 10.1007/s00425-010-1322-7

Contento, A. L., Kim, S. J., and Bassham, D. C. (2004). Transcriptome profiling of the response of *Arabidopsis* suspension culture cells to Suc starvation. *Plant Physiol.* 135, 2330–2347. doi: 10.1104/pp.104.044362

Cooke, J. E., Eriksson, M. E., and Juntilla, O. (2012). The dynamic nature of bud dormancy in trees: environmental control and molecular mechanisms. *Plant Cell Environ.* 35, 1707–1728. doi: 10.1111/j.1365-3040.2012.02552.x

Cramer, G. R., Ergül, A., Grimplet, J., Tillett, R. L., Tattersall, E. A., Bohlman, M. C., et al. (2007). Water and salinity stress in grapevines: early and late changes in transcript and metabolite profiles. *Funct. Integr. Genomics* 7, 111–134. doi: 10.1007/s10142-006-0039-y

Debast, S., Nunes-Nesi, A., Hajirezaei, M. R., Hofmann, J., Sonnewald, U., Fernie, A. R., et al. (2011). Altering Trehalose-6-Phosphate content in transgenic potato tubers affects tuber growth and alters responsiveness to hormones during sprouting. *Plant Physiol.* 156, 1754–1771. doi: 10.1104/pp.111.179903

Degu, A., Hochberg, U., Sikron, N., and Venturini, L. (2014). Metabolite and transcript profiling of berry skin during fruit development elucidates differential regulation between Cabernet Sauvignon and Shiraz cultivars at branching points in the polyphenol pathway. *BMC Plant Biol.* 14, 188. doi: 10.1186/s12870-014-0188-4

Deluc, L. G., Decendit, A., Papstamoulis, Y., Merillon, J.-M., Cushman, J. C., and Cramer, G. R. (2011). Water deficit increases stilbene metabolism in Cabernet Sauvignon berries. *J. Agric. Food Chem.* 59, 289–297. doi: 10.1021/jf1024888

Deluc, L. G., Quilici, D. R., Decendit, A., Grimplet, J., Wheatley, M. D., Schlauch, K. A., et al. (2009). Water Deficit alters differentially metabolic pathways affecting important flavor and quality traits in grape berries of Cabernet Sauvignon and Chardonnay. *BMC Genomics* 10:212 doi: 10.1186/1471-2164-10-212

Díaz-Riquelme, J., Grimplet, J., Martínez-Zapater, J. M., and Carmona, M. J. (2012). Transcriptome variation along bud development in grapevine (*Vitis vinifera* L.). *BMC Plant Biol.* 12:181. doi: 10.1186/1471-2229-12-181

El Kayal, W., Allen, C. C. G., Ju, C. J.-T., Adams, E., King-Jones, S., Zaharia, L. I., et al. (2011). Molecular events of apical bud formation in white spruce, *Picea glauca*. *Plant Cell Environ.* 34, 480–500. doi: 10.1111/j.1365-3040.2010.02257.x

Fang, L. Y., Hou, L., Wang, H., Xin, N., Wang, S., and Li., S. (2014). *Myb14*, a direct activator of *STS*, is associated with resveratrol content variation in berry skin in two grape cultivars. *Plant Cell Rep.* 33, 1629–1640. doi: 10.1007/s00299-014-1642-3

Farmer, E. E., and Mueller, M. J. (2013). ROS-mediated lipid perozidation and RES-activated signaling. *Ann. Rev. Plant Biol.* 64, 429–450. doi: 10.1146/annurev-arplant-050312-120132

Fennell, A. (2004). Freezing tolerance and injury in grapevines. *J. Crop Improvement* 10, 201–235. doi: 10.1300/J411v10n01_09

Fennell, A., and Hoover, E. (1991). Photoperiod influences growth, bud dormancy and cold acclimation in *Vitis labruscana* and *V. riparia*. *J. Amer. Soc. Hort. Sci.* 116, 270–273.

Fennell, A., and Mathiason, K. (2002). "Early acclimation response in grapes (*Vitis*)," in Plant Cold Hardiness, eds P. Li and T. Palva (New York, NY: Kluwer Academic/Plenum Publishers), 93–107.

Forde, B. G., and Lea, P. J. (2007). Glutamate in plants: metabolism regulation and signaling. *J. Expt. Bot.* 58, 2339–2358. doi: 10.1093/jxb/erm121

Garris, A., Clark, L., Owens, C., McKay, S., Luby, J., Mathiason, K., et al. (2009). Mapping of photoperiod-induced growth cessation in the wild grape *V. riparia*. *J. Amer. Soc. Hort. Sci.* 134, 261–272.

Gautier, L., Cope, L., Bolstad, B., and Irizarry, R. A. (2004). Affy-analysis of Affymetrix GeneChip data at the probe level. *Bioinformatics* 20, 307–415. doi: 10.1093/bioinformatics/btg405

Gordon, A. D. (1999). *Classification. 2nd Edn.* Boca Raton, FL; London; New York,NY; Washington, D.C.: Chapman & Hall/RC.

Grimplet, J., Cramer, G. R., Dickerson, J. A., Mathiason, K., Hemert, J. V., and Fennell, A. (2009). VitisNet: "Omics" Integration through Grapevine Molecular Networks. *PLoS ONE* 4, e8365. doi: 10.1371/journal.pone.0008365

Grimplet, J., Hemert, J. V., Carbonell-Bejerano, P., Díaz-Riquelme, J., Dickerson, J. A., Fennell, A., et al. (2012). Comparative analysis of grapevine whole-genome gene predictions, functional annotation, categorization and integration of the predicted gene sequences. *BMC Res. Notes* 5:213.doi: 10.1186/1756-0500-5-213

Hachez, C., Zelazny, E., and Chaumont, F. (2006). Modulating the expression of aquaporin genes in planta: a key to understand their physiological functions? *Biochim. et Biophys. Acta* 8, 1142–1166. doi: 10.1016/j.bbamem.2006.02.017

Hamman, R. A., Dami, I. E., Walsh, T. M., and Stushnoff, C. M. (1996). Seasonal carbohydrate changes and cold hardiness of Chardonnay and Riesling grapevines. *Am. J. Enol. Vitic.* 47, 31–36.

Horvath, D. P., Chao, W. S., Suttle, J. C., Thimmapuram, J., and Anderson, J. V. (2008). Transcriptome analysis identifies novel responses and potential regulatory genes involved in seasonal dormancy transitions of leafy spurge (*Euphorbia esula* L.). *BMC Genomics* 9:536. doi: 10.1186/1471-2164-9-536

Irizarry, R. A., Hobbs, R., Collin, R., Beazer-Barclay, Y. D., Antonellis, K. J., Scherf, U., et al. (2003). Exploration, normalization and summaries of high density oligonucleotide array probe level data. *Biostatistics* 4, 249–264. doi: 10.1093/biostatistics/4.2.249

Jensen, M. K., Kjaersgaard, T., Nielsen, M. M., Galberg, P., Petersen, K., O'Shea, C., et al. (2013). The *Arabidopsis thaliana* NAC transcription factor family: structure-function relationships and determinants of ANAC019 stress signaling. Biochem. *J.* 426, 183–196. doi: 10.1042/BJ20091234

Jiménez, S., Li, Z., Reighard, G. L., and Bielenberg, D. G. (2010). Identification of genes associated with growth cessation and bud dormancy entrance using a dormancy-incapable tree mutant. *BMC Plant Biol.* 10, 25–35. doi: 10.1186/1471-2229-10-25

Johnson, W. E., Li, C., and Rabinovic, A. (2007). Adjusting batch effects in microarray expression data using empirical Bayes methods. *Biostatistics* 8, 118–127. doi: 10.1093/biostatistics/kxj037

Jones, K. S., Paroschy, J., McKersie, B. D., and Bowley, S. R. (1999). Carbohydrate composition and freezing tolerance of canes and buds in *Vitis vinifera*. *J. Plant Physiol.* 155, 101–106. doi: 10.1016/S0176-1617(99)80146-1

Kerr, M. K., Martin, M., and Churchill, G. A. (2000). Analysis of variance for gene expression microarray data. *J. Comput. Biol.* 7, 819–837. doi: 10.1089/10665270050514954

Knudsen, S. (2002). *A Biologist's Guide to the Analysis of DNA Microarray Data.* New York, NY: John Wiley & Sons.

Koike, M., Takezawa, D., Arakawa, K., and Yoshida, S. (1997). Accumulation of 19-kDa Plasma Membrane polypeptide during induction of freezing tolerance in wheat suspension-cultured cells by abscisic acid. *Plant Cell Physiol.* 38, 707–716. doi: 10.1093/oxfordjournals.pcp.a029224

Krochko, J. E., Abrams, G. D., Loewen, M. K., Abrams, S. R., and Cutler, A. J. (1998). (+)-abscisic acid 8'-hydroxylase is a cytochrome P450 monooxygenase. *Plant Physiol.* 118, 849–860. doi: 10.1104/pp.118.3.849

Kuhn, A., Schlauch, K., Lao, R., Halayko, A., Gerthoffer, W. T., and Singer, C. A. (2010). MicroRNA expression in human airway smooth muscle cells: role of miR-25 in regulation of airway smooth muscle phenotype. *Am. J. Respir. Cell Mol. Biol.* 42, 506–513. doi: 10.1165/rcmb.2009-0123OC

Langcake, P., and Pryce, R. J. (1976). The production of resveratrol by *Vitis vinifera* and other members of the Vitaceae as a response to infection or injury. *Physiol. Plant Pathol.* 9, 77–86. doi: 10.1016/0048-4059(76)90077-1

Lau, S., and Deng, X. W. (2010). Plant hormone signaling lightens up: integrators of light and hormone. *Curr. Opin. Plant Biol.* 13, 571–577. doi: 10.1016/j.pbi.2010.07.001

Less, H., and Galilli, G. (2008). Principal transcriptional programs regulating plant amino acid metabolism in response to abiotic stresses. *Plant Physiol.* 147, 316–220. doi: 10.1104/pp.108.115733

Liu, Y., Koornneef, M., and Soppe, W. J. J. (2007). The absend of histone H2B monoubiquitination in the *Arabidopsis* hub1 (rdo4) mutant reveals a role for chromatin remodeling in seed dormancy. *Plant Cell* 19, 433–444. doi: 10.1105/tpc.106.049221

Miller, G., Schlauch, K., Tam, R., Cortes, D., Torres, M. A., Shulaev, V., et al. (2009). The plant NADPH oxidase RBOHD mediates rapid systemic signaling in response to diverse stimuli. *Sci Signal.* 2, ra45. doi: 10.1126/scisignal.2000448

Mullins, M. G., Bouquet, A., and Williams, L. E. (1992). *Biology of the Grapevine.* Cambridge: Cambridge University Press.

Oh, S., Park, S., and Han, K. H. (2003). Transcriptional regulation of secondary growth in *Arabidopsis thaliana*. *J. Exp. Bot.* 54, 2709–2722. doi: 10.1093/jxb/erg304

Olsen, J. E., Junttila, O., Nilsen, J., Eriksson, M. E., Martinussen, I., Olsson, O., et al. (1997). Ectopic expression of oat phytochrome A in hybrid aspen changes critical daylength for growth and prevents cold acclimatization. *Plant J.* 12, 1339–1350. doi: 10.1046/j.1365-313x.1997.12061339.x

Owen, S. J., and Abrams, S. R. (2009). "Measurement of plant hormones by liquid chromatography-mass spectrometry in plant hormones" in *Methods and Protocols*, eds. S. Cutler and D. Bonetta (New York, NY: Humana Press), 495.

Park, M.-R., Yun, K.-Y., Mohanty, B., Herath, V., Xu, F., Wijaya, E., et al. (2010). Supra-optimal expression of the cold-regulated *OsMyb4* transcription factor in transgenic rice changes the complexity of transcriptional network with major effects on stress tolerance and panicle development. *Plant Cell Environ.* 33, 2209–2230. doi: 10.1111/j.1365-3040.2010.02221.x

Pasquali, G., Biricolti, S., Locatelli, F., Baldoni, E., and Mattana, M. (2008). *Osmyb4* expression improves adaptive responses to drought and cold stress in transgenic apples. *Plant Cell Rep.* 27, 1677–1686. doi: 10.1007/s00299-008-0587-9

Paul, M. J., Primavesi, L. F., Jhurreea, D., and Zhang, Y. (2008). Trehalose metabolism and signaling. *Annu. Rev. Plant Biol.* 59, 417–441. doi: 10.1146/annurev.arplant.59.032607.092945

Puranik, S., Pranav, P. S., Srivastava, P. S., and Prasad, M. (2012). NAC proteins: regulation and role in stress tolerance. *Trends Plant Sci.* 17, 369–381. doi: 10.1016/j.tplants.2012.02.004

Resman, L., Howe, G., Jonsen, D., Englund, M., Druart, N., Schrader, J., et al. (2010). Components acting downstream of short day perception regulate differential cessation of cambial activity and associated responses in early and late clones of hybrid poplar. *Plant Physiol.* 154, 1294–1303. doi: 10.1104/pp.110.163907

Rhode, A., Prinsen, E., De Rycke, R., Engler, G., Van Montagu, M., and Boerjan, W. (2002). PtABI3 impinges on the growth and differentiation of embryonic leaves during bud set in poplar. *Plant Cell* 14, 1885–1901. doi: 10.1105/tpc.003186

Rinne, P. L. H., Kaikuranta, P. M., and van der Schoot, C., (2001). The shoot apical meristem restores its symplasmic organization during chilling-induced release from dormancy. *Plant J.*, 26, 249–264. doi: 10.1046/j.1365-313X.2001.01022.x

Ruonala, R., Rinne, P. L. H., Baghour, M., Moritz, T., Tuominen, H., and Kangajärvi, J. (2006). Transitions in the functioning of the shoot apical meristem in birch (*Betula pendula*) involve ethylene. *Plant J.* 46, 628–640. doi: 10.1111/j.1365-313X.2006.02722.x

Ruonala, R., Rinne, P. L. H., Kangasjärvi, J., and van der Schoot, C. (2008). *CENL1* expression in the rib meristem affects stem elongation and the transition to dormancy in *Populus*. *Plant Cell* 20, 59–74. doi: 10.1105/tpc.107.056721

Rushton, P. J., Somssich, I. E., Ringler, P., and Shen, Q. J. (2010). WRKY transcription factors. *Trends Plant Sci.* 15, 247–258. doi: 10.1016/j.tplants.2010.02.006

Ruttink, T., Arend, M., Morreel, K., Storme, V., Rombauts, S., Fromm, J., et al. (2007). A molecular timetable for apical bud formation and dormancy induction in poplar. *Plant Cell* 19, 2370–2390. doi: 10.1105/tpc.107.052811

Santamaría, M. E., Rodríuez, R., Cañal, M. J., and Toorop, P. E. (2011). Transcriptome analysis of chestnut (*Castanea sativa*) tree buds suggests a putative role for epigenetic control of bud dormancy. *Ann. Bot.* 108, 485–498. doi: 10.1093/aob/mcr185

Schnabel, B. J., and Wample, R. L. (1987). Dormancy and cold hardiness in *Vitis vinifera* L. cv. White Riesling as influenced by photoperiod and temperature. *Amer. J. Enol. Vitic.* 34, 265–272.

Solano, R., Stepanova, A., Chao, Q., and Ecker, J. R. (1998). Nuclear events in ethylene signaling: a transcriptional cascade mediated by ETHYLENE-INSENSITIVE3 and ETHYLENE-RESPONSE-FACTOR1. *Genes Dev.* 12, 1703–3714. doi: 10.1101/gad.12.23.3703

Sonnewald, S., and Sonnewald, U. (2014). Regulation of potato tuber sprouting. *Planta* 239, 27–38. doi: 10.1007/s00425-013-1968-z

Sreekantan, L., Mathiason, K., Grimplet, J., Schlauch, K., Dickerson, J., and Fennell, A. (2010). Differential floral development and gene expression in grapevines during long and short photoperiods suggests a role for floral genes in dormancy transitioning. *Plant Mol. Bio.* 73, 191–205. doi: 10.1007/s11103-010-9611-x

Stekel, D. (2003). *Microarray Bioinformatics*. Cambridge; New York, NY; Melbourne, VIC; Madrid; Cape Town; Singapore; Sao Paulo: Cambridge University Press.

Stoev, K. D., Manarov, P. T., and Benchev, I. B. (1960). Sugars and free amino acids during ripeningand dormancy of the grape plant. *Fiziol. Rast.* 7, 145–150.

Stushnoff, C., Remmele, R. L., Esensee, V., and McNeil, M. (1993). "Low temperature induced biochemical mechanisms: implications for cold acclimation and de-acclimation," in *Interacting Stresseson Plants in a Changing Climate*, Vol. 116, eds M. B. Jackson and C. R. Black (Berlin; Heidelberg: Springer-Verlag), 647–657.

Subramanian, A., Kuehn, H., Gould, J., Tamayo, P., and Mesirov, J. P. (2007). GSEA-P: a desktop application for gene set enrichment analysis. *Bioinformatics* 23, 3251–3253. doi: 10.1093/bioinformatics/btm369

Subramanian, A., Tamayo, P., Mootha, V. K., Mukherjee, S., Ebert, B. L., Gillette, M. A., et al. (2005). Gene set enrichment analysis: a knowledge-based approach for interpreting genome-wide expression profiles. *Proc. Natl. Acad. Sci. U.S.A.* 102, 15545–15550. doi: 10.1073/pnas.0506580102

Tattersall, E. A., Grimplet, J., DeLuc, L., Wheatley, M. D., Vincent, D., Osborne, C., et al. (2007). Transcript abundance profiles reveal larger and more complex responses of grapevine to chilling compared to osmotic and salinity stress. *Funct. Integr. Genomics* 7, 317–333. doi: 10.1007/s10142-007-0051-x

Ueno, S., Klopp, C., Leplé, J. C., Derory, J., Noirot, C., Léger, V., et al. (2013). Transcriptional profiling of bud dormancy induction and release in oak by next-generation sequencing. *BMC Genomics* 14, 236–250. doi: 10.1186/1471-2164-14-236

Wake, C. M. F., and Fennell, A. (2000). Morphological, physiological and dormancy responses of three Vitis genotypes to short photoperiod. *Physiol. Plant* 109, 203–210 doi: 10.1034/j.1399-3054.2000.100213.x

Wang, M., Vannozzi, A., Wang, G., Liang, Y.-H., Tornielli, G. B., Zenoni, S., et al. (2014). Genome and transcriptome analysis of the grapevine (*Vitis vinifera* L.) WRKY gene family. *Hort. Res.* 1, 14016. doi: 10.1038/hortres.2014.16

Wang, W., Tang, K., Yang, H. R., Wen, P. F., Zhang, P., Wang, H. L., et al. (2010). Distribution of resveratrol and stilbene synthase in young grape plants (*Vitis vinifera* L. cv. Cabernet Sauvignon) and the effect of UV-C on its accumulation. *Plant Physiol. Biochem.* 48, 142–152. doi: 10.1016/j.plaphy.2009.12.002

Wood, M., Rae, G. M., Wu, R.-M., Walton, E. T., Xue, B., Hellens, R. P., et al. (2013). *Actinidia* DRM1 An intrinsically disordered protein whose mRNA expression is inversely correlated with spring budbreak in kiwifruit. *PLoS ONE* 8:e57354. doi: 10.1371/journal.pone.0057354

You, J., Zhang, L., Song, B., Qi, X., and Chan, Z. (2015). Systematic analysis and identification of stress-responsive genes of the NAC gene family in *Brachypodium distachyon*. *PLoS ONE* 10:e0122027. doi: 10.1371/journal.pone.0122027

Zheng, C., Halaly, T., Kwame Acheampong, A., Takebayashi, Y., Jikumaru, Y., Kamiya, Y., et al. (2015). Abscisic Acid (ABA) regulates grape bud dormancy, and dormancy release stimuli may act through modification of ABA metabolism. *J. Exp. Bot.* 66, 1527–1542. doi: 10.1093/jxb/eru519

Zhong, R., Lee, C., and Ye, Z.-H. (2010). Evolutionary conservation of the transcriptional network regulating secondary cell wall biosynthesis. *Trends Plant Sci.* 15, 625–632. doi: 10.1016/j.tplants.2010.08.007

Zhou, X., He, F., Liu, F., Zheng, X., Di, C., Zhou, S., et al. (2007). "Integration of cold signal transduction pathway related to ABA 8'- hydroxylase in *Arabidopsis*," in *Biotechnology and Sustainable Agriculture 2006 and Beyond*, ed Z. Xu (Dordrecht: Springer), 414–474.

Conflict of Interest Statement: The authors declare that the research was conducted in the absence of any commercial or financial relationships that could be construed as a potential conflict of interest.

17

Water Use Patterns of Four Tropical Bamboo Species Assessed with Sap Flux Measurements

Tingting Mei[1]*[†], Dongming Fang[1†], Alexander Röll[1], Furong Niu[1], Hendrayanto[2] and Dirk Hölscher[1]

[1] Tropical Silviculture and Forest Ecology, Georg-August-Universität Göttingen, Göttingen, Germany, [2] Department of Forest Management, Institut Pertanian Bogor, Bogor, Indonesia

Bamboos are grasses (Poaceae) that are widespread in tropical and subtropical regions. We aimed at exploring water use patterns of four tropical bamboo species (*Bambusa vulgaris, Dendrocalamus asper, Gigantochloa atroviolacea,* and *G. apus*) with sap flux measurement techniques. Our approach included three experimental steps: (1) a pot experiment with a comparison of thermal dissipation probes (TDPs), the stem heat balance (SHB) method and gravimetric readings using potted *B. vulgaris* culms, (2) an *in situ* calibration of TDPs with the SHB method for the four bamboo species, and (3) field monitoring of sap flux of the four bamboo species along with three tropical tree species (*Gmelina arborea, Shorea leprosula,* and *Hevea brasiliensis*) during a dry and a wet period. In the pot experiment, it was confirmed that the SHB method is well suited for bamboos but that TDPs need to be calibrated. *In situ*, species-specific parameters for such calibration formulas were derived. During field monitoring we found that some bamboo species reached high maximum sap flux densities. Across bamboo species, maximal sap flux density increased with decreasing culm diameter. In the diurnal course, sap flux densities in bamboos peaked much earlier than radiation and vapor pressure deficit (VPD), and also much earlier than sap flux densities in trees. There was a pronounced hysteresis between sap flux density and VPD in bamboos, which was less pronounced in trees. Three of the four bamboo species showed reduced sap flux densities at high VPD values during the dry period, which was associated with a decrease in soil moisture content. Possible roles of internal water storage, root pressure and stomatal sensitivity are discussed.

Keywords: calibration, environmental drivers, hysteresis, stem heat balance, thermal dissipation probes, trees, bamboos

Edited by:
Julia Cooke,
The Open University, UK

Reviewed by:
Teresa E. Gimeno,
Institut National de la Recherche
Agronomique, France
Tomonori Kume,
National Taiwan University, Taiwan

*Correspondence:
Tingting Mei
tmei@gwdg.de

[†] These authors have contributed
equally to this work.

INTRODUCTION

Bamboos (Poaceae, Bambuseae) are abundant in the natural vegetation of tropical and subtropical regions. They have been used by people for millennia and are still used as food and construction materials. In addition, a large variety of bamboo usages have been developed in recent decades, for example for pulp, paper, or clothing production (International Network for Bamboo and Rattan [INBAR], 2014). The increasing economic exploitation of bamboos goes along with a considerable expansion of bamboo plantations in some regions (Chen et al., 2009;

Food Agriculture Organization [FAO], 2010), which may lead to changes in ecological processes such as water use patterns (Uchimura, 1994; Komatsu et al., 2010). Some bamboo stands were reported to evaporate more water than tree-dominated forests (Komatsu et al., 2010; Ichihashi et al., 2015), but studies focusing on water use patterns of bamboos are still rare thus far (Pereira and Hosegood, 1962; Dierick et al., 2010; Komatsu et al., 2010; Kume et al., 2010; Ichihashi et al., 2015).

Water use patterns of bamboos and trees differ in several aspects. In contrast to trees, bamboos are monocotyledonous species and lack secondary growth (Zimmermann and Tomlinson, 1972). Therefore, vascular conduits of bamboo xylem have to remain functional throughout the ontogeny of a bamboo culm. Bamboos consequently have great ability to avoid cavitation (Cochard et al., 1994; Cao et al., 2012; Petit et al., 2014); root pressure mechanisms may contribute to repairing embolized conduits at night (Cao et al., 2012). Such features and structural traits of bamboos may also lead to particular water use patterns.

In general, plant water use is driven by micrometeorological factors and can be limited by soil water availability (O'Brien et al., 2004; Bovard et al., 2005; Kume et al., 2007); it is regulated by stomata opening and closing (Jarvis, 1989) and can be influenced by internal water storage mechanisms (Waring and Running, 1978; Goldstein et al., 1998; Carrasco et al., 2014). Xylem sap flux reflects these multiple factors. For some tree species, for example, hysteresis in the diurnal sap flux response to radiation and vapor pressure deficit (VPD) of the air have been reported (Goldstein et al., 1998; O'Brien et al., 2004). Sap flux measurements thus appear suitable to study the water use patterns of bamboos as well as their controlling environmental factors.

Thermal dissipation probes (TDP) are widely used to measure sap flux density (J_s) in trees (Granier, 1985). Several studies suggest calibrating the method before studying new species (Lu et al., 2004; Wullschleger et al., 2011; Vandegehuchte and Steppe, 2013). To our knowledge, only two studies have applied the TDP method on bamboos so far. Both reported an underestimation of bamboo sap flux compared to stem heat balance (SHB) and reference gravimetric measurements (GM) when the TDP method was not calibrated (Dierick et al., 2010; Kume et al., 2010). In contrast, the SHB method (Sakuratani, 1981) was suggested to be well suited for sap flux measurements on bamboos (Dierick et al., 2010). Bamboo culms are hollow; hence heat loss in the form of heat storage inside culms is marginal, so that steady thermal conditions as a main assumption of the method are met (Baker and Van Bavel, 1987).

The aim of this study was to analyze water use patterns of tropical bamboo species and particularly the response of J_s to the principal environmental drivers. First, we calibrated the SHB and the TDP method with reference GM in an experiment on potted culms of *Bambusa vulgaris*. We then measured J_s in the field in four bamboo species including *B. vulgaris* with both the TDP and SHB method, and calibrated the TDP method with the SHB method. Herein, three factors which may influence the quality of the calibration were tested: time step of the data, formula specificity and calibration formula type. After calibration of the TDP method, we applied it to monitor J_s in four bamboo

and three tree species in a common garden in Bogor, Indonesia. Differences in the response of J_s to fluctuations in environmental conditions were assessed. The study intends to contribute to expanding the yet limited knowledge on the eco-hydrological functioning of bamboos.

MATERIALS AND METHODS

Study Sites and Species Selection

The pot calibration experiment was conducted in Guangzhou, China (23°26′13″ N, 113°12′33″E, 13 m asl). The field calibration experiment and monitoring campaign were carried out in a common garden in Bogor, Indonesia (6°33′40″ S, 106°43′27″ E, 182 m asl). Average annual temperature in Bogor is 25.6°C and annual precipitation is 3978 mm. Relatively dry conditions with consecutive rainless days can occur between June and September. During this dry period, monthly precipitation is on average 40% lower than during the wet period (230 vs. 383 mm), and the number of consecutive dry days (rainfall < 1 mm) is twice that of the wet period (8 vs. 4 days, 1989–2008, Van Den Besselaar et al., 2014). During our study period (July 2012–January 2013), differences between dry and wet period were more pronounced, i.e., 155 vs. 489 mm monthly precipitation, 14 vs. 2 consecutive dry days, and 0.29 vs. 0.39 m^{-3} m^{-3} daily soil water content. In Bogor, four bamboo species (*B. vulgaris, Dendrocalamus asper, Gigantochloa atroviolacea, G. apus*) with five culms per species and three tree species (*Gmelina arborea, Shorea leprosula* and *Hevea brasiliensis*, **Table 1**) with five stems per species were selected and their J_s were monitored with the TDP method for 7 months.

TDP Construction and Installation

To measure J_s in trees and bamboos, we used self-made TDP (1 and 2 cm length, respectively). In sensor design and construction, we followed Wang et al. (2012). Each TDP sensor was comprised of a heating (downstream) and a reference (upstream) probe made of steel hypodermic needles. The probes were placed 10 cm apart (vertically). For bamboos and trees, TDP installation depths in culms and stems were 1 and 2 cm, respectively. After installation, each TDP was supplied with a constant current of

TABLE 1 | Structural characteristics of the studied bamboo and tree species (n = 5 per species; mean ± SD).

	Species	DBH (cm)	Bamboo culm wall thickness (cm)	Height (m)
Bamboo	*B. vulgaris*	7.0 ± 0.3	1.3 ± 0.1	17.9 ± 0.8
	G. apus	8.6 ± 0.4	1.2 ± 0.2	16.2 ± 2.7
	D. asper	11.9 ± 1.9	2.4 ± 0.2	21.1 ± 0.9
	G. atroviolacea	8.9 ± 0.6	1.6 ± 0.1	17.0 ± 1.0
Tree	*H. brasiliensis*	27.4 ± 2.3	–	25.2 ± 3.0
	G. arborea	26.3 ± 7.7	–	26.5 ± 2.3
	S. leprosula	20.7 ± 4.8	–	19.2 ± 2.5

Culm wall thickness (derived from five culms per species) and culm height (derived from three cut culms per species) of the studied bamboos.

120 mA; the respective power outputs of 1 and 2 cm length TDP were ~0.1 and ~0.2 W. TDP signals were sampled every 30 s and stored as 10-min averages for the pot calibration experiment and as 1-min averages for all other experiments by data loggers and multiplexers (CR1000, AM16/32, Campbell Scientific Inc., USA).

Calibration of the TDP Method
Pot Calibration Experiment: TDP, SHB, and GM
Five culms of *B. vulgaris* (diameter 5.3–7.3 cm, height 2.2–3.2 m) with trimmed canopies were transplanted into plastic bags (diameter 30 cm, height 25 cm) 6 months before the calibration experiment. One day before the experiment, the five bamboos were transplanted into bigger plastic pots (diameter 50 cm, height 65 cm). The pots were filled with cobblestones and water and were then fully sealed with plastic cover and aluminum foil to prevent evaporation of water from the pots (**Figure 1A**). A scaled syringe tube was attached to each pot and connected into the pot through a U-type tube. At the beginning of the experiment, the water was added into the pot through the syringe tube to a fixed level (5 cm below the pot cover). Subsequently, water was added manually every 30 min to reach the pre-defined level. The weight of the added water was determined gravimetrically (GM). To measure J_s, each bamboo culm was equipped with three pairs of 1 cm length TDP which were evenly installed circumferentially, about 15 cm above the plastic cover. To minimize potential measurement errors induced by circumferential variations of J_s, the thermocouple wires of the three TDP were connected in

parallel to get an average voltage signal for each bamboo culm (Lu et al., 2004). For a second J_s estimate, a SHB gage (SGB50 or SGA70, Dynagage Inc., USA) was installed about 1.5 m above the TDP. Both sensor types were protected by foil and the sensor signals were subsequently recorded as described in Section "TDP Construction and Installation." For the comparison to reference GM, 10-min TDP and SHB derived values were aggregated to half-hourly values.

To assess the performance of TDP and SHB in the pot experiments, J_s derived from TDP and SHB (J_{s_TDP} and J_{s_SHB}, respectively) on daily and 30-min scales were compared to GM derived J_s (J_{s_GM}) with paired *t*-tests. Additionally, the slopes of the respective linear fits between J_{s_TDP}, J_{s_SHB}, and J_{s_GM} were tested for significant differences from one with the test of homogeneity of slopes. The same statistical analyses were applied again later when testing for significant differences between J_{s_TDP} and J_{s_SHB} in the field calibration experiments.

Field Calibration Experiment: TDP and SHB
Five culms per bamboo species (*B. vulgaris*, *D. asper*, *G. atroviolacea*, *G. apus*) were selected for TDP measurements (**Table 1**), three to four of which were additionally measured with SHB for a field calibration of the TDP method. TDP sensors were installed at 1.3 m height, and SHB gages (SGB50, SGA70, Dynagage Inc., USA) were installed about 2.5 m above the TDP. Simultaneous TDP-SHB measurements were conducted for a minimum of 5 days per culm (**Figure 1B**). Heat storage inside bamboo culms is assumed to be negligible, which was confirmed

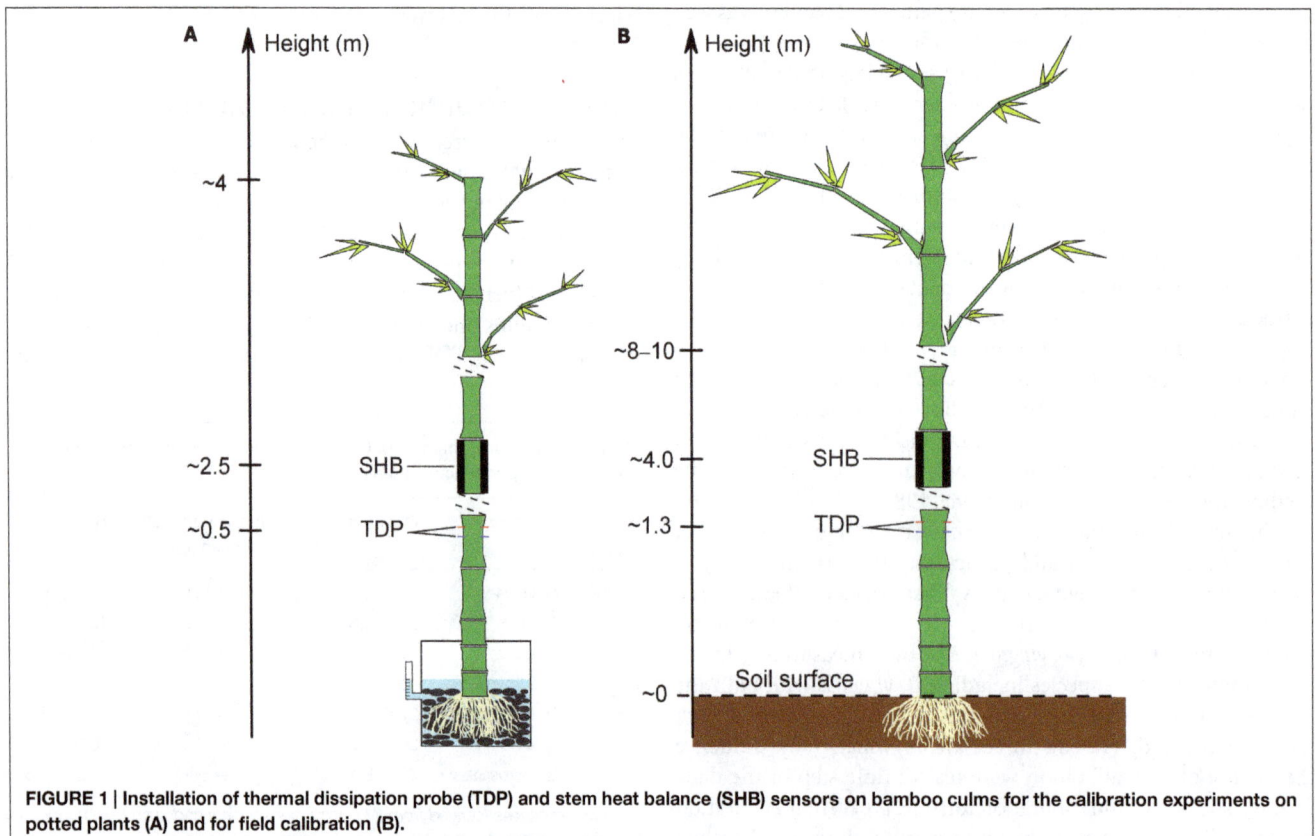

FIGURE 1 | Installation of thermal dissipation probe (TDP) and stem heat balance (SHB) sensors on bamboo culms for the calibration experiments on potted plants (A) and for field calibration (B).

by installing thermocouple wires inside the measured segments of the respective bamboo culms to detect fluctuations in culm temperature (Dierick et al., 2010). The observed fluctuations were marginal, which meant stable thermal conditions as a requirement of the SHB method were met.

Parametrization for TDP Calibration

We derived cross-sectional water conductive areas (A_{TDP}) from the culm wall thickness at the location of TDP sensor installation. In the pot calibration experiment, reference J_s were calculated by dividing water flow rates (g h^{-1}, GM-derived) by A_{TDP}. In the field calibration experiment, reference J_s were taken from the SHB measurements. The reference J_s could subsequently be used to calibrate J_{s_TDP}. Nighttime sap flux values were excluded in both calibration experiments.

In the field calibration, three factors were considered for obtaining a TDP calibration formula from reference (SHB) measurements: time step of the data, formula specificity and calibration formula type. To examine effects of varying time steps, the formulas were built and tested on data at varying intervals (1-, 10-, 30-, and 60-min averages, respectively). The effects of formula specificity were examined by using common (i.e., all bamboo species pooled), species-specific and culm-specific formulas, respectively. Regarding the calibration formula type, two formulas were compared: one was non-linear ($J_s = aK^b$) and generated by deriving new a and b parameters for the original Granier (1985) formula. The second was a linear formula ($J_{s\_SHB} = c \times J_{s\_TDP}$) which was calculated from the linear relationship between J_{s_TDP} and J_{s_SHB}.

To obtain stable calibration formulas, pooled data sets were randomly split in half for calibration and independent validation, respectively (Niu et al., 2015). First, for each time step (1-, 10-, 30-, and 60-min, respectively), a data pool was built. Three culms of each bamboo species were randomly chosen, and for each, 3 days of data were randomly chosen from an initial common dataset. With these data pools, formula specificity was examined. For the common calibration, culms of all four species were selected for calibration. For species-specific and culm-specific calibration, only the data of the respective species or culms was selected. Next, the selected data was randomly split in half, for building the calibration formula and testing it, respectively. When testing the formula, the differences between J_{s_SHB} and calibrated J_{s_TDP} ($J_{s\_TDP\_cali}$, abnormal distribution, $P > 0.05$) were examined with the Wilcoxon Signed-Rank Test (no significant differences at $P > 0.05$). The process of randomly building and testing the formula was iterated 10,000 times. Final calibration formula parameters were derived by averaging the parameters of those iterations which passed the Wilcoxon Signed-Rank Test ($P > 0.05$).

For an evaluation of the performance of the different formulas and the influence of the three factors (time scale, formula specificity and calibration formula type), differences in normalized Root-Mean-Square Errors (nRMSE) were assessed for each culm, species and formula factor, respectively. First, the RMSE for each day was derived with the J_{s_SHB} and $J_{s\_TDP\_cali}$ values, and the nRMSE was calculated by normalizing the RMSE with the observed daily range of J_{s-SHB} (difference between maximum and minimum J_{s_SHB}). Then, the nRMSE were analyzed regarding the three formula factors (data time scale, formula specificity and calibration formula type) by ANOVA (Analysis of variance). Additionally, for each day, $J_{s\_TDP\_cali}$ with each formula type was tested for significant differences from J_{s_SHB} with the Wilcoxon Signed-Rank Test. The rates of passing the Wilcoxon Signed-Rank Test ($P > 0.05$ when no significant difference between TDP and SHB derived values) were assessed for each formula.

Field Study

Monitoring Bamboo and Tree Sap Flux

Four calibrated bamboo species as well as three tree species (G. arborea, S. leprosula, and H. brasiliensis) were monitored with the TDP method for 7 months (July, 2012–January, 2013). Five bamboo culms and five tree trunks per species were selected for the measurements. On bamboos, three pairs of TDP (10 mm in length) were installed evenly around each culm at 1.3 m height and connected in parallel (see TDP Construction and Installation). On trees, two pairs of 20 mm TDP were installed in the trunk 1.3 m above the ground, in the North and South, respectively. J_s for the two sensors were separately derived with the original calibration formula (Granier, 1985) and subsequently averaged to obtain values for each tree. For bamboos, J_s derived with the original formula were calibrated with species-specific calibration parameters (from reference SHB field measurements) to obtain final J_s values.

Environmental Measurements and Analyses

A micrometeorological station was set up in an open area. It was about 100 and 600 m away from the closer measurement sites (D. asper, G. arborea, G. atroviolacea, G. apus, S. leprosula) and farthest sites (B. vulgaris, H. brasiliensis), respectively. Air temperature (Ta, °C) and air relative humidity (RH, %) were measured with a temperature and relative humidity probe (CS215, Campbell) installed in a radiation shield. VPD (kPa) was calculated from Ta and RH. Radiation (J m^{-2} s^{-1}) was measured with a pyranometer (CS300, Campbell). Data were recorded with the previously described data loggers every minute.

In addition to the mentioned micrometeorological variables, soil moisture (SM, m^{-3} m^{-3}) was measured with time domain reflectometry sensors (TDR, CS616, Campbell) at 0–20 cm depth. As the clump of D. asper and the stand of G. arborea were next to each other, one TDR was positioned between them to measure soil moisture. Likewise, one sensor was used for measurements of G. atroviolacea and G. apus. One TDR each were used for the remaining species (S. leprosula, B. vulgaris, H. brasiliensis). TDR measurements ran in parallel to the sap flux field campaign and data were recorded with the described data loggers every minute.

For the day-to-day analysis of influences of fluctuations in environmental conditions (VPD, radiation, SM) on J_s in the studied bamboo and tree species, daily accumulated J_s (kg cm^{-2} d^{-1}) were normalized by setting the highest daily observation of each species to one and the lowest to zero. For a more isolated analysis of potentially limiting influences of soil moisture on J_s, we focused on 'dry period conditions' with consecutive rainless days, which occurred between June and September in the study area. During this period, monthly precipitation was only 32% of monthly wet period precipitation (155 vs. 489 mm), and the number of consecutive dry days (rainfall < 1 mm) was seven times higher than during the wet period (14 vs. 2 days). Dry period conditions are also characterized by higher VPD (average daily VPD > 0.74 kPa on 92% of the days). 0.74 kPa was chosen as the threshold to distinguish between dry and wet period because it constituted the mean maximum ('turning point') in the fitted J_s response functions to VPD in three of the four studied bamboo species (except *D. asper*, see **Figure 4B**).

For the diurnal analysis of influences of fluctuations in environmental conditions on J_s, time lags between J_s and micrometeorological drivers (radiation and VPD) were calculated as the time difference between the respective occurrences of maximal J_s (J_{s_max}) and maximal radiation and VPD. T-tests were used to test time lags for significant differences from 0 min. 30-min J_s values (average values of three selected sunny days) of each species were plotted against radiation and VPD to examine occurrences of hysteresis. The respective areas of hysteresis were compared between bamboos and trees with t-tests.

All data analyses were performed with SAS 9.3 (SAS Institute Inc., Cary, NC, USA, 2013).

RESULTS

Calibration of the TDP Method for Bamboos

Pot Calibration Experiment: TDP, SHB, and GM

In the pot calibration experiment with *B. vulgaris*, SHB yielded similar absolute values of J_s as GM on daily and 30-min scales ($P > 0.05$). The slope of the linear fit between SHB and GM on the 30-min scale was 0.98 ($R^2 = 0.93$, $P < 0.01$). It did not significantly differ from 1 ($P > 0.05$, **Figure 2A**). In contrast to this, TDP estimates, with the original parameters of the calibration formula (Granier, 1985),

differed substantially from GM values at both the daily (60% underestimation of accumulated J_s, $P < 0.01$) and 30-min scale (56% underestimation, $P < 0.01$). The slope of the linear fit between TDP and GM on the 30-min scale was 0.44 ($R^2 = 0.84$, $P < 0.01$). It was significantly different from 1 ($P < 0.01$, **Figure 2A**).

After applying the TDP calibration parameter for *B. vulgaris* derived from the pot experiment ($c = 2.28$), the 30-min J_{s_TDP} were in line with those from GM. The slope was not significantly different from 1 ($P > 0.05$, **Figure 2B**). When applying the calibration parameters derived for *B. vulgaris* from the SHB field calibration experiment ($c = 2.79$), J_{s_TDP} was 19% higher than J_{s-GM} ($P < 0.01$, **Figure 2B**).

Field Calibration Experiment: TDP and SHB

Formula type and data time step had no significant influence on the performance of the calibration formula, but it mattered whether culm- or species-specific or a common calibration formula was used (Appendix Tables A1 and A2 Supplementary Material). Based on the nRMSE and the passing rate of the Wilcoxon test ($P > 0.05$) between calibrated J_{s_TDP} and J_{s_SHB}, culm-specific formulas performed better than species-specific and common formulas. In our study, there was no statistically significant difference between the species-specific and the common calibration parameters (**Table 2**, $P > 0.05$). For two of the four studied bamboo species (*G. apus* and *B. vulgaris*), however, using species-specific formulas slightly improved the quality of predictions as compared to applying the common formula ($P = 0.06$ and 0.07, respectively, **Table 2**). These two bamboo species had lower nRMSE and higher passing rates than *D. asper* and *G. atroviolacea* (Appendix Table A2 in Supplementary Material). The linear calibration parameters of the four bamboo species were significantly different from each other ($P < 0.01$). The linear calibration parameters, the slopes of J_{s_TDP} vs. J_{s_SHB}, were examined with the test of homogeneity of slopes and were found to differ significantly from each other ($t > 0.01$).

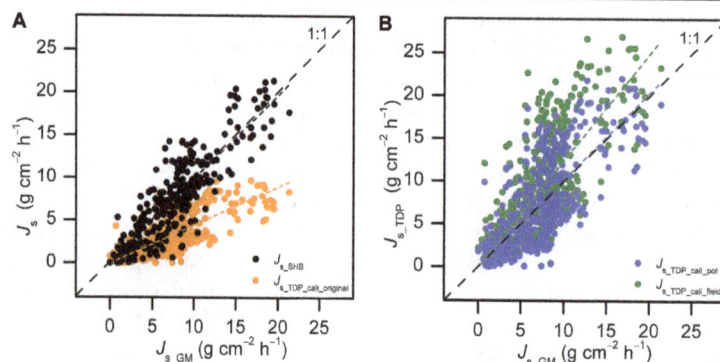

FIGURE 2 | Half-hourly sap flux density (J_s) measured with thermal dissipation probes (TDP) and stem heat balance (SHB) sensors on five potted *Bambusa vulgaris* culms plotted against GM-derived reference sap flux densities (J_{s_GM}) before (A; $J_{s\_TDP\_cali\_original}$: $Y = 0.44X$, $R^2 = 0.84$, $P < 0.01$; J_{s_SHB}: $Y = 0.98X$, $R^2 = 0.93$, $P < 0.01$) and after (B; $J_{s\_TDP\_cali\_field}$: $Y = 1.24X$, $R^2 = 0.84$, $P < 0.01$; $J_{s\_TDP\_cali\_pot}$: $Y = 1.01X$, $R^2 = 0.84$, $P < 0.01$) species-specific calibration and field calibrations of the TDP method. Pooled data from 2 to 5 days of simultaneous TDP, SHB, and gravimetric measurements (GM).

TABLE 2 | Values of the parameter c of different bamboo calibrations (species-specific/common) for TDP sap flux estimates.

Formula specificity	Species	c	nRMSE		
			Species-specific formula	Common formula	P-value
Species	B. vulgaris	2.79 ± 0.13^a	0.10	0.11	0.07
	G. apus	3.32 ± 0.08^b	0.10	0.12	0.06
	D. asper	2.42 ± 0.06^c	0.18	0.18	0.97
	G. atroviolacea	2.53 ± 0.11^d	0.12	0.13	0.81
Common		2.74 ± 0.07^e			

Significant differences between species-specific and common c estimates (Tukey's test, $P < 0.01$) are indicated by superscripted letters. P-values < 0.05 indicate significant differences between Normalized Root-Mean-Square Errors (nRMSE) of species-specific and common formula.

Before calibration, J_{s_TDP} was on average 66 and 63% lower than SHB-derived reference values on the daily and 30-min scales, respectively ($P < 0.01$). This deviation was reduced to 10 and 8% underestimations ($P < 0.01$) when using species-specific calibration parameters (**Table 2**). On average, for $77 \pm 6\%$ of the days that were included in the analysis, the species-specific post-calibration 30-min J_{s_TDP} values were not significantly different from the respective reference J_{s_SHB} (Wilcoxon Signed-Rank test, $P > 0.05$).

Field Study
Monitoring Bamboo and Tree Sap Flux

J_{s_max} in the studied bamboo species (averages from five individuals per species) were 70.5, 21.6, 49.7, and 56.2 g cm^{-2} h^{-1} for B. vulgaris, D. asper, G. apus, and G. atroviolacea, respectively. In trees, corresponding values were 17.7, 10.5, and 23.3 g cm^{-2} h^{-1} for H. brasiliensis, G. arborea, and S. leprosula, respectively. Across bamboo species, J_{s_max} decreased with increasing culm diameter ($R^2 = 0.97$, $P = 0.02$, **Figure 3**).

Environmental Measurements and Analyses

The normalized daily accumulated J_s of all studied species increased with increasing daily integrated radiation. This relationship did not fully hold up for accumulated J_s and average daily VPD. In several species, daily J_s increased with increasing VPD only to a certain VPD threshold (approximately 0.74 kPa, **Figure 4**); after this threshold, accumulated J_s decreased with further increasing VPD. Such conditions of high VPD were characteristic of the dry period. For days with VPD > 0.74 kPa, daily accumulated J_s of most studied species (except in D. asper and G. arborea) declined with decreasing soil moisture content ($R^2 = 0.39$, 0.44, 0.4, 0.52, and 0.55 for B. vulgaris, G. apus, G. atroviolacea, S. leprosula, and H. brasiliensis, respectively; $P < 0.05$, **Figures 5A,B**).

Diurnal peaks in J_s in the studied bamboo species occurred relatively early (on average at about 11 am), which was significantly earlier than the peaks of radiation and VPD (20–82 and 131–206 min, respectively). In the studied tree species, maximal hourly J_s values were observed after the peak of radiation (3–97 min), but still before (51–108 min) VPD peaked. All time lags were significantly different from 0 min ($P < 0.01$; **Table 3**), except for the time lag to radiation for the tree species S. leprosula ($P > 0.05$).

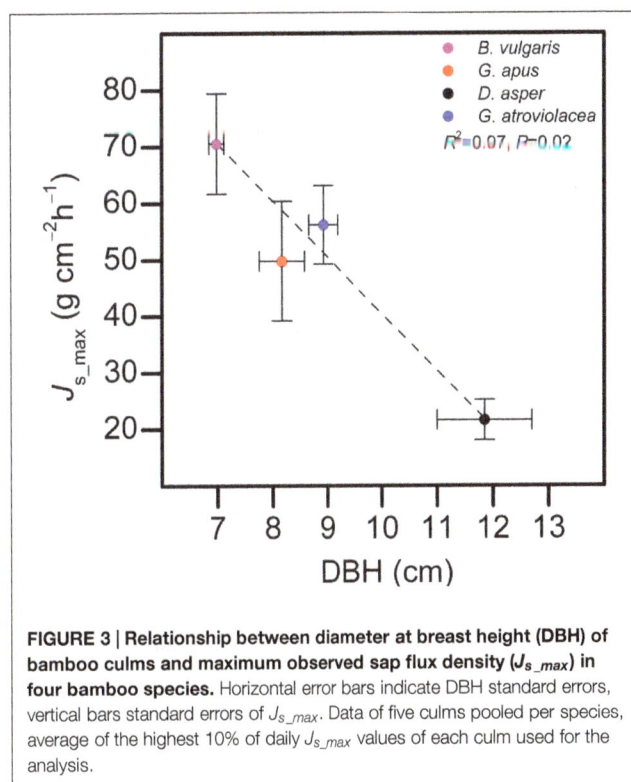

FIGURE 3 | Relationship between diameter at breast height (DBH) of bamboo culms and maximum observed sap flux density (J_{s_max}) in four bamboo species. Horizontal error bars indicate DBH standard errors, vertical bars standard errors of J_{s_max}. Data of five culms pooled per species, average of the highest 10% of daily J_{s_max} values of each culm used for the analysis.

Diurnally, some of the studied species showed pronounced hysteresis of hourly J_s to radiation and VPD. Direction of rotation (i.e., order of observations) was counter-clockwise for radiation (**Figure 6A**) and clockwise for VPD (**Figure 6B**). The area of the hysteresis to VPD was on average 32% larger in bamboos than in trees, while the area of hysteresis to radiation was on average 50% smaller in bamboos ($P < 0.01$).

DISCUSSION

Calibration Experiments

In the pot calibration experiment, SHB yielded similar results as reference GM measurements. Bamboos seem well suited for the SHB method (Dierick et al., 2010) due to their round shape and smooth and barkless surface, which allows for tight contact with the gages. Additionally, the hollow center and thin culm walls

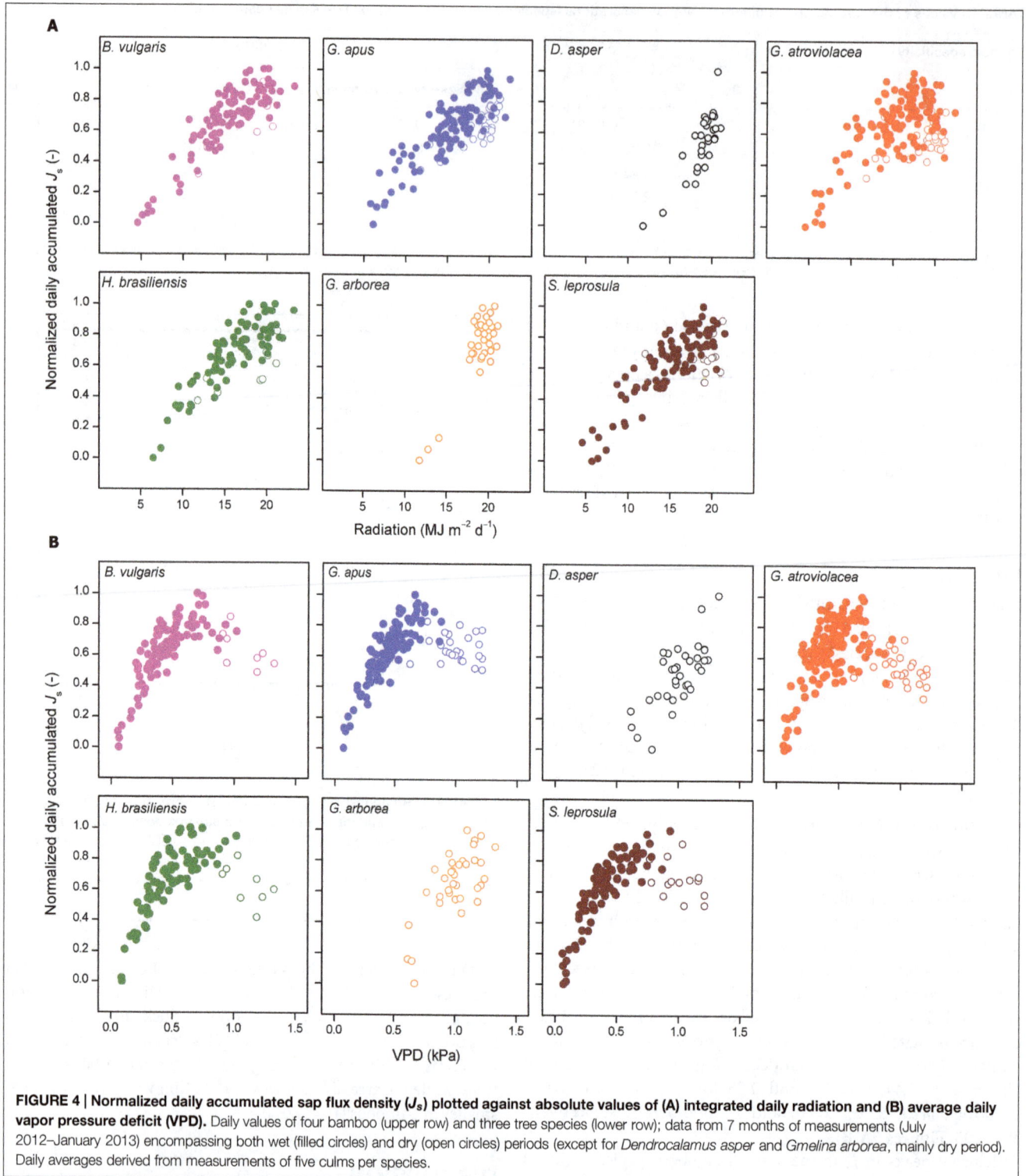

FIGURE 4 | Normalized daily accumulated sap flux density (J_s) plotted against absolute values of (A) integrated daily radiation and (B) average daily vapor pressure deficit (VPD). Daily values of four bamboo (upper row) and three tree species (lower row); data from 7 months of measurements (July 2012–January 2013) encompassing both wet (filled circles) and dry (open circles) periods (except for *Dendrocalamus asper* and *Gmelina arborea*, mainly dry period). Daily averages derived from measurements of five culms per species.

result in relatively low energy losses to heat storage so that the heat balance conditions required for the SHB method are met. 'Zero sap flux' conditions to obtain the heat conductivity of the sheath (K_{sh}, Sakuratani, 1981) as a further requirement of the SHB method are difficult to determine *in situ* due to potential root pressure induced night time sap flux in bamboos (Cao et al.,

2012); however, using K_{sh} derived from field conditions of very low night-time sap flux likely introduced only negligible errors into the calculation of daytime sap flux (Grime and Sinclair, 1999). As we observed very low sap flux over several hours during our experiments (e.g., about $1 \, g \, cm^{-2} \, h^{-1}$ during the pot experiment), our obtained K_{sh} were likely reliable.

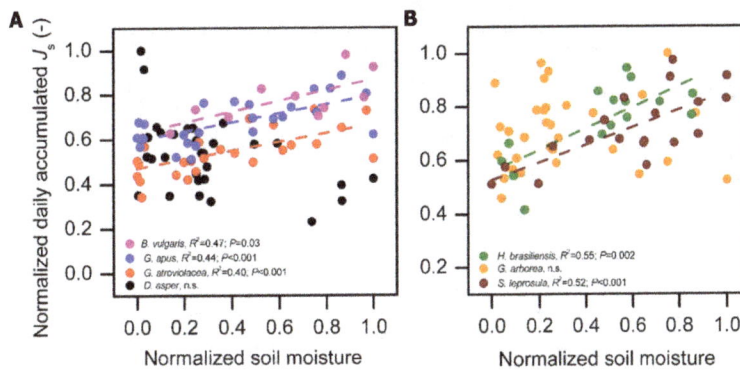

FIGURE 5 | Normalized daily accumulated sap flux density (J_s) of four bamboo species (A) and three tree species (B) in the 'dry period' (characterized with mean daily VPD > 0.74 kPa) plotted against normalized mean daily soil moisture content (SM). There was a significant linear relationship between J_s and SM (P < 0.05) for all species except D. asper and G. arborea. Normalized values do not reach 1.0 for all species in the figure as the normalization was performed by setting the maximum value of the full measurement period of each species (including wet period) to one, while the figure displays only values in dry period. Daily averages derived from measurements on five culms per species, data of at least 10 dry period days per species.

In contrast to SHB, the TDP method was found to substantially underestimate J_s of bamboos in the pot and field calibration experiments. Underestimations by TDP were also reported in two other bamboo species: respective average underestimations of 13% for *B. blumeana* (Dierick et al., 2010) and 31% for *Phyllostachys pubescens* (Moso bamboo, Kume et al., 2010) were reported. Reasons for the observed underestimations could lie in the distinct hydraulic and physiological features of bamboos. Diurnal variations of stem water storage, for example, could affect the accuracy of TDP measurements (Vergeynst et al., 2014). Bamboos have approximately 50% parenchyma in culm walls (Dransfield and Widjaja, 1995), which potentially provides large water reservoirs. The depletion and refilling of the stem during the day and night, respectively, could cause diurnal fluctuations in culm thermal diffusivity. Higher water content during the night could lead to a lower maximum temperature difference (ΔT_{max}) between heated and reference probe under "zero sap flux" conditions. Likewise, lower water content during the day could lead to higher observed ΔT values. As $\Delta T_{max}/\Delta T$ constitutes the basis for calculations of daytime J_s, substantial underestimations of J_s could be introduced when using the original calibration parameters (Granier, 1985; Vergeynst et al., 2014). This hypothesis was assessed further by comparing the linear calibration parameters of *B. vulgaris* from the pot and the field calibration experiment (c = 2.28 and 2.79, respectively). In the pot experiment, the bamboos were always supplied with plenty of water, so that the variability of the culm water content was likely smaller than under field conditions. Effects of varying stem water content on $\Delta T_{max}/\Delta T$ are thus likely much smaller in the pot experiment, which may explain why pot and field calibration experiment yield different parameters for the linear calibration of the same species (*B. vulgaris*). Another potential factor for the divergence could be that the maximum observed J_s in the field (about 70 g cm^{-2} h^{-1}) was much larger than in the pot experiment (about 20 g cm^{-2} h^{-1}). Higher daytime sap flux (and thus transpiration) may cause a quicker depletion of the potential culm water storage, which consequently leads to a higher variability of culm water content between night and day.

We expected the calibration formula type (linear vs. non-linear) and data time step to have an impact on the performance of TDP predictions. However, both were not as important as the factor formula specificity. Even though species-specific calibration formulas generally did not perform significantly better than the common formula, species-specific formulas tended to show slightly better performance (**Table 1**) for two of the studied species (*G. apus* and *B. vulgaris*). Also, the calibration parameters were significantly different among the four studied bamboo species (**Table 2**). Confronting this insight with results from sap flux studies on other bamboo species (Dierick et al., 2010; Kume et al., 2010), differences among species become even more apparent. We thus used the derived species-specific formulas for further analysis. The observed differences among species may be indicative of highly heterogeneous wood anatomical properties among bamboo species. For example,

TABLE 3 | Time lags between diurnal peaks of radiation and VPD and peaks of J_s in studied bamboos and trees.

Species	N	Time lag with radiation (min)	Time lag with VPD (min)
B. vulgaris	5	82 ± 62	171 ± 63
D. asper	5	41 ± 57	206 ± 57
G. apus	4	20 ± 61	131 ± 53
G. atroviolacea	5	64 ± 30	170 ± 35
Bamboo_mean	**19**	**51[A]**	**169[A]**
H. brasiliensis	5	−37 ± 12[a]	51 ± 9
G. aborea	5	−97 ± 87[b]	67 ± 87
S. leprosula	5	−3 ± 25[a]	108 ± 20
Tree_mean	**15**	**−46[B]**	**75[B]**

Positive values indicate a peak of radiation/VPD after the peak of J_s, negative values indicate a peak before J_s; N, culms/trunks per species averaged (mean ± SD). Significant differences in bamboo/tree mean time lags are indicated by different superscripted letters (Tukey's test, P < 0.01). Significant differences between species are indicated by capital letters (P < 0.01).

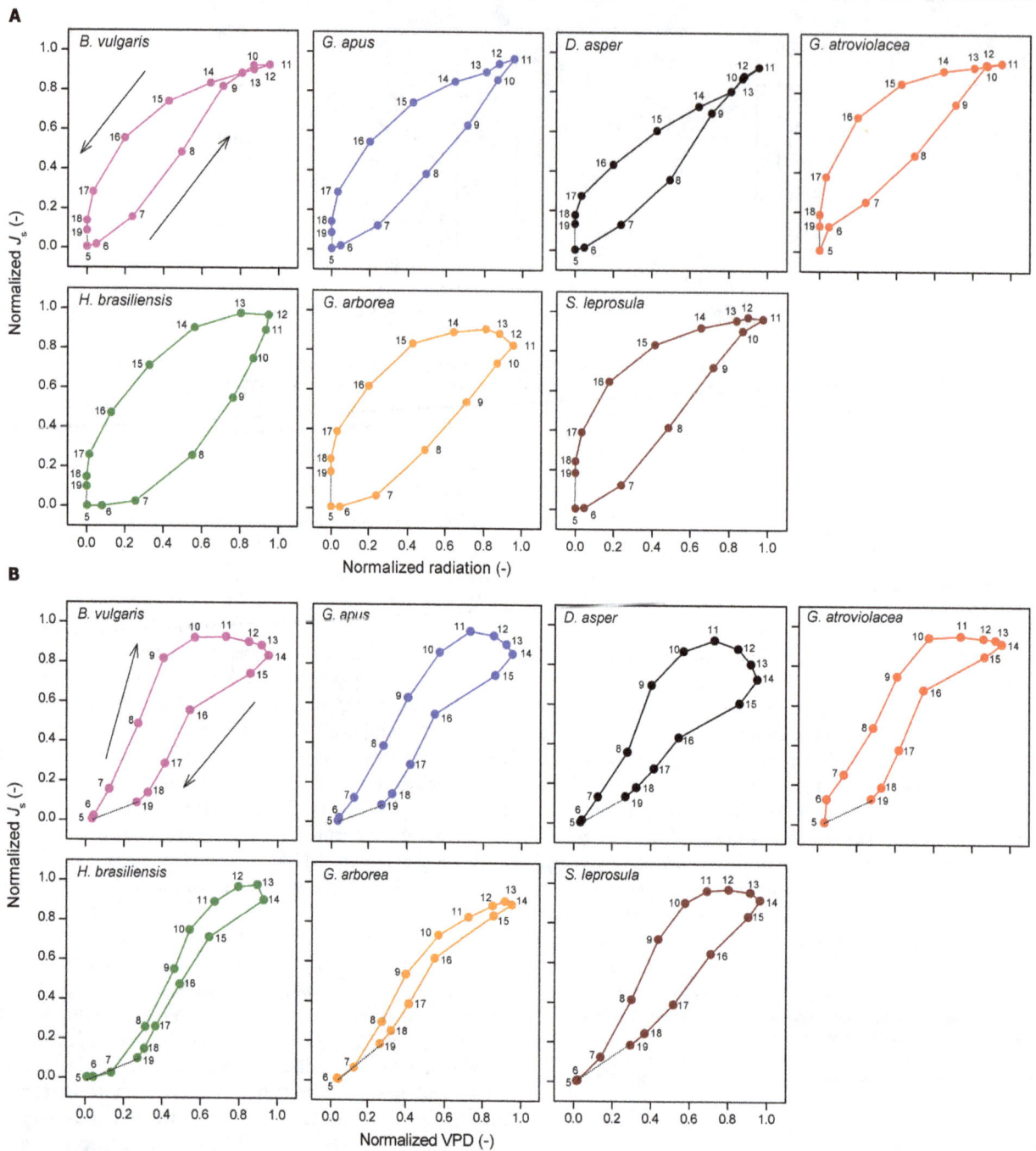

FIGURE 6 | Normalized hourly sap flux density (J_s) plotted against (A) normalized hourly radiation and (B) VPD. Data of four bamboo (upper row) and three tree species (lower row). Hourly averages derived from simultaneous measurements on five culms per species and by averaging the values of three sunny days to minimize influences of weather. The numbers in the sub-figures indicate the respective time of the day.

size and shape of vascular bundles and parenchyma of 15 bamboo species were reported to be highly variable (Rúgolo de Agrasar and Rodríguez, 2003). For two further bamboo species (*Chusquea ramosissima* and *Merostachys claussenii*), it was suggested that differences in number of vascular bundles per unit area (1000 vs. 225 per cm²) and vessel length (~1 m vs.

20 cm) could lead to differences in xylem hydraulic conductivity (Saha et al., 2009). Differences in wood anatomical properties may also lead to heterogeneous heat conductive properties, which potentially affects applicability and accuracy of sap flux measurements and particularly of the TDP method (Wullschleger et al., 2011).

In our study, culm-specific formulas performed better at predicting J_s than species-specific and common calibration formulas (Appendix Tables A1 and A2 in Supplementary Material). This result indicates heterogeneity in conductive properties among culms of the same species. Potential reasons could lie in the age and the ontogeny of individual culms. Even though we carefully selected culms of similar age (approximately 2 years old), the exact age of individual bamboo culms within a given clump is difficult to assess. As all monocot species, bamboos lack secondary growth (Zimmermann and Tomlinson, 1972), so culm diameters are not related to culm age. Additionally, over the ontogeny of a certain culm, events and processes such as conductive circuit failure (drought- or metabolism-related; Cochard et al., 1994; Liese and Weiner, 1996), lignification (Lin et al., 2002) or increasing hydraulic limitations with height (Renninger and Phillips, 2010; Cao et al., 2012) could result in overall reduced hydraulic conductivity and thus lower sap flux densities with increasing culm age. However, these processes remain difficult to assess from the outside of the culm; further studies linking the age and ontogeny of bamboos to (TDP-derived) sap flux and water use patterns are suggested.

Water Use Patterns of Bamboos and Trees

Half-hourly J_{s_max} in the four studied bamboo species ranged from 21.6 to 70.5 g cm^{-2} h^{-1} and were (on average) almost two times greater than in the studied tree species. The observed range for both bamboos and trees falls into the range of J_{s_max} values reported for tropical tree species in a variety of sap flux studies (Meinzer et al., 2001; O'Brien et al., 2004). For *D. asper*, the J_{s_max} (21.6 g cm^{-2} h^{-1}) was similar to values reported for *B. blumeana* culms (25.7 g cm^{-2} h^{-1}, Dierick et al., 2010) and Moso bamboos (approximately 20 g cm^{-2} h^{-1}, Kume et al., 2010) of similar size. Our four studied bamboo species showed significant differences in J_{s_max}, which were negatively correlated with species-specific differences in DBH (**Figure 3**). Consistent with this, in a study on 27 tropical tree species, the negative correlation between J_{s_max} and DBH was also observed (Meinzer et al., 2001). It was assumed to be related to a decline of the leaf area to sapwood area ratio with increasing DBH. This was also observed in a study on *Eucalyptus grandis* trees (Dye and Olbrich, 1993). In our study, we harvested leaves of three bamboo species (*B. vulgaris*, *D. asper*, and *G. apus*) and found that the leaf weight to sapwood area ratio was positively correlated with J_{s_max} ($R^2 = 0.45$, $P < 0.05$). However, studies connecting such anatomical and eco-hydrological properties of bamboos are yet scarce (Saha et al., 2009).

On the day-to-day level, accumulated J_s of both the studied bamboo and tree species were significantly correlated with radiation and VPD (**Figure 4**). During the long wet period, accumulated J_s linearly increased with higher integrated radiation and average daily VPD. Likewise, linear relationships in the day-to-day behavior of J_s to micrometeorological drivers have been reported for some tropical bamboo and several dicot tree species (Dierick and Hölscher, 2009; Köhler et al., 2009).

During the dry period characterized by higher radiation and VPD (13 and 100% higher, respectively) than during the wet period, however, the observed linear relationship to VPD did not hold. Higher average daily VPD ('dry period conditions') led to decreases in accumulated J_s of several studied species (**Figure 4B**). Similar decreases after a certain peak value have been reported for some previously studied tree species (Kubota et al., 2005; Jung et al., 2011), but in most species studied so far, higher average daily VPD leads to increases in accumulated J_s or water use (Wullschleger and Norby, 2001; Tang et al., 2006; Kume et al., 2007; Hernández-Santana et al., 2008; Peters et al., 2010; Horna et al., 2011). This was also reported for Moso bamboo (Komatsu et al., 2010). The observed decreasing accumulated J_s in bamboos under high VPD in our study were related to a reduction of soil moisture in the dry period (for three of the four bamboo and two of the three studied tree species). During the dry period, VPD was generally much higher than during the wet period. Soil moisture may become a limiting factor after several days without rainfall in the dry period. Accumulated J_s decreased strongly and linearly with decreasing soil moisture under 'dry period conditions' (i.e., VPD > 0.74 kPa) for all studied bamboo (except *D. asper*) and tree species (except *G. arborea*, **Figure 5**). Similarly, in a throughfall reduction experiment in Indonesia, declines of monthly J_s of Cacao and *Gliricidia sepium* were found to linearly correlate with reduced soil moisture (Köhler et al., 2010). Such sensitivity of daily J_s to fluctuating soil moisture may be related to a relatively shallow rooting depth (Kume et al., 2007).

Regarding the diurnal course of J_s, the studied bamboo species showed earlier peaks than radiation and VPD, and also earlier than the respective peaks of the studied tree species. In contrast to this, previous studies on tropical trees reported rather small time-lags between peaks of J_s and radiation and VPD, respectively (Dierick and Hölscher, 2009; Köhler et al., 2009; Horna et al., 2011). Pre-noon peaks of J_s have only been described for few species thus far, for example, *Acer rubrum* (Johnson et al., 2011) and oil palms (Niu et al., 2015). The early diurnal peaks of J_s result in substantial hysteresis of J_s particularly to VPD. For another monocot species, oil palm, it has been suggested that such pre-noon peaks of J_s and the resulting large hysteresis to VPD could be indicative of internal trunk water storage and/or root pressure mechanisms (Niu et al., 2015; Röll et al., 2015). Early peaks of J_s could be due to a pre-noon contribution of internal water storage to bamboo transpiration. Likewise, the decoupling of hourly J_s particularly from VPD in the afternoon, i.e., the drop in bamboo J_s (after an early peak) despite further rising VPD, could be connected to the reduced water availability for leaves after the depletion of internal water storage at a certain time of the day. The depletion of stored stem water may be compensated for during the night by root pressure mechanisms (Cao et al., 2012; Yang et al., 2012). Other potential reasons for the diurnally relatively early decline of bamboo J_s and the consequent decoupling of the sap flux response from micrometeorological drivers could be a decline in leaf hydraulic conductance in the afternoon hours, which could contribute to prevent stem water potential loss and subsequent xylem cavitation (Saha et al., 2009; Yang et al., 2012).

CONCLUSION

Adjusting and applying the TDP method for sap flux measurements on four bamboo species pointed to substantial differences in water use patterns between the studied bamboos and three tree species studied. Bamboos had higher J_s, and respective hourly maxima were reached earlier in the day than in tree species. This resulted in strong diurnal hysteresis, particularly to VPD, and in significant time lags between the peaks of J_s in bamboos and the respective peaks of radiation and VPD. Both may point to a strong contribution of internal water storage mechanisms to bamboo transpiration. We found substantial differences in the day-to-day J_s response of most studied bamboo and tree species to fluctuations in environmental conditions between the dry and the wet period. Reduced J_s under conditions of high VPD in the dry period could largely be explained by limiting soil moisture content. The regulation of bamboo water use thus seems to involve mechanisms at the leaf-, culm-, and root- level. However, these mechanisms yet remain to be inter-connected convincingly.

ACKNOWLEDGMENTS

The authors acknowledge the funding from the German Research Foundation (DFG, code number: HO 2119 and CRC 990). DF and FN received scholarships from the China Scholarship Council (CSC). We thank the Departments of Facility and Property, the Ecology Lab and the Agriculture Faculty of Bogor Agriculture University (IPB) for supporting our field work. We especially thank the IPB Faculty of Forestry for close cooperation in our project and our assistants Wahyu Iskandar and Popi Puspita for their work with great enthusiasm in our field experiments. We are very grateful to Diego Dierick for suggestions on experimental methods and to Michael Köhler for suggestions on data processing.

REFERENCES

Baker, J. M., and Van Bavel, C. H. M. (1987). Measurement of mass flow of water in the stems of herbaceous plants. *Plant Cell Environ.* 10, 777–782. doi: 10.1111/1365-3040.ep11604765

Bovard, B. D., Curtis, P. S., Vogel, C. S., Su, H.-B., and Schmid, H. P. (2005). Environmental controls on sap flow in a northern hardwood forest. *Tree Physiol.* 25, 31–38. doi: 10.1093/treephys/25.1.31

Cao, K.-F., Yang, S.-J., Zhang, Y.-J., and Brodribb, T. J. (2012). The maximum height of grasses is determined by roots. *Ecol. Lett.* 15, 666–672. doi: 10.1111/j.1461-0248.2012.01783.x

Carrasco, L. O., Bucci, S. J., Francescantonio, D. D., Lezcano, O. A., Campanello, P. I., Scholz, F. G., et al. (2014). Water storage dynamics in the main stem of subtropical tree species differing in wood density, growth rate and life history traits. *Tree Physiol.* 35, 354–365. doi: 10.1093/treephys/tpu087

Chen, X., Zhang, X., Zhang, Y., Booth, T., and He, X. (2009). Changes of carbon stocks in bamboo stands in China during 100 years. *For. Ecol. Manag.* 258, 1489–1496. doi: 10.1016/j.foreco.2009.06.051

Cochard, H., Ewers, F. W., and Tyree, M. T. (1994). Water relations of a tropical vine-like bamboo (*Rhipidocladum racemiflorum*): root pressures, vulnerability to cavitation and seasonal changes in embolism. *J. Exp. Bot.* 45, 1085–1089. doi: 10.1093/jxb/45.8.1085

Dierick, D., and Hölscher, D. (2009). Species-specific tree water use characteristics in reforestation stands in the Philippines. *Agric. For. Meteorol.* 149, 1317–1326. doi: 10.1016/j.agrformet.2009.03.003

Dierick, D., Hölscher, D., and Schwendenmann, L. (2010). Water use characteristics of a bamboo species (*Bambusa blumeana*) in the Philippines. *Agric. For. Meteorol.* 150, 1568–1578. doi: 10.1016/j.agrformet.2010.08.006

Dransfield, S., and Widjaja, E. (1995). *Plant Resources of South-East Asia No. 7: Bamboos*. Leiden: Backhuys Publishers.

Dye, P. J., and Olbrich, B. W. (1993). Estimating transpiration from 6-year-old Eucalyptus grandis trees: development of a canopy conductance model and comparison with independent sap flux measurements. *Plant Cell Environ.* 16, 45–53. doi: 10.1111/j.1365-3040.1993.tb00843.x

Food Agriculture Organization [FAO] (2010). *Global Forest Resources Assessment 2010?: Main Report, FAO Forestry Paper, 0258-6150; 163*. Rome: Food and Agriculture Organization of the United Nations.

Goldstein, G., Andrade, J. L., Meinzer, F. C., Holbrook, N. M., Cavelier, J., Jackson, P., et al. (1998). Stem water storage and diurnal patterns of water use in tropical forest canopy trees. *Plant Cell Environ.* 21, 397–406. doi: 10.1046/j.1365-3040.1998.00273.x

Granier, A. (1985). Une nouvelle méthode pour la mesure du flux de sève brute dans le tronc des arbres. *Ann. For. Sci.* 42:8. doi: 10.1051/forest:19850204

Grime, V. L., and Sinclair, F. L. (1999). Sources of error in stem heat balance sap flow measurements. *Agric. For. Meteorol.* 94, 103–121. doi: 10.1016/S0168-1923(99)00011-8

Hernández-Santana, V., David, T. S., and Martínez-Fernández, J. (2008). Environmental and plant-based controls of water use in a Mediterranean oak stand. *For. Ecol. Manag.* 255, 3707–3715. doi: 10.1016/j.foreco.2008.03.004

Horna, V., Schuldt, B., Brix, S., and Leuschner, C. (2011). Environment and tree size controlling stem sap flux in a perhumid tropical forest of Central Sulawesi, Indonesia. *Ann. For. Sci.* 68, 1027–1038. doi: 10.1007/s13595-011-0110-2

Ichihashi, R., Komatsu, H., Kume, T., Onozawa, Y., Shinohara, Y., Tsuruta, K., et al. (2015). Stand-scale transpiration of two *Moso bamboo* stands with different culm densities. *Ecohydrology* 8, 450–459. doi: 10.1002/eco.1515

International Network for Bamboo and Rattan [INBAR] (2014). *International Trade in Bamboo and Rattan 2012*. Beijing: International Network for Bamboo and Rattan. Available at: http://www.inbar.int/publications/?did=292 [accessed June 19, 2015].

Jarvis, N. J. (1989). A simple empirical model of root water uptake. *J. Hydrol.* 107, 57–72. doi: 10.1016/0022-1694(89)90050-4

Johnson, D. M., McCulloh, K. A., Meinzer, F. C., Woodruff, D. R., and Eissenstat, D. M. (2011). Hydraulic patterns and safety margins, from stem to stomata, in three eastern US tree species. *Tree Physiol.* 31, 659–668. doi: 10.1093/treephys/tpr050

Jung, E. Y., Otieno, D., Lee, B., Lim, J. H., Kang, S. K., Schmidt, M. W. T., et al. (2011). Up-scaling to stand transpiration of an Asian temperate mixed-deciduous forest from single tree sapflow measurements. *Plant Ecol.* 212, 383–395. doi: 10.1007/s11258-010-9829-3

Köhler, M., Dierick, D., Schwendenmann, L., and Hölscher, D. (2009). Water use characteristics of cacao and Gliricidia trees in an agroforest in Central Sulawesi, Indonesia. *Ecohydrol.* 2, 520–529. doi: 10.1002/eco.67

Köhler, M., Schwendenmann, L., and Hölscher, D. (2010). Throughfall reduction in a cacao agroforest: tree water use and soil water budgeting. *Agric. For. Meteorol.* 150, 1079–1089. doi: 10.1016/j.agrformet.2010.04.005

Komatsu, H., Onozawa, Y., Kume, T., Tsuruta, K., Kumagai, T., Shinohara, Y., et al. (2010). Stand-scale transpiration estimates in a *Moso bamboo* forest: II. Comparison with coniferous forests. *For. Ecol. Manag.* 260, 1295–1302. doi: 10.1016/j.foreco.2010.06.040

Kubota, M., Tenhunen, J., Zimmermann, R., Schmidt, M., Adiku, S., and Kakubari, Y. (2005). Influences of environmental factors on the radial profile of sap flux density in Fagus crenata growing at different elevations in the Naeba Mountains, Japan. *Tree Physiol.* 25, 545–556. doi: 10.1093/treephys/25.5.545

Kume, T., Onozawa, Y., Komatsu, H., Tsuruta, K., Shinohara, Y., Umebayashi, T., et al. (2010). Stand-scale transpiration estimates in a *Moso bamboo* forest: (I) Applicability of sap flux measurements. *For. Ecol. Manag.* 260, 1287–1294. doi: 10.1016/j.foreco.2010.07.012

Kume, T., Takizawa, H., Yoshifuji, N., Tanaka, K., Tantasirin, C., Tanaka, N., et al. (2007). Impact of soil drought on sap flow and water status of evergreen trees in a tropical monsoon forest in northern Thailand. *For. Ecol. Manag.* 238, 220–230. doi: 10.1016/j.foreco.2006.10.019

Liese, W., and Weiner, G. (1996). Ageing of bamboo culms. A review. *Wood Sci. Technol.* 30, 77–89. doi: 10.1007/BF00224958

Lin, J., He, X., Hu, Y., Kuang, T., and Ceulemans, R. (2002). Lignification and lignin heterogeneity for various age classes of bamboo (*Phyllostachys pubescens*) stems. *Physiol. Plant.* 114, 296–302. doi: 10.1034/j.1399-3054.2002.1140216.x

Lu, P., Urban, L., and Zhao, P. (2004). Granier's thermal dissipation probe (TDP) method for measuring sap flow in trees: theory and practice. *ACTA Bot. Sin. Engl. Ed.* 46, 631–646.

Meinzer, F. C., Goldstein, G., and Andrade, J. L. (2001). Regulation of water flux through tropical forest canopy trees: do universal rules apply? *Tree Physiol.* 21, 19–26. doi: 10.1093/treephys/21.1.19

Niu, F., Röll, A., Hardanto, A., Meijide, A., Köhler, M., Hendrayanto, et al. (2015). Oil palm water use: calibration of a sap flux method and a field measurement scheme. *Tree Physiol.* 35, 563–573. doi: 10.1093/treephys/tpv013

O'Brien, J. J., Oberbauer, S. F., and Clark, D. B. (2004). Whole tree xylem sap flow responses to multiple environmental variables in a wet tropical forest. *Plant Cell Environ.* 27, 551–567. doi: 10.1111/j.1365-3040.2003.01160.x

Pereira, H. C., and Hosegood, P. H. (1962). Comparative water-use of softwood plantations and bamboo forest. *J. Soil Sci.* 13, 299–313. doi: 10.1111/j.1365-2389.1962.tb00709.x

Peters, E. B., McFadden, J. P., and Montgomery, R. A. (2010). Biological and environmental controls on tree transpiration in a suburban landscape. *J. Geophys. Res. Biogeosci.* 115:G04006. doi: 10.1029/2009JG001266

Petit, G., DeClerck, F. A. J., Carrer, M., and Anfodillo, T. (2014). Axial vessel widening in arborescent monocots. *Tree Physiol.* 34, 137–145. doi: 10.1093/treephys/tpt118

Renninger, H. J., and Phillips, N. (2010). Intrinsic and extrinsic hydraulic factors in varying sizes of two Amazonian palm species (*Iriartea deltoidea* and *Mauritia flexuosa*) differing in development and growing environment. *Am. J. Bot.* 97, 1926–1936. doi: 10.3732/ajb.1000015

Röll, A., Niu, F., Meijide, A., Hardanto, A., Hendrayanto, Knohl, A., et al. (2015). Transpiration in an oil palm landscape: effects of palm age. *Biogeosciences* 12, 5619–5633. doi: 10.5194/bg-12-5619-2015

Rúgolo de Agrasar, Z., and Rodríguez, M. F. (2003). Culm anatomy of native woody bamboos in argentina and neighboring areas: cross section. *Bamboo Sci. Cult. J. Am. Bamboo Soc.* 17, 28–43.

Saha, S., Holbrook, N. M., Montti, L., Goldstein, G., and Cardinot, G. K. (2009). Water Relations of *Chusquea ramosissima* and *Merostachys clausenii* in Iguazu National Park, Argentina. *Plant Physiol.* 149, 1992–1999. doi: 10.1104/pp.108.129015

Sakuratani, T. (1981). A heat balance method for measuring water flux in the stem of intact plants. *J. Agric. Meteorol.* 37, 9–17. doi: 10.2480/agrmet.37.9

Tang, J., Bolstad, P. V., Ewers, B. E., Desai, A. R., Davis, K. J., and Carey, E. V. (2006). Sap flux–upscaled canopy transpiration, stomatal conductance, and water use efficiency in an old growth forest in the Great Lakes region of the United States. *J. Geophys. Res. Biogeosci.* 111:G02009. doi: 10.1029/2005JG000083

Uchimura, E. (1994). *Invitation to Bamboos: Wonder of Biology.* Tokyo: Kenseisha.

Van Den Besselaar, E. J. M., Klein Tank, A. M. G., Van Der Schrier, G., Abass, M. S., Baddour, O., Van Engelen, A. F. V., et al. (2014). International climate assessment & dataset: climate services across borders. *Bull. Am. Meteorol. Soc.* 96, 16–21. doi: 10.1175/BAMS-D-13-00249.1

Vandegehuchte, M. W., and Steppe, K. (2013). Sap-flux density measurement methods: working principles and applicability. *Funct. Plant Biol.* 40, 213–223. doi: 10.1071/FP12233_CO

Vergeynst, L. L., Vandegehuchte, M. W., McGuire, M. A., Teskey, R. O., and Steppe, K. (2014). Changes in stem water content influence sap flux density measurements with thermal dissipation probes. *Trees* 28, 949–955. doi: 10.1007/s00468-014-0989-y

Wang, H., Zhao, P., Hölscher, D., Wang, Q., Lu, P., Cai, X. A., et al. (2012). Nighttime sap flow of Acacia mangium and its implications for nighttime transpiration and stem water storage. *J. Plant Ecol.* 5, 294–304. doi: 10.1093/jpe/rtr025

Waring, R. H., and Running, S. W. (1978). Sapwood water storage: its contribution to transpiration and effect upon water conductance through the stems of old-growth Douglas-fir. *Plant Cell Environ.* 1, 131–140. doi: 10.1111/j.1365-3040.1978.tb00754.x

Wullschleger, S. D., Childs, K. W., King, A. W., and Hanson, P. J. (2011). A model of heat transfer in sapwood and implications for sap flux density measurements using thermal dissipation probes. *Tree Physiol.* 31, 669–679. doi: 10.1093/treephys/tpr051

Wullschleger, S. D., and Norby, R. J. (2001). Sap velocity and canopy transpiration in a sweetgum stand exposed to free-air CO2 enrichment (FACE). *New Phytol.* 150, 489–498. doi: 10.1046/j.1469-8137.2001.00094.x

Yang, S.-J., Zhang, Y.-J., Sun, M., Goldstein, G., and Cao, K.-F. (2012). Recovery of diurnal depression of leaf hydraulic conductance in a subtropical woody bamboo species: embolism refilling by nocturnal root pressure. *Tree Physiol.* 32, 414–422. doi: 10.1093/treephys/tps028

Zimmermann, M. H., and Tomlinson, P. B. (1972). The vascular system of monocotyledonous stems. *Bot. Gaz.* 133, 141–155. doi: 10.1086/336628

Conflict of Interest Statement: The authors declare that the research was conducted in the absence of any commercial or financial relationships that could be construed as a potential conflict of interest.

18

The Proteomic Response of *Arabidopsis thaliana* to Cadmium Sulfide Quantum Dots, and Its Correlation with the Transcriptomic Response

Marta Marmiroli[1]*, Davide Imperiale[1], Luca Pagano[1], Marco Villani[2], Andrea Zappettini[2] and Nelson Marmiroli[1]

[1] Department of Life Sciences, University of Parma, Parma, Italy, [2] Institute of Materials for Electronics and Magnetism (IMEM-CNR), Parma, Italy

A fuller understanding of the interaction between plants and engineered nanomaterials is of topical relevance because the latter are beginning to find applications in agriculture and the food industry. There is a growing need to establish objective safety criteria for their use. The recognition of two independent *Arabidopsis thaliana* mutants displaying a greater level of tolerance than the wild type plant to exposure to cadmium sulfide quantum dots (CdS QDs) has offered the opportunity to characterize the tolerance response at the physiological, transcriptomic, and proteomic levels. Here, a proteomics-based comparison confirmed the conclusions drawn from an earlier transcriptomic analysis that the two mutants responded to CdS QD exposure differently both to the wild type and to each other. Just over half of the proteomic changes mirrored documented changes at the level of gene transcription, but a substantial number of transcript/gene product pairs were altered in the opposite direction. An interpretation of the discrepancies is given, along with some considerations regarding the use and significance of -omics when monitoring the potential toxicity of ENMs for health and environment.

Keywords: tolerant mutants, engineered nanomaterials (ENM), genotoxicology ecotoxicology, exposure markers, comparative analysis

Edited by:
Manuel González-Guerrero,
Universidad Politécnica de Madrid,
Spain

Reviewed by:
Zhenzhu Xu,
Institute of Botany, Chinese Academy
of Sciences, China
Francisco Cabello-Hurtado,
University of Rennes 1, France

*Correspondence:
Marta Marmiroli
marta.marmiroli@unipr.it

INTRODUCTION

Nanotechnology is regarded as transformative, but its potential long term impact on human health and the environment remains inadequately researched (Colvin, 2003; Royal Society and Royal Academy of Engineering, 2004). Legislative authorities still suffer from a paucity of appropriate data to enable a science-based regulatory framework to be constructed over the release and commercialization of nanomaterials. While much of the focus of nanotechnology has been in the electronics industry and medical research and development, a range of potential applications in agriculture is now opening up, including the incorporation of nanoparticles (NPs) in pesticide formulations, their use as biosensors and devices to aid genetic manipulation and as aids to post-harvest management (Singh Sekhon, 2014; Servin et al., 2015). A wealth of literature over the last decade has addressed the potential toxicity of NPs and enhanced manufactured nanomaterials [ENMs; European Parliament, 2011; Science and Technology Options Assessment (STOA) European Parliament, 2012].

The EU Regulation 1169/2011 (to be applied in December 2016), although attempting a formal definition of ENMs, fails to mention them in the context of food additives, and even the proposed definitions are controversial. Concerns regarding the adequacy of the regulation have been raised by other EU organs [Aschberger et al., 2014; Science and Technology Options Assessment (STOA) European Parliament, 2012].

A general lack of consumer information has been criticized by some parties Friends of the Earth: emerging Tech Project website, 2015 (https://www.foe.org.au/articles/2015-09-25/new-study-raises-further-questions-about-safety-nano-ingredients-food) as has the approach for assessing toxicity and ecotoxicity (Fadeel and Garcia-Bennett, 2010; Saez et al., 2010; Sigg et al., 2014). In a recent report the OECD emphasized the importance of improved toxicity test for assessing ENMs environmental dispersion and impact on human health [Organisation for Economic Co-operation and Development (OECD), 2014]. The potential risks arising from a lack of legislation have been flagged by Abbott et al. (2012) and Hodge et al. (2014). The consensus regarding ENMs is that hard data are still required to clarify the nature and implications of their interaction with biological matrices. Meanwhile, some methods aimed at improving the performance and reducing the toxicity of medical NPs, such as incorporating biocompatible coating materials, modifying their surface to mitigate toxicity and building in biodegradability have been proposed.

Quantum dots (QDs) are crystalline NPs, first synthesized in the early 1980s for use in the electronics industry (Brus, 1984). Cadmium sulfide quantum dots (CdS QDs) have a high surface charge and reactivity and are very stable (Favero et al., 2006). Their biological activity has been studied using both a plant and a yeast model (Marmiroli et al., 2014, 2015), applying both a mutant-based and a genome-wide transcriptomics approach. Two Arabidopsis thaliana mutants have emerged as showing an enhanced level of tolerance; in the first, the mutagenized genes encoded an unknown chloroplast-localized protein, a cytoplasm-localized calmodulin binding protein and a member of the MYB class of transcription factors, while in the second, the candidate genes encoded an O-glycosyl hydrolase localizing to the endomembrane and a chloroplast-localizing ATP binding protein. The contrasting genetic basis of tolerance in the two mutants was taken to imply that CdS QDs tolerance can be achieved by the activation of non-identical master switches. A transcriptomic analysis of wild type (wt) A. thaliana plants exposed to Cd^{2+} ions revealed that a completely different gene set was activated, meaning that the pathway leading to CdS QD tolerance must be unrelated to that determining the response to Cd^{2+} stress.

In contrast to a wealth of transcriptomic data sets, the understanding of the proteomic response to ENM exposure is rather limited. In general, the statistical correlation between transcript and protein abundance in eukaryotic cells is poor (Gygi et al., 1999; Hajduch et al., 2010), a phenomenon ascribed largely to the major role played by post-transcriptional modification (Maier et al., 2009). The aim of the current study was to supplement the documented transcriptomic and phenotypic responses of A. thaliana to CdS QD exposure (Marmiroli et al.,

2014) with a robust set of proteomic data, collected using a double liquid chromatography separation system well-proven to resolve the complex protein mixture present in a plant matrix (Marmiroli et al., 2013).

MATERIALS AND METHODS

Plant Material

A. thaliana, accession Landsberg erecta (L. Heyn) mutants atnp01 and atnp02 were isolated by screening of 378 mutant lines obtained from the (Nottingham) European Arabidopsis Stock Centre (uNASC; http://arabidopsis.info/), for resistance to CdS QDs as described by Marmiroli et al. (2014). The same paper, reports also a genetic and molecular characterization of the two mutants.

Physical Properties of the CdS QDs

The CdS QDs utilized during all the experiments had a bulk density of 4.82 g cm^{-3} and a diameter of about 5 nm, they were synthesized following Villani et al. (2012). Cadmium represented ~78% of the dry weight. The CdS QDs were the same batch used into the previous transcriptomic work (Marmiroli et al., 2014).

Seed Germination, Growth, and Treatments

Twenty-five seeds of A. thaliana wild type (wt) and atnp01 and atnp02 were sawn on Petri dishes containing Murashige and Skoog (MS) nutrient medium (Duchefa Biochemie, Haarlem, Netherlands) containing 1% w/v sucrose and solidified with 0.8% w/v agar, then placed in the dark, under controlled conditions in a growth chamber. After germination, seedlings were grown at 24°C, with relative humidity of 30%, and under a 16-h photoperiod (light intensity 120 μM m^{-2} s^{-1} photosynthetic photon flux) in the same MS medium in the absence of CdS QDs for 14 days. Seedlings were transferred to MS medium containing 80 mg L^{-1} CdS QDs (treatment) or 0 mg L^{-1} (control) and grown for a further 21 days, as above. Plantlets were then removed from the medium, carefully washed with distilled H_2O and used for protein extraction.

Protein Samples Preparation

Crude proteins of wt and the two mutant lines in untreated and treated (80 mg L^{-1} CdS QDs) conditions were extracted by using $MgSO_4$-gbased buffer. A total amounts of 1 g of frozen plants for wt and mutants and for both treatments were finely ground in liquid nitrogen with a ceramic mortar and pestle, adding SiO_2 sand (Sigma-Aldrich, St. Louis, MO, USA), to encourage breakage of the cell walls. The fine powder obtained was resuspended in 50 mM Tris [tris (hydroxymethyl) aminomethane] HCl pH 7.8, 10 mM $MgSO_4$, 0.1% [v/v] β mercaptoethanol and 0.1% [v/v] Protease Inhibitor Cocktail (Sigma-Aldrich). The crude mixture was sonicated for 10 min at 35 kHz (Transsonic T460, Elma Schmidbauer GmbH, Singen, Germany) and then the solution was placed in ice for 40 min. After 10 min more of sonication, the sample was centrifuged in a precooled rotor spun at 16,000×g for 5 min at 4°C (Micorfuge

22R Centrifuge, Beckman Coulter, Fullerton, CA, USA). The pellet, containing the larger cellular residues and SiO_2 sand, was discarded and the supernatant centrifuged at $16.000 \times g$ for 30 min at $4°C$. The upper phase was pipetted into other 15 ml tubes and stored at $-20°C$ for further analysis. Three biological replicates were produced for crude protein extracts from wt and mutants.

Protein Quantification

Proteins were quantified using the Quick Start Bradford Protein Assay (BioRad, Hercules, CA, USA); The protein-dye formed was detected at 595 nm with spectrophotometric assay (Uvikon 931, Kontron Instruments) with a standard curve from different dilutions of BSA (Bovine Serum Albumin; Sigma-Aldrich). The BSA dilutions and sample dilutions were prepared in a suitable Chromatofucusing (CF) Start buffer for the next step of two-dimensional liquid chromatography (2D-LC).

PD10 Desalting Column

Protein extracts were desalted and equilibrated using PD-10 Desalting Workmade disposable columns (GE-Healthcare Biosciences, Uppsala, Sweden) containing prepacked Sephadex G-25 Medium with exclusion limit of 5000 Da. PD-10 column equilibration was performed by using ~25 ml of CF Start buffer (Eprogen, Downers Grove, IL, USA) and the sample was then eluted with 3.5 ml of CF start buffer. The capacity of the system allows the loading of up to 2.5 ml of sample, with a range of loading capacity between 0.5 and 5 mg of protein per sample.

Two Dimensional Liquid Chromatography

Three milligram of protein extract were separated by 2D-LC for each sample. Separation was performed with ProteomeLab™ PF2D by Beckman Coulter equipped with: HPCF-1D column 250×2.1 mm internal diameter, 300 Å pore size and HPRP C18 column 4.66 mm length $\times$ 3.3 mm internal diameter, $1.5\,\mu$m particle size (Eprogen). Proteins are separated in the first dimension by high-performance chromatofocusing (HPCF), performed on an HPCF-1D column. With this technique, proteins bound to a strong anion exchanger followed by elution with a continuously decreasing pH (8.5–4.0) gradient. The pH gradient was generated in the column by two buffers: Start Buffer (SB) and Eluent Buffer (EB; Eprogen). The calibration of both buffers was an important step: SB and EB were sonicated for 5 min and then their pH was adjusted to 8.5 and 4.0 respectively using either a saturated solution (50 mg/ml) of iminodiacetic acid (Sigma-Aldrich) if the buffer was too basic, or 1M NH_4OH (J.T. Baker, Deventer, Holland) if the buffer was too acidic. The column was first equilibrated to the initial pH 8.5 using CF Start buffer at a flow rate of 0.2 ml min^{-1} for 3 h. After this step, 5 ml of sample were injected the column for the first dimension CF. Twenty minutes from sample injection, the first dimension pump switches to the CF Eluent buffer (pH 4) at a flow rate of 0.2 ml min^{-1}. The interaction of the column filling with the CF Eluent buffer produced a gradually decreasing pH gradient that traveled through the column as a retained front. The pH gradient affected the proteins net charge and their adsorption/desorption to the positively-charged matrix of the column, causing protein

separation in the effluent. The pH of the mobile phase was monitored on-line by a post-detector pH flow cell. The proteins were eluted based on their isoeletric point (pI), measured the absorbance at 280 nm, and collected in a 96 deep-well plate by a fraction collector according to pre-determined pH decrements of 0.4 pH units during the gradient, or in 1 ml volumes when the pH did not change. At the 115th min the most acidic proteins were recovered by washing the column with 1M NaCl 30% n-propanol [v/v] for 15 min. The column was finally washed with water for 45 min; the CF separation took of total of $\sim$185 min.

The eluent from the 1st dimension was injected into the 2nd dimension, a high-performance reversed-phase chromatography (HPRP) based on protein hydrophobicity. HPRP was carried out in a C18 column. The mobile phase consisted of A: 0.1% TFA (Trifluoroacetic Acid; J.T. Baker) in water and B: 0.08% TFA in Acetonitrile (J.T. Baker). Separation was performed at 0.75 ml min^{-1} with an increasing gradient of B. During the first 2 min 100% of solvent A was pumped into the column; in the next 35 min the gradient was created in the column by switching the flow from 0 to 100% solvent B; this is followed by 100% B for 4 min and 100% A for 9 min. In order to obtain a better resolution, the separation was done at $50°C$. The eluent from the second dimension was monitored by a second high performance UV/VIS detector at 214 nm, that provided a more universal and sensitive detection of proteins via peptide bonds. Fractions were immediately collected in eppendorf tubes for MS analysis by using an automated fraction collector.

Protein Identification

Matrix-assisted laser desorption/ionization time-of-flight mass spectrometry (MALDI-TOF/MS) analysis was carried out using a 4800 Plus MALDI-TOF/TOF™ (AB SCIEX, Framingham, MA, USA) equipment. Eluted fractions were evaporated to a final individual volume of 10 μl, using a Speed Vac Concentrator 5301 (Eppendorf AG, Barkhausenweg, Hamburg, Germany). Protein digestion was performed by incubating each fraction in 25 mM NH_4HCO_3 and 2 mM DTT (DL-Dithiothreitol) in a water bath at $60°C$ for 1 h. The alkylation of the reduced sulfhydryl groups was carried out adding 1 mM Iodoacetamide, at $25°C$, for 30 min in the dark, and then 10 μL of Trypsin (125 μg ml^{-1}) in 50 mM NH_4HCO_3 were added. Digestion was carried out at $37°C$ for 24 h. The samples digested were then purified and concentrated with a ZipTipC18 using the procedure recommended by the manufacturer (Millipore Corporation, Billerica, MA, USA). Then 1 μL of each purified peptide was spotted directly onto a stainless steel MALDI target plate with 1 μL of a saturated solution of α-cyano-4-hydroxycinnamic acid in 0.1% TFA:ACN (2:1, v/v). The solution was allowed to dry at room temperature and a spot was produced. Positively charged ions were analyzed in reflectron mode. The spectra were obtained by random scanning of the sample surface with an ablation laser. A mass range of 10,000–100,000 Da was used, and about 400 laser shots were averaged to improve the signal-to-noise ratio. Calibration was performed by a ProteoMass Protein MALDI-MS Calibration kit (Sigma-Aldrich). Two technical replicates for each spectrum were analyzed by MS, and peptides common to all of the resolved spectra were considered for protein identification.

Statistical and Bioinformatics Analysis

ProteoVue software (Eprogen) was utilized to convert chromatographic intensities from the 2D-LC of each pH fraction into a band intensity format. This produced a highly detailed map with the dimensions of hydrophobicity and pI. The 2D-LC maps could be viewed in several colored formats where the color intensity was proportional to the relative intensity of each chromatographic peak. DeltaVue software (Eprogen) was utilized for the differential analysis of corresponding fractions from two different sample sets. This software compared chromatogram peaks corresponding to the same protein in the two samples, allowing quantification by a subtractive analysis. A differential map was produced by point-to-point subtraction and it is viewed between the two original sample sets. Mass spectra were analyzed using the mMass software package (http://www.mmass.org/; ver. 5.5, by Martin Strohalm) and the peak list for each mass spectra were obtained. Peptide mass fingerprinting analysis was carried out with the Mascot program (http://www.matrixscience.com). Proteins were identified by searching against Swiss-Prot database of *A. thaliana* (thale cress). The following parameters were used for database search: mass accuracy below 100 ppm, maximum of one missed cleavages by trypsin, carbamidomethylation of cysteine as fixed modifications, oxidation of methionine as variable modifications. The search was based on the monoisotopic masses of the peptides. For mass-spectrometry (MS) analyses, three technical replicates for each spectrum were performed. For proteins identification, only peptides in common to all the resolved spectra were considered.

The gene loci found in the UniProt were searched in TAIR database (https://www.arabidopsis.org/) for the corresponding *A. thaliana* proteins names, description, and GO annotations.

Heat maps of selected proteins were generated by TreeView v1.60 software. Gene Ontology (GO) analysis (Harris et al., 2004) visualized with pie charts were generated by VirtualPlant v1.3 (http://virtualplant.bio.nyu.edu/cgi-bin/vpweb/virtualplant.cgi) applying a p (calculated according to Bonferroni test) cutoff value of 0.05. Venn diagrams were generated by Venny 2.0 (http://bioinfogp.cnb.csic.es/tools/venny/index.html). The correlation between mRNA and protein levels was calculated using the Pearson product moment correlation coefficient in Microsoft Excel 2010.

RESULTS

Proteomic Data Management and Visualization

The proteomic separation identified around 600 proteins in the extracts of wt plants and of each of the two mutants exposed and not to CdS QDs. Coupling treated and not treated results for wt, atnp01 and atnp02 three subset of about 1200 proteins were found. The use of DeltaVue software led to the elaboration of a "differential map" for each genetic comparison, where each "band" corresponded to a unique protein and where each virtual band's intensity was proportional to the protein's relative abundance, measured against its abundance in the non-treated control sample. In order to assess which of the intensity ratios were statistically significant, their $\log_{10}$'s were grouped into frequency categories, producing a normal distribution; only those proteins associated with a ratio differing from the mean by at least two standard deviations ($\pm$) were taken forward for identification, following the strategy outlined by Marmiroli et al. (2013). On this basis almost 200 proteins were selected, but of these, only 130 were abundant enough to be subjected to MALDI-TOF/MS. The identification of some of the proteins using mass fingerprinting was not possible due to low scores, so finally 88 proteins were identified with any statistical confidence. The sets of differentially expressed proteins are listed in **Table 1**, and a global heat map is presented in **Figure 1B**: the chosen calibrator was the treated wt plant, because this was found to most clearly highlight the differences between the set of samples, while also allowed direct comparisons to be made with established transcriptomic data (Marmiroli et al., 2014; **Figure 1A**).

Venn diagrams featuring the differentially represented, both over- and under-represented, proteins in both mutants compared to the wt in both the treated and untreated situation are presented in **Figure 2**. There were 35 over-represented proteins in the treated atnp01 mutant, and 47 in the atnp02 mutant; of these, 26 were in common between the two comparisons. The respective frequencies of under-represented proteins were 44 in atnp01, 40 in atnp02, and 31 common to both mutants. In the comparison between treated wt and atnp01 plants, nine proteins having the same abundance.

In the comparisons involving non-treated plants, there were 44 over-represented proteins in each mutant, of which 35 were in common. With respect to the set of under-represented proteins: 44 for atnp01, 40 for atnp02, of which 35 in common. Inspection of the data revealed that seven of the over-represented and 11 of the under-represented proteins did not change in abundance either as a consequence of the treatment or as a result of a genetic difference, 17 over- and 17 under-represented were ascribable to genetic differences and eight over- and 17 under-represented ones to the CdS QDs exposure. The atnp01 mutation affected eight proteins (two over-, six under-represented), while the atnp02 mutation affected six proteins (one over-, five under-represented). The CdS QDs treatment altered the expression level of 12 proteins in atnp01 (four over-, eight under-represented) and 14 in atnp02 (11 over-, three under-represented).

Functional Analysis of Differentially Expressed Proteins

A GO analysis was conducted to assign functionality to the set of differentially expressed proteins (Supplementary Figures S1–S4). The most frequently encountered GO class was biological process, followed by molecular function and cellular components. For both the mutants, the over- and the under-represented proteins were classified within the biological process category as involving a cellular process, a metabolism or a response to stimuli.

The over-represented proteins in atnp01 concerned metabolic and cellular processes, response to stimuli and regulation (Supplementary Figure S1), the cellular components interested being extracellular parts, cell parts and organelles. The molecular function of relevance were catalytic, binding but also electron

TABLE 1 | atnp01 and atnp02 proteins influenced by mutations and by exposure to 80 mg L^{-1} CdS QDs, separated by 2D-LC and identified by MALDI-TOF/MS and *in silico* analysis.

Frac.[a]	RT[b]	Protein name (UniProt database)[c]	Accession no.[d]	Gene[e]	Locus[f]	Probe[g]	Mass[h]	pI[i]	Score[j]	Expect[k]	Match[l]	Cov.[m] (%)
14	20.15	12S seed storage protein CRC	CRU3_ARATH	CRU3	At4g28520	253767_AT	58,235	6.99	54	0.047	18	23
27	24.50	14-3-3-like protein GF14 nu	14337_ARATH	GRF7	At3g02520	258489_AT	29,920	4.74	39	1.5	4	18
17	19.32	2S seed storage protein 1	2SS1_ARATH	AT2S1	At4g27140	253904_AT	19,013	5.70	50	0.16	9	31
11	20.00	2S seed storage protein 3	2SS3_ARATH	AT2S3	At4g27160	253895_AT	18,762	7.86	51	0.11	11	38
24	14.91	4-alpha-glucanotransferase DPE2	DPE2_ARATH	DPE2	At2g40840	245094_AT	110,562	5.54	49	0.18	14	14
7	20.18	Actin-2	ACT2_ARATH	ACT2	At3g18780	257749_AT	42,078	5.37	62	0.008	11	37
29	34.09	Alanine-tRNA ligase	SYA_ARATH	ALATS	At1g50200	262468_AT	111,275	6.05	45	0.38	13	21
13	20.43	Arogenate dehydratase/prephenate dehydratase 6	AROD6_ARATH	ADT6	At1g08250	261758_AT	45,059	6.11	41	0.96	5	28
8	24.56	ATP synthase subunit beta, chloroplastic	ATPB_ARATH	atpB	AtCg00480	245014_AT	53,957	5.38	45	0.38	7	19
28	26.67	ATP-dependent DNA helicase Q-like 4A	RQL4A_ARATH	RECQL4A	At1g10930	17661_AT	134,654	6.82	45	0.39	12	14
22	16.14	Auxilin-like protein 1	AUL1_ARATH	AUL1	At1g75310	261117_AT	163,087	4.92	60	0.016	37	22
8	23.62	Auxin transport protein BIG	BIG_ARATH	BIG	At3g02260	259128_AT	574,541	5.65	54	0.047	30	7
15	19.30	Beta-amylase 3, chloroplastic	BAM3_ARATH	BAM3	At4g17090	245346_AT	61,713	6.59	40	1.2	20	41
8	26.25	BTB/POZ domain-containing protein At5g67385	Y5738_ARATH	At5g67385	At5g67385	-	68,086	8.16	34	36	3	4
27	18.30	Calcium-binding protein CML31 (probable)	CML31_ARATH	CML31	At2g36180	263903_AT	16,375	4.21	47	0.25	7	53
14	13.37	Calcium-binding protein CML42	CML42_ARATH	CML42	At4g20780	254487_AT	21,191	4.59	55	0.043	7	34
21	17.08	Calcium-binding protein CML45 (Probable)	CML45_ARATH	CML45	At5g39670	249417_AT	24,104	4.93	33	6.9	5	34
4	17.01	Calcium-dependent protein kinase 23	CDPKN_ARATH	CPK23	At4g04740	255306_AT	59,073	6.09	60	0.001	11	23
22	14.95	Calmodulin-like protein 1	CML1_ARATH	CML1	At2g15680	265494_AT	16,3087	4.92	46	0.34	31	18
29	34.09	CLIP-associated protein	CLASP_ARATH	CLASP	At2g20190	265315_AT	160,107	6.72	46	0.34	31	22
2	22.23	Cyclic nucleotide-gated ion channel 6 (Probable)	CNGC6_ARATH	CNGC6	At2g23980	266520_AT	86,299	9.38	45	0.42	12	19
24	15.87	Defensin-like protein 121 (Putative)	DF121_ARATH	LCR55	At3g20997	-	8969	7.51	42	0.85	5	67
14	19.90	Defensin-like protein 192	DF192_ARATH	ATTI7	At1g47540	262431_AT	11,063	6.34	44	0.45	8	55
28	21.11	Defensin-like protein 37	DEF37_ARATH	EDA21	At4g13235	-	9708	9.3	40	1.2	3	36
2	24.91	Defensin-like protein 90	DEF90_ARATH	At1g54445	At1g54445	-	9175	8.44	40	1.4	3	57
18	15.50	DNA damage-binding protein 1b	DDB1B_ARATH	DDB1B	At4g21100	254452_AT	121,178	4.97	44	0.45	6	44
26	23.93	E3 ubiquitin-protein ligase ARI9 (probable)	ARI9_ARATH	ARI9	At2g31770	263463_AT	64,529	4.96	45	0.44	11	19
24	17.90	ELM2 domain-containing protein	D7KQK8_ARAL	At1g13880	At1g13880	259423_AT	48,428	4.18	50	0.15	14	40
14	21.12	F-box/kelch-repeat protein At1g48625	FBK20_ARATH	At1g48625	At1g48625	-	48,704	8.9	51	0.11	11	32
29	34.12	F-box/kelch-repeat protein At3g43710 (putative)	FBK72_ARATH	At3g43710	At3g43710	-	44,064	7.91	40	1.3	8	15
5	23.07	Galacturonosyltransferase 2 (putative)	GAUT2_ARATH	GAUT2	At2g46480	265477_AT	62,393	6.98	34	5.2	4	16
3	25.23	GDP-mannose 4,6 dehydratase 1	GMD1_ARATH	GMD1	At5g66280	247094_AT	40,939	6.61	41	0.94	6	26
23	20.37	GDP-mannose 4,6 dehydratase 2	GMD2_ARATH	MUR1	At3g51160	252121_AT	40,939	6.61	53	0.07	5	22
22	15.87	GDSL esterase/lipase ESM1	ESM1_ARATH	ESM1	At3g14210	257008_AT	44,374	7.59	44	0.49	6	18
19	16.55	Glucan endo-1,3-beta-glucosidase	E13A_ARATH	BGL2	At3g57260	251625_AT	37,338	4.60	44	0.60	9	25
10	23.05	Glutaredoxin-C14	GRC14_ARATH	GRXC14	At3g62960	251197_AT	11,561	7.66	44	0.45	4	66
12	17.40	Glutathione S-transferase DHAR3	DHAR3_ARATH	DHAR3	A5g16710	246454_AT	28,724	7.59	35	4.3	12	41
14	18.51	Glutathione S-transferase U24	GSTUO_ARATH	GSTU24	A1g17170	262518_AT	25,461	5.85	56	0.031	17	47

(Continued)

TABLE 1 | Continued

Frac.[a]	RT[b]	Protein name (UniProt database)[c]	Accession no.[d]	Gene[e]	Locus[f]	Probe[g]	Mass[h]	pI[i]	Score[j]	Expect[k]	Match[l]	Cov.[m] (%)
22	14.51	Glycine cleavage system H protein 2	GCSH2_ARATH	GDH2	At2g35120	266517_AT	17,203	4.97	49	0.18	8	31
27	22.01	Heat shock 70 kDa protein 10, mitochondrial	HSP7J_ARATH	HSP70-10	At5g09590	250502_AT	73,174	3.5	36	3.5	5	9
28	27.65	Homeobox-leucine zipper protein ATHB-7	ATHB7_ARATH	ATHB-7	At2g46680	266327_AT	30,663	5.52	42	0.9	12	50
24	18.36	LOB domain-containing protein 5	LBD5_ARATH	LBD5	At1g36000	260187_AT	14,477	6.73	34	4.9	8	44
19	19.62	Lysine-specific histone demethylase 1	LDL1_ARATH	LDL1	At1g62830	262668_AT	93,767	4.84	64	0.005	10	17
26	24.89	Mechanosensitive ion channel protein 6	MSL6_ARATH	MSL6	At1g78610	263127_AT	97,057	8.73	42	0.8	13	16
22	15.98	Methyl-CpG-binding domain-containing protein 9	MBD9_ARATH	MBD9	At3g01460	258950_AT	243,929	5.34	54	0.059	32	17
2	24.12	N-(5'-phosphoribosyl)anthranilate isomerase 1	PAI1_ARATH	PAI1	At1g07780	259770_S_AT	29,847	8.63	35	3.7	5	32
28	24.68	NF-X1-type zinc finger protein NFXL2	NFXL2_ARATH	NFXL2	At5g05660	250767_AT	105,547	8.63	43	0.67	14	18
27	24.20	Nudix hydrolase 21	NUD21_ARATH	NUDT21	At1g73540	245777_AT	22,793	8.39	52	0.084	4	43
2	21.30	Oleosin GRP-17	GRP17_ARATH	GRP17	At5g07530	250637_AT	53,330	10.34	36	3.5	15	36
4	18.20	Pathogenesis-related protein 1	PR1_ARATH	At2g14610	At2g14610	266385_AT	17,676	8.96	48	0.20	8	35
18	17.80	Pathogenesis-related protein 5	PR5_ARATH	At1g75040	At1g75040	259925_AT	25,252	4.54	53	0.040	16	43
27	23.98	Pectate lyase 9 (probable)	PLY9_ARATH	At3g24230	At3g24230	257243_AT	50,638	8.65	48	0.22	11	31
28	19.63	Pentatricopeptide repeat-containing protein At1g12775	PPR39_ARATH	At1g12775	At1g12775	261194_AT	73,474	6.15	44	0.46	10	28
2	24.33	Pentatricopeptide repeat-containing protein At1g62590	PPR90_ARATH	At1g62590	At1g62590	265106_S_AT	71,899	7.97	53	0.061	6	9
24	15.36	Pentatricopeptide repeat-containing protein At1g62670	PPR91_ARATH	At1g62670	At1g62670	260792_F_AT	110,562	5.54	50	0.15	10	10
27	19.43	Pentatricopeptide repeat-containing protein At1g63330	PP101_ARATH	At1g63330	At1g63330	265106_S_AT	63,778	6.83	36	2.9	6	10
27	21.94	Pentatricopeptide repeat-containing protein At2g18940	PP163_ARATH	At2g18940	At2g18940	266951_F_AT	93,433	8.36	36	3.6	13	20
14	24.38	Pentatricopeptide repeat-containing protein At2g45350	PP202_ARATH	CRR4	At2g45350	245129_F_AT	70,052	6.78	46	0.37	5	10
27	21.44	Pentatricopeptide repeat-containing protein At3g29290	PP262_ARATH	EMB2076	At3g29290	256613_F_AT	63,160	9.01	38	2.1	5	12
24	27.45	Pentatricopeptide repeat-containing protein At3g57430	PP285_ARATH	PCMP-H81	At3g57430	251631_AT	100,466	6.94	37	2.7	15	15
26	25.63	Pentatricopeptide repeat-containing protein At4g26680	PP338_ARATH	At4g26680	At4g26680	253979_AT	60,010	9.45	47	0.28	12	28
24	13.03	Pentatricopeptide repeat-containing protein At5g61400	PP440_ARATH	At5g61400	At5g61400	247548_AT	75,760	7.91	37	2.5	10	14
26	26.77	Peptide methionine sulfoxide reductase B7	MSRB7_ARATH	MSRB7	At4g21830	254385_S_AT	15,789	8.54	51	0.11	4	36
4	14.49	Phosphomethylpyrimidine synthase	THIC_ARATH	THIC	At2g29630	266673_AT	72,568	6.00	46	0.31	9	22
26	26.02	Prefoldin subunit 4 (probable)	PFD4_ARATH	AIP3	At1g08780	264778_AT	14,985	4.55	41	1	5	54
14	13.68	Pre-mRNA-splicing factor SLU7-A	SLU7A_ARATH	At1g65660	At1g65660	264633_AT	62,281	5.73	41	1.1	12	21
4	11.96	Probable beta-D-xylosidase 7	BXL7_ARATH	At1g78060	At1g78060	262181_AT	84,751	8.3	66	0.003	15	22

(Continued)

TABLE 1 | Continued

Frac.[a]	RT[b]	Protein name (UniProt database)[c]	Accession no.[d]	Gene[e]	Locus[f]	Probe[g]	Mass[h]	pI[i]	Score[j]	Expect[k]	Match[l]	Cov.[m] (%)
1	12.50	Proline-rich extensin-like protein EPR1	EPR1_ARATH	EPR1	At2g27380	265644_AT	81,946	10.61	60	0.009	12	19
29	28.05	Prolyl 4-hydroxylase 7 (probable)	P4H7_ARATH	P4H7	At3g28480	257844_AT	36,117	6.32	43	0.64	7	20
16	18.93	Proteasome subunit alpha type-1-A	PSA1A_ARATH	PAF1	At5g42790	249161_AT	30,685	4.99	45	0.4	7	27
29	34.48	Proteasome subunit alpha type-3	PSA3_ARATH	PAG1	At2g27020	266312_AT	27,645	5.93	62	0.008	9	24
19	21.88	Protein FLX-like 2	FLXL2_ARATH	FLXL2	At1g67170	–	39,831	6.93	48	0.21	7	23
15	19.32	Protein KTI12 homolog	KTI12_ARATH	DLR1	At1g13870	259450_AT	34,063	5.97	44	0.73	9	33
29	27.96	Protein MODIFIER OF SNC1 1	MOS1_ARATH	MOS1	At4g24680	254143_AT	153,766	9.39	53	0.067	21	19
7	20.45	Protein phosphatase 2C 13 (probable)	P2C13_ARATH	At1g48040	At1g48040	260722_AT	42,649	4.85	52	0.088	10	25
24	19.84	Protein TONNEAU 1b	TON1B_ARATH	TON1B	At3g55005	251816_AT	29,338	5.06	37	3	14	46
22	17.02	Protein VERNALIZATION INSENSITIVE 3	VIN3_ARATH	VIN3	At5g57380	–	71,128	5.88	39	1.6	16	26
27	31.71	Proton pump-interactor 3A	PPI3A_ARATH	PPI3A	At5g36780	–	66,923	6.07	33	6.2	7	14
2	24.88	Ribulose bisphosphate carboxylase large chain	RBL_ARATH	rbcL	AtCg00490	245015_AT	53,435	5.88	49	0.16	10	23
2	22.76	Ribulose bisphosphate carboxylase small chain 1A	RBS1A_ARATH	RBCS-1A	At1g67090	264474_S_AT	20,488	7.59	101	1.6e-07	12	68
3	22.72	Ribulose bisphosphate carboxylase small chain 2B	RBS2B_ARATH	RBCS-2B	At5g38420	264474_S_AT	20,622	7.59	101	0.00001	13	59
3	23.95	Ribulose bisphosphate carboxylase small chain 3B	RBS3B_ARATH	RBCS-3B	At5g38410	264474_S_AT	20,556	8.22	54	0.048	8	53
26	25.17	RING-H2 finger protein ATL38	ATL38_ARATH	ATL38	At2g34990	267417_AT	34,890	6.71	43	0.64	8	36
3	24.77	Rop guanine nucleotide exchange factor 1	ROGF1_ARATH	ROPGEF1	At4g38430	252975_S_AT	61,380	5.44	40	1.3	8	20
26	26.52	Transcription factor GTE1	GTE1_ARATH	GTE1	At2g34900	257352_AT	43,586	6.04	45	0.37	7	26
26	25.09	WPP domain-interacting tail-anchored protein 1	WIT1_ARATH	WIT1	At5g11390	250363_AT	79,228	4.7	39	1.6	12	13
12	21.17	WRKY transcription factor 61 (probable)	WRK61_ARATH	WRKY61	At1g18860	261429_AT	53,224	6.48	35	4.1	10	18
12	17.80	γ-interferon responsive lysosomal thiol (GILT) reductase family protein	F4JRI7_ARATH	At4g12960	At4g12960	254777_AT	27,196	7.05	38	3.7	14	44

[a]fraction, [b]progressive peak number as given by 2D-LC, [c]putative protein identification, [d]accession number for the closest match in the UniProt database, [e]gene, [f]locus, [g]probe, [h]predicted mass value, [i]pI of the closest match in the database, [j]score, [K]expected value of the database research, [l]matches, [m]percentage of coverage of the matching peptide sequence tags derived by applying MASCOT algorithm.

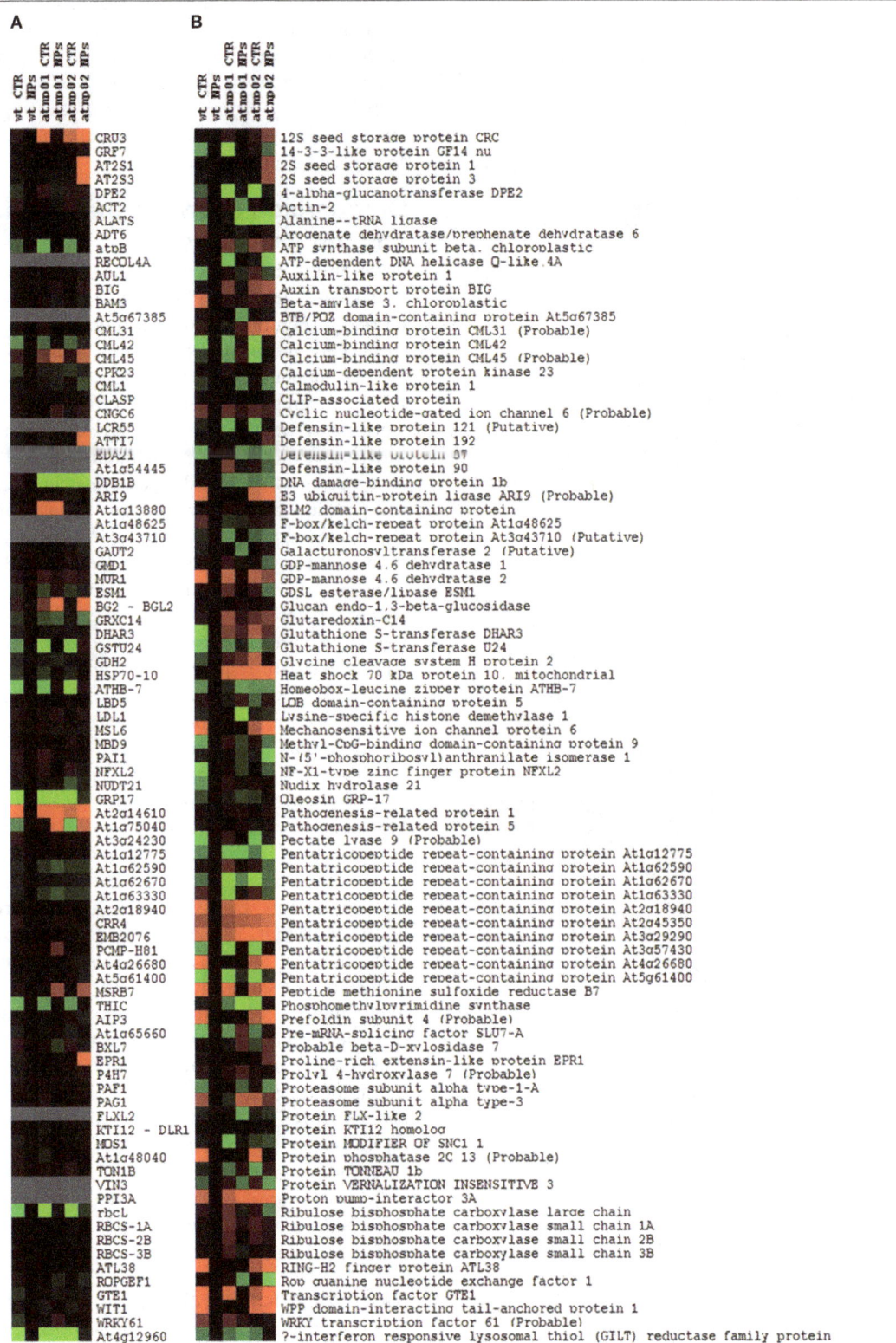

FIGURE 1 | Heat maps of *A. thaliana* wt and mutant lines atnp01 and atnp02 not treated and treated with 80 mg L⁻¹ CdS QDs drawn with TreeView software. Heat map of the transcriptomic data, the probe "wt treated" was used as calibrator (black column). Up-regulated genes compared to the calibrator are shown in shades of red and down-regulated genes in shades of green **(A)**. Heat map of the proteomic data, "wt treated" was used as calibrator (black column). Proteins more abundant in the sample compared to the calibrator are shown in shades of red, and those less abundant in the sample compared to the calibrator in shades of green **(B)**.

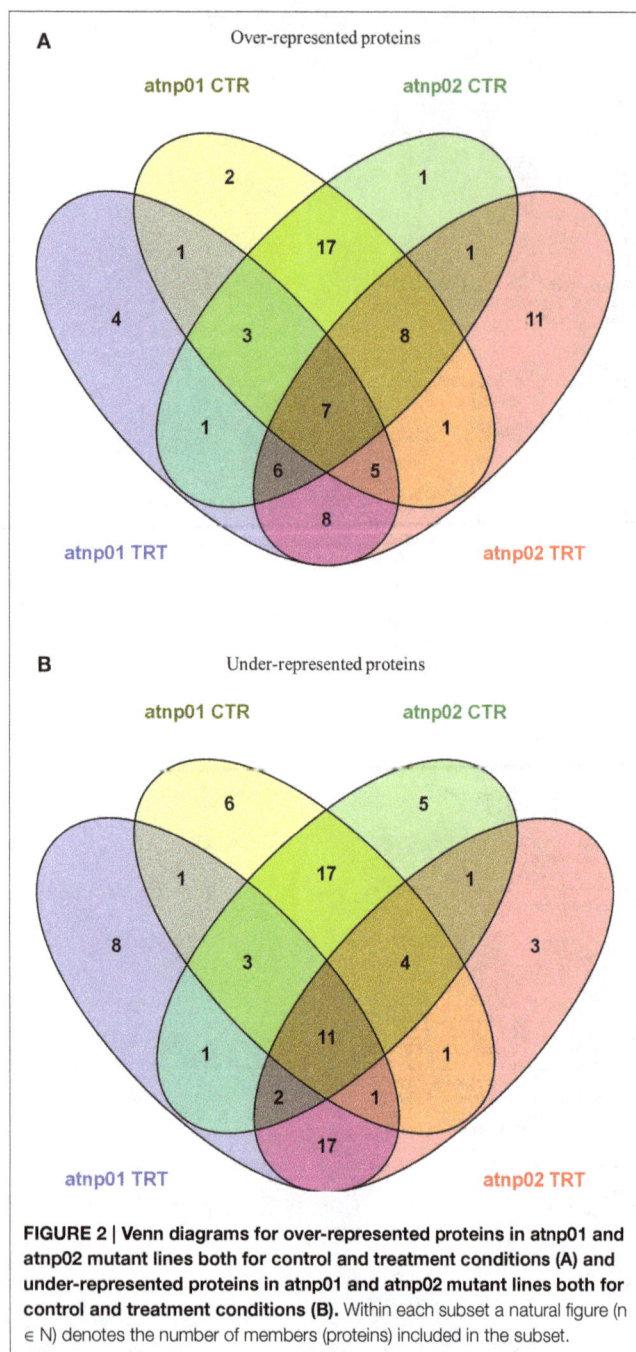

FIGURE 2 | Venn diagrams for over-represented proteins in atnp01 and atnp02 mutant lines both for control and treatment conditions (A) and under-represented proteins in atnp01 and atnp02 mutant lines both for control and treatment conditions (B). Within each subset a natural figure (n ∈ N) denotes the number of members (proteins) included in the subset.

activity, binding and transport (Supplementary Figure S3). In the mutant atnp02 the under-represented concerned proteins of the cell metabolism and developmental process, but also cellular components organization and response to stimuli (Supplementary Figure S4). Cell part and organelles were the more affected with molecular function in the class of catalytic activity and binding as majority.

Comparison Between the Transcriptome and the Proteome

Based on the transcriptome description provided by Marmiroli et al. (2014), 78 of the 88 proteins were assignable an encoding transcript (**Figure 1**). The Pearson product moment correlation coefficients (r) for transcript and protein representation for the two mutants were, respectively, 0.126 and 0.197. **Figure 3** shows a comparison between protein and transcript over-represented with respect to gene product, but under-represented with respect to transcript (column 4 and 5). In atnp01, 46 of the proteins (59%) exhibited a matching level of transcript and protein ("concurrent" gene products), while 16 were over-represented even though their RNA was underexpressed, and 16 behaved in the opposite manner; these 32 gene products were termed "non-concurrent." In atnp02, there were 44 (57%) concurrent and 34 non-concurrent proteins, of which 12 (15%) were over-represented with respect to gene product, but under-represented with respect to transcript, and 22 (28%) *vice versa*. In atnp02, 57% were concurrent, while of the non-concurrent ones, 15%, the reverse holds for the 28% of the proteins (**Table 2**). Since the studied protein set was so much smaller than the number of relevant transcripts, reported in Marmiroli et al. (2014) (88 vs. 456), a correlation analysis based on either Pearson's P or Kendall's τ was considered to be unsuitable. In order to recognize an association between transcript and protein abundance induced by the Cd QDs, the behavior of the two mutants was compared: either the direction of change of the transcript abundance matched that of the protein in both mutants, or it did not. Thus, two broad groups were defined comprising a constant, invariable, or a variable, at times unquestionably opposite, behavior, which describes the trend of protein production rate against the backdrop of transcript. The final column in **Figure 3** depicts the general cascade from transcript to gene product induced by the treatment in the two mutants. In all, for 71% of the gene products, protein representation reflected the behavior of the matching transcript, while for the remaining 29%, there was no apparent relationship; in 3% of the cases, a particularly high transcript expression was matched by a particularly low level of protein representation or *vice versa* (**Table 2**).

Identification of Specific Proteins

In Supplementary Tables S1, S2 is reported all the bibliography relevant to each protein mentioned in this sub-heading.

The proteome of both mutants differed from that of the wt, both when the plants were growth under control conditions and when they were exposed to CdS QDs. In the absence of the stress treatment, just two proteins were specifically over-represented in atnp01 and one in atnp02 (**Figure 2A**). One

carriers and antioxidants activity. The over-represented proteins in atnp02 concerned metabolic and cellular processes and as for atnp01 response to stimuli (Supplementary Figure S2). Also for the cellular components and molecular functions the similarities were remarkable (Supplementary Figure S2).

The proteomic response of the two mutants evidenced, in the condition of treatment for atnp01, under-represented proteins in the biological process metabolisms, cellular response to stimuli, cellular components organization with at this level a predominance of cell part and organelle (Supplementary Figure S3). The molecular function involved were: catalytic

Protein	Gene	Locus	Comparisons between transcriptomic and proteomic data in atnp01 Trt	Comparisons between transcriptomic and proteomic data in atnp02 Trt	Comparisons atnp01/ atnp02
12S seed storage protein CRC	CRU3	At4g28520			yellow
14-3-3-like protein GF14 nu	GRF7	At3g02520		blue	orange
2S seed storage protein 1	AT2S1	At4g27140			orange
2S seed storage protein 3	AT2S3	At4g27160			yellow
4-alpha-glucanotransferase DPE2	DPE2	At2g40840		blue	yellow
Actin-2	ACT2	At3g18780			yellow
Alanine--tRNA ligase	ALATS	At1g50200			yellow
Arogenate dehydratase/prephenate dehydratase 6	ADT6	At1g08250			pink
ATP synthase subunit beta, chloroplastic	atpB	AtCg00480			orange
ATP-dependent DNA helicase Q-like 4A	RECQL4A	At1g10930			yellow
Auxilin-like protein 1	AUL1	At1g75310			yellow
Auxin transport protein BIG	BIG	At3g02260	light blue		yellow
Beta-amylase 3, chloroplastic	BAM3	At4g17090			yellow
BTB/POZ domain-containing protein At5g67385	At5g67385	At5g67385			yellow
Calcium-binding protein CML31 (Probable)	CML31	At2g36180	light blue		yellow
Calcium-binding protein CML42	CML42	At4g20780	light blue		yellow
Calcium-binding protein CML45 (Probable)	CML45	At5g39670	blue		yellow
Calcium-dependent protein kinase 23	CPK23	At4g04740			yellow
Calmodulin-like protein 1	CML1	At2g15680			yellow
CLIP-associated protein	CLASP	At2g20190	blue		yellow
Cyclic nucleotide-gated ion channel 6 (Probable)	CNGC6	At2g23980			yellow
Defensin-like protein 121 (Putative)	LCR55	At3g20997			yellow
Defensin-like protein 192	ATTI7	At1g47540			yellow
Defensin-like protein 37	EDA21	At4g13235			yellow
Defensin-like protein 90	At1g54115	At1g54115			
DNA damage-binding protein 1b	DDB1B	At4g21100			yellow
E3 ubiquitin-protein ligase ARI9 (Probable)	ARI9	At2g31770	blue		yellow
ELM2 domain-containing protein	At1g13880	At1g13880			yellow
F-box/kelch-repeat protein At1g48625	At1g48625	At1g48625	light blue		yellow
F-box/kelch-repeat protein At3g43710 (Putative)	At3g43710	At3g43710	light blue		yellow
Galacturonosyltransferase 2 (Putative)	GAUT2	At2g46480			yellow
GDP-mannose 4,6 dehydratase 1	GMD1	At5g66280			yellow
GDP-mannose 4,6 dehydratase 2	MUR1	At3g51160			yellow
GDSL esterase/lipase ESM1	ESM1	At3g14210			yellow
Glucan endo-1,3-beta-glucosidase	BGL2	At3g57260			yellow
Glutaredoxin-C14	GRXC14	At3g62960			yellow
Glutathione S-transferase DHAR3	DHAR3	At5g16710	blue		yellow
Glutathione S-transferase U24	GSTU24	At1g17170	light blue		yellow
Glycine cleavage system H protein 2	GDH2	At2g35120	light blue		yellow
Heat shock 70 kDa protein 10, mitochondrial	HSP70-10	At5g09590	blue		orange
Homeobox-leucine zipper protein ATHB-7	ATHB-7	At2g46680	blue		orange
LOB domain-containing protein 5	LBD5	At1g36000	blue		yellow
Lysine-specific histone demethylase 1	LDL1	At1g62830	blue		yellow
Mechanosensitive ion channel protein 6	MSL6	At1g78610	blue		yellow
Methyl-CpG-binding domain-containing protein 9	MBD9	At3g01460	blue		yellow
N-(5'-phosphoribosyl)anthranilate isomerase 1	PAI1	At1g07780	blue		yellow
NF-X1-type zinc finger protein NFXL2	NFXL2	At5g05660			yellow
Nudix hydrolase 21	NUDT21	At1g73540	light blue		yellow
Oleosin GRP-17	GRP17	At5g07530			yellow
Pathogenesis-related protein 1	At2g14610	At2g14610			yellow
Pathogenesis-related protein 5	At1g75040	At1g75040			yellow
Pectate lyase 9 (Probable)	At3g24230	At3g24230	blue		yellow
Pentatricopeptide repeat-containing protein At1g12775	At1g12775	At1g12775			yellow
Pentatricopeptide repeat-containing protein At1g62590	At1g62590	At1g62590	blue		yellow
Pentatricopeptide repeat-containing protein At1g62670	At1g62670	At1g62670			yellow
Pentatricopeptide repeat-containing protein At1g63330	At1g63330	At1g63330			yellow
Pentatricopeptide repeat-containing protein At2g18940	At2g18940	At2g18940	light blue		yellow
Pentatricopeptide repeat-containing protein At2g45350	CRR4	At2g45350			yellow
Pentatricopeptide repeat-containing protein At3g29290	EMB2076	At3g29290			yellow
Pentatricopeptide repeat-containing protein At3g57430	PCMP-H81	At3g57430			orange
Pentatricopeptide repeat-containing protein At4g26680	At4g26680	At4g26680			orange
Pentatricopeptide repeat-containing protein At5g61400	At5g61400	At5g61400			yellow
Peptide methionine sulfoxide reductase B7	MSRB7	At4g21830	blue		orange
Phosphomethylpyrimidine synthase	THIC	At2g29630			yellow
Prefoldin subunit 4 (Probable)	AIP3	At1g08780			yellow
Pre-mRNA-splicing factor SLU7-A	At1g65660	At1g65660			yellow
Probable beta-D-xylosidase 7	BXL7	At1g78060			yellow
Proline-rich extensin-like protein EPR1	EPR1	At2g27380			yellow
Prolyl 4-hydroxylase 7 (Probable)	P4H7	At3g28480			yellow
Proteasome subunit alpha type-1-A	PAF1	At5g42790			yellow
Proteasome subunit alpha type-3	PAG1	At2g27020			yellow
Protein FLX-like 2	FLXL2	At1g67170			yellow
Protein KTI12 homolog	DLR1	At1g13870			yellow
Protein MODIFIER OF SNC1 1	MOS1	At4g24680	blue		yellow
Protein phosphatase 2C 13 (Probable)	At1g48040	At1g48040	blue	blue	yellow
Protein TONNEAU 1b	TON1B	At3g55005			yellow
Protein VERNALIZATION INSENSITIVE 3	VIN3	At5g57380	light blue		yellow
Proton pump-interactor 3A	PPI3A	At5g36780			yellow
Ribulose bisphosphate carboxylase large chain	rbcL	AtCg00490			yellow
Ribulose bisphosphate carboxylase small chain 1A	RBCS-1A	At1g67090			yellow
Ribulose bisphosphate carboxylase small chain 2B	RBCS-2B	At5g38420	blue		pink
Ribulose bisphosphate carboxylase small chain 3B	RBCS-3B	At5g38410	light blue		yellow
RING-H2 finger protein ATL38	ATL38	At2g34990			yellow
Rop guanine nucleotide exchange factor 1	ROPGEF1	At4g38430			yellow
Transcription factor GTE1	GTE1	At2g34900			yellow
WPP domain-interacting tail-anchored protein 1	WIT1	At5g11390			yellow
WRKY transcription factor 61 (Probable)	WRKY61	At1g18860		blue	orange
γ-interferon responsive lysosomal thiol (GILT) reductase family protein	At4g12960	At4g12960			yellow

FIGURE 3 | Visualization as "heat map" of the comparison between transcriptomic data and proteomic data. In column 4 and 5, white rectangles indicate concurrency between transcript level and protein abundance, light blue indicates that high level of transcript has a low protein abundance, blue indicates that low level of transcript has high protein abundance. In the last column, yellow rectangles are for consistent behavior between the two mutants in the transcriptomic-proteomic cascade, orange is for contrasting behavior, specifically, pink is for a markedly opposite trend.

TABLE 2 | Comparison between transcriptomic and proteomic data.

		atnp01 (%)	atnp02 (%)	Comparisons atnp01/atnp02 (%)
% Concurrent		58.97	56.41	70.51
% Non-concurrent	High transcript low protein	20.51	15.38	29.49
	Low transcript high protein	20.51	28.21	

of the former was an alanine-tRNA ligase expressed in the mitochondria and the chloroplasts, which forms part of the response to both salinity and Cd^{2+} stress; the other was N-(5'-phosphoribosyl) anthranilate isomerase 1, an enzyme which catalyzes a step in the tryptophan synthesis pathway, and is active in guard cell chloroplasts. The sole atnp02-specific over-represented protein of unknown function was a member of the pentatricopeptide repeat superfamily active in the mitochondria (Supplementary Table S1). There were six atnp01-specific under-represented proteins in the non-stressed plants (Supplementary Table S2). These comprised (1) a calcium-binding protein CML31 localizing to the nucleus, (2,3) two pentatricopeptide repeat-containing proteins of unknown function expressed in the mitochondria, (4) a pentatricopeptide repeat-containing protein member of the PCMP-E subfamily involved in RNA editing in the chloroplast, (5) an alpha type 3 proteasome subunit active in both the cytosol and various organelles, and involved in glycolysis, photorespiration, proteolysis, the hyperosmotic response, the response to various abiotic stress agents (including Cd^{2+}) and water transport, and (6) a proton pump-interactor 3A, which may be responsible for the regulation of plasma membrane ATPase activity and proton transport. There were five under-represented proteins specific to atnp02 (Supplementary Table S2). These comprised (1) a calcium-binding protein CML42 involved in protein binding and trichome branching, (2) a glucan endo-1,3-beta-glucosidase which participates in a MAPK cascade and in a variety of other processes, and localizes within the apoplast, cell wall, chloroplast and vacuole, (3) a homeobox-leucine zipper protein ATHB-7 thought to be a transcription factor acting in a signal transduction pathway mediating the drought response, and (4,5) the two pathogenisis-related proteins PR1 and PR5, present in the apoplast, cell wall and extracellular regions.

Exposure to CdS QDs resulted in the specific over-representation of four proteins in atnp01 and 11 in atnp02 (**Figure 2A**; Supplementary Table S1). The former set comprised (1) a calcium-binding protein CML45 of unknown function, (2) a putative defensin expressed extracellularly, (3) a KTI12 homolog expressed in both the cytoplasm and the nucleus, and involved in 5-carbamoylmethyluridine metabolism and also in the regulation of transcription and in tRNA modification and, (4) a WRKY transcription factor involved in the response to Zn^{2+}. The 11 atnp02-specfic proteins were as follows: (1) a plastid-localized arogenate dehydratase involved in phenylalanine synthesis, in anthocyanin accumulation in response to UV irradiation and in the vernalization response,

(2) a probable E3 ubiquitin-protein ligase, (3) an F-box/kelch-repeat protein of unknown function, (4) a mechanosensitive ion channel protein responsible for ion transmembrane transport, (5) a mitochondrion localizing member of the tetratricopeptide repeat-like superfamily, (6) a mitochondrion localizing methionine sulfoxide reductase involved in oxidation-reduction, protein repair and the response to singlet oxygen, (7) a plastid localizing phosphomethylpyrimidine synthase involved in pathogen detection, glucosinolate and maltose metabolism and several other processes, (8) a prefolding subunit 4, expressed in the cytosol and the nucleus, (9) a membrane RING-H2 finger protein associated with Zn^{2+} binding, (10) a nuclear GTE1 transcription factor involved in the regulation of germination, and (11) a nuclear WPP domain-interacting tail-anchored protein 1 involved in lateral root development and nucleocytoplasmic transport. Overall, a somewhat larger number of proteins was over-represented in atnp02 than in atnp01.

Exposure to CdS QDs resulted in the specific under-representation of eight proteins in atnp01 and three in atnp02 (**Figure 2B**; Supplementary Table S2). The former set comprised: (1) an extracellular 2S seed storage and lipid binding protein, (2) a chloroplastic ATP synthase subunit β, involved in photosynthesis and aspects of the biotic and abiotic stress response, (3) BIG, a cytosolic protein, involved in auxin polar transport, auxin-activated signaling, inflorescence morphogenesis, lateral root formation and development, and the anti-fungal response, (4) a putative galacturonosyltransferase 2, responsible for carbohydrate and pectin synthesis and cell wall organization, (5) a chloroplast nudix hydrolase 21, (6) a structural constituent of the extensin-like EPR1 involved in cell wall modification, seed lipid storage, embryo development, dormancy and germination, and sugar-mediated signaling, and (7,8) two RuBisCO small subunits (RBCS2B and RBCS3B). The three specifically under-represented atnp02 proteins were: (1) esterase/lipase ESM1 involved in photosynthesis, starch synthesis, pest/pathogen defense, (2) an extracellular pectate lyase, and (3) a mitochondrion localizing pentatricopeptide repeat-containing protein acting as an adenylate cyclase.

Among the proteins showing altered representation in both mutants (either in the control plants and/or in the CdS QDs exposed ones), seven were over-represented and 11 under-represented (**Figures 2A,B**; Supplementary Tables S1, S2). The former group comprised: (1) a 12S CRC protein responsive to abscisic acid and associated with lipid storage, protein ubiquitination, germination, seed maturation and sugar-mediated signaling, (2) a calcium-dependent protein kinase 23 involved in abscisic acid-activated signaling, intracellular signal transduction, protein phosphorylation and the response to Cd^2, (3) glutaredoxin C14, (4) DHAR3—a chloroplast-localizing dehydroacorbate reductase involved in protein glutathionylation and toxin catabolism, (5) a mitochondrion localizing glycine cleavage system H protein 2, (6) the nuclear pre-mRNA splicing factor SLU7-A, and (7) the RuBisCO small subunit RBCS1A. The 11 down-regulated proteins were: (1) a chloroplast and cytosolic 4- alpha glucanotransferase DPE2, involved in the sensing of the circadian rhythm, polysaccharide and starch metabolism and cell wall organization, (2) actin2, a cytosolic protein involved in anthocyanin accumulation, cellulose metabolism, the response to various abiotic stresses and water transport, (3) a nuclear ATP-dependent DNA helicase Q-like 4A involved in DNA recombination, repair and replication, and the cellular responses to DNA damage and low temperature stress, (4) a BTP/POZ domain-containing plasma membrane protein responsible for protein ubiquitination, (5) a CLIP-associated protein involved in anthocyanin accumulation, cellulose metabolism, polysaccharide and cell wall synthesis and root hair elongation, (6) a nuclear DNA damage binding protein 1b involved in DNA repair, cell division and embryo and reproductive structure development, (7) a nuclear F-box/kelch-repeat protein belonging to the galactose oxidase/kelch repeat superfamily, (8) a pentatricopeptide repeat-containing protein required for the $5'$ processing of nad9 and cox3 mRNAs in the mitochondria, (9,10) two mitochondrial pentatricopeptide repeat-containing proteins of unknown function, and (11) an SNC1 modifier involved in the regulation of gene expression, glucuronoxylan metabolism, and nuclear-transcribed mRNA catabolism.

Exposure to CdS QDs resulted in an increase in the number of over and under-represented proteins in both mutants (**Figures 2A,B**; Supplementary Tables S1, S2). The over-represented proteins were eight: (1) a chloroplast-localized beta-amylase 3 involved in maltose and starch synthesis and the response to low temperature, (2) a cyclic nucleotide-gated ion channel 6, (3) a GDP-mannose 4,6-dehydratase 2, involved in de novo GDP-L-fucose synthesis and GDP-mannose and glucose metabolism, (4) a nuclear LOB domain-containing protein 5 of unknown function, (5) a chloroplast pentatricopeptide repeat-containing protein involved in chloroplast RNA and mRNA processing, (6) a probable beta-D-xylosidase 7 involved in carbohydrate metabolism, (7) a probable prolyl 4-hydroxylase 7, involved in oxidation-reduction, (8) the cytoskeletal protein TONNEAU 1-B, which is probably involved in cortical cytoskeleton organization and microtubule organization. The set of under-represented proteins in both mutants comprised 17 proteins: (1) a chloroplast auxilin-like protein 1, which binds to certain heat shock proteins and is associated with protein folding, (2) a calmodulin-like protein 1 localizing to the mitochondria and the plasma membrane, (3,4) two extracellular defensin-like proteins involved in anti-fungal defense, embryo sac development and transition metal ion transport, (5) a GDP-mannose 4,6-dehydratase 1, (6) a glutathione S-transferase U 24 involved in fatty acid beta-oxidation as well as in protein and toxin catabolism, (7) a 70 kDa heat shock protein 10 involved in protein folding, peroxide neutralization and the response to various abiotic stresses (including Cd^{2+}), (8) a nuclear lysine-specific histone demethylase 1 involved in histone H3-K4 methylation, histone deacetylation, oxidation-reduction and the regulation of transcription, (9) a methyl-CpG-binding domain-containing protein 9 involved in cell wall organization, the regulation of transcription, embryo development, the sensing of photoperiod, flowering and secondary shoot formation, (10) a nuclear NF-X1-type zinc finger protein NFXL2 involved in sensing the circadian rhythm, floral development and the regulation of transcription, (11) oleosin GRP-17, a lipid-binding protein involved in lipid storage, cell wall modification and pollen development, (12) a proteasome subunit alpha type 1-A endowed with endopeptidase and peptidase activity and involved in fatty acid oxidation, protein catabolism and the response to As stress, (13) an FLX-like 2 protein of unknown function expressed in the guard cells, (14) the nuclear protein vernalization insensitive 3, which forms part of the low temperature-induced regulation of gene expression, (15) a RuBisCO large subunit, (16) the ROP guanine nucleotide exchange factor 1 involved in anthocyanin accumulation in response to UV irradiation, polysaccharide synthesis, the regulation of hormone levels and pollen tube growth, root hair elongation and root morphogenesis and (17) an extracellular gamma-interferon responsive lysosomal thiol (GILT) reductase with catalytic activity.

DISCUSSION

The most frequently reported toxicity problem associated with ENMs is oxidative stress (Pujalté et al., 2011; Ma et al., 2015). When taken up, they can drive down the cellular content of antioxidants and/or increase its production of reactive oxygen species (ROS; Maysinger and Lovric, 2007; Mahmoudi et al., 2011; Santos et al., 2012). A better understanding of the properties of these materials, along with technical improvements in their synthesis, should provide the means to reduce their hazard: examples are the use of biocompatible coating materials and the exploitation of surface functionalization, which both help mask the particles' surface reactivity (Lynch et al., 2014; Burello and Worth, 2015). The toxicity of CdS QDs has been related to not just their small size but also their high surface charge and reactivity, photolytic activity, shape, composition, and mechanical stability (Favero et al., 2006; Maysinger and Lovric, 2007). Toxicity tests based on conventional pharmacokinetic and/or pharmacodynamic approaches (Holford, 2007; Steele and Austin, 2009) may be inadequate to identify the full range of potential hazards posed by CdS QDs. This realization explains the present application of a genotoxicological approach.

Transposon mutagenesis has succeeded in identifying two A. thaliana mutants (atnp01 and atnp02) able to tolerate a level of CdS QDs sufficient to strongly compromise the growth of a wt plant (Marmiroli et al., 2014). Comparing the transcriptomes of these mutants with that of the wt has provided a ready means to define which genes which were up- or down-regulated in one or both of the mutant(s), both in non-stressed conditions and when the plants were exposed to CdS QDs (Marmiroli et al., 2014). Here, the comparisons have been extended to the protein level, by exploiting platforms able to identify not just specific gene products but also some of their post-translationally modified forms. Combining these data with those acquired from other omics platforms is the aspiration of current system biology strategies, which aim to define the complex pathways and networks involved in response to different external stimuli (Chen and Harmon, 2006; Jorrín-Novo et al., 2015; Wang et al., 2015). Of particular note are the two proteins DRL and ELM, the encoding transcripts for which were both abundant in atnp01 plants whether or not the Cd QD treatment was imposed; despite this, both proteins were only slightly over-represented (**Figure 1**) confirming their epistatic role (Marmiroli et al., 2014). The two mutants differed quite markedly at the proteomic level: while atnp01 has a mixed change in its proteins abundance, ready to cope with general stress situations, the mutations affecting atnp02 were more closely related to the response to oxidative stress. Many of the proteins altered in their level of expression in atnp01 were concerned with DNA transcription, lipid binding and the auxin response; in contrast, in atnp02, although there was also an effect on some proteins involved in DNA transcription, a range of other functions were also modified, including protein metabolism, cell wall formation and photosynthesis. Note that the oxidative stress response is triggered by excessive amounts of ROS, which not only induces changes in DNA transcription, but also triggers the metabolism of proteins, starch and sugars (Desikan et al., 2001; Mittler et al., 2004; Fujita et al., 2006; Foyer and Noctor, 2011). In both mutants, there was an over-representation of proteins associated with the oxidative stress response and an under-representation of those associated with DNA and RNA processing and with cell development.

An over-representation of lytic proteins and an under-representation of stress-related and hormone-regulated proteins was an unexpected feature of the CdS QD treatment. Characteristic of an oxidative stress response was the up-regulation of sugar metabolism, a disturbance in phytohormone levels and the prominence of glutathione/ascorbate cycle related enzymes (Couée et al., 2006; Foyer and Noctor, 2011; Villiers et al., 2011). There was overall little commonality between the two mutants with respect to either which proteins were over- or which were under-represented, compared with the WT (Supplementary Figure S5). This pointed to possibly divergent phenotypic traits as a result of the over-represented proteins in respect to possibly convergent traits as a result of the under-represented proteins in atnp01 and atnp02. Nevertheless, the numbers of altered proteins expressed in the two mutants in plants not exposed to CdS QDs were rather similar to one another, even though the proteins differed so widely in type, function and cell localization (Supplementary Figure S5).

A growing body of literature has confirmed that transcription levels in eukaryotes are poorly correlated with the levels of their encoded products (Griffin et al., 2002; Lan et al., 2012). This uncoupling is assumed to reflect the action of a number of cellular phenomena, notably the influence of RNA secondary structure, the activity of regulatory proteins and regulatory siRNAs, codon bias and codon adaptation, ribosomal density, and protein half-life (Gygi et al., 1999; Hajduch et al., 2010). The Pearson r correlation coefficients for atnp01 and atnp02 were, respectively, 0.126 and 0.197, levels which confirm the anticipated poor correlation between transcriptome and proteome. As an alternative means of linking the two data

sets, a qualitative rather than a quantitative view was taken of the relationship between each transcript/protein pair (**Figure 3**). The criterion adopted highlighted the direction rather than the extent to which the amount of a particular couple of cognate transcript and protein was affected. The number of up- and down-regulated genes which, in this sense, matched the behavior of their encoded protein was quite similar in the two mutants (46 and 44), of which 31 were represented in both mutants (Supplementary Figure S6). There were 16 genes in atnp01 associated with an increased expression of transcript but a decreased representation of protein, and 12 behaving in this manner in atnp02. The frequency of genes responding in the opposite direction (low transcript/high protein abundance) was 16 in atnp01 and 22 in atnp02. Overall, therefore, about 59% of differentially represented proteins in atnp01 and 56% of those in atnp02 behaved in a concurrent manner (**Table 2**), a frequency which is quite consistent with the outcome of cognate studies in other eukaryotes (Hajduch et al., 2010). It is thus possible to argue that the mutants' responses were split into two almost equally-sized parts: one was a shared response, and the other was specific to the mutant. For 70% of the reprogrammed genes, at a certain level of transcript corresponded the same amount of protein, either over or under-represented in the two mutants. On the other hand, without referring to the nature of the type of change within a single mutant, for the remaining genes, a difference in direction of regulation between the protein and its transcript was observed (**Figure 3**; **Table 2**). The outcome of the combined analysis of the transcriptomic and proteomic data implies that a significant level of translational and/or post-translational regulation must have been taking place, presumably triggered by the CdS QD treatment. Moreover, they differed in their response to the treatment, in fact there was general protein requirement to be met in order to achieve resistance to CdS QDs that both mutants should achieve.

The use of plants as test organism to investigate the environmental and biological effect of ENM exposure, coupled with the exploitation of tolerant (or hypersensitive) mutants, provides a convenient means to discriminate between non-essential and essential molecular functions. The substantial number of concurrent transcripts and proteins which were regulated by the stress treatment provides the necessary sequence information which can be used in risk assessment through the construction of exposition and effect markers.

REFERENCES

Abbott, K. W., Marchant, G. E., and Corley, E. A. (2012). Soft law oversight mechanisms for nanotechnology. *Jiurimetrics* 59, 279–312.

Aschberger, K., Rauscher, H., Crutzen, H., Rasmussen, K., Christensen, F. M., Sokull-klüttgen, B., et al. (2014). *Considerations on Information Needs for Nanomaterials in Consumer Products Consumer Products in the EU.* Luxembourg City: Publications Office of the European Union.

Brus, L. E. (1984). Electron–electron and electron-hole interactions in small semiconductor crystallites: the size dependence of the lowest excited electronic state. *J. Chem. Phys.* 80, 4403. doi: 10.1063/1.447218

Burello, E., and Worth, A. P. (2015). A rule for designing safer nanomaterials: do not interfere with the cellular redox equilibrium. *Nanotoxicology* 9, 116–117. doi: 10.3109/17435390.2013.828109

Chen, S., and Harmon, A. C. (2006). Advances in plant proteomics. *Proteomics* 6, 5504–5516. doi: 10.1002/pmic.200600143

Colvin, V. L. (2003). The potential environmental impact of engineered nanomaterials. *Nat. Biotechnol.* 21, 1166–1170. doi: 10.1038/nbt875

Couée, I., Sulmon, C., Gouesbet, G., and El Amrani, A. (2006). Involvement of soluble sugars in reactive oxygen species balance and responses to oxidative stress in plants. *J. Exp. Bot.* 57, 449–459. doi: 10.1093/jxb/erj027

Desikan, R., A-H-Mackerness, S., Hancock, J. T., and Neill, S. J. (2001). Regulation of the Arabidopsis transcriptome by oxidative stress. *Plant Physiol.* 127, 159–172. doi: 10.1104/pp.127.1.159

European Parliament (2011). Regulation (EU) No 1169/2011 of the European Parliament and of the Council of 25 October 2011 on the Provision of Food Information to Consumers, Amending Regulations (EC) No 1924/2006 and (EC) No 1925/2006 of the European Parliament and of the Council, and Repealing Commission Directive 87/250/EEC, Council Directive 90/496/EEC, Commission Directive 1999/10/EC, Directive 2000/13/EC of the European Parliament and of the Council, Commission Directives 2002/67/EC and 2008/5/EC and Commission Regulation (EC) No 608/2004. OJ L 304/18-63. (Brussel)

Fadeel, B., and Garcia-Bennett, A. E. (2010). Better safe than sorry: understanding the toxicological properties of inorganic nanoparticles manufactured for biomedical applications. *Adv. Drug Deliv. Rev.* 62, 362–374. doi: 10.1016/j.addr.2009.11.008

Favero, P. P., de Souza-Parise, M., Fernandez, J. L. R., Miotto, R., and Ferraz, A. C. (2006). Surface properties of CdS nanoparticles. *Braz. J. Phys.* 36, 1032–1034. doi: 10.1590/S0103-97332006000600062

Foyer, C. H., and Noctor, G. (2011). Ascorbate and glutathione: the heart of the redox hub. *Plant Physiol.* 155, 2–18. doi: 10.1104/pp.110.167569

Fujita, M., Fujita, Y., Noutoshi, Y., Takahashi, F., Narusaka, Y., Yamaguchi-Shinozaki, K., et al. (2006). Crosstalk between abiotic and biotic stress responses: a current view from the points of convergence in the stress signaling networks. *Curr. Opin. Plant Biol.* 9, 436–442. doi: 10.1016/j.pbi.2006.05.014

Griffin, T. J., Gygi, S. P., Ideker, T., Rist, B., Eng, J., Hood, L., et al. (2002). Complementary profiling of gene expression at the transcriptome and proteome levels in *Saccharomyces cerevisiae*. *Mol. Cell. Proteomics* 1, 323–333. doi: 10.1074/mcp.M200001-MCP200

Gygi, S. P., Rochon, Y., Franza, B. R., and Aebersold, R. (1999). Correlation between protein and mRNA abundance in yeast. *Mol. Cell. Biol.* 19, 1720–1730. doi: 10.1128/MCB.19.3.1720

Hajduch, M., Hearne, L. B., Miernyk, J. A., Casteel, J. E., Joshi, T., Agrawal, G. K., et al. (2010). Systems analysis of seed filling in Arabidopsis: using general linear modeling to assess concordance of transcript and protein expression. *Plant Physiol.* 152, 2078–2087. doi: 10.1104/pp.109.152413

Harris, M. A., Clark, J., Ireland, A., Lomax, J., Ashburner, M., Foulger, R., et al. (2004). The Gene Ontology (GO) database and informatics resource. *Nucleic Acids Res.* 32, D258–D261. doi: 10.1093/nar/gkh036

Hodge, G. A., Maynard, A. D., and Bowman, D. M. (2014). Nanotechnology: rhetoric, risk and regulation. *Sci. Public Policy* 41, 1–14. doi: 10.1093/scipol/sct029

Holford, N. H. G. (2007). "Pharmacokinetics & pharmacodynamics: rational dosing & the ime course of drug action," in *Basic and Clinical Pharmacology, 10th Edn.*, ed B. G. Katzung (San Francisco, CA: McGraw Hill), 34–49.

Jorrín-Novo, J. V., Pascual, J., Sánchez-Lucas, R., Romero-Rodríguez, M. C., Rodríguez-Ortega, M. J., Lenz, C., et al. (2015). Fourteen years of plant proteomics reflected in Proteomics: moving from model species and 2DE-based approaches to orphan species and gel-free platforms. *Proteomics* 15, 1089–1112. doi: 10.1002/pmic.201400349

Lan, P., Li, W., and Schmidt, W. (2012). Complementary proteome and transcriptome profiling in phosphate-deficient arabidopsis roots reveals multiple levels of gene regulation. *Mol. Cell. Proteomics* 11, 1156–1166. doi: 10.1074/mcp.M112.020461

Lynch, I., Weiss, C., and Valsami-Jones, E. (2014). A strategy for grouping of nanomaterials based on key physico-chemical descriptors as a basis for safer-by-design NMs. *Nano Today* 9, 266–270. doi: 10.1016/j.nantod.2014.05.001

Ma, C., White, J. C., Dhankher, O. P., and Xing, B. (2015). Metal-based nanotoxicity and detoxification pathways in higher plants. *Environ. Sci. Technol.* 49, 7109–7122. doi: 10.1021/acs.est.5b00685

Mahmoudi, M., Lynch, I., Ejtehadi, M. R., Monopoli, M. P., Bombelli, F. B., and Laurent, S. (2011). Protein-nanoparticle interactions: opportunities and challenges. *Chem. Rev.* 111, 5610–5637. doi: 10.1021/cr100440g

Maier, T., Güell, M., and Serrano, L. (2009). Correlation of mRNA and protein in complex biological samples. *FEBS Lett.* 583, 3966–3973. doi: 10.1016/j.febslet.2009.10.036

Marmiroli, M., Imperiale, D., Maestri, E., and Marmiroli, N. (2013). The response of Populus spp. to cadmium stress: chemical, morphological and proteomics study. *Chemosphere* 93, 1333–1344. doi: 10.1016/j.chemosphere.2013.07.065

Marmiroli, M., Pagano, L., Pasquali, F., Zappettini, A., Tosato, V., Bruschi, C. V., et al. (2015). A genome-wide nanotoxicology screen of *Saccharomyces cerevisiae* mutants reveals the basis for cadmium sulphide quantum dot tolerance and sensitivity. *Nanotoxicology* 1–10. doi: 10.3109/17435390.2015.1019586. [Epub ahead of print].

Marmiroli, M., Pagano, L., Savo Sardaro, M. L., Villani, M., and Marmiroli, N. (2014). Genome-wide approach in *Arabidopsis thaliana* to assess the toxicity of cadmium sulfide quantum dots. *Environ. Sci. Technol.* 48, 5902–5909. doi: 10.1021/es404958r

Maysinger, D., and Lovric, J. (2007). "Quantum dots and other fluorescent nanoparticles: quo vadis in the cell?" in *Bio-Applications of Nanoparticles*, ed W. C. W. Chan (New York, NY: Springer; Landes Bioscience), 156–167. doi: 10.1007/978-0-387-76713-0_12

Mittler, R., Vanderauwera, S., Gollery, M., and Van Breusegem, F. (2004). Reactive oxygen gene network of plants. *Trends Plant Sci.* 9, 490–498. doi: 10.1016/j.tplants.2004.08.009

Organisation for Economic Co-operation and Development (OECD) (2014). "Ecotoxicology and environmental fate of manufactured nanomaterials: test guidelines," in *Expert Meeting Report, Series on the Safety of Manufactured Nanomaterials, No. 40. (ENV/JM/MONO(2014)1)* (Paris).

Pujalté, I., Passagne, I., Brouillaud, B., Tréguer, M., Durand, E., Ohayon-Courtès, C., et al. (2011). Cytotoxicity and oxidative stress induced by different metallic nanoparticles on human kidney cells. *Part. Fibre Toxicol.* 8:10. doi: 10.1186/1743-8977-8-10

Royal Society and Royal Academy of Engineering (2004). *Nanoscience and Nanotechnologies, Opportunities and Uncertainties*. Plymouth: Latimer Trend Ltd.

Saez, G., Moreau, X., De Jong, L., Thiéry, A., Dolain, C., Bestel, I., et al. (2010). Development of new nano-tools: towards an integrative approach to address the societal question of nanotechnology? *Nano Today* 5, 251–253. doi: 10.1016/j.nantod.2010.06.002

Santos, A. R., Miguel, A. S., Fevereiro, P., and Oliva, A. (2012). "Evaluation of cytotoxicity of 3-Mercaptopropionic acid-modified quantum dots on Medicago sativa cells and tissues," in *Nanoparticles in Biology and Medicine, Methods and Protocols*, ed M. Soloviev (New York, NY: Springer; Springer Protocols), 435–449. doi: 10.1007/978-1-61779-953-2_36

Science and Technology Options Assessment (STOA) European Parliament (2012). *NanoSafety - Risk Governance of Manufactured Nanoparticles, Final Report*, eds T. Fleischer, J. Jahnel, and S. B. Seitz (IP/A/STOA/FWC/2008-096/LOT5/C1/SC3) (Brussels).

Servin, A., Elmer, W., Mukherjee, A., De la Torre-Roche, R., Hamdi, H., White, J. C., et al. (2015). A review of the use of engineered nanomaterials to suppress plant disease and enhance crop yield. *J. Nanopart. Res.* 17, 92. doi: 10.1007/s11051-015-2907-7

Sigg, L., Behra, R., Groh, K., Isaacson, C., Odzak, N., Piccapietra, F., et al. (2014). Chemical aspects of nanoparticle ecotoxicology. *Chim. Int. J. Chem.* 68, 806–811. doi: 10.2533/chimia.2014.806

Singh Sekhon, B. (2014). Nanotechnology in agri-food production: an overview. *Nanotechnol. Sci. Appl.* 7, 31–53. doi: 10.2147/NSA.S39406

Steele, G., and Austin, T. (2009). "Preformulation investigation using small amounts of compound as an hide to candidate drug selection and early development," in *Pharmaceutical Preformulation and Formulation, 2nd Edn.*, ed M. Gibson (New York, NY: Informa Healthcare USA, Inc.), 17–128.

Villani, M., Calestani, D., Lazzarini, L., Zanotti, L., Mosca, R., and Zappettini, A. (2012). Extended functionality of ZnO nanotetrapods by solution-based coupling with CdS nanoparticles. *J. Mater. Chem.* 22, 5694. doi: 10.1039/c2jm16164h

Villiers, F., Ducruix, C., Hugouvieux, V., Jarno, N., Ezan, E., Garin, J., et al. (2011). Investigating the plant response to cadmium exposure by proteomic and metabolomic approaches. *Proteomics* 11, 1650–1663. doi: 10.1002/pmic.201000645

Wang, X.-C., Li, Q., Jin, X., Xiao, G.-H., Liu, G.-J., Liu, N.-J., et al. (2015). Quantitative proteomics and transcriptomics reveal key metabolic processes associated with cotton fiber initiation. *J. Proteomics* 114, 16–27. doi: 10.1016/j.jprot.2014.10.022

Conflict of Interest Statement: The authors declare that the research was conducted in the absence of any commercial or financial relationships that could be construed as a potential conflict of interest.

19

Differentiated Responses of Apple Tree Floral Phenology to Global Warming in Contrasting Climatic Regions

Jean-Michel Legave [1*], Yann Guédon [2], Gustavo Malagi [3], Adnane El Yaacoubi [4] and Marc Bonhomme [5]

[1] INRA, Unité Mixte de Recherche 1334 Amélioration Génétique et Adaptation des Plantes Méditerranéennes et Tropicales, Montpellier, France, [2] CIRAD, Unité Mixte de Recherche 1334 et Inria, Virtual Plants, Montpellier, France, [3] Faculdade de Agronomia, Universidade Federal de Pelotas, Pelotas, Brazil, [4] Faculté des Sciences, Université Moulay Ismail, Meknès, Morocco, [5] Unité Mixte de Recherche 547, INRA et Université Blaise Pascal, PIAF, Clermont-Ferrand, France

Edited by:
Sergio Rossi,
Université du Québec à Chicoutimi,
Canada

Reviewed by:
Ignacio García-González,
Universidade de Santiago de
Compostela, Spain
Rebecca Darbyshire,
University of Melbourne, Australia

*Correspondence:
Jean-Michel Legave
legave@supagro.inra.fr

The responses of flowering phenology to temperature increases in temperate fruit trees have rarely been investigated in contrasting climatic regions. This is an appropriate framework for highlighting varying responses to diverse warming contexts, which would potentially combine chill accumulation (CA) declines and heat accumulation (HA) increases. To examine this issue, a data set was constituted in apple tree from flowering dates collected for two phenological stages of three cultivars in seven climate-contrasting temperate regions of Western Europe and in three mild regions, one in Northern Morocco and two in Southern Brazil. Multiple change-point models were applied to flowering date series, as well as to corresponding series of mean temperature during two successive periods, respectively determining for the fulfillment of chill and heat requirements. A new overview in space and time of flowering date changes was provided in apple tree highlighting not only flowering date advances as in previous studies but also stationary flowering date series. At global scale, differentiated flowering time patterns result from varying interactions between contrasting thermal determinisms of flowering dates and contrasting warming contexts. This may explain flowering date advances in most of European regions and in Morocco vs. stationary flowering date series in the Brazilian regions. A notable exception in Europe was found in the French Mediterranean region where the flowering date series was stationary. While the flowering duration series were stationary whatever the region, the flowering durations were far longer in mild regions compared to temperate regions. Our findings suggest a new warming vulnerability in temperate Mediterranean regions, which could shift toward responding more to chill decline and consequently experience late and extended flowering under future warming scenarios.

Keywords: fruit tree, flowering, chill period, heat period, warming vulnerability, multiple change-point models

INTRODUCTION

Phenological events are highly responsive to temperature (Menzel and Fabian, 1999) and the abundance of information on plant phenology outlined substantial responses to global warming (Rutishauser et al., 2009). Most studies focused on bud phenology in natural vegetation and exhibited flowering advances as a main warming response (Abu-asab et al., 2001). Concerning fruit trees, flowering advances were highlighted in the European warming context for apple, pear and cherry trees (Chmielewski et al., 2004; Guédon and Legave, 2008; Eccel et al., 2009), hazelnut tree (Črepinšek et al., 2012), and olive tree (Garcia-Mozo et al., 2009). This was also observed in the Northeastern American context for apple tree (Wolfe et al., 2005) and in various parts of Asia for cherry tree (Miller-Rushing et al., 2007), apple tree (Fujisawa and Kobayashi, 2010), chestnut tree (Guo et al., 2013), and citrus species (Fitchett et al., 2014). Similar flowering advances have been scarcely reported in the Southern Hemisphere for apple and pear trees (Grab and Craparo, 2011).

Moreover, several studies dealing with warming responses in various perennial plants grown in temperate conditions found stationary or delayed bud phenology, in spite of temperature increases (Doi and Katano, 2008; Gordo and Sanz, 2009; Schwartz and Hanes, 2010; Yu et al., 2010). In fact, the timing of flowering is controlled by multiple and complex determinisms related to temperature at different periods of the year (Cook et al., 2012; Guo et al., 2013). Most temperate trees, including fruit species, are dormant in autumn and winter. Since the work of Lang et al. (1987), it has been widely accepted that among the different phases of bud dormancy, endodormancy corresponds to the growth suspension of the meristematic activity. The dormant buds require exposure to chill temperatures in order to overcome the endodormancy phase, followed by exposure to heat temperatures to resume growth during an ecodormancy phase and to initiate flowering in spring (Campoy et al., 2011). One likely warming impact during endodormancy is a delay in the fulfillment of chill requirements and consequently a delay in the time at which perennial plants become receptive to heat temperatures (Yu et al., 2010; Luedeling et al., 2013). This may explain unexpected phenological changes like those observed in walnut trees grown in California for which the vegetative buds (high chill requirements) shifted to late leaf-out since 1994 (Pope et al., 2013). Inversely, the flowering advances, that have dominated climate-warming responses thus far, were explained by increasing temperatures during ecodormancy leading to a more rapid fulfillment of heat requirements, as shown for apple trees in Europe (Legave et al., 2013) and in Japan (Fujisawa and Kobayashi, 2010). A comprehensive assessment of divergent responses to warming in temperate perennial plants must thus include the potential impacts on the fulfillment of both chill and heat requirements (Schwartz and Hanes, 2010). The sequential chill-growth model was therefore commonly used for analyzing flowering times in temperate fruit trees (Eccel et al., 2009; Darbyshire et al., 2014). When the fulfillment of chill requirements is inadequate, as is currently the case in mild climates, a typical symptom is the extended duration of the flowering phase (Atkinson et al., 2013). However, less

attention has been paid to change in flowering duration in response to climate warming (Miller-Rushing et al., 2007; Legave et al., 2013). Moreover, in the case of fruit trees, nearly all the studies have reported warming responses in only one location or a few locations submitted to similar climatic contexts (Chmielewski et al., 2004; Fujisawa and Kobayashi, 2010; Grab and Craparo, 2011; Črepinšek et al., 2012), whereas it has been demonstrated that a given species can have contrasting responses in different locations (Primack et al., 2009). As an illustration, a large spatially-distributed lilac data set in North America demonstrated that the floral phenology has progressively changed from advances in flowering in northern regions to delays in flowering in southern regions (Zhang et al., 2007). In fact, there is evidence that more field studies are needed to determine the extent to which phenological shifts are occurring on large geographical scales (Primack et al., 2009).

Another key question is the use of appropriate statistical methods for analyzing flowering date and temperature series. The statistical analysis of such series is not standardized and various methods were used including linear regression (Fujisawa and Kobayashi, 2010; Grab and Craparo, 2011), multiple change-point models (Guédon and Legave, 2008) and segmented regression models (Pope et al., 2013). Compared to our previous study (Guédon and Legave, 2008), we extended in this study the statistical modeling framework in order to test not only piecewise constant models but also piecewise linear models that include simple linear regression models when no change point can be detected. We were thus able in this way to identify both abrupt changes and linear trends in phenological series.

Our objectives here were (i) to propose a statistical modeling framework for analyzing flowering date and temperature series with minimum a priori assumptions (ii) to identify on this basis differentiated flowering changes on a large geographical scale in apple tree and (iii) to understand how changing temperature conditions can lead to differentiated flowering changes. These complementary objectives included changes both of the flowering time and the flowering duration. Apple tree offers a relevant study plant because of its worldwide cultivation and relatively high chill requirements (Hauagge and Cummins, 1991a; Ghariani and Stebbins, 1994) which can result in divergent responses to change in temperature conditions (Schwartz and Hanes, 2010).

MATERIALS AND METHODS

Flowering and Temperature Data
Collection of Flowering Date Series

A collaborative international network on apple tree phenology has been established between research institutes in six countries. We selected 10 locations, seven in Western Europe, one in Northern Morocco and two in Southern Brazil (**Table 1**; **Figure 1**). The eight locations in the Northern Hemisphere are located across a large latitudinal range (from 34 to 50°N) with a corresponding large range of climatic conditions during the dormancy and flowering phases, from a cold continental climate in Europe (Bonn, Gembloux, Conthey, Trento) to a mild climate in Northern Morocco (Ain Taoujdate). This includes

TABLE 1 | Description of the collected data.

World region *Location*	Latitude/ Longitude	Elevation (m.a.s.l.)	Climatic conditions *Climatic influence*	Temperature data period	Phenological data			Collaborative institute
					Period	Cultivar	Stage (BBCH)	
Western Europe			Temperate					
Bonn, Germany	50.62/6.98	160	*Continental*	1959–2013	1958–2013	Golden D.	61,65	INRES
Gembloux, Belgium	50.57/4.68	138	*Continental*	1964–2013	1984–2013	Golden D.	61	CRA-W
Angers, France	47.47/-0.63	38	*Oceanic*	1963–2013	1963–2013	Golden D.	61	INRA France
Conthey, Switzerland	46.22/7.30	504	*Continental*	1970–2013	1970–2013	Golden D.	65	Agroscope
					1975–2013	Gala	65	
Trento, Italy	46.07/11.12	419	*Continental*	1983–2013	1983–2013	Golden D.	61,65	CRA-FRF
Forli, Italy	44.22/12.03	34	*Mediterranean*	1970–2013	1970–2013	Golden D.	61,65	CRA-FRF
Nîmes, France	43.73/4.50	52	*Mediterranean*	1966–2013	1974–2013	Golden D.	61,65	Ctifl
					1979–2013	Gala	61,65	
					1980–2013	Fuji	61,65	
Northern Africa			Mild					
Aïn Taoujdate, Morocco	00.00/ 6.22	499	*Mediterranean*	1973–2013	1984–2013	Golden D.	61,65	INRA Morocco
Southern Brazil			Mild					
Caçador, Santa Catarina	−26.78/−51.02	960	*Subtropical*	1961–2013	1984–2013	Golden D.	61,65	EPAGRI
					1982–2013	Gala	61,65	
					1982–2013	Fuji	61,65	
Sao Joaquim, Santa Catarina	−28.29/−49.93	1353	*Subtropical*	1955–2013	1972–2013	Golden D.	61,65	EPAGRI
					1972–2013	Gala	61,65	
					1976–2003	Fuji	61,65	

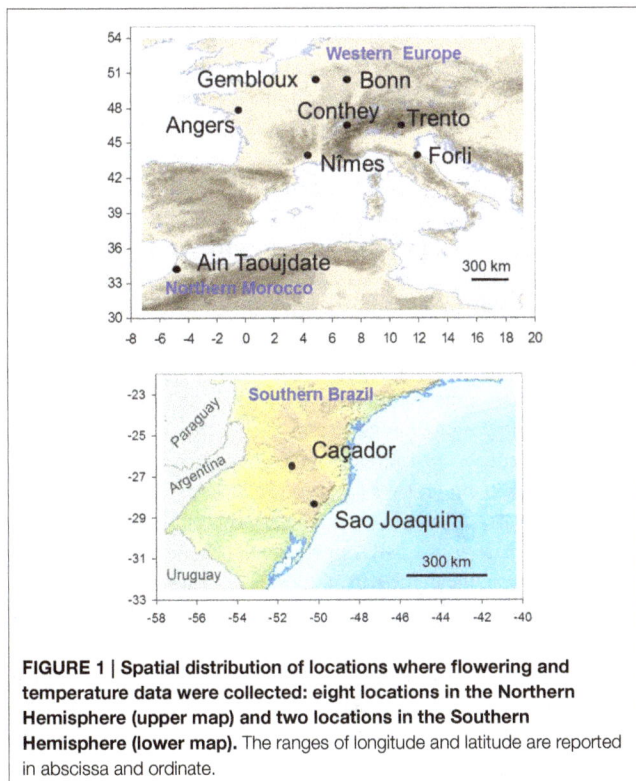

FIGURE 1 | Spatial distribution of locations where flowering and temperature data were collected: eight locations in the Northern Hemisphere (upper map) and two locations in the Southern Hemisphere (lower map). The ranges of longitude and latitude are reported in abscissa and ordinate.

European locations with intermediate climates such as oceanic (Angers) and Mediterranean (Forli, Nîmes). While situated at high elevation to favor apple cropping, the two Brazilian locations

in the Southern Hemisphere are clearly characterized by mild climates during the dormancy and flowering phases (mean temperature up to 11°C).

Within this extensive geographical area, flowering dates were recorded for the beginning of the bloom phase (~10% of flowers open) and the full bloom (~50% of flowers open, first petals may have fallen). These dates correspond to stages 61 and 65 of the international BBCH code, respectively. Experienced observers recorded them using similar observation procedures on adult trees grown in long-term orchards. At each location, the flowering dates were assessed at least twice weekly on several trees of a given cultivar. In the mild conditions of Morocco and Brazil where flowering duration is extended (see Results) and flowering intensity is frequently weak due to floral abortions (Oukabli et al., 2003; Petri and Leite, 2004), the observers were trained to collect accurate data for comparison with those collected in temperate conditions. New trees were observed periodically at all locations as trees aged, whereas new observers were trained by the preceding ones.

To compare long-term flowering series between different locations, we chose cultivars grown worldwide. We therefore collected numerous data for Golden Delicious for which records were available at all 10 locations. In addition, records for Gala and Fuji were collected since these cultivars were frequently grown in Southern Brazil, but also in Europe. These three cultivars were characterized by nearly the same high chill requirements (Hauagge and Cummins, 1991a) and concomitant flowering times both in Southern Brazil and Europe (cross pollination in orchard). The collection of different varietal series in a given location was thus considered as a way to repeat the statistical analysis to reveal a strong phenological change in the location, and not as a way to study the genotype × location interactions. Our data set consists of 30 flowering date series including series for the stages 61 and 65 (16 of them corresponded to the temperate conditions and 14 to the mild conditions). Each series was defined by a location, a cultivar and a flowering stage, including a total of 1121 measurements. Most series were complete aside from some missing data (not interpolated) in some series. The longest series contains 56 years in Bonn (Golden Delicious, stages 61 and 65) and the shortest contains 25 years in Caçador (Golden Delicious, stages 61 and 65; **Table 1**). The consistency of collected data was assessed by the fact that the flowering dates were consistently related to the geographical characteristics (latitude, elevation) and temperature conditions of the locations. Moreover, the flowering duration between the dates of stages 61 and 65 was assessed at all locations where the two dates were recorded. This included 13 series of flowering durations (six for the temperate conditions and seven for the mild conditions) ranging from 56 years in Bonn (Golden Delicious) to 25 years in Caçador (Golden Delicious).

Collection of Temperature Series

For characterizing the relationships between flowering and temperatures, we analyzed series of mean temperatures during two successive periods respectively determining for the fulfillment of chill and heat requirements. Annual chill accumulation (CA) period and subsequent heat accumulation (HA) period have thus been defined. Based on previous results concerning the bud dormancy dynamics (Malagi et al., 2015) and the relationships between flowering and temperatures (Legave et al., 2013; El Yaacoubi et al., 2014) in apple tree, the CA period ranged from October to January for the European and Moroccan locations (Northern Hemisphere) and from April to July for the Brazilian locations (Southern Hemisphere). The HA period ranged from February to April for the European locations and from August to October for the Brazilian locations. We chose a shorter HA period for the Moroccan location (March to mid-April), because previous works using Partial Least Squares regression clearly suggested this period as a major period of heat requirement fulfillment in Morocco (El Yaacoubi et al., 2014).

The mean temperature series were constituted from the average minimum and maximum daily temperatures collected from weather stations located near the orchards where the flowering dates were recorded (no more than 10 km). The daily temperatures were provided for each location and checked by the corresponding research institute. The French partner performed a complementary global check for this study. The few missing data were estimated by linear interpolation. All the series started before the end of the 1980s, the instant at which marked increases of temperature have been frequently recorded at the world scale, particularly in Europe (Jones and Moberg, 2003). When temperature series longer than the corresponding flowering date series were available, we collected the longest possible temperature series; this was the case for Gembloux, Nîmes, Ain Taoujdate, Caçador, and Sao Joaquim (**Table 1**).

Statistical Modeling
Definition of Piecewise Constant and Piecewise Linear Models

Multiple change-point models were used to delimit segments within a flowering date or temperature series of length T, for which the data characteristics were homogeneous within each segment while markedly differing from one segment to another. We made the assumption of homoscedastic Gaussian multiple change-point models, either piecewise constant or piecewise linear models. In the first case, the slope is assumed to be zero and the only within-segment parameter is the intercept (which is also the segment mean in this case) whereas in the second case, the within-segment parameters are the intercept and the slope. In both cases, the variance is assumed to be common to the segments. This homoscedasticity assumption is justified by the data characteristics but also by the fact that the series were rather short (between 25 and 56 years). The two associated models are denoted by $M_{constant}$ (for piecewise constant) and M_{linear} (for piecewise linear). Piecewise linear models are somewhat related to the segmented regression models used by Pope et al. (2013). Segmented regression or broken-line models are regression models where the regression function is piecewise linear, i.e., made of straight lines connected at change points (Muggeo, 2003). The regression function is thus continuous, but first derivatives are discontinuous. In our case, the regression function is not constrained to be continuous.

For the $M_{constant}$ model, we suppose that some $J - 1$ instants $\tau_1 < \cdots < \tau_{J-1}$ (with the convention $\tau_0 = 0$ and $\tau_J = T$) exist such that the mean is constant between two successive change points and the variance is assumed to be constant,

$$\text{if } \tau_j \leq t < \tau_{j+1}, \quad \begin{cases} E(X_t) = \alpha_j, \\ \text{Var}(X_t) = \sigma^2. \end{cases}$$

These two families of models enable to test and combine two assumptions: change point of sufficient amplitude separating two phases and linear trend (within phase or for the whole series in the case of no change point).

We adopted a retrospective or off-line inference approach whose objective was to infer the number of segments J, the instants of the $J - 1$ change points $\tau_1, \ldots, \tau_{J-1}$, the J within-segment intercepts α_j, the global variance σ^2 and the J within-segment slopes β_j (for M_{linear} model). For the selection of the number of segments J, we used the modified Bayesian information criterion (mBIC) proposed by Zhang and Siegmund (2007) and specifically dedicated to Gaussian homoscedastic multiple change-point models. The principle of this kind of penalized likelihood criterion consists in making a trade-off between an adequate fitting of the model to the data and a reasonable number of parameters to be estimated. Jeffreys' rules of thumb (Kass and Raftery, 1995) suggest that a difference of mBIC of at least $2 \log(100) = 9.2$ is needed to deem the model with the higher mBIC substantially better. For the optimal segmentation of the series into J segments, we applied the dynamic programming algorithm proposed by Auger and Lawrence (1989). This optimal segmentation defines the optimal change points and relies on the estimation of within-segment and global variance parameters; see details on these statistical methods for multiple change-point models in Supplementary Material, Appendix S1.

Comparison Between the Selected Piecewise Constant Model and the Selected Piecewise Linear Model

For many flowering date series, we obtained two models that were not discernible according to mBIC: the 2-segment piecewise constant model and the simple linear model (i.e., 1-segment piecewise linear model). This situation is illustrated by the Forli series (**Figure 2A**) for which the difference of mBIC is <1. This can be explained by the similar orders of magnitude for the change-point amplitude and the global standard deviation in the case of the 2-segment piecewise constant model. We thus extracted the residual series from the linear function and we found that the residual series was not stationary but that a change point can be identified in 1988 (this was the change point of the selected 2-segment piecewise constant model), between two increasing linear trends (for this, we selected the best piecewise linear model for the residual series using mBIC; **Figure 2B**). The Ain Taoujdate series illustrates another situation where the 2-segment piecewise constant model can be identified using piecewise linear models (the selected model in this family was a 2-segment model and the two estimated slopes were not significantly different from 0; **Figure 3**). Finally, the Sao Joaquim series for Golden Delicious illustrates the case of very short segments at one end of the series (**Figure 4**). In this situation, we chose to not consider these very short segments that cannot be interpreted in our application context.

Assessment of the Segmentation Assumption

It is often of interest to quantify the uncertainty concerning change point instants. Let $L_J(\mathbf{s}, \mathbf{x}; \hat{\theta})$ denote the likelihood of the segmentation $\mathbf{s}$ in J segments of the observed series $\mathbf{x}$ where θ denotes the set of within-segment and global variance parameters. In the case of a single change point ($J = 2$), the posterior probability of entering the second segment at τ_1 is given by:

$$L_2(\mathbf{s}(\tau_1), \mathbf{x}; \hat{\theta}) / \sum_{\mathbf{s}} L_2(\mathbf{s}, \mathbf{x}; \hat{\theta}),$$

where each segmentation $\mathbf{s}$ defines a unique change point. In our case of short series, the dynamic programming algorithm for computing the top N most probable segmentations proposed by Guédon (2013) was used to compute the $T - 1$ possible segmentations and the associated likelihood and then to extract the change-point posterior distribution. In this particular case of a single change point, this posterior distribution therefore summarizes the possible segmentations. In particular, the posterior probability of the optimal segmentation $\mathbf{s}^*$ given by:

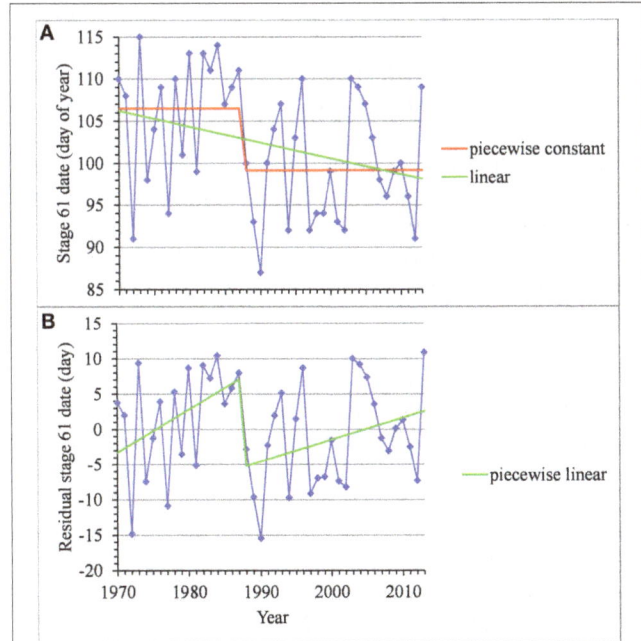

FIGURE 3 | Segmentation of the Ain Taoujdate BBCH 61 stage date series using a 2-segment piecewise constant model and a 2-segment piecewise linear model.

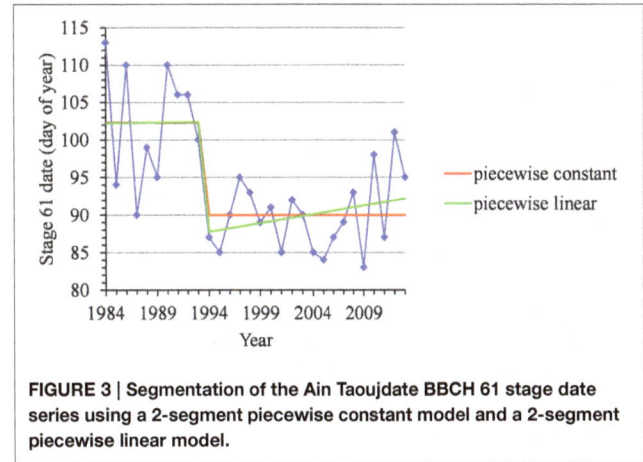

FIGURE 2 | (A) Segmentation of the Forli BBCH 61 stage date series using a 2-segment piecewise constant model and estimation of a linear model. (B) Segmentation of the residual series deduced from the estimated linear model using a 2-segment piecewise linear model.

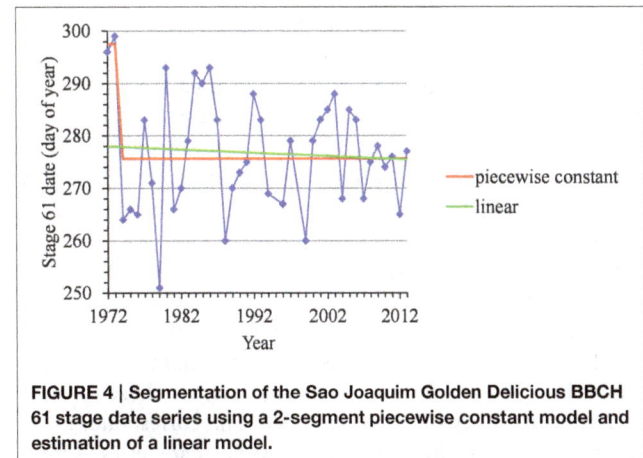

FIGURE 4 | Segmentation of the Sao Joaquim Golden Delicious BBCH 61 stage date series using a 2-segment piecewise constant model and estimation of a linear model.

$$P(\mathbf{s}^*|\mathbf{x}; 2) = L_2(\mathbf{s}^*, \mathbf{x}; \hat{\theta})/\sum_{\mathbf{s}} L_2(\mathbf{s}, \mathbf{x}; \hat{\theta}),$$

which is the mode of the change-point posterior distribution, can be used to assess the segmentation assumption.

More generally, the posterior probability of the optimal segmentation given by:

$$P(\mathbf{s}^*|\mathbf{x}; J) = L_J(\mathbf{s}^*, \mathbf{x}; \hat{\theta})/\sum_{\mathbf{s}} L_J(\mathbf{s}, \mathbf{x}; \hat{\theta}),$$

can be computed using the dynamic programming algorithm for computing the top N most probable segmentations in our case of short series segmented into a few segments (up to $J = 3$). The assessment of multiple change-point models thus relies on two posterior probabilities:

• Posterior probability of the optimal segmentation $\mathbf{s}^*$ for a fixed number of segments J $P(\mathbf{s}^*|\mathbf{x}; J)$ i.e., weight of the optimal segmentation among all the possible segmentations for a fixed number of segments.

• Posterior probability of the J-segment model M_J, $P(M_J|\mathbf{x})$ deduced from the mBIC computed for a collection of multiple change-point models for $J = 1, \ldots, J_{max}$ i.e., weight of the J-segment model among all the possible models between 1 and J_{max} segments; see Supplementary Material, Appendix S1.

RESULTS

In this study, we systematically favored longitudinal analyses of the various series (flowering dates and durations, mean temperatures during the CA and HA periods) in order to identify phenological patterns with minimum a priori assumptions. We also chose to not build simple regression models on the basis of these longitudinal data since this would rely on an oversimplified view of the influence of the temperatures on the flowering process regarding the biological bases and current functional models of bud phenology (Cook et al., 2012; Guo et al., 2013; Darbyshire et al., 2014; Pope et al., 2014).

Flowering Time

The dates of stages 61 and 65 appeared highly correlated, meaning that flowering durations fluctuate around quite constant values for a given series and that the changes in dates of stages 61 and 65 are markedly synchronous for a given series (see Section Flowering Duration). We thus focused the analysis on the 15 series of stage 61 dates for which data were collected at nine locations. We also analyzed the 2 series of stage 65 dates in Conthey for which only stage 65 dates were collected over long periods (44 and 39 years for Golden Delicious and Gala, respectively; **Table 1**).

Combining model selection criterion (mBIC) and residual analysis in the case of piecewise linear models, we found that the assumption of a piecewise constant model was better supported than the assumption of a piecewise linear model. In order to ease comparison between locations and cultivars, we chose to focus on 2-segment piecewise constant models. This corresponds to the models selected by mBIC for seven flowering date series: Angers

(Golden Delicious), Forli (Golden Delicious), Trento (Golden Delicious), Gembloux (Golden Delicious), Conthey (Golden Delicious and Gala), and Ain Taoujdate (Golden Delicious). This was a well-supported alternative model for four other flowering date series: Nîmes (Gala and Fuji), Caçador (Golden Delicious and Gala) according to the posterior probability of the 2-segment model (**Table 2**). It should be noted that in our context of short series (length between 25 and 56), the number of segments given by mBIC should only be considered as indicative. We chose to discard 2-segment piecewise constant models selected by mBIC for Sao Joaquim (Golden Delicious—**Figure 4**—and Gala) since this corresponds to very short segments at the beginning of the series (2 and 1 years respectively) that cannot be reliably interpreted in our context (**Table 3**). Two-segment piecewise constant models are well-defined if the single change point of sufficient amplitude with respect to the global segment standard deviation separates two sufficiently long segments. It should be noted that the 3-segment model selected by mBIC for Bonn includes a short 4-year segment (between 1958 and 1961) at the beginning of the series. Since this range of years was not represented in other series, it was difficult to interpret this first segment. The 2-segment model retained for comparison of locations (**Table 2**) was simply this optimal 3-segment model where the first two segments were merged (**Table 3**; Figure S1 in Supplementary Material). No change point can be detected for Sao Joaquim (Fuji) and Caçador (Fuji). In the case of 2-segment models the change point is located between 1987 and 1989 for most of the flowering date series, which is consistent with our previous analyzes (Guédon and Legave, 2008), but with the notable exceptions of Ain Taoujdate (change point in 1994; **Table 2**). For the flowering series starting at the beginning of the 1980s with a change point detected at the end of the 1980s, Nîmes (Fuji), Trento (Golden Delicious), Gembloux (Golden Delicious), Caçador (Golden Delicious, Gala), the rather short length of the first segment (between 3 and 6 years) makes the mean estimated for this segment less reliable and, consequently, the change-point amplitude. This explains the difference in change-point amplitude for Nîmes between Fuji and the other two cultivars, as well as the difference between Gembloux (Golden Delicious) and Bonn (Golden Delicious; **Table 2**) for which the climatic conditions were rather similar (**Table 1**).

For each flowering date series, the uncertainty concerning the instant of the change point is low for most locations (**Figure 5**) except for Nîmes (Golden Delicious) for which the posterior probability of the segmentation is the lowest among the segmentation in 2-segments (**Table 2**). Moreover, this series is the only one for which the change-point amplitude is lower than the global standard deviation in the case of a 2-segment piecewise constant model (**Table 2**). Hence, the segmentation in 2-segments is not well defined in this case. This can be illustrated by the segmentation in 2- and 3-segments of this flowering date series where the segmentation in 3-segments highlights a change toward later flowering dates since 2003 (**Figure 6**). It should be noted that in the case of Nîmes, the mBIC favors the constant model (i.e., no change point) regardless of the cultivar (**Tables 2, 3**).

TABLE 2 | Segmentations of flowering date series (BBCH 61 stage for all locations except Conthey—BBCH 65 stage) using piecewise constant models (2 or 1 segment when the 2-segment model was irrelevant): observation period, change-point instant and amplitude, global standard deviation, optimal segmentation posterior probability, model posterior probability, mBIC model, average flowering duration, correlation coefficient between BBCH 61 and 65 stage dates.

Location	Cultivar	Observation period	Change point		Standard deviation	Posterior probability		mBIC model	Average flowering duration	61–65 stage date correlation coef.
			Instant	Amplitude		Segmentation	Model			
Angers	Golden D.	1963–2013	1989	−7.81	6.89	0.33	0.81*			
Nîmes	Golden D.	1974–2013	1989	−5.39	7.05	0.15	0.1	1	2.71	0.99
Nîmes	Gala	1979–2013	1989	−6.7	6.59	0.25	0.2	1	2.68	0.99
Nîmes	Fuji	1983–2013	1988	−11.2	7.51	0.24	0.27	1	3.15	0.96
Forli	Golden D.	1970–2013	1988	−7.35	6.87	0.36	0.62*		3.95	0.97
Trento	Golden D.	1983–2013	1988	−13.75	5.96	0.32	0.52*		3.58	0.98
Gembloux	Golden D.	1984–2013	1987	−15.44	5.63	0.64	0.92*			
Bonn	Golden D.	1958–2013	1989	−9.33	7.96	0.33	0.12	3	4.45	0.95
Conthey (65)	Golden D.	1970–2013	1988	−8.3	5.45	0.43	0.78*			
Conthey (65)	Gala	1975–2010	1988	−9.88	5.25	0.52	0.71*			
Ain Taoujdate	Golden D.	1984–2013	1994	−12.35	5.97	0.73	0.73*		14.3	0.87
Sao Joaquim	Golden D.	1972–2013	–	–	11.05	1	0.37	2	9.33	0.92
Sao Joaquim	Gala	1972–2013	–	–	9.95	1	0.21	2	10.24	0.89
Sao Joaquim	Fuji	1976–2013	–	–	10.33	1	0.98*		7.92	0.95
Caçador	Golden D.	1984–2013	1988	−13.29	9.49	0.19	0.27	1	10.96	0.91
Caçador	Gala	1982–2013	1988	−14.75	8.8	0.47	0.33	3	10.78	0.91
Caçador	Fuji	1982–2013	–	–	11.2	1	0.52*		13.53	0.78

In the model posterior probability column, an asterisk indicates that the model is the one given by mBIC. If this is not the case, the model given by mBIC is indicated in the next column (mBIC model).

TABLE 3 | Optimal segmentations of flowering date series (BBCH 61 stage) using piecewise constant models for series where the optimal J-segment model according to the mBIC was not retained for the comparison in Table 2: observation period, change-point instant and amplitude, global standard deviation, optimal segmentation posterior probability, model posterior probability.

Location	Cultivar	Observation period	Change point 1		Change point 2		Standard deviation	Posterior probability	
			Instant	Amplitude	Instant	Amplitude		Segmentation	Model
Nîmes	Golden D.	1974–2013	–	–			7.44	1	0.87
Nîmes	Gala	1979–2013	–	–			7.12	1	0.77
Nîmes	Fuji	1983–2013	–	–			8.17	1	0.68
Bonn	Golden D.	1958–2013	1962	13.81	1989	−11.12	7.21	0.28	0.49
Sao Joaquim	Golden D.	1972–2013	1974	−21.87			10.06	0.58	0.47
Sao Joaquim	Gala	1972–2013	1975	−29.16			8.87	0.72	0.73
Caçador	Golden D.	1984–2013	–	–			10.53	1	0.54
Caçador	Gala	1982–2013	1988	−13.42	2012	−17.25	7.82	0.27	0.56

Flowering Duration

We did not detect any change point or linear trend in the flowering duration series and we thus analyzed them as simple frequency distributions without considering the year indexing (**Table 2**). We first grouped samples corresponding to the different cultivars in a given location (Golden Delicious, Gala and Fuji for Nîmes, Sao Joaquim and Caçador, respectively) for which the frequency distributions were not significantly different according to the Kruskal Wallis test (ANOVA by ranks for these frequency distributions defined on a small set of values). The cumulative frequency distribution functions of the seven samples (**Figure 7**) highlights a clear order for flowering duration (from the shortest to the longest): (1) Nîmes, 2.82 days on average; (2)

Forli, Trento and Bonn, between 3.6 and 4.45 days (these three samples are not significantly different according to the Kruskal Wallis test); (3) Sao Joaquim, 9.18 days; (4) Caçador, 11.82 days; (5) Ain Taoujdate, 14.3 days.

Temperature During the CA and HA Periods vs. Flowering Date and Duration

The flowering date was not significantly correlated with the mean temperature during the CA period (defined for the Northern Hemisphere) for most of the European locations (**Table 4**). The only exceptions were Trento and Conthey for which we found slightly significant negative correlations. The situation was very different for the Brazilian locations for which we found strongly

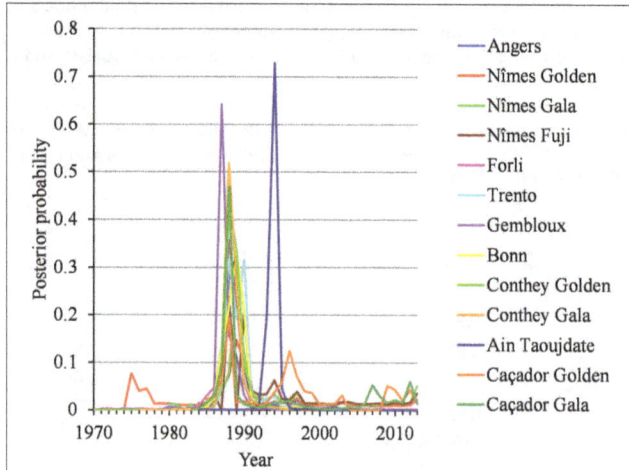

FIGURE 5 | Two-segment piecewise constant models estimated on the basis of BBCH 61 stage date series for Angers (Golden Delicious), Nîmes (Golden Delicious, Gala, and Fuji), Forli (Golden Delicious), Trento (Golden Delicious), Gembloux (Golden Delicious), Bonn (Golden Delicious), Ain Taoujdate (Golden Delicious), and Caçador (Golden Delicious and Gala) and BBCH 65 stage date series for Conthey (Golden Delicious and Gala): posterior change-point distributions.

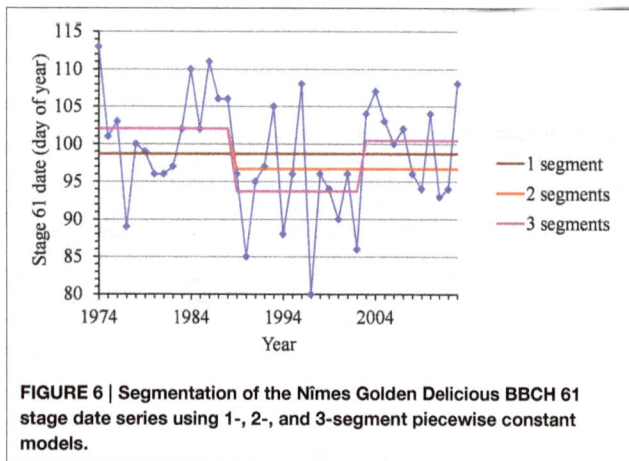

FIGURE 6 | Segmentation of the Nîmes Golden Delicious BBCH 61 stage date series using 1-, 2-, and 3-segment piecewise constant models.

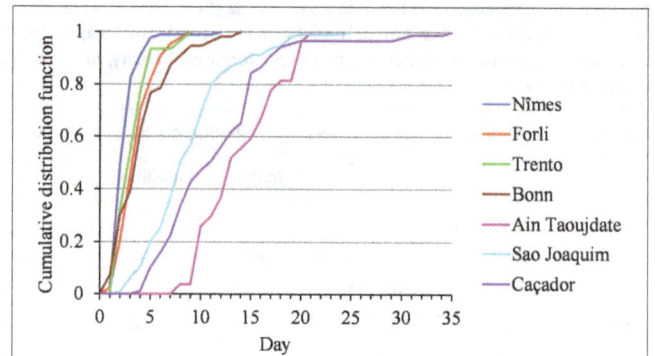

FIGURE 7 | Flowering duration for Nîmes (Golden Delicious, Gala, and Fuji samples pooled), Forli, Trento, Bonn, Ain Taoujdate; Sao Joaquim (Golden Delicious, Gala, and Fuji samples pooled) and Caçador (Golden Delicious, Gala, and Fuji samples pooled): comparison of cumulative frequency distribution functions.

significant positive correlations (except for Caçador, cultivar Fuji) between the flowering date and the mean temperature during the CA period (defined for the Southern Hemisphere), which means that the warmer the austral CA period is, the later the flowering date will be. The Moroccan situation seems to be closer to the Brazilian situations than to the European ones but we cannot be conclusive in this case because the correlation coefficient was not significantly different from 0 (**Table 4**). We thus conducted a longitudinal analysis of the series of mean temperatures during the CA period using the methodology previously applied to the flowering date series. We found various patterns:

- Two stationary segments for Nîmes (1974–2013, change point in 1988; **Figure 8A**) and Forli (1970–2013, change point in 1993), but with a rather small change-point amplitude

with respect to the residual standard deviation in this latter case.

- Stationary series for Trento (1983–2013), Ain Taoujdate (1973–2013) and Sao Joaquim (1955–2013).
- Slightly positive slope for Angers (1963–2013), Gembloux (1964–2013), Bonn (1959–2013), Conthey (1970–2013) and Caçador (1961–2013).

We found strongly significant negative correlations for the European locations between the flowering date and the mean temperature during the HA period (defined for Europe), which means that the warmer the Northern HA period is, the earlier the flowering date will be (**Table 5**). The situation was very different for the Brazilian locations for which the flowering date was not significantly correlated with the mean temperature during the HA period defined in Brazil (**Table 5**). The only exception was Sao Joaquim, cultivar Golden Delicious, for which we found a slightly significant negative correlation. For the Moroccan location we found a significant negative correlation between the flowering date and the mean temperature during the HA period. This relationship appears to be closer to the relationships found for the European locations than to the ones found for the Brazilian locations (**Table 5**). Because of the significant correlations between the flowering date and the mean temperature during the HA period found for the European and Moroccan locations, we conducted a longitudinal analysis of the series of mean temperatures during this period (**Table 5**) using the methodology previously applied to the flowering date series. We found a change point at the end of the 1980s in all the European temperature series when applying the piecewise constant model (see an illustration in **Figure 8B** for Nîmes) and the change-point amplitude was around 1.3°C for most locations (it should be noted that for Gembloux, the change-point amplitude estimated on the complete series up to 1966 is far more reliable than the one estimated on the series corresponding to the flowering date range of years since, in this latter case, the first segment was very short—four years). We found a change point of similar amplitude but in 1993 in the Moroccan location

TABLE 4 | Correlation coefficients between flowering date (BBCH 61 stage for all locations except Conthey—BBCH 65 stage) and mean temperature during the CA period (**significant at 1% level; *significant at 5% level; n.s., non-significant).

Location	Observation Period	Correlation coefficient		
		Golden D.	Gala	Fuji
Angers	1963–2013	−0.2 n.s.		
Nîmes	1974–2013	0 n.s.	−0.05 n.s.	−0.05 n.s.
Forli	1970–2013	−0.05 n.s.		
Trento	1983–2013	−0.3 n.s.		
Gembloux	1984–2013	−0.19 n.s.		
Bonn	1959–2013	−0.22 n.s.		
Conthey	1970–2013	−0.4**	−0.36*	
Ain Taoujdate	1984–2013	0.19 n.s.		
Sao Joaquim	1972–2013	0.69**	0.63**	0.72**
Caçador	1982–2013	0.83**	0.61**	0.06 n.s.

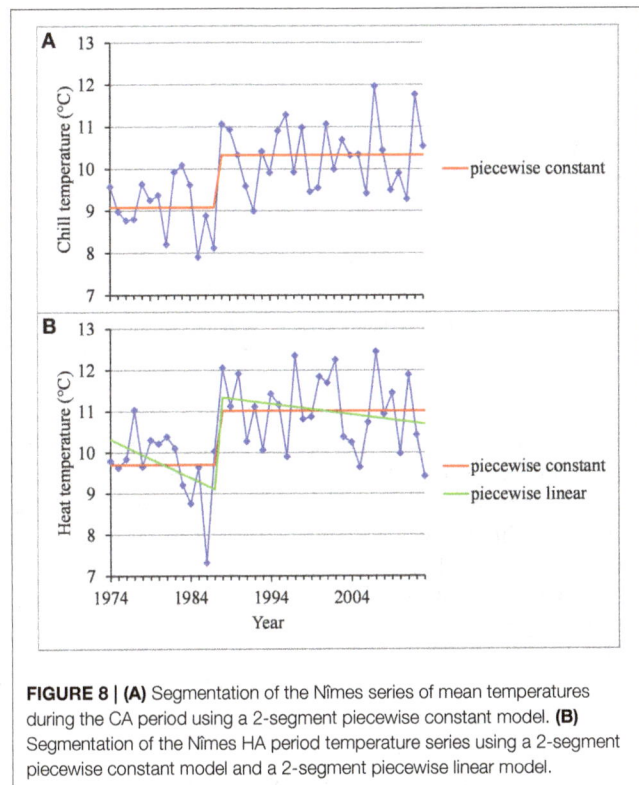

FIGURE 8 | (A) Segmentation of the Nîmes series of mean temperatures during the CA period using a 2-segment piecewise constant model. **(B)** Segmentation of the Nîmes HA period temperature series using a 2-segment piecewise constant model and a 2-segment piecewise linear model.

when applying a piecewise constant model (**Table 5**), although the assumption of a simple linear model was better supported (**Figure 9**). In contrast, the series of mean temperatures during the HA period were stationary for the Brazilian locations (**Table 5**).

We also found non-significant correlations between the flowering durations and the mean temperatures, both during the CA and HA periods at all the seven locations where these correlations were analyzed (especially in Morocco and Brazil; results not shown).

DISCUSSION

Differentiated Flowering Date Series

The analyses showed clear advances of apple flowering time since the end of the 1980s in most locations of Western Europe and a few years later in Northern Morocco, whereas the flowering dates remained stationary in Southern Brazil. Our study thus provides a new overview in space and time of the flowering time changes in apple tree highlighting contrasting behaviors (advance or stationarity) contrary to previous studies that only reported flowering advances (Chmielewski et al., 2004; Wolfe et al., 2005; Guédon and Legave, 2008; Eccel et al., 2009; Fujisawa and Kobayashi, 2010; Grab and Craparo, 2011; Darbyshire et al., 2013). Nevertheless, the flowering advances have been found to a lesser amplitude in the Southern Hemisphere (Australia, Southern Africa) than in the Northern Hemisphere (Europe, Japan, United States) as outlined by Darbyshire et al. (2013). Such lesser changes in flowering date in the Southern Hemisphere are consistent with the stationary series mainly found in Southern Brazil.

While there was a strong consensus concerning abrupt flowering advances in Europe, a notable exception was the French Mediterranean location (Nîmes) for which the flowering dates were stationary. This was clearly demonstrated for Golden Delicious for which the flowering advance reported by Guédon and Legave (2008) up to 2002 became an alternative not strongly supported when records up to 2013 were included in the series.

When all our findings are taken into account, the overview of flowering date changes reveals that flowering advances or stationary flowering dates may be detected in temperate as well as in mild regions: (i) advances for most European series, the Moroccan series and one Brazilian series; and (ii) stationarity for the French Mediterranean series and most Brazilian series. A simple relationship between the flowering date change pattern and a geographical trait such as the elevation or the localization in the Northern or Southern Hemisphere is no longer valid.

Relationships Between Temperatures and Flowering Dates

The correlation analysis gave evidence of contrasting climatic determinisms of flowering time in relation to the localization in either temperate or mild regions. In Western Europe, our analysis showed that the higher the HA is the earlier the flowering date will be, whereas the CA was not related to the flowering time. These results emphasize that flowering time in recent past have been strongly determined by the HA in the temperate conditions of Western Europe. This was similarly showed for apple tree in other temperate conditions like those of Northeastern America and Japan (Wolfe et al., 2005; Fujisawa and Kobayashi, 2010).

In Southern Brazil, a main influence of CA on flowering time was emphasized since the correlation analysis mainly showed that the lower the CA is the later the flowering date will be, as would be expected in mild conditions (Atkinson et al., 2013). Although a similar determinism of flowering time was highlighted in Northern Morocco (Oukabli et al., 2003), this appeared less evident on the basis of our correlation analysis. That may be explained by an additional influence of the HA, which was shown

TABLE 5 | Segmentations of series of mean temperatures during the HA period using piecewise constant models (2 or 1 segment when the 2-segment model was irrelevant): recording period, change-point instant and amplitude, global standard deviation, optimal segmentation posterior probability, model posterior probability, mBIC model, correlation coefficients between flowering date (BBCH 61 stage for all locations except Conthey–BBCH 65 stage) and mean temperature during the HA period (**significant at 1% level; *significant at 5% level; n.s., non-significant).

Location	Recording period	Change point		Standard deviation	Posterior probability		mBIC model	Correlation coefficient		
		Instant	Amplitude		Segmentation	Model		Golden D.	Gala	Fuji
Angers	1963–2013	1987	1.35	0.94	0.63	0.16	3	−0.77**		
Nîmes	1974–2013	1988	1.31	0.88	0.57	0.37	3	−0.7**	−0.72**	−0.76**
	1966–2013	1988	1.26	0.9	0.56	0.45*				
Forli	1970–2013	1988	0.91	0.92	0.26	0.42	1	−0.8**		
Trento	1983–2013	1988	1.72	0.89	0.54	0.81*		−0.78**		
Gembloux	1984–2013	1988	2.84	1.19	0.48	0.46*		−0.81**		
†	1964–2013	1989	1.49	1.16	0.29	0.57*				
Bonn	1959–2013	1989	1.34	1.26	0.27	0.59*		−0.84**		
Conthey	1970–2013	1988	1.31	1.06	0.31	0.7*		−0.78**	−0.74**	
Ain Taoujdate	1984–2013	1993	1.32	0.86	0.33	0.75*		−0.47**		
	1973–2013	1993	1.21	1.08	0.14	0.52*				
Sao Joaquim	1972–2013	-	-	0.76	1	0.91*		−0.34*	−0.27 n.s.	−0.12 n.s.
	1955–2013	-	-	0.79	1	0.97*				
Caçador	1982–2013	-	-	0.69	1	0.74*		−0.23 n.s.	−0.2 n.s.	−0.05 n.s.
	1961–2013	-	-	0.8	1	0.76*				

In the model posterior probability column, an asterisk indicates that the model is the one given by mBIC. If this is not the case, the model given by mBIC is indicated in the next column (mBIC model). We analyzed the mean temperatures during the HA period over the range of years corresponding to the flowering dates and in certain cases (Nîmes, Gembloux, Ain Taoujdate, Sao Joaquim, Caçador) over an extended range of years.
†*Gembloux (1964–2013): Posterior probability of 0.29 for 1989 but of 0.27 for 1988.*

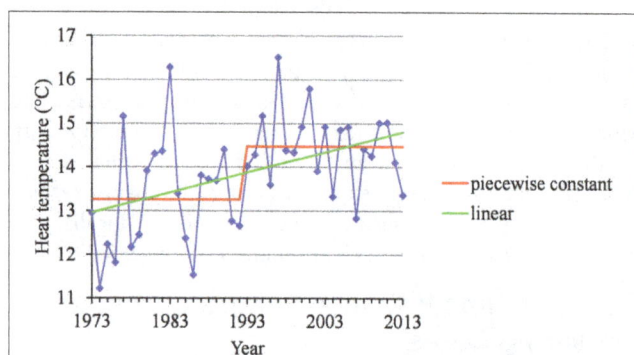

FIGURE 9 | Segmentation of the Ain Taoujdate series of mean temperatures during the HA period using a 2-segment piecewise constant model and estimation of a linear model.

by the correlation analysis in Northern Morocco. Nevertheless, our results globally showed that flowering time would be mainly determined by the CA in mild conditions of both Southern Brazil and Northern Morocco. Such determinism is consistent with the high genetic heritabilities of the chill requirement trait that were found from apple progenies grown in mild conditions like those of Southern Africa (Labuschagné et al., 2002). In addition severe symptoms of inadequate chill requirements, including delays of floral budbreak and flowering time were observed in high-chill genotypes selected from Golden Delicious progenies. Therefore, it was generally accepted in apple tree that budbreak time in mild regions was an accurate biomarker of the fulfillment of chill

requirements (Hauagge and Cummins, 1991b). This supposes a short time between the fulfillment of chill requirements and floral budbreak in mild conditions, which was demonstrated in Southern Brazil by comparison with the temperate conditions of Southern France (Malagi et al., 2015).

The analysis of temperature series accounts for contrasting warming contexts in relation to the localization in the Northern or Southern Hemisphere. In Western Europe, a marked warming during the HA period was detected at the end of the 1980s, similarly to the change-point instant generally found for the flowering date. Such a concomitance was also detected in Northern Morocco at the beginning of the 1990s. Heat increase can thus explain flowering advances in Western Europe where the HA would mainly determine flowering time and in Northern Morocco where the influence of HA would be also involved. In Southern Brazil, the stationarity of temperatures during both the CA and HA periods can explain the stationarity of most flowering date series. The absence of warming in Southern Brazil over the last four decades could be attributed to the relatively high elevation of the studied locations and to the globally lower temperature increases since the 1970s in the Southern Hemisphere compared to the Northern Hemisphere (Jones and Moberg, 2003).

As a consequence of these relationships, flowering date advances and stationary flowering dates result from interactions between contrasting thermal determinisms (temperate vs. mild regions) and warming contexts (Northern vs. Southern Hemisphere). A particularly interesting result is the stationarity of flowering dates both in the Brazilian and the French

Mediterranean regions. In Southern Brazil no thermal influences could have been exercised in the absence of warming, whereas the French Mediterranean region would have been progressively submitted to two opposite thermal influences: an increase in HA compensated by a decrease in CA, respectively linked to marked warming during the HA and CA periods. While less marked, the decrease of CA tended to delay the flowering dates. This appears mainly due to certain years since 1988, including 1988, 1996, 2007, and 2012, that were characterized by relatively high temperatures during the CA period (**Figure 8A**). This may explain the absence of correlation up to 2013 between the flowering date and the temperature during the CA period in this Mediterranean region.

Stationary flowering dates resulting from two opposite warming impacts, such as in the French Mediterranean region, may thus cast doubt on the relevance of flowering advances in apple tree as a suitable indicator of recent warming in early spring in Europe (Menzel et al., 2011). This indicator would be even more questionable in the near future in the entire Mediterranean region, as suggested by a similar temperature increase during the CA period since 1993 in the Italian Mediterranean region of Forli near the Adriatic coastline (see Section Temperature During the CA and HA Periods vs. Flowering Date and Duration).

There are still many unknowns concerning the relationships between the flowering dates and temperatures from bud dormancy to flowering time. This requires new researches on the timings of fulfillment of the chill and heat requirements and better knowledge of temperature thresholds on flowering time (Guo et al., 2013; Luedeling et al., 2013; Pope et al., 2014).

Specificity of the Flowering Duration

As expected (Atkinson et al., 2013), our analyses highlighted contrasting flowering duration patterns between temperate (short durations) and mild conditions (far longer durations). An additional specificity of flowering duration patterns is the lack of detection of a change point or linear trend at all locations, contrary to the flowering time patterns. This suggests different climatic determinisms for the flowering duration and the flowering time. In temperate conditions, a likely climatic determinism of flowering duration is the HA rate during the blooming phase, which is consistent with a significant increase in the average flowering duration from the Mediterranean location of Nîmes (2.8 days) to the colder continental location of Bonn (4.5 days). In mild conditions inversely, longer durations of the blooming phase are currently attributed to insufficient CAs (Campoy et al., 2011). Nevertheless, our results suggest more complex thermal determinisms to explain extended flowering durations in mild conditions since non-significant correlations were found between the flowering duration and the mean temperature during the CA period as well as during the HA period in both the Moroccan and Brazilian locations (see Section Temperature During the CA and HA Periods vs. Flowering Date and Duration).

Strengths of the Statistical Modeling Approach

Homoscedastic 2-segment piecewise constant models, where the variance was assumed to be common to the two segments, played

a key role in this study. This homoscedasticity assumption can be assessed by examining the standard deviations empirically estimated for the two segments in the case of sufficiently long flowering date or series of mean temperatures during the HA period (see Supplementary Materials, Tables S1, S2). In our context of potentially short segments, homoscedastic models are more robust than heteroscedastic models where changes in variance can induce artifactual short segments. One shortcoming of homoscedastic models with respect to heteroscedastic models used in our previous study (Guédon and Legave, 2008) was the inability to compute posterior change-point distributions. This is no longer the case since, for a single change point, the posterior change-point distribution can be computed using the dynamic programming algorithm for computing the top N most probable segmentations proposed by Guédon (2013). More generally, the posterior probability of the optimal segmentation can be computed using this dynamic programming algorithm in our case of short series segmented in a few segments. The other main methodological improvement consists in the systematic comparison of homoscedastic piecewise constant models with homoscedastic piecewise linear models. The fact that simple linear models corresponding to the linear trend assumption were systematically compared with piecewise constant models and the ability to identify piecewise constant models using piecewise linear models (see the Ain Taoujdate example in **Figure 3**) strengthen the reliability of our results with respect to our previous study.

Implications of Flowering Changes on Climatic Adaptation

Early flowering dates might increase the risk of spring frost damages, as pointed out for a long time by Cannell and Smith (1986) in Great Britain. More recently, spring frosts in southern regions of Great Britain have been seen to be decreasing both in frequency and severity (Sunley et al., 2006). Likewise, investigations into the frost risk for apple tree in Northern Italy (Eccel et al., 2009) showed that the risk was lower than in the past, and suggested that it will remain stable or decrease slightly in the future. Thus, the frost risk during the fruit tree flowering phase in Europe will probably be more open to debate in the future context of climate warming because of regional differences in both the magnitude of flowering advances and the frequency of negative temperatures. In particular, our results suggest that the French Mediterranean region may be rarely subject to frost risk at present time because flowering advance is decreasing. Inversely, this risk might remain a true concern for growers in continental regions because of constant flowering advances and also relatively short flowering durations (as shown in Bonn in Germany).

Mediterranean and oceanic regions in Europe might be more affected in the future by excessive delays in chill fulfillment, particularly the French Mediterranean region where our study showed an increase in temperature from October to January since the end of the 1980s. Declining chill will become a limiting cropping factor (Atkinson et al., 2013) and a new warming vulnerability in many European Mediterranean regions, particularly for the apple tree characterized by high chill requirements for most commercial cultivars. At a certain point, CA might shift from sub-optimal, just inducing a delayed phenology as illustrated in our study by the Nîmes location, to below the requirement leading to irregular or insufficient yields. At the orchard scale, this could cause in the future phenological disorders similar to those observed in the Mediterranean mild conditions of Northern Morocco (Oukabli et al., 2003), such as insufficient flowering synchronization between non self-fertile cultivars (all apple cultivars), which require cross-pollination, and excessive duration of the fruit maturity phase, that may be consequences of late and extended flowering.

CONCLUSION

The collection of flowering date series recorded in contrasting climatic locations and their analysis based on multiple change-point models proved to be appropriate for identifying differentiated flowering patterns in apple tree. The same methodology applied to the corresponding temperature series provided complementary results, making possible to establish a comprehensive overview of the relationships between the flowering phenology and the warming context at the world scale. On the one hand, we showed that the different patterns of flowering time change are not closely related to the localization of apple trees in temperate or mild regions or in the Northern or Southern Hemisphere. On the other hand, contrasting temperature influences during successive CA and HA periods on flowering time appeared to be mainly in relation with the localization in temperate or mild conditions, while contrasting warming contexts appeared to be in relation with the localization in the Northern or Southern Hemisphere. This overview was completed by information on the relationships between flowering duration and temperature, which appeared to be different and more complex than those concerning the flowering time. Because continuous warming will change

the relationships between phenology and temperature, a new warming vulnerability is expected in the more or less long term in Europe especially in Mediterranean regions where apple tree is a native crop.

AUTHOR CONTRIBUTIONS

JL conceived the study, conducted the data collection, contributed to the data analysis and edited the manuscript; YG contributed to the study conception, conducted the statistical analysis and edited the manuscript; GM and AE supplied data for Brazil and Morocco respectively, contributed to the data analysis and manuscript approval; MB contributed to the study conception, the data analysis, and manuscript approval.

ACKNOWLEDGMENTS

The authors thank INRA Angers (France), Vincent Mathieu (Ctifl, France), Daniela Giovannini, and Marco Fontanari (CRA-FRF, Italy), Danilo Christen (Agroscope C-W, Switzerland), Michael Blanke (INRES-Horticultural Science, Germany), Robert Oger and Viviane Planchon (CRA-W, Belgium), Ahmed Oukabli (INRA, Morocco), Frederico Denardi, and Gabriel Leite (Epagri, Brazil) for providing the data used in the analyzes. The authors also thank Pierre-Éric Lauri, Evelyne Costes (INRA, France), and Isabelle Farrera (Montpellier SupAgro, France) for their fruitful comments on the manuscript.

REFERENCES

Abu-asab, M. S., Peterson, P. M., Shetler, S. G., and Orli, S. S. (2001). Earlier plant flowering in spring as a response to global warming in the Washington, DC, area. *Biodivers. Conserv.* 10, 597–612. doi: 10.1023/A:1016667125469

Atkinson, C. J., Brennan, R. M., and Jones, H. G. (2013). Declining chilling and its impact on temperate perennial crops. *Env. Exp. Bot.* 91, 48–62. doi: 10.1016/j.envexpbot.2013.02.004

Auger, I. E., and Lawrence, C. E. (1989). Algorithms for the optimal identification of segment neighborhoods. *Bull. Math. Biol.* 51, 39–54. doi: 10.1007/BF02458835

Campoy, J. A., Ruiz, D., and Egea, J. (2011). Dormancy in temperate fruit trees in a global warming context: a review. *Sci. Hortic.* 130, 357–372. doi: 10.1016/j.scienta.2011.07.011

Cannell, M. G. R., and Smith, R. I. (1986). Climatic warming, spring budburst and frost damage on trees. *J. Appl. Ecol.* 23, 177–191. doi: 10.2307/2403090

Chmielewski, F. M., Müller, A., and Bruns, E. (2004). Climate changes and trends in phenology of fruit trees and field crops in Germany, 1961–2000. *Agric. For. Meteorol.* 121, 69–78. doi: 10.1016/S0168-1923(03)00161-8

Cook, B. I., Wolkovich, E. M., and Parmesan, C. (2012). Divergent responses to spring and winter warming drive community level flowering trends. *Proc. Natl. Acad. Sci. U.S.A.* 109, 9000–9005. doi: 10.1073/pnas.1118364109

Črepinšek, Z., Štampar, F., Kajfež-Bogataj, L., and Solar, A. (2012). The response of *Corylus avellana* L. phenology to rising temperature in north-eastern Slovenia. *Int. J. Biometeorol.* 56, 681–694. doi: 10.1007/s00484-011-0469-7

Darbyshire, R., Webb, L., Goodwin, I., and Barlow, E. W. R. (2013). Evaluation of recent trends in Australian pome fruit spring phenology. *Int. J. Biometeorol.* 57, 409–421. doi: 10.1007/s00484-012-0567-1

Darbyshire, R., Webb, L., Goodwin, I., and Barlow, E. W. R. (2014). Challenges in predicting climate change in pome fruit phenology. *Int. J. Biometeorol.* 58, 1119–1133. doi: 10.1007/s00484-013-0705-4

Doi, H., and Katano, I. (2008). Phenological timings of leaf budburst with climate change in Japan. *Agric. For. Meteorol.* 148, 512–516. doi: 10.1016/j.agrformet.2007.10.002

Eccel, E., Rea, R., Caffarra, A., and Crisci, A. (2009). Risk of spring frost to apple production under future climate scenarios: the role of phenological adaptation. *Int. J. Biometeorol.* 53, 273–286. doi: 10.1007/s00484-009-0213-8

El Yaacoubi, A., Malagi, G., Oukabli, A., Hafidi, M., and Legave, J. M. (2014). Global warming impact on floral phenology of fruit tree species in Mediterranean region. *Sci. Hortic.* 180, 243–253. doi: 10.1016/j.scienta.2014.10.041

Fitchett, J. M., Grab, S. W., Thompson, D. I., and Roshan, G. (2014). Increasing frost risk associated with advanced citrus flowering dates in Kerman and Shiraz, Iran: 1960-2010. *Int. J. Biometeorol.* 58, 1811–1815. doi: 10.1007/s00484-013-0778-0

Fujisawa, M., and Kobayashi, K. (2010). Apple (*Malus pumila* var. domestica) phenology is advancing due to rising air temperature in northern Japan. *Glob. Chang. Biol.* 16, 2651–2660. doi: 10.1111/j.1365-2486.2009.02126.x

Garcia-Mozo, H., Orlandi, F., Galan, C., Fornaciari, M., Romano, B., Ruiz, L., et al. (2009). Olive flowering phenology variation between different cultivars in Spain and Italy: modeling analysis. *Theor. Appl. Climatol.* 95, 385–395. doi: 10.1007/s00704-008-0016-6

Ghariani, K., and Stebbins, R. L. (1994). Chill requirements of apple and pear cultivars. *Fruit Var. J.* 48, 215–222.

Gordo, O., and Sanz, J. J. (2009). Long-term temporal changes of plant phenology in the Western Mediterranean. *Glob. Chang. Biol.* 15, 1930–1948. doi: 10.1111/j.1365-2486.2009.01851.x

Grab, S., and Craparo, A. (2011). Advance of apple and pear tree full bloom dates in response to climate change in the southwestern Cape, South Africa: 1973–2009. *Agric. For. Meteorol.* 151, 406–413. doi: 10.1016/j.agrformet.2010.11.001

Guédon, Y. (2013). Exploring the latent segmentation space for the assessment of multiple change-point models. *Comput. Stat.* 28, 2641–2678. doi: 10.1007/s00180-013-0422-9

Guédon, Y., and Legave, J. M. (2008). Analyzing the time-course variation of apple and pear tree dates of flowering stages in the global warming context. *Ecol. Modell.* 219, 189–199. doi: 10.1016/j.ecolmodel.2008.08.010

Guo, L., Dai, J., Ranjitkar, S., Xu, J., and Luedeling, E. (2013). Response of chestnut phenology in China to climate variation and change. *Agric. For. Meteorol.* 180, 164–172. doi: 10.1016/j.agrformet.2013.06.004

Hauagge, R., and Cummins, J. N. (1991a). Phenotypic variation of length of bud dormancy in apple cultivars and related Malus species. *J. Am. Soc. Hortic. Sci.* 116, 100–106.

Hauagge, R., and Cummins, J. N. (1991b). Relationships among indices for the end of bud dormancy in apple cultivars and related *Malus* species. *J. Am. Soc. Hortic. Sci.* 116, 95–99.

Jones, P. D., and Moberg, A. (2003). Hemispheric and large-scale surface air temperature variations: an extensive revision and an update to 2001. *J. Clim.* 16, 206–223. doi: 10.1175/1520-0442(2003)016<0206:HALSSA>2.0.CO;2

Kass, R. E., and Raftery, A. E. (1995). Bayes factors. *J. Am. Stat. Assoc.* 90, 773–795. doi: 10.1080/01621459.1995.10476572

Labuschagné, I. F., Louw, J. H., Schmidt, K., and Sadie, A. (2002). Genetic variation in chilling rquirement in apple progeny. *J. Am. Soc. Hort. Sci.* 127, 663–672.

Lang, G. A., Early, J. D., Martin, G. C., and Darnell, R. L. (1987). Endo-, para-, and ecodormancy: physiological terminology and classification for dormancy research. *HortScience* 22, 371–377.

Legave, J. M., Blanke, M., Christen, D., Giovannini, D., Mathieu, V., and Oger, R. (2013). A comprehensive overview of the spatial and temporal variability of apple bud dormancy release and blooming phenology in Western Europe. *Int. J. Biometeorol.* 57, 317–331. doi: 10.1007/s00484-012-0551-9

Luedeling, E., Kunz, A., and Blanke, M. M. (2013). Identification of chilling and heat requirements of cherry trees - a statistical approach. *Int. J. Biometeorol.* 57, 679–689. doi: 10.1007/s00484-012-0594-y

Malagi, G., Sachet, M. R., Citadin, I., Herter, F. G., Bonhomme, M., Regnard, J. L., et al. (2015). The comparison of dormancy dynamics in apple trees grown under temperate and mild Winter climates imposes a renewal of classical approaches. *Trees* 29, 1365–1380. doi: 10.1007/s00468-015-1214-3

Menzel, A., and Fabian, P. (1999). Growing season extended in Europe. *Nature* 397, 659. doi: 10.1038/17709

Menzel, A., Seifert, H., and Estrella, N. (2011). Effects of recent warm and cold spells on European plant phenology. *Int. J. Biometeorol.* 55, 921–932. doi: 10.1007/s00484-011-0466-x

Miller-Rushing, A. J., Katsuki, T., Primack, R. B., Ishii, Y., Lee, S. D., and Higuchi, H. (2007). Impact of global warming on a group of related species and their hybrids: cherry tree (Rosaceae) flowering at Mt. Takao, Japan. *Am. J. Bot.* 94, 1470–1478. doi: 10.3732/ajb.94.9.1470

Muggeo, V. M. R. (2003). Estimating regression models with unknown breakpoints. *Stat. Med.* 22, 3055–3071. doi: 10.1002/sim.1545

Oukabli, A., Bartolini, S., and Viti, R. (2003). Anatomical and morphological study of apple (*Malus x domestica* Borkh) flower buds growing under inadequate winter chill. *J. Hortic. Sci. Biotechnol.* 78, 580–585.

Petri, J. L., and Leite, G. B. (2004). Consequences of insufficient winter chill on apple tree bud-break. VIIIth international symposium on temperate zone fruits in the tropics and subtropics, Florianopolis (Brazil). *Acta Hort.* 872, 53–60. doi: 10.17660/ActaHortic.2004.662.4

Pope, K. S., Da Silva, D., Brown, P. H., and Dejong, T. M. (2014). A biologically based approach to modeling spring phenology in temperate deciduous trees. *Agric. For. Meteorol.* 198–199, 15–23. doi: 10.1016/j.agrformet.2014.07.009

Pope, K. S., Dose, V., Da Silva, D., Brown, P. H., Leslie, C. A., and Dejong, T. M. (2013). Detecting nonlinear response of spring phenology to climate change by Bayesian analysis. *Glob. Chang. Biol.* 19, 1518–1525. doi: 10.1111/gcb.12130

Primack, R. B., Ibanez, I., Higuchi, H., Don Lee, S., Miller Rhushing, A. J., Wilson, A. M., et al. (2009). Spatial and interspecific variability in phenological responses to warming temperatures. *Biol. Conserv.* 142, 2569–2577. doi: 10.1016/j.biocon.2009.06.003

Rutishauser, T., Schleip, C., Sparks, T., Nordli, Ø., Menzel, A., Wanner, H., et al. (2009). Temperature sensitivity of Swiss and British plant phenology from 1753 to 1958. *Clim. Res.* 39, 179–190. doi: 10.3354/cr00810

Schwartz, M. D., and Hanes, J. M. (2010). Continental-scale phenology: warming and chill. *Int. J. Climatol.* 30, 1595–1598. doi: 10.1002/joc.2014

Sunley, R. J., Atkinson, C. J., and Jones, H. G. (2006). Chill unit models and recent changes in the occurrence of winter chill and spring frost in the United Kingdom. *J. Hortic. Sci. Biotechnol.* 8, 949–958. doi: 10.1080/14620316.2006.11512181

Wolfe, D. W., Schwartz, M. D., Lakso, A. N., Otsuki, Y., Pool, R. M., and Shaukis, N. J. (2005). Climate change and shifts in spring phenology of three horticultural woody perennials in northeastern USA. *Int. J. Biometeorol.* 49, 303–309. doi: 10.1007/s00484-004-0248-9

Yu, H., Luedeling, E., and Xu, J. (2010). Winter and spring warming result in delayed spring phenology on the Tibetan Plateau. *Proc. Natl. Acad. Sci. U.S.A.* 107, 1–6. doi: 10.1073/pnas.1012490107

Zhang, N. R., and Siegmund, D. O. (2007). A modified Bayes information criterion with applications to the analysis of comparative genomic hybridization data. *Biometrics* 63, 22–32. doi: 10.1111/j.1541-0420.2006.00662.x

Zhang, X., Tarpley, D., and Sullivan, J. T. (2007). Diverse responses of vegetation phenology to a warming climate. *Geophys. Res. Lett.* 34, 1–5. doi: 10.1029/2007gl031447

Conflict of Interest Statement: The authors declare that the research was conducted in the absence of any commercial or financial relationships that could be construed as a potential conflict of interest.

20

Evidence of Phytotoxicity and Genotoxicity *in Hordeum vulgare* L. Exposed to CeO$_2$ and TiO$_2$ Nanoparticles

Alessandro Mattiello[1], Antonio Filippi[1], Filip Pošćić[1], Rita Musetti[1], Maria C. Salvatici[2], Cristiana Giordano[2,3], Massimo Vischi[1], Alberto Bertolini[1] and Luca Marchiol[1]*

[1] Department of Agriculture and Environmental Sciences, University of Udine, Udine, Italy, [2] Centro di Microscopie Elettroniche "Laura Bonzi", Istituto di Chimica dei Composti OrganoMetallici, Consiglio Nazionale delle Ricerche, Firenze, Italy, [3] Tree and Timber Institute, Istituto Per La Valorizzazione del Legno e delle Specie Arboree-CNR, Firenze, Italy

Edited by:
Nelson Marmiroli,
University of Parma, Italy

Reviewed by:
Margarita Sánchez-Domínguez,
Centro de Investigación en Materiales
Avanzados, S.C., Mexico
Jozsef Dudas,
Medical University of Innsbruck,
Austria

*Correspondence:
Luca Marchiol
marchiol@uniud.it

Engineered nanoscale materials (ENMs) are considered emerging contaminants since they are perceived as a potential threat to the environment and the human health. The reactions of living organisms when exposed to metal nanoparticles (NPs) or NPs of different size are not well known. Very few studies on NPs–plant interactions have been published, so far. For this reason there is also great concern regarding the potential NPs impact to food safety. Early genotoxic and phytotoxic effects of cerium oxide NPs (nCeO$_2$) and titanium dioxide NPs (nTiO$_2$) were investigated in seedlings of *Hordeum vulgare* L. Caryopses were exposed to an aqueous dispersion of nCeO$_2$ and nTiO$_2$ at, respectively 0, 500, 1000, and 2000 mg l^{-1} for 7 days. Genotoxicity was studied by Randomly Amplified Polymorphism DNA (RAPDs) and mitotic index on root tip cells. Differences between treated and control plants were observed in RAPD banding patterns as well as at the chromosomal level with a reduction of cell divisions. At cellular level we monitored the oxidative stress of treated plants in terms of reactive oxygen species (ROS) generation and ATP content. Again nCeO$_2$ influenced clearly these two physiological parameters, while nTiO$_2$ were ineffective. In particular, the dose 500 mg l^{-1} showed the highest increase regarding both ROS generation and ATP content; the phenomenon were detectable, at different extent, both at root and shoot level. Total Ce and Ti concentration in seedlings was detected by ICP-OES. TEM EDSX microanalysis demonstrated the presence of aggregates of nCeO$_2$ and nTiO$_2$ within root cells of barley. nCeO$_2$ induced modifications in the chromatin aggregation mode in the nuclei of both root and shoot cells.

Keywords: barley, cerium oxide nanoparticles, titanium oxide nanoparticles, genotoxicity, oxidative stress

INTRODUCTION

It is estimated that by 2020 about six million people will be employed worldwide in industries that use nanotechnologies, which will have the potential to produce goods for a market of more than 3,000 billion dollars (Roco, 2011). There is therefore a tumultuous development of new materials justified by a rapid growth of technological and commercial applications. Model simulations demonstrated that flows of engineered nanomaterials (ENMs) are able to reach several natural ecosystems (Colman et al., 2013). Cerium oxide nanoparticles (NPs; nCeO$_2$) and titanium oxide NPs (nTiO$_2$) are among the top ten most produced ENMs by mass (Keller et al., 2013) and used in cosmetics industries, in solar cells, paints, cements, coatings, in agriculture and the food industry (Gogos et al., 2012; Piccinno et al., 2012; Parisi et al., 2015). nCeO$_2$ and nTiO$_2$ were included in the list of ENMs of priority for immediate testing by the Organization for Economic Cooperation and Development (OECD, 2010). From point sources (e.g., discharges of wastewaters from industries or landfills), such materials will tend to accumulate in sediments and soils, exposing the organisms inhabiting these environments to potential risks (Liu and Cohen, 2014).

Plants are able to assimilate metal nanoparticles (MeNPs) largely depending on the type of plant and the size of MeNPs (Rico et al., 2011). In addition, the primary particle size of MeNPs is relevant for their bioavailability and therefore their toxicity (Van Hoecke et al., 2009), also raising questions on the potential for MeNPs exposure of crops and food safety (Hong et al., 2013). Experimental evidences were reported by Zhang et al. (2011) which studied the nCeO$_2$ uptake and translocation in cucumber, reporting an higher Ce assimilation in plants treated with 7 nm Ce than 25 nm ones. Clément et al. (2013) reported similar results for nTiO$_2$ on rapeseed plantlets treated with 14–25 nm particles. Another functional property that influences the MeNPs plant assimilation is the agglomeration/aggregation status that in turn is influenced directly by the zeta-potential (Navarro et al., 2008). Song et al. (2013a) demonstrated a negative correlation between nTiO$_2$ agglomeration/aggregation and assimilation in tomato. A similar behavior could be hypothesized also for nCeO$_2$.

Although there are potential positive applications of ENMs in agriculture (Parisi et al., 2015), studies on the toxicity of MeNPs have shown early negative consequences on crops due to genotoxic and phytotoxic effects (Miralles et al., 2012; Gardea-Torresdey et al., 2014). From an ecological point of view, this raises questions about potential risks due to the input of MeNPs in the food chain. Early plant MeNPs toxicity can be measured observing seed germination, root elongation, DNA mutations (López-Moreno et al., 2010a; Atha et al., 2012) or changes in biochemical parameters (Rico et al., 2013; Schwabe et al., 2013).

The aims of this study were to determine the early phytotoxic and genotoxic effects of nCeO$_2$ and nTiO$_2$ on barley (*Hordeum vulgare* L.) plants. The FAO ranks barley fourth in the top five cereals in the world ordered based on production tonnage (FAOSTAT, 2014) and the cereal is one of the major crops grown worldwide for human and animal consumption. Suspensions of nCeO$_2$ and nTiO$_2$ were prepared at 0, 500, 1000, and 2000 mg l^{-1}. Phytotoxicity of NPs was determined through percentage of germination and root elongation, ATP and ROS generation in root and leaf cells. Genotoxicity was investigated by the mitotic index and RAPDs. Ce and Ti uptake and translocation within seedling tissues were determined by inductively coupled plasma-optical emission spectroscopy (ICP-OES), while nCeO$_2$ and nTiO$_2$ within plant cells were detected by transmission electronic microscope and energy dispersive x-ray spectrometer (TEM–EDX).

MATERIALS AND METHODS

Nanoparticles Characterization

The cerium[IV] oxide (nCeO$_2$) and titanium[IV] oxide anatase (nTiO$_2$) powders with a nominal average particle size <25 nm were purchased from Sigma–Aldrich (Milwaukee, WI, USA). The specific surface area of the nCeO$_2$ and nTiO$_2$ was measured by the Brunauer–Emmett–Teller (BET) method by using the Surface Area and Pore Size Analyzer SA 3100 plus (Beckman Coulter, USA).

The nCeO$_2$ and nTiO$_2$ powders were suspended in deionized water at a concentration of 1000 mg l^{-1} and sonicated at 60°C for 30 min. The suspensions were characterized for Z-average size, measured as hydrodynamic diameter, zeta potential, via electrophoretic mobility, and polydispersity index (PDI), calculated from the signal intensity, by the dynamic light scattering (DLS) method using the Nano ZS90 (Malvern Instruments, UK). The nCeO$_2$ and nTiO$_2$ powder suspensions at three different concentrations (500, 1000, and 2000 mg l^{-1}) were prepared in MilliQ® water by sonication for 30 min at room temperature and then stirred for 15 min. The range of concentrations (0, 500, 1000, and 2000 mg L^{-1}) was chosen according to Yang and Watts (2005), Lin and Xing (2007), and López-Moreno et al. (2010a).

Seed Germination and Root Elongation

Caryopses of *H. vulgare* L. var. Tunika were provided by S.I.S. Società Italiana Sementi (San Lazzaro di Savena, Bologna, Italy). The caryopses were sterilized by orbital agitation with 70% ethanol for 2 min and then with 5% sodium hypochlorite plus some drops of Tween 80 for 30 min. They were rinsed six times with sterilized MilliQ® water. All caryopses were transferred in sterile conditions into 15 mm Petri dishes containing filter paper (Ø 90 mm Whatman No. 1) soaked with 8 ml of MilliQ® water (control treatment) or 8 ml of nCeO$_2$ or nTiO$_2$ suspensions at different concentrations. The Petri dishes were taped and placed in the dark at 21°C for 3 days. The germination percentage was calculated as the ratio of germinated seeds out the total seeds of each Petri dish. A second set of caryopses were treated for 7 days in the same conditions to evaluate root elongation and Ce and Ti uptake. The seedlings were photographed and Image J software (Schneider et al., 2012) was used to measure roots length. Root elongation was calculated as the average or the sum of all roots emerged from each seed. The experiments were performed in triplicate.

Mitotic Index

The germinated seedlings with actively growing roots (2.5 cm in length) were placed in the nCeO$_2$ and nTiO$_2$ suspensions (0, 500, 1000, 2000 mg l^{-1}) for 24 h. After treatment the root tips were fixed in 3:1 alcohol : acetic acid and then, kept in 70% ethanol at 4°C. The root tips were rinsed in deionized water for 5 min, hydrolyzed in 1N HCl for 8 min at 60°C, rinsed in deionized water for 5 min, stained in leuco-basic-fuchsine for 45 min and washed in tap water for 5 min. The root tips were then transferred to 45% acetic acid for 1 to 5 min, root caps were removed, and the roots were dissected to release the meristematic cells. Ten tips per treatment were evaluated and each treatment was replicated three times, for a total of about 10,000 cell observations. The mitotic index was evaluated in Feulgen stained preparations as the percentage of dividing cells out of the total number of cells scored.

Random Amplified Polymorphic DNA (RAPD) Analysis

The genotoxicity of $nCeO_2$ and $nTiO_2$ was investigated by observing the band profile after a random amplified polymorphic DNA (RAPD) assay on six replicates per treatment obtained from seedlings exposed as for mitotic index experiment. Plant genomic DNA was extracted from root tips using the DNeasy Plant Mini Kit (QIAGEN®) according to manufacturer's protocol. PCR reactions were performed with 30 ng of genomic DNA as a template using six primer pairs: OPA04 (AATCGGGCTG), OPA10 (GTGATCGCAG), OPB01 (GTTTCGCTCC), OPB03 (CATCCCCCTG), OPB12 (CCTTGACGCA), and OPB20 (GGACCCTTAC). The PCR conditions consisted of an initial Taq polymerase activation at 95°C for 5 min, followed by 45 cycles of denaturation (95°C, 1 min), annealing (35°C, 1 min), and extension (72°C, 1 min) with a final extension for 10 min at 72°C. The PCR products were subjected to electrophoresis on 1.6% agarose in TBE 0.5%, for 2 h at 60 V/cm stained with GelRed® and photographed for band scoring.

Evaluation of ATP Content

ATP content was determined by means of the luciferin–luciferase luminometric assay (Lundin, 1984). Root and shoot of each seedling were separated, frozen with liquid nitrogen and ground to a fine powder. Tissue powder (100 ± 20 mg FW) was suspended in 1 ml of 50 mM Tris-HCl (pH 7.5), 0.05% (w/v) Triton X-100 and immediately kept at 95°C for 3 min to inactivate any possible hydrolytic activity. After cooling, samples were centrifuged to obtain the cellular soluble fraction in the supernatant. The sample assays were performed in a 96-well plate with ATPlite Luminescence ATP Detection Assay System, (PerkinElmer) according to manufacturer's protocol. Aliquots (20 μl) of soluble fraction were mixed with 20 μl of ATPlite buffer in 130 μl of 50 mM Tris-HCl (pH 7.5) and 5 mM $MgCl_2$. Signals were detected by a Multilabel Counter (WALLAC, model 1420, PerkinElmer, Waltham, MA, USA). Actual ATP concentration of each experiment (expressed as nmol ATP g^{-1} f. w.) was calculated by a calibration curve obtained with commercially purchased ATP (Sigma, USA) in a 8–100 nM range.

Reactive Oxygen Species (ROS) Determination

The generation of ROS was monitored by the method of Wang and Joseph (1999), using 2′,7′-dichlorodihydrofluorescein diacetate (H_2DCFDA) as a probe. Tissue powder (0.5 g f. w.) obtained from both roots and shoots was extracted in 2.5 ml cold acetone and incubated for 4 h at 4°C. After centrifugation at 1000 g for 10 min, the pellet was homogenized in 1 ml of 50 mM Tris-HCl (pH 7.5), 0.4 M sucrose and 1 mM EDTA by

Turrax device. The sample was again centrifuged for 15 min and the supernatant stored at –80°C until analysis. Aliquots of sample (20 μl) were incubated in 96-well microplate with 5 μM H_2DCFDA and 180 μl of 50 mM Tris-HCl (pH 7.5). Detection was performed by fluorimetric assay using Multilabel Counter (WALLAC, model 1420, PerkinElmer) with orbital shaking and reading for 1.75 h at 5 min intervals with excitation filter set at 485 ± 10 nm and the emission filter set at 530 ± 10 nm. Values of relative fluorescence (RFU) were expressed as RFU mg^{-1} protein. Protein concentration was estimated by the Bradford (1976) method.

Cerium and Titanium in Seedling Tissues

The seedlings were washed by agitation with 0.01 M HNO_3 for 30 min and rinsed three times by agitation with MilliQ® water for 15 min. The seedling roots and shoots were then oven-dried at 105°C for 24 h and 0.5 g material was digested using 10 ml of HNO_3 in a microwave oven (CEM, MARS Xpress) according to the USEPA 3052 method (USEPA, 1995). After mineralization, the plant extracts were filtered (0.2 μm PTFE), diluted and analyzed. Total content of Ce and Ti was determined by an ICP-OES (Varian Inc., Vista MPX). The accuracy of the analytical procedure adopted for ICP-OES analysis was checked by running standard solutions every 20 samples. Yttrium was used as the internal standard.

TEM Observations and X-ray Microanalysis

The morphology of NPs was assessed by direct observation of suspension of $nCeO_2$ or $nTiO_2$ NPs under the TEM. Drops of suspensions (prepared as described above) were placed on carbon–formvar coated nickel grids, dried at room temperature and observed under a Philips CM 10 (FEI, Eindhoven, The Netherlands) TEM, operating at 80 kV.

For microscopic analyses *in planta*, tissues from seedlings treated with nCO_2 or $nTiO_2$ at 1000 and 2000 mg l^{-1} were sampled as in the root elongation experiment were sampled. Roots and shoots were excised, cut into small portions (2 mm × 3 mm) and fixed for 2 h at 4°C in 0.1% (w/vol) buffered sodium phosphate and 3% (w/v) glutaraldehyde at pH 7.2. They were then post-fixed with 1% osmium tetroxide (w/v) in the same buffer for 2 h, dehydrated in an ethanol series, and embedded in Epon/Araldite epoxy resin (Electron Microscopy Sciences, Fort Washington, PA, USA). For conventional TEM observations, serial ultrathin sections from embedded leaf tissues were cut with a diamond knife, mounted on uncoated 200 mesh copper grids (Electron Microscopy Sciences, Fort Washington, PA, USA), stained in uranyl acetate and lead citrate, and then observed under TEM as reported above.

For X-ray microanalysis, unstained ultrathin sections were placed on formvar/carbon-coated 200 mesh nickel grids and dried at room temperature. The nature of NPs observed in plant tissues was determined by a TEM (PHILIPS CM 12, FEI, Eindhoven, The Netherlands) equipped with an EDS-X-ray microanalysis system (EDAX, AMETEK, Mahwah NJ, USA, software EDAX Genesis). The images were recorded by a Megaview G2 CCD Camera (Olympus; software iTEM FEI, Analysis Image Processing).

Data Analysis

One-way analysis of variance (ANOVA) was conducted to test differences in the plants' behavior. Tukey's Multiple Comparison test at 0.05 p level were used to compare means. Statistical analyses were performed using the SPSS program (SPSS Inc., Chicago, IL, USA, ver. 17). Principal Coordinate Analysis (PCoA) was computed based on the binary genetic distance option in GenAlEx v. 6.501 software (Peakall and Smouse, 2012). Graphics were produced using CoPlot (CoHort ver. 6.204, Monterey, CA, USA).

RESULTS

Nanoparticles Characterization

The specific surface values obtained by BET measurements were 46.1 $m^2 g^{-1}$ for $nCeO_2$ and 61.6 $m^2 g^{-1}$ for $nTiO_2$. The Z-average sizes of the $nCeO_2$ and $nTiO_2$ suspended in deionized water were 174 ± 1.2 nm and 925 ± 105 nm, respectively, these values result remarkable higher respect the declared producer dimensions. The zeta potentials were 0.027 ± 0.064 mV for $nCeO_2$ and 19.9 ± 0.55 mV $nTiO_2$. These parameter values put in evidence their instability, in fact for both NP types are included in the range of the NP instability (–30 mV ÷ +30 mV) and justify the differences between the declared dimension and the measured ones. The PDI of $nCeO_2$ and $nTiO_2$ were 0.339 ± 0.011 and 0.841 ± 0.173, respectively. These values indicate a narrow dimensional distribution of $nCeO_2$ respect to $nTiO_2$.

Caryopses Germination and Root Elongation

Effects of $nCeO_2$ and $nTiO_2$ on caryopses germination and root growth are shown in **Table 1**. Since there was not a statistically significant effect of concentrations for $nCeO_2$ and $nTiO_2$, our results demonstrate that, even at the highest level of concentration, caryopses germination is not affected by $nCeO_2$ or $nTiO_2$ (**Table 1**). At the end of our experiment the barley seedlings had reached coleoptile emergence. At this stage typically has between six and seven seminal roots (Knipfer and Fricke, 2011). In our experiment the number of seminal roots was not affected by $nCeO_2$ and $nTiO_2$ (**Table 1**). On the contrary, in

both cases the development of root tissues was influenced in a similar manner by the treatments. In fact, there was a significant effect of both $nCeO_2$ ($p < 0.05$) and $nTiO_2$ ($p < 0.05$) on the average length of the seminal roots. *Post hoc* comparison tests indicated that root elongation in seedlings treated with 500 mg l^{-1} $nCeO_2$ and $nTiO_2$ was significantly lower than controls (–24.5 and –14.8%, respectively). At higher $nCeO_2$ and $nTiO_2$ concentrations we would have expected to see a further reduction in the development of seminal roots. However, this did not occur since the average length of seminal roots was similar to controls (**Table 1**).

Cerium and Titanium in Plant Tissues

Although without visible symptoms of phytotoxicity, the concentration of total Ce and Ti in the tissues of barley seedlings showed (i) a dose-response and (ii) a different magnitude of accumulation between Ce and Ti. **Table 2** shows the concentration of Ce and Ti in the fractions of barley seedlings. As expected Ce and Ti accumulated much more within root tissues than in the shoot ($p < 0.05$). Ce concentration in the roots increased significantly ($p < 0.05$) as the concentration of $nCeO_2$ in the growth medium increased (**Table 2**). A statistically significant effect of treatments in Ce accumulation in the shoots ($p < 0.001$) was verified. Mean comparisons showed differences among the treatments. Ce concentration in shoots did not significantly differ between the 500 and 1000 mg l^{-1} Ce treatment (38.3 and 98.1 mg Ce kg^{-1} DW, respectively), whereas at 2000 mg $nCeO_2$ L^{-1} a Ce concentration of 622 mg Ce kg^{-1} DW was observed in the shoots, which is significantly different from other values (**Table 2**).

Titanium concentrations in barley roots and shoots were one–two orders of magnitude lower compared to Ce. However, also in this case a statistically significant dose dependent increase was also observed. With the lowest $nTiO_2$ treatment (500 mg l^{-1}) Ti concentration in roots was negligible and no Ti was detected in shoots (**Table 2**). At the intermediate $nTiO_2$ treatment (1000 mg l^{-1}) the root tissues had 37.2 mg Ti kg^{-1} DW which is significantly lower ($p = 0.0001$) than 413 mg Ti kg^{-1} DW found at highest $nTiO_2$ treatment (**Table 2**). Finally, we verified that also Ti concentration in the shoots also responded positively to the treatments ($p < 0.001$). The mean Ti concentration detected in barley shoots were 7.83 mg kg^{-1} DW and 26.2 mg kg^{-1} DW for 1000 and 2000 mg $nTiO_2$ l^{-1}, respectively (**Table 2**).

TABLE 1 | Germination percentage of seeds, number of seminal roots and root length in barley seedlings treated with 0, 500, 100, and 2000 mg l^{-1} of $nCeO_2$ and $nTiO_2$.

Treatment	$nCeO_2$			$nTiO_2$		
	Germination (%)	Seminal roots (n)	Root length (mm)	Germination (%)	Seminal roots (n)	Root length (mm)
Ctrl	87 ± 1.76 a	5.2 ± 0.18 a	52.7 ± 4.13 a	88 ± 1.20 a	6.6 ± 0.34 a	53.3 ± 3.03 a
500 mg l^{-1}	83 ± 2.03 a	5.5 ± 0.22 a	39.8 ± 2.24 b	87 ± 1.76 a	6.1 ± 0.27 a	45.4 ± 2.85 b
1000 mg l^{-1}	80 ± 2.08 a	5.2 ± 0.26 a	45.8 ± 17.8 ab	85 ± 1.45 a	6.5 ± 0.22 a	53.9 ± 3.13 a
2000 mg l^{-1}	79 ± 1.86 a	4.9 ± 0.25 a	43.8 ± 1.72 ab	87 ± 1.76 a	6.4 ± 0.13 a	58.5 ± 2.97 a

Values are mean ± SE (n = 3). Different letters indicate statistical difference between treatments at Tuckey's test (p < 0.05).

TABLE 2 | Concentration of total Ce and Ti in roots and shoots of barley seedlings treated with 0, 500, 100, and 2000 mg l⁻¹ of $nCeO_2$ and $nTiO_2$.

Treatment	Ce roots (mg kg⁻¹ DW)	Ce coleoptile (mg kg⁻¹ DW)	Ti roots (mg kg⁻¹ DW)	Ti coleoptile (mg kg⁻¹ DW)
Ctrl	<d.l.	<d.l.	<d.l.	<d.l.
500 mg l⁻¹	579 ± 168 b	38.3 ± 5.77 b	<d.l.	<d.l.
1000 mg l⁻¹	5262 ± 1751 b	98.1 ± 40.2 b	35.2 ± 17.3 b	7.83 ± 3.3 b
2000 mg l⁻¹	20,714 ± 5722 a	622 ± 95.1 a	412 ± 127 a	26.2 ± 8.71 a

Values are mean ± SE (n = 3). Different letters indicate statistical difference between treatments at Tuckey's test (p < 0.05).

Ce and Ti Nano-aggregates in Plant Tissues

The morphology of $nCeO_2$ and $nTiO_2$ NPs is visible in **Figures 1A,B**, respectively. Transmission electron microscopy analysis demonstrated that CeO_2 particles exhibited an approximate equi-axes shape with sharp edges (**Figure 1A**), while particle sharp edges are less evident in TiO_2. To assess the possible uptake of $nCeO_2$ or $nTiO_2$ from the culture medium to the root tissues and the translocation to the different parts of the plantlets, we performed ultrastructural analyses on roots and shoot tissues. Several clusters of NPs were found in cortical parenchymal tissues of roots, both in the case of $nCeO_2$ (**Figure 2A**) and $nTiO_2$ treatment, at all concentrations. Clusters were also observed in the xylem, even if in to lesser extent (**Figure 2B**). EDS-X ray microanalysis allowed the identification of the clusters as aggregates of Ce and Ti nanoparticles.

No NPs were detected in the shoots of $nCeO_2$ or $nTiO_2$ treated plantlets. The ultrastructure of all observed tissues appeared preserved. No necrosis or damage to membranes, nor cell modifications were detected. In general, the cell compartments were not significantly affected by treatments, except for the nuclei of parenchymal cells of root and shoot of seedlings treated with $nCeO_2$ (1000 and 2000 mg l⁻¹), which showed compact chromatin (**Figures 3A–D**).

ATP and ROS

The evaluation of ATP concentration aimed to evidence the energetic status in different fractions of barley seedlings exposed to $nCeO_2$ and $nTiO_2$. The different concentrations of $nCeO_2$ induced a statistically significant effect (**Figure 4**), with a trend of values peaking at 500 and 1000 mg l⁻¹, in root and lowering at 2000 mg l⁻¹ in shoot samples. The highest $nCeO_2$ (2000 mg l⁻¹) reached a low concentration of ATP in roots, statistically comparable to control samples. On the contrary, $nTiO_2$ induce no significant changes of ATP concentration, since different $nTiO_2$ doses were similar to the controls in both roots and shoots (**Figure 4**).

The measurement of ROS was performed as marker for oxidative stress. Similarly to ATP content, $nCeO_2$ were able to induce an increase of a ROS formation at all the concentrations assayed (**Figure 5**), in comparison with the control, although no statistically significant differences were observed. Also for this parameter, a trend with a peak at 500 mg l⁻¹ was present in both roots and shoots. In the case of $nTiO_2$ (**Figure 5**), the treatments did not show any difference, if compared with the control in roots, whereas a decrease of ROS level was observed at the higher dose (2000 mg l⁻¹) in shoots.

Mitotic Index and RAPDs

The mitotic index was significantly reduced by $nCeO_2$ 2000 mg l⁻¹ (from 4 ± 1.2% in the control to 2.4 ± 1.2%). Instead, the $nCeO_2$ 500 and 1000 mg l⁻¹ treatments with mean values of 4 ± 1.3% were very similar to the control (**Figure 6A**). The treatments with $nTiO_2$ with values of 6.2 ± 3.2%, 4.6 ± 3.2%, 4.9 ± 2.5% for the concentration at 500, 1000, and 2000 mg l⁻¹, respectively, were not significantly different from the control (4.9 ± 2.8%; **Figure 6A**).

The six primers used for the RAPD analysis amplified for a total of 40 representative bands in controls with a variable number of 3 to 9 (9, 5, 6, 3, 9, 8, bands, respectively, for OPA04, OPB01, OPA10, OPB20, OPB12, and OPB03). Amplification was highly reproducible since the same RAPD profile was observed within control replicates. A concentration effect was observed for the $nCeO_2$ treatments on the RAPD profiles. The same banding pattern as controls was obtained for the $nCeO_2$ 500 mg l⁻¹ treatment, whereas new profiles at 1000 mg l⁻¹ were observed and three additional bands appeared and eight disappeared. Even greater variability was observed at $nCeO_2$ 2000 mg l⁻¹ with a total of 20 differences (appearing and disappearing bands) in treated plants (**Figures 6B,C**). The results were summarized by Principal Coordinates Analysis (PCoA), with almost 94% of the total variability explained by the two axes (**Figure 6D**). The overlap of the control and 500 mg l⁻¹ treatments is notable, while the treatments at 1000 and 2000 mg l⁻¹ are well separated in different quadrants. The band polymorphism in the different replicates at the higher concentration (2000 mg l⁻¹) can be noticed by the point cloud (**Figure 6D**). In a similar way to what was observed fro the mitotic index, the $nTiO_2$ treatments at each concentration have no effect on the RAPD profiles (**Figure 6D**).

DISCUSSION

Since plant nanotoxicology is a new field of investigation, specific ecotoxicological methods for the estimation of toxicity of ENPs have not yet been developed (Jośko and Oleszczuk, 2014). According to OECD guidelines, the acute effects of MeNPs on plant physiology are currently investigated by adapting the methods already used for traditional contaminants (Kühnel and Nickel, 2014). Evidence of MeNPs plant uptake and toxicity are still scarce and contradictory (Etheridge et al., 2013). This is likely because, compared to their bulk counterparts, MeNPs show particular properties, which are subjected to transformations (e.g., redox reactions, aggregation or agglomeration, and dissolution) according to different environmental factors. These

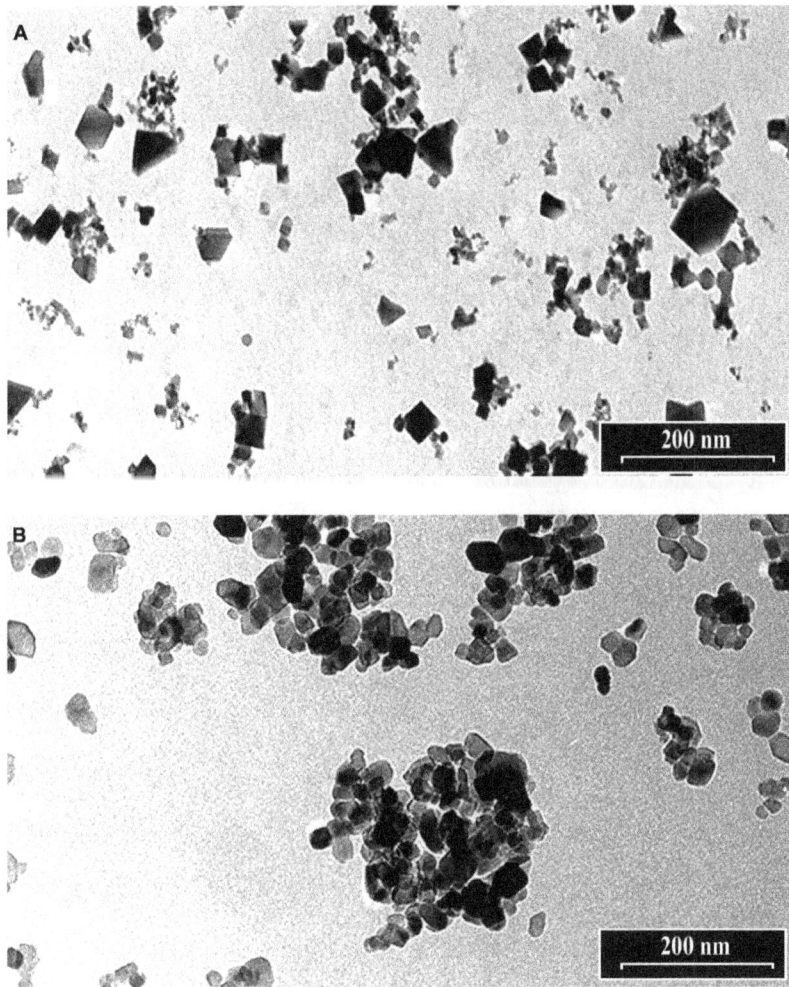

FIGURE 1 | Transmission electron microscope images of 1000 mg l^{-1} suspensions of (A) nCeO$_2$ and (B) nTiO$_2$.

changes might modify the ecotoxicological properties of MeNPs and thus, their interactions with the biota (Nowack et al., 2012; Maurer-Jones et al., 2013). However, despite these limitations, the experimental results obtained so far offer early indications on MeNPs phytotoxicity (Li et al., 2015; Rico et al., 2015a). Our data suggests that also in very simple experimental conditions, nTiO$_2$, as expected taking into account their intrinsic properties, forms bigger agglomerates with a wider dimensional distribution than nCeO$_2$.

nCeO$_2$ and nTiO$_2$ Affects Seed Germination and Seedling Development

Previous studies carried out in controlled conditions reported that the toxicity of MeNPs in the early stages of plant growth is likely due to the following factors: (i) chemical and physical properties which influence the release of ions or the aggregation of particles in more stable forms and (ii) the size and shape of the particles, which determine the specific surface area of MeNPs (Yang and Watts, 2005; Lin and Xing, 2007).

In agreement with Rico et al. (2015b), we found that germination of barley was unaffected by 500–2000 mg l^{-1} nCeO$_2$. This is in contrast with the results provided by López-Moreno et al. (2010a) who reported that suspensions of 2000 mg l^{-1} nCeO$_2$ significantly reduced seed germination in maize, cucumber, tomato, and soybean. Possible explanations could be the greater Ce tolerance of barley to the treatment if compared to other species and/or to the very small size of Ce NPs they used (7 nm). Another explanation could be related to the chemical and physical properties of nCeO$_2$, in particular his zeta potential value. This parameter is the cause of the agglomeration behavior of the nCeO$_2$ that brings to a low bioavailability and the absence of phytotoxic effects on the treated seeds regards the germination percentage.

Another important issue that plays a role on seed/NP interaction, is the methodology adopted for seed treatment. In fact, following Lin and Xing (2007), we prepared the barley seeds for germination trials by soaking them in distilled water before starting treatments, whereas López-Moreno et al. (2010a)

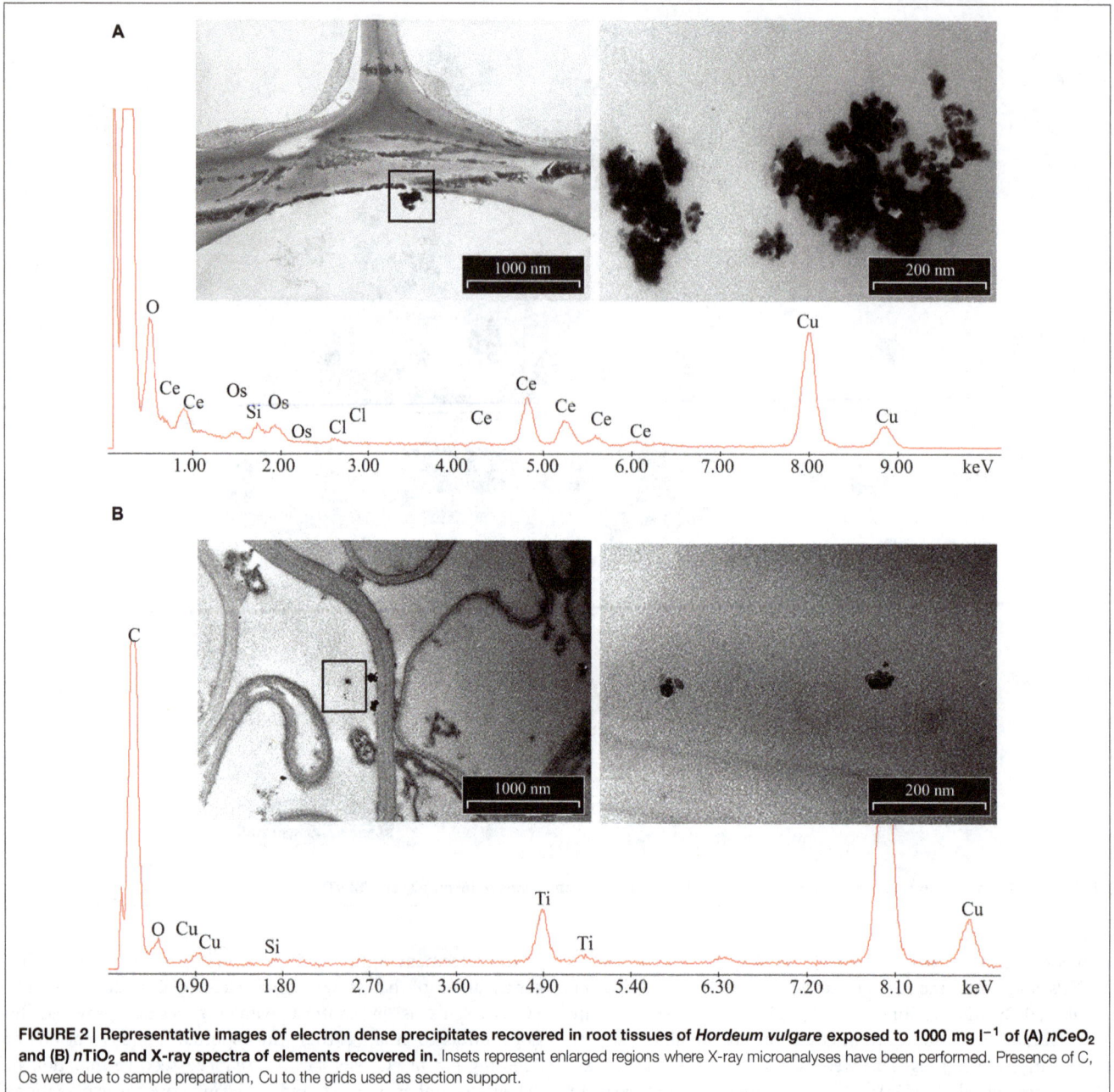

FIGURE 2 | Representative images of electron dense precipitates recovered in root tissues of *Hordeum vulgare* exposed to 1000 mg l⁻¹ of (A) *n*CeO₂ and (B) *n*TiO₂ and X-ray spectra of elements recovered in. Insets represent enlarged regions where X-ray microanalyses have been performed. Presence of C, Os were due to sample preparation, Cu to the grids used as section support.

soaked the seeds directly in the nCeO$_2$ suspensions. This different experimental approach could result in a different exposure of germinating seeds to nCeO$_2$.

As regards Ti, there is a substantial agreement in literature on the fact that suspensions of nTiO$_2$ do not affect seed germination, with few exceptions, as reported by Zheng et al. (2005) and Feizi et al. (2012). Our results are in accordance with those reported by other authors on rice, lettuce, radish, cucumber, tomato, and pea (Boonyanitipong et al., 2011; Wu et al., 2012; Song et al., 2013a; Fan et al., 2014).

Besides the germination percentage, we observed a negative influence of the treatments with nCeO$_2$ and nTiO$_2$ on root elongation in barley seedlings. However, this did not occur in

seedlings treated with nCeO$_2$ at the highest concentration, in which the root length was very similar to controls. In addition, in this case the literature reports contradictory evidence. López-Moreno et al. (2010a) reported that the root growth in maize and cucumber seedlings was significantly promoted by nCeO$_2$ (up to 4000 mg l^{-1}) whereas the same treatments resulted in a negative effect on root development in alfalfa and tomato. An inhibitory effect of nTiO$_2$ on root elongation in cucumber was reported by Mushtaq (2011). A decrease in the number of secondary lateral roots in pea seedlings was verified by Fan et al. (2014), whereas Boonyanitipong et al. (2011) did not record any effect on root length in rice seedlings exposed to nTiO$_2$. In our case, the different effect of the nCeO$_2$ and nTiO$_2$ on the root

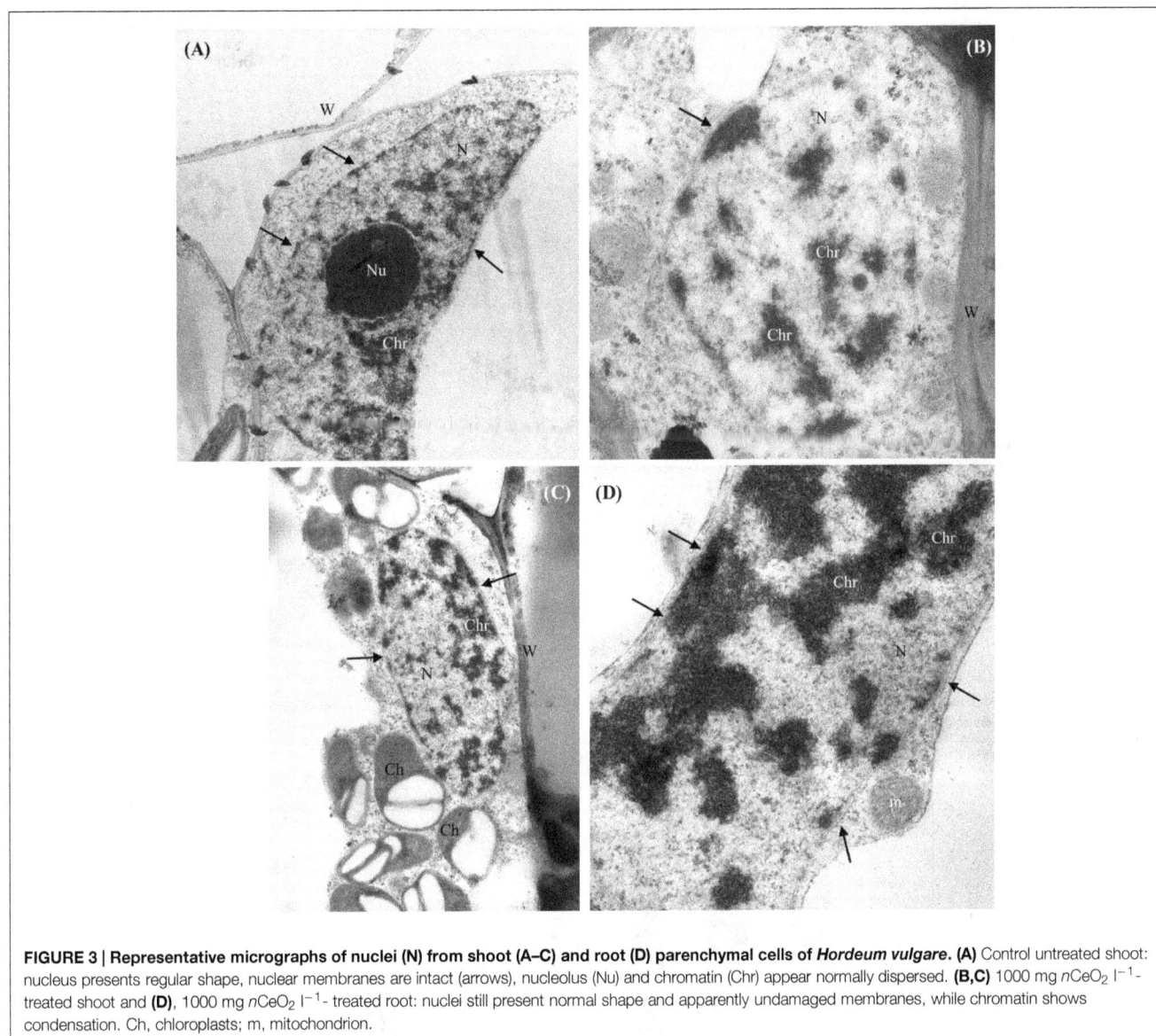

FIGURE 3 | Representative micrographs of nuclei (N) from shoot (A–C) and root (D) parenchymal cells of *Hordeum vulgare*. (A) Control untreated shoot: nucleus presents regular shape, nuclear membranes are intact (arrows), nucleolus (Nu) and chromatin (Chr) appear normally dispersed. **(B,C)** 1000 mg nCeO$_2$ l^{-1}-treated shoot and **(D)**, 1000 mg nCeO$_2$ l^{-1}- treated root: nuclei still present normal shape and apparently undamaged membranes, while chromatin shows condensation. Ch, chloroplasts; m, mitochondrion.

elongation is likely due to their different grade of agglomeration demonstrated by the z-average size and PDI values of nTiO$_2$ that results significantly higher than nCeO$_2$.

It might happen that the quantification analysis of trace metals in plant roots is disturbed by external contamination. In this case, the concentration of the element in the plant tissues could be significantly overestimated due to a fraction of metal, which is not taken up but simply adsorbed onto the external root surface. In our experiment, a concentration of Ce about 60 times greater than Ti, was found in barley root tissues. This substantial difference indicates that the procedures for preparation of the samples were conducted properly; otherwise, we would also have very high concentrations also for Ti.

Our results showed that the exposure of *H. vulgare* to nCeO$_2$, which are smaller and less aggregated than nTiO$_2$, resulted in a greater total Ce concentration in roots compared to Ti. In can therefore be assumed that, for some still unknown reasons, the

model of root uptake of the two elements could differ, depending in part on the intrinsic properties of solubility and agglomeration properties of nCeO$_2$ and nTiO$_2$. On the other hand, this is in agreement with the findings by Zhang et al. (2011), who verified that cucumber roots absorbed higher amounts of 7 nm nCeO$_2$ than 25 nm ones. On the other, some studies pointed out the possibility of interactions between the root metabolism and MeNPs. Lin and Xing (2007) demonstrated that root exudates such as proteins, phenolic acids, and aminoacids have a role in the adsorption of ZnO NPs to the root surface of perennial ryegrass. More recently, Schwabe et al. (2015) observed that root uptake of dissolved Ce$^{(III)}$ was promoted by the dissolution of nCeO$_2$ at the medium-root interface in hydroponically growth sunflower and maize. A further confirmation about the role of root exudates on the adsorption of MeNPs was provided by Lv et al. (2015) and Ma et al. (2015), respectively, for nCeO$_2$ and nZnO, respectively. However, Lv et al. (2015) reported that a

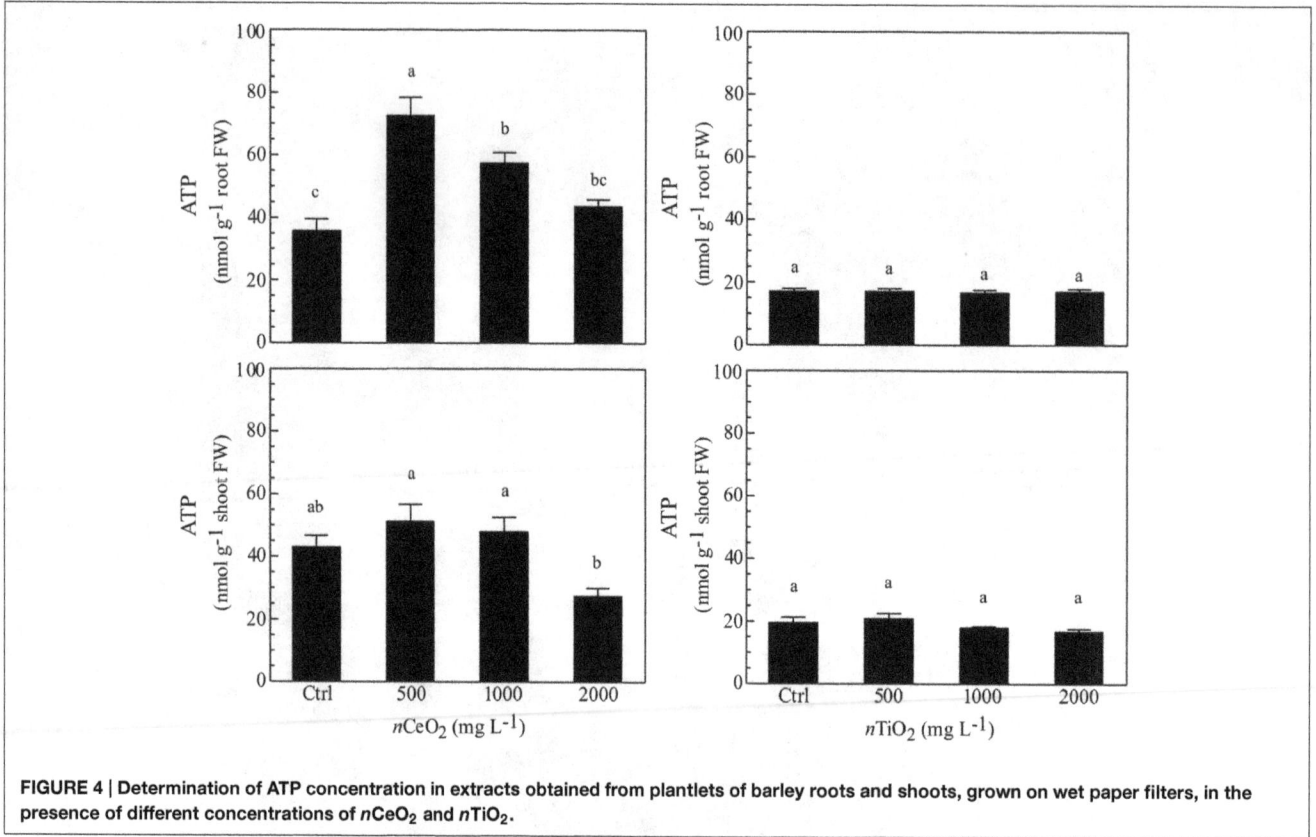

FIGURE 4 | Determination of ATP concentration in extracts obtained from plantlets of barley roots and shoots, grown on wet paper filters, in the presence of different concentrations of nCeO$_2$ and nTiO$_2$.

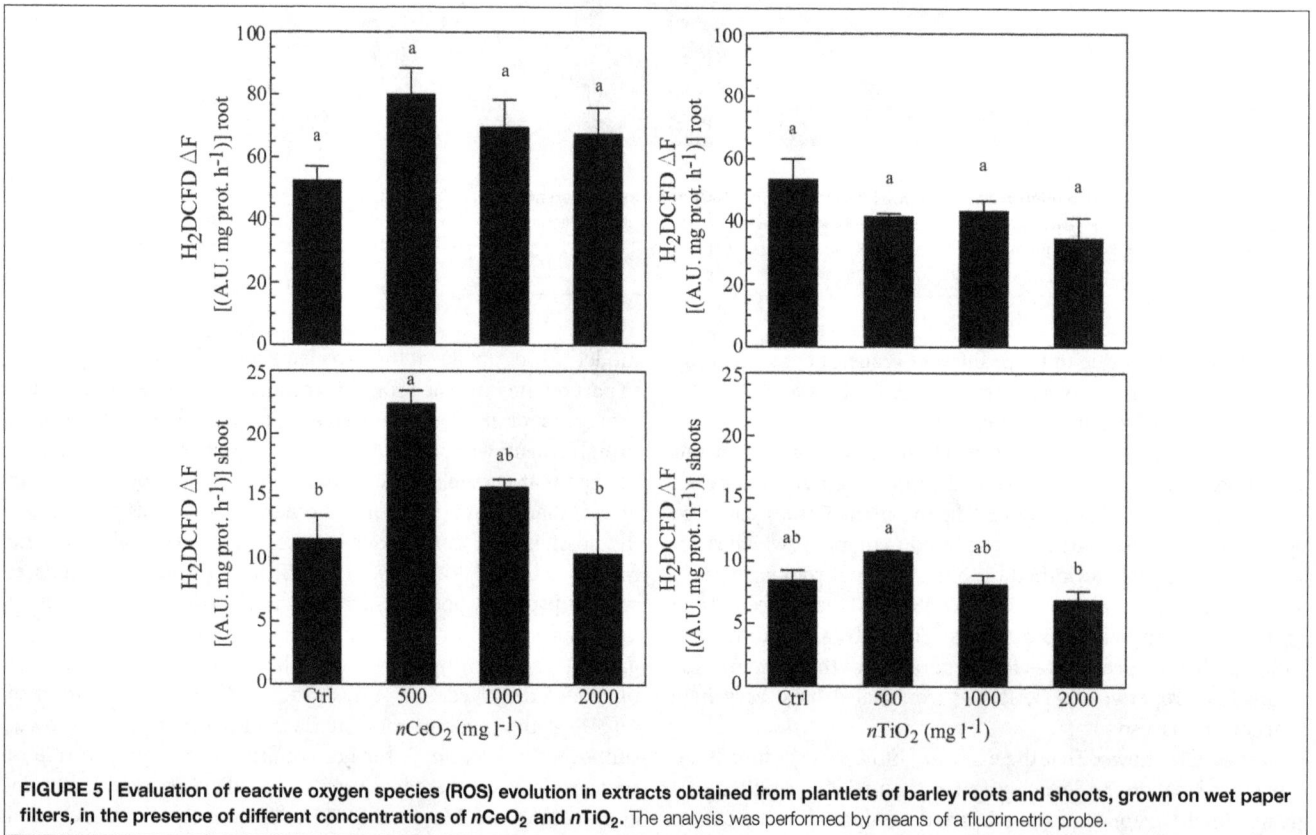

FIGURE 5 | Evaluation of reactive oxygen species (ROS) evolution in extracts obtained from plantlets of barley roots and shoots, grown on wet paper filters, in the presence of different concentrations of nCeO$_2$ and nTiO$_2$. The analysis was performed by means of a fluorimetric probe.

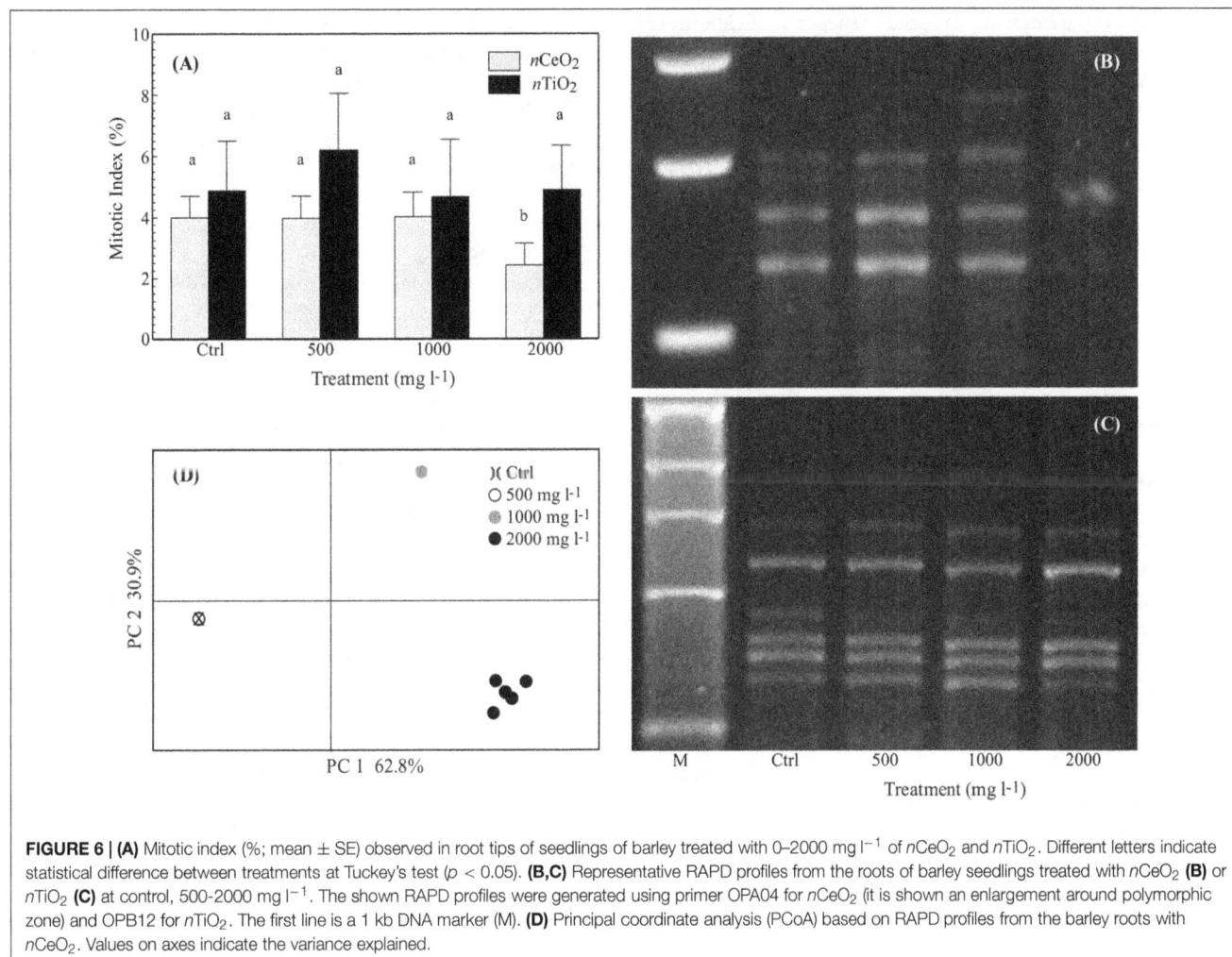

FIGURE 6 | (A) Mitotic index (%; mean $\pm$ SE) observed in root tips of seedlings of barley treated with 0–2000 mg l^{-1} of nCeO$_2$ and nTiO$_2$. Different letters indicate statistical difference between treatments at Tuckey's test ($p < 0.05$). **(B,C)** Representative RAPD profiles from the roots of barley seedlings treated with nCeO$_2$ **(B)** or nTiO$_2$ **(C)** at control, 500-2000 mg l^{-1}. The shown RAPD profiles were generated using primer OPA04 for nCeO$_2$ (it is shown an enlargement around polymorphic zone) and OPB12 for nTiO$_2$. The first line is a 1 kb DNA marker (M). **(D)** Principal coordinate analysis (PCoA) based on RAPD profiles from the barley roots with nCeO$_2$. Values on axes indicate the variance explained.

possible access of nZnO to the root tissues could be through the root apex or the meristematic zone to the lateral root system where the Casparian strip is not yet developed.

Root-to-shoot translocation of nCeO$_2$ has been previously described in soybean (Priester et al., 2012), tomato (Wang et al., 2012), cucumber (Zhao et al., 2013), and cotton (Van Nhan et al., 2015) after treatments with nCeO$_2$ suspensions. Different observations have been made on nCeO$_2$ root-to-shoot translocation in graminaceous crops. Schwabe et al. (2013) reported that wheat does not translocate nCeO$_2$ into the aerial tissues, whereas Rico et al. (2013, 2015a) reported the translocation of nCeO$_2$ from roots to rice grains and maize kernels, respectively. According to Rico et al. (2015b), we report evidence of Ce translocation from roots to the aerial part of barley. As regards Ti uptake and translocation, fewer data are available in the literature compared. However, our data are consistent with the findings reported by Song et al. (2013b) on tomato seedlings exposed to Ti at concentrations ranging from 50 to 5000 mg l^{-1}.

Finally, we reported that root length in barley seedlings treated with 500 mg nCeO$_2$ l^{-1} was significantly shorter than controls. This apparent dose-effect was not confirmed at higher nCeO$_2$

concentrations, since the root length was similar to that of controls. Similar evidence was reported by López-Moreno et al. (2010a). According to Nascarella and Calabrese (2012) and Bell et al. (2014), such unexpected results might be interpreted as a hormetic effect of nCeO$_2$ on root elongation in barley seedlings.

Plant Stress Induced by Nanoparticle Treatments

Within the plants, NPs may interact with the host cells, causing different effects, ranging from cell death (if the host is sensitive) to not relevant cell modifications (in the case of host tolerance), depending on their type, shape, and concentration (Rico et al., 2011; Gardea-Torresdey et al., 2014). The microscopic observations on barley seedlings indicated that both nCeO$_2$ and nTiO$_2$, at the used concentrations used, were able to enter the root tissues, being detected in the parenchymal cells and xylem vessels. Even though we did not observe Ce and Ti crystalline aggregates in the shoots, ICP analyses suggested a root-to-shoot mobilization of Ce and Ti ions. At histological level the accumulation of such elements induced limited injuries. On the contrary, important differences in the effects of treatments were obtained at nuclear level, where only the nCeO$_2$ treatments induced visible modifications in the chromatin aggregation in the nuclei of root and shoot parenchymal cells.

Condensed chromatin and fragmented nuclei are described as part of the programmed cell death (PCD), occurring in response to different environmental stimuli and stresses, induced by pathogens (Lam et al., 2001) and by diverse abiotic factors (White, 1996; Kratsch and Wise, 2000) including the exposition to nanomaterials (Shen et al., 2010). PCD plays an important role in mediating plant adaptation to the environment. In cells that undergo programmed death, chromatin condenses into masses with sharp margins, and DNA is hydrolyzed into a series of fragments (Gladish et al., 2006). Dynamic compaction of chromatin is an important step in the DNA-damage response, because it activates DNA-damage-repair signaling (Burgess et al., 2014) in response to injuries.

The hypothesis of Ce-induced DNA damage in treated seedlings finds further support in the results obtained with the RAPD test. RAPD can potentially detect a broad range of DNA damage and mutations, so it is suitable for studying MeNPs genotoxicity (Atienzar and Jha, 2006). The RAPD modified patterns at high concentrations of nCeO$_2$ (1000–2000 mg l^{-1}) indicated a genotoxic effect, which could directly influence the cell cycle. This is further confirmed by the reduced mitotic index recorded in the samples treated with nCeO$_2$ 2000 mg l^{-1}, which clearly demonstrated the negative effect of high nCeO$_2$ concentrations on the cell cycle. Our results are in agreement with López-Moreno et al. (2010b), who demonstrated nCeO$_2$ genotoxicity on soybean plants subjected to treatments similar to those reported in our work.

It is still far too early to conclude if the observed effects were direct or indirect consequences of the treatments, since nCeO$_2$ were not found in the nucleus. As it is known that increasing oxidative stress leads to DNA damage, the higher presence of ROS in treated samples could cause modification in RAPD patterns. However, as our analysis on ROS indicated a peak at 500 mg nCeO$_2$ l^{-1}, it can be rationalized that lower concentrations triggered an initial oxidative signal, while only higher nCeO$_2$ doses were able to induce damage at nuclear level. The oxidative stress peak at 500 mg l^{-1} dose and could be rationalized by the well-known SOD mimetic activity attributable to nCeO$_2$, which could cause a dismutation of superoxide anions into H$_2$O$_2$. Since a similar pattern is also found for ATP measured in nCeO$_2$ treated tissues, it is suggested that the oxidative burst induced by the more effective dose of nCeO$_2$ could be associated to a stimulation of cellular respiration and a consequent increase in ATP production. This could be due to a defense response signal or an increased requirement for energy (Vranová et al., 2002).

On the contrary, the nTiO$_2$ treatments did not influence either the mitotic index or RAPD pattern. This is in contrast to Moreno-Olivas et al. (2014) who observed nTiO$_2$-induced genotoxicity in hydroponically cultivated zucchini. As the size of nTiO$_2$ they reported is comparable to that used in our work, the different results obtained

can be explained by (i) the different cultivation systems (Petri dishes vs. full nutrient solution in hydroponics) and (ii) the nTiO$_2$ concentration used by Moreno-Olivas et al. (2014; 10-fold smaller). The latter potentially prevents the formation of big NP agglomerates, making them more bioavailable.

CONCLUSION

Although investigations into the effects of NPs in plants continue to increase, there are still many unresolved issues and challenges, in particular at the biota-nanomaterial interface (Nowack et al., 2015). In this multidisciplinary work, we studied the phytotoxic and genotoxic impact of nCeO$_2$ and nTiO$_2$ cerium and titanium oxide NP suspensions on the early growth of barley. Seed germination was not affected by the nCeO$_2$ and nTiO$_2$ suspensions, indicating that nCeO$_2$ and nTiO$_2$ are not allowed to enter the seed coatings. However, we verified that the concentration of Ce and Ti in the seedling fractions, as well as the root-to-shoot translocation, were dose-dependent. Then, we found signals of genotoxicity (RAPD banding patterns and mitotic index) and phytotoxicity in root cells (oxidative stress and chromatin modifications) resulting in a shortage of root elongation.

The different magnitude of bioaccumulation of Ce and Ti suggests a different uptake mechanism, likely due to the different behavior of nCeO$_2$ and nTiO$_2$. Recent studies have shown that plant toxic effects of nanomaterials are not merely due to the particle size and concentration of a suspension. Phytotoxicity of metal oxide NPs is related both to the direct adsorption of particles onto the root structures and to the aptitude of the metal ion to dissolve, possibly mediated by binding molecules produced by plants in the medium-root interface.

Our study had not the objective to investigate the details of the mechanisms by which the NPs entering within the roots. However, we verified the presence of both nCeO$_2$ and nTiO$_2$ into the root cells where an increase in oxidative stress occurred. More research needs to be conducted to verify whether germination can be affected by smaller nCeO$_2$ and nTiO$_2$. In addition, we need to understand if modification of the physical–chemical properties of NPs at the root interface can foster the plant uptake of Ce and Ti forms.

AUTHOR CONTRIBUTIONS

AM conducted the experiments. AF and AB provided the biochemical parameters. FP performed out the ICP and RAPD analysis. RM made TEM observation and observed MeNPs distribution *in planta*. CG and MS carried out TEM–EDAX observations. MV contributed to the mitotic index. LM designed, coordinated the study, performed statistical analysis, and prepared the figures. All authors were involved in manuscript writing. All authors contributed to the revision of the manuscript.

ACKNOWLEDGMENTS

This work was in part supported by a project funded by DISA – Department of Agriculture and Environmental Sciences, University of Udine, through Grant n.64 dd. 08-09.2014 (RANDOLPH – Relazioni tra nanoparticelle metalliche e piante superiori). The authors acknowledge technical assistance in the use of BET by Anastasios Papadiamantis and access to the lab facilities of the Environmental Nanoscience research group at Birmingham University (UK). The authors also thank Francesco Bertolini and Nicola Zorzin for their relevant contribution to the work.

REFERENCES

Atha, D. H., Wang, H., Petersen, E. J., Cleveland, D., Holbrook, D., Jaruga, P., et al. (2012). Copper oxide nanoparticle mediated DNA damage in terrestrial plant models. *Environ. Sci. Technol.* 46, 1819–1827. doi: 10.1021/es202660k

Atienzar, F. A., and Jha, A. N. (2006). The random amplified polymorphic DNA (RAPD) assay and related techniques applied to genotoxicity and carcinogenesis studies: a critical review. *Mutat. Res.* 613, 76–102. doi: 10.1016/j.mrrev.2006.06.001

Bell, I. R., Ives, J. A., and Jonas, W. D. (2011). Nonlinear effects of nanoparticles: biological Variability from Hormetic Doses, small particle sizes, and dynamic adaptive interactions. *Dose Response* 12, 202–232. doi: 10.2203/dose-response.13-025.Bell

Boonyanitipong, B., Kositsup, B., Kumar, P., Baruah, S., and Dutta, J. (2011). Toxicity of ZnO and TiO2 nanoparticles on germinating rice seed Oryza sativa L. *Int. J. Biosci. Biochem. Bioinform.* 1, 282–285. doi: 10.7763/IJBBB.2011.V1.53

Bradford, M. (1976). A rapid and sensitive method for the quantitation of microgram quantities of protein utilizing the principle of protein-dye binding. *Anal. Biochem.* 72, 248–254. doi: 10.1016/0003-2697(76)90527-90523

Burgess, R. C., Burman, B., Kruhlak, M. J., and Misteli, T. (2014). Activation of DNA damage response signaling by condensed chromatin. *Cell Rep.* 9, 1703–1717. doi: 10.1016/j.celrep.2014.10.060

Clément, L., Hurel, C., and Marmier, N. (2013). Toxicity of TiO2 nanoparticles to cladocerans, algae, rotifers and plants – Effects of size and crystalline structure. *Chemosphere* 90, 1083–1090. doi: 10.1016/j.chemosphere.2012.09.013

Colman, B. P., Arnaout, C. L., Anciaux, S., Gunsch, G. K., Hochella, M. F. Jr., Kim, B., et al. (2013). Low concentrations of silver nanoparticles in biosolids cause adverse ecosystem responses under realistic field scenario. *PLoS ONE* 8:e57189. doi: 10.1371/journal.pone.0057189

Etheridge, M. L., Campbell, S. A., Erdman, A. G., Haynes, C. L., Wolf, S. M., and McCullough, J. (2013). The big picture on nanomedicine: the state of investigational and approved nanomedicine products. *Nanomed. Nanotechnol. Biol. Med.* 9, 1–14. doi: 10.1016/j.nano.2012.05.013

Fan, R., Huang, Y. C., Grusak, M. A., Huang, C. P., and Sherrier, D. J. (2014). Effects of nano-TiO(2) on the agronomically-relevant Rhizobium-legume symbiosis. *Sci. Total Environ.* 466–467, 503–512. doi: 10.1016/j.scitotenv.2013.07.032

FAOSTAT (2014). *Food and Agricultural Commodities Production: Commodities by Regions.* Available at: http://faostat3.fao.org/faostat-gateway/go/to/browse/ranking/commodities_by_regions/E [accessed Sept 4, 2015]

Feizi, H., Moghaddam, P. R., Shahtahmassebi, N., and Fotovat, A. (2012). Impact of bulk and nanosized titanium dioxide (TiO2) on wheat seed germination and seedling growth. *Biol. Trace Elem. Res.* 146, 101–106. doi: 10.1007/s12011-011-9222-7

Gardea-Torresdey, J. L., Rico, C. M., and White, J. C. (2014). Trophic transfer, transformation and impact of engineered nanomaterials in terrestrial environments. *Environ. Sci. Technol.* 48, 2526–2540. doi: 10.1021/es4050665

Gladish, D. K., Xu, J., and Niki, T. (2006). Apoptosis-like programmed cell death occurs in procambium and ground meristem of Pea (Pisum sativum) root tips exposed to sudden flooding. *Ann. Bot.* 97, 895–902. doi: 10.1093/aob/mcl040

Gogos, A., Knauer, K., and Bucheli, T. D. (2012). Nanomaterials in plant protection and fertilization: current state, foreseen applications, and research priorities. *J. Agric. Food Chem.* 60, 9781–9792. doi: 10.1021/jf302154y

Hong, J., Peralta-Videa, J. R., and Gardea-Torresdey, J. L. (2013). "Nanomaterials in agricultural production: benefits and possible threats?" in *Sustainable Nanotechnology and the Environment: Advances and Achievements*, eds N. Shamim and V. K. Sharma, *ACS Symposium Series* (Washington, DC: American Chemical Society).

Jośko, I., and Oleszczuk, P. (2014). Phytotoxicity of nanoparticles—problems with bioassay choosing and sample preparation. *Environ. Sci. Pollut. Res.* 21, 10215–10224. doi: 10.1007/s11356-014-2865-0

Keller, A., McFerran, S., Lazareva, A., and Suh, S. (2013). Global life cycle releases of engineered nanomaterials. *J. Nanopart. Res.* 15, 1692. doi: 10.1007/s11051-013-1692-4

Knipfer, T., and Fricke, W. (2011). Water uptake by seminal and adventitious roots in relation to whole-plant water flow in barley (Hordeum vulgare L.). *J. Exp. Bot.* 62, 717–733. doi: 10.1093/jxb/erq312

Kratsch, H. A., and Wise, R. R. (2000). The ultrastructure of chilling stress. *Plant Cell Environ.* 23, 337–350. doi: 10.1046/j.1365-3040.2000.00560.x

Kühnel, D., and Nickel, C. (2014). The OECD expert meeting on ecotoxicology and environmental fate — Towards the development of improved OECD guidelines for the testing of nanomaterials. *Sci. Tot. Environ.* 472, 347–353. doi: 10.1016/j.scitotenv.2013.11.055

Lam, E., Kato, N., and Lawton, M. (2001). Programmed cell death, mitochondria and the plant hypersensitive response. *Nature* 411, 848–853. doi: 10.1038/35081184

Li, K.-E., Chang, Z.-Y., Shen, C.-X., and Yao, N. (2015). "Toxicity of nanomaterials to plants," in *Nanotechnology and Plant Sciences*, eds M. H. Siddiqui, M. H. Al-Whaibi, and F. Mohammad (Cham: Springer International Publishing), 101–124.

Lin, D., and Xing, B. (2007). Phytotoxicity of nanoparticles: inhibition of seed germination and root growth. *Environ. Pollut.* 150, 243–250. doi: 10.1016/j.envpol.2007.01.016

Liu, H. H., and Cohen, Y. (2014). Multimedia environmental distribution of engineered nanomaterials. *Environ. Sci. Technol.* 48, 3281–3292. doi: 10.1021/es405132z

López-Moreno, M. L., de La Rosa, G., Hernández-Viezcas, J. A., Peralta-Videa, J. R., and Gardea-Torresdey, J. L. (2010a). X-ray Absorption Spectroscopy (XAS) corroboration of the uptake and storage of CeO2 nanoparticles and assessment of their differential toxicity in four edible plant species. *J. Agric. Food Chem.* 58, 3689–3693. doi: 10.1021/jf904472e

López-Moreno, M. L., de la Rosa, G., Hernandez-Viezcas, J. A., Castillo-Michel, H., Botez, C. E., Peralta-Videa, J. R., et al. (2010b). Evidence of the differential biotransformation and genotoxicity of ZnO and CeO2 nanoparticles on soybean (Glycine max) plants. *Environ. Sci. Technol.* 44, 7315–7320. doi: 10.1021/es903891g

Lundin, A. (1984). *Extraction and Automatic Luminometric Assay of ATP, ADP and AMP.* New York: Academic Press.

Lv, J., Zhang, S., Luo, L., Zhang, J., Yang, K., and Christie, P. (2015). Accumulation, speciation and uptake pathway of ZnO nanoparticles in maize. *Environ. Sci. Nano* 2, 68–77. doi: 10.1039/c4en00064a

Ma, Y., Zhang, P., Zhang, Z., He, X., Zhang, J., Ding, Y., et al. (2015). Where does the transformation of precipitated ceria nanoparticles in hydroponic plants take place? *Environ. Sci. Technol.* 49, 10667–10674. doi: 10.1021/acs.est.5b02761

Maurer-Jones, M. A., Gunsolus, I. L., Murphy, C. J., and Haynes, C. L. (2013). Toxicity of engineered nanoparticles in the environment. *Anal. Chem.* 85, 3036–3049. doi: 10.1021/ac303636s

Miralles, P., Church, T. L., and Harris, A. T. (2012). Toxicity, Uptake, and Translocation of engineered nanomaterials in vascular plants. *Environ. Sci. Technol.* 46, 9224–9239. doi: 10.1021/es202995d

Moreno-Olivas, F., Gant, V. U. Jr., Johnson, K. L., Peralta-Videa, J. R., and Gardea-Torresdey, J. L. (2014). Random amplified polymorphic DNA reveals that TiO2 nanoparticles are genotoxic to Cucurbita pepo. *J. Zhejiang Univ. Sci. A* 15, 618–623. doi: 10.1631/jzus.A1400159

Mushtaq, Y. K. (2011). Effect of nanoscale Fe3O4, TiO2 and carbon particles on cucumber seed germination. *J. Environ. Sci. Health A* 46, 1732–1735. doi: 10.1080/10934529.2011.633403

Nascarella, M. A., and Calabrese, E. J. (2012). A method to evaluate hormesis in nanoparticle dose-response. *Dose Response* 10, 344–354. doi: 10.2203/dose-response.10-025

Navarro, E., Baun, A., Behra, R., Hartmann, N. B., Filser, J., Miao, A.-J., et al. (2008). Environmental behavior and ecotoxicity of engineered nanoparticles to algae, plants, and fungi. *Ecotoxicology* 17, 372–386. doi: 10.1007/s10646-008-0214-0

Nowack, B., Baalousha, M., Bornhöft, N., Chaudhry, O., Cornelis, G., Cotterill, J., et al. (2015). Progress towards the validation of modelled environmental concentrations of engineered nanomaterials by analytical measurements. *Environ. Sci. Nano* 2, 421–428. doi: 10.1039/c5en00100e

Nowack, B., Ranville, J. F., Diamond, S., Gallego-Urea, J. A., Metcalfe, C., Rose, J., et al. (2012). Potential scenarios for nanomaterial release and subsequent alternation in the environment. *Environ. Toxicol. Chem.* 31, 50–59. doi: 10.1002/etc.726

OECD (2010). *List of Manufactured Nanomaterials and List of Endpoints for Phase One of the Sponsorship Programme for the Testing of Manufactured Nanomaterials: Revision.* Paris: OECD Environment, Health and Safety Publications Series on the Safety of Manufactured Nanomaterials.

Parisi, C., Vigani, M., and Rodríguez-Cerezo, E. (2015). Agricultural nanotechnologies: what are the current possibilities? *Nano Today* 10, 124–127. doi: 10.1016/j.nantod.2014.09.009

Peakall, R., and Smouse, P. E. (2012). GenAlEx 6.5: genetic analysis in Excel. Population genetic software for teaching and research – an update. *Bioinformatics* 28, 2537–2539. doi: 10.1093/bioinformatics/bts460

Piccinno, F., Gottschalk, F., Seeger, S., and Nowack, B. (2012). Industrial production quantities and uses of ten engineered nanomaterials in Europe and the world. *J. Nanopart. Res.* 14, 1109. doi: 10.1007/s11051-012-1109-9

Priester, J. H., Ge, Y., Mielke, R. E., Horst, A. M., Moritz, S. C., Espinosa, K., et al. (2012). Soybean susceptibility to manufactured nanomaterials with evidence for food quality and soil fertility interruption. *Proc. Natl. Acad. Sci. U.S.A.* 109, E2451. doi: 10.1073/pnas.1205431109

Rico, C., Morales, M., Barrios, A., McCreary, R., Hong, J., Lee, W., et al. (2013). Effect of cerium oxide nanoparticles on the quality of rice (*Oryza sativa* L.) grains. *J. Agric. Food Chem.* 61, 11278–11285. doi: 10.1021/jf404046v

Rico, C. M., Majumdar, S., Duarte-Gardea, M., Peralta-Videa, J. R., and Gardea-Torresdey, J. L. (2011). Interaction of nanoparticles with edible plants and their possible implications in the food chain. *J. Agric. Food Chem.* 59, 3485–3498. doi: 10.1021/jf104517j

Rico, C. M., Peralta-Videa, J. R., and Gardea-Torresdey, J. L. (2015a). "Chemistry, biochemistry of nanoparticles, and their role in antioxidant defense system in plants," in *Nanotechnology and Plant Sciences*, eds M. H. Siddiqui, M. H. Al-Whaibi, and F. Mohammad (Cham: Springer International Publishing), 1–18.

Rico, C. M., Peralta-Videa, J. R., and Gardea-Torresdey, J. L. (2015b). Differential effects of cerium oxide nanoparticles on rice, wheat, and barley roots: a Fourier Transform Infrared (FT-IR) microspectroscopy study. *Appl. Spectrosc.* 69, 287–295. doi: 10.1366/14-07495

Roco, M. M. (2011). The long view of nanotechnology development: the National Nanotechnology Initiative at 10 years. *J. Nanopart. Res.* 13, 427–445. doi: 10.1007/s11051-010-0192-z

Schneider, C. A., Rasband, W. S., and Eliceiri, K. W. (2012). NIH Image to ImageJ: 25 years of image analysis. *Nat. Methods* 9, 671–675. doi: 10.1038/nmeth.2089

Schwabe, F., Schulin, R., Limbach, L. K., Stark, W., and Burge, D. (2013). Influence of two types of organic matter on interaction of CeO$_2$ nanoparticles with plants in hydroponic culture. *Chemosphere* 91, 512–520. doi: 10.1016/j.chemosphere.2012.12.025

Schwabe, F., Tanner, S., Schulin, R., Rotzetter, A. C., Stark, W. J., Von Quadt, A., et al. (2015). Dissolved cerium contributes to uptake of Ce in presence of differently sized CeO$_2$-nanoparticles by three crop plants. *Metallomics* 7, 466–477. doi: 10.1039/c4mt00343h

Shen, C. S., Zhang, Q.-F., Li, J., Bi, F.-C., and Yan, N. (2010). Induction of programmed celle death in *Arabidopsis* and rice by single-wall carbon nanotubes. *Am. J. Bot.* 97, 1602–1609. doi: 10.3732/ajb.1000073

Song, U., Jun, H., Waldman, B., Roh, J., Kim, Y., Yi, J., et al. (2013a). Functional analysis of nanoparticle toxicity: a comparative study of the effects of TiO2 and Ag on tomatoes (Lycopersicon esculentum). *Ecotoxicol. Environ. Safe.* 93, 60–67. doi: 10.1016/j.ecoenv.2013.03.033

Song, U., Shin, M., Lee, G., Roh, J., Kim, Y., and Lee, E. J. (2013b). Functional Analysis of TiO2 Nanoparticle toxicity in three plant species. *Biol. Trace Elem. Res.* 155, 93–103. doi: 10.1007/s12011-013-9765-x

USEPA (1995). *EPA Method 3052: Microwave Assisted Acid Digestion of Siliceous and Organically Based Matrices*, 3rd Edn. Washington, DC: Test Methods for Evaluating Solid Waste.

Van Hoecke, K., Quik, J. T., Mankiewicz-Boczek, J., De Schamphelaere, K. A., Elsaesser, A., Van der Meeren, P., et al. (2009). Fate and effects of CeO$_2$ nanoparticles in aquatic ecotoxicity tests. *Environ. Sci. Technol.* 15, 4537–4546. doi: 10.1021/es9002444

Van Nhan, L., Ma, C., Rui, Y., Liu, S., Li, X., Xing, B., et al. (2015). Phytotoxic mechanism of nanoparticles: destruction of chloroplasts and vascular bundles and alteration of nutrient absorption. *Sci. Rep.* 5, 11618. doi: 10.1038/srep11618

Vranová, E., Inzé, D., and Van Breusegem, F. (2002). Signal transduction during oxidative stress. *J. Exp. Bot.* 53, 1227–1236. doi: 10.1093/jexbot/53.372.1227

Wang, H., and Joseph, J. A. (1999). Quantifying cellular oxidative stress by dichlorofluorescein assay using microplate reader. *Free Radic. Biol. Med.* 27, 612–616. doi: 10.1016/S0891-5849(99)00107-0

Wang, Q., Ma, X., Zhang, W., Pei, H., and Chen, Y. (2012). The impact of cerium oxide nanoparticles on tomato (*Solanum lycopersicum* L.) and its implications for food safety. *Metallomics* 4, 1105–1112. doi: 10.1039/C2MT20149F

White, E. (1996). Life, death, and the pursuit of apoptosis. *Genes Dev.* 10, 1–15. doi: 10.1101/gad.10.1.1

Wu, S. G., Huang, L., Head, J., Chen, D. R., Kong, I. C., and Tang, Y. J. (2012). Phytotoxicity of metal oxide nanoparticles is related to both dissolved metals ions and adsorption of particles on seed surfaces. *J. Pet. Environ. Biotechnol.* 3, 126. doi: 10.4172/2157-7463.1000126

Yang, L., and Watts, D. J. (2005). Particle surface characteristics may play an important role in phytotoxicity of alumina nanoparticles. *Toxicol. Lett.* 158, 122–132. doi: 10.1016/j.toxlet.2005.03.003

Zhang, Z., He, X., Zhang, H., Ma, Y., Zhang, P., Ding, Y., et al. (2011). Uptake and distribution of ceria nanoparticles in cucumber plants. *Metallomics* 3, 816–822. doi: 10.1039/c1mt00049g

Zhao, L., Youping, S., Hernandez-Viezcas, J. A., Servin, A., Hong, J., Genhua, N., et al. (2013). Influence of CeO2 and ZnO nanoparticles on cucumber physiological markers and bioaccumulation of Ce and Zn: a life cycle study. *J. Agric. Food Chem.* 61, 11945–11951. doi: 10.1021/jf404328e

Zheng, L. F., Lu, S., and Liu, C. (2005). Effect of nano-TiO$_2$ on strength of naturally aged seeds and growth of spinach. *Biol. Trace Elem. Res.* 104, 83–91. doi: 10.1385/BTER:104:1:083

Conflict of Interest Statement: The authors declare that the research was conducted in the absence of any commercial or financial relationships that could be construed as a potential conflict of interest.

21

Seed germination strategies: an evolutionary trajectory independent of vegetative functional traits

Gemma L. Hoyle[1], Kathryn J. Steadman[2], Roger B. Good[3,4], Emma J. McIntosh[1], Lucy M. E. Galea[1] and Adrienne B. Nicotra[1]*

[1] Department of Evolution, Ecology and Genetics, Research School of Biology, Australian National University, Canberra, ACT, Australia, [2] School of Pharmacy and Queensland Alliance for Agriculture and Food Innovation, The University of Queensland, QLD, Australia, [3] Australian National Botanic Gardens, Canberra, ACT, Australia, [4] Fenner School of the Environment, Australian National University, Canberra, ACT, Australia

Edited by:
Julia Cooke,
The Open University, UK

Reviewed by:
Gerald Moser,
Justus-Liebig-University Giessen,
Germany
Zhenzhu Xu,
Institute of Botany Chinese Academy
of Sciences, China

*Correspondence:
Adrienne B. Nicotra,
Department of Evolution, Ecology and
Genetics, Research School of Biology,
Australian National University, Bldg.
116, Canberra, ACT 2601, Australia
adrienne.nicotra@anu.edu.au

1. Seed germination strategies vary dramatically among species but relatively little is known about how germination traits correlate with other elements of plant strategy systems. Understanding drivers of germination strategy is critical to our understanding of the evolutionary biology of plant reproduction.

2. We present a novel assessment of seed germination strategies focussing on Australian alpine species as a case study. We describe the distribution of germination strategies and ask whether these are correlated with, or form an independent axis to, other plant functional traits. Our approach to describing germination strategy mimicked realistic temperatures that seeds experience in situ following dispersal. Strategies were subsequently assigned using an objective clustering approach. We hypothesized that two main strategies would emerge, involving dormant or non-dormant seeds, and that while these strategies would be correlated with seed traits (e.g., mass or endospermy) they would be largely independent of vegetative traits when analysed in a phylogenetically structured manner.

3. Across all species, three germination strategies emerged. The majority of species postponed germination until after a period of cold, winter-like temperatures indicating physiological and/or morphological dormancy mechanisms. Other species exhibited immediate germination at temperatures representative of those at dispersal. Interestingly, seeds of an additional 13 species "staggered" germination over time. Germination strategies were generally conserved within families. Across a broad range of ecological traits only seed mass and endospermy showed any correlation with germination strategy when phylogenetic relatedness was accounted for; vegetative traits showed no significant correlations with germination strategy. The results indicate that germination traits correlate with other aspects of seed ecology but form an independent axis relative to vegetative traits.

Keywords: alpine plants, climate change, dormancy, endosperm, germination strategy, phylogenetic regression

Introduction

The timing of seed germination dictates a seedling's seasonal exposure to potentially lethal environmental factors, and thus has strong fitness consequences (Simons and Johnston, 2000; Donohue, 2005). However, for much of the world's flora the particular mechanisms that regulate seasonal emergence patterns are unknown. These mechanisms may include a combination of environmental germination requirements and seed dormancy. Given the importance of germination timing these traits are likely to evolve in correlated suites with other key functional traits. However, it is unclear whether germination strategy is correlated with other axes of plant strategy (e.g., seed mass or leaf mass per unit area), or indeed constitutes an additional independent axis. In the context of a rapidly changing climate, understanding the germination strategies of native species from threatened communities moves from being a question of evolutionary and ecological interest, to an urgent matter for conservation and management goals.

Seed mass declines with increasing latitude (Moles and Westoby, 2004) and has been shown to be correlated with a range of traits, including early seedling survival in low light, growth form, and dispersal syndrome (Leishman et al., 1995; Westoby et al., 1996; Moles et al., 2007). However, little is known about whether other reproductive traits, including germination strategies, correlate with other seed traits, with leaf or whole plant traits, or whether they might form another independent axis entirely.

Seed dormancy mechanisms are regarded as the principle means by which seeds can control the timing of germination and thus are expected to be under strong selective pressure. Dormant seeds sense and respond to their environment (Vleeshouwers et al., 1995) in order to avoid a germination response to temperature or rainfall that would not support subsequent seedling growth (Tielborger et al., 2012). Dormancy may result from physical, physiological or developmental/morphological mechanisms, or combinations thereof (Baskin and Baskin, 2001). In understanding dormancy, however, documenting the presence of a dormancy mechanism is just the first step: understanding the role of that dormancy mechanism in controlling the timing of germination is a crucial step. Studies of germination strategy frequently bypass or terminate dormancy through the application of chemical agents for logistical reasons (Cohn et al., 1989; Foley, 1992), but doing so reveals little about when it is alleviated naturally, *in situ*. In contrast, investigating germination strategies under ecologically relevant experimental conditions that mimic seasonal temperature regimes and seed moisture content can alleviate dormancy in a way that reveals much more about innate germination strategies (Baskin and Baskin, 2004; Albrecht and McCarthy, 2006; Hoyle et al., 2008a).

Climatically extreme environments, such as alpine and high montane regions, are characterized by spatially variable and temporally unpredictable conditions, particularly low temperatures and short growing seasons (Bliss, 1971; Körner, 2003). Conditions for seedling establishment may not be favorable for all species immediately after seeds are dispersed or even during the subsequent growing season, nor will species be equally equipped to cope with winter conditions during early seedling establishment. Therefore, seeds of species found in alpine environents are expected to vary in germination traits and strategies (Wagner and Simons, 2009). Seed dormancy was once considered relatively rare among alpine plant species (Amen, 1966), however, more recent studies indicate physiological dormancy may control the timing of germination in alpine systems more often than previously thought (Densmore and Zasada, 1983; Cavieres and Arroyo, 2000; Shimono and Kudo, 2005; Mondoni et al., 2009; Sommerville et al., 2013). Thus, the alpine flora provides an ideal context in which to assess the evolution of germination strategy.

When the ecology of plant species is described from the perspective of functional traits it becomes apparent that some traits form correlated suites that are robust whether considered at an absolute scale or in a phylogenetically controlled design. For example, the leaf economic spectrum (LES) describes a continuum of strategies ranging from slow growth and high cost leaves to rapid growth potential, short lived leaves and high photosynthetic rates (Reich et al., 1997; Wright et al., 2004). The LES can be viewed in an extended form as a whole plant strategy for water or carbon use (Reich, 2014). But in some cases traits sort more effectively into suites that form independent axes. For example, seed mass, mature plant height and leaf mass per unit area show little intercorrelation (Westoby, 1998). It remains to be determined how other aspects of seed ecology, particularly germination strategy, correlate with functional traits such as seed mass and vegetative characteristics.

The present study investigated germination strategies and ecological correlates thereof in 54 Australian species from 16 families, including 10 endemics and include a range of species from grassland specialists, to widely distributed species also found in bogs, fens and shrublands. We hypothesized that species would exhibit one of two possible germination strategies: pre-winter germination of non-dormant seeds or postponed germination via dormancy mechanisms and that these would show conservation among families. Further, we asked whether germination strategy would be correlated with other ecological traits such that species that germinated immediately would also show traits of opportunistic growth (e.g., higher leaf mass per unit area, shorter mature heights and/or smaller seed mass), or whether germination strategy would comprise an independent axis to these functional traits.

Materials and Methods

Study Site

The Australian Alps are located in southeast Australia and cover approximately 25000 km^2. Seed collections were made at altitudes ranging from 1605 to 2212 m a.s.l. in the New South Wales portion of Kosciuszko National Park and including high elevation frost hollows as well as true alpine sites above treeline. Seeds were collected from herbfields and grasslands incorporating a range of bog and fen habitats and a mix of specialist and generalist species distributed along moisture gradients. More than half of Kosciuszko National Park's annual rainfall (1800–3100 mm) falls as winter snow and persists for at least 4 months. Data collected by the Bureau of Meteorology at Charlotte Pass (Kosciuszko Chalet; 36.43°S, 148.33°E, 1755 m a.s.l.) in Kosciuszko National Park between 1968 and 2014 indicate that air temperatures are commonly below zero during winter months and average between 15 and 20°C during summer (Bureau of Meteorology, 2010, http://www.bom.gov.au. **Figure 1**).

With a view to uncovering temperature conditions close to those that seeds experience post-dispersal, daily maximum and minimum soil temperature data were collected within the study site (Seaman's Hut; 36.27°S and 148.17°E, 2030 m a.s.l.) using ibutton data loggers ($n = 10$, Embedded Data Systems, USA) placed 4 cm below the soil surface at the base of vegetation from 17 January to 17 December 2012. We placed loggers at 4 cm to avoid surface disturbance but still be representative of conditions to which seeds are exposed in soil; this depth is intermediate to what has been used in prior alpine soil temperature monitoring exercises (Scherrer and Körner, 2011; Pauli et al., 2015). As suspected, average soil temperature did not drop below freezing during winter (**Figure 1**). In summer, soil temperature under vegetation is known to track ambient temperature, whereas temperatures of bare soil will exceed ambient by up to 15°C (Soil Conservation Service unpublished records, 1960s–1970s).

Seed Collecting and Germination

Mature seeds of 54 species from 16 families and 37 genera were collected between January and April 2009, 2010, and 2011 (see **Table 1** for full names and authorities. Vouchers were lodged at the Australian National Herbarium, Canberra). In total the species represented more than a quarter of the Australian angiosperm flora found in alpine regions (Costin et al., 2000, see **Table 1**), though many of these species extend to below treeline as well. The viability of all collections was estimated prior to sowing in experimental germination conditions using the tetrazolium chloride (TZ) staining technique (International Seed Testing Association, 2003). For more details on collection and processing see Appendix A in Supplementary Material.

Phenology of seed germination was investigated by mimicking temperature regimes that alpine seeds experience *in situ*, post-dispersal, in an artificially shortened progression of seasons. Seeds were imbibed throughout the experiment thus results indicate potential for germination when water is not limiting. Air and soil temperature data (**Figure 1**) were used to guide and validate incubator temperature regimes, though the logistical contraints of working with incubators precluded incorporation of the fluctuations inherent to natural conditions. Germination tests used eight replicates of 25 seeds per collection, sown into 9 cm diameter plastic Petri dishes containing 1% plain water-agar. Petri dishes were sealed using Parafilm to avoid agar desiccation, before being placed in germination incubators (Thermoline Scientific, Melbourne, NSW, Australia). Half of the

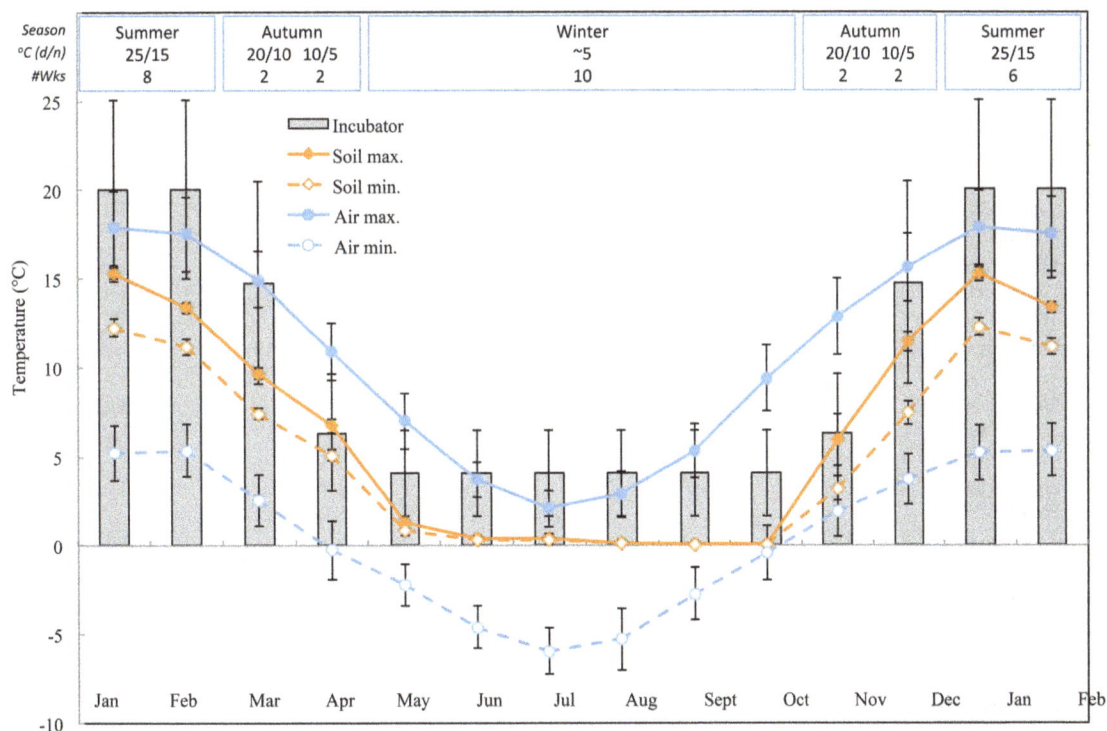

FIGURE 1 | Monthly maximum and minimum air and soil temperature data in the Australian Alps (mean ± STDEV). Air temperatures are averages from the BOM database, soil temperatures were measured with iButtons (see methods). Bars indicate experimental incubator temperatures (mean ± STDEV). The equivalent time of year (season), incubator temperature regime (°C, 12/12 h, light/dark) and duration of treatment (weeks) as indicated in the top panel.

TABLE 1 | Study species, family and authority, growth form, and collections details, percentage seed viability (TZ-estimated, ± s.e.), final percentage germination achieved (± s.e.), statistical difference between viability and germination (G < V, tested only in cases where germination was less than viability).

Family	Species	Growth form	Date collected	Elev. (m a.s.l.)	Viability (%, SE)		Germination (%, SE)		G < V	Cluster	Accession no.
Apiaceae	*Aciphylla glacialis* (F.Muell.) Benth.	Herb	11-Mar-09	2164	100	0	95	3.8	ns	Postponed	CANB 782521
	Aciphylla glacialis (F.Muell.) Benth.	Herb	17-Feb-10	1937	85	4.8	91	2.1		Postponed	CANB 792173
	Aciphylla simplicifolia (F.Muell.) Benth.	Herb	12-Mar-09	1698	100	0	81	5.9	*	Postponed	CANB 782524
	Aciphylla simplicifolia (F.Muell.) Benth.	Herb	16-Feb-10	1743	85	15	80	3	ns	Postponed	CANB 792163
	Diplaspis nivis Van den Borre and Henwood	Herb	28-Jan-10	1960	64	16	88	3.1		Postponed	CANB 786949
	Gingidia algens (F.Muell.) J. W. Dawson[†]	Herb	16-Feb-10	1740	97	3.3	94	1.2	ns	Postponed	CANB 792164
	Oreomyrrhis ciliata Hook.f.	Herb	24-Mar-09	1748	92	8.3	94	2.6		Staggered (A2)	CANB 783452
	Oreomyrrhis ciliata Hook.f.	Herb	27-Jan-10	1748	100	0	100	0		Staggered (A3)	CANB 792149
	Oreomyrrhis eriopoda (DC.) Hook.f.	Herb	24-Feb-09	1743	97	2.8	93	3.1	ns	Staggered (A2)	CANB 783437
	Oreomyrrhis eriopoda (DC.) Hook.f.	Herb	3-Feb-10	1744	83	12	54	9	ns	Postponed	CANB 792151
	Oreomyrrhis pulvinifica F.Muell.	Herb	4-Feb-10	1733	77	8.8	88	2.9		Staggered (A3)	CANB 786952
	Schizeilema fragoseum (F.Muell.) Domin	Herb	31-Mar-11	2092	67	6.7	1	1	***	too low to score	CANB 798462
Asteraceae	*Brachyscome obovata* G. L. R. Davis	Herb	2-Mar-11	2090	77	5.5	65	3.1	ns	Immediate	CANB 797826
	Brachyscome stolonifera G. L. Davis[†]	Herb	5-Feb-10	1941	70	15	22	6.4	*	Staggered (B2)	CANB 792161
	Celmisia costiniana M. Gray and Given	Herb	2-Mar-11	2046	61	3.2	90	3.5		Staggered (B2)	CANB 798295
	Craspedia jamesii J. Everett and Joy Thomps.	Herb	1-Feb-11	1730	70	10	78	3.6		Immediate	CANB 797734
	Craspedia lamicola J. Everett and Joy Thomps.	Herb	3-Mar-11	2067	93	3.3	93	2.1	ns	Immediate	CANB 798305
	Craspedia leucantha F.Muell.[†]	Herb	25-Feb-09	1996	93	3.7	96	1.6		Immediate	CANB 783441
	Erigeron bellidioides (Hook.f.) S. J. Forbes and D. I. Morris	Herb	29-Jan-10	1730	83	6.7	95	1.9		Immediate	CANB 792150
	Erigeron nitidus S. J. Forbes	Herb	17-Feb-10	2007	70	5.8	93	4.3		Immediate	CANB 792172
	Erigeron setosus (Benth.) M.Gray[†]	Herb	4-Feb-10	1928	57	1.5	100	0		Immediate	CANB 792158
	Leucochrysum alpinum (F.Muell.) R. J. Dennis and N. G. Walsh	Mat/ Herb	11-Mar-09	2212	77	3.3	100	0		Immediate	CANB 782523
	Ozothamnus cupressoides Puttock and D. J. Ohlsen	Shrub	9-Apr-09	1605	82	19	21	13.8	*	Staggered (A3)	CANB 782558
Campanulaceae	*Wahlenbergia ceracea* Lothian	Herb	3-Feb-10	1744	87	7.2	78	10.6	ns	Staggered (A2)	CANB 792185

(Continued)

TABLE 1 | Continued

Family	Species	Growth form	Date collected	Elev. (m a.s.l.)	Viability (%, SE)		Germination (%, SE)		G<V	Cluster	Accession no.
Caryophyllaceae	*Colobanthus affinis* (Hook.) Hook.f.	Herb	25-Mar-09	2069	90	5.8	95	3.9		Immediate	CANB 782529
	Colobanthus affinis (Hook.) Hook.f.	Herb	5-Jan-10	1736	93	3.3	100	0		Immediate	CANB 786943
	Scleranthus biflorus (J. R. Forst. and G. Forst.) Hook.f.	Mat/ Cushion	10-Mar-09	1751	77	6.5	82	3.6		Immediate	CANB 782512
Cyperaceae	*Carex cephalotes* F.Muell.	Sedge	4-Jan-09	1613	77	5.5	78	5.5		Postponed	CANB 771067
	Carex cephalotes F.Muell.	Sedge	5-Jan-10	1640	96	3.7	93	2.3	ns	Postponed	CANB 771067
	Carex echinata Murray	Sedge	21-Jan-09	1684	83	11	98	1.5		Postponed	CANB 783404
	Carex echinata Murray	Sedge	3-Feb-10	1742	83	3.3	98	2.1		Postponed	CANB 792155
	Carpha nivicola F.Muell.	Sedge	3-Feb-10	1730	86	8.8	84	5.9	ns	Postponed	CANB 792153
	Oreobolus pumilio R.Br.	Sedge	7-Apr-09	1945	80	5.5	81	8.9		Postponed	CANB 782540
	Uncinia flaccida S. T. Blake^	Sedge	3-Mar-11	2047	97	3.3	13	15.7	***	Postponed	CANB 798306
Droseraceae	*Drosera arcturi* Hook.	Herb	27-Jan-10	1752	97	3.3	98	1.2		Postponed	CANB 786948
Ericaceae	*Epacris petrophila* Hook.f.	(Sub) shrub	26-Mar-09	1745	71	5.8	93	3.2		Postponed	CANB 782533
	Pentachondra pumila (J. R. Forst. and G. Forst.) R.Br.^	Mat/Sub -shrub	17-Feb-10	2096	85	8.3	27	4.3	***	Postponed	CANB 792168
	Richea continentis B. L. Burtt	(Sub) shrub	8-Apr-09	2078	83	3.8	95	1.1		Postponed	CANB 782545
Gentiananaceae	*Gentianella muelleriana* subsp. *alpestris* (L. G. Adams) Glenny[†]	Herb	24-Mar-09	1748	62	9.6	74	3.8		Postponed	CANB 783451
	Gentianella muelleriana subsp. *alpestris* (L. G. Adams) Glenny[†]	Herb	30-Mar-11	2109	83	3.8	1	1	***	too low to score	CANB 798460
Juncaceae	*Juncus falcatus* E.Mey.	Rush	10-Mar-09	1742	64	4.7	54	15.8	ns	Postponed	CANB 782513
	Luzula acutifolia subsp. *nana* Edgar[†]	Rush	25-Feb-09	1932	73	3.3	68	4.9	ns	Immediate	CANB 783440
	Luzula acutifolia subsp. *nana* Edgar[†]	Rush	16-Feb-11	1957	93	3.3	93	1	ns	Immediate	CANB 797746
Liliaceae	*Astelia alpina* var. *novae-hollandiae* Skottsb.^	Herb	4-Feb-10	1880	77	3.3	76	12.9	ns	Postponed	CANB 792157
	Astelia psychrocharis F.Muell.[†]	Herb	4-Feb-10	1830	63	8.8	80	3.5		Staggered (A2)	CANB 792156
	Herpolirion novae-zelandiae Hook.f.	Herb	3-Mar-11	1758	53	3.3	0	0	***	too low to score	CANB 797834
Poaceae	*Agrostis muelleriana* Vickery	Grass	2-Mar-11	2041	83	3.3	79	2.9	ns	Staggered (B2)	CANB 798298
	Deschampsia cespitosa (L.) P.Beauv.	Grass	24-Feb-09	1743	83	8.8	96	2.4		Staggered (A2)	CANB 783436
	Poa costiniana Vickery	Grass	30-Mar-11	2109	73	6.7	75	4		Immediate	CANB 798459
	Poa hiemata Vickery	Grass	2-Mar-11	2042	83	8.8	65	4.3	ns	Immediate	CANB 798300
	Rytidosperma alpicola (Vickery) Connor and Edgar	Grass	5-Feb-09	1747	77	3.5	78	5.6		Immediate	CANB 783601
	Rytidosperma alpicola (Vickery) Connor and Edgar	Grass	2-Mar-11	2041	100	0	88	5.4	ns	Immediate	CANB 798297

(Continued)

TABLE 1 | Continued

Family	Species	Growth form	Date collected	Elev. (m a.s.l.)	Viability (%, SE)		Germination (%, SE)		G<V	Cluster	Accession no.
	Rytidosperma nudiflorum (P.Morris) Connor and Edgar	Grass	15-Feb-11	1925	97	3.3	96	1.6	ns	Immediate	CANB 797742
Plantaginaceae	*Plantago glacialis* B. G. Briggs, Carolin and Pulley	Herb	16-Feb-11	1953	93	3.7	82	3.4	ns	Immediate	CANB 797747
Ranunculaceae	*Psychrophila introloba* (F.Muell.) W. A. Weber	Herb	5-Feb-10	1941	96	3.7	98	1.2		Postponed	CANB 792160
	Ranunculus acrophilus B. G. Briggs[†]	Herb	6-Jan-10	2039	57	27	76	4.7		Immediate	CANB 786947
	Ranunculus clivicola B. G. Briggs	Herb	29-Jan-10	1742	75	6.9	95	2		Staggered (B2)	CANB 792145
	Ranunculus dissectifolius F.Muell. ex Benth.[†]	Herb	24-Mar-09	1747	68	17	84	3.4		Postponed	CANB 783454
	Ranunculus dissectifolius F.Muell. ex Benth.[†]	Herb	27-Jan-10	1752	48	15	86	6		Postponed	CANB 792148
	Ranunculus graniticola Melville	Herb	6-Jan-10	1734	81	9.2	89	4.3		Staggered (A2)	CANB 786945
	Ranunculus gunnianus Hook.	Herb	5-Jan-10	1943	97	3.3	81	1.2	**	Postponed	CANB 786944
	Ranunculus muelleri Benth.	Herb	28-Jan-10	1955	74	16	96	2.4		Staggered (A2)	CANB 786950
Scrophulariaceae	*Euphrasia alsa* F.Muell.[†]	Herb	16-Feb-11	2047	93	6.7	2.3	1.3	***	too low to score	CANB 797754
Stackhousiaceae	*Stackhousia pulvinaris* F.Muell.	Mat	8-Apr-09	2085	67	8.3	4	1.6	***	too low to score	CANB 782543
Thymalaceae	*Pimelea axiflora* subsp. *alpina* (Benth.) Threlfall	(Sub) shrub	4-Jan-09	1616	47	11	23	4.8	ns	Postponed	CANB 771063

*Cluster refers to germination strategy, see **Figure 2**. Herbarium accession number in final column.*
[†]*Indicates endemic species (Costin et al., 2000).*
[^]*Indicates species that were cycled through the experiment for a second "year."*
*P < 0.05, 0.01, and 0.001 shown as *, **, and *** respectively.*

dishes were wrapped in aluminum foil to exclude light and provide an indication of light requirements for germination. Each of the four replicates per light treatment per collection was placed on a different shelf within the same incubator and location within that shelf was re-arranged weekly. All replicates of each collection were moved through the following simulated thermoperiods: 8 weeks at 25/15°C (day/night temperature) → 2 weeks at 20/10°C → 2 weeks at 10/5°C → 10 weeks at constant 4–5°C → 2 weeks at 10/5°C → 2 weeks at 20/10°C → 6 weeks at 25/15°C. Thus, one cycle of the entire experiment mimicked temperatures reminiscent of summer, early autumn, late autumn, winter, early spring, late spring, and summer, and lasted 32 weeks in total. The abbreviated year was considered reasonable because accessions generally had ceased further germination in a given treatment prior to being shifted to the next "season". A 12/12 h light/dark photoperiod was provided throughout by fluorescent tubes (> 90 µmol m^{-2} s^{-1} at the center of each shelf), and in spring, summer and autumn temperature regimes light coincided with the warm temperature period of the day. Germination, defined as radicle emergence by more than 1 mm,

was scored every 7 days, and germinated seeds and seedlings were subsequently removed from the dishes. Germination of foil-wrapped plates was assessed as above but under very low green light.

Following one experimental cycle (32 weeks) most collections were terminated (see supplemental info for exceptions), having exhibited zero germination for at least 2 weeks at this time. Following termination, all remaining intact seeds were dissected with a scalpel under a microscope. Seeds with a firm, fresh endosperm and embryo were deemed viable and seeds empty of an embryo were deducted from the total when calculating percentage germination for each collection. We calculated an index of the light requirements for germination for each species as a fraction of germination in the dark treatment at the time that cumulative germination percentiles of 50 and 95% were reached in the alternating light/dark treatment (see below).

Ecological Correlates of Germination Strategy

We compiled a database of species' functional traits and geographic measures from published data to assess correlations with germination strategy. The database included the following continuous traits: average seed mass, light requirement for germination (see above), specific leaf area (fresh area/dry mass), individual leaf area, individual leaf dry weight, and height at maturity. The minimum, average and maximum collection elevations for the species were determined from all herbarium records in the Australian Virtual Herbarium, and the elevation at which the specific seed collections used in the study was made was also included as continuous traits. Growth form was scored as one of three categories: graminoid, herb, or shrub. We also included a binary trait: whether the seeds were endospermic or not.

The trait data were collected from a variety of sources including the Australian Virtual Herbarium (http://avh.chah.org.au/), the Australian National Herbarium specimen information register (https://www.anbg.gov.au/cgi-bin/anhsir), *Kosciuszko Alpine Flora* (Costin et al., 2000), the *New South Wales Flora* (Harden, 1990), and *The comparative internal morphology of seeds* (Martin, 1946). Seed mass and specific leaf area were obtained from individual field collections used in the germination trials. Seed mass was obtained by weighing four lots of 10 seeds. SLA was determined based on five fully expanded leaves from separate individuals per species. These were collected in the field, scanned on a flatbed scanner, dried at 60°C to constant weight and weighted to the nearest 0.000 g. Where insufficient seed was available, or specific leaf area was not available, data were obtained from other ANBG collections of the same species or from other data sources (S. Venn, *unpublished data*). Whether or not a seed was endospermic was determined from drawings of the seed in the ANGB seed bank records in consultation with ANBG seed bank staff. The full database and sources is available on request from the authors.

Data Analysis

Germination patterns were assessed visually and using a cluster analysis in PC-ORD for Windows, version 6 (MJM Software Design, Gleneden Beach, OR, USA). The analysis used a Euclidean distance matrix and a nearest neighbor joining algorithm. Data points for the cluster analysis were the time (weeks) to specific cumulative germination percentiles (t20, t40, t50, t60, t80, and t95%) for each collection (averaged across the 4 dishes). Thus, a collection that germinated early in the experiment would have short times to percentiles up to t80% for example, whereas a collection that postponed germination until temperatures increased following the simulated winter would have long times to these percentiles. More complex germination patterns, such as slow, steady germination or bursts of germination, could also be detected, and would be indicated by evenly spaced times or times clustered at the start and end of the experiment, respectively. This method was used as a way to impose an objective assessment of germination pattern, independent of temperature *per se*.

A one-way analysis of variance (ANOVA) was carried out for the replicate plates of each collection to assess whether there was a significant difference between percentage viability (as determined by the TZ test), and final percentage germination. Further general linear models were used to assess whether the clusters resulting

from the cluster analysis differed in viability or final germination percentage.

Associations between ecological traits and germination strategy clusters were tested in two ways. Firstly, we tested for mean differences between clusters using ANOVA with cluster as a fixed effect and ecological correlates as the response variables, or using replicated G-tests for endospermy and growth form. Secondly, because germination and other functional traits are highly likely to be influenced by evolutionary history, we assessed these correlations using phylogenetic regressions. A phylogeny was constructed for the study species using Phylocom Phylomatic and the bladj packing in R (Webb et al., 2008; R Core Team, 2013). We assessed the presence of phylogenetic structure in variables using the *phylosig* or *phylo.d* (discrete or binary variables) function of the *Phytools* package in R (Revell, 2012). Phylogenetic regressions were conducted using the *pgls* function in *Caper* (Orme et al., 2013). Analyses were conducted in a structured way reflecting the clustering revealed in PC-Ord (see below). Initially we included four germination strategies (postponed, immediate, and staggered categories within each of the preceding), then three (lumping all staggered species and comparing to postponed and intermediate) and finally we assessed differences between the two broadest strategies, postponed and intermediate.

Results

Seed Viability Compared to Final Germination

Overall, seed viability and final germination percentage were high. Mean TZ-estimated viability across all collections was $80 \pm 1.7\%$, with more than two thirds of collections exhibiting more than 75% viability (**Table 1**). There was no difference in the quality of collections across the 3 years (mean viability in 2009, 2010, and 2011 was 79 ± 2.8, 80 ± 2.7, and $81 \pm 3.7\%$, respectively), indicating that the banking of seeds collected in 2009 did not lessen their viability.

Germination that exceeded or was not significantly lower than TZ-estimated viability was achieved for 45 of the 54 species (One-Way ANOVA: $P > 0.05$, **Table 1**). Of the remaining nine species, final germination of four species was very low (< 5%), despite good viability (*ca.* 55–94%), and thus we were unable to draw any conclusions about the germination strategies of these species. Failure to germinate could reflect either that our treatments did not alleviate dormancy or that the seeds were not fully mature at collection. Cut tests at the end of the experiment indicated that, in all cases, the majority of un-germinated seeds contained a healthy, imbibed embryo. Interestingly, 34 species achieved germination that was greater than the TZ-estimated viability (**Table 1**), suggesting that the TZ test commonly underestimated seed viability.

Germination Clusters

Cluster analysis revealed two major clusters (A and B), with sub-structure in each (**Figure 2**). For the majority of species ($n = 21$ species from 10 families), germination was postponed until during a period of winter temperatures (5°C day/night) or until temperatures were raised to 10/5°C following winter

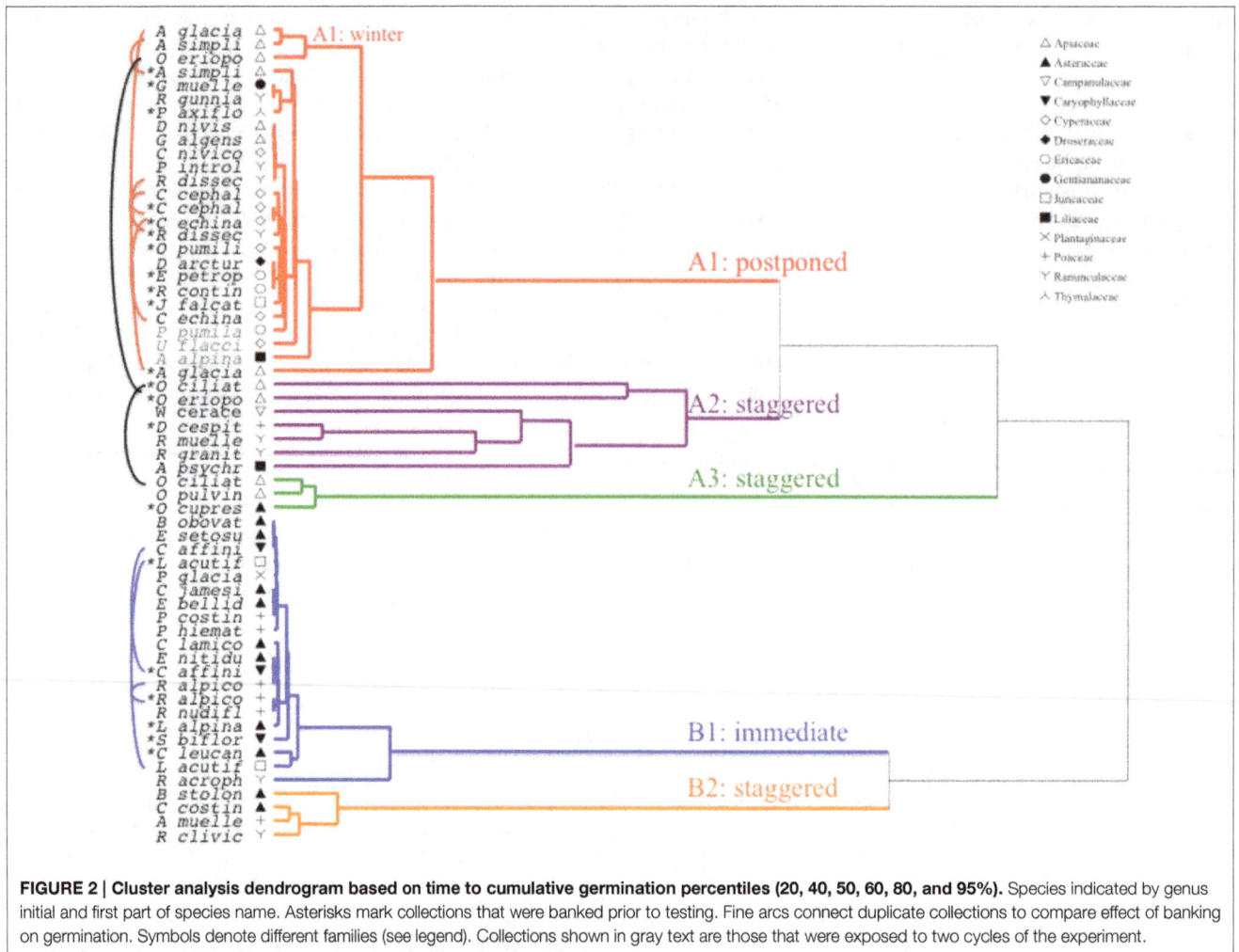

FIGURE 2 | Cluster analysis dendrogram based on time to cumulative germination percentiles (20, 40, 50, 60, 80, and 95%). Species indicated by genus initial and first part of species name. Asterisks mark collections that were banked prior to testing. Fine arcs connect duplicate collections to compare effect of banking on germination. Symbols denote different families (see legend). Collections shown in gray text are those that were exposed to two cycles of the experiment.

(A1 and A1:winter, **Figures 2**, **3A**). Alternatively, germination began immediately at 25/15°C ($n = 17$ species, six families, seen in cluster B1, **Figures 2**, **3B**). For the majority of these species, germination rate was fast; 50% of the total germination (t_{50}) occurred in less than 3 weeks (Supplementary Figure 1).

Within each of these broad classes, however, there were clear sub-clusters reflecting germination that was staggered over time i.e., a proportion of the seed lot germinated before exposure to winter temperatures while the remainder germinated after winter. Staggered germination occurred in 13 species from six families and occurred primarily when daytime temperatures exceeded 10°C (**Figures 3A,B**). These staggered germinators were broken into three groups in the cluster analysis. Cluster A2 with seven collections showed substantial germination both before and after the cold temperature period. Where germination ceased prior to the temperatures being changed we concluded that the lack of further pre-winter germination was not driven by temperature (A3). In contrast, germination of much of the A2 group appeared to be halted only when temperatures were reduced to 10/5°C. Cluster B2 contained a further four species which exhibited substantial germination early in the experiment and went on to achieve a relatively small proportion of their

total germination after temperatures were returned to above 5°C (**Figure 3B**).

Compared to the relatively minor variation among species within clusters A1 and B1, the staggered clusters also show much greater differentiation among species. Individual curves for all species are shown in Supplementary Figures 1i–vii. For more specific detail on germination strategy see Appendix B in Supplementary Material.

Ecological Correlates of Germination Strategy

Of all the ecological correlates that we considered, only those directly associated with seed traits were significantly correlated with germination strategy regardless of whether we assessed the correlation at the level of two major clusters (postponed vs. immediate germination), or at the level of three or four clusters, or whether we accounted for phylogenetic structure or not (**Table 2**). Species with postponed germination had heavier seeds and were more likely to be endospermic compared to those of species that germinated immediately (**Figure 4**). Species with staggered germination were intermediate in seed mass and endospermy. However, not all seed characteristics were correlated with germination strategy: there was no correlation

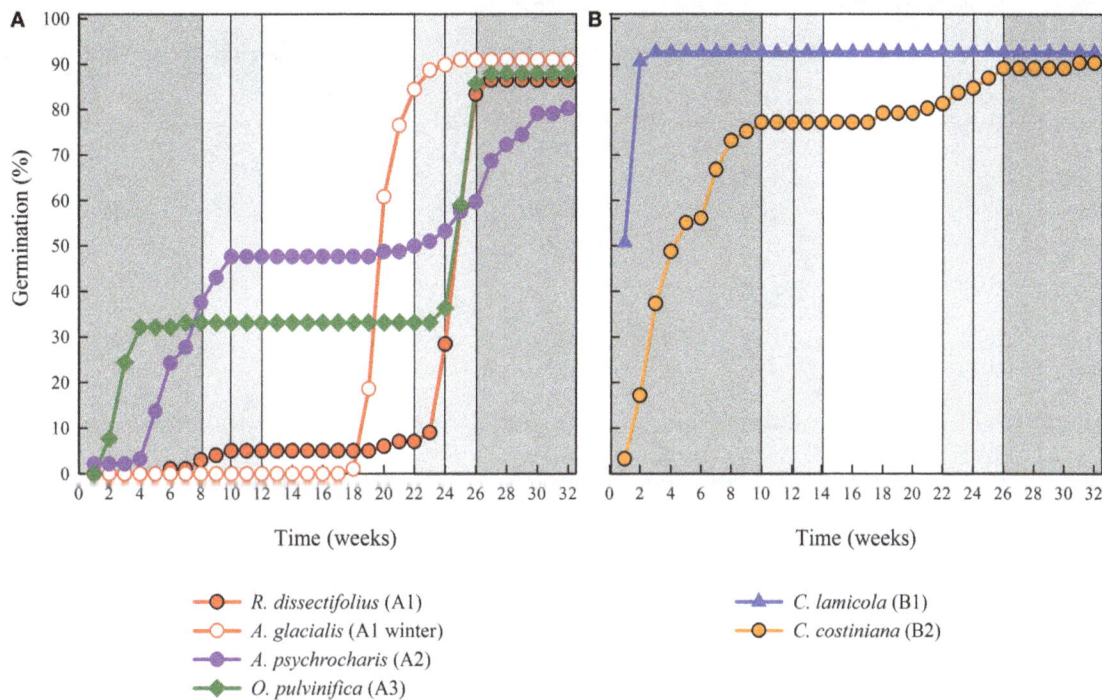

FIGURE 3 | Cumulative percentage germination of one representative species from each germination cluster. (A) cluster A, species that postponed germination, **(B)** cluster B, species that germinated immediately. Note that staggered germination patterns are nested within the above clusters. Incubator temperature regimes are represented by shading (see **Figure 1**).

between light requirements for seed germination and strategy. Contrary to our expectations, species with higher specific leaf area (potentially indicative of higher growth rates) were not more likely to exhibit immediate germination. Likewise, species with higher average elevation or smaller elevation ranges were not more likely to exhibit postponed germination.

Discussion

Germination in alpine habitats was historically deemed to be environmentally controlled, with winter snow insulating seeds against potential germination cues and rendering dormancy unnecessary (Billings and Mooney, 1968). However, our results support more recent evidence of dormancy mechanisms being common among alpine species both in Australia (Hoyle et al., 2013; Sommerville et al., 2013) and elsewhere around the world (Cavieres and Arroyo, 2000; Schutz, 2002; Shimono and Kudo, 2005; Mondoni et al., 2009; Schwienbacher et al., 2011). Dormancy would appear to play a significant role in controlling the *in situ* timing of germination of many Australian alpine species, acting to delay, postpone, or slow the rate of germination and thus potentially conferring risk-averse regeneration strategies in a harsh and variable environment. Perhaps most striking of our germination results was the proportion of species for which germination strategy varied within a seed collection for a given species suggesting that both dormant and non-dormant characteristics were exhibited within the same seed collection of these species. Notably,

germination strategies were generally highly conserved within families and also correlated with elements of seed anatomy: mass and endospermy. Germination strategies were, however, independent of other ecologically significant functional traits of the mature plant.

Variation in Germination Strategy

Seeds of nearly half of the species studied appeared unable to germinate until they were exposed to a cool, wet period (constant $5°C$), suggesting that cold stratification alleviated a physiological dormancy mechanism. Postponing germination until the following spring may enable seedlings to avoid establishing over or before the harsh winter, while also optimizing the short forthcoming growing season. In contrast the opportunistic germination exhibited by the species that germinated immediately may provide a selective advantage when the risk of winter seedling mortality is low, by enabling plants to flower earlier the following spring or at a larger size (Donohue, 2002).

To date, there has been little published evidence of intra-specific variability in dormancy mechanisms within alpine seed collections such as would explain the staggered germination strategies observed here. Germination staggered over time may also be explained by varying levels of seed maturity among individuals in the population at the time of collection, and/or could reflect a dimorphic strategy within a single plant associated with position on the plant or timing of development.

Evolutionary bet-hedging is often evoked to explain the diversified strategy in seed germination characteristics (Simons,

TABLE 2 | Incidence of phylogenetic structure in, and correlates of, germination strategy.

Trait	Lambda or D value	Phylogenetic structure	PGLS phyolenetic correlation with germination		
			4 Clusters	3 Clusters	2 Clusters
Germination strategy[a]	1.532	0.000	na	na	na
Seed mass	1.527	0.000	**0.1098**	**0.008036**	**0.1104**
Endospermic/Non[b]	−9.030	0.000	0.003366	**0.01715**	0.004963
Height	1.070	0.022	0.1276	0.4391	0.01948
SLA	< 0.0001	1.000	0.7886	0.2082	0.1998
Surface area	< 0.0001	1.000	0.9141	0.4042	0.7755
Leaf dry weight	1.783	0.020	0.6367	0.2903	0.6017
Average elevation	< 0.0001	1.000	0.6521	0.5852	0.5954
Range	1.227	0.013	0.5775	0.8864	0.63
Light requirement 50%	1.532	0.000	0.1738	0.5999	0.2943
Light requirement 95%	1.007	0.180	0.8247	0.5494	0.9611
Growth form (woody/non)	−0.690	1.000	0.1135	0.8466	0.1499

Bold indicates significance in non-phylogenetic analyses, yellow cells indicate phylogenetic structure or significance in PGLS regressions.
[a]Germination strategy shows phylogenetic structure at all levels of clustering.
[b]Phylogenetic structure for endospermy and woodiness assessed with Phylo-D for binary value. Probability values for D reflect probablity of pattern resulting from random (no phylogenetic) structure.

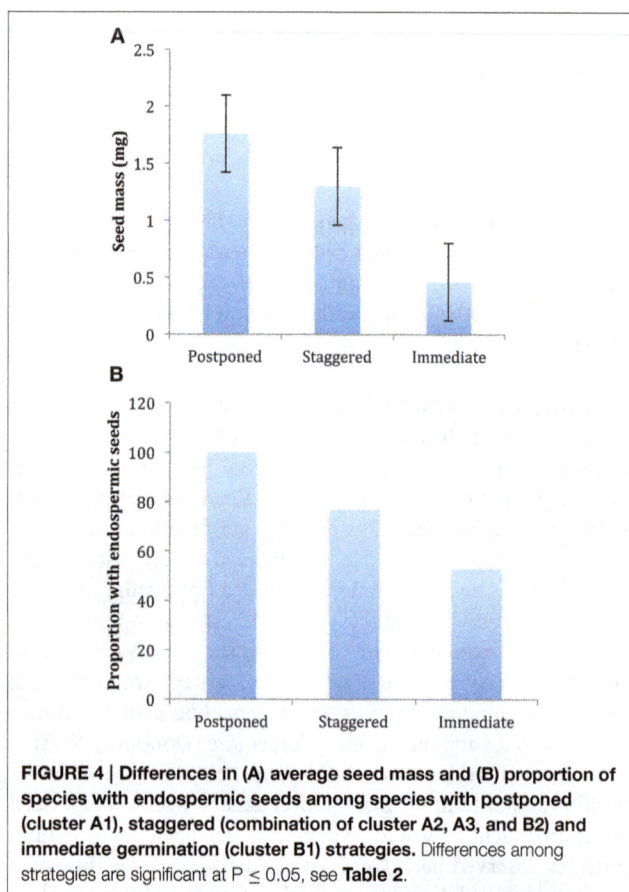

FIGURE 4 | Differences in (A) average seed mass and (B) proportion of species with endospermic seeds among species with postponed (cluster A1), staggered (combination of cluster A2, A3, and B2) and immediate germination (cluster B1) strategies. Differences among strategies are significant at P ≤ 0.05, see **Table 2**.

(e.g., by reducing competition among siblings), then it does not constitute bet-hedging. Likewise, if the apparent diversification reflects differences among, not within, individuals then it is not bet-hedging (Starrfelt and Kokko, 2012). Investigations into alpine seedling dispersal patterns and tolerance of frost and snow melt, paired with studies of determinants of germination strategy variation within species, may help explain germination phenology of the species exhibiting a staggered strategy and may indicate whether they are likely to represent bet-hedging strategies.

Ecological Correlates of Germination Strategy

Our examination of the relationship between germination strategy and other ecologically important traits demonstrated that germination traits were correlated with other seed traits but not with vegetative traits. Sommerville et al. (2013) found that Australian alpine herbs with non-endospermic seeds were more likely to be non-dormant at dispersal. Our work supports that finding and further indicates that smaller seeds are more likely to be among the immediate germinators. Although Sommerville et al. (2013) did not posit why endospermy may be associated with dormancy, we suggest that placement of reserves in cotyledons may improve early growth rate and establishment in immediate germinating species. Our results did not indicate any correlation between light requirement and dormancy, although light requirements did vary among species. There are a variety of other seed traits that buffer extinction risk in variable environments, for example seed dispersal mechanisms and seed longevity in the soil seed bank (see Tielborger et al., 2012 for a review). At this stage we cannot say whether those traits would show correlations with germination strategy or not.

We further asked whether germination strategy would be correlated with ecologically important non-reproductive traits, such as specific leaf area, or plant height (Westoby, 1998;

2011), such as that demonstrated by the collections with staggered germination. If, however, a diversified strategy results in both increased average fitness and a reduction in variance

Wright et al., 2004). In particular we posited that the immediate germinating species might show trait values indicative of faster growth rates (using SLA as a proxy) as might be necessary for establishment in the short period between dispersal and the onset of winter. However, we found no such correlations. Whether physiological and growth traits of seedlings at early establishment stages might show different patterns than do adult vegetative traits is a question worthy of consideration, but beyond the scope of the present work.

Finally, we examined whether species from higher elevations where growing seasons are generally shorter might be more likely to exhibit dormancy but there was no association between either elevation range, collection elevations, or the average elevation of collection for the species as a whole and germination strategy. Together these results indicate that while the evolution of germination strategy is closely linked to ecologically important seed traits, it comprises a largely independent axis to vegetative or distributional characteristics.

Implications for Alpine Plant Communities in a Changing Climate

Australian alpine ecosystems, like alpine areas around the world, are under threat from climate change combined with changes in fire frequency, land use patterns, influx of invasive species, and the impacts of increased human visitation. Warming associated with climate change is occurring more rapidly above the treeline than at lower elevations, and alpine areas are predicted to continue to experience above average warming in the future (Kullman, 2004). The capacity for continued regeneration via seed under novel conditions is likely to play a significant role in the response of alpine plants to climate and associated changes. In addition, altered disturbance regimes are likely to require an increased role of restoration and rehabilitation in the region. The insights provided here are therefore crucial to conserving and managing alpine systems under change; they can help inform our predictions of how study species may be affected, form the basis of seed propagation plans for these species, and be used to guide future investigation into alpine seed germination strategies with identifiable conservation and management outcomes.

Changes in temperature are likely to affect the dormancy status of seeds both pre- and post-dispersal, consequently altering and/or disrupting germination strategies and resulting in shifts in current germination phenology (Hoyle et al., 2008b). For alpine plants, climate warming may lead to a shift from spring to autumn emergence, driven primarily by changes in seed dormancy status and resulting in major implications for species currently adapted to emergence in spring (Mondoni et al., 2012). Further research into how changes in temperatures that seeds experience post-dispersal, reduced duration of snow cover, and increased frequency of extreme temperature events will affect germination are required in order to predict how species with dormancy will respond to climate change. For those alpine species that germinate in autumn, prior to frost and snowfall, our results raise the question of how seedlings of these species will cope with the predicted reduction in precipitation and increased frequency of a-seasonal frost events.

Finally, our results raise important questions regarding variation within species, particularly in association with predicting the impact of climate change or selecting which seeds to utilize for restoration. Germination strategies may vary within species depending on population, flowering time, seed mass, parental investment, climate, elevation, and/or maternal environment and this variation may reflect genetic, epigenetic, and/or environmental factors (Cochrane et al., 2015). Increasing elevation accounts for variation in germinability and/or dormancy status of several alpine species (Billings and Mooney, 1968; Dorne, 1981; Cavieres and Arroyo, 2000) but see (Hoyle et al., 2014). Reduced dormancy status can result from early termination of seed development resulting from environmental stresses (Steadman et al., 2006; Hoyle et al., 2008b). Examination of variation in germination strategies within species may reveal the potential for species to cope with, and adapt to, changes in climate, thus, playing an important role in determining future survival and species distributions and providing important insight for management and conservation.

Conclusions

Our approach has revealed widespread occurrence of dormancy in the Australian alpine flora and demonstrates that germination strategies vary within a seed collection, as well as between collections. Further, we have shown that while dormancy strategy is phylogenetically conserved and correlated with seed mass and endospermy, it is largely independent of vegetative traits and range characteristics. While one cannot extrapolate directly from our results how climate and associated changes will affect the alpine flora, these results can form the basis for design of propagation plans and further experimentation on the impact of changing climatic regimes. In particular, we advocate further research into understanding the causes of within-species variation in germination strategy and the role this may play in the ability of species to withstand predicted climate change.

Data Accessibility

Full germination data and trait database will be lodged on DRYAD and with the Atlas of Living Australia.

Acknowledgments

We acknowledge funding from the Australian Research Council (LP0991593) and the Australian National Botanic Garden and the Friends thereof. We thank J. McAuliffe, S. Fethers, and S. Lawatsch of the ANBG. We also thank M. Holloway-Phillips, M. Cardillo, and S. Cunningham for input on earlier versions of the manuscript and for analytical advice and S. Venn for access to trait data.

References

Albrecht, M. A., and McCarthy, B. C. (2006). Seed germination and dormancy in the medicinal woodland herbs *Collinsonia canadensis* L. (Laminaceae) and *Dioscorea villosa* L. (Dioscoreaceae). *Flora* 201, 24–31. doi: 10.1016/j.flora.2005.04.001

Amen, R. D. (1966). The extent and role of seed dormancy in alpine plants. *Q. Rev. Biol.* 41, 271–281. doi: 10.1086/405055

Baskin, C. C., and Baskin, J. M. (2001). *Seeds: Ecology, Biogeography and Evolution of Dormancy and Germination*. London; California, CA: Academic Press.

Baskin, C. C., and Baskin, J. M. (2004). Germinating seeds of wildflowers, an ecological perspective. *Horticult. Technol.* 14, 467–473.

Billings, W. D., and Mooney, H. A. (1968). Ecology of arctic and alpine plants. *Biol. Rev. Camb. Philos. Soc.* 43, 481–529. doi: 10.1111/j.1469-185X.1968.tb00968.x

Bliss, L. C. (1971). Arctic and alpine plant life cycles. *Annu. Rev. Ecol. Syst.* 2, 405–438. doi: 10.1146/annurev.es.02.110171.002201

Bureau of Meteorology. (2010). *Climate Statistics at Charlotte Pass (Kosciusko Chalet), New South Wales, Australia between 1968 and 2009 [Online]*. Available online at: http://www.bom.gov.au/climate/data-services/

Cavieres, L. A., and Arroyo, M. T. K. (2000). Seed germination response to cold stratification period and thermal regime in *Phacelia secunda* (Hydrophyllaceae). *Plant Ecol.* 149, 1–8. doi: 10.1023/A:1009802806674

Cochrane, A., Yates, C. J., Hoyle, G. L. and Nicotra, A. B. (2015). Will among-population variation in seed traits improve the chance of species persistence under climate change? *Glob. Ecol. Biogeogr.* 24, 12–24. doi: 10.1111/geb.12234

Cohn, M. A., Jones, K. L., Chiles, L. A., and Church, D. F. (1989). Seed dormancy in red rice VII. Structure-activity studies of germination stimulants. *Plant Physiol.* 89, 879–882. doi: 10.1104/pp.89.3.879

Costin, A., Gray, M., Totterdell, C., and Wimbush, D. (2000). *Kosciuszko Alpine Flora*. Collingwood, VIC: CSIRO Publishing.

Densmore, R., and Zasada, J. (1983). Seed dispersal and dormancy patterns in northern willows: ecological and evolutionary significance. *Can. J. Bot.* 61, 3207–3216. doi: 10.1139/b83-358

Donohue, K. (2002). Germination timing influences natural selection on life-history characteristics in *Arabidopsis thaliana*. *Ecology* 83, 1006–1016. doi: 10.1890/0012-9658(2002)083[1006:GTINSO]2.0.CO;2

Donohue, K. (2005). Seeds and seasons: interpreting germination timing in the field. *Seed Sci. Res.* 15, 175–187. doi: 10.1079/SSR2005208

Dorne, A. J. (1981). Variation in seed germination inhibition of *Chenopodium bonus-henricus* in relation to altitude of plant growth. *Can. J. Bot.* 59, 1893–1901. doi: 10.1139/b81-249

Foley, M. E. (1992). Effect of soluble sugars and gibberellic-acid in breaking dormancy of excised Wild Oat (*Avena fatua*) embryos. *Weed Sci.* 40, 208–214.

Harden, G. J. (1990). *Flora of New South Wales*. Sydney, NSW: Royal Botanic Gardens.

Hoyle, G. L., Cordiner, H., Good, R. B., and Nicotra, A. B. (2014). Effects of reduced winter duration on seed dormancy and germination in six populations of the alpine herb *Aciphyllya glacialis* (Apiaceae). *Conserv. Physiol.* 2:cou015. doi: 10.1093/conphys/cou015

Hoyle, G. L., Daws, M. I., Steadman, K. J., and Adkins, S. W. (2008a). Mimicking a semi-arid tropical environment achieves dormancy alleviation for seeds of Australian native Goodeniaceae and Asteraceae. *Ann. Bot.* 101, 701–708. doi: 10.1093/aob/mcn009

Hoyle, G. L., Steadman, K. J., Daws, M. I., and Adkins, S. W. (2008b). Pre-and post-harvest influences on seed dormancy status of an Australian Goodeniaceae species, *Goodenia fascicularis*. *Ann. Bot.* 102, 93–101. doi: 10.1093/aob/mcn062

Hoyle, G. L., Venn, S. E., Steadman, K. J., Good, R. B., McAuliffe, E. J., Williams, E. R., et al. (2013). Soil warming increases plant species richness but decreases germination from the alpine soil seed bank. *Glob. Change Biol.* 19, 1549–1561. doi: 10.1111/gcb.12135

International Seed Testing Association. (2003). *Working Sheets on Tetrazolium Testing*. Bassersdorf: ISTA.

Körner, C. (2003). "Life under snow: protection and limitation," in *Alpine Plant Life: Functional Plant Ecology of High Mountain Ecosystems* (Berlin; Heidelberg; New York, NY: Springer-Verlag), 47–62.

Kullman, L. (2004). The changing face of the alpine world. *Glob. Change Newsl.* 57, 12–14.

Leishman, M., Westoby, M., and Jurado, E. (1995). Correlates of seed size variation: a comparison among five temperate floras. *J. Ecol.* 83, 517–530. doi: 10.2307/2261604

Martin, A. C. (1946). The comparative internal morphology of seeds. *Am. Midl. Nat.* 36, 513–660. doi: 10.2307/2421457

Moles, A. T., Ackerly, D. D., Tweddle, J. C., Dickie, J. B., Smith, R., Leishman, M. R., et al. (2007). Global patterns in seed size. *Glob. Ecol. Biogeogr.* 16, 109–116. doi: 10.1111/j.1466-8238.2006.00259.x

Moles, A. T., and Westoby, M. (2004). Seedling survival and seed size: a synthesis of the literature. *J. Ecol.* 92, 372–383. doi: 10.1111/j.0022-0477.2004.00884.x

Mondoni, A., Daws, M. I., Belotti, J., and Rossi, G. (2009). Germination requirements of the alpine endemic *Silene elisabethae* Jan: effects of cold stratification, light and GA3. *Seed Sci. Technol.* 37, 79–87. doi: 10.15258/sst.2009.37.1.10

Mondoni, A., Rossi, G., Orsenigo, S., and Probert, R. J. (2012). Climate warming could shift the timing of seed germination in alpine plants. *Ann. Bot.* 110, 155–164. doi: 10.1093/aob/mcs097

Orme, D., Freckleton, R., Thomas, G., Petzoldt, T., Fritz, S., Isaac, N., et al. (2013). *caper: Comparative Analyses of Phylogenetics and Evolution in R*. R package version 0.5.2.

Pauli, H., Gottfried, M., Lamprecht, A., Niessner, S., Rumpf, S., Winkler, M., et al. (2015). *The GLORIA Field Manual – Standard Multi-Summit approach, Supplementary Methods and Extra Approaches*. Vienna: GLORIA-Coordination, Austrian Academy of Sciences & University of Natural Resources and Life Sciences.

R Core Team. (2013). *R: A Language and Environment for Statistical Computing, [Online]*. Vienna: R Foundation for Statistical Computing. Available online at: http://www.R-project.org/

Reich, P. B. (2014). The world-wide 'fast-slow' plant economics spectrum: a traits manifesto. *J. Ecol.* 102, 275–301. doi: 10.1111/1365-2745.12211

Reich, P. B., Walters, M. B., and Ellsworth, D. S. (1997). From tropics to tundra: global convergence in plant functioning. *Proc. Natl. Acad. Sci. U.S.A.* 94, 13730–13734. doi: 10.1073/pnas.94.25.13730

Revell, L. J. (2012). phytools: an R package for phylogenetic comparative biology (and other things). *Methods Ecol. Evol.* 3, 217–223. doi: 10.1111/j.2041-210X.2011.00169.x

Scherrer, D., and Körner, C. (2011). Topographically controlled thermal-habitat differentiation buffers alpine plant diversity against climate warming. *J. Biogeogr.* 38, 406–416. doi: 10.1111/j.1365-2699.2010.02407.x

Schutz, W. (2002). Dormancy characteristics and germination timing in two alpine *Carex* species. *Basic Appl. Ecol.* 3, 125–134. doi: 10.1078/1439-1791-00090

Schwienbacher, E., Navarro-Cano, J. A., Neuner, G., and Erschbamer, B. (2011). Seed dormancy in alpine species. *Flora* 206, 845–856. doi: 10.1016/j.flora.2011.05.001

Shimono, Y., and Kudo, G. (2005). Comparisons of germination traits of alpine plants between fellfield and snowbed habitats. *Ecol. Res.* 20, 189–197. doi: 10.1007/s11284-004-0031-8

Simons, A. M. (2011). Modes of response to environmental change and the elusive empirical evidence for bet hedging. *Proc. R. Soc. B Biol. Sci.* 278, 1601–1609. doi: 10.1098/rspb.2011.0176

Simons, A. M., and Johnston, M. O. (2000). Variation in seed traits of *Lobelia inflata* (Campanulaceae): sources and fitness consequences. *Am. J. Bot.* 87, 124–132. doi: 10.2307/2656690

Sommerville, K. D., Martyn, A. J., and Offord, C. A. (2013). Can seed characteristics or species distribution be used to predict the stratification requirements of herbs in the Australian Alps? *Bot. J. Linn. Soc.* 172, 187–204. doi: 10.1111/boj.12021

Starrfelt, J., and Kokko, H. (2012). Bet-hedging - a triple trade-off between means, variances and correlations. *Biol. Rev.* 87, 742–755. doi: 10.1111/j.1469-185X.2012.00225.x

Steadman, K. J., Eaton, D. M., Plummer, J. A., Ferris, D. G., and Powles, S. B. (2006). Late-season non-selective herbicide application reduces *Lolium rigidum* seed numbers, seed viability and seedling fitness. *Aust. J. Agric. Res.* 57, 133–141. doi: 10.1071/AR05122

Tielborger, K., Petru, M., and Lampei, C. (2012). Bet-hedging germination in annual plants: a sound empirical test of the theoretical foundations. *Oikos* 121, 1860–1868. doi: 10.1111/j.1600-0706.2011.20236.x

Vleeshouwers, L. M., Bouwmeester, H. J., and Karssen, C. M. (1995). Redefining seed dormancy: an attempt to integrate physiology and ecology. *J. Ecol.* 83, 1031–1037. doi: 10.2307/2261184

Wagner, I., and Simons, A. M. (2009). Divergence in germination traits among arctic and alpine populations of *Koenigia islandica*: light requirements. *Plant Ecol.* 204, 145–153. doi: 10.1007/s11258-009-9578-3

Webb, C. O., Ackerly, D. D., and Kembel, S. W. (2008). Phylocom: software for the analysis of phylogenetic community structure and trait evolution. *Bioinformatics* 24, 2098–2100. doi: 10.1093/bioinformatics/btn358

Westoby, M. (1998). A leaf-height-seed (LHS) plant ecology strategy scheme. *Plant Soil* 199, 213–227. doi: 10.1023/A:1004327224729

Westoby, M., Leishman, M., and Lord, J. (1996). Comparative ecology of seed size and dispersal. *Philos. Trans. R. Soc. Lond. B Biol. Sci.* 1309–1318. doi: 10.1098/rstb.1996.0114

Wright, I. J., Reich, P. B., Westoby, M., Ackerly, D. D., Baruch, Z., Bongers, F., et al. (2004). The worldwide leaf economics spectrum. *Nature* 428, 821–827. doi: 10.1038/nature02403

Conflict of Interest Statement: The authors declare that the research was conducted in the absence of any commercial or financial relationships that could be construed as a potential conflict of interest.

Permissions

List of Contributors

Erica M. Waters and Maxine A. Watson
Department of Biology, Indiana University, Bloomington, IN, USA

Torben Lübbe, Bernhard Schuldt and Christoph Leuschner
Plant Ecology and Ecosystems Research, Albrecht von Haller Institute for Plant Sciences, University of Göttingen, Göttingen, Germany

Madjelia C. E. Dao
Département Productions Forestières, Institut de l'Environnement et de Recherches Agricoles, Ouagadougou, Burkina Faso

Sergio Rossi, Denis Walsh and Hubert Morin
Département des Sciences Fondamentales, Université du Québec à Chicoutimi, Chicoutimi, QC, Canada

Daniel Houle
Direction de la Recherche Forestière, Forêt Québec, Ministère des Forêts de la Faune et des Parcs, Québec, QC, Canada
Ouranos, Consortium Sur la Climatologie Régionale et l'Adaptation aux Changements Climatiques, Montréal, QC, Canada

Pei-Jian Shi and Fu-Sheng Wang
Co-Innovation Centre for Sustainable Forestry in Southern China, Bamboo Research Institute, Nanjing Forestry University, Nanjing, China

Pei-Jian Shi
Key Laboratory of Vegetation Restoration and Management of Degraded Ecosystems, South China Botanical Garden, Chinese Academy of Sciences, Guangzhou, China

Jian-Guo Huang and Li-Hong Zhai
Key Laboratory of Vegetation Restoration and Management of Degraded Ecosystems, South China Botanical Garden, Chinese Academy of Sciences, Guangzhou, China
Provincial Key Laboratory of Applied Botany, South China Botanical Garden, Chinese Academy of Sciences, Guangzhou, China,

Cang Hui
Department of Mathematical Sciences, Centre for Invasion Biology, Stellenbosch University, Matieland, South Africa
Mathematical and Physical Biosciences, African Institute for Mathematical Sciences, Cape Town, South Africa

Henri D. Grissino-Mayer
Department of Geography, The University of Tennessee, Knoxville, TN, USA

Jacques C. Tardif
Centre for Forest Interdisciplinary Research, University of Winnipeg, Winnipeg, MB, Canada

Bai-Lian Li
Ecological Complexity and Modelling Laboratory, Department of Botany and Plant Sciences, University of California, Riverside, Riverside, CA, USA

Adrien Taudiere, Anne-Christine Monnet, Jean-Michel Bellanger and Franck Richard
UMR 5175, CEFE – CNRS – Université de Montpellier – Université Paul Valéry Montpellier – EPHE –INSERM, Montpellier, France

François Munoz
UM2, UMR AMAP, Montpellier, France
French Institute of Pondicherry, Pondicherry, India

Annick Lesne
CNRS, LPTMC UMR 7600, Université Pierre et Marie Curie-Paris 6, Sorbonne Universités, Paris, France
CNRS, IGMM UMR 5535, Université de Montpellier, Montpellier, France

Marc-André Selosse
CNRS, Muséum National d'Histoire Naturelle, UMR 7205, Origine, Structure et Evolution de la Biodiversité, Paris, France,

Pierre-Arthur Moreau
Département de Botanique, Faculté des Sciences Pharmaceutiques et Biologiques, Université Lille, Lille, France

Wataru Ishizuka
Forestry Research Institute, Hokkaido Research Organization, Bibai, Japan

Kiyomi Ono and Toshihiko Hara
Institute of Low Temperature Science, Hokkaido University, Sapporo, Japan

Susumu Goto
Graduate School of Agricultural and Life Sciences, The University of Tokyo, Tokyo, Japan

Yu-Kun Hu, Xu Pan, Wen-Bing Li, Wen-Hong Dai, Tao Xiao, Ling-Yun Chen, Wei Xiong, Meng-Yao Zhou, Yao-Bin Song and Ming Dong
Key Laboratory of Hangzhou City for Ecosystem Protection and Restoration, College of Life and Environmental Sciences, Hangzhou Normal University, Hangzhou, China,

Yu-Kun Hu, Guo-Fang Liu, Shuang-Li Tang, Ya-Lin Zhang and Ming Dong
State Key Laboratory of Vegetation and Environmental Change, Institute of Botany, Chinese Academy of Sciences, Beijing, China

Xu Pan
Institute of Wetland Research, Chinese Academy of Forestry, Beijing, China

Meng Xu, Guoan Wang, Xiaoliang Li, Xiaolin Li, Peter Christie and Junling Zhang
College of Resources and Environmental Sciences, China Agricultural University, Beijing, China

Xiaobu Cai
Tibet Agricultural and Animal Husbandry College, Tibet University, Linzhi, China

Glenn T. Howe, Palitha Dharmawardhana and Steven H. Strauss
Department of Forest Ecosystems and Society, Oregon State University, Corvallis, OR, USA

David P. Horvath
Biosciences Research Laboratory, United States Department of Agriculture-Agricultural Research Service, Fargo, ND, USA

Todd C. Mockler and Palitha Dharmawardhana
Department of Botany and Plant Pathology, Oregon State University, Corvallis, OR, USA,

Henry D. Priest and Todd C. Mockler
Donald Danforth Plant Science Center, Saint Louis, MO, USA

Henry D. Priest
Division of Biology and Biomedical Sciences, Washington University in Saint Louis, Saint Louis, MO, USA

Adam Kimak and Markus Leuenberger
Climate and Environmental Physics, Physics Institute, University of Bern, Bern, Switzerland
Oeschger Centre for Climate Change Research, University of Bern, Bern, Switzerland

Zoltan Kern
Institute for Geological and Geochemical Research, Research Centre for Astronomy and Earth Sciences, Hungarian Academy of Sciences (MTA), Budapest, Hungary

Georg von Arx and Patrick Fonti
Landscape Dynamics Research Unit, Swiss Federal Institute for Forest, Snow and Landscape Research WSL, Birmensdorf, Switzerland

Alberto Arzac
Departamento de Biología Vegetal y Ecología, Facultad de Ciencia y Tecnología, Universidad del País Vasco, Leioa, Spain

José M. Olano
Departamento de Ciencias Agroforestales, Escuela Universitaria de Ingenierías Agrarias, Instituto Universitario de Investigación en Gestión Forestal Sostenible-Universidad de Valladolid, Soria, Spain

Jean-Luc Maeght
Institut de Recherche pour le Développement, UMR 242/iEES – Paris (IRD-UPMC-CNRS-UPEC-UDD-INRA), Bondy, France

Jean-Luc Maeght and Alexia Stokes
INRA, UMR-AMAP, Montpellier, France

Santimaitree Gonkhamdee and Supat Isarangkool Na Ayutthaya
Khon Kaen University, Faculty of Agriculture, Khon Kaen, Thailand
Corentin Clément
International Water Management Institute, Vientiane, Laos

Alain Pierret
Institut de Recherche Pour le Développement, UMR IEES-Paris – Department of Agricultural Land Management (DALaM), Vientiane, Laos

Xiaoping Wang, Xiyu Yang, Siyu Chen, Qianqian Li, Wei Wang, Chunjiang Hou, Xiao Gao, Li Wang and Shucai Wang
Key Laboratory of Molecular Epigenetics of Ministry of Education, Northeast Normal University, Changchun, China

Xiliang Song, Zhenzhu Xu, Xiaomin Lv and Yuhui Wang
State Key Laboratory of Vegetation and Environmental Change, Institute of Botany, Chinese Academy of Science, Beijing, China, 2 University of Chinese Academy of Sciences, Beijing, China

Guangsheng Zhou
State Key Laboratory of Vegetation and Environmental Change, Institute of Botany, Chinese Academy of Science, Beijing, China
Chinese Academy of Meteorological Sciences, China Meteorological Administration, Beijing, China

Arnab Mukherjee, Erving Morelius, Carlos Tamez, Susmita Bandyopadhyay, Jose R. Peralta-Videa and Jorge L. Gardea-Torresdey
Environmental Science and Engineering, The University of Texas at El Paso, El Paso, TX, USA
University of California Center for Environmental Implications of Nanotechnology, The University of Texas at El Paso, El Paso, TX, USA

Genhua Niu and Youping Sun
Texas A&M AgriLife Research Center at El Paso, El Paso, TX, USA

Jason C. White
Department of Analytical Chemistry, The Connecticut Agricultural Experiment Station, New
Haven, CT, USA

Jose R. Peralta-Videa and Jorge L. Gardea-Torresdey
Department of Chemistry, The University of Texas at El Paso, El Paso, TX, USA

Anne Y. Fennell, Vedbar Khadka, Lekha Sreekantan and Katherine L. Mathiason
Northern Plains BioStress Laboratory, Plant Science Department, South Dakota State University, Brookings, SD, USA

Grant R. Cramer and Karen A. Schlauch
Department of Biochemistry and Molecular Biology, University of Nevada, Reno, Reno, NV, USA

Satyanarayana Gouthu and Laurent G. Deluc
Department of Horticulture, Oregon State University, Corvallis, OR, USA,

Jerome Grimplet and Lekha Sreekantan
Instituto de Ciencias de la Vid y del Vino (CSIC, Universidad de La Rioja, Gobierno de La Rioja), Logroño, Spain

Tingting Mei, Dongming Fang, Alexander Röll, Furong Niu and Dirk Hölscher
Tropical Silviculture and Forest Ecology, Georg-August-Universität Göttingen, Göttingen, Germany

Hendrayanto
Department of Forest Management, Institut Pertanian Bogor, Bogor, Indonesia

Marta Marmiroli, Davide Imperiale, Luca Pagano and Nelson Marmiroli
Department of Life Sciences, University of Parma, Parma, Italy

Marco Villani and Andrea Zappettini
Institute of Materials for Electronics and Magnetism (IMEM-CNR), Parma, Italy

Jean-Michel Legave
INRA, Unité Mixte de Recherche 1334 Amélioration Génétique et Adaptation des Plantes Méditerranéennes et Tropicales, Montpellier, France

Yann Guédon
CIRAD, Unité Mixte de Recherche 1334 et Inria, Virtual Plants, Montpellier, France

Gustavo Malagi
Faculdade de Agronomia, Universidade Federal de Pelotas, Pelotas, Brazil

Adnane El Yaacoubi
Faculté des Sciences, Université Moulay Ismail, Meknès, Morocco

Marc Bonhomme
Unité Mixte de Recherche 547, INRA et Université Blaise Pascal, PIAF, Clermont-Ferrand, France

Alessandro Mattiello, Antonio Filippi, Filip Pošćić, Rita Musetti, Massimo Vischi, Alberto Bertolini and Luca Marchiol
Department of Agriculture and Environmental Sciences, University of Udine, Udine, Italy

Maria C. Salvatici and Cristiana Giordano
Centro di Microscopie Elettroniche "Laura Bonzi", Istituto di Chimica dei Composti OrganoMetallici, Consiglio Nazionale delle Ricerche, Firenze, Italy

Cristiana Giordano
Tree and Timber Institute, Istituto Per La Valorizzazione del Legno e delle Specie Arboree-CNR, Firenze, Italy

Gemma L. Hoyle, Emma J. McIntosh, Lucy M. E. Galea and Adrienne B. Nicotra
Department of Evolution, Ecology and Genetics, Research School of Biology, Australian National University, Canberra, ACT, Australia

Kathryn J. Steadman
School of Pharmacy and Queensland Alliance for Agriculture and Food Innovation, The University of Queensland, QLD, Australia

Roger B. Good
Australian National Botanic Gardens, Canberra, ACT, Australia
Fenner School of the Environment, Australian National University, Canberra, ACT, Australia